Encyclopedia of
Plant Physiology

New Series   Volume 5

Editors
A. Pirson, Göttingen
M. H. Zimmermann, Harvard

Photosynthesis I

Photosynthetic Electron Transport and Photophosphorylation

Edited by

A. Trebst and M. Avron

Contributors

R. S. Alberte J. Amesz D. I. Arnon T. Bakker-Grunwald
R. J. Berzborn N. K. Boardman H. Bothe W. L. Butler
C. Carmeli G. A. Corker W. A. Cramer R. A. Dilley B. A. Diner
G. Forti J. M. Galmiche R. Giaquinta H. Gimmler
J. H. Golbeck N. E. Good Z. Gromet-Elhanan D. O. Hall
G. Hauska G. E. Hoch S. Izawa G. Jacobi A. T. Jagendorf
P. Joliot W. Junge S. Katoh B. Kok R. Kraayenhof
D. W. Krogmann S. Lien W. Lockau R. Malkin S. Malkin
D. Mauzerall R. E. McCarty B. A. Melandri K. Mühlethaler
N. Nelson R. J. Radmer K. K. Rao H. Rottenberg B. Rumberg
A. San Pietro P. V. Sane S. Schuldiner N. Shavit
J. P. Thornber W. Urbach E. C. Weaver J. S. C. Wessels

Springer-Verlag Berlin Heidelberg New York 1977

Professor Dr. A. TREBST
Abteilung für Biologie, Ruhr-Universität Bochum
Postfach 102148, 4630 Bochum-Querenburg

Professor Dr. M. AVRON
Department of Biochemistry, The Weizmann Institute of Science
Rehovot/Israel

With 128 Figures

ISBN 3-540-07962-9 Springer-Verlag Berlin Heidelberg New York
ISBN 0-387-07962-9 Springer-Verlag New York Heidelberg Berlin

Library of Congress Cataloging in Publication Data. Main entry under title: Photosynthesis. (Encyclopedia of plant physiology; new ser., v. 5). Includes bibliographical references. Contents: 1. Photosynthetic electron transport and photophosphorylation. 1. Photosynthesis–Collected works. I. Trebst, Achim. II. Avron, Mordhay. III. Series. QK711.2.E5 vol. 5 [QK882] 581.1′08s [581.1′3342] 76-30357.

© by Springer-Verlag Berlin · Heidelberg 1977
Printed in Germany

Typesetting, printing and bookbinding: Universitätsdruckerei H. Stürtz AG, Würzburg.
2131/3130–543210

Preface

As editor of the two-part Volume V on photosynthesis in RUHLAND's Encyclopedia, the forerunner of this series published in 1960, I have been approached by the editors of the present volume to provide a short preface. The justification for following this suggestion lies in the great changes which have been taking place in biology in the two decades between these publications, changes which are reflected in the new editorial plan.

Twenty years ago it appeared convenient and formally easy to consider photosynthesis as a clearly separated field of research, which could be dealt with under two major headings: one presenting primarily photochemical and biochemical principles, the other physiological and environmental studies. Such a partition, however, as far as aims and opinions of the authors were concerned, resulted in a rather heterogeneous volume.

Today, the tendency in experimental biology is towards a merger of previously distinct disciplines. Biochemists and biophysicists have developed their methods to such an extent that, over and above the analysis of individual reaction sequences, work on the manifold interrelationships among cellular activities has become increasingly possible. Joining them in growing numbers are the physiologists and ecologists with their wealth of information on activity changes in vivo and on the variability and efficiency of the organisms concerned. Furthermore, biochemists, biophysicists and physiologists also now share a lively interest in ultrastructure research, the results and implications of which, through continually improving methodology, have generated important stimuli for the work in the field of cell function.

This general trend towards cell biology is also apparent in photosynthesis research. Therefore, for the new Encyclopedia, it was originally felt worth considering whether or not to present photosynthesis explicitly in terms of its multifunctional and regulatory principles. This, however, would have meant exceeding the planned two volumes, and might also have given premature coverage to some still emerging concepts. Above all, one must not forget that a great arsenal of experience and ideas is now at hand, which must first be harnessed to provide a thorough insight into the function and elaborate structure of the thylakoid system. Moreover, the thylakoid functions deserve particular attention, as being integrated with other energy-conserving processes into a unifying concept of bioenergetics.

Consequently it seemed justifiable to devote "Photosynthesis I" to the membrane-associated part of photosynthesis, that is, to the photochemical primary reactions, electron transport and photophosphorylation. In order to understand the magnitude of this theme more clearly, one should remember that the authors of the old Encyclopedia still proceeded on the basis of a single photoreaction and, 17 years ago, important partners of the photosynthetic electron transport chain were unknown. Of course, essential features of photophosphorylation had already been revealed at that time, but its possible molecular mechanisms could not be proposed. To overcome the problem of evaluating the enormous amount of information accumulated since then, the editors have approached prominent scientists who have been major contributors in their field, to provide general reviews for each of the main sections. These introductory articles, suitable for the general reader, are followed by a group of succinct essays on limited topics written by specialists. A meticulous recording of the progress over the last two decades, as perhaps expected from the dictionary definition of the word encyclopedia, did not seem practical, a thorough presentation of the actual trends being preferred. Nevertheless, the reader will be thankful, that Professor D.I. ARNON, indefatigable advocate of the field of photophosphorylation ever since its discovery, has presented an introductory insight into his field related to the framework of photosynthesis as a whole. Although this article extends to 1975, the great turnover of facts and ideas is duly characterized by its title "History".

The forthcoming volume "Photosynthesis II" (Editors M. GIBBS and E. LATZKO) addresses itself to the problems of carbon metabolism and other reactions more or less directly resulting from photosynthetic electron transport. Since this second volume deals primarily with processes occurring predominantly in the chloroplast stroma, the well-tried methods of enzymology are here applicable and principles of regulation are therefore much easier to point out and to discuss. Regulatory mechanisms within the electron transport itself are certainly to be expected, and the reader will find relevant indications, but the time appeared not yet right for this viewpoint to serve as the leitmotiv for Volume I.

Whereas 20 years ago the conception of the chloroplast as a biochemical unit, fully competent for the whole process of photosynthesis, had scarcely been established, the interconnections of this same unit with cellular functions outside of it are a main subject of today's research. This theme, together with other intracellular transport and exchange processes, is already included in Volume 3 of this encyclopedia (Transport III, edited by C.R. STOCKING and U. HEBER). The development of the chloroplast and its photosynthetic apparatus, a chapter on which is to be found in the present volume, will be treated extensively in a later volume on plant cell development, together with problems of plastid genetics. Also in contrast to the old encyclopedia, the ecological aspects of photosynthesis will be fully covered in a group of volumes entitled "Ecophysiology of Plants."

The editors of the series hope, with this partitioning of the material, to have given the optimal representation of the current state of photosynthesis research without allowing subject overlapping, inherent in modern physiology, to extend beyond reasonable boundaries.

Finally, also on behalf of the editors of this volume, I wish to express my thanks to all the authors who agreed to contribute their expertise, time and efforts, and to the publisher whose continued readiness has made the rapid publication of "Photosynthesis I" a reality.

Göttingen, March 1977   A. PIRSON

Contents

I. History

II. Electron Transport

1. General

1b. Electron Transport in Chloroplasts
J.H. GOLBECK, S. LIEN, and A. SAN PIETRO (With 4 Figures)

2. Porphyrins, Chlorophyll, and Photosynthesis
D. MAUZERALL (With 2 Figures)

3. Light Conversion Efficiency in Photosynthesis
R.J. RADMER and B. KOK (With 3 Figures)

4. P-700
G.E. HOCH (With 2 Figures)

5. Chlorophyll Fluorescence: A Probe for Electron Transfer and Energy Transfer
W.L. BUTLER (With 2 Figures)

6. Electron Paramagnetic Resonance Spectroscopy
E.C. WEAVER and G.A. CORKER (With 1 Figure)

7. Primary Electron Acceptors
R. MALKIN (With 4 Figures)

8. Oxygen Evolution and Manganese
B.A. DINER and P. JOLIOT (With 2 Figures)

9. Ferredoxin
D.O. HALL and K.K. RAO (With 3 Figures)

10. Flavodoxin
H. BOTHE

11. Flavoproteins
G. FORTI

12. Cytochromes
W.A. CRAMER (With 2 Figures)

13. Plastoquinone
J. AMESZ (With 5 Figures)

III. Energy Conservation

1. Photophosphorylation
A.T. JAGENDORF (With 7 Figures)

2. Proton and Ion Transport Across the Thylakoid Membranes
H. ROTTENBERG (With 3 Figures)

3. Bound Nucleotides and Conformational Changes in Photophosphorylation
N. SHAVIT (With 1 Figure)

9. Acid Base ATP Synthesis in Chloroplasts
S. SCHULDINER (With 2 Figures)

10. Energy-Dependent Conformational Changes
R. KRAAYENHOF (With 1 Figure)

11. Uncoupling of Electron Transport from Phosphorylation in Chloroplasts
N. E. GOOD (With 1 Figure)

12. Energy Transfer Inhibitors of Photophosphorylation in Chloroplasts
R. E. MCCARTY (With 2 Figures)

13. Photophosphorylation in vivo
H. GIMMLER

Contents

IV. Structure and Function

V. Algal and Bacterial Photosynthesis

1. Eukaryotic Algae
W. URBACH

2. Blue-Green Algae
D.W. KROGMANN (With 1 Figure)

3. Electron Transport and Photophosphorylation in Photosynthetic Bacteria
Z. GROMET-ELHANAN (With 1 Figures)

List of Contributors

R.S. ALBERTE
Department of Biology and
Molecular Biology Institute
University of California
Los Angeles, CA 90024/USA

J. AMESZ
Department of Biophysics
Huygens Laboratory
University of Leiden
Wassenaarseweg 18
Leiden, The Netherlands

D.I. ARNON
Department of Cell Physiology
University of California
Berkeley, CA 94720/USA

T. BAKKER-GRUNWALD
University of Colorado
Medical Center, School of Medicine
Department of Physiology
4200 East Ninth Avenue
Denver, CO 80262/USA

R.J. BERZBORN
Lehrstuhl für Biochemie der Pflanzen,
Abt. Biologie
Ruhr-Universität Bochum
Postfach 10 21 48
4630 Bochum
Federal Republic of Germany

N.K. BOARDMAN
CSIRO, Division of Plant Industry
P.O. Box 1600
Canberra, City, A.C.T. 2601/Australia

H. BOTHE
Botanisches Institut
Universität Heidelberg
Hofmeisterweg 4
6900 Heidelberg
Federal Republic of Germany

W.L. BUTLER
Department of Biology
University of California
San Diego
La Jolla, CA 92093/USA

C. CARMELI
Department of Biochemistry
Tel-Aviv University
Ramat-Aviv, Tel-Aviv/Israel

G.A. CORKER
Thomas J. Watson Res. Center
P.O. Box 218
Yorktown Heights
NY 10598/USA

W.A. CRAMER
Department of Biological Sciences
Purdue University
W. Lafayette, IN 47907/USA

R.A. DILLEY
Department of Biological Sciences
Purdue University
W. Lafayette, IN 47907/USA

B.A. DINER
Institut de Biologie Physico-Chimique
13, rue Pierre et Marie Curie
75005 Paris/France

G. FORTI
Istituto Botanico
Via Giuseppe Colombo 60
Milano/Italy

J.M. GALMICHE
Centre d'Etudes Nucléaires de Saclay
Département de Biologie
91190 GIF-sur-YVETTE/France

R. GIAQUINTA
E.I. Du Pont De Nemours &
Company Central Research &
Development Department
Experimental Station
Wilmington, DE 19898/USA

H. GIMMLER
Botanisches Institut der Universität
Mittlerer Dallenbergweg 64
8700 Würzburg
Federal Republic of Germany

J.H. GOLBECK
Martin Marietta Laboratories
1450 South Rolling Road
Baltimore, MD 21227/USA

N.E. GOOD
Department of Botany and
Plant Pathology
Michigan State University
East Lansing, MI 48824/USA

Z. GROMET- ELHANAN
Biochemistry Department
The Weizmann Institute of Science
Rehovot/Israel

D.O. HALL
Department of Plant Sciences
University of London King's College
68 Half Moon Lane
London SE24 9JF/Great Britain

G. HAUSKA
Universität Regensburg
FB Biologie und Vorkl. Medizin
Universitätsstraße 31
8400 Regensburg
Federal Republic of Germany

G.E. HOCH
Department of Biology
University of Rochester
Rochester, NY 14627/USA

S. IZAWA
Department of Biology
Wayne State University
Detroit, MI 48202/USA

G. JACOBI† (deceased 25.6.1976)
Lehrstuhl für Biochemie der Pflanzen
Universität Göttingen
Untere Karspüle 2
3400 Göttingen
Federal Republic of Germany

A.T. JAGENDORF
Section of Genetics
Development and Physiology
Plant Science Building
Cornell University
Ithaca, NY 14853/USA

P. JOLIOT
Institut de Biologie Physico-Chimique
13, rue Pierre et Marie Curie
75005 Paris/France

W. JUNGE
Max-Volmer-Institut für
Physikalische Chemie und
Molekularbiologie
Technische Universität Berlin
Strasse des 17. Juni 135
1000 Berlin 12/W. Germany

S. KATOH
Department of Pure and
Applied Sciences
College of General Education
University of Tokyo
Komaba, Meguro-ku
Tokyo 153/Japan

B. KOK
Martin Marietta Laboratories
1450 South Rolling Road
Baltimore, MD 21227/USA

R. KRAAYENHOF
Biological Laboratory
Free University
De Boelelaan 1087
Amsterdam/The Netherlands

D.W. KROGMANN
Department of Biochemistry
Purdue University
W. Lafayette, IN 47907/USA

S. LIEN
Department of Plant Sciences
Indiana University
Bloomington, IN 47401/USA

W. LOCKAU
Lehrstuhl für Biochemie der Pflanzen,
Abt. Biologie
Ruhr-Universität Bochum
Postfach 10 21 48
4630 Bochum
Federal Republic of Germany

R. MALKIN
Department of Cell Physiology
251 Hilgard Hall
University of California
Berkeley, CA 94720/USA

S. MALKIN
Biochemistry Department
The Weizmann Institute of Science
Rehovot/Israel

D. MAUZERALL
Rockefeller University
New York, NY 10021/USA

R.E. McCarty
Section of Biochemistry
Molecular and Cell Biology
Cornell University
Ithaca, NY 14853/USA

B.A. Melandri
Institute of Botany
University of Bologna
Via Irnerio 42
40126 Bologna/Italy

K. Mühlethaler
Institut für Zellbiologie
HPT-C5.1
ETHZ Hönggerberg
8093 Zurich/Switzerland

N. Nelson
Department of Biology
Technion-Israel Institute
of Technology
Haifa/Israel

R.J. Radmer
Martin Marietta Laboratories
1450 South Rolling Road
Baltimore, MD 21227/USA

K.K. Rao
Department of Plant Sciences
University of London King's
College
68 Half Moon Lane
London SE24 9JF/Great Britain

H. Rottenberg
Biochemistry Department
The George S. Wise Center
for Life Sciences
Tel-Aviv University
Ramat-Aviv
Tel-Aviv/Israel

B. Rumberg
Max-Volmer-Institut für
Physikalische Chemie und Molekularbiologie der Technischen Universität
1000 Berlin 12/W. Germany

A. San Pietro
Department of Plant Sciences
Indiana University
Bloomington, IN 47401/USA

P.V. Sane
Biology and Agriculture
Division
Bhabha Atomic Research Centre
Bombay 400085/India

S. Schuldiner
Department Molecular Biology
The Hebrew University-Hadassah
Medical School
Jerusalem/Israel

N. Shavit
Biology Department
Ben Gurion University
of the Negev
Beer-Sheva/Israel

J.P. Thornber
Department of Biology and
Molecular Biology Institute
University of California
Los Angeles, CA 90024/USA

W. Urbach
Botanisches Institut der
Universität Würzburg
Mittlerer Dallenbergweg 64
8700 Würzburg
Federal Republic of Germany

E.C. Weaver
Department of Biological Sciences
San José State University
San José, CA 95192/USA

J.S.C. Wessels
Philips Research Laboratories
Eindhoven/The Netherlands

List of Abbreviations

9-AA	9-aminoacridine	EPR	Electron paramagnetic resonance
ADP	Adenosine diphosphate		
AMP	Adenosine monophosphate	Fab	Antigen binding fragment of an antibody
ANS	8-anilinonaphthalene-1-sulfonate	FAD	Flavin adenine dinucleotide
ATP	Adenosine triphosphate	FCCP	Carbonylcyanide-p-trifluoro-methoxy-phenyl-hydrazone
CCP (CCCP, m-Cl-CCP)	Carbonyl cyanide m-chloro-phenylhydrazone		
CDIS	Cyclohexyl-morpholinoethyl carbodiimide metho-p-toluenesulfonate	Fd	Ferredoxin
		FMN	Flavin mononucleotide
		Fe-S	Sulfur-iron (=non heme iron)
CDP	Cytidine diphosphate	Fp	"Flavoprotein"=ferredoxin-NADP$^+$-reductase
CF$_1$	Coupling factor 1		
CP-I, CP-II	Chlorophyll protein complex I and II	GDP	Guanosine diphosphate
		GTP	Guanosine triphosphate
DABS	p-(Diazonium)-benzene-sulfonic acid	HES	High energy state
		HOQNO	2-n-heptyl-4-hydroxy-quinoline-N-oxide
DAD	Diaminodurene=2,3,5,6-tetramethyl-p-phenylenediamine	IDP	Inosine diphosphate
		KCN	Potassium cyanide
DBMIB	Dibromothymoquinone=2,5-dibromo-3-methyl-6-isopropyl-p-benzoquinone	NAD$^+$	Nicotinamide-adenine dinucleotide oxidized
DCCD	Dicyclohexyl-carbodiimide	NADH	Nicotinamide-adenine dinucleotide reduced
DCMU	3-(3′,4′-Dichlorophenyl)-1,1-dimethylurea	NADP$^+$	Nicotinamide-adenine dinucleotide phosphate oxidized
DCPIP (DPIP)	Dichlorophenolindophenol	NADPH	Nicotinamide-adenine dinucleotide phosphate reduced
DEAE	Diethylaminoethyl		
d.l. (DLE)	Delayed luminescence	NEM	N-ethylmaleimide
DNA	Deoxyribonucleic acid	NMR	Nuclear magnetic resonance
DPC	Diphenylcarbazide	P	Inorganic phosphate
DSPD	Disalicylidenepropane-diamine	PCy (PC)	Plastocyanine
		PD	p-phenylenediamine
DTNB	Dithio-tris-nitrobenzoic-acid (Ellman's reagent)	P/e$_2$	Ratio of ATP to 2 electrons
		PMF	Proton motive force
DTT	Dithiothreitol	PMS	N-methylphenazonium metho-sulfate
EDAC	1-Ethyl-3-(3′-dimethyl-aminopropyl)-carbodiimide		
EDTA	Ethylenediaminetetracetic acid	P/O	Ratio of ATP to oxygen
		PPi	Inorganic pyrophosphate

PPNR	Photosynthetic pyridine nucleotide reductase	SCP	Subchloroplast particles
		SDS	Sodium dodecylsulfate
PQ	Plastoquinone	TMPD	N,N,N′N′-Tetramethyl-p-phenylenediamine
PSI, PSII	Photosystem I and II		
Q	Quencher; primary electron acceptor of photosystem II	TNBS	Trinitrobenzene sulfonate
		UTP	Uridine triphosphate
RCBchl	Reaction center bacterio-chlorophyll	X_e	Hypothetical high energy intermediate or state

Introduction

A. Trebst and M. Avron

Photosynthesis is (still) defined as the assimilation of CO_2 in the light to form carbohydrates and oxygen. However, as A. Pirson has pointed out already in his introduction to the volumes on photosynthesis of the first edition of the Encyclopedia, this definition may change, if the primary products of the light reactions (and photosynthesis proper) should prove to be used up to a large extent not only in carbohydrate biosynthesis, but also in other metabolic pathways. The shifting of emphasis in photosynthesis research to the characterization of these primary products of the light reactions, the mechanisms and kinetics of their formation and turnover, as anticipated in the first edition of the Encyclopedia, has indeed occurred. The theory of van Niel served as a guide for the very successful early photosynthetic research period. Even though it is no longer tenable that the water-splitting reaction is common to all photosynthetic organisms, plants and bacteria, his principle concept of the homology of the primary light reactions in bacterial and plant photosynthesis and of carbon metabolism in photo- and chemosynthesis has in fact been strengthened.

It has now been established that in plant photosynthesis light-driven electron transport, coupled to phosphorylation, produces oxygen, NADPH and ATP. The latter two are in turn consumed in, and drive the CO_2-reduction-cycle. The generation of reducing equivalents and of ATP in the light might be regarded as photosynthesis proper, though the electron transport system as well as the energy-conserving system share many properties with other energy-conserving and transducing mechanism.

The two phases in photosynthesis seem to be separated also in space in the chloroplast. The primary photochemical acts, electron flow and photophosphorylation, are bound to the inner membrane system (thylakoid membranes of chloroplasts), whereas the carbon-fixing enzymes are localized in the matrix of the chloroplast. Of course, as we are beginning to understand, in addition to NADPH and ATP several regulatory processes, link electron flow and carbon fixation, such as the light-driven magnesium extrusion from the inner thylakoid space into the matrix space, the alkalinization of the matrix compartment due to light-driven proton uptake into the inner thylakoid space and the light-dependent activation and inactivation of some carbon metabolism enzymes (as will be discussed in the second photosynthesis volume of this series).

Several important features of the chemistry and physics of the primary light reactions in photosynthesis, of carbon assimilation and of the physiology of photosynthesis are well covered in chapters of the first edition of the Encyclopedia, written in 1959. By that time the path of carbon had been essentially discovered, principal methods such as fluorescence-, absorbance- and flash-light-spectroscopy had been developed, functionally active chloroplasts and thylakoid membrane fractions had been prepared and the biochemistry of the primary reactions, i.e. photosynthetic $NADP^+$-reduction and oxygen evolution and of cyclic and non-cyclic photophosphorylation, had been identified.

Thus the basis for modern photosynthesis had been laid by 1960, but many startling discoveries came shortly thereafter. A short list of some of the major advances in our understanding of electron flow and phosphorylation which appeared during the time between the first and this present edition could serve to illustrate the impact of great and unexpected discoveries in the last fifteen years: operation of two light reactions in sequence, the new electron carriers plastoquinone, plastocyanin and ferredoxin, the coupling factor, light-driven proton uptake, post-illumination ATP synthesis, dark phosphorylation by an acid base transition, the electric field across the membrane, vectorial electron flow, and step-wise four-charge accumulation in the oxygen evolution path. One might equally well point out some of the major problems which have not been resolved in the last fifteen years: the mechanism and intermediates of oxygen evolution still remain largely a mystery and the mechanism through which an energy-rich state, generated by the electron flow system, drives an ATP synthetase awaits the break-through of tomorrow.

A comparison of the various schemes in the first edition of 1960 with the master scheme of Figure 1 demonstrates the long path traversed from the information available in 1960 and illustrates the progress which has been made since. This master scheme also attempts to combine the essence of the various schemes presented in this text by the different authors, who emphasize other, sometimes conflicting, points.

The two light reactions in sequence with reaction centers P-700 and P-680 in photosystems I and II, undergoing redox changes, drive the photooxidation of water to oxygen and the reduction of $NADP^+$. A manganese-containing complex is oxidized by photosystem II, and after accumulating four charges reacts with water to yield oxygen. The primary electron acceptor of photosystem II (Q) reduces plastoquinone with the uptake of protons from the matrix space. Plastoquinone also serves as an electron buffer interconnecting several photosystem I and II units. Furthermore, by reduction of plastoquinone on the outside and oxidation of plastohydroquinone on the inside of the thylakoid membrane, protons are pumped across. Cytochrome b-559 might participate in the reduction of plastoquinone in non-cyclic, cytochrome b-563 in cyclic electron flow, cytochrome f in the oxidation of plastohydroquinone. Plastocyanin acts as the primary electron donor for photosystem I which, via a primary electron acceptor, possibly a non-heme-iron absorbing at 430 nm, reduces ferredoxin. $NADP^+$ reduction requires further a flavin-containing enzyme, the ferredoxin-$NADP^+$ reductase. Artificial acceptors for photosystem I, such as viologens, do not require ferredoxin. Artificial electron acceptors like phenylenediimine or silicomolybdate may be reduced close to photosystem II, and so constitute a ferricyanide Hill reaction driven by photosystem II only. Artificial electron donors donate electrons either into photosystem I, like DCPIP, or into photosystem II, like diphenylcarbazide (DPC). The point of inhibition by several compounds is indicated, DSPD inhibiting at ferredoxin, KCN at plastocyanin, DBMIB at plastoquinone, DCMU following the primary quencher level, and NH_2OH serving both as an inhibitor and an electron donor at the donating side to photosystem II.

Photosynthetic electron flow from water to $NADP^+$ is coupled stoichiometrically to ATP formation (non-cyclic photophosphorylation). Hill reactions involving photosystem II only or photoreduction of $NADP^+$ or methylviologen at the expense

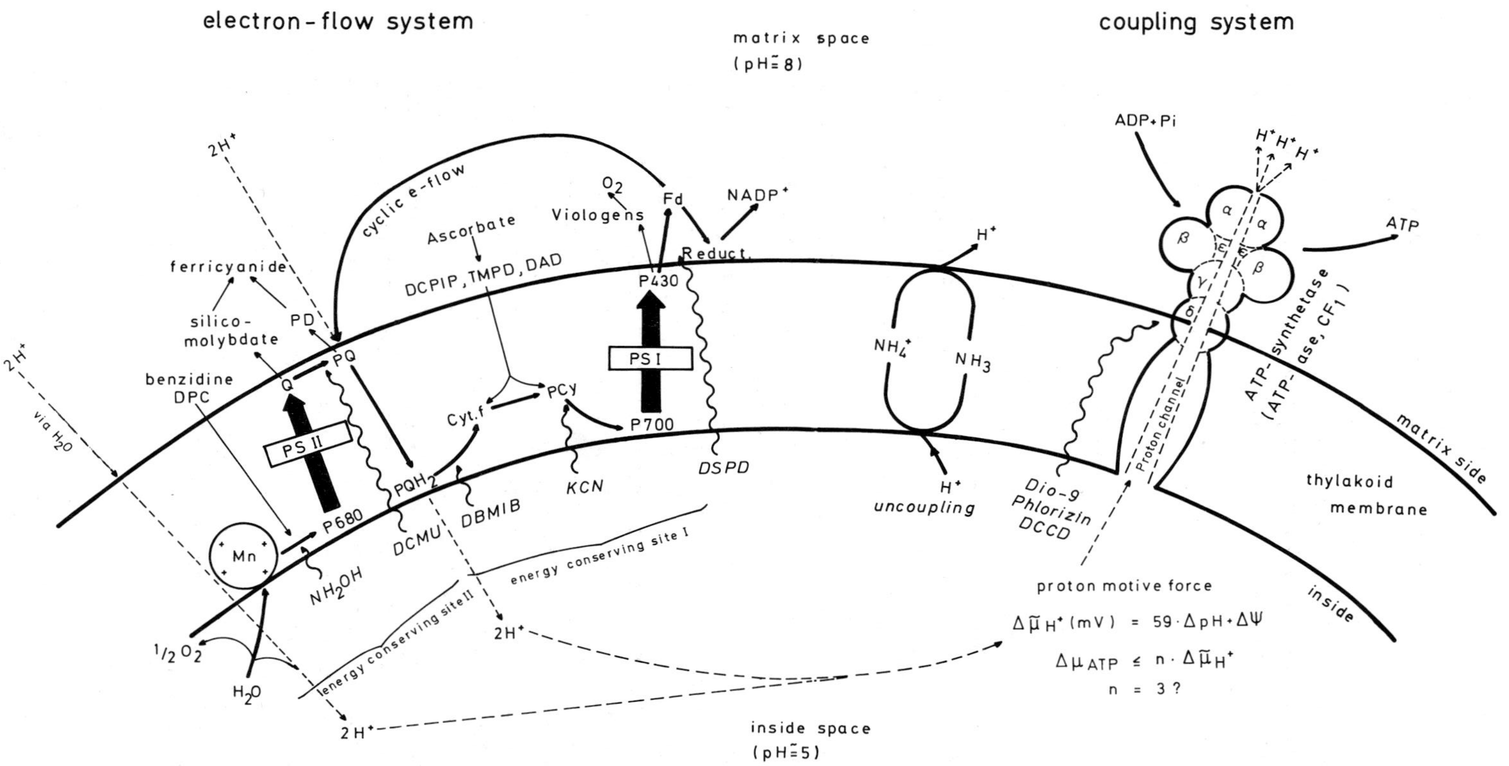

electron-flow system

coupling system

matrix space (pH̄ = 8)

matrix side

thylakoid membrane

inside

ATP

ATP-synthetase (ATP-ase, CF₁)

$H^+ H^+ H^+$

proton channel

ADP + Pi

Dio-9
Phlorizin
DCCD

H^+

NH_3

NH_4^+

H^+ uncoupling

proton motive force

$\Delta\tilde{\mu}_{H^+}(mV) = 59 \cdot \Delta pH + \Delta\Psi$

$\Delta\mu_{ATP} \leq n \cdot \Delta\tilde{\mu}_{H^+}$

n = 3 ?

inside space (pH̄ = 5)

NADP⁺

Fd

Reduct.

P430

DSPD

O_2

Viologens

PS I

P700

KCN

$2H^+$

Ascorbate

TMPD, DAD

DCPIP,

PCY

Cyt.f

DBMIB

energy conserving site I

cyclic e-flow

PQH₂

DCMU

2 H⁺

PD

PQ

Q

P680

energy conserving site II

ferricyanide

silico-molybdate

benzidine
DPC

NH_2OH

Mn

H_2O

$\frac{1}{2} O_2$

2 H⁺

2 H⁺

via H₂O

Fig. 1. Master scheme for photosynthetic electron flow coupled to ATP formation. According to a chemiosmotic mechanism of coupling the electron flow system is separated in the membrane from the ATP synthetase. Non-cyclic electron flow generates NADPH, $1/2 O_2$ and 4 protons inside the thylakoid membrane. It includes two energy-conserving sites defined as an electrogenic photosystem + a proton-translocating electron neutral redox reaction across the membrane. 3 protons (n=3 in the PMF) are assumed to drive the formation of 1 ATP in the coupling system consisting of a base piece and the coupling factor CF_1 with 5 different subunits. This yields a stoichiometry of 1.33 ATP per NADPH. In cyclic photophosphorylation in vivo ferredoxin reacts back with PQ. It includes one energy-conserving site. *P-680*, *P-700*: reaction centers of photosystem II and I; *Q*: quencher; *PQ*: plastoquinone; *PQH₂*: plastohydroquinone; *PCy*: plastocyanin; *Fd*: ferredoxin; *reduct*: ferredoxin-NADP⁺ oxidoreductase. Artificial donors and acceptors: *DPC*: diphenylcarbazid; *DCPIP*: dichlorophenolindophenol; *TMPD*: N-tetramethyl-p-phenylenedia-mine; *DAD*: diaminodurol; *PD*: p-phenylenediamine. e-flow inhibitors: *DCMU*: dichlorophenyldimethylurea; *DBMIB*: dibromomethyl-isopropyl-p-benzoquinone. Energy-transfer inhibitors: *DCCD*: dicyclohexylcarbodiimid, also closing the proton conducting channel

of an artificial electron donor via photosystem I only are coupled to ATP formation with half the stoichiometry of a Hill reaction involving both photosystems.

In cyclic electron flow ferredoxin is connected to plastoquinone with the cooperation of cytochrome b-563 leading to (cyclic) photophosphorylation coupled to photosystem I only.

An artificial lipophilic hydrogen-carrying redox carrier, reduced by photosystem I and reoxidized by cytochrome f, plastocyanin or P-700 may be used as a cofactor of artificially induced cyclic photophosphorylation requiring photosystem I only.

Photosynthetic electron flow is vectorial across the thylakoid membranes leading to an electrical membrane potential and a pH gradient across the membrane in the light. Two loops according to the chemiosmotic theory of coupling may be recognized, both comprised of an electrogenic photosystem and a proton-translocating redox reaction across the membrane. The two proton-translocating steps, releasing protons inside (water and plastohydroquinone oxidation) may constitute the two energy-conserving sites: the coupling of the electron transport system to the generation of an energy-rich state. The electrochemical gradient, or proton motive force, generated in the light by the electron transport system, consisting of a (small) membrane potential and a (large) pH gradient (inside pH below 5, outside 8), is dissipated by driving ATP formation via the coupling system of the ATP synthetase. The latter consists of a base piece (with a proton channel through the membrane) and the coupling factor CF_1 containing five different subunits, located on the matrix side of the membrane. Dio 9 and phlorizin are energy-transfer inhibitors, acting on the ATPase. Ammonia, other amines and ionophores such as gramicidin are uncouplers, releasing photosynthetic control by acting as proton carriers across the membrane.

The photosynthetic system provides unique opportunities for probing the energy-transducing system. Under normal circumstances isolated chloroplasts constitute a highly efficient ATP-synthesizing machine which is essentially irreversible, i.e. no ATP is hydrolyzed in the light or in the dark. However, after proper tampering, the system will hydrolyze ATP, catalyze ATP-Pi exchange, ATP-dependent proton accumulation, and ATP-dependent reverse electron flow. When the ATP synthetase is deprived of one of its substrates, the system charges itself to a "high-energy state" which can be observed via the large proton concentration gradients that are maintained across the thylakoid vesicles. Such a state was shown to yield in a post-illumination phase large amounts of ATP. A similar state can be generated in the dark by rapid transition of chloroplasts equilibrated at an acid pH to a basic pH. These and other types of ingenious methods have added immeasurably to our present understanding of energy interconversion in general, and in the photosynthetic system in particular.

The large amount of detailed information, summarized in the scheme of Figure 1 and even more in the following chapters should not confuse readers from other fields. One has to dissociate the emerging picture from the perhaps confusing specific tools and methodology of present day photosynthesis research, needed to obtain the data. Present thinking on the mechanism of electron flow, energy transduction and energy conservation, of membrane topology and of comparative biochemistry of the individual constituents of chloroplasts, mitochondria and bacterial chromatophores points to major similarities among these systems. Photosynthesis thus becomes part of a general mechanism and of a unifying concept in bioenergetics.

I. History

Photosynthesis 1950–75:
Changing Concepts and Perspectives

D.I. Arnon

A. Introduction

In the two centuries that have elapsed since the discovery of photosynthesis, no period has surpassed the last 25 years in richness of discovery and conceptual advances. It so happens that the progress made in the last quarter of a century can be assessed with considerable accuracy, for at midcentury, in July 1950, the Society for Experimental Biology held a Symposium on *Carbon Dioxide Fixation and Photosynthesis* at Sheffield, England, and the Proceedings of that Symposium (published in 1951) provide a reliable record of some of the main experimental and conceptual realities that characterized photosynthesis research up to the mid-twentieth century. This article will use the Sheffield Symposium as a point of departure for tracing progress in the next quarter of a century and will give special emphasis to the experimental and conceptual advances in the areas of photosynthetic electron transport and photophosphorylation in chloroplasts. No attempt will be made to cover all of the vast literature that has accumulated in these areas in the last 25 years. Such an attempt would be neither practical nor desirable; individual topics are covered in separate chapters of this volume. The emphasis here will be on developments that led to changes in concepts and perspectives.

B. Photosynthesis Research at Midcentury

I. Investigations with Whole Cells

The Sheffield Symposium included reports of research with whole cells and with isolated chloroplasts. With whole cells, the two most active areas of research at midcentury were the maximum quantum efficiency and the identity of the early products of photosynthesis as observed in experiments with $^{14}CO_2$.

Few questions in the history of photosynthesis have received more intensive theoretical and experimental study and led to more controversy than the efficiency with which photosynthetic cells convert absorbed light energy into chemical energy. At midcentury, quantum efficiency measurements were thought to be of crucial importance to the elucidation of the mechanism of photosynthesis (NISHIMURA et al., 1951). The Sheffield Symposium reflected the conflicting views of WARBURG and his former student, EMERSON. WARBURG et al. (1951) reported on manometric techniques with which they obtained a requirement of between 3 and 5 quanta per molecule of carbon dioxide consumed or oxygen produced by *Chlorella* cells. NISHIMURA et al. (1951) stressed that they and other investigators obtained values of 8 to 10 quanta per molecule of CO_2 consumed or O_2 evolved and attributed the higher efficiency reported from Warburg's laboratory to errors in manometric techniques.

In contrast to the disagreement over the quantum efficiency of photosynthesis, the Sheffield Symposium marked the end of another controversy that had arisen earlier from the use of $^{14}CO_2$ as a tracer for the detection of the first products of carbon assimilation. It was at Sheffield that GAFFRON et al. (1951) withdrew their previous objections and confirmed that phosphoglycerate was the first product of carbon dioxide assimilation — a finding enlarged upon then (and reported earlier) by CALVIN et al. (1951).

It is interesting to note that the pursuit of the problem of quantum efficiency of photosynthesis and the emergence of phosphoglycerate as the first product of CO_2 assimilation soon led to two diametrically opposed concepts of the nature of CO_2 assimilation by plants. But before dealing with these, we will consider the research with isolated chloroplasts that was presented at the Sheffield Symposium.

II. Investigations with Isolated Chloroplasts

There were two papers dealing with isolated chloroplasts, one by HILL (1951) and the other by FRENCH and MILNER (1951). FRENCH and MILNER reported on improvements in experimental methods for the investigation of the oxygen-evolving capacity of isolated chloroplasts in the presence of an artificial hydrogen acceptor, as discovered earlier by HILL (1937, 1939). The paper of HILL (1951) was, in retrospect, one of the high points of the Sheffield Symposium, for it provided a recapitulation of the photosynthetic activity of isolated chloroplasts as it was known at midcentury by the investigator whose own work about a decade earlier had opened the modern period of research in this area.

HILL (1951) wrote: "If we break the green cell, it is possible to separate the fluid containing the chloroplast and chloroplast fragments from the tissue. This green juice can no longer assimilate carbon dioxide, but in the case of many plants the insoluble material, for a time at least, is still capable of giving oxygen in light ... This evolution of oxygen takes place in the presence of soluble substances contained in the plant juice or better by the addition of reagents which can act as hydrogen acceptors ... The cell preparations then though not showing photosynthesis, can still convert light into a form of chemical energy ... We may very well call this the *chloroplast reaction*." (original italics.)

The chloroplast reaction, now appropriately known as the Hill reaction, established that the photoproduction of oxygen by chloroplasts was basically independent of CO_2 assimilation. This evidence provided strong support for the concept, developed by VAN NIEL (1941) from studies of bacterial photosynthesis, that the source of oxygen in plant photosynthesis is water. The importance of the Hill reaction was that it provided compelling evidence for this conclusion, whereas (contrary to a widely held belief) other evidence on this point from experiments with $^{18}O_2$ was not decisive (BROWN and FRENKEL, 1953; METZNER, 1975).

Apart from providing direct experimental evidence that water was the ultimate source of the oxygen liberated in photosynthesis, the Hill reaction also provided a method for investigating the nature of the photochemically generated reducing power that accompanied oxygen evolution by chloroplasts. This reducing power could be measured as the oxidation-reduction potential of the reduced hydrogen acceptor, AH_2, formed during the Hill reaction [Eq. (1)].

$$A + H_2O \xrightarrow[\text{chloroplasts}]{hv} AH_2 + 1/2 O_2. \tag{1}$$

At midcentury, the reducing power known to be generated by isolated chloroplasts was insufficient for the reduction of CO_2. As summarized by HILL (1951), the hydrogen acceptors (ferric oxalate, 2,6-dichlorophenolindophenol, benzoquinone, ferricyanide) effective in oxygen evolution were "substances of a type with oxidation-reduction potentials well below the oxygen electrode and could not possibly yield molecular oxygen unless energy in some form is supplied". Hill postulated "that oxygen must be consumed again in order to give the necessary additional energy ... This means first the reduction of a substance by chloroplasts giving oxygen and then a reoxidation of the reduced substance by part of the oxygen together with a simultaneous reduction of the necessary equivalent of carbon dioxide." Thus, Hill envisaged a dismutation reaction as being necessary to raise the initially inadequate reducing power generated by chloroplasts to a potential sufficiently negative to cope with the reduction of CO_2. This conclusion was in accord with the shortly thereafter published findings of WESSELS and HAVINGA (1952) that, of the large number of compounds tested, only those with a standard oxidation-reduction potential greater than about $+40$ mV could function as hydrogen acceptors in the Hill reaction.

The inability of isolated chloroplasts to fix CO_2 was confirmed by experiments with $^{14}CO_2$ (BROWN and FRANCK, 1948; BENSON and CALVIN, 1950). It seemed reasonable, therefore, to conclude that the chloroplast was an "incomplete machine" which "can evolve O_2 from H_2O in light provided an external substance 'B' is present ... [and that] this requirement for 'B' can be attributed to a deficiency in the enzymatic mechanism operative in various steps by which CO_2 is converted into the normal assimilation product" (VAN NIEL, 1956). A few years earlier, when the oxygen evolution by chloroplasts was linked (via $NADP^+$) to CO_2 assimilation by a malic enzyme of cytoplasmic origin, ARNON (1951) similarly concluded that "isolated chloroplasts lack either the appropriate hydrogen carriers or enzymes, or both, required for the reduction of carbon dioxide by the hydrogen derived from the photolysis of water, and that in the intact cell these factors are found chiefly or wholly outside the chloroplasts." In sum, it seemed well-established that the chloroplast was indeed a "system much simpler than that required for photosynthesis" and was the site of only "the light absorbing and water splitting reactions of the overall photosynthetic process" (LUMRY et al., 1954).

The new, more restricted view of the role of chloroplasts in photosynthesis gained ready acceptance in the middle of this century and displaced the older concept that chloroplasts are the site of complete photosynthesis—a concept that had prevailed without challenge for almost a hundred years. The basis for the older concept was the conviction that oxygen evolution and CO_2 assimilation were inseparable events. For example, SACHS (1887), one of the most distinguished plant physiologists of his time, drew on his own research to conclude that "the most definite proof that ... the chlorophyll body [chloroplast] itself is the organ which decomposes carbon dioxide and consequently assimilates this organic substance, is afforded by the fact ... that the first recognizable products of assimilation [starch] appear not in any haphazard place in the green cell, but in the chlorophyll body itself." Similarly, another leading plant physiologist (PFEFFER, 1900) concluded that "the actual CO_2 assimilation takes place entirely in the chloroplastid" because, as ENGELMANN (1881) and others had demonstrated, "by means of the delicate bacterium method it may be shown that isolated chloroplastids occasionally continue to evolve oxygen in the light ... if placed in an iso-osmotic sugar solution" (PFEFFER, 1900). Clearly, from the new perspective afforded by the Hill reaction, oxygen evolution by chloroplasts could no longer be regarded as evidence of concurrent CO_2 assimilation. Thus, at midcentury, the concept of CO_2 assimilation by chloroplasts was abandoned.

C. Research Past Midcentury: Some Major Advances

This partial account of photosynthesis research at midcentury highlights four areas in which major advances were made within a few years after the Sheffield Symposium. Subsequent investigations with whole cells led to a cogent concept of CO_2 assimilation in photosynthesis, and subsequent investigations with isolated chloroplasts changed the concensus about their role in photosynthesis in three respects: (1) Chloroplasts were found to be the sites of CO_2 assimilation; (2) they were found capable of photochemically generating the strong reducing power required for CO_2 assimilation, and (3) they were found to be the sites of light-induced ATP formation. We will now review these developments in more detail.

D. CO_2 Assimilation: Experiments with Whole Cells

I. The Photolyte Concept

WARBURG continued to pursue his work on the quantum efficiency of photosynthesis with the conviction that research in this direction would lead to the finding of a unique photochemical mechanism for CO_2 assimilation. One of Warburg's primary aims was to establish for photosynthesis the validity of Einstein's law of photochemical equivalence, according to which a photochemical reaction involves primarily the absorption of one quantum by a molecule of "photolyte," i.e., the substance that becomes reactive after photon absorption. Warburg's earlier findings that the evolution of one molecule of oxygen required four collaborative quantum absorption acts appeared to be in conflict with Einstein's law of photochemical equivalence. In fact, it was on these grounds that HENRI (1926) suggested long ago that oxygen evolution by whole cells (i.e. complete photosynthesis) is unsuited for measurement of the quantum efficiency of the primary photochemical act in photosynthesis. WARBURG (1926) responded by postulating that CO_2 molecules remain adsorbed on the chloroplast surface ("Grenzfläche") until successive step-wise reduction by light-activated chlorophyll converts them to glucose and liberates oxygen. By invoking such a mechanism, Warburg continued to regard the oxygen evolution that accompanies CO_2 assimilation as a valid measurement of primary photochemical reactions.

Soon after the Sheffield Symposium, BURK and WARBURG (1951) reported a major advance in the elucidation of the photosynthetic energy conversion process, which they divided into two parts: (1) a one-quantum light reaction which liberates oxygen and converts a bound species of CO_2 into carbohydrate; and (2) a dark oxidative reaction which provides the rest of the total energy needed for CO_2 assimilation. Warburg viewed this one-quantum light reaction as the long-sought primary reaction that conformed to Einstein's law of photochemical equivalence and in his last major paper on photosynthesis (WARBURG et al., 1969) represented this one-quantum light reaction as:

$$1 \text{ Photolyte} + 1\,N \cdot h\nu = 1\,H_2CO + 1\,O_2 + 1 \text{ chlorophyll} \tag{2}$$

where photolyte is defined as "one molecule of carbonic acid combined with one molecule of chlorophyll" and where one quantum of light splits the photolyte, thereby liberating chlorophyll, oxygen, and "reduced carbonic acid," i.e. carbohydrate.

According to this concept, the light reaction proper [Eq. (2)] supplies only one-third of the total energy required for CO_2 assimilation. The remaining energy, equivalent to about two quanta (75 kcal), is needed to form the photolyte and is provided by a dark reaction [Eq. (3)], i.e. the oxidation of two-thirds of the carbohydrate product (H_2CO) by the oxygen formed in Eq. (2).

$$2/3\,H_2CO + 2/3\,O_2 = 2/3\,H_2CO_3 + 75 \text{ kcal}. \tag{3}$$

Thus, the net yield per one quantum is $1/3\,H_2CO$ and $1/3\,O_2$. Accordingly, three quanta would be required to liberate one molecule of O_2 [Eq. (4)].

$$1\,H_2CO_3 + 3\,N \cdot h\nu = 1\,H_2CO + 1\,O_2. \tag{4}$$

Eq. (4) signifies an astonishing energy conversion efficiency of about 90%. WARBURG et al. (1969) maintained that this extraordinary efficiency of photosynthesis occurs only under optimal conditions (which they described) when all of the chlorophyll is present as the photolyte, i.e. as the chlorophyll–carbonic acid complex. Lower efficiencies of photosynthesis, they reasoned, resulted from light absorption by a portion of chlorophyll that had not been converted to the photolyte and was therefore photochemically inactive.

Warburg's concept of the bioenergetics of photosynthesis has not gained wide acceptance. On the biochemical side, his concept was resisted because the photolyte was not isolable and could therefore not be characterized chemically. Biochemists could not accept a mechanism for a one-step transformation of a compound at the level of carbonic acid to carbohydrate without any evidence for intermediates or for hitherto unrecognized enzymes. Moreover, as discussed later, investigations of $^{14}CO_2$ assimilation by photosynthetic cells soon produced evidence in favor of a carbon assimilation cycle which relied on specific enzyme systems that were components of the photosynthetic apparatus of plants. Furthermore, work with isolated chloroplast preparations yielded evidence that one-quantum light reactions in photosynthesis are concerned not with CO_2 assimilation and terminal events in oxygen evolution but with intermediate light-induced electron transfer steps.

II. The Photosynthetic Carbon Cycle

An entirely different concept of CO_2 assimilation in photosynthesis has emerged from experiments in Calvin's laboratory, where the discovery of phosphoglycerate and other labeled compounds was followed by the discovery of phosphate esters of ribulose and sedoheptulose as early products of photosynthesis (BENSON, 1951; BENSON et al., 1952) and the concept of a photosynthetic carbon cycle (BASSHAM et al., 1954). The concept of a photosynthetic carbon cycle (now also known as the Calvin cycle, the Calvin–Benson cycle, or the reductive pentose phosphate cycle) was greatly strengthened by the work of WEISSBACH et al. (1956), JAKOBY et al. (1956), and RACKER (1957), who documented the presence in photosynthetic tissues of two new enzymes, phosphoribulokinase and ribulose diphosphate carboxylase (carboxydismutase).

The reductive pentose phosphate cycle consists of three phases (Fig. 1). The first, carboxylative phase involves a phosphorylation of ribulose 5-phosphate to ribulose 1,5-diphosphate, which in turn combines with a molecule of CO_2 and is cleaved into two molecules of 3-phosphoglycerate. The second phase consists of a reversal of two reactions of glycolysis. In one, each of the two molecules of phosphoglycerate is phosphorylated to 1,3-diphosphoglycerate by phosphoglycerate kinase and in the other, 1,3-diphosphoglycerate is reduced to glyceraldehyde 3-phosphate by glyceraldehyde-3-phosphate dehydrogenase. In the third, regenerative phase of the cycle, ribulose-5-phosphate is regenerated from glyceraldehyde-3-phosphate by enzymes of the pentose phosphate cycle and the remaining glyceraldehyde-3-phosphate is converted to starch. Six turns of the cycle are needed to form one molecule of glucose that is polymerized to starch.

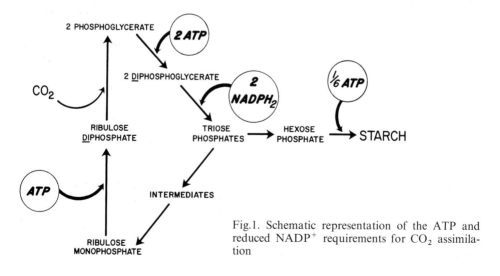

Fig.1. Schematic representation of the ATP and reduced $NADP^+$ requirements for CO_2 assimilation

The reductive pentose phosphate cycle is now accepted as the major pathway for carbon assimilation in green plants. In some groups of plants, known as C_4 plants, the reductive pentose phosphate cycle is supplemented by special enzymatic mechanisms which provide for additional fixation of CO_2 into malate and aspartate, which by decarboxylation reactions supply more CO_2 to the reductive pentose phosphate cycle (see review by BLACK, 1973). The reductive pentose phosphate cycle is also found in photosynthetic bacteria, but in certain types of these organisms the major pathway for CO_2 assimilation is the reductive carboxylic acid cycle that is particularly well suited to the synthesis of amino acids (EVANS et al., 1966; SIREVÅG and ORMEROD, 1970; SIREVÅG, 1974; cf. also BEUSCHER and GOTTSCHALK, 1972; TABITA et al., 1974).

III. Energy Requirements for Photosynthetic CO_2 Assimilation

All of the component steps of the carbon cycles are enzymatic reactions that in a direct sense are independent of light. Indeed, the entire reductive pentose phosphate cycle has also been identified in CO_2 assimilation by autotrophic, non-photosynthetic bacteria [see, for example, TRUDINGER (1956) and AUBERT et al. (1957)]. As pointed out by STANIER (1961), insofar as the light reactions of photosynthesis are concerned, the main contribution of the reductive pentose phosphate cycle was the identification of the quantitative energy demands for photosynthetic CO_2 assimilation, by revealing which of the component enzymatic reactions require an input of energy-rich chemical intermediates that must be formed at the expense of light energy. As summarized in Figure 1, the cycle shows that the conversion of 1 mol of CO_2 to the level of hexose phosphate requires 3 mol of ATP and 2 mol of reduced $NADP^+$. The need for light energy in photosynthesis could thus be traced to those photochemical reactions that generate ATP and NADPH, two energy-rich products that jointly constitute the assimilatory power required for CO_2 assimilation.

This identification of the energy requirements for CO_2 assimilation vindicated the brilliant hypothesis of RUBEN (1943) — for which at the time there was no experimental evidence — that photosynthetic CO_2 assimilation is a dark process, dependent on photochemically generated ATP and reduced pyridine nucleotides. As shown below, before its final vindication, Ruben's hypothesis did not fare well in the succeeding years: it was deemed incorrect on both theoretical and experimental grounds.

E. Evidence for CO_2 Assimilation by Isolated Chloroplasts

In the early 1950s, when it appeared that isolated chloroplasts did not assimilate CO_2, photosynthesis came to be regarded, like fermentation in the days of Pasteur, as a process that could not be separated from the structural and functional complexity of whole cells. Thus, RABINOWITCH (1953) concluded that "the task of separating it [photosynthesis] from other life processes in the cell and analyzing it into its essential chemical reactions has proved to be more difficult than was anticipated. The photosynthetic process like certain other groups of reactions in living cells seems to be bound to the structure of the cell. It cannot be repeated outside that structure."

This conclusion, however, soon proved untenable. In 1954, a reinvestigation of photosynthesis in isolated chloroplasts by different methods yielded evidence for a light-dependent assimilation of CO_2 (ARNON et al., 1954b). Isolated spinach chloroplasts, unaided by other enzyme systems, were found on exposure to light to fix CO_2 at an almost constant rate for at least an hour. There was approximate correspondence between the oxygen evolved and the CO_2 fixed, as would be expected from the photosynthetic quotient in green plants, $O_2/CO_2 = 1$ (ALLEN et al., 1955; ARNON, 1955).

Both soluble and insoluble products resulted from the fixation of $^{14}CO_2$ by chloroplasts (ALLEN et al., 1955). Among the soluble products the following were identified: phosphate esters of fructose, glucose, ribulose, sedoheptulose, dihydroxy-actone, and glyceric acid; glycolic, malic, and aspartic acids; alanine, glycine, and free dihydroxyacetone, and glucose. The insoluble product of $^{14}CO_2$ fixation was identified as starch. The ability of whole isolated chloroplasts to carry out the synthesis of starch from $^{14}CO_2$ and water without the aid of external enzyme systems and organic substrates provided incontrovertible evidence for concluding that chloroplasts are indeed, as was asserted but not documented in earlier periods (Sect. B. II), the sites of both O_2 evolution and CO_2 assimilation (ARNON, 1955).

The ability of isolated chloroplasts to fix CO_2 was almost completely lost when they were treated with a dilute salt solution or with water (WHATLEY et al., 1956). This osmotic shock treatment disrupted the chloroplasts and gave "broken chloroplasts" or "grana" (the water-insoluble green membrane fraction that was the site of the Hill reaction) and a straw-colored chloroplast extract ("stroma") that contained the water-soluble components, including the CO_2-fixing enzymes that were subsequently identified and characterized (LOSADA et al., 1960; TREBST, et al., 1960). The "broken chloroplasts" without the stroma fraction lacked the requisite enzyme for CO_2 assimilation, but their capacity for CO_2 assimilation was restored by adding the chloroplast extract (WHATLEY et al., 1956). The ease with which the soluble CO_2-fixing enzymes and, as shown later, other soluble components were lost into the aqueous phase during the isolation of chloroplasts may account for many of the previous failures to observe CO_2 fixation by isolated chloroplasts.

I. Rates of CO_2 Assimilation by Isolated Chloroplasts

Because of the earlier negative results, special experimental safeguards were deemed necessary to establish that chloroplasts alone, uncontaminated by other organelles or enzyme systems, were capable of a total synthesis of carbohydrates from CO_2, with light as the only energy source. The chloroplasts were washed and, to eliminate a possible source of chemical energy and metabolites, their isolation was performed not in isotonic sugar solutions (HILL, 1939) but in isotonic sodium chloride (ARNON et al., 1954b; ALLEN et al., 1955). In comparison with the parent leaves, the washed "salt" chloroplasts gave low rates of CO_2 assimilation—a situation reminiscent of the first reconstruction of other complex cellular processes in vitro, e.g. fermentation (BÜCHNER, 1897a, b; RUBNER, 1913), protein synthesis (ZAMECNIK and KELLER, 1954) and polymerization of DNA (KORNBERG, 1960). Crucial for the documentation of complete CO_2 assimilation by isolated chloroplasts were not high rates but consistency and reproducibility of results which yielded the same intermediate and final products as did photosynthesis by intact cells.

KAHN and VON WETTSTEIN (1961), on examining by light and electron microscopy spinach chloroplasts isolated in 0.35 M NaCl, found that these preparations included chloroplasts that retained all of their native structural characteristics and also chloroplasts that retained normal lamellar organization but lost their "limiting membranes" (envelopes) and stroma. More recently, when CO_2 assimilation by isolated chloroplasts ceased to be a matter of dispute and such special experimental safeguards as the use of NaCl and washing of chloroplasts were no longer needed, much higher rates of CO_2 assimilation, comparable to those in intact cells, were obtained with chloroplasts isolated by modified procedures that protected the integrity of the outer chloroplast envelopes (WALKER, 1964, 1965; JENSEN and BASSHAM, 1966; KALBERER et al., 1967).

F. Investigations of Light Reactions of Photosynthesis: Experimental Advantages of Chloroplasts Over Whole Cells

A changed research perspective emerged from the finding that isolated chloroplasts photoassimilate CO_2. Once the complete photosynthetic capacity of isolated chloroplasts was experimentally established, it was possible to concentrate on them rather than on whole cells in the search for those photochemical reactions that generate the assimilatory power needed for the conversion of CO_2 into organic compounds. It was clear that the assimilatory power must include ATP and a reductant with a reducing power at least equal to that of NADPH, and it was also clear that both ATP and the reductant must be formed photochemically prior to CO_2 assimilation. But kinetic experiments with whole cells provided no direct evidence for this conclusion. For example, an investigation of the photoassimilation of phosphorus by *Scenedesmus* cells with the aid of carrier-free ^{32}P showed that the shortest exposure to light gave the lowest incorporation of ^{32}P into ATP; the highest incorporation of ^{32}P into ATP occurred not in the light but after a brief exposure of the cells to $KH_2{}^{32}PO_4$ in the dark (GOODMAN et al., 1953). The first compound to be labeled in the light proved to be not the expected ATP but again 3-phosphoglycerate (GOODMAN et al., 1953).

Other investigations of direct photoassimilation of phosphorus by intact cells gave results that were at best suggestive (see review, ARNON, 1956). In short, experiments with whole cells proved, for reasons discussed below, incapable of yielding evidence for photoreduction of pyridine nucleotides and for a special, light-induced phosphorylation. The occurrence of these light reactions in photosynthesis was discovered not in whole cells but in isolated chloroplasts.

Since neither ATP nor reduced $NADP^+$ was added to isolated chloroplasts that fixed CO_2, it was clear that these energy-rich compounds were being photochemically generated from their respective precursors within the chloroplasts. The advantages of isolated chloroplasts for investigations of these photochemical reactions are substantial. Chloroplasts cannot respire — they lack the terminal respiration enzyme, cytochrome oxidase (James and Das, 1957; Lundegårdh, 1961). This feature insures that the photochemically generated ATP will not be confused with the ATP formed by respiration, a possibility that cannot be excluded with certainty in intact cells. Intact cells contain only catalytic amounts of the precursors AMP, ADP, and $NADP^+$ and these, because of permeability barriers, could not be increased by external additions. By contrast, in experiments with isolated and fragmented chloroplasts, it was possible to supply these normally catalytic substances in subtrate amounts and to determine chemically their light-induced conversion to ATP and reduction to NADPH, respectively. These features of isolated chloroplasts proved to be of decisive experimental advantage in the discovery of light-induced formation of ATP and reducing power required for CO_2 assimilation.

G. Discovery of Photosynthetic Phosphorylation

As already mentioned, the first comprehensive hypothesis for a role of ATP in photosynthesis was formulated by Ruben (1943). Shortly thereafter, Emerson et al. (1944) proposed on the basis of experiments with intact *Chlorella* cells that the "sole function of light energy in photosynthesis is the formation of 'energy-rich' phosphate bonds." This proposal was strongly opposed by Rabinowitch (1945), not only because he found the supporting evidence inadequate, but also because he considered the idea theoretically unsound. "What good can be served," he argued, "by converting light quanta (even those of red light, which amount to about 43 kcal per Einstein) into 'phosphate quanta' of only 10 kcal per mol? This appears to be a start in the wrong direction — toward *dissipation* rather than toward accumulation of energy." (Original italics.)

These theoretical objections received experimental support when the first experiments with the sensitive ^{32}P technique to test the ability of isolated chloroplasts to form ATP (on illumination) gave negative results (Aronoff and Calvin, 1948). The most plausible model for ATP formation in photosynthesis became one that envisaged a collaboration between chloroplasts and mitochondria. Chloroplasts would, in that scheme, reduce NAD^+ photochemically and mitochondria would reoxidize it with oxygen and form ATP via oxidative phosphorylation (Vishniac and Ochoa, 1952). This model posed a serious physiological problem. Photosynthesis in saturating light can proceed at a rate almost 30 times greater than the rate of respiration. It was difficult to see, therefore, how the respiratory mechanisms of mitochondria could cope with the ATP requirement in photosynthesis.

In 1954, work with the same spinach chloroplast preparations that fixed CO_2 led to the discovery that they were also able to convert light energy into chemical energy and trap it in the pyrophosphate bonds of ATP (Arnon et al., 1954a, b). Several unique features distinguished this photosynthetic phosphorylation (photophosphorylation), as the process was named, from substrate-level phosphorylation in fermentation and oxidative phosphorylation in respiration: (1) ATP formation occured in the chlorophyll-containing lamellae and was independent of other enzyme systems or organelles; (2) no energy-rich substrate, other than absorbed photons, served as a source of energy; (3) no oxygen was produced or consumed; (4) ATP formation was not accompanied by a measurable electron transport involving any external electron donor or acceptor (ARNON, et al., 1954a, b; ARNON et al., 1954). The light-induced ATP formation could be expressed by Eq. (5).

$$n\,\text{ADP} + n\,\text{P}_i \xrightarrow{\ hv\ } n\,\text{ATP}. \tag{5}$$

When photophosphorylation in chloroplasts was followed by evidence of a similar phenomenon in cell-free preparations of such diverse types of photosynthetic organisms as photosynthetic bacteria (FRENKEL, 1954) and algae (THOMAS and HAANS, 1955; PETRACK and LIPMANN, 1961), it became evident that photophosphorylation is not peculiar to plants containing chloroplasts but is a major ATP-forming process in nature that supplies ATP for the biosynthetic reactions in all types of photosynthesis.

A question arose initially whether one fundamental property of photophosphorylation in chloroplasts, i.e. independence from chemical substrates, applied also to bacterial photophosphorylation. FRENKEL (1954) found that photophosphorylation in a cell-free preparation from *Rhodospirillum rubrum* became dependent on a substrate (α-ketoglutarate) when the chlorophyll-containing particles were washed. However, in later experiments, FRENKEL (1956) and other investigators (KAMEN and NEWTON, 1957; ANDERSON and FULLER, 1958; GELLER and LIPMANN, 1960) found that the role of α-ketoglutarate and other organic acids in the bacterial system was regulatory and not that of a substrate. When this basic point was clarified, the fundamental similarity of photophosphorylation in chloroplasts and in bacterial systems was no longer in doubt.

Photophosphorylation was discovered in isolated whole chloroplasts, to which no cofactors (other than ascorbate) were added to increase the rate of ATP formation. Without the addition of cofactors, the integrity of the chloroplast structure was essential for ATP formation; chloroplast fragments, which gave good rates of the Hill reaction, had only feeble photophosphorylation activity (ARNON et al, 1954b; ARNON et al., 1956). The rates of photophosphorylation in whole chloroplasts, even though much higher than in chloroplast fragments, were, like those of CO_2 assimilation (Sect. E.I.), still low enough to give rise to questions about the quantitative importance of photophosphorylation in photosynthesis (RABINOWITCH, 1957). With further improvements in experimental methods (which included the addition of such cofactors as menadione (Sect. J.) and the use of broken chloroplasts with lowered permeability barriers), rates of photophosphorylation increased 170 times (ALLEN et al., 1958) and, with phenazine methosulfate, even more (JAGENDORF and AVRON, 1958) over those originally described (ARNON et al., 1954b).

The improved rates of photophosphorylation were equal to, or greater than, the maximum known rates of carbon assimilation in intact leaves. It appeared, therefore, that isolated chloroplasts retain, without substantial loss, the enzymic apparatus for photophosphorylation—a conclusion in harmony with evidence that the phosphorylating system was tightly bound in the water-insoluble lamellar portion of the chloroplasts.

H. The Concept of a Light-Induced Electron Flow

Once the main features of photophosphorylation were firmly established, the next objective was to explain its mechanism, particularly its absolute dependence on illumination (Arnon et al., 1954b; Arnon et al., 1954). On the one hand, photo-phosphorylation was independent of such classical manifestations of photosynthesis as oxygen evolution and CO_2 assimilation; on the other hand, it seemed unlikely that light was involved in the formation of ATP itself, a reaction universally occurring in all cells independently of photosynthesis. Light energy, therefore, had to be used in photophosphorylation before ATP synthesis and in a manner unrelated to CO_2 assimilation or oxygen evolution. The most probable mechanism for such a role seemed to be a light-induced electron flow (Arnon, 1959, 1961).

It is often difficult for the student of photosynthesis today to realize that before the discovery of photophosphorylation the concept of a light-induced elec-tron transport had no substantial basis in photosynthesis research. The idea that photon energy is used in photosynthesis to transfer electrons rather than cumber-some atoms had a few proponents at various times, for example Katz (1949) and Levitt (1953, 1954) but, as the literature before the late 1950s shows, it did not become a viable concept in photosynthesis—it was merely one of several speculative ideas based on model systems. The situation changed with the discovery of light-induced ATP formation. ATP is formed in nonphotosynthetic cells at the expense of energy released by electron transport. The idea that ATP may also be formed in photosynthesis through a special light-induced electron flow mechanism in chloroplasts now had a high probability that could be experimentally tested.

After abandoning early attempts to link photophosphorylation with a hypotheti-cal photolysis of water and recombination of [H] and [OH] (Arnon, 1955), a hypothesis was put forward that a chlorophyll molecule, on absorbing a quantum of light, becomes excited and promotes an electron to an outer orbital with a higher energy level (Arnon, 1959, 1961). This high-energy electron is then transferred to an adjacent electron acceptor molecule, a catalyst (A) with a strongly electronegative oxidation-reduction potential. The transfer of an electron from excited chlorophyll to this first acceptor is the energy conversion step proper and terminates the photochemical phase of the process. By transforming a flow of photons into a flow of electrons, it constitutes a mechanism for generating a strongly electronegative reductant at the expense of the excitation energy of chlorophyll.

Once the strongly electronegative reductant is formed, no further input of energy is needed. Subsequent electron transfers within the chloroplast liberate energy, since they constitute an electron flow from the electronegative reductant to electron acceptors (thought to include chloroplast cytochromes) with more elec-tropositive oxidation-reduction potentials (Arnon, 1959, 1961). Several of the exer-gonic electron transfer steps, particularly those involving cytochromes, were thought to be coupled with phosphorylation. At the end of one cycle, an electron is returned to the electron-deficient chlorophyll molecule and the quantum absorption process is repeated. A mechanism of this kind would account for the observed lack of any net oxidation-reduction change and would not need any external electron donor or acceptor. Because of the envisaged cyclic pathway traversed by the emitted electron, the process was named *cyclic photophosphorylation* (ARNON 1959, 1961).

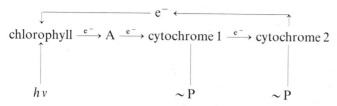

A cyclic electron flow that is driven by light and that liberates chemical energy, used for the synthesis of the pyrophosphate bonds of ATP, is unique to photosynthetic cells. The idea has been discussed elsewhere that cyclic photophosphorylation may be a primitive manifestation of photosynthetic activity—an activity that is the common denominator of plant and bacterial photosynthesis (ARNON, 1959, 1961).

I. Noncyclic Photophosphorylation

As noted above (Sect. B.II), the prevailing concept at midcentury was that the Hill reaction could not generate a strong reductant; only compounds with standard oxidation-reduction potentials more positive than 40 mV were known to be photoreduced. A new concept of the reducing power of chloroplasts emerged when three laboratories independently found that isolated chloroplasts could photoreduce pyridine nucleotides, despite their strongly electronegative oxidation-reduction potential ($E_m = -320$ mV, at pH 7). VISHNIAC and OCHOA (1951) and TOLMACH (1951) observed that chloroplasts could photoreduce either $NADP^+$ or NAD^+; ARNON (1951) observed only the photoreduction of $NADP^+$. The newly discovered capacity of isolated chloroplasts for pyridine nucleotide photoreduction was low at first, but the situation changed within a few years. SAN PIETRO and LANG (1956) found that long incubation (60 min) of high concentrations of NAD^+ (38 μmol) with high concentrations of grana (4.9 mg chlorophyll) (or low concentrations of grana supplemented with a chloroplast extract) yielded an accumulation of large amounts of NADH; $NADP^+$ was reduced to a lesser extent than NAD^+. ARNON et al. (1957) isolated from an extract of spinach chloroplasts a soluble protein, named the "TPN-reducing factor," which when added to low concentrations of grana ("broken chloroplasts") catalyzed the photoreduction of substrate amounts of $NADP^+$; NAD^+ was photoreduced at a much lower rate. This lack of agreement about which pyridine nucleotide was preferentially photoreduced by chloroplasts was resolved when SAN PIETRO and LANG (1958) isolated from leaves and chloroplasts a "soluble enzyme" ["photosynthetic pyridine nucleotide reductase" (PPNR)] which initially catalyzed the photoreduction of either $NADP^+$ or NAD^+ but which on further purification catalyzed only the reduction of $NADP^+$ (KEISTER et al., 1960). Thus, the preferential photoreduction of $NADP^+$ by chloroplasts ceased to be an issue. The mechanism of $NADP^+$ reduction is discussed in Section N, in which it is noted how both the TPN-reducing factor and the PPNR became recognized as synonyms for ferredoxin, an electron carrier protein that interacts with the enzyme ferredoxin-NADP$^+$ reductase.

At first there was no experimental evidence linking photophosphorylation to the photoreduction of NADP$^+$ by chloroplasts. In fact, these two photochemical activities of chloroplasts were once thought to be antagonistic (ARNON et al., 1956). It was, therefore, wholly unexpected when a second type of photophosphorylation was discovered in 1957, which provided direct experimental evidence for a coupling between photoreduction of NADP$^+$ and the synthesis of ATP (ARNON et al., 1958). Here, in contrast to cyclic photophosphorylation, ATP formation was stoichiometrically coupled with a light-driven transfer of electrons from water to NADP$^+$ (or to a nonphysiological electron acceptor such as ferricyanide) and a concomitant evolution of oxygen [Eq. (6)]. Moreover, the light-induced, thermodynamically "uphill" electron transport in chloroplasts was greatly increased by the addition of the phosphate acceptor ADP (in the presence of P$_i$), i.e. the rate of electron transport was at its maximum when it was coupled (as it would be under physiological conditions) to the synthesis of ATP (ARNON et al., 1958, 1959). Thus, photosynthetic electron transport in chloroplasts, like respiratory electron transport in mitochondria, was found to be under *phosphorylation control*. The phosphorylation control of photosynthetic electron transport was soon confirmed (AVRON et al., 1958; DAVENPORT, 1959) and extended by the discovery that ammonium ions effectively uncouple electron transport from photophosphorylation in chloroplasts (KROGMANN et al., 1959).

$$NADP^+ + H_2O + ADP + P_i \xrightarrow{hv} NADPH + 1/2\,O_2 + ATP + H^+. \qquad (6)$$

The electron flow concept postulated for cyclic photophosphorylation was also applicable to the new type of photophosphorylation (ARNON, 1959, 1961). It was envisaged that a chlorophyll molecule excited by a captured photon transfers an electron to NADP$^+$ (or to ferricyanide). It was postulated that electrons thus removed from chlorophyll are replaced by electrons from water (OH$^-$, at pH 7) with a resultant evolution of oxygen. In this manner, light would induce an electron flow from OH$^-$ to NADP$^+$ and a coupled phosphorylation. Because of the unidirectional or noncyclic nature of this electron flow, this process was named *noncyclic photophosphorylation* (ARNON, 1960, 1961).

J. Role of Cyclic Photophosphorylation: Early Views

Prior to the discovery of noncyclic photophosphorylation, cyclic photophosphorylation was regarded as the only source of ATP for CO$_2$ assimilation (ARNON et al., 1956). But this view could no longer be sustained when noncyclic photophosphorylation was found to account for all three products of the light phase of photosynthesis: ATP, NADPH, and O$_2$. Three possibilities now had to be considered with respect to cyclic photophosphorylation: (1) that it produced ATP for processes other than CO$_2$ assimilation, (2) that it contributed ATP for CO$_2$ assimilation to supplement the insufficient ATP generated by noncyclic photophosphorylation, and (3) that it was an artifact and that noncyclic photophosphorylation was the only physiological source of photosynthetic ATP.

ARNON et al. (1958) favored possibilities (1) and (2). They suggested a physiological role for cyclic photophosphorylation as a source of photochemically generated ATP when CO_2 assimilation is, for one reason or another, diminished or stopped altogether. In higher plants, this situation might arise during the well-known midday closure of stomata (STÅLFELT, 1955; HEATH and ORCHARD, 1957). Thus, cyclic photophosphorylation would be a source of ATP for diverse metabolic purposes that might or might not include CO_2 assimilation.

When the observed stoichiometry of noncyclic photophosphorylation (ATP/NADPH $= 1$) was compared with the ATP/NADPH ratio of 1.5 required for the conversion of CO_2 to carbohydrates (Sect. D.III), the need for extra ATP in CO_2 assimilation was apparent. Evidence that the extra ATP could be supplied by cyclic photophosphorylation came from experiments on CO_2 assimilation with reconstituted chloroplasts that were supplied with catalytic amounts of ADP and $NADP^+$ under three conditions: (1) when the photochemical phase was limited to noncyclic photophosphorylation, (2) when the photochemical phase was limited to cyclic photophosphorylation, and (3) when the photochemical phase included both cyclic and noncyclic photophosphorylation (TREBST et al., 1959).

Under condition (1) or (2), CO_2 assimilation was limited almost entirely to the formation of phosphoglycerate. Sugar phosphates were the predominant products of CO_2 assimilation only in condition (3). These results were interpreted as having been caused by a shortage of ATP in condition (1) and of NADPH in condition (2). Only in condition (3), when cyclic and noncyclic photophosphorylation operated concurrently, was a balance maintained between ATP and NADPH that made possible the formation of sugar phosphates, i.e. complete CO_2 assimilation (TREBST et al., 1959).

Direct as this evidence was, it did not prevent later doubts about the role of cyclic photophosphorylation in chloroplasts. In the above experiments on CO_2 assimilation, the balance between noncyclic and cyclic photophosphorylation needed to bring about the formation of sugar phosphates was achieved by adding to the reconstituted chloroplast system one of the then-known catalysts of cyclic photophosphorylation, FMN (WHATLEY et al., 1955), menadione (vitamin K_3) (ARNON et al., 1955), or a substance foreign to chloroplasts, phenazine methosulfate (JAGENDORF and AVRON, 1958). Each of these three substances was found earlier to be an effective catalyst that greatly increased the rate of cyclic photophosphorylation under anaerobic conditions. FMN and menadione (a component of vitamin K_1), being natural components of chloroplasts, were tentatively assumed to be physiological catalysts of cyclic photophosphorylation (ARNON, 1960), but this assumption was open to question in view of the even greater catalytic effectiveness of phenazine methosulfate.

To recapitulate, the capacity of isolated chloroplasts to carry on cyclic and noncyclic photophosphorylation become well established and there was experimental evidence that both types might be needed for CO_2 assimilation. However, the identity of the physiological catalyst of cyclic photophosphorylation was uncertain and remained so until the discovery of the role of ferredoxin (Sects. N and O.I).

K. Physical Separation of Light and Dark Phases of Photosynthesis in Chloroplasts

It is clear from the foregoing account that in the late 1950s there was sufficient evidence to identify the light phase of photosynthesis with the cyclic and noncyclic photophosphorylations that generate the assimilatory power (consisting of ATP and NADPH) to be used in the "dark" phase of photosynthesis for the conversion of CO_2 to carbohydrates and other products of carbon assimilation. The validity of this concept was directly substantiated through a physical separation of the light and dark phases of photosynthesis by fractionating isolated chloroplasts (TREBST et al., 1958). The light phase was completed first by the complete chloroplast system, in the absence of CO_2, and resulted in an evolution of oxygen accompanied by an accumulation of substrate amounts of NADPH and ATP in the reaction mixture. The green lamellar portion of the chloroplasts (grana) was then discarded and CO_2 was next supplied to the stroma portion in the dark.

In the presence of substrate amounts of NADPH and ATP, the chlorophyll-free extract was able to fix $^{14}CO_2$ in the dark. Only feeble $^{14}CO_2$ fixation occurred in the dark without the two components of assimilatory power. The dark fixation of CO_2 by the chlorophyll-free extract supplemented with assimilatory power was comparable with fixation in the light by the complete chloroplast system, to which $^{14}CO_2$ was supplied at the beginning of the illumination period. Here, prior accumulation of assimilatory power was not needed, because it was formed continuously in the light and used at once for CO_2 fixation. Very little CO_2 fixation occurred in the complete chloroplast system in the dark.

The products of CO_2 assimilation were found to be the same whether CO_2 assimilation occurred during continuous illumination or in the dark, at the expense of assimilatory power generated during a preceding light period. The products included hexose and pentose mono- and di-phosphates, phosphoglyceric acid, dihydroxyacetone phosphate, and small amounts of phosphoenolpyruvate and malate.

L. Ferredoxins in Chloroplasts and Bacteria

Further progress in elucidating the role of light in chloroplast reactions came from investigations which led to the recognition of the key role in photosynthesis of the iron-sulfur proteins, ferredoxins. The name ferredoxin was introduced by MORTENSON et al. (1962) to describe a nonheme-iron-containing protein which they isolated from *Clostridium pasteurianum*. In *C. pasteurianum* and in other nonphotosynthetic anaerobic bacteria in which it was later found, ferredoxin appeared to function as an electron carrier, either between molecular hydrogen (activated by hydrogenase) and various electron acceptors or in the breakdown of compounds that, like pyruvate, generate strong reducing power.

The isolation of ferredoxin from *C. pasteurianum*—an anaerobic bacterium devoid of chlorophyll and normally living in the soil at a depth to which sunlight does not penetrate—at first had no bearing on photosynthesis or any other photo-chemical process. A connection between ferredoxin and photosynthesis was estab-lished when TAGAWA and ARNON (1962) crystallized *C. pasteurianum* ferredoxin and found it to mediate the photoreduction of $NADP^+$ by spinach chloroplasts. In this reaction, *Clostridium* ferredoxin replaced a native chloroplast protein that had an unusual history, given in detail elsewhere (ARNON, 1969; BUCHANAN and ARNON, 1970) but one that can be discussed here only briefly.

Beginning in 1952, three proteins were successively and independently isolated from chloroplasts, each under a different name, to denote a different function. DAVENPORT et al. (1952) isolated the "methemoglobin-reducing factor"; ARNON et al. (1957) isolated from an aqueous extract of spinach chloroplasts a "TPN^+-reducing factor," required for the photoreduction of $NADP^+$; and SAN PIETRO and LANG (1958) isolated a "photosynthetic pyridine nucleotide reductase" which they characterized as the enzyme required for the photoreduction of $NADP^+$ and NAD^+ by chloroplasts.

By 1960, it became clear that "methemoglobin-reducing factor," "TPN-reduc-ing factor," and "photosynthetic pyridine nucleotide reductase" were different names for the same protein. Because the protein under its various names was isolated from, and effective in, chloroplasts, it was thought at first to be a unique chloroplast protein. However, in 1961 the association of the protein with chloro-plasts ceased to be unique when K. TAGAWA and M. NOZAKI (unpublished data from this laboratory) and LOSADA et al. (1961) isolated a protein with similar properties from the photosynthetic bacterium *Chromatium*. Although *Chromatium* cells do not have chloroplasts, do not photoreduce $NADP^+$, and do not evolve oxygen, the "pyridine nucleotide reductase" isolated from these bacterial cells was able to replace the native protein of spinach chloroplasts in mediating the photoreduction of $NADP^+$ and the concomitant evolution of oxygen. The full implications of this finding—namely, that the *Chromatium* protein, the chloroplast protein, and the *Clostridium* ferredoxin constituted a new group of electron carriers with sufficient chemical similarities to warrant the common designation of ferre-doxins—became clear only a year later, when the new nomenclature was proposed (TAGAWA and ARNON, 1962).

I. Chemical and Physical Properties of Ferredoxins

The similarities found initially between the clostridial and chloroplast (spinach) ferredoxins were that the two proteins contained iron, were free of heme and flavin groups, had low molecular weights, underwent reversible reduction that could be measured by a decrease in their respective absorption peaks, and—what was considered especially important—both proteins had a reducing power about equal to that of hydrogen gas, i.e. their oxidation-reduction potentials (at pH 7) were close to $-420\,\text{mV}$ (TAGAWA and ARNON, 1962, 1968).

Thus, ferredoxins were recognized as a group of electron carrier proteins that transfer to appropriate enzyme systems some of the most "reducing" electrons in cellular metabolism. In recent years the intensive study of the properties of ferredoxins and related proteins has added much to our knowledge of their chemical and physical properties. These will now be briefly recapitulated. They are discussed in detail elsewhere (ARNON, 1969; BUCHANAN and ARNON, 1970; MALKIN, 1973; ORME-JOHNSON, 1973, see chapter II,9).

Ferredoxins are now recognized as a subdivision of a broad category of iron-sulfur proteins present in all living cells. Photosynthetic cells contain ferredoxins whose properties include: An equal number of iron and acid-labile sulfur (S*) atoms, low molecular weight, a strongly negative oxidation-reduction potential, and (in the reduced state and at low temperatures) a characteristic electron paramagnetic resonance (EPR) spectrum with the principal signal at $g = 1.94$. Photosynthetic cells contain two types of soluble ferredoxins, distinguishable by different absorption spectra (that undergo reversible changes on oxidation-reduction): a plant or chloroplast type and a bacterial type (ARNON, 1965, 1969).

In the oxidized state, all ferredoxins are colored proteins, characterized by a bacterial or a chloroplast-type absorption spectrum. The bacterial type, found in ferredoxins from the different types of photosynthetic bacteria (*Rhodospirillum rubrum*, *Chromatium*, and *Chlorobium*) and the nonphotosynthetic anaerobes (e.g. *Clostridium pasteurianum*), has a single absorption peak in the visible region (at 380 nm to 390 nm) and a peak in the UV region at about 280 nm with a shoulder at 300 nm (Fig. 2). The chloroplast-type ferredoxin has the usual protein peak in the UV region around 280 nm and peaks at 463 nm, 420 nm, and 330 nm in the near UV region. On reduction, the absorbance peaks in the visible region disappear (Fig. 3); there is little change in the UV region. In the reduced state, the absorption spectra of all types of ferredoxins are similar to one another.

X-ray diffraction studies of the molecular structure of an 8 iron-8 sulfur ferredoxin from a nonphotosynthetic fermentative bacterium have shown that its iron and sulfur atoms exist in two tetrameric clusters, each with 4 iron, 4 inorganic sulfur, and 4 cysteine sulfur atoms, as shown in Figure 4 (ADMAN et al., 1973). [To denote this two-cluster structure, the designation $(Fe_4S_4^*)_2$ is used.] It seems likely that all 8 iron-8 sulfur bacterial ferredoxins have the 2-cluster $(Fe_4S_4^*)_2$ structure, whereas the 2 iron-2 sulfur chloroplast ferredoxin and the 4 iron-4 sulfur

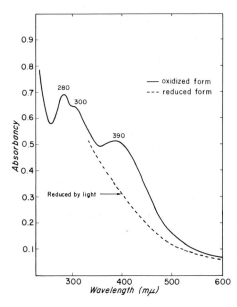

Fig. 2. Absorption spectra of *Chromatium* ferredoxin

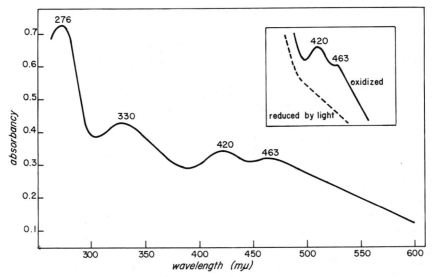

Fig. 3. Absorption spectra of spinach ferredoxin

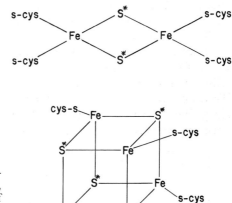

Fig. 4. *Top:* Proposed model of a 2Fe-2S* cluster of ferredoxin (after BLOMSTROM et al., 1964). *Bottom:* Model of a 4Fe-4S* cluster of bacterial ferredoxin derived from X-ray diffraction data (after ADMAN et al., 1973)

bacterial ferredoxins have the $Fe_2S_2^*$ or $Fe_4S_4^*$ 1-cluster structure. A proposed model of the 2Fe-2S* cluster, characteristic of the chloroplast-type ferredoxin, is shown in Figure 4.

The number of electrons transferred by plant and bacterial ferredoxins has been a point of considerable interest. It now appears that a ferredoxin containing a single iron-sulfur cluster transfers one electron and that ferredoxins containing two iron-sulfur clusters transfer one electron at a time from each of the two clusters.

The chloroplast-type ferredoxin is found in all oxygen-evolving cells, including blue-green algae that do not have chloroplasts (MITSUI and ARNON, 1971; ARNON, 1965). A crystalline chloroplast-type ferredoxin from the blue-green alga *Nostoc*

Fig. 5. Crystalline ferredoxin of *Nostoc muscorum* (MITSUI and ARNON, 1971)

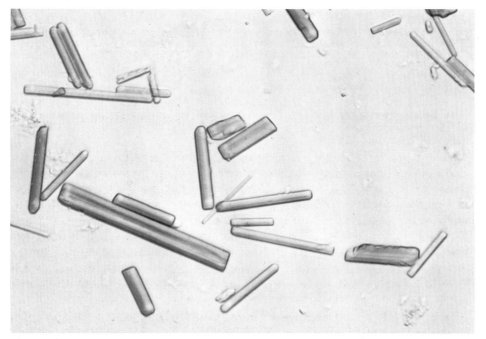

Fig. 6. Crystalline ferredoxin of *Chromatium* (BACHOFEN and ARNON, 1966)

Table 1. Some properties of soluble ferredoxins from photosynthetic organisms[a]

Property	Photosynthetic bacteria				Chloroplast
	Chlorobium	Chromatium	Rhodospirillum rubrum		Spinach
	Ferredoxin	Ferredoxin	Ferredoxin I	Ferredoxin II	Ferredoxin
Color	Brown	Brown	Brown	Brown	Red
Molecular weight	7,000	10,000	8,800	14,500	12,000
Oxidation-reduction potential (mV at pH 7)	—	−490	—	−430	−420
Iron (atoms/molecule)	8	8	8	4	2
Labile sulfide (atoms/molecule)	8	8	8	4	2
Absorption maximum, ultraviolet (nm)	280	280	280	280	280, 330
Absorption maximum, visible (nm)	385	385	385	385	463, 420
Principal EPR signals in reduced state (g value)	1.94	1.94	1.94	1.94	1.96

[a] Dashed spaces indicate that no determination has as yet been made

is shown in Figure 5 and a crystalline bacterial-type from *Chromatium* is shown in Figure 6.

Table 1 summarizes the iron and labile sulfide content, molecular weight, and certain other properties of plant and bacterial iron-sulfur proteins.

M. Role of Ferredoxin in Noncyclic Photophosphorylation

The elucidation of the nature of photosynthetic ferredoxins led to significant advances in the understanding of the photochemical events in photosynthesis. Insofar as the mechanism of $NADP^+$ reduction by chloroplasts is concerned, progress was greatly aided by the finding that hydrogen gas, in the presence of added hydrogenase, could substitute for light in the reduction of $NADP^+$ by chloroplasts (TAGAWA and ARNON, 1962). Using the hydrogen-hydrogenase system as the source of reducing power, $NADP^+$ was reduced by a reconstituted system that required only two chloroplast components, ferredoxin and a flavoprotein enzyme, which was also isolated in a crystalline form (SHIN et al., 1963). These experiments demonstrated that the role of chloroplast ferredoxin was that of an electron carrier protein (replaceable by ferredoxins from other sources) that could not react directly with $NADP^+$. Ferredoxin transferred electrons to the flavoprotein enzyme that served as the specific ferredoxin-$NADP^+$ reductase of chloroplasts (SHIN and AR-

NON, 1965). The mechanism of $NADP^+$ reduction was resolved into (1) a photo-chemical reduction of ferredoxin followed by two "dark" steps, (2) reoxidation of ferredoxin by ferredoxin-$NADP^+$ reductase, and (3) reoxidation of the reduced ferredoxin-$NADP^+$ reductase by $NADP^+$ (SHIN and ARNON, 1965).

Illuminated chloroplasts $\xrightarrow{e^-}$ ferredoxin

$\xrightarrow{e^-}$ ferredoxin-$NADP^+$ reductase $\xrightarrow{2e^-}$ $NADP^+$

A further advance in the understanding of the mechanism of $NADP^+$ reduction by chloroplasts was the finding that the ferredoxin-$NADP^+$ reductase forms stoi-chiometric 1/1 complexes with ferredoxin and $NADP^+$ (SHIN and SAN PIETRO, 1968; FOUST et al., 1969; NELSON and NEUMANN, 1969).

The affinity of ferredoxin-$NADP^+$ reductase for $NADP^+$ was much greater than for NAD^+. The $NADP^+$ concentration required to give half-maximal velocity of the reaction was found to be about 400 times smaller than that for NAD^+ (SHIN and ARNON, 1965). This great difference in affinities accounts for the apparent specificity of the purified enzyme toward $NADP^+$ and for the preferential reduction of $NADP^+$ by chloroplasts.

The evidence for ferredoxin as the terminal electron acceptor in the photochem-ical events that lead to $NADP^+$ reduction was subject to a rigid test. The evolution of oxygen by chloroplasts is uniquely dependent on light and it occurs only in the presence of a suitable electron acceptor. Thus, ferredoxin supplied in substrate amounts should support the production of oxygen by illuminated chloroplasts in the absence of any other electron acceptor.

Such evidence was indeed obtained (ARNON et al., 1964). Of special interest was the ratio between the ferredoxin reduced and the oxygen produced. Four molecules of ferredoxin were reduced for each molecule of oxygen produced. These findings were in accord with earlier evidence that the oxidation-reduction of ferre-doxin involved a transfer of one electron (WHATLEY et al., 1963). The coupling of oxygen evolution to photoreduction of ferredoxin provided strong support for the important role assigned to ferredoxin in the photochemical reactions of photo-synthesis. The evidence for that became even stronger when ferredoxin was also found to catalyze noncyclic photophosphorylation. ATP formation was linked to photoreduction of ferredoxin in the expected theoretical ratio of two molecules of ferredoxin reduced for each molecule of ATP formed. Here, as in oxygen evolution, the oxidized form of ferredoxin was the terminal electron acceptor. $NADP^+$ was not essential; its omission did not affect ATP formation (ARNON et al., 1964). The new equation for noncyclic photophosphorylation could now be written as Eq. (7):

$$4\ \text{Ferredoxin}_{\text{oxidized}} + 2\ \text{ADP} + 2\ \text{P}_i + 2\ \text{H}_2\text{O}$$
$$\xrightarrow{h\nu}\ 4\ \text{Ferredoxin}_{\text{reduced}} + 2\ \text{ATP} + \text{O}_2 + 4\ \text{H}^+. \tag{7}$$

Although Eq. (7) became the true equation for noncyclic photophosphorylation it is still experimentally more convenient and economical to use ferredoxin in only catalytic amounts by coupling reaction (7) with a reoxidation of reduced ferredoxin by $NADP^+$ (Eq. 8).

$$4 \text{ Ferredoxin}_{\text{reduced}} + 2\,\text{NADP}^+ + 2\,\text{H}^+$$

$$\xrightarrow{\text{Ferredoxin-NADP}^+\text{ reductase}} \quad 4 \text{ Ferredoxin}_{\text{oxidized}} + 2\,\text{NADPH}. \tag{8}$$

The sum of reactions (7) and (8) gives the earlier equation for noncyclic photophosphorylation (Eq. 6).

Under aerobic conditions and in the absence of NADP^+, the photoreduction of ferredoxin [Eq. (7)] may be coupled with its chemical reoxidation by molecular oxygen [Eq. (9)] (ARNON et al., 1964, 1967).

$$4 \text{ Ferredoxin}_{\text{reduced}} + O_2 + 4\,\text{H}^+ \longrightarrow 4 \text{ Ferredoxin}_{\text{oxidized}} + 2\,H_2O. \tag{9}$$

Reactions (7) and (9) were observed in the presence of catalase, which is normally found in the spinach chloroplasts used (ARNON et al., 1964, 1967). However, when catalase is removed, the photoreduction of substrate amounts of ferredoxin is accompanied by the accumulation of hydrogen peroxide (TELFER et al., 1970).

The sum of Eqs. (7) and (9) represents an oxygen-linked noncyclic photophosphorylation [Eq. (10)] that is catalyzed by ferredoxin.

$$\text{Sum (7)} + \text{(9)}: \quad 2\,\text{ADP} + 2\,P_i \xrightarrow[\text{ferredoxin, } O_2]{h\nu} 2\,\text{ATP}. \tag{10}$$

Reaction (10) is superficially similar to cyclic photophosphorylation in that it yields only ATP, but it differs from true cyclic photophosphorylation in being dependent on a continuous production and consumption of oxygen. It is, therefore, known as oxygen-dependent noncylic or *pseudocyclic photophosphorylation*, a name coined in an earlier period, prior to our knowledge of ferredoxin, when other catalysts were used in lieu of ferredoxin in reactions (8), (9), and (10) (ARNON et al., 1961; TREBST and ECK, 1961).

Reduced ferredoxin has a stronger affinity for ferredoxin-NADP^+ reductase and NADP^+ than for oxygen (TAGAWA et al., 1963 b). This property of ferredoxin ensures that NADP^+-linked noncyclic photophosphorylation will continue to generate ATP and NADP in an aerobic environment.

Ever since the discovery of oxygen production by green plants (PRIESTLEY, 1772), a continuing objective of research in photosynthesis has been the isolation of the first stable chemically defined molecular species that is formed by trapping and converting light energy into chemical energy. As summarized in Figure 7, investigations with whole cells successively identified products of CO_2 assimilation,

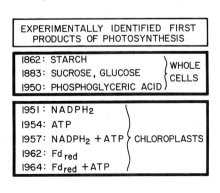

Fig. 7. Chronology of the first stable product(s) of photosynthesis

starch, soluble sugars, and phosphoglycerate, as the first products of photosynthesis. Investigations with isolated chloroplasts identified the first stable, energy-rich products of photosynthesis that are formed prior to, and are essential for, the conversion of CO_2 into organic compounds: NADP, ATP, and reduced ferredoxin.

N. Ferredoxin as the Physiological Catalyst of Cyclic Photophosphorylation

Work with spinach ferredoxin has revealed that its role is not limited to noncyclic photophosphorylation. It will be recalled (Sect. J) that under anaerobic conditions cyclic photophosphorylation in chloroplasts depended on catalysts whose physiological role was at best uncertain. Moreover, experiments with labeled oxygen revealed that the catalysts of cyclic photophosphorylation, FMN and menadione (but not phenazine methosulfate), also catalyzed an exchange between O_2 and H_2O (NAKAMOTO et al., 1960; KRALL et al., 1961). Such an exchange could give rise to a pseudocyclic type of photophosphorylation.

There were also differences between cyclic photophosphorylation in chloroplasts and bacterial chromatophores that cast doubt on the physiological role of menadione or FMN as catalysts of this process. Neither was required for cyclic photophosphorylation in freshly prepared bacterial chromatophores (NEWTON and KAMEN, 1957; ANDERSON and FULLER, 1958). Furthermore, with these exogenous catalysts, cyclic photophosphorylation in chloroplasts was not sensitive to such characteristic inhibitors of cyclic bacterial photophosphorylation and mitochondrial oxidative phosphorylation as antimycin A (BALTSCHEFFSKY, 1960).

A possible explanation for the need of exogenous catalysts in cyclic photophosphorylation by chloroplasts was that, during isolation, these organelles lost an endogenous soluble constituent that they require for this process. This possibility was verified when evidence was obtained for an anaerobic cyclic photophosphorylation in chloroplasts that was dependent on catalytic amounts of ferredoxin and on no other catalyst of photophosphorylation (TAGAWA et al., 1963b). When catalyzed by ferredoxin, cyclic photophosphorylation in chloroplasts for the first time became sensitive to inhibition by low concentrations of antimycin A and oligomycin and resembled in this respect cyclic photophosphorylation in bacteria and oxidative phosphorylation in mitochondria (TAGAWA et al., 1963b, c; ARNON, 1969; BÖHME et al., 1971).

Thus, ferredoxin was found to catalyze two distinctly different types of photophosphorylation, cyclic and noncyclic. They could be readily distinguished by their opposite responses to monochromatic light and to inhibitors. Noncyclic photophosphorylation exhibited a sharp "red drop" at the longer wavelengths of monochromatic light, coming to an almost complete halt at 714 nm. By contrast, cyclic photophosphorylation showed a marked "red rise," i.e. its rate increased greatly under far-red illumination (TAGAWA et al., 1963c; ARNON et al., 1967).

Differential sensitivity to antimycin A and to other inhibitors provided another sharp distinction between ferredoxin-catalyzed cyclic and noncyclic photophosphorylation. Low concentrations of antimycin A, oligomycin, and other inhibitors which sharply inhibited cyclic photophosphorylation had no effect on noncyclic photophosphorylation. By contrast, low concentrations of 3-(3,4-dichlorophenyl)-1,1-dimethyl urea (DCMU) and o-phenanthroline, which markedly inhibited noncyclic photophosphorylation, actually stimulated cyclic photophosphorylation (ARNON et al., 1967; ARNON, 1969).

Several other lines of evidence pointed to ferredoxin as the physiological catalyst of cyclic photophosphorylation in chloroplasts: (1) as expected of a true catalyst, ferredoxin stimulated cyclic photophosphorylation at low concentrations (100 μM), comparable on a molar basis to those of the other known catalysts of the process; (2) when light intensity was restricted, ferredoxin catalyzed ATP formation more effectively than any other catalyst; (3) in adequate light, cyclic photophosphorylation by ferredoxin produced ATP at a rate comparable with the maximum rates of photosynthesis in vivo (ARNON et al., 1967; ARNON, 1969).

The optimal concentration of ferredoxin for cyclic photophosphorylation, although still within a catalytic range (ca. 100 μM), was about an order of magnitude higher than that (ca. 10 μM) needed to catalyze $NADP^+$ reduction and its coupled noncyclic photophosphorylation. This difference in concentration was sometimes used as an argument against ferredoxin being the physiological catalyst of cyclic photophosphorylation (see, for example, AVRON and NEUMANN, 1968). However, in more recent work, discussed below, the optimum concentration for ferredoxin-catalyzed cyclic photophosphorylation was found to be about 10 μM, the same as that for $NADP^+$ reduction and the coupled noncyclic photophosphorylation.

To recapitulate, low concentrations of ferredoxin were found to catalyze two distinct types of photophosphorylation in chloroplasts, cyclic and noncyclic. Their contrasting responses to inhibitors pointed to differences in some intermediate electron transfer steps and their contrasting responses to far-red monochromatic light pointed to their dependence, at least in part, on different chloroplast pigment systems.

O. Stoichiometry and Regulation of Ferredoxin-Catalyzed Photophosphorylations

I. Cyclic Photophosphorylation in the Presence of Oxygen

Cyclic photophosphorylation in chloroplasts has usually been investigated under anaerobic conditions. Historically, this ATP formation in the absence of oxygen played an important part in overcoming the early resistance to the concept that light-induced ATP synthesis by chloroplasts was totally independent of oxidative phosphorylation by mitochondria that consume oxygen (ARNON, 1955, 1956). Although the distinction between photophosphorylation and oxidative phosphorylation is no longer in dispute, anaerobic conditions have continued to be widely used in studies of cyclic photophosphorylation in chloroplasts for another reason: anaerobicity helped to distinguish cyclic photophosphorylation from the pseudocyclic type. Reduced ferredoxin can be reoxidized by air and may catalyze a pseudocyclic photophosphorylation which, like the true cyclic type, yields only ATP but

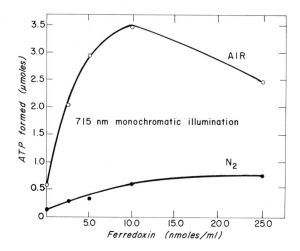

Fig. 8. Effect of ferredoxin on cyclic photophosphorylation under aerobic and anaerobic conditions (ARNON and CHAIN, 1975)

in reality is a variety of noncyclic photophosphorylation that depends on continuous production and consumption of oxygen (Sect. M).

Putting these historical and experimental reasons aside, the compelling consideration remains that, in vivo, oxygen is never excluded from the immediate environment of the thylakoids in which cyclic photophosphorylation occurs. Indeed, early evidence indicated that lower concentrations of ferredoxin were needed to catalyze cyclic photophosphorylation (at 708 nm) in the presence of air than were needed under anaerobic conditions (TAGAWA et al., 1963a, c). A ferredoxin-catalyzed cyclic photophosphorylation under aerobic conditions was also observed by BOTHE (1969) in a cell-free preparation of a blue-green alga. Recently, therefore, an investigation was carried out to characterize more fully cyclic photophosphorylation in the presence of air (ARNON and CHAIN, 1975).

Figure 8 shows that, at 715 nm, ferredoxin was a far more effective catalyst of cyclic photophosphorylation when the reaction mixture contained dissolved air rather than dissolved N_2. The optimum ferredoxin concentration in air was about 10 µM, i.e. an order of magnitude less than that used earlier under argon, and about the same as that needed to catalyze noncyclic photophosphorylation (ARNON et al., 1967). Much higher concentrations were less effective, probably because they facilitated a reoxidation of reduced ferredoxin by oxygen, rather than by the cyclic electron transport chain of chloroplasts. Similar results were obtained in parallel experiments with green (554-nm) monochromatic illumination (Fig. 9). Here again, low concentrations of ferredoxin were found to catalyze cyclic photophosphorylation more effectively under aerobic than under anaerobic conditions but—and this was crucial—only when the light-induced electron flow from water was severely inhibited by 3-(3,4-dichlorophenyl)-1,1-dimethylurea (DCMU). Thus, unlike pseudocyclic photophosphorylation, aerobic cyclic photophosphorylation at 554 nm was stimulated, rather than inhibited, by DCMU.

In short, ferredoxin-catalyzed cyclic photophosphorylation functioned best when light-induced electron flow from water (and hence noncyclic photophosphorylation and its pseudocyclic variant) was materially restricted either by the use of 715-nm illumination or by combining 554-nm illumination with the addition

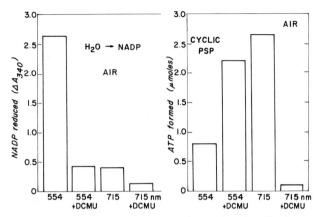

Fig. 9. Comparison of noncyclic electron transport (*left*) with ferredoxin-catalyzed cyclic photophosphorylation (*right*) (ARNON and CHAIN, 1975)

of DCMU. The extent to which these conditions restricted electron flow from water is reflected in the parallel inhibition of the photoreduction of NADP$^+$ by water (Fig. 9).

Another observation was that under the aerobic conditions of these experiments a light-induced electron flow from water, though severely restricted, could not be dispensed with entirely. When this remaining "trickle" electron flow from water was totally suppressed, as by adding DCMU to chloroplasts illuminated with 715-nm light, cyclic photophosphorylation stopped (Fig. 9). This observation is contrary to the generally held view that DCMU does not inhibit cyclic photophosphorylation (see review by AVRON and NEUMANN, 1968) but is in agreement with the findings of KAISER and URBACH (1973) that in far-red light endogenous cyclic photophosphorylation in isolated chloroplasts is inhibited by DCMU.

In the presence of oxygen, the trickle of electrons from water appeared to maintain the proper oxidation-reduction balance or "poising" that is required for cyclic photophosphorylation. The need for poising of ferredoxin-catalyzed cyclic photophosphorylation in chloroplasts had become apparent in earlier experiments under conditions different from those used here (TAGAWA et al., 1963a; GRANT and WHATLEY, 1967). The general importance of poising for cyclic photophosphorylation in both chloroplasts and bacterial chromatophores has been reviewed and stressed by AVRON and NEUMANN (1968).

A comparison of the effectiveness of ferredoxin, menadione, and phenazine methosulfate as catalysts of cyclic photophosphorylation in the presence of air is shown in Figure 10. With equal 715- or 554-nm illumination (in the presence of DCMU), and at equal concentrations, ferredoxin was a decisively more effective catalyst of cyclic photophosphorylation than either menadione or phenazine methosulfate. The great superiority of low concentrations of ferredoxin over other catalysts in catalyzing the conversion of light energy into phosphate bond energy is consistent with the role assigned to ferredoxin as the physiological catalyst of cyclic photophosphorylation.

The conclusion that the aerobic ferredoxin-catalyzed cyclic photophosphorylation is physiological in nature is also supported by its sensitivity to low concentra-

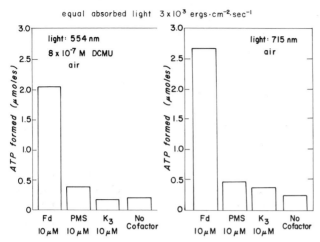

Fig. 10. Comparison of ferredoxin with other cofactors of cyclic photophosphorylation (Arnon and Chain, 1975)

tions of dibromothymoquinone (Arnon and Chain, 1975). Dibromothymoquinone is an antagonist of plastoquinone and inhibits chloroplast thylakoid reactions in which plastoquinone is involved (Trebst et al., 1970). Recently, Hauska et al. (1974) concluded from dibromothymoquinone inhibition experiments (under argon) that ferredoxin is the probable cofactor of cyclic photophosphorylation in vivo, in which plastoquinone serves as the natural energy-conserving site. A similar conclusion now seems applicable to ferredoxin-catalyzed cyclic photophosphorylation in the presence of air.

II. Concurrent Cyclic and Noncyclic Photophosphorylations and the ATP Requirements for CO_2 Assimilation

In the work on ferredoxin-catalyzed aerobic cyclic photophosphorylation discussed so far, the poising was maintained by decreasing the electron flow from water by means foreign to photosynthesis in nature: far-red monochromatic illumination or shorter wavelength illumination combined with an addition of DCMU. In seeking a physiological regulation mechanism, Arnon and Chain (1975) tested a hypothesis that poising of cyclic photophosphorylation in vivo is accomplished by its concurrent operation with noncyclic photophosphorylation. During CO_2 assimilation, electrons from water would be used (via ferredoxin and $NADP^+$) for the reduction of 1,3-diphosphoglycerate. Thus, an overreduction of the thylakoid milieu would be avoided and proper poising would be maintained, and cyclic photophosphorylation would operate concurrently with the noncyclic type. With cyclic and noncyclic photophosphorylation operating concurrently, the ATP/NADPH (or ATP/e_2) ratio would become greater than 1—a result that would help to resolve a current controversy over the stoichiometry of noncyclic photophosphorylation and the role of cyclic photophosphorylation in CO_2 assimilation. The basis of the controversy is as follows:

When noncyclic photophosphorylation was discovered, a P/e_2 ratio (or, with reference to the oxygen produced, a P/O ratio) equal to one was obtained with either $NADP^+$ or its nonphysiological substitute, ferricyanide, as the terminal electron acceptor (Sect. I). A P/e_2 ratio equal to one for noncyclic photophosphorylation was confirmed by a later study in this laboratory (DEL CAMPO et al., 1968) and by many other investigators (JAGENDORF, 1958; AVRON and JAGENDORF, 1959; STILLER and VENNESLAND, 1962; TURNER et al., 1962). However, higher P/e_2 (or P/O) ratios were reported by other investigators and led, on the basis of certain assumptions (which, owing to limitations of space, cannot be examined here), to proposals that noncyclic photophosphorylation produces 2 ATPs per pair of electrons transferred (WINGET et al., 1965; IZAWA and GOOD, 1968; SAHA and GOOD, 1970; HALL et al., 1971; REEVES and HALL, 1973; WEST and WISKICH, 1973). With these P/e_2 ratios greater than 1, only a negligible amount of the ATP was thought to come from a possibly concurrent cyclic or pseudocyclic photophosphorylation (IZAWA and GOOD, 1968; REEVES and HALL, 1973).

The P/e_2 ratio of noncyclic photophosphorylation is of considerable importance for the assessment of the role of cyclic photophosphorylation in photosynthetic CO_2 assimilation, which requires, as stated earlier, three molecules of ATP and two molecules of NADPH for each molecule of CO_2. A P/e_2 ratio of noncyclic photophosphorylation greater than one could account for all the ATP needs of CO_2 assimilation. Conversely, with a P/e_2 ratio of one, a contribution of ATP from cyclic or pseudocyclic photophosphorylation would be required for CO_2 assimilation.

The P/e_2 ratios of noncyclic photophosphorylation were recently investigated on the premise that in the presence of air the ratios might include a concurrent contribution of ATP not from pseudocyclic but from cyclic photophosphorylation. Noncyclic photophosphorylation with ferricyanide as the terminal electron acceptor (and in the absence of ferredoxin) again gave a P/e_2 ratio of 1; but with the physiological acceptor $NADP^+$ (and ferredoxin present) the P/e_2 ratio was about 1.5 (ARNON and CHAIN, 1975). Evidence that the extra ATP originated from cyclic rather than from pseudocyclic photophosphorylation came from inhibition studies with antimycin A. Antimycin A, within a concentration range that inhibited cyclic photophosphorylation and did not inhibit noncyclic (including pseudocyclic) photophosphorylation, suppressed the extra ATP formed by cyclic photophosphorylation, thereby restoring to noncyclic photophosphorylation its intrinsic P/e_2 ratio of one (Fig. 11).

It appears, therefore, that, in the presence of $NADP^+$, noncyclic electron flow poises the chloroplast system to permit the concurrent operation of ferredoxin-catalyzed cyclic photophosphorylation. The same concentration of ferredoxin serves both types of photophosphorylation. The concurrent operation of cyclic and noncyclic photophosphorylation would prevail during CO_2 assimilation and could provide the needed extra ATP ($P/e_2 > 1$).

ALLEN (1975) has recently proposed that the extra ATP needed for CO_2 assimilation comes from ferredoxin-catalyzed pseudocyclic photophosphorylation. This seems an unlikely possibility as much higher concentrations of ferredoxin are required for pseudocyclic photophosphorylation than for concurrent cyclic photophosphorylation under aerobic conditions (ARNON et al., 1967; ALLEN, 1975; ARNON and CHAIN, 1975).

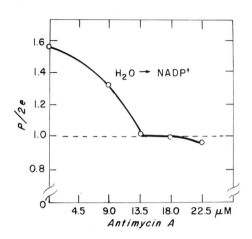

Fig. 11. Effect of antimycin A on concurrent cyclic and noncyclic photophosphorylation (ARNON and CHAIN, 1975)

III. Regulation of Cyclic Photophosphorylation

During photosynthetic CO_2 assimilation, cyclic and noncyclic photophosphorylation are envisaged as operating concurrently, but when CO_2 assimilation is curtailed or stopped altogether cyclic photophosphorylation may operate by itself as a general source of cellular ATP (cf. Sect. J). A curtailment or cessation of CO_2 assimilation would result in an accumulation of NADPH. Hence, the possibility was recently tested that NADPH might have a regulatory effect on the operation of ferredoxin-catalyzed cyclic photophosphorylation in isolated chloroplasts (ARNON and CHAIN, 1975).

Figure 12 shows that under aerobic conditions the addition of NADPH greatly stimulated ferredoxin-catalyzed cyclic photophosphorylation (sensitive to antimycin A) in 554-nm light. Unexpectedly, no DCMU was required for poising (cf. Fig. 9). It appeared that NADPH, like DCMU, diminished the electron flow from water. Such a novel role of NADPH was tested further by measuring its effect on the photoreduction of C-550 (KNAFF and ARNON, 1969b), a primary indicator

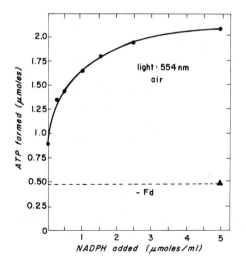

Fig. 12. Effect of NADPH on ferredoxin-catalyzed cyclic photophosphorylation (ARNON and CHAIN, 1975)

Fig. 13. Effect of preincubation with and without NADPH on subsequent photoreduction of C-550 and photooxidation of cytochrome b-599 at low temperature (ARNON and CHAIN, 1975)

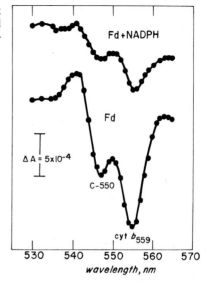

of photosystem II activity (BUTLER, 1973). Figure 13 (upper curve) shows that preincubation of chloroplasts with NADPH (in the presence of ferredoxin) in the dark diminished the subsequent photoreduction of C-550 (and photooxidation of cytochrome b-559). No such diminution was produced by NADPH without ferredoxin.

It appears, therefore, that when CO_2 assimilation is curtailed and NADPH accumulates, the back reaction of NADPH and ferredoxin with C-550 and possibly cytochrome b-559 provides a regulatory mechanism which, by diminishing electron pressure from water, maintains proper poising for the operation of cyclic photophosphorylation as a source of ATP for diverse cellular needs.

IV. Ferredoxin-Catalyzed Cyclic Photophosphorylation in vivo

The elucidation of the properties of ferredoxin-catalyzed cyclic photophosphorylation in isolated chloroplasts provided new tests for the operation of cyclic photophosphorylation in vivo and for its role as a source of ATP for CO_2 assimilation and other endergonic activities. Previously, the most widely used test for the operation of cyclic photophosphorylation was its resistance to inhibition by DCMU under white light or short-wavelength monochromatic illumination. The new tests, applicable to endogenous cyclic photophosphorylation in intact chloroplasts or in whole cells, involved the use of antimycin A (or oligomycin) and of far-red monochromatic illumination that cannot support noncyclic photophosphorylation. Another, recently added, special test for endogenous, ferredoxin-catalyzed cyclic photophosphorylation, valid only under far-red monochromatic illumination and particularly in the presence of air, was sensitivity to inhibition by DCMU (ARNON and CHAIN, 1975).

MIGINIAC-MASLOW (1971) found that antimycin A (at concentrations too low to inhibit $NADP^+$-linked noncyclic photophosphorylation) inhibited, under either N_2 or air, the endogenous photophosphorylation of intact spinach chloroplasts.

An endogenous photophosphorylation, inhibited by antimycin A, was observed in intact spinach chloroplasts, capable of high rates of CO_2 assimilation, by Schürmann et al. (1972). More recently, Kaiser and Urbach (1973) described in intact spinach chloroplasts an endogenous photophosphorylation (in air) at 721 nm that was inhibited by DCMU.

There are several lines of evidence that endogenous, ferredoxin-catalyzed cyclic photophosphorylation contributes ATP to CO_2 assimilation by intact chloroplasts. One line of evidence has to do with shortening of the lag or induction period in CO_2 assimilation by chloroplasts. A lag period that preceded the onset of linear CO_2 fixation was observed when light-dependent CO_2 assimilation by isolated chloroplasts was discovered (Arnon et al., 1954b; Allen et al., 1955). The lag period could be shortened either by preillumination or by the addition of certain phosphorylated intermediates of the reductive pentose phosphate cycle (Bamberger and Gibbs, 1965; Baldry et al., 1966; Bucke et al., 1966; Jensen and Bassham, 1966; Walker, 1973). In more recent experiments with isolated intact chloroplasts, the lag period in CO_2 assimilation was markedly shortened, either by the addition of ATP or by preillumination with monochromatic illumination (720 nm) that could support only cyclic photophosphorylation (Schürmann et al., 1972). That cyclic photophosphorylation contributed ATP, which shortened the induction phase of CO_2 assimilation and favored the formation of intermediates of the reductive pentose phosphate cycle, was concluded by Miginiac-Maslow and Champigny (1974) from work with isolated chloroplasts and by Klob et al. (1973) from work with *Chlorella* cells. (The latter authors limited the contribution of ATP from cyclic photophosphorylation to this phase of CO_2 assimilation.)

The inhibitors of ferredoxin-catalyzed cyclic photophosphorylation also had a marked effect on the pattern of products of CO_2 assimilation by intact chloroplasts: more phosphoglycerate and less triose phosphate (accompanied by higher total CO_2 fixation) was found in the presence of antimycin A (Champigny and Gibbs, 1969; Schacter et al., 1971; Miginiac-Maslow and Champigny, 1974). A similar shift in the pattern of products of CO_2 assimilation by intact spinach chloroplasts, as a result of the addition of either antimycin A or oligomycin, i.e., an increase in phosphoglycerate and a decrease in sugar phosphates, was observed by Schürmann et al. (1972). This shift in the pattern of products found in antimycin-inhibited chloroplasts was reversed by the addition of exogenous ATP. Sufficient ATP appeared to have penetrated the chloroplast envelope to increase the ratio of ATP to NADPH and thereby to increase the percentage of $^{14}CO_2$ that was fixed as sugar phosphates (Schürmann et al., 1972). (The antimycin-linked increase in total CO_2 fixation was correlated with high light intensity.)

These findings again support the conclusion that in intact chloroplasts the ATP requirements for the conversion of CO_2 to carbohydrates are not fully met by noncyclic photophosphorylation but must be supplemented by the ATP produced by cyclic photophosphorylation. It is noteworthy that the same conclusion was reached from earlier experiments with reconstituted ("broken") chloroplast systems, in which nonphysiological catalysts of cyclic photophosphorylation were used but which, nonetheless, yielded qualitatively similar evidence on the contribution of cyclic photophosphorylation to CO_2 assimilation (Sect. J).

Apart from photosynthetic CO_2 assimilation in which cyclic photophosphorylation acts as a source of additional ATP, endogenous cyclic photophosphorylation may, as discussed earlier (Sect. J), operate by itself and serve as a source of ATP for endergonic cellular processes which, like protein synthesis, require ATP but are independent of reducing power generated (along with ATP) by noncyclic photophosphorylation. For example, evidence has been obtained that intact chloroplasts incorporate amino acids and synthesize proteins at the expense of ATP produced by endogenous cyclic photophosphorylation (RAMIREZ et al., 1968; ELLIS, 1975).

Evidence has also been accumulating for a physiological role of cyclic photophosphorylation in intact cells. URBACH and SIMONIS (1964) and SIMONIS (1964, 1966) observed a light-dependent incorporation of ^{32}P by intact algal cells in far-red light (708 nm) or in the presence of inhibitors that inhibit noncyclic but not cyclic photophosphorylation. NULTSCH (1966, 1969) found that light-induced changes in rates of movement (photokinesis) of blue-green algae depend on cyclic photophosphorylation as the main source of energy. Of special interest is the growing evidence that nitrogen fixation in blue-green algae is utilizing the ATP formed in these cells by cyclic photophosphorylation (FAY, 1970; BOTHE and LOOS, 1972; LYNE and STEWART, 1973).

Other evidence for the operation of cyclic photophosphorylation in intact cells comes from investigations of the photoassimilation of organic compounds and light-dependent ion uptake. *Chlorella* cells photoassimilate glucose at 711 nm, a wavelength that can support cyclic, but not noncyclic, photophosphorylation or CO_2 assimilation. Moreover, the light-dependent glucose assimilation had a striking sensitivity to inhibition by antimycin A (TANNER et al., 1966; TANNER et al., 1967). Extensive evidence for the role of cyclic photophosphorylation as an energy source for active potassium influx in *Hydrodictyon africanum* cells has been presented by RAVEN (1971). Other work in this and related areas of photophosphorylation in vivo has recently been reviewed by SIMONIS and URBACH (1973).

P. Other Ferredoxin-Dependent Reactions in Photosynthetic Cells

As discussed so far, the key role of ferredoxin in plant photosynthesis rests on evidence that this iron-sulfur protein is essential for the light-induced generation of reducing power and ATP required for CO_2 assimilation. Only brief mention can be made here of other important functions that reduced ferredoxin performs in chloroplasts, blue-green algae, and photosynthetic bacteria; these are discussed more extensively elsewhere (BUCHANAN and ARNON, 1970).

Ferredoxin serves as the electron donor for the reduction of nitrite to ammonia and of sulfite to sulfide (RAMIREZ et al., 1966; JOY and HAGEMAN, 1966; NEYRA and HAGEMAN, 1974; SCHMIDT and TREBST, 1969). Reduced ferredoxin provides reducing power for nitrogen fixation by blue-green algae and photosynthetic bacteria (see review, BUCHANAN and ARNON, 1970). Another aspect of the role of photoreduced ferredoxin in plant photosynthesis concerns its regulation of CO_2 assimilation by activating two regulatory enzymes, fructose 1,6-diphosphatase and sedoheptulose 1,7-diphosphatase (BUCHANAN et al., 1967, 1971; SCHÜRMANN and BUCHANAN, 1975). Through its control of the levels of fructose-6-phosphate and fructose 1,6-diphosphate, the ferredoxin-1,6-diphosphatase system also provides

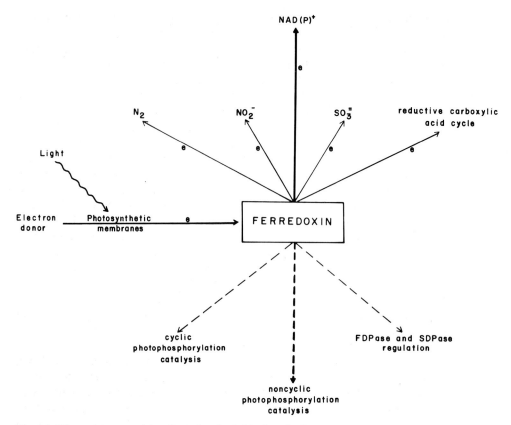

Fig. 14. Diagram summarizing the role of soluble ferredoxin

a mechanism for regulation of the activity of ribulose diphosphate carboxylase in chloroplasts (BUCHANAN and SCHÜRMANN, 1973a, b).

In photosynthetic bacteria, but not in chloroplasts, ferredoxin is directly involved in CO_2 assimilation. Its strong reducing power drives the energetically difficult carboxylations of acyl-CoA intermediates and results in the biosynthesis of α-ketocarboxylic acids, e.g. pyruvate from acetyl-CoA and α-ketoglutarate from succinyl-CoA. These carboxylations form the basis for a new pathway, the previously cited reductive carboxylic acid cycle, for CO_2 assimilation in bacterial photosynthesis (Sect. D.II.).

A diagrammatic summary of the known functions of ferredoxin in photosynthetic cells is shown in Figure 14.

Q. Multiple Ferredoxins: Soluble and Bound

Until recently, several characteristic properties readily distinguished ferredoxin from all of the other photosynthetic electron carriers. Whether isolated from chloroplasts or from whole cells, ferredoxin was a single, soluble, well-characterized protein that could be readily restored to the photosynthetic apparatus without detectable impairment of any ferredoxin-dependent reaction. All of the properties and activities of ferredoxin discussed so far were discovered in studies with such single, soluble ferredoxins. By contrast, the other chloroplast electron carriers are tightly bound to the thylakoids and their removal from the photosynthetic apparatus usually requires a physical or chemical disruption of the photosynthetic membrane system; restoration of an isolated membrane-bound carrier to its original membrane site of activity is often uncertain and in many cases impossible. Another—and from the standpoint of electron transport, unique—property of soluble ferredoxin was its strong reducing potential ($E_m = -420$ mV) that is 100 mV more electronegative than that of $NADP^+$ and is close to that of hydrogen gas (TAGAWA and ARNON, 1962). Thus, soluble ferredoxin (in its reduced state) emerged as the strongest, chemically characterized reductant that is photochemically generated by, and isolable from, chloroplasts. The existence of stronger reductants in chloroplasts has been suggested on the basis of observations that chloroplasts can photoreduce nonphysiological dyes, some of which have polarographically measured oxidation-reduction potentials that are more negative than that of ferredoxin (KOK et al., 1965; ZWEIG and AVRON, 1965; BLACK, 1966). However, without evidence that reductants more negative than soluble ferredoxin existed in chloroplasts, these suggestions remained speculative. The reported isolation of a stronger reductant from chloroplasts, e.g. a "ferredoxin-reducing substance" (YOCUM and SAN PIETRO, 1969, 1970) could not be confirmed (TSUJIMOTO et al., 1973).

Several recent findings have altered this perspective on ferredoxin. To begin with, evidence was found that some organisms contain more than one soluble ferredoxin. The first photosynthetic organism in which two ferredoxins were found was *Rhodospirillum rubrum* (SHANMUGAM et al., 1972). The two ferredoxins, one of which was formed only when the cells were grown photosynthetically, were thought at first to be present in a bound form, but more recent work has established that both ferredoxins are components of the soluble protein fraction of the isolated chromatophores (YOCH et al., 1975). The known coexistence in one organism of two soluble ferredoxins is still not universal: it has been documented in blue-green algae (HUTSON and ROGERS, 1975; HASE et al., 1975) and in the nonphotosynthetic nitrogen-fixing bacteria *Azotobacter vinelandii* (YOCH and ARNON, 1972) and *Bacillus polymyxa* (YOCH and VALENTINE, 1972; STOMBAUGH et al., 1973; YOCH, 1973).

In addition, little is known as yet about the significance and possible functional differences between the different soluble ferredoxin components of photosynthetic cells. In an initial study with *R. rubrum,* ferredoxin I (found only in photosynthetically grown cells) was found to be an 8Fe-8S* protein that was three times as effective in coupling the reducing power generated by illuminated chloroplasts to *R. rubrum* nitrogenase activity as was ferredoxin II, the 4Fe-4S* protein that was formed whether the cells were grown in the light or in the dark (YOCH and ARNON, 1975; YOCH et al., 1975).

After the removal of the two soluble ferredoxins, the washed chromatophores still contained a large amount of tightly bound iron-sulfur protein(s), which, upon reduction, exhibited the characteristic low-temperature electron paramagnetic resonance (EPR) signals (YOCH and ARNON, 1975). Similar EPR signals were observed in chromatophores from several species of photosynthetic bacteria (DUTTON and LEIGH, 1973; EVANS et al., 1974; HIGUTI et al., 1974; PRINCE et al., 1974). It appears, therefore, that chromatophores contain two kinds of iron-sulfur proteins, one consisting of soluble ferredoxins and the other consisting of iron-sulfur proteins that are tightly membrane-bound.

Limited as our present knowledge of multiple soluble ferredoxins is, the pattern of one or more soluble ferredoxin(s) accompanying iron-sulfur proteins that are tightly bound to the photosynthetic membranes is definitely a general feature of the photosynthetic apparatus of plants and photosynthetic bacteria. The pattern was first uncovered in isolated chloroplasts when studies were initiated to test whether soluble ferredoxin was a primary electron acceptor in photosynthesis, a view that stemmed from the conclusion that photoreduced ferredoxin was the strongest reductant generated by the action of light on chloroplasts (TAGAWA and ARNON, 1962).

Evidence for a primary electron acceptor role for soluble ferredoxin was first sought by spectrophotometric measurements of its photoreduction at cryogenic temperatures (77 K), at which electron transfer by ordinary chemical reactions is improbable. This approach, however, proved to be technically unworkable because of the large absorption by chlorophyll and cytochromes at the absorption peaks of ferredoxin (420 nm and 460 nm). MALKIN and BEARDEN (1971) then investigated the problem by low-temperature EPR spectroscopy, which permitted measurement of the EPR spectrum of reduced ferredoxin without interference by other components of chloroplasts.

By means of EPR spectroscopy, MALKIN and BEARDEN (1971) detected in spinach chloroplasts the photoreduction at cryogenic temperatures of a new bound species of ferredoxin. Despite similarities in EPR spectra, the photoreduction of the bound ferredoxin was independent of the soluble ferredoxin, since it also occurred in washed, broken chloroplast fragments from which soluble ferredoxin had been removed. The occurrence of bound ferredoxins in spinach chloroplasts has been confirmed and extended to other chloroplasts and to green and blue-green algae (EVANS et al., 1972; EVANS et al., 1973).

The bound ferredoxin of chloroplasts which exhibited upon photoreduction at low temperature EPR signals with g values of 1.86, 1.94, and 2.05 was assigned by BEARDEN and MALKIN (1975) the role of the primary electron acceptor of photosystem I (Sect. S). KE and BEINERT (1973) suggested that this bound ferredoxin may be identical with P-430, a chloroplast component that HIYAMA and KE (1971) had earlier identified spectrophotometrically as the primary electron acceptor of photosystem I.

Other work has indicated that chloroplasts contain not one but at least two iron-sulfur centers, which may reflect either the presence of two different proteins or of two iron-sulfur centers located on the same protein. The midpoint oxidation-reduction potentials of the two iron-sulfur centers were found to be strongly electronegative: about -530 mV and about -580 mV (KE et al., 1973; EVANS et al., 1974). An unresolved discrepancy exists between these values and that found by KE (1972) for the potential of P-430 (-470 mV).

Although the subject of the primary electron acceptor cannot be pursued here in detail, it must be noted that earlier agreement about bound ferredoxin as the primary electron acceptor of photosystem I has been challenged by MCINTOSH et al. (1975) and EVANS et al. (1975), who assign this function to other component(s) of chloroplasts. Regardless of the final resolution of the identity of the primary acceptor of this photosystem, there now seems to be wide agreement on two points: (1) the membrane-bound ferredoxin of chloroplasts generates, under the influence of light, a more negative potential than that of soluble ferredoxin; and (2) the bound ferredoxin (or its equivalent) serves as an electron donor for the soluble ferredoxin-NADP$^+$ system.

Evidence on the second point was sought by isolation of bound ferredoxin from spinach chloroplasts (MALKIN et al., 1974). Unexpectedly, the isolated, bound ferredoxin was more similar to a bacterial than to a plant (chloroplast) type ferredoxin (Sect. L). In its biological activity, the protein, like soluble plant and bacterial ferredoxin, was reduced by illuminated chloroplasts. However, it differed from these soluble ferredoxins in being unable to interact with ferredoxin-NADP$^+$ reductase. Thus, the photoreduced protein could not catalyze the reduction of NADP$^+$, a reaction for which ferredoxin-NADP$^+$ reductase is required, but it was able to reduce cytochrome c, a reaction for which ferredoxin-NADP$^+$ reductase is not required. The protein has as yet to be fully characterized; at present there is no evidence that it acts as a reductant for soluble ferredoxin. In short, an electron transfer from a reduced, bound iron-sulfur protein to soluble ferredoxin is yet to be demonstrated.

R. Photosynthetic Electron Carriers

The new knowledge of the importance and versatility of ferredoxin in mediating photosynthetic electron transport (Fig. 14) illustrates one of the major advances in photosynthesis research during the past quarter century: the discovery of electron carriers that are components of the photosynthetic electron transport system(s) that were unknown at midcentury. Apart from ferredoxin, the electron carriers include four other protein components, three of which are cytochromes: a c-type cytochrome, known as cytochrome f, with an α-absorption peak at 554 nm and a midpoint oxidation-reduction potential (E_m) at pH 7 of $+360$ mV (DAVENPORT and HILL, 1952); cytochrome b_6 with an α-absorption peak at 563 nm and an E_m of -60 mV (HILL and SCARISBRICK, 1951; HILL, 1954); and cytochrome b-559 with an α-absorption peak at 559 nm and several unusual properties, which include the coexistence, in freshly prepared chloroplasts, of a high-potential form ($E_m = +370$ mV) and a low-potential form ($E_m = +70$ mV) (LUNDEGÅRDH, 1962; BOARDMAN and ANDERSON, 1967; ARNON et al., 1968; BENDALL, 1968; WADA and ARNON, 1971; KNAFF and MALKIN, 1973). The other protein electron carrier in chloroplasts is plastocyanin, a copper-containing protein discovered by KATOH (1960) in *Chlorella* and by KATOH et al. (1961) in leaf tissue and chloroplasts. The E_m of plastocyanin is $+390$ mV.

The most prominent nonprotein electron carriers in chloroplasts are quinones. Our knowledge of the occurrence of quinones in the photosynthetic apparatus began with the demonstration that the vitamin K_1 of leaves is localized in chloroplasts (DAM et al., 1948). This finding was first documented by bioassay and was later confirmed by direct chemical methods (KEGEL and CRANE, 1962). Aside from the finding of the naphthoquinone, vitamin K_1, the early work of Koffler and that of Crane and co-workers has led to the recognition of seven benzoquinones as normal constituents of chloroplasts: plastoquinones A, B, C, and D and α-, β-, and γ-tocopherolquinone (see review, ARNON and CRANE, 1965).

The extensive research on chloroplast cytochromes, plastocyanin, and plastoqui-nones in photosynthetic electron transport and photophosphorylation has been reviewed by BENDALL and HILL (1968), HIND and OLSON (1968), BOARDMAN (1970), BISHOP (1971), and TREBST (1974).

S. Two Photosystems in Plant Photosynthesis: Origins of a Concept

One of the most popular concepts that emerged in the last quarter century is that of two photosystems in plant photosynthesis. The subject in its modern phase had its origin in measurements by EMERSON et al. (1957) of quantum efficiency of complete photosynthesis in *Chlorella* cells. In far-red light (at wavelengths longer than 680 nm) they observed a drop in quantum efficiency which was counteracted by the addition of supplementary monochromatic illumination of shorter wave-lengths. EMERSON and CHALMERS (1958) suggested that, at the long wavelengths of light that are absorbed only by chlorophyll a, the full efficiency of photosynthesis depends on a second light reaction, i.e. a simultaneous absorption of light of shorter wavelengths by "accessory" pigments: chlorophyll b (and carotenoids) in *Chlorella* and carotenoids and phycobilins in other algae. Cells illuminated simultaneously by long- and short-wavelength light beams exhibited an "enhance-ment" effect: a rate of photosynthesis higher than the sum of the rates obtained from each light beam separately.

Another hypothesis that greatly influenced the concept of two light reactions was that of HILL and BENDALL (1960), who attempted to explain the experimental evidence for noncyclic photophosphorylation in chloroplasts. They suggested that one light reaction oxidizes cytochrome f and reduces $NADP^+$ and a second light reaction reduces cytochrome b_6 and oxidizes water. Their proposal envisaged that the ATP formed in noncyclic photophosphorylation results from a dark, thermody-namically favorable, electron transfer from reduced cytochrome b_6 to oxidized cytochrome f. The hypothesis of HILL and BENDALL (1960) gained support from the experiments of LOSADA et al. (1961), who separated noncyclic photophosphory-lation by isolated chloroplasts into two light reactions: one light reaction photoox-idized water, yielded oxygen, and reduced dichlorophenol indophenol (DCIP), and a second light reaction gave a photoreduction of $NADP^+$ by reduced DCIP and a coupled photophosphorylation. A schematic representation of the two reac-tions joined together is shown in Figure 15. The photoreduction of $NADP^+$ by reduced DCIP, without an accompanying phosphorylation, was demonstrated ear-lier by VERNON and ZAUGG (1960).

DUYSENS et al. (1961) and DUYSENS and AMESZ (1962) arrived at a concept of two light reactions from measurements (at 420 nm) of the oxidation-reduction changes of cytochrome f in the red alga *Porphyridium cruentum*. The cytochrome was oxidized by long-wavelength light and reduced by an added beam of short-wavelength light. The reduction, but not the oxidation, of cytochrome f was abo-lished by inhibitors like DCMU.

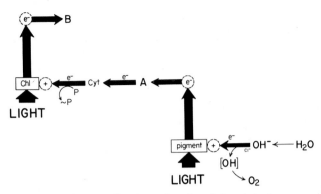

Fig. 15. Diagram depicting the separation of noncyclic photophosphorylation into two reactions: photooxidation of water coupled with a photoreduction of an intermediate electron carrier A and photoreduction of $NADP^+$ (B) by A^- coupled with phosphorylation. The role of A in these experiments was played by dichlorophenol indophenol (LOSADA et al., 1961)

On the basis of this and related evidence, DUYSENS and AMESZ (1962) proposed that photosynthesis in higher plants and algae consists of two light reactions operating in series: a long-wavelength, photosystem I, which oxidizes cytochrome f and photoreduces $NADP^+$, and a short-wavelength, photosystem II, which reduces cytochrome f and produces oxygen by dehydrogenation of water.

These, and other studies that followed, led to the wide acceptance of the concept that plant photosynthesis includes two distinct photochemical acts (photosystems I and II) that are joined by a series of dark electron transfer steps. Aside from the many later elaborations, the essential features of the concept (often referred to as the series scheme or the Z scheme) can still be represented by the simplified diagram in Figure 15 which shows: (1) a pigment system, identified with photosystem II, that generates a strong oxidant (capable of oxidizing water) and a weak reductant, A; (2) a pigment system, identified with photosystem I, which generates a weak oxidant and the strong reductant, B (NADPH); and (3) an electron transport chain that joins the reductant produced by photosystem II with the oxidant produced by photosystem I, thereby releasing energy that is used to form ATP.

The electron transport chain is currently depicted as including not only cytochromes but also plastoquinone and plastocyanin (Sect. R). The pigment assembly of each photosystem is now thought to include a small number of special reaction center chlorophyll molecules that undergo photooxidation and act as traps for excitation energy collected by the much greater number of "light-harvesting" chlorophyll and accessory pigment molecules (see for example, VERNON et al., 1971). The best known and most readily demonstrable is the reaction center chlorophyll of photosystem I, P-700, discovered by KOK (1956) (see also MARSHO and KOK, 1971). The reaction center chlorophyll of photosystem II has been designated chlorophyll a_{II}, or P-680 (DÖRING and WITT, 1972; FLOYD et al., 1971; BUTLER, 1972).

Insofar as monochromatic light is concerned, the important point that emerged is that both photosystems I and II can be driven by short-wavelength light (ca. $\lambda < 700$ nm), but only photosystem I can be driven effectively by far-red light (ca. $\lambda > 700$ nm).

T. Two Photosystems: Facts, Hypotheses, and Dogma

HILL and BENDALL (1960) characterized their proposal for two light reactions as a "hypothesis in barest outline" and DUYSENS and AMESZ (1962) characterized their scheme as a "working hypothesis." Yet, within a few years, the series scheme evolved into what RAVEN (1970) called "the central dogma of photosynthesis." Apart from the general repugnance of dogma to any branch of science, there are specific and cogent reasons, some of which have been put forward by MYERS (1974) and need not be repeated here, for again viewing the series scheme as a working hypothesis that is subject to change as new facts are uncovered.

As already discussed, work with isolated chloroplasts has led to the identification of cyclic and noncyclic photophosphorylation as the two photochemical processes that provide ATP and reduced ferredoxin—the only two products formed at the expense of light energy that are needed for CO_2 assimilation and related enzymatic events in the dark phase of photosynthesis. Thus, a unified view of the light events in photosynthesis should relate cyclic and noncyclic photophosphorylation to the concept of two photosystems.

The work with monochromatic light, which showed that ferrodoxin-catalyzed cyclic photophosphorylation proceeded most effectively in long-wavelength light that could not support oxygen evolution, permitted a ready identification of cyclic photophosphorylation with photosystem I (TAGAWA et al., 1963c). In fact, cyclic photophosphorylation was identified as the only physiological part of plant photosynthesis that functioned well in far-red light and exhibited a "red rise" (Sect. N), in marked contrast to the $NADP^+$-linked noncyclic photophosphorylation and oxygen evolution which showed a marked "red drop" (ARNON et al., 1967).

The dependence of $NADP^+$-linked noncyclic photophosphorylation on photosystem II is clearly beyond dispute but, as discussed below, there is some controversy whether noncyclic photophosphorylation depends on photosystem II alone or whether it depends on a collaboration between photosystems II and I. There is general agreement that two light reactions are involved. Measurements of the quantum requirements for the photoreduction of $NADP^+$ by water in short-wavelength light show a requirement of two quanta per electron, indicative of two light reactions, as opposed to a requirement of one quantum per electron (indicative of one light reaction) in the photoreduction of $NADP^+$ by reduced DCIP in far-red light where only photosystem I is involved (HOCH and MARTIN, 1963; SAUER and BIGGINS, 1965; SAUER and PARK, 1965; McSWAIN and ARNON, 1968; AVRON and BEN-HAYYIM, 1969).

There now are basically two hypotheses that relate photosystems I and II to cyclic and noncyclic photophosphorylation. They will be discussed briefly as the "two-light-reactions" and the "three-light-reactions" hypotheses.

I. The Two-Light-Reactions Hypothesis

The two-light-reactions hypothesis is currently by far the more popular hypothesis and is invoked most often in interpretation of many observations which cannot be discussed here in detail but which will be found in other chapters in this volume and in several reviews (HIND and OLSON, 1968; BOARDMAN, 1970; BISHOP, 1971; MYERS, 1971; TREBST, 1974). The basic premise of this hypothesis is that $NADP^+$-linked noncyclic photophosphorylation involves the collaboration of a photosystem II light reaction, responsible for the photooxidation of water, with a photosystem I light reaction, responsible for the photoreduction of ferredoxin (i.e. $NADP^+$). There are several variations of this hypothesis with respect to the identity and sequence of the electron carriers that join the two light reactions. Plastoquinone and plastocyanin are generally included, but there is now a divergence of views with respect to the cytochromes, which were the cornerstones of the original proposals (Sect. S). Cytochrome b_6 is no longer identified with noncyclic electron transport—it is now assigned to cyclic electron transport (ARNON et al., 1965; WEIKARD, 1968; KNAFF, 1972; BÖHME and CRAMER, 1972). Some investigators have included cytochrome b-559 in the noncyclic electron transport chain (e.g. CRAMER and BUTLER, 1967) but others have placed it on a side pathway (e.g. BOARDMAN et al., 1971), leaving cytochrome f as the only cytochrome electron carrier remaining in the noncyclic electron transport chain. However, WITT (1975), reversing an earlier assignment (WITT et al., 1961), has now removed cytochrome f from the noncyclic electron transport chain and depicts that chain as containing no cytochrome components at all.

Several lines of evidence are put forward in support of this hypothesis. They include: (1) findings on the effect of Mg^{2+} on fluorescence yield that led to postulations that Mg^{2+} controls the partition of energy between photosystems II and I (e.g. MURATA, 1969); (2) experiments with algal mutants that lack one or more components of the photosynthetic apparatus (cf. BISHOP, 1973); (3) fractionation of chloroplasts into fragments enriched in photosystem II or photosystem I activity (cf. BOARDMAN, 1970); and (4) enhancement effects in $NADP^+$ photoreduction by isolated chloroplasts (cf. MYERS, 1971).

As regards cyclic electron transport and photophosphorylation, the conventional series hypothesis views them as depending on the same photosystem I that functions in noncyclic electron transport and photophosphorylation.

II. The Three-Light-Reactions Hypothesis

The main feature which distinguishes the three-light-reactions hypothesis from the previous one is a provision for a third light reaction, identified with photosystem I and cyclic photophosphorylation, that can function in parallel to the two light reactions that are identified with noncyclic photophosphorylation. The three-light-reactions hypothesis has two variants: (1) one which envisages two photosystem I reactions and one photosystem II light reaction and (2) one which envisages two photosystem II reactions and one photosystem I light reaction.

Variant (1) is consistent with the work of PARK and SANE (1971) on the relationship between structure and function in chloroplast lamellae. These investigators isolated the stroma lamellae from chloroplasts and found that they contain only photosystem I activity, i.e., the long-wavelength photosystem that can perform cyclic photophosphorylation. The isolated grana fraction, on the other hand, contained both photosystem I and photosystem II activity. PARK and SANE (1971) suggested that there may be two kinds of photosystem I: one operating in the stroma lamellae, and a second kind present in the grana region. This suggestion may explain the findings of ARNTZEN et al. (1972) that the photosystem II fraction from the grana region could functionally recombine only with the photosystem I fraction isolated from the grana region, not with the photosystem I fraction isolated from the stroma lamellae.

The second variant of the three-light-reactions hypothesis attributes the photoreduction of $NADP^+$ by water and the accompanying noncyclic photophosphorylation to two photosystem II light reactions that operate in series; it identifies photosystem I as a third light reaction that operates in parellel to photosystem II and is responsible for cyclic electron flow and photophosphorylation (KNAFF and ARNON, 1969c; ARNON et al., 1971). This hypothesis is supported by several lines of evidence derived from fractionation of chloroplasts and algal membranes; studies of the role of P-700, measurements of enhancement, and studies of cytochrome b-559.

ARNON et al. (1970) prepared two kinds of chloroplast fragments, one fragment with photosystem II activity capable (in the presence of plastocyanin) of photoreducing $NADP^+$ with water but low in P-700 and functional cytochrome f and another fragment that had only photosystem I activity and was enriched in P-700. In preparing these chloroplast fragments, ARNON et al. (1970) used the methods of HUZISIGE et al. (1969), who also formulated a scheme of three light reactions on the basis of somewhat different considerations (HUZISIGE and TAKIMOTO, 1974).

The results were consistent with the findings of RURAINSKI et al. (1970, 1971) and RURAINSKI and HOCH (1972) on the role of P-700 in noncyclic electron flow. With steady-state relaxation spectroscopy, they obtained evidence for the participation of P-700 in cyclic electron flow but not in the photoreduction of $NADP^+$ by water. They observed that magnesium ions increased the light-induced electron flow from water to $NADP^+$ but decreased the light.induced electron flow through P-700. They concluded that, contrary to the conventional two-light-reactions series scheme, the photoreduction of $NADP^+$ with water and the photoinduced electron flow through P-700 are not in series but are parallel photoacts.

Unlike the conventional two-light-reactions hypothesis, the three-light-reactions hypothesis regards the photoreduction of $NADP^+$ by artificial electron donors not as reflecting the involvement of a segment of the overall electron transport chain from water to $NADP^+$, as was initially proposed by LOSADA et al. (1961) but as an artificial, unidirectional electron transport that involves a segment of the cyclic electron pathway (ARNON et al., 1965). Further evidence for this point of view was obtained by McSWAIN et al. (1976) with *Nostoc* membrane fragments, noted for their high rates of $NADP^+$ reduction with either water or artificial photosystem I electron donors (ARNON et al., 1974). In the photoreduction of $NADP^+$ by water, magnesium ions were required for the preservation of activity of membrane fragments that had been stored in the dark as well as for the subsequent light reaction. Neither effect of magnesium was observed in the photoreduction of $NADP^+$ by reduced DCIP or other photosystem I electron donors. Moreover, at high concentrations magnesium severely inhibited the photoreduction of $NADP^+$ by $DCIPH_2$ and similar photosystem I donors but did not affect the photoreduction of $NADP^+$ by water. In short, the data did not support the concept that the photoreduction of $NADP^+$ by $DCIPH_2$, a reaction that was inhibited by magnesium ions, represented a segment of a complete electron transport chain used in the photoreduction of $NADP^+$ by water, a reaction that was stimulated by magnesium ions (MCSWAIN et al., 1976).

As already stated, the conventional two-light-reaction hypothesis was supported by evidence, reported by several investigators, for enhancement in the photoreduction of $NADP^+$ with water by isolated chloroplasts (see review, MYERS, 1971). However, in the work of this laboratory, under a wide range of experimental conditions, no such enhancement effect was observed with isolated chloroplasts (MCSWAIN and ARNON, 1968, 1972). By contrast, by the same experimental methods, a marked and consistent enhancement effect was observed in oxygen evolution by isolated chloroplasts engaged in CO_2 assimilation (MCSWAIN and ARNON, 1968). Furthermore, the enhancement associated with CO_2 assimilation was abolished by the addition of exogenous ATP (MCSWAIN and ARNON, 1972). These results suggested that enhancement was the result of a joint contribution by cyclic and noncyclic photophosphorylation of ATP for CO_2 assimilation (Sect. O.II) and was not related to noncyclic electron flow from water to $NADP^+$. That enhancement may be due to chemical products of light reactions was already implied in the findings of MYERS and FRENCH (1960) who observed enhancement effects in cells to which two light beams were applied even a few seconds apart — a period too long for purely photic interactions.

An entirely different line of evidence in support of the concept of two light reactions in photosystem II concerns cytochrome b-559. The conventional two-light-reactions hypothesis was based on the premise that chloroplast cytochromes are photoreduced by photosystem II and photooxidized by photosystem I (HILL and BENDALL, 1960). It was, therefore, wholly unexpected when cytochrome b-559 (high potential form) was found to undergo photooxidation at low temperature by photosystem II light but not by photosystem I light (KNAFF and ARNON, 1969a). Photooxidation at low temperature suggested that cytochrome b-559 might be a primary electron donor in photosystem II, but this possibility seemed unlikely on thermodynamic grounds. It was, therefore, postulated that cytochrome b-559 is situated between two photosystem II light reactions (KNAFF and ARNON, 1969c; ARNON et al., 1971; KNAFF and MCSWAIN, 1971).

According to this concept, the first light reaction of photosystem II oxidizes water and reduces Component 550 (C-550) while the second light reaction of photosystem II oxidizes plastocyanin and reduces ferredoxin. C-550 and plastocyanin are joined by an electron transport chain that includes (but is not limited to) plastoquinone and cytochrome b-559 and is coupled to the noncyclic phosphorylation site (KNAFF and ARNON, 1971). C-550 is a new photoreactive component of chloroplasts that undergoes photoreduction by water only in short-wavelength (i.e., photosystem II) light (KNAFF and ARNON, 1969b; ERIXON and BUTLER, 1971; BOARDMAN et al., 1971).

An alternative explanation was proposed by ERIXON and BUTLER (1971), namely, that photooxidation of cytochrome b-559 by photosystem II occurs only under special nonphysiological conditions (such as low temperature or Tris treatment) that stop the normal path of electron transport from water to the oxidant of photosystem II. However, this explanation recently had to be abandoned when experiments by TSUJIMOTO et al. (1976) with *Nostoc* membrane fragments demonstrated a photooxidation of cytochrome b-559 at physiological temperatures and, particularly, during active electron transport from water to $NADP^+$ (Fig. 16).

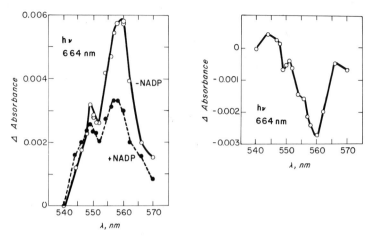

Fig. 16. *Left:* Light-induced absorption spectra in the presence and absence of electron transport from water to NADP$^+$. *Right:* Difference spectrum showing the shift toward oxidation of cytochromes as a result of electron transport from water to NADP$^+$ (TSUJIMOTO et al., 1976)

In other experiments, these membrane fragments were found, like chloroplasts, to be capable of photooxidizing cytochrome b-559 at cryogenic temperatures only with photosystem II light (APARICIO et al., 1974).

The shortcomings of the conventional two-light-reactions hypothesis have been discussed above. There are also, however, experimental findings which at present cannot be explained by the three-light-reactions hypothesis. In sum, it appears that, while the concepts of two photosystems, and of cyclic and noncyclic photophosphorylation, are well documented, the interrelations among them are still open to different interpretations by several hypotheses, each of which is supported by some experimental evidence. It seems, therefore, best to regard the entire subject as a still-open field of inquiry, in which final judgments must await future developments.

U. Concluding Remarks

During the last quarter century our knowledge of photosynthesis has advanced not only in the areas surveyed here but also in areas such as the structure of the photosynthetic apparatus and mechanisms of phosphorylation, that it was not possible to include in this article. Progress in photosynthesis research, as in other fields of scientific inquiry, can be measured by the degree to which knowledge based on important and verifiable facts has displaced conjectures. From this perspective, the last quarter century has been a period of striking advances that greatly enlarged our earlier knowledge of this photochemical process that sustains life on our planet. Indeed, historians of science may count the last 25 years among the golden years of photosynthesis research.

References

Adman, E.T., Sieker, L.C., Jensen, L.H.: J. Biol. Chem. **248**, 3987–3996 (1973)

Allen, J.F.: Nature (London) **256**, 599–600 (1975)

Allen, M.B., Arnon, D.I., Capindale, J.B., Whatley, F.R., Durham, L.J.: J. Am. Chem. Soc. **77**, 4149–4155 (1955)

Allen, M.B., Whatley, F.R., Arnon, D.I.: Biochim. Biophys. Acty **27**, 16–23 (1958)

Anderson, I.C., Fuller, R.C.: Arch. Biochem. Biophys. **76**, 168–179 (1958)

Aparicio, P.J., Ando, K., Arnon, D.I.: Biochim. Biophys. Acta **357**, 246–251 (1974); errata, **368**, 459 (1974)

Arnon, D.I.: Nature (London) **167**, 1008–1010 (1951)

Arnon, D.I.: Science **122**, 9–16 (1955)

Arnon, D.I.: Ann. Rev. Plant Physiol. **7**, 325–354 (1956)

Arnon, D.I.: Nature (London) **184**, 10–21 (1959)

Arnon, D.I.: In: Encyclopedia of Plant Physiology. Ruhland, W. (ed.). Berlin-Heidelberg-New York: Springer, 1960, Vol. V, pp. 773–829

Arnon, D.I.: In: Light and Life. McElroy, W.D., Glass, B. (eds.). Baltimore: Johns Hopkins, 1961, pp. 489–566

Arnon, D.I.: Science **149**, 1460–1469 (1965)

Arnon, D.I.: Naturwissenschaften **56**, 295–305 (1969)

Arnon, D.I., Allen, M.B., Whatley, F.R.: Proc. 8th Congr. Intern. Botan., Paris, July, 1954. Report Commun. Sect. **11, 12**, 1–2, 1954a

Arnon, D.I., Allen, M.B., Whatley, F.R.: Nature (London) **174**, 394–396 (1954b)

Arnon, D.I., Allen, M.B., Whatley, F.R.: Biochim. Biophys. Acta **20**, 449–461 (1956)

Arnon, D.I., Chain, R.K.: Proc. Nat. Acad. Sci. U.S. **72**, 4961–4965 (1975)

Arnon, D.I., Chain, R.K., McSwain, B.D., Tsujimoto, H.Y., Knaff, D.B.: Proc. Nat. Acad. Sci. U.S. **67**, 1404–1409 (1970)

Arnon, D.I., Crane, F.L.: In: Biochemistry of Quinones. Morton, R.A. (ed.). New York: Academic Press, 1965, pp. 433–458

Arnon, D.I., Knaff, D.B., McSwain, B.D., Chain, R.K., Tsujimoto, H.Y.: Photochem. Photobiol. **14**, 397–425 (1971)

Arnon, D.I., Losada, M., Whatley, F.R., Tsujimoto, H.Y., Hall, D.O., Horton, A.A.: Proc. Nat. Acad. Sci. U.S. **47**, 1314–1344 (1961)

Arnon, D.I., McSwain, B.D., Tsujimoto, H.Y., Wada, K.: Biochim. Biophys. Acta **357**, 231–245 (1974); errata, **368**, 459 (1974)

Arnon, D.I., Tsujimoto, H.Y., McSwain, B.D.: Proc. Nat. Acad. Sci. U.S. **51**, 1274–1282 (1964)

Arnon, D.I., Tsujimoto, H.Y., McSwain, B.D.: Nature (London) **207**, 1367–1372 (1965)

Arnon, D.I., Tsujimoto, H.Y., McSwain, B.D.: Nature (London) **214**, 562–566 (1967)

Arnon, D.I., Tsujimoto, H.Y., McSwain, B.D., Chain, R.K.: In: Comparative Biochemistry and Biophysics of Photosynthesis. Shibata, K., Takamiya, A., Jagendorf, A.T., Fuller, R.C. (eds.). State College, Pa: Univ. Park Press, 1968, pp. 113–132

Arnon, D.I., Whatley, F.R., Allen, M.B.: J. Am. Chem. Soc. **76**, 6324–6329 (1954)

Arnon, D.I., Whatley, F.R., Allen, M.B.: Biochim. Biophys. Acta **16**, 607–608 (1955)

Arnon, D.I., Whatley, F.R., Allen, M.B.: Nature (London) **180**, 182–185 (1957); errata p. 1325

Arnon, D.I., Whatley, F.R., Allen, M.B.: Science **127**, 1026–1034 (1958)

Arnon, D.I., Whatley, F.R., Allen, M.B.: Biochim. Biophys. Acta **32**, 47–57 (1959)

Arntzen, C.J., Dilley, R.A., Peters, G.A., Shaw, E.R.: Biochim. Biophys. Acta **256**, 85–107 (1972)

Aronoff, S., Calvin, M.: Plant Physiol. **23**, 351–358 (1948)

Aubert, J.P., Milhaud, G., Millet, J.: Ann. Inst. Pasteur **92**, 515–528 (1957)

Avron, M., Ben-Hayyim, G.: In: Progress in Photosynthesis Research. Metzner, H. (ed.). Tübingen: Laupp, 1969, Vol. III, pp. 1185–1196

Avron, M., Jagendorf, A.T.: J. Biol. Chem. **234**, 1315–1320 (1959)

Avron, M., Krogmann, D.W., Jagendorf, A.T.: Biochim. Biophys. Acta **30**, 144–153 (1958)

Avron, M., Neumann, J.: Ann. Rev. Plant Physiol. **19**, 137–166 (1968)

Bachofen, R., Arnon, D.I.: Biochim. Biophys. Acta **120**, 259–265 (1966)

Baldry, C.W., Walker, D.A., Bucke, C.: Biochem. J. **101**, 642–646 (1966)

Baltscheffsky, H.: Svensk Kem. Tidskrift **72**, 310–325 (1960)
Bamberger, E., Gibbs, M.: Plant Physiol. **40**, 919–926 (1965)
Bassham, J.A., Calvin, M., Benson, A.A., Kay, L.D., Harris, A.Z., Wilson, A.T.: J. Am. Chem. Soc. **76**, 1760–1770 (1954)
Bearden, A.J., Malkin, R.: Quart. Rev. Biophys. **7**, 131–177 (1975)
Bendall, D.S.: Biochem. J. **109**, 46P (1968)
Bendall, D.S., Hill, R.: Ann. Rev. Plant Physiol. **19**, 167–186 (1968)
Benson, A.A.: J. Am. Chem. Soc. **73**, 2971 (1951)
Benson, A.A., Bassham, J.A., Calvin, M., Hall, A.G., Hirsch, H.E., Kawaguchi, S., Lynch, V., Tolbert, N.E.: J. Biol. Chem. **196**, 703–716 (1952)
Benson, A.A., Calvin, M.: Ann. Rev. Plant Physiol. **1**, 25–42 (1950)
Beuscher, N., Gottschalk, G.: Z. Naturforsch. **27B**, 967–973 (1972)
Bishop, N.I.: Ann. Rev. Biochem. **40**, 197–226 (1971)
Bishop, N.I.: Current Topics Photobiol. Photochem. **8**, 65–96 (1973)
Black, C.C.: Biochim. Biophys. Acta **120**, 332–340 (1966)
Black, C.C., Jr.: Ann. Rev. Plant Physiol. **24**, 253–386 (1973)
Blomstrom, D.E., Knight, E., Jr., Phillips, W.D., Weiher, J.F.: Proc. Nat. Acad. Sci. U.S. **51**, 1085–1092 (1964)
Boardman, N.K.: Ann. Rev. Plant Physiol. **21**, 115–140 (1970)
Boardman, N.K., Anderson, J.M.: Biochim. Biophys. Acta **143**, 187–203 (1967)
Boardman, N.K., Anderson, J.M., Hiller, R.G.: Biochim. Biophys. Acta **234**, 126–136 (1971)
Böhme, H., Cramer, W.A.: Biochim. Biophys. Acta **283**, 302–315 (1972)
Böhme, H., Reimer, S., Trebst, A.: Z. Naturforsch. **26B**, 341–352 (1971)
Bothe, H.: Z. Naturforsch. **24B**, 1574–1582 (1969)
Bothe, H., Loos, E.: Arch. Mikrobiol. **86**, 241–254 (1972)
Brown, A.H., Franck, J.: Arch. Biochem. **16**, 55–60 (1948)
Brown, A.H., Frenkel, A.W.: Ann. Rev. Plant Physiol. **4**, 23–58 (1953)
Buchanan, B.B., Arnon, D.I.: Advan. Enzymol. **33**, 119–176 (1970)
Buchanan, B.B., Kalberer, P.P., Arnon, D.I.: Biochem. Biophys. Res. Commun. **29**, 74–79 (1967)
Buchanan, B.B., Schürmann, P.: J. Biol. Chem. **248**, 4956–4964 (1973a); errata, **248**, 8616 (1973a)
Buchanan, B.B., Schürmann, P.: In: Current Topics in Cellular Regulation. Horecker, B.L., Stadtman, E.R. (eds.). New York: Academic Press, 1973b, Vol. VII, pp. 1–20
Buchanan, B.B., Schürmann, P., Kalberer, P.P.: J. Biol. Chem. **246**, 5952–5959 (1971)
Büchner, E.: Ber. Deut. Chem. Ges. **30**, 117–124 (1897a)
Büchner, E.: Ber. Deut. Chem. Ges. **30**, 1110–1113 (1897b)
Bucke, C., Walker, D.A., Baldry, C.W.: Biochem. J. **101**, 636–641 (1966)
Burk, D., Warburg, O.: Z. Naturforsch. **6B**, 12–22 (1951)
Butler, W.L.: Biophys. J. **12**, 851–857 (1972)
Butler, W.L.: Acc. Chem. Res. **6**, 177–184 (1973)
Calvin, M., Bassham, J.A., Benson, A.A., Lynch, V.H., Ouellet, C., Schou, L., Stepka, W., Tolbert, N.E.: In: Carbon Dioxide Fixation and Photosynthesis. Soc. Exp. Biol. Symp. V. New York: Academic Press, 1951, pp. 284–305
Campo, F.F. del, Ramirez, J.M., Arnon, D.I.: J. Biol. Chem. **243**, 2805–2809 (1968)
Champigny, M.-L., Gibbs, M.: In: Progress in Photosynthesis Research. Metzner, H. (ed.). Tübingen: Laupp, 1969, pp. 1534–1537
Cramer, W.A., Butler, W.L.: Biochim. Biophys. Acta **143**, 332–339 (1967)
Dam, H., Hjorth, E., Kruse, I.: Physiol. Plantarum **1**, 379–381 (1948)
Davenport, H.E.: Nature (London) **184**, 524–526 (1959)
Davenport, H.E., Hill, R.: Proc. Roy. Soc., Ser. B **139**, 327–345 (1952)
Davenport, H.E., Hill, R., Whatley, F.R.: Proc. Roy. Soc., Ser. B **139**, 346–358 (1952)
Döring, G., Witt, H.T.: In: Proc. 2nd Intern. Congr. Photosynthesis. Forti, G., Avron, M., Melandri, A. (eds.). The Hague: Dr. Junk, 1972, pp. 39–45
Dutton, P.L., Leigh, J.S.: Biochim. Biophys. Acta **314**, 178–190 (1973)
Duysens, L.N.M., Amesz, J.: Biochim. Biophys. Acta **64**, 243–260 (1962)
Duysens, L.N.M., Amesz, J., Kamp, B.M.: Nature (London) **190**, 510–511 (1961)
Ellis, R.J.: In: Membrane Biogenesis: Mitochondrial, Chloroplast, and Bacterial. Tzagoloff, A. (ed.). New York: Plenum, 1975, pp. 247–278

Emerson, R.L., Chalmers, R.V.: Physiol. Soc. Am. Bull. **11**, 51–56 (1958)

Emerson, R.L., Chalmers, R.V., Cederstrand, C.: Proc. Nat. Acad. Sci. U.S. **43**, 133–143 (1957)

Emerson, R.L., Stauffer, J.F., Umbreit, W.W.: Am. J. Botany **31**, 107–120 (1944)

Engelmann, T.W.: Botan. Z. **39**, 441–448 (1881)

Erixon, K., Butler, W.L.: Photochem. Photobiol. **14**, 427–433 (1971)

Evans, M.C.W., Buchanan, B.B., Arnon, D.I.: Proc. Nat. Acad. Sci. U.S. **55**, 928–934 (1966)

Evans, M.C.W., Lord, A.V., Reeves, S.G.: Biochem. J. **138**, 177–183 (1974)

Evans, M.C.W., Reeves, S.G., Cammack, R.: FEBS Lett. **49**, 111–114 (1974)

Evans, M.C.W., Reeves, S.G., Telfer, A.: Biochem. Biophys. Res. Commun. **51**, 593–596 (1973)

Evans, M.C.W., Shira, C.K., Bolton, J.R., Cammack, R.: Nature (London) **256**, 668–670 (1975)

Evans, M.C.W., Telfer, A., Lord, A.V.: Biochim. Biophys. Acta **267**, 530–537 (1972)

Fay, P.: Biochim. Biophys. Acta **216**, 353–356 (1970)

Floyd, R.A., Chance, B., de Vault, D.: Biochim. Biophys. Acta **226**, 103–112 (1971)

Foust, G.P., Mayhew, S.G., Massey, V.: J. Biol. Chem. **244**, 964–970 (1969)

French, C.S., Milner, H.W.: In: Carbon Dioxide Fixation and Photosynthesis. Soc. Exp. Biol. Symp. V. New York: Academic Press, 1951, pp. 232–250

Frenkel, A.W.: J. Am. Chem. Soc. **76**, 5568–5569 (1954)

Frenkel, A.W.: J. Biol. Chem. **222**, 823–834 (1956)

Gaffron, H., Fager, E.W., Rosenberg, L.L.: In: Carbon Dioxide Fixation and Photosynthesis. Soc. Exp. Biol. Symp. V. New York: Academic Press, 1951, pp. 262–283

Geller, D., Lipmann, F.: J. Biol. Chem. **235**, 2478–2484 (1960)

Goodman, M., Bradley, D.F., Calvin, M.: J. Am. Chem. Soc. **75**, 1962 (1953)

Grant, B.R., Whatley, F.R.: In: Biochemistry of Chloroplasts. Goodwin, T.W. (ed.). New York: Academic Press, 1967, pp. 505–521

Hall, D.O., Reeves, S.G., Baltscheffsky, H.: Biochem. Biophys. Res. Commun. **43**, 359–366 (1971)

Hase, T., Wada, K., Matsubara, H.: J. Biochem. (Tokyo) **78**, 605–610 (1975)

Hauska, G., Reimer, S., Trebst, A.: Biochim. Biophys. Acta **357**, 1–12 (1974)

Heath, O.V.S., Orchard, B.: Nature (London) **180**, 180–181 (1957)

Henri, V.: Naturwissenschaften **14**, 165–167 (1926)

Higuti, T., Shiga, T., Kakuno, T., Horio, T.: J. Biochem. **75**, 1363–1371 (1974)

Hill, R.: Nature (London) **139**, 881–882 (1937)

Hill, R.: Proc. Roy. Soc., Ser. B. **127**, 192–210 (1939)

Hill, R.: In: Carbon Dioxide Fixation and Photosynthesis. Soc. Exp. Biol. Symp. V. New York: Academic Press, 1951, pp. 222–231

Hill, R.: Nature (London) **174**, 501–503 (1954)

Hill, R., Bendall, F.: Nature (London) **186**, 136–137 (1960)

Hill, R., Scarisbrick, R.: New Phytologist **50**, 98–111 (1951)

Hind, G., Olson, J.M.: Ann. Rev. Plant Physiol. **19**, 249–282 (1968)

Hiyama, T., Ke, B.: Proc. Nat. Acad. Sci. U.S. **68**, 1010–1013 (1971)

Hoch, G., Martin, I.: Arch. Biochem. Biophys. **102**, 430–438 (1963)

Hutson, K.G., Rogers, L.J.: Biochem. Soc. Transact. **3**, 377–379 (1975)

Huzisige, H., Takimoto, N.: Plant Cell Physiol. **15**, 1099–1113 (1974)

Huzisige, H., Usiyama, H., Kikuti, T., Azi, T.: Plant Cell Physiol. **10**, 441–455 (1969)

Izawa, S., Good, N.E.: Biochim. Biophys. Acta **162**, 380–391 (1968)

Jagendorf, A.T.: Brookhaven Symp. Biol. **11**, 236–258 (1958)

Jagendorf, A.T., Avron, M.: J. Biol. Chem. **231**, 277–290 (1958)

Jakoby, E.B., Brummond, D.O., Ochoa, S.: J. Biol. Chem. **218**, 811–822 (1956)

James, W.O., Das, V.S.R.: New Phytologist **56**, 325–343 (1957)

Jensen, R.G., Bassham, J.A.: Proc. Nat. Acad. Sci. U.S. **56**, 1095–1101 (1966)

Joy, K.W., Hageman, R.H.: Biochem. J. **100**, 263–273 (1966)

Kahn, A., von Wettstein, D.: J. Ultrastructure Res. **5**, 557–574 (1961)

Kaiser, W., Urbach, W.: Ber. Deut. Botan. Ges. **86**, 213–226 (1973)

Kalberer, P.P., Buchanan, B.B., Arnon, D.I.: Proc. Nat. Acad. Sci. U.S. **57**, 1542–1549 (1967)

Kamen, M., Newton, J.W.: Biochim. Biophys. Acta **25**, 462–474 (1957)

Katoh, S.: Nature (London) **186**, 533–534 (1960)

Katoh, S., Suga, I., Shiratori, I., Takamiya, A.: Arch. Biochem. Biophys. **94**, 136–141 (1961)

Katz, E.: In: Photosynthesis in Plants. Franck, J., Loomis, W.E. (eds.). Ames Ia: Iowa St. College Press, 1949, p. 287

Ke, B.: Arch. Biochem. Biophys. **152**, 70–77 (1972)

Ke, B., Beinert, H.: Biochim. Biophys. Acta **305**, 689–693 (1973)

Ke, B., Hansen, R.E., Beinert, H.: Proc. Nat. Acad. Sci. U.S. **70**, 2941–2945 (1973)

Kegel, L.P., Crane, F.L.: Nature (London) **194**, 1282 (1962)

Keister, D., San Pietro, A., Stolzenbach, F.E.: J. Biol. Chem. **235**, 2989–2996 (1960)

Klob, W., Kandler, O., Tanner, W.: Plant Physiol. **51**, 825–827 (1973)

Knaff, D.B.: FEBS Lett. **23**, 92–94 (1972)

Knaff, D.B., Arnon, D.I.: Proc. Nat. Acad. Sci. U.S. **63**, 956–962 (1969a)

Knaff, D.B., Arnon, D.I.: Proc. Nat. Acad. Sci. U.S. **63**, 963–969 (1969b)

Knaff, D.B., Arnon, D.I.: Proc. Nat. Acad. Sci. U.S. **64**, 715–722 (1969c)

Knaff, D.B., Arnon, D.I.: Biochim. Biophys. Acta **226**, 400–408 (1971)

Knaff, D.B., McSwain, B.D.: Biochim. Biophys. Acta **245**, 105–108 (1971)

Knaff, D.B., Malkin, R.: Arch. Biochem. Biophys. **159**, 555–562 (1973)

Kok, B.: Biochim. Biophys. Acta **22**, 399–401 (1956)

Kok, B., Rurainski, J., Owens, O.V.H.: Biochim. Biophys. Acta **109**, 347–356 (1965)

Kornberg, A.: Nobel Prize, 1959. Stockholm: Norsted, 1960

Krall, A.B., Good, N.E., Mayne, B.C.: Plant Physiol. **36**, 44–47 (1961)

Krogmann, D.W., Jagendorf, A.T., Avron, M.: Plant Physiol. **34**, 272–277 (1959)

Levitt, L.S.: Science **118**, 696 (1953)

Levitt, L.S.: Science **120**, 33–35 (1954)

Losada, M., Trebst, A.V., Arnon, D.I.: J. Biol. Chem. **235**, 832–839 (1960)

Losada, M., Whatley, F.R., Arnon, D.I.: Nature (London) **190**, 606–610 (1961)

Lumry, R., Spikes, J.D., Eyring, H.: Ann. Rev. Plant Physiol. **5**, 271–340 (1954)

Lundegårdh, H.: Nature (London) **192**, 243–248 (1961)

Lundegårdh, H.: Physiol. Plantarum **15**, 390–398 (1962)

Lyne, R.L., Stewart, W.D.P.: Planta (Berl.) **109**, 27–38 (1973)

Malkin, R.: In: Iron-Sulfur Proteins. Lovenberg, W. (ed.). New York: Academic Press, 1973, Vol. II, pp. 1–26

Malkin, R., Aparicio, P.J, Arnon, D.I.: Proc. Nat. Acad. Sci. U.S. **71**, 2362–2366 (1974)

Malkin, R., Bearden, A.J.: Proc. Nat. Acad. Sci. U.S. **68**, 16–19 (1971)

Marsho, T.V., Kok, B.: Methods Enzymol. **23**, 515–523 (1971)

McIntosh, A.R., Chu, M., Bolton, J.R.: Biochim. Biophys. Acta **376**, 308–314 (1975)

McSwain, B.D., Arnon, D.I.: Proc. Nat. Acad. Sci. U.S. **61**, 989–996 (1968)

McSwain, B.D., Arnon, D.I.: Biochem. Biophys. Res. Commun. **49**, 68–75 (1972); errata, **49**, 1709 (1972)

McSwain, B.D., Tsujimoto, H.Y., Arnon, D.I.: Biochim. Biophys. Acta **413**, 313–322 (1976)

Metzner, H.: J. Theor. Biol. **51**, 201–231 (1975)

Miginiac-Maslow, M.: Biochim. Biophys. Acta **234**, 353–359 (1971)

Miginiac-Maslow, M., Champigny, M.-L.: Plant Physiol. **53**, 856–862 (1974)

Mitsui, A., Arnon, D.I.: Physiol. Plantarum **25**, 135–140 (1971)

Mortenson, L.E., Valentine, R.C., Carnahan, J.E.: Biochem. Biophys. Res. Commun. **7**, 448–452 (1962)

Murata, N.: Biochim. Biophys. Acta **189**, 171–181 (1969)

Myers, J.: Ann. Rev. Plant Physiol. **22**, 289–312 (1971)

Myers, J.: Plant Physiol. **54**, 420–426 (1974)

Myers, J., French, C.S.: Plant Physiol. **35**, 963–969 (1960)

Nakamoto, T., Krogmann, D.W., Mayne, B.: J. Biol. Chem. **235**, 1843–1845 (1960)

Nelson, W., Neumann, J.: J. Biol. Chem. **244**, 1932–1936 (1969)

Newton, J.W., Kamen, M.D.: Biochim. Biophys. Acta **25**, 462–474 (1957)

Neyra, C.A., Hageman, R.H.: Plant Physiol. **54**, 480–483 (1974)

Niel, C.B. van: Advan. Enzymol. **1**, 263–328 (1941)

Niel, C.B. van: In: The Microbe's Contribution to Biology. Kluyver, A.J., van Niel, C.B. (eds.). Cambridge Ma.: Harvard Univ., 1956, p. 83

Nishimura, M.S., Whittingham, C.P., Emerson, R.: In: Carbon Dioxide Fixation and Photosynthesis. Soc. Exp. Biol. Symp. V. New York: Academic Press, 1951, pp. 176–210

Nultsch, W.: In: Currents in Photosynthesis. Thomas, J.B., Goedheer, J.C. (eds.). Rotterdam: Donker, 1966, pp. 421–427

Nultsch, W.: Photochem. Photobiol. **10**, 119–123 (1969)
Orme-Johnson, W.H.: Ann. Rev. Biochem. **42**, 159–204 (1973)
Park, R.B., Sane, P.V.: Ann. Rev. Plant Physiol. **22**, 395–430 (1971)
Petrack, B., Lipmann, F.: In: Light and Life. McElroy, W.D., Glass, B. (eds.). Baltimore: Johns Hopkins, 1961, pp. 621–630
Pfeffer, W.: Physiology of Plants. Oxford: Clarendon Press, 1900
Priestley, J.: Phil. Trans. **62**, 147–264 (1772)
Prince, R.C., Leigh, J.S., Jr., Dutton, P.L.: Biochem. Soc. Transact. **2**, 950–954 (1974)
Rabinowitch, E.I.: Photosynthesis and Related Processes. New York: Interscience, 1945, p. 228
Rabinowitch, E. I.: Sci. Am. 80 (1953)
Rabinowitch, E.I.: In: Research in Photosynthesis. Gaffron, H. (ed.). New York: Interscience, 1957, p. 345
Racker, E.: Arch. Biochem. Biophys. **69**, 300–310 (1957)
Ramirez, J.M., Campo, F.F. del, Arnon, D.I.: Proc. Nat. Acad. Sci. U.S. **59**, 606–612 (1968)
Ramirez, J.M., Campo, F.F. del, Paneque, A., Losada, M.: Biochim. Biophys. Acta **118**, 58–71 (1966)
Raven, J.A.: Nature (London) **227**, 1170–1171 (1970)
Raven, J.A.: J. Exp. Botany **22**, 420–433 (1971)
Reeves, S.G., Hall, D.O.: Biochim. Biophys. Acta **314**, 66–78 (1973)
Ruben, S.: J. Am. Chem. Soc. **65**, 279–282 (1943)
Rubner, M.: Die Ernährungsphysiologie der Hefezelle bei alkoholischer Gärung. Leipzig: Veit, 1913, p. 59
Rurainski, H.J., Hoch, G.E.: In: Proc. 2nd Intern. Congr. Photosynthesis. Forti, G., Avron, M., Melandri, A. (eds.). The Hague: Dr. Junk, 1972, pp. 1283–1291
Rurainski, H.J., Randles, J., Hoch, G.E.: Biochim. Biophys. Acta **205**, 254–262 (1970)
Rurainski, H.J., Randles, J., Hoch, G.E.: FEBS Lett. **13**, 98–100 (1971)
Sachs, J.: Lectures on the Physiology of Plants. Oxford: Clarendon Press, 1887, p. 299
Saha, S., Good, N.E.: J. Biol. Chem. **245**, 5017–5021 (1970)
San Pietro, A., Lang, H.M.: Science **124**, 118–119 (1956)
San Pietro, A., Lang, H.M.: J. Biol. Chem. **231**, 211–229 (1958)
Sauer, K., Biggins, J.: Biochim. Biophys. Acta **102**, 55–72 (1965)
Sauer, K., Park, R.B.: Biochemistry **4**, 2791–2798 (1965)
Schacter, B.Z., Gibbs, M., Champigny, M.-L.: Plant Physiol. **48**, 443–446 (1971)
Schmidt, A., Trebst, A.: Biochim. Biophys. Acta **180**, 529–535 (1969)
Schürmann, P., Buchanan, B.B.: Biochim. Biophys. Acta **376**, 189–192 (1975)
Schürmann, P., Buchanan, B.B., Arnon, D.I.: Biochim. Biophys. Acta **267**, 111–124 (1972)
Shanmugam, K.T., Buchanan, B.B., Arnon, D.I.: Biochim. Biophys. Acta **256**, 477–486 (1972)
Shin, M., Arnon, D.I.: J. Biol. Chem. **240**, 1405–1412 (1965)
Shin, M., San Pietro, A.: Biochem. Biophys. Res. Commun. **33**, 38–42 (1968)
Shin, M., Tagawa, K., Arnon, D.I.: Biochem. Z. **338**, 84–96 (1963)
Simonis, W.: Ber. Deut. Botan. Ges. **77**, 5013 (1964)
Simonis, W.: In: Currents in Photosynthesis. Thomas, J.B., Goedheer, J.C. (eds.), Rotterdam: Donker, 1966, pp. 217–223
Simonis, W., Urbach, W.: Ann. Rev. Plant Physiol. **24**, 89–114 (1973)
Sirevåg, R.: Arch. Microbiol. **98**, 3–18 (1974)
Sirevåg, R., Ormerod, J.G.: Biochem. J. **120**, 399–408 (1970)
Stålfelt, M.G.: Physiol. Plantarum **8**, 572–593 (1955)
Stanier, R.Y.: Bacteriol. Rev. **25**, 1–17 (1961)
Stiller, M., Vennesland, B.: Biochim. Biophys. Acta **60**, 562–579 (1962)
Stombaugh, N.A., Burris, R.H., Orme-Johnson, W.H.: J. Biol. Chem. **248**, 7951–7956 (1973)
Tabita, R.F., McFadden, B., Pfennig, N.: Biochim. Biophys. Acta **341**, 187–194 (1974)
Tagawa, K., Arnon, D.I.: Nature (London) **195**, 537–543 (1962)
Tagawa, K., Arnon, D.I.: Biochim. Biophys. Acta **153**, 602–613 (1968)
Tagawa, K., Tsujimoto, H.Y., Arnon, D.I.: Nature (London) **199**, 1247–1252 (1963a)
Tagawa, K., Tsujimoto, H.Y., Arnon, D.I.: Proc. Nat. Acad. Sci. U.S. **49**, 567–572 (1963b)
Tagawa, K., Tsujimoto, H.Y., Arnon, D.I.: Proc. Nat. Acad. Sci. U.S. **50**, 544–549 (1963c)
Tanner, W., Loos, E., Kandler, O.: In: Currents in Photosynthesis. Thomas, J.B., Goedheer, J.C. (eds.). Rotterdam: Donker, 1966, pp. 243–250

Tanner, W., Zinecker, U., Kandler, O.: Z. Naturforsch. **22b**, 358–359 (1967)

Telfer, A., Cammack, R., Evans, M.C.W.: FEBS Lett. **10**, 21–24 (1970)

Thomas, J.B., Haans, J.M.: Biochim, Biophys. Acta **18**, 286–288 (1955)

Tolmach, L.J.: Nature (London) **167**, 946–949 (1951)

Trebst, A.V.: Ann. Rev. Plant Physiol. **25**, 423–458 (1974)

Trebst, A.V., Eck, H.: Z. Naturforsch. **16B**, 455–461 (1961)

Trebst, A.V., Harth, E., Draber, W.: Z. Naturforsch. **25B**, 1157–1159 (1970)

Trebst, A.V., Losada, M., Arnon, D.I.: J. Biol. Chem. **234**, 3055–3058 (1959)

Trebst, A.V., Losada, M., Arnon, D.I.: J. Biol. Chem. **235**, 840–844 (1960)

Trebst, A.V., Tsujimoto, H.Y., Arnon, D.I.: Nature (London) **182**, 351–355 (1958)

Trudinger, P.A.: Biochem. J. **64**, 273–286 (1956)

Tsujimoto, H.Y., Chain, R.K., Arnon, D.I.: Biochem. Biophys. Res. Commun. **51**, 917–923 (1973)

Tsujimoto, H.Y., McSwain, B.D., Hiyama, T., Arnon, D.I.: Biochim. Biophys. Acta **423**, 303–312 (1976)

Turner, J.F., Black, C.C., Gibbs, M.: J. Biol. Chem. **237**, 577–579 (1962)

Urbach, W., Simonis, W.: Biochem. Biophys. Res. Commun. **17**, 39–45 (1964)

Vernon, L.P., Shaw, E.R., Ogawa, T., Raveed, D.: Photochem. Photobiol. **14**, 343–357 (1971)

Vernon, L.P., Zaugg, W.S.: J. Biol. Chem. **235**, 2728–2733 (1960)

Vishniac, W., Ochoa, S.: Nature (London) **167**, 768–770 (1951)

Vishniac, W., Ochoa, A.: J. Biol. Chem. **198**, 501–506 (1952)

Wada, K., Arnon, D.I.: Proc. Nat. Acad. Sci. U.S. **68**, 3064–3068 (1971)

Walker, D.A.: Biochem. J. **92**, 22C–23C (1964)

Walker, D.A.: Plant Physiol. **40**, 1157–1161 (1965)

Walker, D.A.: New Phytologist **72**, 209–235 (1973)

Warburg, O.: Naturwissenschaften **14**, 167–168 (1926)

Warburg, O., Burk, D., Schade, A.L.: In: Carbon Dioxide Fixation and Photosynthesis. Soc. Exp. Biol. Symp. V. New York: Academic Press, 1951, pp. 306–311

Warburg, O., Krippahl, G., Lehman, A.: Am. J. Botany **56**, 961–971 (1969)

Weikard, J.: Z. Naturforsch. **23B**, 235–238 (1968)

Weissbach, A., Horecker, B.L., Hurwitz, J.: J. Biol. Chem. **218**, 795–810 (1956)

Wessels, J.S.C., Havinga, E.: Recueil des Travaux Chimiques des Pays-Bas **71**, 809–812 (1952)

West, K.R., Wiskich, J.T.: Biochim. Biophys. Acta **292**, 197–205 (1973)

Whatley, F.R., Allen, M.B., Arnon, D.I.: Biochim. Biophys. Acta **16**, 605–606 (1955)

Whatley, F.R., Allen, M.B., Rosenberg, L.L., Capindale, J.B., Arnon, D.I.: Biochim. Biophys. Acta **20**, 462–468 (1956)

Whatley, F.R., Tagawa, K., Arnon, D.I.: Proc. Nat. Acad. Sci. U.S. **49**, 266–270 (1963)

Winget, G.D., Izawa, S., Good, N.E.: Biochem. Biophys. Res. Commun. **21**, 438–443 (1965)

Witt, H.T.: In: Bioenergetics of Photosynthesis. Govindjee (ed.). New York: Academic Press, 1975, pp. 493–554

Witt, H.T., Müller, A., Rumberg, B.: Nature (London) **192**, 967–969 (1961)

Yoch, D.C.: Arch. Biochem. Biophys. **158**, 633–640 (1973)

Yoch, D.C., Arnon, D.I.: J. Biol. Chem. **247**, 4514–4520 (1972)

Yoch, D.C., Arnon, D.I.: J. Bacteriol. **121**, 743–745 (1975)

Yoch, D.C., Arnon, D.I., Sweeney, W.V.: J. Biol. Chem. **250**, 8330–8336 (1975)

Yoch, D.C., Valentine, R.C.: J. Bacteriol. **110**, 1211–1213 (1972)

Yocum, C.S., San Pietro, A.: Biochem. Biophys. Res. Commun. **36**, 614–620 (1969)

Yocum, C.S., San Pietro, A.: Arch. Biochem. Biophys. **140**, 152–157 (1970)

Zamecnik, P.C., Keller, E.B.: J. Biol. Chem. **209**, 337–354 (1954)

Zweig, G., Avron, M.: Biochem. Biophys. Res. Commun. **19**, 397–400 (1965)

II. Electron Transport

1. General

1a. Physical Aspects of Light Harvesting, Electron Transport and Electrochemical Potential Generation in Photosynthesis of Green Plants

W. Junge

A. Introduction

This article reviews the primary events in photosynthesis of green plants. Emphasis lies on those aspects which were accessible to the physicist's approach. The biological object of this report is the isolated chloroplast, with its envelope broken, but its inner membrane system largely intact.

Various physical techniques were applied to study the molecular structure and function of the photosynthetic apparatus. These range from magnetic resonance spectroscopy to X-ray scattering. It is only natural with this highly pigmented system, that spectroscopy in the visible spectral region proved especially valuable (for an introduction into molecular spectroscopy, see CLAYTON, 1970; for spectrophotometric instrumentation in photosynthesis studies, see RÜPPEL and WITT, 1969; BUTLER, 1972; JUNGE, 1976).

The inner membrane system of chloroplasts transforms the energy of light quanta into other forms of energy which are useful to drive the secondary processes of photosynthesis. Light is absorbed by antennae pigments and transferred into photochemical reaction centers. There it drives electron transport from water to $NADP^+$ and, coupled to this, it charges the internal phase of the functional vesicles with protons. The reducing power of NADPH is used directly for the secondary processes, while the electrochemical potential difference of the proton is first converted into ATP (see JAGENDORF, this vol. Chap. III.1) which then is used for CO_2-fixation.

Antennae function, electron transport and electrochemical potential generation are tightly coupled to each other. They all have components linked to the very first photochemical events which occur in the time domain between pico- and nanoseconds.

The functional units of the electrochemical events are the inner vesicles of chloroplasts. According to morphological studies (cf. MÜHLETHALER, this vol. Chap. IV.1) these are thylakoid disks interconnected by intergrana lamellae. A typical disk from mature spinach contains at least 10^5 chlorophyll molecules, hence its membrane is covered by at least 200 electron transport chains. One electron transport chain has two distinct photochemical reaction centers, which are supplied with quanta by an ensemble of about 300 antennae chlorophylls each. The electron transport chains are not isolated from each other but communicate at three levels: They exchange electromagnetic energy at the level of the antennae. They exchange electrons at the level of the electron carriers. They share one common pool of electrochemical energy at the level of the thylakoid membrane. This is illustrated in Figure 1.

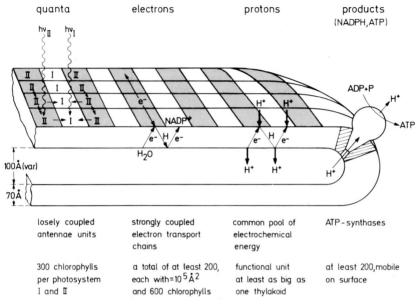

Fig. 1. Model for the primary events of photosynthesis. The thylakoid membrane is covered with at least 200 units with about 600 chlorophylls each plus a complete set of electron carriers. These units interact at the level of the antennae pigments, at the one of the electron carriers and via a common pool of electrochemical energy of the proton. The distribution of photosystems I and II over the membrane is arbitrary

Each of the three subsequent chapters is headed by a summary which states the facts and open questions. Specific references follow in the specialized sections. The presentation is rather synthetic than historic. For the sequence of events the reader may refer to the article by ARNON (this vol. Chap. I) and to the first edition of this Encyclopedia (vol. V/1). The citation policy has emerged under the realization that our present knowledge of the primary processes originates from the joined efforts of very many workers rather than from singular contributions of a few. This lengthened the reference list considerably, even though some of the earlier papers were omitted, if later and more detailed ones gave reference to them.

In only hope that the presentation is balanced in those fields where I did not work, and not too unbalanced where I did.

B. Antennae

In intact, well coupled chloroplasts each electron transport chain may turn over once every 15 msec. If a single chlorophyll molecule were to drive the reaction, the supply rate of light quanta would not suffice, even if the pigment was exposed to bright sunlight. An average chlorophyll a molecule absorbs one quantum in 100 msec under bright sunlight, one in 1 sec under diffuse daylight and only 1 in 10 sec on a cloudy day (for data on skylight, see VALLEY, 1965). To match the rather low absorption rate of quanta to the higher possible rates of chemical reactions in photosynthesis each photochemical reaction center is endowed with antennae pigments (light harvesting pigments, accessory pigments). In green plants these are mainly chlorophylls and carotenoids.

Quanta absorbed by the antennae pigments are funnelled into reaction centers. The efficiency of energy transfer is high. This implies that the probability for the transfer of a quantum between two neighboring pigments is higher than the probability for any competing process as there are fluorescence emission, formation of metastable states, wasteful photochemistry and radiationless deactivation (see Fig. 2). As these processes may occur within nanoseconds, the transfer through the whole antennae system of a reaction center must occur in a much shorter interval. Rapid transfer of energy occurs via dipolar coupling between pigments in resonance which are rather tightly packed.

Within the antennae different types of chlorophyll a are discernable according to their spectra. These pigments differ in the degree of interaction with their neighbors, only, chemically they are identical. One can only speculate about the physiological relevance of this. Two possible advantages are: (1) Chlorophylls with different peak wavelength use a larger proportion of the sunlight spectrum. (2) Resonant energy transfer is directed towards pigments absorbing at longer wavelength. Hence the variety of chlorophylls may serve to shorten the diffusion

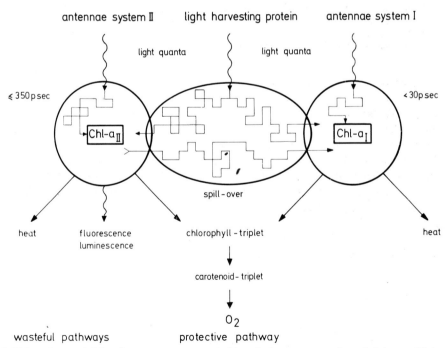

Fig. 2. Tripartite model for the antennae system. Absorbed quanta are funneled by multiple resonant energy transfer steps into the reaction centers or, if not used, flow away via wasteful pathways. The antennae systems closely associated with reaction centers I and II, resp., exchange energy via light-harvesting chlorophyll proteins

path for quanta through the antennae into the trap, which, in both photosystems, lies at the red end of the absorption spectrum.

The size of the antennae system (about 300 chlorophylls per reaction center) is designed to drive electron transport at full velocity even on a cloudy day. In consequence, under bright sunlight, much more quanta are absorbed than can be processed by the reaction centers. These quanta flow away via wasteful pathways (see Fig. 2). One of these is the formation of the triplet state of chlorophyll a. In the presence of the ubiquitous oxygen triplet excited chlorophyll is destroyed. To avoid photodestruction of the antennae system, there is a protective pathway. The energy is rapidly transferred into carotenoid triplets from where it is dissipated as heat.

The physical principles underlying the variety of chlorophylls, rapid resonant energy transfer and the protective reaction within the antennae are fairly well understood (see below). On the other hand there are still important white fields:

1. The molecular arrangement of the antennae within the thylakoid membrane, the size of aggregates, the pigment-protein assembly, rigidity or variability of the structure.

2. The degree and the mechanism of the functional interaction between antennae patches which serve different reaction canters.

3. The physiological relevance of the well-established variability of this interaction.

I. Physically Different Types of Chlorophylls in Chloroplasts

The major pigments in photosynthesis are chlorophylls. The chemical structure of chlorophyll a is illustrated in Figure 3. The absorption spectrum of chlorophyll a in ethanol is illustrated in Figure 4. Absorption in the visible spectral region mainly results from electronic transitions between orbitals of π-symmetry (π-π*-transition). These orbitals extend over the whole porphyrin ring, as illustrated by shading in Figure 3.

Four subtransitions can be discriminated which, according to molecular orbital calculations (GOUTERMAN et al., 1963; WEISS, 1972), differ in their polarization with respect to the molecular coordinate system. If one says "a transition is polarized in the x-direction" this is to indicate that the pigment absorbs light of the respective wavelength by preference if its electric vector is polarized parallel to the x-axis of the molecular coordinate system. The convention for the axes x and y of chlorophylls is illustrated in Figure 3, the attribution of the absorption bands to these axes is given in Figure 4.

The absorption spectrum of chloroplasts extends into regions where diluted solutions of chlorophyll a and b do not absorb. Part of the complexity is due to the presence of carotenes and xanthophylls. But even if the contribution of these pigments to the spectrum is eliminated a remarkable complexity above a wavelength of 670 nm still remains. Solvent extraction, however, revealed only one type of chlorophyll a and chlorophyll b, respectively. This has led to the suggestion (CLAYTON, 1966) that the apparent variety of pigments is due to physically different states of chlorophylls in the thylakoid membrane. Four approaches led to a better understanding of this phenomenon:

Fig. 3. Chemical structure of chlorophyll a.
Shaded portions indicate the area across
which the π-electrons are delocalized. (From
SAUER, 1975)

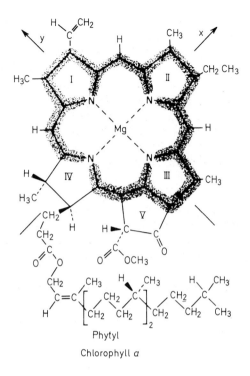

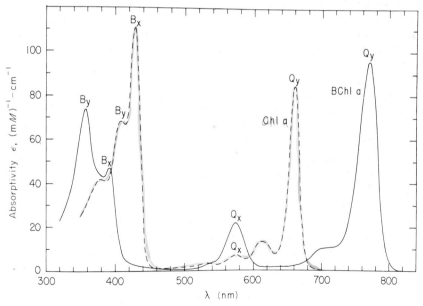

Fig. 4. Absorption spectra of chlorophyll a and bacteriochlorophyll a in ethanolic solution.
The polarization of the respective transitions within the molecular coordinate system is given
with reference to Figure 2. A synonymous term for the major band in the blue is Soret
band. (From SAUER, 1975)

1. Computer deconvolution of chloroplast absorption spectra into gaussian or lorentzian absorption profiles of hypothetical chlorophylls absorbing at different wavelength (Brown, 1963; Smith and French, 1963; French et al., 1971; Cederstrand et al., 1966; Gulyaev and Litvin, 1970; Butler and Hopkins, 1970).

2. Chemical and mechanical fractionation of chloroplasts into subunits with reduced pigment contents (Boardman and Anderson, 1964; Vernon et al., 1966; Sane et al., 1970).

3. Studies on the action spectra of both photosystems (e.g. Ried, 1972).

4. Model studies on chlorophylls in different solvents and in different states of interaction (see below).

Computer deconvolution of a complex spectrum into say gaussian components is not unequivocal for principal reasons. Moreover, it requires ad hoc assumptions. Nevertheless, analyses of chloroplast spectra yielded some insight into the complexity of chlorophyll a in the thylakoid membrane. Absorption bands peaking at 660, 670, 678, 685, 690, 705 and 720 nm were identified (for a review see Govindjee and Govindjee, 1975).

Fractionation of chloroplasts into subunits revealed that the antennae pigments are unevenly distributed among particles with photosystem I activity and photosystem II activity, respectively. Photosystem I has a preference for red shifted chlorophyll a while photosystem II for chlorophyll b.

Studies on chlorophyll a in various solvents have shown that the microenvironment may shift the peak wavelength of the Q band (red band) of chlorophyll a by up to 80 nm towards longer wavelength. Two types of interaction are responsible for this:

1. Pigment-Solvent Interaction

If chlorophyll is embedded in a solid or a fluid medium it is exposed to very high electric field strength ($> 10^6$ Vcm^{-1}) resulting from the polar or dipolar properties of its microenvironment. This distorts its electronic orbitals and in consequence shifts its absorption bands (solvatochroism, cf. Liptay, 1969). More specific interactions between the central magnesium atom in chlorophyll and solvent molecules which are electron donors have been reported (Seely, 1965; Cotton et al., 1974). Solvent effects produce only slight shifts of the Q_y band of chlorophyll a which range from 660–670 nm.

2. Pigment-Pigment Interaction

If strongly absorbing organic dye molecules are concentrated to mutual distances of less than about 50 Å their resonance interaction may exceed the interaction with the ambient solvent molecules. Shifts and eventual splitting of absorption bands result. These effects can be theoretically understood in terms of the molecular exciton theory (e.g. Hochstrasser and Kasha, 1964). For very close aggregates, however, the simple dipole–dipole model of the exciton theory has to be extended by taking overlap of the π-electron orbitals of neighboring chlorophylls into account. The bandshifts due to chlorophyll–chlorophyll aggregation are rather large. Peak wavelength ranging from 660 nm to 740 nm were reported (for a review see Norris et al., 1975).

There is no doubt that the extension of the chlorophyll a absorption in vivo into the red spectral region up to 720 nm reflects aggregation. However, the question is open as to what extent chlorophyll a is just self-aggregated (for models, see KATZ and NORRIS, 1973), aggregated within proteins (for an X-ray analysis of a pigment-protein complex from bacteria, see FENNA and MATHEWS, 1975) or aggregated by interaction with lipids (for model studies, see TRURNITT and COLMANO, 1959; STEINEMANN et al., 1971).

II. Resonant Energy Transfer

A quantum of light absorbed by one of the antennae pigments is transferred through the antennae system until it is trapped by a photochemical reaction center. In order to compete with wasteful processes the energy transfer through the antennae has to be very rapid, more rapid for instance than reemission of the quantum as fluorescence. The fluorescence lifetime of chlorophyll a is rather short, already, 5 nsec in "good solvents" (BRODY and RABINOVITCH, 1957).

Resonant energy transfer between pigments is more rapid the stronger they interact with each other via their resonant transition dipoles. The degree of interaction depends on the mutual distance and the relative orientation of the chromophores. For strongly absorbing pigments such as chlorophylls, resonant energy transfer starts beating out fluorescence already at distances of about 100 Å between neighboring pigments (FÖRSTER, 1946; for a review on experimental data and theories, see KNOX, 1975).

Two extreme cases of energy transfer have to be distinguished: At weaker interaction, when the transfer time is long in comparison with the time for intramolecular vibrational relaxation (10^{-12} sec), the transfer is unidirectional as illustrated in the left of Figure 5. The excited energy donor molecule relaxes to the vibrational ground state before transferring its energy to the acceptor. After the transfer the acceptor molecule relaxes vibrationally and drops out of resonance with the donor. Backtransfer can occur only, if the excited state of the acceptor is thermally uplifted into a higher vibrational level, which is improbable, although not impossible. (Such a process causes the emission of delayed fluorescence in photosystem II,

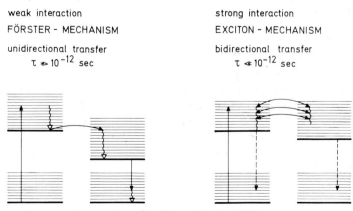

Fig. 5. Two extreme cases of resonant energy transfer. (From JUNGE, 1976)

see below.) This case of incoherent energy transfer which involves considerable energy dissipation between subsequent transfer steps was theoretically first described by Förster (1946, 1947, 1960). It proceeds from pigments at shorter toward those absorbing at longer wavelength. Its velocity depends strongly, by the reciprocal of the sixth power, on the distance between pigments. It has been argued that the irreversible transfer of energy from the antennae into the photochemical trap is governed by a Förster-transfer mechanism.

The other extreme mechanism for resonance energy transfer is characterized by transfer times which are very short in comparison with the times for vibrational relaxation (10^{-12} sec). The transfer is bidirectional, as illustrated in the right of Figure 5. Before vibrational relaxation occurs a quantum may be transferred repeatedly between the resonant vibrational levels of the donor and the acceptor. At the extremes the energy is delocalized over a whole aggregate of strongly interacting pigment molecules. The theoretical description of the delocalized "exciton" in cristals goes back to Davidov (1962). It was adapted for dimers and higher oligomers of organic pigments by McRae and Kasha (1958), Tinoco (1963) and Hochstrasser and Kasha (1964).

The presently available data are insufficient to decide which type of coupling governs energy transfer through the antennae in green plants. A probable assumption is that the transfer occurs via the exciton mechanism within larger chlorophyll–protein aggregates ("pebbles", see below) while the transfer between these units proceeds via the Förster-mechanism (Knox, 1975).

Alternatively chlorophylls might be arranged in quasicristalline arrays wherein electrons and holes (electron deficiencies) could move independently from each other until they are finally trapped. This mechanism which is analogous to the ones in semiconductors requires a rather perfect lattice structure. It is improbable that such a structure of chlorophylls exists throughout the antennae system of green plants.

III. Distinctive Properties of Antennae Systems I and II

The antennae serving photosystems I and II, respectively, differ in their *chemical composition*. Carotenoids and more pronouncedly chlorophyll b are enriched in photosystem II. This became evident from fragmentation of chloroplasts (Boardman and Anderson, 1964; Vernon et al., 1966; Sane et al., 1970) and from the respective action spectra of photosystem I (Müller et al., 1963; Vidaver and French, 1965) and photosystem II (Joliot, 1965) (see Fig. 6).

The antennae systems I and II differ also in the *physical state of their chlorophyll a molecules*. This is evident from inspection of the action spectra of both photosystems in Figure 6. The major differences are in the blue spectral region between 450 and 500 nm (due to different proportions of carotenoids and chlorophyll b) and at the red end. The red portion of the spectrum above 660 nm is entirely due to chlorophyll a absorption. It is obvious that chlorophyll a molecules in photosystem I are red-shifted in comparison to those in photosystem II. The respective half efficiencies for excitation are at 697 nm and 688 nm, about 10 nm apart. One consequence of this red shift is that energy transfer between different systems is more likely to occur from II to I than vice versa.

Fig. 6. Action spectra for photosystem I and photosystem II activity (RIED, 1972)

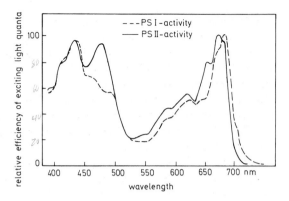

The *total transfer time* of a quantum through the antennae into the trap is at variance between the two systems by more than one order of magnitude. The transfer times were not measured directly. Instead upper limits were derived from the actual lifetime of fluorescence. For both photosystems the actual lifetime of fluorescence is much shorter than the intrinsic lifetime (τ^0_{fluor}) of chlorophyll a (15 nsec, BRODY and RABINOVITCH 1957) or the lifetime which is observed in "good" solvents (5 nsec). The actual rate of fluorescence decay ($1/\tau_{actual}$) equals the sum of the rates of all processes which discharge the first excited singlet state of chlorophylls in chloroplasts:

$$1/\tau_{actual} = 1/\tau^0_{fluor} + 1/\tau_{triplet} + 1/\tau_{radless} + 1/\tau_{trap}.$$

Therefrom it follows that the actual lifetime of fluorescence represents an upper limit for the time which characterizes any process including the transfer of the quantum into the trap.

Under saturating illumination of chloroplasts, when the traps are closed the following actual lifetimes of fluorescence were observed for the two photosystems:

Photosystem II: 1.9 nsec (MÜLLER et al., 1969), 1.7 nsec (BORISOV and IL'INA, 1973)
Photosystem I: < 30 psec, instrument limited (BORISOV and IL'INA, 1973), 10 psec (SEIBERT and ALFANO, 1974)

Since the traps were closed the above actual lifetimes characterize all wasteful events. The high quantum yield of photosynthesis implies that the transfer into the respective traps occurs much more rapidly. In fact, under low light, when the traps are open, MÜLLER et al. (1969) observed an actual lifetime of fluorescence from photosystem II of only 0.35 nsec.

It is probable that the very rapid transfer of quanta into the traps involves several resonant energy transfer steps between chlorophyll molecules of the respective antennae system. On a modified random walk basis BORISOV and IL'INA (1973) estimated that an average quantum in an antennae unit of 300 chlorophylls takes about 125 steps before falling into a trap. If so, the less than 30 psec required for the whole walk in photosystem I imply that the average transfer step needs $2.5 \cdot 10^{-13}$ sec, only. This would imply that chlorophylls interact via the exciton mechanism. The data, however, are equally well compaible with an interval-walk of the quantum, with many rapid transfer steps between closely coupled chlorophylls plus a few slower ones (Förster-mechanism) between less closely coupled pigments (cf. Sauer, 1975).

The *fluorescence yield* largely differs between the two photosystems. The major portion of fluorescence in green plants and algae results from photosystem II (Duysens, 1963). This reflects the fact that the rates of wasteful processes are higher by almost two orders of magnitude in photosystem I (see the above data on the "closed trap" fluorescence). The mechanism of the extremely rapid energy dissipation in photosystem I is not as yet quite understood (see Clayton, 1972). Fluorescence from photosystem II has two distinct components, a constant one ("dead fluorescence") and another one ("life fluorescence") the yield of which reacts on the state of the trap (Clayton, 1969). The dependence of life fluorescence on the redox state of the photochemically active chlorophyll in reaction center II was used to evaluate the kinetic parameters of the center and of its electron donors and acceptors (cf. Papageorgiou, 1975).

The evaluation of fluorescence data for conclusions as to electron transport around the reaction center is complicated by the existence of several effectors. The major influence acting on the fluorescence yield is the primary electron acceptor, named therefore the quencher Q (Duysens and Sweers, 1963). If this component is oxidized, and if it is ready to accept an electron from the reaction center chlorophyll, the trap is open and the fluorescence yield is low. If Q is reduced the yield goes up.

The fluorescence yield is also varied by the state of chlorophyll a_{II} (Butler, 1972) and by the state of endogenous plastoquinone (Delosme, 1967; Joliot and Joliot, 1973).

In addition to the above "prompt" fluorescence delayed light emission was observed from photosystem II with decay times ranging up to seconds (cf. Lavorel, 1975). This results from the same excited singlet state of antennae chlorophyll a as the prompt fluorescence. As proposed by Arthur and Strehler (1957) luminescence reflects the reversal of the primary photochemical act, the recombination of an electron-hole pair to yield the excited singlet state of the reaction center chlorophyll. Since the forward reaction from the excited singlet into the charge separated state involves energy dissipation (see Sect. C), the production of delayed light requires energy input, for instance by thermal collision. The longevity of delayed light reflects the low probability of collisions which transfer enough energy. The probability of delayed fluorescence is increased if an electric potential is generated across the thylakoid membrane with positive polarity inside (Barber and Kraan, 1970). This was interpreted to indicate that the primary charge separation in photosystem II is directed across the thylakoid membrane (Wraight and Crofts, 1971; Crofts et al., 1971).

IV. Size and Interaction of the Antennae Systems

The first indirect determination of the *size* of the total antennae system of both light reactions goes back to Emerson and Arnold's (1932) observation that the amount of oxygen evolved after excitation with a single short flash of light is one molecule of O_2 on 2,500 chlorophylls. Since four oxidizing equivalents are required to produce one dioxygen from water each of these four equivalents is associated with the action of about 600 chlorophylls which are grouped around two photosystems.

It is almost generally accepted that the antennae units of photosystems II and I are not segregated from each other but communicate by transfer of excitation. Quantitative estimates for the degree of interaction are still at variance with each other.

The *transfer* of energy *between units serving reaction center II* was studied via the oxygen yield in dependence on the energy of exciting flashes. If the trap of the donor unit was closed transfer probabilities of 0.55 (JOLIOT et al., 1968) and 0.3 (WANG and MYERS, 1973) were determined. Fluorescence studies revealed that the transfer probability increases with the Mg^{2+}-concentration in the suspending medium of broken chloroplasts (BRIANTAIS et al., 1973; S. MALKIN, 1974). It is still open as to why studies on the dependence of the luminescence yield seemed to indicate the total absence of energy transfer between antennae units of photosystem II (WRAIGHT, 1972).

There is, as yet, no indication for energy transfer between antennae units serving photosystem I (JOLIOT et al., 1968).

There is evidence for *spill-over* of energy from antennae units serving photosystem II to those of photosystem I. Spill-over was first postulated to account for the almost constant quantum yield of oxygen evolution under excitation varying between 600 and 680 nm (MYERS, 1963). On the other hand the observation of enhancement phenomena (EMERSON, 1957) and the distinct action spectra of the two photosystems (see Fig. 6) demonstrated that spill-over could not be too efficient. While some studies seemed to indicate the total absence of energy transfer between different units (JOLIOT et al., 1968) it was later shown that its efficiency depends critically on the concentration of divalent cations, especially Mg^{2+} (MURATA, 1969), and to lesser extent on monovalent ones (WYDRZYNSKI et al., 1975). The difficulties in deciding whether energy transfer goes from photosystem II to photosystem I or vice versa were discussed by SUN and SAUER (1972). At least for low temperatures the question is settled in favor of photosystem II to I transfer. BUTLER and KITAJIMA (1975) were able to derive semiquantitative estimates for the transfer probabilities by interpreting their fluorescence data in terms of a simple tripartite model for the distribution of quanta among antennae system II, antennae system I and "light harvesting protein". This configuration is illustrated in Figure 2. At excitation with 633 nm-light the transfer probabilities from II to I ranged between 0.065 (trap II open) and 0.23 (trap II closed) in the presence of Mg^{2+} (5 mM) and between 0.12 (trap open) to 0.28 (trap closed) in the absence of Mg^{2+}.

Mg^{2+} is one of the major ions which electrically balance the light driven proton uptake into the internal phase of thylakoids (DILLEY and VERNON, 1965; SCHRÖDER et al., 1972; HIND et al., 1974). Certain transients in photosystem II fluorescence after the onset of illumination was attributed to the regulation of spill-over by the generation of Mg^{2+}-gradient across the thylakoid membrane (KRAUSE, 1973, 1974; BARBER et al., 1974a, b).

The physiological advantage of controlling the distribution of quanta between the two photosystems by the ionic gradients across the thylakoid membrane still awaits quantitative interpretation.

Qualitatively it is obvious that spill-over increases the efficiency of the electron transport chain at intermediate light intensities. Approaching steady state the plastoquinone-pool becomes largely reduced and therewith less effective as a kinetic buffer for the fluctuations of quanta falling into trap II. At the same time a Mg^{2+}-gradient is built up which tends to enhance spill-over. If, then, antennae system II is hit by a quantum twice, the second quantum is not necessarily lost, but with a probability of about 25% may be used by photosystem I, where it acts to reopen trap II which was just closed by the first hit.

V. Protective Reactions

During their lifetime chloroplasts operate on the same set of chlorophyll molecules. In contrast to this, chlorophyll in vitro if illuminated is irreversibly destroyed in the presence of the ubiquitous oxygen. Its oxidation occurs from the triplet state. The triplet state is formed from the excited singlet state with relatively low probability by a spin flip process (intersystem crossing). Despite of its low probability this may be harmful under continous illumination.

Based on the observation that mutants of bacteria deficient in carotenoids lose their antennae system under strong illumination, it has been suggested that carotenoids may somehow protect triplet excited chlorophyll from oxydation (GRIFFITHS et al., 1955). Flash photometry led to the identification of the protective reaction. Triplet excited chlorophyll a transfers its excitation into the triplet state of carotenoids from where it is dissipated mainly by interaction with the triplet ground state of dioxygen.

This concept is based on the following observations: Under excitation of isolated chloroplasts with intense flashes absorption changes were detected with three negative bands in the blue plus a broad positive band at 520 nm (ZIEGER et al., 1961; WOLFF and WITT, 1969). No correlated absorption changes were observed in the red (BUCHWALD and WOLFF, 1971). This spectrum closely resembles the one for the formation of the carotenoid triplet state in vitro (CHESSIN et al., 1966). The changes became observable only, if the photochemical traps were closed (WOLFF and WITT, 1969). This suggested that the formation of carotenoid triplets is an alternative pathway to the photochemical use of light energy. The carotenoid triplet state is generated very rapidly (rise time 20 nsec, MATHIS, 1970; K. WITT and WOLFF, 1970). Its decay rate depends on the oxygen pressure (MATHIS and GALMICHE, 1967). Under atmospheric conditions the decay time is short enough (3 μsec) to cope with all quanta falling into the antennae even under bright sunlight. A quantitative estimate for the relative efficiencies of three wasteful processes in photosystem II yielded the following figures (WOLFF, personal communication): fluorescence, 6%, formation of carotenoid triplets, 10–20%, radiationless deactivation, 78%.

VI. Structure

In mature chloroplasts the membrane area available for each chlorophyll molecule is about 200 Å^2 (WOLKEN and SCHWERTZ, 1953; THOMAS et al., 1956). This is just the area required to cover the membrane with a homogenous layer of porphyrin rings.

Structural models were derived with the antennae chlorophylls forming a layer on top of a more or less homogenous layer of lipids. While WEIER and BENSON (1967) based their model on electron microscopy, KREUTZ (1970) mainly relied on his X-ray small-angle scattering data plus the then available information on fluorescence polarization. Both approaches suffered from a relatively poor resolution in plane of the membrane plus the principal difficulty in assigning specific membrane components to the observed electron density profiles. Hence most of the finer details of these models are hypothetical.

Chemical fractionation of chloroplasts revealed at least three different structural units for the antennae. It was shown that each reaction center is associated with a separate pack of antennae chlorophylls (BOARDMAN and ANDERSON, 1964; VERNON et al., 1966; SANE et al., 1970). In addition to these antennae units II and I "light harvesting chlorophyll-protein complexes" were isolated which have no photochemical activity by themselves but serve as antennae for photosystem I and II (THORNBER and HIGHKIN, 1974). One light-harvesting chlorophyll-protein complex from a green bacterium was cristallized and analyzed by X-ray diffraction at a resolution of 2.8 Å. It proved to be a trimer of three polypeptide chains with seven porphyrin rings each, wrapped into the polypeptide chain in rather irregular fashion (FENNA et al., 1974; FENNA and MATHEWS, 1975). Together with further evidence this suggested that the antenna rather than being a two-dimensional lattice of chlorophyll is a "pebble-mosaik" of chlorophyll proteins (see SAUER, 1975).

The position of chlorophylls at the outer or the inner side of the thylakoid membrane is open. There is indication that some chlorophylls are accessible from outside to specific antibodies (RADUNZ et al., 1971).

Information on the orientation of antennae chlorophylls with respect to the membrane was obtained from studies on linear dichroism and fluorescence polarization with macroscopically aligned thylakoid membranes. While earlier attempts to orient chloroplasts by exposure to high electric field strength and hydrodynamic shear (SAUER and CALVIN, 1962; SAUER, 1965), did not produce high degrees of orientation, sedimentation of chloroplasts on glass plates (BRETON and ROUX, 1971) and exposure to high magnetic field strength (GEACINTOV et al., 1971) proved to be more efficient. Polarized absorption and fluorescence studies revealed preferential orientation of antennae transition moments to the membrane (for the orientation of transition moments within the molecular coordinate system, see Fig. 3). The Q_y-band transition moments of those antennae chlorophylls which absorb above 680 nm are oriented in the plane of the membrane, while the transition moments of the Soret band (B-bands, see Fig. 4) are partly in the plane and partly out of plane. Carotenoids are by preference in the plane (BRETON and ROUX, 1971; BRETON et al., 1973; GEACINTOV et al., 1971, 1972). By linear dichroism in photoselection it was shown that there is no preferential orientation of the in plane oriented transition moments of photosystem I within the plane of the membrane. This implies that antennae system I behaves as an absorber with circular symmetry (JUNGE and ECKHOF, 1974; JUNGE, 1974).

C. Electron Transport

Photosynthetic electron transport from water to NADP$^+$ serves two purposes:
it produces reducing equivalents in form of NADPH and molecular oxygen and
it generates an electrochemical potential of the proton across the thylakoid membrane, which is used for the production of the energy carrier ATP.

Two photoreactions operate in series to lift up electrons from the low potential
donor, water, to the high potential acceptor NADP$^+$. The transfer is mediated
by several components. Their functional sequence is illustrated in Figure 7. Most
of the electron pathways are well characterized, others await further elucidation
(broken lines in Fig. 7). According to their function the *electron carriers* may
be grouped into the following classes (for specific references, see Sects. C.I and
C.II):

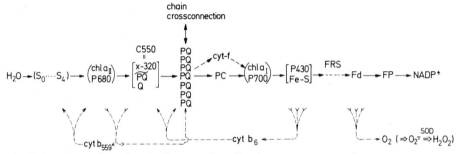

Fig. 7. Sequence of carriers in linear and cyclic electron transport. Synonymous terms are
written stacked in brackets. *Broken lines:* pathways which are not fully identified. *FRS:*
ferredoxin reducing substance, *SOD:* superoxiddismutase

Two photochemically active chlorophyll complexes, chlorophyll a$_I$ and chlorophyll a$_{II}$, transduce energy of light quanta into the energy of redox couples. Excitation of the photochemically active complex leads to the oxidation of a chlorophyll
a dimer in the respective reaction center. The transfer of an electron to the primary
acceptor crosses the thylakoid membrane and thus charges the membrane electrically.

The primary acceptors, X-320, a "bound" plastoquinone, and P-430, which
might be a "bound" iron-sulfur protein are tightly coupled to their respective chlorophyll a dimer to secure extremely rapid and practically irreversible transfer of
electrons. This is required in order to compete with the rapid wasteful processes
which deactivate singlet excited chlorophylls.

Plastoquinone transfers electrons from photosystem II to photosystem I. It
serves three special purposes. Being higher abundant than the other components
of the electron transport chain it provides an electron buffer, which provides
"smooth" operation of the chain even under larger fluctuations around an even
distribution of light quanta among the two photosystems. The second function
is to interconnet several electron transport chains. This improves the reliability
of the system under conditions where some of the photochemical reaction centers
may be damaged. If, for instance, one of the centers II fails to operate the reaction
center I attached to it will still be served with electrons from another center II,
which under saturating light would just perform an extra turnover. The third

function of plastoquinone is to transfer hydrogen across the membrane and thus to contribute to the generation of the electrochemical potential of the proton across the membrane (cf. Sect. D). The oxidation of plastohydroquinone by plasto-cyanin is the rate-limiting step of the whole chain and the site of control of the electron transfer rate by the electrochemical gradient and therewith by the ATP/ADP ratio.

Plastocyanin, ferredoxin and the flavoprotein are necessary components in that the electron transfer is interrupted if they are extracted or blocked. Probably their function is merely catalytic, to enhance the velocity of the electron transport.

Cytochrome b-559, cytochrome b-6, cytochrome f and the ferredoxin-reducing-substance are still ill defined in their functional role and their place in the electron transport chain. There is evidence that cytochrome b-559 and cytochrome f are not always in the chain. In dependence on the state of other components they apparently alter their kinetic (cytochrome f) and thermodynamic (cytochrome b-559) properties. It is conceivable that their role is regulatory, to avoid overreduction or -oxidation in the chain. Cytochrome b-6 mediates cyclic electron flow around reaction center I. The precise pathway and the relative efficiency of this has still to be elucidated. The position of the ferredoxin-reducing-substance illustrated in Figure 7 is questionable (see below).

C-550, finally, is probably not directly involved in the transfer of electrons, but, closely associated with the primary acceptor of reaction center II, it reflects its redox state as an indicator.

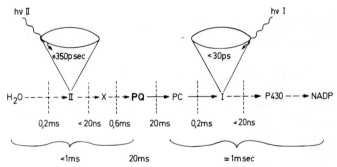

Fig. 8. Characteristic time constants of the energy transfer into the reaction centers and for the electron transport (for details, see text)

The *kinetic parameters* of the electron transport chain are illustrated in Figure 8 (for references, see Sects. C.I and C.II). The transfer of quanta into the traps and the primary charge separation in both reaction centers are very rapid reactions (at least nanosecond range). The adjacent reactions are much slower (millisecond range), although by order of magnitude faster than the rate limiting step, the oxidation of plastohydroquinone by plastocyanin (20 msec). This is the step upon which the control of the reaction velocity by the ATP/ADP-level is executed (external pH = 8).

The *energetics* of the electron transport chain according to the Z-scheme of HILL and BENDALL (1960) is illustrated in Figure 9. The abzissa follows the sequence of electron carriers, while the ordinate scale represents the "isotropic midpoint

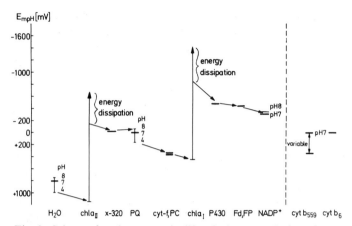

Fig. 9. Scheme for the energetic lift of electrons during electron transport. The isotropic midpoint potential of the electron carriers is given for pH 7. For plastoquinone and water they are given at other pH-values, also (for details, see text)

potentials" of these components in millivolts. For a review on redox-potentiometry with biological systems, see DUTTON and WILSON (1974). Electron flow, if not photochemically driven, proceeds from negative towards positive midpoint potential.

When processed through the chain, an electron is lifted from a potential of $+810$ mV up to -310 mV. This process is driven by two photochemical reactions with an energy input of at least 1,800 mV each (at excitation with light at 700 nm) which are followed by several dissipative and energy conserving reaction steps, respectively. Even if the photochemical reactions occurred in perfect thermodynamic equilibrium, about 30% of the light energy is lost, due to the temperature difference between the radiation field and the ambience (DUYSENS, 1966; ROSS and CALVIN, 1967; KNOX, 1969). The then available free energy of about 2,500 mV is only half conserved while the other half is dissipated as heat. In Figure 9 dissipative reaction steps are indicated by downward directed arrows. There are two regions where part of the energy can be conserved in form of an electrochemical potential difference of the proton across the membrane. These are indicated by vertical bars at H_2O and PQ. While the midpoint potentials of the other components of the electron transport chain are almost independent of the ambient electrochemical potential of the proton, those of water, $NADP^+$ and plastoquinone are not. This is because their oxidoreduction is coupled to the release and binding, respectively, of protons. Under steady illumination the electrochemical potential of the proton in the internal phase of thylakoids is by about 240 mV (equivalent to 4 pH-units) higher than the one in the external phase. Since the reduction of plastoquinone occurs at the outer side of the membrane, and its oxidation at the inner (see Sect. D) its midpoint potential differs by the same amount between these two reaction sites. The drop in the free energy of the electron during transfer on plastoquinone across the membrane, however, is about poised by the gain of electrochemical potential of the proton.

Functional neighborhood of electron transport components as illustrated in Figure 7 does not imply their *topological neighborhood within the membrane*. It is debatable, whether photosystems I and II cover the thylakoid membrane in an alternating pattern as on chess board or if they are arranged in accumulated multi-system I and multi-system II units. Even in the latter case they may be functionally cross-connected via "light harvesting protein" and plastoquinone for rapid exchange of quanta and electrons.

It is almost certain from fractionation studies that the intergrana lamellae, which interconnect thylakoid stacks, carry photosystem I activity, only (JACOBI and LEHMANN, 1969; SANE et al., 1970; ARNTZEN et al., 1972). The ratio of chlorophyll a_I, which is inaccessible to the linear electron transport from photosystem II, to accessible chlorophyll a_I is 1:3 (HAEHNEL, 1976). Thus the cyclic pathway around photosystem I which is illustrated in Figure 7 may be topologically segregated from the linear one.

In contrast to most of the functional interrelations, *mechanistic details of the electron transport are unknown*: For instance, which of the electron transfer steps are jumps within a more or less rigid matrix of carriers, which do not require translational or rotational motion of the latter; whether plastoquinone, which translocates hydrogen across the membrane and probably in the plane of the membrane, is acting as a mobile carrier; whether plastocyanin, which at least partly resides loosely bound at the inner side of the membrane (HAUSKA et al., 1971), mediates cross-talk between electron transport chains in grana stacks and photosystems I in the intergrana lamellae by migrating back and forth. Progress in this field certainly necessitates much more information on the fine structure of the thylakoid membrane.

I. Photochemical Reactions

Two distinct photochemical reactions were identified by spectroscopic techniques and by fractionation of photosynthetic membranes from green plants and algae. In both photosystems light causes the oxidation of a chlorophyll a-dimer. The active dimers are denoted chlorophyll a_I and chlorophyll a_{II}, in the following.

Chlorophyll a_I (or P-700) is characterized by certain absorption changes which appear after illumination (e.g. by a flash) of chloroplasts. Their negative maxima are at 700 nm (KOK, 1957), 438 nm (RUMBERG and WITT, 1964) and 682 nm (DÖRING et al., 1968). A broad positive peak appears around 810 nm (KE, 1973; INOUE et al., 1973; MATHIS and VERMEGLIO, 1975). This photoinduced difference spectrum resembles the one for the photooxidation of chlorophyll a to yield the radical cation in vitro (BORG et al., 1970). The attribution of these absorption changes to a univalent photooxidation of chlorophyll a is firmly established by three lines of evidence: (1) redox titration [midpoint potential $E_{m, pH\,7} = +450$ mV, KOK (1961)], (2) redox kinetics (RUMBERG and WITT, 1964) and (3) electron spin resonance (WARDEN and BOLTON, 1973, 1974; KE et al., 1974b). Prior inconsistencies between the number of spins and the amount of chlorophyll a bleached (BEINERT and KOK, 1964) were eliminated in the former esr-studies.

The narrow linewidth of the epr-signal led to the conclusion that the unpaired electron in the radical cation of chlorophyll a_I is delocalized over two porphyrin rings (NORRIS et al., 1971; confirmed in Endor-technique NORRIS et al., 1974). This suggested that chlorophyll a_I is a dimer of two chlorophyll a molecules.

Indirect evidence for this came from the circular dichroism of the difference spectrum (Phillipson et al., 1972) and from the existence of a double band in the red spectral region (Döring et al., 1968, see above).

The details of the difference spectrum of chlorophyll a_I are not well understood as yet. The interpretation is difficult because of the unknown absorption spectrum of the dimer and the possible influence of the redox state of chlorophyll a_I on the absorption spectrum of the antennae chlorophylls in its neighborhood. For the extinction coefficient of chlorophyll a_I, see Hiyama and Ke (1972).

The internal structure of the chlorophyll a_I dimer is unknown. Model studies on chlorophyll aggregation in vitro revealed that chlorophyll water adducts ($-chl-H_2O-chl-H_2O-chl-$) share properties (far red absorption, spin delocalization) with the chlorophyll a_I dimer (for a review, see Katz and Norris, 1973). The structure of these adducts was analyzed chemically (Cotton et al., 1970) by X-ray technique (Katz et al., 1968) and by infrared spectroscopy (Ballschmiter and Katz, 1969). A model was derived with one molecule of water between two chlorophyll a molecules, interacting by its oxygen with the central magnesium atom of one chlorophyll and by its hydrogens with the two $C=O$-groups at ring V of the other one. The validity of this model for the in vivo dimer, however, was challenged for its lack of symmetry (Fong, 1974).

After excitation of chloroplasts with a short flash of light chlorophyll a_I is oxidized in less than 20 nsec (K. Witt and Wolff, 1970). The trapping of a quantum in photosystem I, however, occurs in less than 30 psec. The sequence of events covering the three decades in between these time marks is unknown for chlorophyll a_I.

Up to now picosecond-spectroscopy has been restricted to reaction centers from bacteria. One reason lies in the relative ease by which highly enriched and optically "clear" preparations are available. The following sequence of events for the reaction center pigment (P) of *Rhodopseudomonas sphaeroides* was proposed (Rockley et al., 1975; see also Kaufmann et al., 1975):

$$P + h\nu \longrightarrow P^+_{singlet} \longrightarrow P^F \longrightarrow P^+$$
$$\text{less than 8 psec} \qquad \text{246 psec}$$

The state P^F, according to its difference spectrum was tentatively identified with the biradical of the photochemically active dimer, $bchl^- - bchl^+$.

It is probable that the photooxidation of chlorophyll a_I occurs as rapidly as the equivalent reaction in bacteria. The reaction does not require thermal activation as documented by its occurrence at low temperature (Witt et al., 1961; Floyd et al., 1971; K. Witt, 1973b).

The reduction of oxidized chlorophyll a_I has three kinetically distinct components: 10 μsec, 200 μsec and 20 msec (Haehnel et al., 1971; Haehnel and Witt, 1972). The reaction partner in the most rapid reaction is unknown, the one in the intermediate reaction is plastocyanin, and the one in the slowest reaction is plastoquinone reacting slowly via plastocyanin (Haehnel, 1974).

Chlorophyll a_{II} (or P-680) was identified by negative directed changes of absorption at 687 nm and 433 nm (Döring et al., 1969) plus a positive band around 825 nm (Mathis and Vermeglio, 1975). Two lines of evidence indicate that this difference spectrum is attributable to the photooxidation of chlorophyll a: (1)

its similarity with the respective difference spectrum in vitro (BORG et al., 1970) and (2) e.p.r.-experiments revealing the photoproduction of oxidized chlorophyll a in photosystem II (MALKIN and BEARDEN, 1973; KE et al., 1974a; VAN GORKOM et al., 1974). Again, the relatively narrow width of the e.s.r. spectrum indicated that chlorophyll a_{II} is a dimer of two chlorphyll a molecules across which the unpaired electron of the radical cation is delocalized.

The velocity of the photooxydation of chlorophyll a_{II} has not been resolved, yet. Indirect arguments led to the conclusions that it occurs in less than 20 nsec (WITT, 1971). The argument was based on the rise time of the electric potential across the membrane (less than 20 nsec, WOLFF et al., 1969) and on the contribution of chlorphyll a_{II} to the generation of the electric potential (SCHLIEPHAKE et al., 1968).

The reduction kinetics of chlorophyll a_∞ is biphasic. The half-life time of one component is 200 μsec (DÖRING et al., 1969), that of the other is 35 μsec (GLÄSER et al., 1974). The same life times were observed in delayed fluorescence (ZANKEL, 1971) which reflects the lifetime of the pair chlorophyll a_{II}^+ Q^-, with Q, the primary electron acceptor (LAVOREL, 1968). The electron donors responsible for these two phases of chlorophyll a_{II} reduction are not identified chemically.

The relative abundance of chlorophyll a_I, counting only those dimers which are accessible to linear electron transport from photosystem II, is lower than the abundance of chlorophyll a_{II}. Their ratio is 0.85/1 (HAEHNEL, 1976).

II. Non-Photochemical Components

The *water-oxidizing enzyme* system is chemically uncharacterized except for the participation of manganese (KESSLER, 1957). Its function is to accumulate four oxidizing equivalents before interacting with two molecules of water to produce di-oxygen, four electrons and four protons. The accumulation of electron holes became evident from studies on the oxygen yield under a sequence of exciting flashes starting from the dark adapted state (JOLIOT et al., 1969; KOK et al., 1970). Whether or not the storage sites for oxidizing equivalents are on manganese as in the model of RENGER (1970, 1976) is open.

Cytochrome b-559 has been isolated (GAREWAL and WASSERMAN, 1974). Its midpoint potential in situ is variable between $+350$ mV and 0 V (FAN and CRAMER, 1970). Its functional role under physiological conditions is unknown. Direct spectroscopic observation of its oxidoreduction has so far been restricted to unphysiological conditions (low temperature, modified chloroplasts). Reduction of cytochrome b-559 and subsequent re-oxidation by chlorophyll a_{II} was demonstrated at low temperature (MATHIS and VERMEGLIO, 1975). According to indirect evidence it also occurs at room temperature (KOK et al., 1974). Other authors reported that it mediates electron transport between chlorophyll a_{II} and plastoquinone (BÖHME and CRAMER, 1971).

C-550 is chemically uncharacterized. Its absorption changes kinetically match those of the primary electron acceptor in photosystem II, X-320 (AMESZ et al., 1974; VAN GORKOM, 1974; HAVEMAN et al., 1975). It was proposed that the absorption changes of C-550 are due to a shift in the absorption bands of pheophytin a in response to the redox state of X-320 (VAN GORKOM, 1974). Hence, C-550 is probably no electron carrier, but just accidentially serves as an indicator, courtesy of the chloroplast to the student of photosynthesis.

X-320 was first identified by its characteristic absorption changes in the ultra-violet spectral region (STIEHL and WITT, 1968, 1969). It was shown that it reacts as an electron acceptor for chlorophyll a_{II} which is reduced in less than 30 μsec (unresolved) (STIEHL and WITT, 1969). As it reacts even at low temperature it was identified with the primary electron acceptor (K. WITT, 1973a). The chemical reaction underlying to the absorption changes of X-320 was identified with the univalent reduction of plastoquinone to yield the semi-quinone anion. This was based on the similarity between the in vivo spectrum of X-320 (STIEHL and WITT, 1969; AMESZ et al., 1974; Van Gorkom, 1974) with the difference spectrum of the above reaction in vitro (BENSASSON and LAND, 1973). In fluorescence studies the primary electron acceptor of photosystem II was named the quencher Q (DUYSENS and SWEERS, 1963). The kinetics of Q-reduction as revealed by fluorescence (FORBUSH and KOK, 1968) parallel the one of X-320 in absorption spectroscopy. The midpoint potential of X-320 can be inferred from the one of Q: $-35\,\mathrm{mV}$ (BUTLER, 1966).

Plastoquinone was early identified as a necessary component for electron transport in green plants and algae (BISHOP, 1959). Its midpoint potential is about 0 V at pH 7. The difference spectra for its reduction to yield the monoreduced plastosemiquinone (BENSASSON and LAND, 1973) and the double reduced plastohydroquinone (HENNINGER and CRANE, 1964) are known from in vitro studies. Similar absorption changes were detected in chloroplasts from green plants (KLINGENBERG et al., 1962) and algae (AMESZ, 1964). Plastoquinone is assumed to be reduced with a half-rise-time of 0.6 msec by photosystem II (STIEHL and WITT, 1968, 1969), its oxidation by photosystem I occurs in 20 msec. The abundance of plastoquinone is about seven molecules per reaction center II. Under steady illumination about four of these are kept in the reduced state (STIEHL and WITT, 1969). The reduction into hydroquinone being divalent, this implies that plastoquinone provides a storage capacity for eight electrons between chlorophyll a_{II} and the rate limiting reaction step. The storage capacity between this site and chlorophyll a_I is only three electrons (STIEHL and WITT, 1969; MARSHO and KOK, 1970). Studies on the linear electron flow with successive blocking of reaction centers II led to the conclusion that at least ten electron transport chains are interconnected at the level of plastoquinone (SIGGEL et al., 1972). The oxidation of plastohydroquinone by plastocyanin is the major site of control of the electron transport by the internal pH (RUMBERG et al., 1969; SIGGEL, 1974).

Cytochrome b-6 has been isolated (STUART and WASSARMAN, 1975). Its midpoint potential at pH 7 is around 0 V (HILL and BENDALL, 1967; KNAFF and ARNON, 1969; BÖHME and CRAMER, 1973). Its relative abundance is about one molecule per electron transport chain. Kinetic analysis of its absorption changes revealed that it mediates cyclic electron flow around photosystem I (HIND and OLSON, 1966; WEIKARD, 1968; KNAFF and ARNON, 1969). It is reduced by photosystem I at a half-rise time of 1.3 msec and oxidized by the same photosystem at 35 msec (DOLAN and HIND, 1974).

Cytochrome f was first characterized by HILL and SCARISBRICK (1951). It was reported that its absorption changes were observable at low temperature (DUYSENS et al., 1961; WITT, et al., 1961). Cytochrome f is reduced by photosystem II while oxidized by photosystem I (DUYSENS and AMESZ, 1962). The half-rise time of the oxidation is about 200 μsec (HILDRETH, 1968; DOLAN and HIND, 1974). The half-rise time of the reduction by photosystem II is 17 msec according to HAEHNEL (1973) but, at variance from this, biphasic at 7.3 msec and 85 msec according to DOLAN and HIND (1974). It has been argued that cytochrome f functions in the main chain. The argument was based on the above kinetic data and on its midpoint potential, which is +350 mV (HILL and BENDALL, 1967). It has been shown recently, however, that only a minor fraction (about 15%) of the electrons flowing from photosystem II into photosystem I pass via cytochrome f, at least under flash excitation of the respective studies (HAEHNEL, 1973, 1974). This led to the suggestion that cytochrome f acts on a side branch of the main chain. A side chain action finds support from the relatively low abundance of cytochrome f, which is about 0.5/1 in relation to chlorophyll a_I or plastocyanin (PLESNICAR and BENDALL, 1970; HAEHNEL, 1974).

Plastocyanin is a copper protein (KATOH, 1960, 1971) which accepts two electrons per molecule. Its relative abundance in relation to chlorophyll a_I is about 1/1 (PLESNICAR and BENDALL, 1970; HAEHNEL, 1974). It is oxidized by chlorophyll a_I, probably at 200 μsec, and reduced with a half-rise time of about 20 msec by plastoquinone (HAEHNEL, 1974). Its midpoint potential is +350 mV (KATOH et al., 1962). Based on mutant studies it has been postulated that plastocyanin mediates electron flow between the two photosystems in the sequence plastoquinone–cytochrome f–plastocyanin–chlorophyll a_I (GORMAN and LEVINE, 1966). This was strengthened by the observation that extraction and reconstitution of plastocyanin abolished and restored the oxidation of cytochrome f by P-700 (SIEDOW et al., 1973). Several lines of evidence suggest that plastocyanin is the only component between plastoquinone and chlorophyll a_I: (1) the above studies, which place cytochrome f on a side path (HAEHNEL, 1973, 1974); (2) further independent kinetic studies revealed that there is only one component reacting between plastoquinone and chlorophyll a_I (BOUGES-BOQUET, 1975); (3) the capacity of three electrons for the donor pool of photosystem I (STIEHL and WITT, 1969; MARSHO and KOK, 1970) is fully accounted for by plastocyanin (two electrons) and chlorophyll a_I (one electron). When taken together, the physical and the biochemical evidence seems to indicate that plastocyanin acts in parallel to cytochrome f. However these two components are not independent from each other.

P-430 was identified first by absorption changes in the visible and in the near UV (HIYAMA and KE, 1971) which meet the expectation for those of the primary electron acceptor of chlorophyll a_I (KE, 1973). Although the optical absorption spectrum of P-430 deviates from the expected one for an iron sulphur protein KE et al. (1974 b) from parallel studies (e.p.r. and optical) suggested that P-430 might be "bound" ferredoxin (MALKIN and BEARDEN, 1971; EVANS et al., 1972) which is reduced by photosystem I. The midpoint potential of P-430, −470 mV KE et al. (1974 b) from parallel studies (e.p.r. and optical) suggested that P-430 a_I and soluble ferredoxin.

Another *ferredoxin-reducing substance* was isolated from chloroplasts (YOCUM and SAN PIETRO, 1972; TREBST, 1972; REGITZ and OETTMEIER, 1972) which was claimed to mediate electron flow between chlorophyll a꜀ and soluble ferredoxin. Its relatively low midpoint potential of −550 mV, however, makes this role questionable in view of the higher one of P-430.

Ferredoxin is the next intermediate. This iron-sulphur protein has a midpoint potential of -420 mV (ARNON, 1965; BUCHANAN and ARNON, 1971). It is oxidized by the flavoprotein ferredoxin-NADP$^+$-reductase, which finally reduces NADP$^+$.

D. Electrochemical Potential Generation

Physical and biochemical studies confirmed the hypothesis of MITCHELL (1961, 1966) that the electron transport chain crosses the thylakoid membrane twice, as illustrated in the upper part of Figure 10. In a single turnover both reaction centers transfer one electron from the inner to the outer side of the membrane. This charges the membrane's electric capacitance in less than 20 nsec (see Fig. 11). Although originating as a localized dipole field in each reaction center, the electric field delocalizes rapidly due to redistribution of ions in both aqueous phases.

Both photochemical reactions are linked in series by an electrically neutral reaction step, the transfer of electrons plus protons on plastoquinone from the outer to the inner side of the thylakoid membrane. The oxydation of water and of plastohydroquinone respectively by carriers which do not bind protons on reduction releases one proton per electron at each of these sites into the internal phase. The reduction of plastoquinone and of NADP$^+$ (plus PGA in the secondary processes) leads to the uptake of one proton per electron at each site from the external phase. Thus the electric potential, although originating from rapid electron

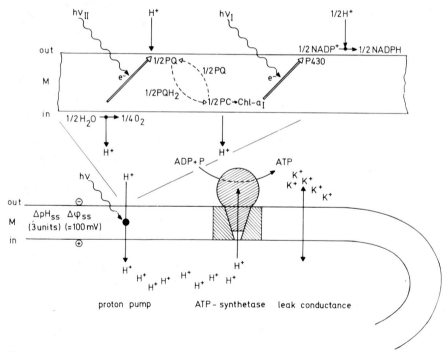

Fig. 10. Model for the generation and the use of an electrochemical potential difference of the proton across the thylakoid membrane (JUNGE, 1976; for details, see text)

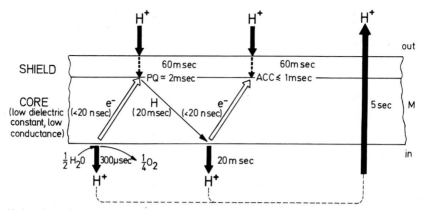

Fig. 11. Half-rise times for the electric potential generation *(open arrows)* and for the release and uptake of protons in either aqueous phase *(filled arrows)*; (for details see text)

transfer, is ultimately equivalent to the translocation of two protons per electron from outside into the internal phase. The two sites of proton release can be identified with the "coupling sites" between electron transport and ATP-synthesis.

The kinetic parameters of the electrochemical potential generation are illustrated in Figure 11. While proton release into the internal phase kinetically matches the respective redox reactions, proton uptake from the external phase is delayed against the reduction of plastoquinone and of the terminal acceptor respectively. This delay was attributed to a proton diffusion barrier shielding the outer side of the dielectric core of the membrane against the external aqueous phase. The structure of this shield and its physiological relevance are unknown. While a proton gradient relaxes at 60 msec across the shield it relaxes in about 5 sec across the core of the membrane.

The vectorial electron transport scheme in Figures 10 and 11 does not include any contribution of eventual proton translocating proteins to the electric potential generation. There is no evidence for such a contribution from stoichiometric and kinetic correlation of proton and electron transfer. The only possible hint for such special proteins are the high values determined for the quantum yield of proton translocation. These values are contradictory with the generally accepted proton over electron stoichiometries. The discrepancy is unsettled as yet.

Under steady illumination of chloroplasts in the absence of phosphorylation about 500 protons are accumulated in the internal phase per electron transport chain (600 chlorophylls). This gives rise to an acidification by about 3 pH-units of the internal phase (pH 5) against the external one (pH 8). The steady state level of the electric potential is still under discussion, its order of magnitude is 50 mV. More than 99% of the protons accumulated inside are electrically compensated by extruded magnesium-cations and inwardly accumulated chloride-anions. About 99% of the protons are bound by buffering groups at the inner side of the membrane.

Major open questions as to the mechanism of the electrochemical potential generation are:

1. The inconsistency of the proton over electron stoichiometries with the quantum yield for proton uptake.

2. The mechanism of the extremely rapid electron transfer crossing the membrane core in each reaction center.

3. The mechanism by which plastoquinone translocates hydrogen across the membrane (migration, collision).

4. The chemical and structural nature of the diffusion barrier for protons at the outer side of the membrane and its relevance for photophosphorylation.

I. The Generation of an Electric Potential

Due to the submicroscopic dimensions of thylakoids the electric potential generated across the membrane was not measurable by straightforwardly interpretatable electric techniques.

The spectroscopic detection of the electric potential via *electrochromic band shifts* of membrane bound pigments is distinguished from the other techniques by its high time resolution and by its applicability to the unperturbed biological system. Elektrochromism denotes the influence of high electric field strength on the band spectra of organic dyes in condensed phases. Band shifts in the order of some tenth of a nanometer are typical if a dye is exposed to electric field strength in the range between 10^5 and 10^6 V/cm, as frequently encountered across ultrathin biological membranes (100 mV across 100 Å are 10^5 V/cm). Electrochromic effects are theoretically well understood (LABHARD, 1961; LIPTAY, 1969) and experimentally established for chloroplast bulk pigments in vitro (MALLEY et al., 1968; KLEUSER and BÜCHER, 1969; SCHMIDT et al., 1971). Some absorption changes observed on flash excitation of chloroplasts were attributed to an electrochromic response of bulk pigments to a photoinduced electric field across the membrane. The argument was first based on their kinetic properties (JUNGE and WITT, 1968) but later corroborated by the similarity between their difference spectrum (EMRICH et al., 1969) with the one for electrochromic effects of these pigments in vitro (SCHMIDT et al., 1971). Further support came from the observation of similar absorption changes in bacterium chromatophores, which were flash excited or exposed to a diffusion potential across their membrane (JACKSON and CROFTS, 1969) or to an externally applied electric field (BORISEVITCH et al., 1975). The linearity of the electrochromic response of chloroplast bulk pigments to the electric potential across the membrane is empirically established by comparison with signals observed with the electric induction technique (WITT and ZICKLER, 1974). Although, a thorough theoretical understanding of the "molecular voltmeter" in the membrane of thylakoids is still lacking (for a discussion of complication, see SCHMIDT and REICH, 1972; CROFTS et al., 1974), the empirical basis seems sound enough to justify even quantitative kinetic studies on the voltage and the current across the membrane (e.g. SCHMID and JUNGE, 1975). The major disadvantage of the electrochromic technique lies in its inapplicability to slow transients of the voltage. This is caused by the possible interference of changes in the light scattering properties of chloroplasts, which may mimic absorption changes and be mistaken for electrochroism (for a discussion, see THORNE et al., 1975).

When an aqueous suspension of chloroplasts is excited with a flash of nonsaturating energy, membranes which are nearer to the light source will recieve more quanta than those which are further away and hence shaded by the former. In consequence, the light induced electric potential across the membrane will differ in dependence on the distance of the membrane from the light source. This by *electrostatic induction* is measurable by macroscopic electrodes placed at different depth in an aqueous suspension of chloroplasts (FOWLER and KOK, 1972, 1974; WITT and ZICKLER, 1973). The major disadvantage of this effect for the evaluation of the electric phenomena at the thylakoid membrane is its sensitivity to ion flow around thylakoids rather than to the one across their membranes.

Although their tip radii are much greater than the thickness of the inner phase of thylakoids, *microelectrodes* were used to probe the electric potential across the membrane (BULYCHEV et al., 1971; VREDENBERG, 1974). Almost unexpectedly their response was reasonable with no indication for major changes in the activity of the thylakoid membrane after impalement by the electrode (VREDENBERG and TONK, 1975).

Ion selective electrodes were used to measure the redistribution of ions between the two aqueous phases separated by the thylakoid membrane in response to a membrane potential (DILLEY and VERNON, 1965; SCHRÖDER et al., 1972; HIND et al., 1974). To avoid possible misinterpretation caused by the interference of active transport systems, the technique was refined to be used with synthetic ions, which are "unknown" to any biological transport system (LIBERMAN and SKULACHEV, 1970).

The electric potential across the thylakoid membrane is generated by two *electrogenic steps*. These were identified with the electron transport in reaction center I and reaction center II (SCHLIEPHAKE et al., 1968). That it is the primary photochemical electron transfer which crosses the membrane is suggested by the following observations: (1) The electric potential is rapidly generated even at low temperatures, which slow down thermally activated reactions (MATHIS and VERMEGLIO, 1975; AMESZ and DE GROOTH, 1975). (2) At room temperature the electric potential rises within less than 20 nsec (unresolved) after excitation with a short flash of light (WOLFF et al., 1969). (3) The reverse reaction of the primary photochemical act, which produces delayed luminescence, is sensitive to an artificially induced diffusion potential across the thylakoid membrane (BARBER and KRAAN, 1970; WRAIGHT and CROFTS, 1971).

As yet, there is no indication of other electrogenic reactions in chloroplasts, than the two photochemical events. The polarity of the electric potential (positive inside) led to the conclusion that the photochemically active dimers, chlorophyll a_I and a_{II}, respectively, are located at the inner side, their primary acceptors at the outer side of the thylakoid membrane. Kinetic correlation of the electric potential with the electron transport practically excludes that the photochemically active chlorophyll-dimers extend into the dielectric core of the membrane (JUNGE, 1974).

The electric potential, although generated as a *localized dipole field* in the reaction centers, *delocalizes* over the membrane. In the time-domain of milliseconds it is delocalized over a unit with at least 10^5 chlorophyll molecules, the minimum size of one thylakoid. This was concluded from the observation that one molecule of a pore forming agent, gramicidin, on 10^5 chlorophyll molecules accelerates the decay of electric potential down to the zero level (JUNGE and WITT, 1968). The time domain of the delocalization which is caused by ion redistribution around thylakoids is in the range below micro-seconds (JUNGE, 1974).

The absolute value of the electric potential generated by a single turnover of both reaction centers is not precisely known. A gross estimate was based on reasonable assumptions as to the membrane's electric capacitance and on the surface density of reaction centers. It yielded a value of 50 mV (SCHLIEPHAKE et al., 1968). After the onset of continous illumination of high intensity the electric potential may rise up to about 200 mV and then decline to a lower value (see WITT, 1971, 1975). The steady state value of the electric potential under continous illumination is under discussion. Figures ranging between 10 and 100 mV have been reported in literature (BARBER, 1972; SCHRÖDER et al., 1972; GRÄBER and WITT, 1974; VREDENBERG and TONK, 1975). For a discussion of methodical problems in the determination of the steady state potential, see RUMBERG this vol. Chap. III.8.

The major open question as to the rapid electric potential generation has to do with the unknown structure that conducts electrons across the membrane in both reaction centers. The thickness of the dielectric core of the membrane across which the primary charge separation takes place, is not known. It will likely approach the total membrane thickness which is 70 Å according to X-ray (KREUTZ, 1970) and electron microscopy (MÜHLETHALER, 1972). It is also unknown, whether this thickness is homogenous over the membrane or if it is reduced at the reaction centers. The only known fact is the location of the photochemically active chlorophyll dimers and their primary acceptors at opposite sides of the membrane core (see above), the chlorophyll a_I dimer with its porphyrin rings oriented rather parallel to the membrane (JUNGE and ECKHOF, 1973, 1974; BRETON et al., 1975). One can only speculate about the structure which, for rapid electronic conduction, bridges the gap between the dimers and their electron acceptors. Carotenoids by overlap of their π-electron system were discussed as possible electron conductors across membranes (DARTNALL, 1946; JAHN, 1962; ILANI and BERNS, 1973). Model studies on photovoltaic effects in artificial membranes with chlorophylls and carotenoids support this suggested action (MANGEL et al., 1975). Alternatively it is conceivable that a proteinacous moeity might act as an injection semiconductor between the photochemical dimers and their acceptors (TRIBUTSCH, 1972). Also, chlorophylls might act in double function, as antennae collecting energy for the reactive dimer and as electron channels by overlap of their π-electron system with the one of the dimer.

II. Proton Translocation

On illumination of broken chloroplasts JAGENDORF and HIND (1963) observed *alkalization* of the suspending medium. Up to 300 protons disappeared from the external phase per electron transport chain (600 chlorophylls) (NEUMANN and JAGENDORF, 1964). While the pH-changes in the external phase were directly observable with a glass electrode, the ones in the internal phase required indirect measuring techniques. By several indirect methods it was confirmed that the internal phase of thylakoids goes acidic:

HAGER (1969) attributed activity changes of the internally trapped violaxanthin deepoxidase to a light induced acidification of the internal phase. RUMBERG and SIGGEL (1969) assumed that the slowing down of the chlorophyll a_I-reduction after the onset of illumination is due to the internal acidification and calibrated this effect quantitatively. Several authors used the distribution of permeant weak acids between the two aqueous phases in response to the pH-difference between them. CROFTS (1967) discussed the distribution of ammonium as induced by a light generated pH-difference. ROTTENBERG et al. (1971, 1972) introduced acridine dyes, where the read-out of the distribution is convenient, as their fluorescence is quenched, if inside. PORTIS and McCARTY (1973) determined the distribution by radioactive tracers and a special centrifugation technique. AUSLÄNDER and JUNGE (1975) used a permeant pH-indicating dye together with a non-permeant macromolecular buffer to record pH-changes in the internal phase at high time resolution.

There is still some discussion as to whether or not some of these indirect methods are valuable for quantitative studies (see BAMBERG et al., 1973; cf. RUMBERG and SIGGEL, 1969; see FIOLET et al., 1974; cf. ROTTENBERG, GRUNWALD and AVRON, 1971, 1972). However, there is general agreement that the internal phase is acidified by about three pH-units against the external one under continuous illumination. This holds if the external pH is at 8 (then $pH_{in} = 5$).

How are protons translocated across the thylakoid membrane? One *possible mechanism* is illustrated in Figure 10. It goes back to Mitchell's hypothesis (1961, 1966) that alternating electron-hydrogen transport across the membrane generates an electrochemical potential difference of the proton, which is used by a proton translocating enzyme for the synthesis of ATP. In this scheme proton translocation is a natural consequence of vectorial electron transport in both reaction centers plus the involvement of protons in the redox-reactions of water, plastoquinone and the terminal electron acceptor. This model neglects any contribution of special proton translocating proteins to the generation of the electric potential. Proton translocating proteins were suggested to act in mitochondria and chromatophores (CHANCE et al., 1970) as well as in chloroplasts (SIGGEL, 1974; MITCHELL, 1975). One such example is experimentally confirmed in *Halobacterium halobium* (OESTERHELT and STOECKENIUS, 1973).

Any discrimination between the relative contributions of electron-hydrogen-loops and proton translocating proteins necessitates stoichiometric and kinetic resolution of all sites at either side of the thylakoid membrane which interact with protons. This requires methods with high time resolution. In this respect *pH-indicating dyes* are superior to any glass electrode even to those modified for higher resolution (cf. SCHWARTZ, 1968; FOWLER and KOK, 1974). From the first applications of pH-indicating dyes to biological membranes (CHANCE and MELA, 1966) there was discussion of possible artifacts attributable to binding changes, solvatochromic effects and even redox reactions (COST and FRENKEL, 1967; MITCHELL et al., 1968; FIOLET and VAN DE VLUGT, 1975). Appropriate controls for the elimination of these artifacts were given by JUNGE and AUSLÄNDER (1974). Another complication arises from the distribution of these dyes between the external, the internal and the membrane phase. However, it was shown, that a discrimination between the dye's response to changes in either aqueous phase is feasible by means of permeant and non-permeant buffers (AUSLÄNDER and JUNGE, 1975).

Studies with pH-indicating dyes under flash excitation of chloroplasts supported the vectorial electron-hydrogen transport scheme illustrated in Figures 10 and 11.

Two *sites of proton uptake* from the external phase were detected and, by their action spectra, attributed to either photochemical reaction center (SCHLIEP-HAKE et al., 1968). By variation of artificial electron acceptors they were attributed to the reduction of the terminal acceptor and of plastoquinone (JUNGE and AUSLÄN-DER, 1974; AUSLÄNDER et al., 1974). Their proton-over-electron stoichiometries are $1 H^+/e^-$ for the PQ site and $1 H^+/e^-$ if oxygen via benzylviologen was the terminal electron acceptor ($0.5 H^+/e^-$ for $NADP^+$, $0 H^+/e^-$ for ferricyanide).

Two *sites of proton release* into the internal phase with a stoichiometry of $1 H^+/e^-$ each were identified and attributed to the oxydation of water and of plastohydroquinone at the inner side of the membrane (JUNGE and AUSLÄNDER, 1974; FOWLER and KOK, 1974). While these stoichiometries were recorded after the internally liberated protons had flown back into the external phase, they were later confirmed for the internal phase by the acridine fluorescence technique (GRÄBER and WITT, 1975) and by the permeant dye neutral red (AUSLÄNDER and JUNGE, 1975).

The *kinetic parameters* of the four protolytic reactions were resolved. Proton release into the internal phase occurred at 300 µsec (half-rise time), in agreement with the expected time for the oxidation of water, and for the other site at 18 msec, in agreement with the oxidation of plastohydroquinone (AUSLÄNDER and JUNGE, 1975). Proton uptake from the external phase, however, was considerably delayed against the respective redox reactions. It occurred at 60 msec (AUSLÄNDER and JUNGE, 1974). It was demonstrated that the half-rise-time of the external alkalization could be accelerated by mechanical or chemical disruption of chloroplasts. If, in addition, proton carrying agents were present proton uptake was accelerated until it almost matched the reduction of plastoquinone and the terminal acceptor (AUSLÄNDER and JUNGE, 1974). This led to the suggestion that the sites for the reduction of these components, which are located at the outer side of the membrane core, are shielded against the external aqueous phase by a diffusion barrier for protons. The structural and chemical properties of the shield are unknown. It had to be postulated for kinetic reasons that reduced plastoquinone is protonized rapidly from a proton reservoir beneath the shielding layer which only slowly refills with protons from the external aqueous phase (AUSLÄNDER and JUNGE, 1974). The implication of such proton sinks in the membrane for the mechanism of photophosphorylation is open (for a discussion, see WILLIAMS, 1975).

The above results gave no evidence for the involvement of any proton translocat-ing protein. Instead proton translocation appeared as a consequence of the vectorial electron-hydrogen transport. However, as the above experiments were restricted to flash excited chloroplasts, the question arises as to whether such proton translo-cating proteins come in under continuous illumination. This is a question as to the proton over electron *stoichiometry under continuous light.* While some authors reported stoichiometries of $2 H^+/e^-$, in agreement with the above flash studies (IZAWA and HIND, 1967; RUMBERG et al., 1968; SCHWARTZ, 1968), other authors reported lower (GOULD and IZAWA, 1974; HOPE and CHOW, 1974) and higher stoichiometries (KARLISH and AVRON, 1967; LYNN and BROWN, 1967; DILLEY, 1970). The stoichiometries different from 2 were subjected to methodical criticism. SCHRÖDER et al. (1972) pointed out that values apparently lower than 2, if recorded with a glass electrode, might be due to the slow response of this device, with necessarily ignores the first short transient of proton efflux after cessation of illumination. They were able to define the conditions under which this rapid transient occurs. JAGENDORF (1975) criticized the approach of those authors, who obtained stoichiometries higher than 2, for comparing the steady-state rate of electron flow with the initial rate of proton uptake. In consequence, the most probable proton over electron stoichiometry of the linear electron transport chain is 2, if the terminal electron acceptor is proton-binding. Thus there is no serious objection against the above mechanism of the proton pump resulting from steady-state proton over electron stoichiometries.

Biochemical evidence for the vectorial electron transport scheme illustrated in Figures 10 and 11 came from biochemical attempts to localize the electron carriers at either side of the membrane (for reviews, see TREBST, 1974; ARNTZEN and BRIANTAIS, 1975) and from studies with artificial electron carriers with different lipid solubility (see TREBST, 1974).

Two independent experiments on the *quantum yield of proton translocation* yielded results contradictory to the above stoichiometry. At excitation with far red light, stimulating by preference photosystem I, DILLEY and VERNON (1967) observed the uptake of 5 H$^+$/$h\nu$, while HEATH (1972) reported 6.7 H$^+$/$h\nu$. In the light of an electron over quantum stoichiometry of 1 e$^-$/$h\nu$ for each reaction center, these figures imply H$^+$/e$^-$-ratios of 5 and 6.7, respectively, for photosystem I, in contrast to the above results which revealed a ratio of 1 H$^+$/e$^-$ for each reaction center. As yet, no serious argument has been raised against the calibration techniques in the above studies on the quantum yield. Hence the discrepancy has still to be resolved.

Some remarks on *electric compensation and buffering* of the inwardly translocated protons.

Under continous illumination the electric potential is in the order of 50 mV (see Sect. D.1). This is equivalent to the displacement of two elementary charges across the membrane area covered by one electron transport chain (SCHLIEPHAKE et al., 1968). Under continous illumination more than 100 protons are translocated by the same membrane element. This implies that about 98% of the inwardly translocated protons are electrically compensated by the efflux of cations and the influx of anions. In intact chloroplasts the ionic species which contribute most to this compensation are Mg^{2+} (KRAUSE, 1973, 1974; BARBER et al., 1974; HIND et al., 1974) and Cl$^-$ (DILLEY and VERNON, 1965; HIND et al., 1974).

Taking the estimates for the internal volume of thylakoids into account (about 50 l per mol chlorophyll, REINWALD, 1970; SCHULDINER et al., 1972), an internal pH of 5, as induced under continous illumination, implies that the internal volume contains about 0.3 *free* protons per electron transport chain (600 chlorophylls). Hence, more than 99% of the protons accumulated in the internal phase are bound to buffering groups. The chemical nature of these groups is as yet unknown.

Acknowledgements. I am very much indebted to Mrs. I. COLUMBUS for technical assistance, to Mrs. B. SANDER for the graphs and to Drs. G. RENGER and U. SIGGEL for critical comments. Financial support from the Deutsche Forschungsgemeinschaft is gratefully acknowledged.

References

Amesz, J.: Biochim. Biophys. Acta **79**, 257–265 (1964)

Amesz, J., DeGrooth, B.G.: Biochim. Biophys. Acta **376**, 298–307 (1975)

Amesz, J., Pulles, M.P.J., DeGrooth, B.G., Kerkhof, P.L.M.: Proc. 3rd Intern. Congr. Photosynthesis. Avron, M. (ed.). Amsterdam: Elsevier, 1975, pp. 307–314

Arnon, D.I.: Science **149**, 1460–1469 (1965)

Arntzen, C.J., Briantais, J.M.: In: Bioenergetics of Photosynthesis. Govindjee (ed.). New York: Academic Press 1975, pp. 51–113

Arntzen, C.J., Dilley, R.A., Peters, G.A., Shaw, E.R.: Biochim. Biophys. Acta **256**, 85–107 (1972)

Arthur, W.E., Strehler, B.L.: Arch. Biochem. Biophys. **70**, 507–526 (1957)

Ausländer, W., Heathcote, P., Junge, W.: FEBS Lett. **47**, 229–235 (1974)

Ausländer, W., Junge, W.: Biochim. Biophys. Acta **357**, 285–298 (1974)

Ausländer, W., Junge, W.: FEBS Lett. **59**, 310–315 (1975)

Ballschmiter, K., Katz, J.J.: Biochim. Biophys. Acta **180**, 347–359 (1969)

Bamberger, E.S., Rottenberg, H., Avron, M.: Europ. J. Biochem. **34**, 557–563 (1973)

Barber, J.: FEBS Lett. **20**, 251–254 (1972)

Barber, J., Kraan, G.P.B.: Biochim. Biophys. Acta **197**, 49–59 (1970)

Barber, J., Telfer, A., Mills, J., Nicolson, J.: In: Proc. 3rd Intern. Congr. Photosynthesis. Avron, M. (ed.). Amsterdam: Elsevier, 1975, pp. 53–63

Barber, J., Telfer, A., Nicolson, J.: Biochim. Biophys. Acta **357**, 161–165 (1974)

Beinert, H., Kok, B.: Biochim. Biophys. Acta **88**, 278–288 (1964)

Bensasson, R., Land, E.G.: Biochim. Biophys. Acta **325**, 175–181 (1973)

Bishop, N.I.: Biochemistry **45**, 1696–1702 (1959)

Boardman, N.K., Anderson, J.M.: Nature (London) **203**, 166–167 (1964)

Böhme, H., Cramer, W.A.: FEBS Lett. **15**, 349–351 (1971)

Böhme, H., Cramer, W.A.: Biochim. Biophys. Acta **325**, 275–283 (1973)

Borg, D.C., Fajer, J., Fetton, R., Dolphin, D.: Proc. Nat. Acad. Sci. U.S. **67**, 813–820 (1970)

Borisevitch, G.P., Kononenko, A.A., Venediktov, R.S., Verknoturov, V.N., Rubin, A.B.: Biofizika **20**, 250–253 (1975)

Borisov, A.Yu., Il'ina, M.D.: Biochim. Biophys. Acta **305**, 364–371 (1973)

Bouges-Boquet, B.: Biochim. Biophys. Acta **396**, 382–391 (1975)

Breton, J., Michel-Villaz, M., Paillotin, G.: Biochim. Biophys. Acta **314**, 42–56 (1973)

Breton, J., Roux, E.: Biochem. Biophys. Res. Commun. **45**, 557–563 (1971)

Breton, J., Roux, E., Whitmarsh, J.: Biochem. Biophys. Res. Commun. **64**, 1274–1277 (1975)

Briantais, J.M., Vernotte, C., Moya, I.: Biochem. Biophys. Acta **325**, 530–538 (1973)

Brody, S.S., Rabinovitch, E.: Science **125**, 555–558 (1957)

Brown, J.S.: Photochem. Photobiol. **2**, 159–173 (1963)

Buchanan, B.B., Arnon, D.I.: Methods Enzymol. **23**, 413–440 (1971)

Buchwald, H.E., Wolff, C.: Z. Naturforsch. **26 B**, 51–53 (1971)

Bulychev, A.A., Andrianov, V.K., Kurella, G.A., Litvin, F.F.: Soviet Plant Physiol. (Engl. transl.) **18**, 204–210 (1971)

Butler, W.: Current Topics Bioenerg. **1**, 49–73 (1966)

Butler, W.: Proc. Nat. Acad. Sci. U.S. **69**, 3420–3422 (1972a)

Butler, W.: Methods Enzymol. **24**, 3–25 (1972b)

Butler, W., Hopkins, D.W.: Photochem. Photobiol. **12**, 439–450 (1970)

Butler, W., Katajima, M.: Biochim. Biophys. Acta **396**, 72–85 (1975)

Cederstrand, C.N., Rabinovitch, E., Govindjee: Biochim. Biophys. Acta **126**, 1–12 (1966)

Chance, B., McCray, J.A., Bunkenburg, J.: Nature (London) **225**, 705–708 (1970)

Chance, B., Mela, L.: J. Biol. Chem. **241**, 4588–4596 (1966)

Chessin, M., Livingston, T.G., Truscott, T.G.: Trans. Farad. Soc. **62**, 1519–1524 (1966)

Clayton, R.K.: Photochem. Photobiol. **5**, 669–678 (1966)

Clayton, R.K.: Biophys. J. **9**, 60–76 (1969)

Clayton, R.K.: Light and Living Matter. The Physical Part. New York: McGraw, 1970, Vol. I

Clayton, R.K.: Proc. Nat. Acad. Sci. U.S. **69**, 44–49 (1972)

Cost, K., Frenkel, A.W.: Biochemistry **6**, 663–667 (1967)

Cotton, T.M., Ballschmiter, K., Katz, J.J.: J. Chromatography **8**, 546–549 (1970)
Cotton, T.M., Trifunac, A.D., Ballschmiter, K., Katz, J.J.: Biochim. Biophys. Acta **368**, 181–198 (1974)
Crofts, A.R.: J. Biol. Chem. **242**, 3352–3359 (1967)
Crofts, A.R., Prince, R.C., Holmes, N.G., Crowthers, D.: In: Proc. 3rd Intern. Congr. Photosynthesis. Avron, M. (ed.). Amsterdam: Elsevier, 1975, pp. 1131–1146
Crofts, A.R., Wraight, C.A., Fleischman, D.E.: FEBS Lett. **15**, 89–100 (1971)
Dartnall, H.J.A.: Nature (London) **162**, 122 (1948)
Davidov, A.S.: Theory of Molecular Excitons. New York: McGraw-Hill, 1962
Delosme, R.: Biochim. Biophys. Acta **143**, 108–128 (1967)
Dilley, R.A.: Arch. Biochem. Biophys. **137**, 270–283 (1970)
Dilley, R.A., Vernon, L.P.: Arch. Biochem. Biophys. **111**, 365–375 (1965)
Dilley, R.A., Vernon, L.P.: Proc. Nat. Acad. Sci. U.S. **57**, 395–400 (1967)
Döring, G., Bailey, J.L., Kreutz, W., Weikard, J., Witt, H.T.: Naturwissenschaften **55**, 219–220 (1968)
Döring, G., Renger, G., Vater, J., Witt, H.T.: Z. Naturforsch. **24B**, 1139–1143 (1969)
Dolan, E., Hind, G.: Biochim. Biophys. Acta **357**, 380–385 (1974)
Dutton, P.L., Wilson, F.: Biochim. Biophys. Acta **346**, 165–212 (1974)
Duysens, L.N.M.: Proc. Roy. Soc., Ser. B. **157**, 301–313 (1963)
Duysens, L.N.M., Amesz, J.: Biochim. Biophys. Acta **64**, 243–260 (1962)
Duysens, L.N.M., Amesz, J., Kamp, B.M.: Nature (London) **190**, 510–511 (1961)
Duysens, L.N.M.: Brookhaven Symp. Biol. **19**, 71–80 (1966)
Duysens, L.N.M., Sweers, H.E.: In: Studies on Microalgae and Photosynthetic Bacteria. Miyachi, S. (ed.). Tokyo: Univ. Tokyo, 1963, pp. 353–372
Emerson, R.: Science **125**, 746 (1957)
Emerson, R., Arnold, W.: J. Gen. Physiol. **15**, 391–420 (1932)
Emrich, H.M., Junge, W., Witt, H.T.: Z. Naturforsch. **24B**, 1144–1146 (1969)
Evans, N.C.W., Telfer, A., Lord, A.V.: Biochim. Biophys. Acta **267**, 530–537 (1972)
Fan, H.N., Cramer, W.A.: Biochim. Biophys. Acta **260**, 200–207 (1970)
Fenna, R.E., Mathews, B.W.: Nature (London) **258**, 573–577 (1975)
Fenna, R.E., Mathews, B.W., Olson, J.M., Shaw, E.K.: J. Mol. Biol. **84**, 231–240 (1974)
Fiolet, J.W., Bakker, E.P., VanDam, K.: Biochim. Biophys. Acta **368**, 432–445 (1974)
Fiolet, J.W., Van de Vlugt, F.C.: FEBS Lett. **53**, 287–291 (1975)
Floyd, R.A., Chance, B., DeVault, D.: Biochim. Biophys. Acta **226**, 103–112 (1971)
Förster, T.: Naturwissenschaften **33**, 166–175 (1946)
Förster, T.: Z. Naturforsch. **2B**, 147–182 (1947)
Förster, T.: In: Comparative Effects of Radiation. Burton, M., Kirby, J.S. (eds.). New York: Wiley, 1960, pp. 300–341
Fong, F.: J. Theor. Biol. **46**, 407–420 (1974a)
Fong, F.: Proc. Nat. Acad. Sci. U.S. **71**, 3692–3696 (1974b)
Forbush, B., Kok, B.: Biochim. Biophys. Acta **162**, 243–253 (1968)
Fowler, C.F., Kok, B.: 6th Intern. Congr. Photobiol. Abstr. 417, Bochum (1972)
Fowler, C.F., Kok, B.: Biochim. Biophys. Acta **357**, 299–307 (1974a)
Fowler, C.F., Kok, B.: Biochim. Biophys. Acta **357**, 308–318 (1974b)
French, C.S., Brown, J.S., Lawrence, M.C.: Carnegie Inst. Wash. Yearbook **60**, 487–495 (1971)
Garewal, H.S., Wasserman, A.R.: Biochemistry **13**, 4072–4079 (1974)
Geacintov, N., VanNostrand, F., Becker, J.F., Tinkel, J.B.: Biochim. Biophys. Acta **267**, 65–79 (1972)
Geacintov, N., VanNostrand, F., Pope, M., Tinkel, J.B.: Biochim. Biophys. Acta **226**, 486–491 (1971)
Gläser, M., Wolff, C., Buchwald, H.E., Witt, H.T.: FEBS Lett. **42**, 81–85 (1974)
Gorkom, H.J. van: Biochim. Biophys. Acta **347**, 439–442 (1974)
Gorkom, H.J. van, Tamminga, J.J., Haveman, J.: Biochim. Biophys. Acta **347**, 417–438 (1974)
Gorman, D.S., Levine, R.P.: Plant Physiol. **41**, 1643–1647 (1966)
Gould, N., Izawa, S.: Biochim. Biophys. Acta **333**, 509–524 (1974)
Gouterman, M., Wagniere, G.H.: J. Mol. Spectrosc. **11**, 118–127 (1963)
Govindjee, Govindjee, R.: In: Bioenergetics of Photosynthesis. Govindjee (ed.). New York: Academic Press, 1975, pp. 1–50

Gräber, P., Witt, H.T.: Biochim. Biophys. Acta **333**, 389–392 (1974)
Gräber, P., Witt, H.T.: FEBS Lett. **59**, 184–189 (1975)
Griffiths, M., Sistrom, W.R., Cohen-Bazire, G., Stannier, R.Y.: Nature (London) **176**, 1211–1214 (1955)
Gulyaev, V.A., Litvin, F.F.: Biofizika (engl. transl.) **15**, 701–712 (1970)
Haehnel, W.: Biochim. Biophys. Acta **305**, 618–631 (1973)
Haehnel, W.: In: Proc. 3rd Intern. Congr. Photosynthesis. Avron, M. (ed.). Amsterdam: Elsevier, 1975, pp. 557–567
Haehnel, W.: Biochim. Biophys. Acta (1976)
Haehnel, W., Döring, G., Witt, H.T.: Z. Naturforsch. **26B**, 1171–1174 (1971)
Haehnel, W., Witt, H.T.: In: Proc. 2nd Intern. Congr. Photosynthesis. Forti, G., Avron, M., Melandri, A. (eds.). The Hague: Dr. Junk, 1972, pp. 469–476
Hager, A.: Planta (Berl.) **89**, 224–243 (1969)
Hauska, G.A., McCarty, R.E., Berzborn, R.J., Racker, E.: J. Biol. Chem. **246**, 3524–3531 (1971)
Haveman, J., Mathis, P., Vermeglio, A.: FEBS Lett. **58**, 259–261 (1975)
Heath, R.L.: Biochim. Biophys. Acta **256**, 645–655 (1972)
Henninger, M.D., Crane, F.D.: Plant Physiol. **39**, 598–602 (1964)
Hildreth, W.W.: Biochim. Biophys. Acta **153**, 197–202 (1968)
Hill, R., Bendall, F.: Nature (London) **186**, 136–137 (1960)
Hill, R., Bendall, F.: In: Biochemistry of Chloroplasts. Goodwin, T.W. (ed.). London: Academic Press, 1967, Vol. II, pp. 559–564
Hill, R., Scarisbrick, R.: New Phytologist **50**, 48–103 (1951)
Hind, G., Nakatani, H.Y., Izawa, S.: Proc. Nat. Acad. Sci. U.S. **71**, 1484–1488 (1974)
Hind, G., Olson, J.M.: Brookhaven Symp. Biol. **19**, 188–194 (1966)
Hiyama, T., Ke, B.: Arch. Biochem. Biophys. **147**, 99–108 (1971)
Hiyama, T., Ke, B.: Biochim. Biophys. Acta **267**, 160–171 (1972)
Hochstrasser, R.M., Kasha, N.: Photochem. Photobiol. **3**, 317–331 (1964)
Hope, A.B., Chow, W.S.: In: Membrane Transport in Plants. Zimmermann, U., Dainty, J. (eds.). Berlin-Heidelberg-New York: Springer, 1974, pp. 256–263
Ilani, A., Berns, D.S.: Biophysik **9**, 209–224 (1973)
Inoue, Y., Ogawa, T., Shibata, K.: Biochim. Biophys. Acta **305**, 483–487 (1973)
Izawa, S., Hind, G.: Biochim. Biophys. Acta **143**, 377–390 (1967)
Jackson, J.B., Croft, A.R.: FEBS Lett. **4**, 185–189 (1969)
Jacobi, G., Lehmann, H.: In: Progress in Photosynthesis Research. Metzner, H. (ed.). Tübingen: Laupp, 1969, pp. 159–173
Jagendorf, A.T.: In: Bioenergetics of Photosynthesis. Govindjee (ed.). New York: Academic Press, 1975, pp. 413–492
Jagendorf, A.T., Hind, G.: In: Photosynth. Mech. Green Plants, Nat. Acad. Sci. U.S. Nat. Res. Council **1145**, 599–610 (1963)
Jahn, T.L.: J. Theor. Biol. **2**, 129–138 (1962)
Joliot, P.: Biochim. Biophys. Acta **102**, 116–134 (1965)
Joliot, P., Barbieri, G., Chabaud, R.: Photochem. Photobiol. **10**, 309–329 (1969)
Joliot, P., Joliot, A.: Biochim. Biophys. Acta **305**, 302–316 (1973)
Joliot, P., Joliot, A., Kok, B.: Biochim. Biophys. Acta **153**, 635–652 (1968)
Junge, W.: In: Proc. 3rd Intern. Congr. Photosynthesis. Avron, M. (ed.). Amsterdam: Elsevier, 1975, pp. 273–286
Junge, W.: In: Biochem. Plant Pigments. Goodwin, T.W. (ed.). New York: Academic Press, 1975, vol. II, pp. 233–333
Junge, W., Ausländer, W.: Biochim. Biophys. Acta **333**, 59–70 (1974)
Junge, W., Eckhof, A.: FEBS Lett. **36**, 207–212 (1973)
Junge, W., Eckhof, A.: Biochim. Biophys. Acta **357**, 103–117 (1974)
Junge, W., Witt, H.T.: Z. Naturforsch. **23B**, 244–254 (1968)
Karlish, S.J.D., Avron, M.: Nature (London) **216**, 1107–1109 (1967)
Katoh, S.: Nature (London) **186**, 533–534 (1960)
Katoh, S.: Methods Enzymol. **23**, 408–413 (1971)
Katoh, S., Shiratori, I., Takamiya, A.: J. Biochem. (Tokyo) **51**, 32–40 (1962)
Katz, J.J., Ballschmiter, K., Garcia-Morin, M., Strain, H.H., Uphaus, R.A.: Proc. Nat. Acad. Sci. U.S. **60**, 100–107 (1968)

Katz, J.J., Norris, J.R.: Current Topics Bioenerg. **5**, 41–75 (1973)
Kaufmann, K.J., Dutton, P.L., Netzel, T.L., Leigh, J.S., Rentzepis, P.M.: Science **188**, 1301–1304 (1975)
Ke, B.: Arch. Biochem. Biophys. **152**, 70–77 (1972)
Ke, B.: Biochim. Biophys. Acta **301**, 1–33 (1973)
Ke, B., Sane, S., Shaw, E., Beinert, H.: Biochim. Biophys. Acta **347**, 36–48 (1974a)
Ke, B., Sugahara, K., Shaw, E.R., Hansen, R.E., Hamilton, W.D., Beinert, H.: Biochim. Biophys. Acta **368**, 401–408 (1974b)
Kessler, E.: Planta (Berl.) **49**, 435–454 (1957)
Kleuser, D., Bücher, M.: Z. Naturforsch. **24B**, 1371–1374 (1969)
Klingenberg, M., Müller, A., Schmidt-Mende, P., Witt, H.T.: Nature (London) **194**, 379–380 (1962)
Knaff, D.B., Arnon, D.I.: Proc. Nat. Acad. Sci. U.S. **63**, 956–962 (1969)
Knox, R.S.: Biophys. J. **9**, 1351–1362 (1969)
Knox, R.S.: In: Bioenergetics of Photosynthesis. Govindjee (ed.). New York: Academic Press, 1975, pp. 183–224
Kok, B.: Acta Botan. Neerl. **6**, 316–337 (1957)
Kok, B.: Biochim. Biophys. Acta **48**, 527–535 (1961)
Kok, B., Forbush, B., McGloin, M.: Photochem. Photobiol. **11**, 457–475 (1970)
Kok, B., Radmer, R., Fowler, C.F.: In: Proc. 3rd Intern. Congr. Photosynthesis. Avron, M. (ed.). Amsterdam: Elsevier, 1975, pp. 485–496
Krause, G.M.: Biochim. Biophys. Acta **292**, 715–728 (1973)
Krause, G.M.: Biochim. Biophys. Acta **333**, 301–313 (1974)
Kreutz, W.: Advan. Botan. Res. **3**, 53–169 (1970)
Labhard, H.: Helv. Chim. Acta **44**, 447–456 (1961)
Lavorel, J.: Biochim. Biophys. Acta **153**, 727–730 (1968)
Lavorel, J.: In: Bioenergetics of Photosynthesis. Govindjee (ed.). New York: Academic Press, 1975, pp. 223–317
Liberman, E.A., Skulachev, V.P.: Biochim. Biophys. Acta **216**, 30–42 (1970)
Liptay, W.: Angew. Chem. **81**, 195–231 (1969)
Lynn, W., Brown, R.: J. Biol. Chem. **242**, 426–432 (1967)
Malkin, R., Bearden, A.J.: Proc. Nat. Acad. Sci. U.S. **68**, 16–19 (1971)
Malkin, R., Bearden, A.J.: Proc. Nat. Acad. Sci. U.S. **70**, 294–297 (1973)
Malkin, S.: In: Proc. 3rd Intern. Congr. Photosynthesis. Avron, M. (ed.). Amsterdam: Elsevier, 1975, pp. 199–204
Malley, M., Feher, G., Mauzerall, D.: J. Molec. Spectrosc. **25**, 544–547 (1968)
Mangel, M., Berns, D.S., Ilani, A.: J. Membrane Biol. **20**, 171–180 (1975)
Marsho, T.V., Kok, B.: Biochim. Biophys. Acta **223**, 240–250 (1970)
Mathis, P.: C.R. Acad. Sci. Paris **271**, 1094–1096 (1970)
Mathis, P., Galmiche, J.M.: C.R. Acad. Sci. Paris **264**, 1903–1906 (1967)
Mathis, P., Vermeglio, A.: Biochim. Biophys. Acta **369**, 371–381 (1975)
McRae, E.G., Kasha, M.: J. Chem. Phys. **28**, 721–722 (1958)
Mitchell, P.: Nature (London) **191**, 144–148 (1961)
Mitchell, P.: Biol. Rev. **41**, 445–502 (1966)
Mitchell, P.: In: Electron Transfer Chains and Oxidative Phosphorylation. Quagliarello, E. et al. (eds.). Amsterdam: North Holland, 1975, pp. 305–316
Mitchell, P., Moyle, J., Smith, L.: Europ. J. Biochem. **4**, 9–19 (1968)
Mühlethaler, K.: Chem. Phys. Lipids **8**, 259–264 (1972)
Müller, A., Fork, D.C., Witt, H.T.: Z. Naturforsch. **18B**, 142–145 (1963)
Müller, A., Lumry, R., Walker, M.S.: Photochem. Photobiol. **9**, 113–126 (1969)
Murata, N.: Biochim. Biophys. Acta **189**, 171–181 (1969)
Myers, J.: Natl. Acad. Sci. U.S. (Publ. 1145) 301–317 (1963)
Neumann, J., Jagendorf, A.T.: Arch. Biochem. Biophys. **107**, 109–119 (1964)
Norris, J.R., Scheer, H., Druyan, M.E., Katz, J.J.: Proc. Nat. Acad. Sci. U.S. **71**, 4897–4900 (1974)
Norris, J.R., Scheer, H., Katz, J.J.: Ann. N.Y. Acad. Sci. **244**, 260–280 (1975)
Norris, J.R., Uphaus, R.A., Crespi, H.L., Katz, J.J.: Proc. Nat. Acad. Sci. U.S. **68**, 625–628 (1971)
Oesterhelt, D., Stoeckenius, W.: Proc. Nat. Acad. Sci. U.S. **70**, 2853–2857 (1973)

Papageorgiou, G.: In: Bioenergetics of Photosynthesis. Govindjee (ed.). New York: Academic Press, 1975, pp. 319–412

Phillipson, K.D., Sato, V.L., Sauer, K.: Biochem. **11**, 4591–4595 (1972)

Plesničar, M., Bendall, D.S.: Biochim. Biophys. Acta **216**, 192–199 (1970)

Portis, A.R., McCarty, R.: Arch. Biochem. Biophys. **156**, 621–625 (1973)

Radunz, A., Schmid, G.H., Menke, W.: Z. Naturforsch. **266**, 435–446 (1971)

Regitz, G., Oettmeier, W.: In: Proc. 2nd Intern. Congr. Photosynthesis. Forti, G., Avron, M., Melandri, A. (eds.). The Hague: Dr. Junk, 1972, pp. 499–506

Reinwald, E.: Thesis, Technische Universität Berlin, 1970

Renger, G.: Z. Naturforsch. **25 B**, 966–971 (1970)

Renger, G.: In: Topics in Current Chemistry, vol. 67 (1976), in press

Ried, A.: In: Proc. 2nd Intern. Congr. Photosynthesis. Forti, G., Avron, M., Melandri, A. (eds.). The Hague: Dr. Junk, 1972, pp. 763–772

Rockley, M.G., Windsor, M.W., Cogdell, R.J., Parson, W.W.: Proc. Nat. Acad. Sci. U.S. **72**, 2251–2255 (1975)

Ross, R.T., Calvin, M.: Biophys. J. **7**, 595–614 (1967)

Rottenberg, H., Grunwald, T., Avron, M.: FEBS Lett. **13**, 41–44 (1971)

Rottenberg, H., Grunwald, T., Avron, M.: Europ. J. Biochem. **25**, 54–63 (1972)

Rüppel, H., Witt, H.T.: Methods Enzymol. **16**, 316–379 (1969)

Rumberg, B., Reinwald, E., Schröder, H., Siggel, U.: Naturwissenschaften **55**, 77–79 (1968)

Rumberg, B., Reinwald, E., Schröder, H., Siggel, U.: In: Progress in Photosynthesis Res. Metzner, H. (ed.). Tübingen: Laupp, 1969, Vol. III, pp. 1374–1382

Rumberg, B., Siggel, U.: Naturwissenschaften **56**, 130–132 (1969)

Rumberg, B., Witt, H.T.: Z. Naturforsch. **19 B**, 693–706 (1964)

Sane, P.V., Goodchild, D.J., Park, R.D.: Biochim. Biophys. Acta **216**, 162–178 (1970)

Sauer, K.: Biophys. J. **5**, 337–348 (1965)

Sauer, K.: In: Bioenergetics of Photosynthesis. Govindjee (ed.). New York: Academic Press, 1975, pp. 115–181

Sauer, K., Calvin, M.: J. Molec. Biol. **4**, 451–466 (1962)

Schliephake, W., Junge, W., Witt, H.T.: Z. Naturforsch. **23 B**, 1571–1578 (1968)

Schmid, R., Junge, W.: Biochim. Biophys. Acta **394**, 76–92 (1975)

Schmidt, S., Reich, R.: Ber. Bunsengesellschaft **76**, 1202–1208 (1972)

Schmidt, S., Reich, R., Witt, H.T.: Naturwissenschaften **58**, 414–415 (1971)

Schröder, H., Muhle, H., Rumberg, B.: In: Proc. 2nd Intern. Congr. Photosynthesis. Forti, G., Avron, M., Melandri, A. (eds.). The Hague: Dr. Junk, 1972, pp. 919–930

Schuldiner, S., Rottenberg, H., Avron, M.: Europ. J. Biochem. **25**, 64–70 (1972)

Schwartz, M.: Nature (London) **219**, 915–919 (1968)

Seely, G.R.: Spectrochim. Acta **21**, 1847–1856 (1965)

Seibert, M., Alfano, R.R.: Biophys. J. **14**, 269–283 (1974)

Siedow, J.N., Curtis, V.A., San Pietro, A.: Arch. Biochem. Biophys. **158**, 889–897 (1973)

Siggel, U.: In: Proc. 3rd Intern. Congr. Photosynthesis. Avron, M. (ed.). Amsterdam: Elsevier, 1975, pp. 645–654

Siggel, U., Renger, G., Stiehl, H.H., Rumberg, B.: Biochim. Biophys. Acta **256**, 328–335 (1972)

Smith, J.H.C., French, C.S.: Ann. Rev. Plant Physiol. **14**, 181–224 (1963)

Steinemann, A., Alamuti, N., Brodmann, W., Marshall, O., Läuger, P.: J. Membrane Biol. **4**, 284–294 (1971)

Stiehl, H.H., Witt, H.T.: Z. Naturforsch. **23 B**, 220–224 (1968)

Stiehl, H.H., Witt, H.T.: Z. Naturforsch. **24 B**, 1588–1598 (1969)

Stuart, A.L., Wassarman, A.R.: Biochim. Biophys. Acta **376**, 561–572 (1975)

Sun, A.S.K., Sauer, K.: Biochim. Biophys. Acta **256**, 409–427 (1972)

Thomas, J.B., Minnaert, K., Elbers, P.D.: Acta Botan. Neerl. **5**, 314–321 (1956)

Thornber, J.P., Highkin, H.R.: Europ. J. Biochem. **41**, 109–116 (1974)

Thorne, S.W., Horvath, G., Kahn, A., Boardman, K.: Proc. Nat. Acad. Sci. U.S. **72**, 3858–3862 (1975)

Tinoco, M.: Rad. Res. **20**, 133–139 (1963)

Trebst, A.: In: Proc. 2nd Intern. Congr. Photosynthesis. Forti, G., Avron, M., Melandri, A. (eds.). The Hague: Dr. Junk, 1972, pp. 399–417

Trebst, A.: Ann. Rev. Plant Physiol. **25**, 423–458 (1974)

Tributsch, H.: Photochem. Photobiol. **16**, 261–269 (1972)
Trurnitt, H.J., Colmano, G.: Biochim. Biophys. Acta **31**, 434–447 (1959)
Valley, S.L.: Handbook of Geophysics and Space Environment. New York: McGraw-Hill, 1965
Vernon, L.P., Shaw, E., Ke, B.: J. Biol. Chem. **241**, 4101–4109 (1966)
Vidaver, W., French, C.S.: Plant Physiol. **40**, 7–12 (1965)
Vredenberg, W.J.: In: Proc. 3rd Intern. Congr. Photosynthesis. Avron, M. (ed.). Amsterdam: Elsevier, 1975, pp. 929–939
Vredenberg, W.J., Tonk, W.J.M.: Biochim. Biophys. Acta **387**, 580–587 (1975)
Wang, R.T., Myers, J.: Photochem. Photobiol. **17**, 321–332 (1973)
Warden, J.T., Bolton, J.R.: J. Am. Chem. Soc. **95**, 6435–6436 (1973)
Warden, J.T., Bolton, J.R.: Acc. Chem. Res. **7**, 189–195 (1974)
Weier, T.E., Benson, A.A.: Am. J. Botany **54**, 389–402 (1967)
Weikard, J.: Z. Naturforsch. **23B**, 235–238 (1968)
Weiss, C., Jr.: J. Mol. Spectr. **44**, 37–80 (1972)
Williams, R.P.J.: FEBS Lett. **53**, 123–125 (1975)
Witt, H.T.: Quart. Rev. Biophys. **4**, 365–477 (1971)
Witt, H.T.: In: Bioenergetics of Photosynthesis. Govindjee (ed.). New York: Academic Press, 1975, pp. 493–554
Witt, H.T., Müller, A., Rumberg, B.: Nature (London) **192**, 967–969 (1961)
Witt, H.T., Zickler, A.: FEBS Lett. **37**, 307–310 (1973)
Witt, H.T., Zickler, A.: FEBS Lett. **39**, 205–208 (1974)
Witt, K.: FEBS Lett. **38**, 116–118 (1973a)
Witt, K.: FEBS Lett. **38**, 112–115 (1973b)
Witt, K., Wolff, C.: Z. Naturforsch. **25B**, 387–388 (1970)
Wolff, C., Buchwald, H.E., Rüppel, H., Witt, K., Witt, H.T.: Z. Naturforsch. **24B**, 1038–1041 (1969)
Wolff, C., Witt, H.T.: Z. Naturforsch. **24B**, 1031–1037 (1969)
Wolken, J.J., Schwertz, F.A.: J. Gen. Physiol. **37**, 111–120 (1953)
Wraight, C.A.: Biochim. Biophys. Acta **283**, 247–258 (1972)
Wraight, C.A., Crofts, A.R.: Europ. J. Biochem. **19**, 386–397 (1971)
Wydrzynski, T., Gross, E.L., Govindjee: Biochim. Biophys. Acta **376**, 151–161 (1975)
Yocum, C.S., San Pietro, A.: In: Proc. 2nd Intern. Congr. Photosynthesis. Forti, G., Avron, M., Melandri, A. (eds.). The Hague: Dr. Junk, 1972, pp. 477–489
Zankel, K.L.: Biochim. Biophys. Acta **245**, 373–385 (1971)
Zieger, G., Müller, A., Witt, H.T.: Z. Physik. Chem. **29**, 13–24 (1961)

1 b. Electron Transport in Chloroplasts

J.H. Golbeck, S. Lien, and A. San Pietro

A. General

The result of photosynthetic electron transport in green plants is the evolution of molecular oxygen and the formation of the assimilatory power necessary for the conversion of carbon dioxide into cellular material at the expense of light energy. The currently accepted representation of photosynthetic electron transport as the cooperative interaction of two light reactions originated with Hill and Bendall (1960). A modified schematic representation of their hypothesis is presented in Figure 1. Their formulation was proposed primarily to account for three major experimental observations: First, the decline in efficiency of photosynthesis at long wavelengths ($\lambda > 685$ nm) and the synergistic effect of shorter wavelengths on far red illumination (Emerson and Lewis, 1943; Blinks, 1957). Secondly, the presence in green tissues of two cytochromes, cytochrome f and b_6

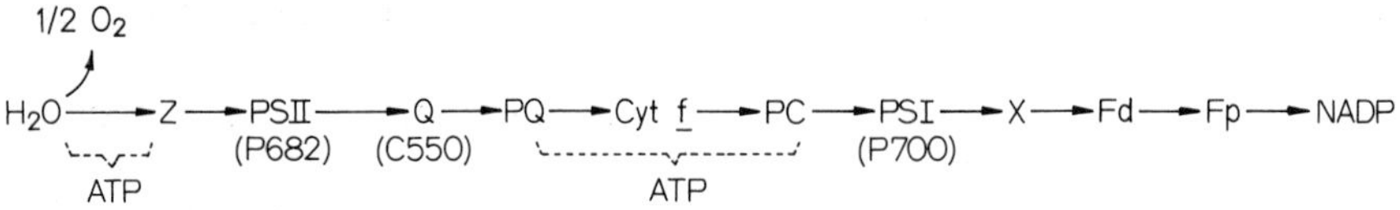

Fig. 1. Schematic representation of photosynthetic electron transport. *Z*: primary electron donor of photosystem II; *Q*: quencher of chlorophyll a fluorescence; *C-550*: compound with a difference absorbance maximum at 550 nm; *PQ*: plastoquinone; *PC*: plastocyanin; *P-700*: primary electron donor to photosystem I; *Fp*: flavoprotein

(564) whose characteristic potentials ($E_{m, 7}$) differed by about 0.4 V and their light-induced absorbance changes (Hill, 1965). Lastly, the stimulation of electron flow to $NADP^+$ when ATP formation occurred concurrently (Arnon et al., 1958; Davenport, 1959).

The two light reactions function in tandem partially and collaboratively to allow for electron flow to proceed against the thermochemical gradient (left to right in Fig. 1). The resultant of either light reactions is electron flow against the thermochemical gradient equivalent to a stabilized net change in characteristic potential of about 0.8–0.9 V: for photosystem II, the origin and terminus are about $+0.8$ V and 0 V, respectively; for photosystem I, the corresponding values are close to $+0.4$ V and -0.5 V. The overall net change in characteristic potential for the cooperative interaction of the two light systems is about 1.2–1.3 V.

A wealth of evidence (Govindjee, 1975; Radmer and Kok, 1975; Trebst, 1974; Avron, 1975) in support of this hypothesis has appeared in the last 15 years and it is primarily biochemical, biophysical or genetic in nature. The biochemical approach of disruption and reconstitution (associated generally with purification and characterization of the component under study) has been particularly noteworthy in the cases of ferredoxin, ferredoxin-NADP$^+$ oxidoreductase, plastoqui-none(s) and cytochrome 552 (of Euglena). This approach has also been success-fully with the coupling factor (CF$_1$) of chloroplasts and plastocyanin using sub-chloroplast particles. In addition, the immunological approach has provided sub-stantive information with regard to component position within the chloroplast membrane; e.g. the coupling factor, cytochrome f and plastocyanin.

The fragmentation of chloroplasts (by physical means or with detergents) and the subsequent isolation of sub-chloroplast particles enriched in one of the two photosystems has been a most useful approach. As a consequence, it is clear that the components of the photosynthetic electron transport pathway can clearly be assigned to one or the other of the two photosystems. This type of information, in association with electron microscopic studies, should provide for increased clarity of the alignment of the two photosystems within the membrane architecture. Fur-ther, the availability of reaction center preparations from higher plants and algae (similar to those available from photosynthetic bacteria) is most important if we are to understand this process completely.

The genetic approach focuses on the isolation of mutant strains lacking, or defective in, a single component of the overall electron transport pathway. This approach has been used most successfully in the past with mutant strains of algae but is now being applied to higher plants (see review by Miles, 1975). A basic requirement for successful application of this technique is the availability of an efficient, and positive, screening procedure to select for the mutant in question. This approach has been used successfully with many of the components of the electron transport pathway; e.g. cytochrome f, P-700, etc.

In this review, we have chosen to consider the photosynthetic electron transfer pathway in terms of the electron transfer processes associated with each of the two photosystems. The components of the overall pathway are not treated individu-ally but rather solely in terms of their association with the photosystems. Of necessity, a great wealth of interesting literature has been overlooked or scarcely recognized but is available in other sections of this volume and elsewhere (GOV-INDJEE, 1975; RADMER and KOK, 1975; TREBST, 1974; AVRON, 1975).

B. Photosystem II

The photosystem II of higher plants and algae is associated with water oxidation from which molecular oxygen is evolved and electrons are provided for subsequent reductive processes mediated by the interphotosystem redox carrier chain, PS I and low potential redox carriers (such as Fd and Fd-NADP$^+$ reductase) leading to the generation of NADPH. As in PS I, the photochemically active pigment-enzyme complex of PS II is membrane-bound and requires a high degree of struc-tural and organizational integrity. The electron transport reactions of PS II are initiated by primary photochemical reactions involving specialized chlorophyll a molecules which function as the trap for excitation energy absorbed by the asso-ciated network of light harvesting pigment molecules. The molar ratio of trap (T$_{II}$) to bulk pigment molecules associated functionally with PS II is estimated to be one per two to three hundred, based on theoretical considerations (DUYSENS, 1964; KNOX, 1968).

According to models proposed for PS II reactions by various investigators (BUTLER et al., 1973; VAN GORKOM and DONZE, 1973; GOVINDJEE and PAPAGEORGIOU, 1971) the trap chlorophyll a of PS II, T_{II}, is visualized to exist as a reaction center complex, $Z\,T_{II}\,Q$, with a primary donor, Z, and a primary electron acceptor, Q. (In a strict sense, according to the mechanism of the primary photochemistry of PS II, through reactions 1 and 2b below, the trap, T_{II}, itself rather than Z is the primary electron donor.) When the reaction center complex is in the proper redox state, $Z\,T_{II}\,Q$ (i.e. Z reduced and Q oxidized), the trapping of an exciton by T_{II} creates a photochemical product which can be expressed in the form of $Z^+\,T_{II}\,Q^-$ (see Scheme 1). Eq. (1) is simply the creation of an excited trap.

Scheme 1

$$Z\,T_{II}\,Q \longrightarrow Z\,T_{II}^*\,Q \tag{1}$$

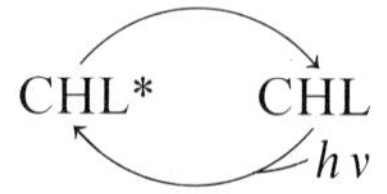

$$\text{CHL}^* \quad \text{CHL}$$
$$h\,v$$

and

$$Z\,T_{II}^*\,Q \xrightarrow{\ \text{Sensitization}\ } Z^+\,T_{II}\,Q^- \tag{2a}$$

$$Z\,T_{II}^*\,Q \longrightarrow Z\,T_{II}^+\,Q^- \longrightarrow Z^+\,T_{II}\,Q^- \tag{2b}$$

$$Z\,T_{II}^*\,Q \longrightarrow Z^+\,T_{II}^-\,Q \longrightarrow Z^+\,T_{II}\,Q^- \tag{2c}$$

Three mechanisms for the formation of the charge-separated product $(Z^+\,T_{II}\,Q^-)$ are visualized in Eq. (2a), (2b), and (2c). The first mechanism [Eq. (2a)] was favored by DÖRING and WITT (1972) on the basis of flash-induced absorption changes of P-680. In this formulation, T_{II} does not engage directly in the redox reaction with Z or Q but acts solely as a sensitizer, the details of the sensitization reaction are unknown. In the mechanisms described by Eq. (2b) and (2c), the trap molecule, T_{II}, is directly involved in the redox reaction. Because of the tendency of an excited molecule to donate rather than accept an electron the mechanism represented by Eq. (2c) is not considered very likely. On the other hand, the mechanism depicted by Eq. (2b) is consistent with recent observations on light-induced absorbance or fluorescence changes at low temperature and has been adopted by a number of investigators. As indicated in Figure 2, the events following formation of the charge-separated product $(Z^+\,T_{II}\,Q^-)$ can be described by two secondary dark oxidation-reduction reactions. Through reaction I, the oxidized form of the primary electron donor of PS II, Z^+, receives an electron (ultimately from H_2O) to regenerate Z. On the other hand, regeneration of the oxidized form of the primary acceptor, Q, proceeds through reaction II; i.e. the interaction of Q^- with components of the secondary redox carrier pool (A) linking the two photosystems. It should be mentioned that in the normal function of PS II, the overall rate of reaction I is faster than that of reaction II. Thus, under steady-state conditions, the reaction center complex is partitioned mainly between states $Z\,T_{II}\,Q$ and $Z\,T_{II}\,Q^-$ (plus a very small fraction as $Z^+\,T_{II}\,Q^-$); i.e. the reactions proceed mainly through a path involving the lower cycle in Figure 2. However, a significant amount of $Z^+ T_{II} Q$ may accumulate under conditions that favor a slowing down of reaction I relative to reaction II.

According to this general formulation of PS II reactions, it is convenient to analyze the electron transport system of PS II in three subsegments: (1) The oxidizing side; (2) the reaction center proper; and (3) The reducing side.

I. The Oxidizing Side of PS II

Reactions involved in the regeneration of Z (from $Z^+ T_{II} Q^-$ or $Z^+ T_{II} Q$) and the eventual oxidation of water comprise the steps on the oxidizing side of PS II and are collectively designated as reaction I in Figure 2. Among the partial reactions of photosynthesis, the coupling between the reaction center complex and the water oxidizing complex (reaction I in Fig. 2) is the most labile. Various treatments which leave PS I associated activities intact (and in some cases even an active photophosphorylation capacity) generally destroy oxygen evolution. The enzymatic constituents involved in reaction I are not known; however, certain cofactors are known to be involved.

A chloride requirement for the Hill reaction with isolated grana preparations was recognized some years ago (WARBURG and LÜTTGENS, 1946; GORHAM and CLENDENNING, 1952). The site of chloride action has recently been localized on the oxidizing side of PS II (HIND et al., 1969; IZAWA et al., 1969; HEATH and HIND, 1969). The exact mechanism of chloride action is not known; however, the effect of chloride depletion can be readily reversed by its inclusion in the aqueous medium.

The result of manganese deficiency indicated that it might be involved in the activity of PS II (KESSLER, 1957; KESSLER et al., 1957). The manganese content of well-washed chloroplast preparations from photosynthetically competent plant and algal sources is approximately 5–8 atoms of manganese per 400 chlorophyll molecules (CHENIAE and MARTIN, 1971a). These "bound" atoms of manganese

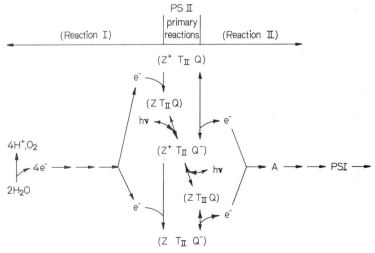

Fig. 2. Electron transfer reactions associated with photosystem II

are neither readily released from chloroplasts upon washing with a variety of metal complexing agents, nor do they exchange readily in the dark with Mn^{2+} or MnO_4^- (CHENIAE and MARTIN, 1966). A variety of treatments of higher plant chloroplasts or algae, including mild-thermal shock (heating at 50° C for a few min), washing with Tris at alkaline pH, washing with chaotropic agents and hydroxylamine extraction (CHENIAE and MARTIN, 1966; YAMASHITA and BUTLER, 1968a, 1968b, 1969; LOZIER et al., 1971; BENNOUN and JOLIOT, 1969), will inactivate primarily the oxidizing side of PS II. These treatments also result in varying degrees of manganese depletion. In some cases it is possible to reactivate the oxygen evolving capacity of the treated preparation. CHENIAE and MARTIN (1971b, 1972) demonstrated that reactivation of the oxygen evolving activity of Mn-depleted algae requires a light-dependent incorporation of manganese catalyzed by PS II. Neither the exact site(s) nor mechanism of this in vivo insertion of manganese into a native protein (enzyme?) is known. Extensive studies on both the inhibition of the oxygen-evolving activity and the manganese content of chloroplasts subjected to the Tris-washing treatment reveal that: (1) The residual oxygen-evolving activity is not related linearly to the amount of manganese retained by the chloroplasts (YAMASHITA et al., 1971; YAMASHITA and TOMITA, 1974); (2) Up to one-third of the manganese is not essential for oxygen evolution since chloroplasts inactivated by Tris-washing but containing at least two-thirds or more of the manganese can be fully reactivated simply by dark incubation with reduced DPIP (YAMASHITA et al., 1971; BLANKENSHIP et al., 1975); (3) The remaining two-thirds of the total manganese (4–5 Mn/400 chl) appears to be essential for the oxygen-evolving activity and can exist in two different states; namely, an EPR-detectable form and an EPR non-detectable form. Conversion of the EPR non-detectable form into an EPR-detectable form is associated with loss of activity. Upon reactivation by the above-mentioned procedure the Mn returns to the EPR non-detectable (and probably more tightly bound) form (BLANKENSHIP et al., 1975); and (4) More exhaustive extraction of Mn by Tris/acetone treatment results in a preparation which cannot be reactivated by incubation with reduced DPIP. Under these conditions, light and externally added Mn^{2+}, together with other agents such as Ca^{2+}, DTT (Dithiothreitol) and BSA (bovine serum albumin), are required for effective reactivation of oxygen evolution (YAMASHITA and TOMITA, 1974). On the basis of these in vitro manganese extraction and reinsertion experiments, BLANKENSHIP et al. (1975) concluded that "the incorporation of Mn^{2+} is a two-step process; a light-driven transport of manganese into the thylakoid and a dark binding to the site responsible for activity". Finally, it should be mentioned that a heat-stable, acid-resistant and highly water-soluble factor has been reported to be involved on the oxidizing side of PS II in a cell-free particulate preparation of *Phormidium luridium* (TEL-OR and AVRON, 1975); reconstitution of Hill activity was achieved by incubation of the "Hill factor" with inactive particles. Further, the Hill factor was reported to be a low molecular weight substance ($<1,000$ daltons) and to contain (bound) manganese. Therefore a detailed characterization of the chemical environment of the manganese by EPR analysis and the elucidation of the mechanism involved in the reassociation of this "manganese catalyst" with the photochemically active pigment complex of PS II should be highly informative.

As mentioned earlier in this section, various mild treatments lead either to inactivation of the water-oxidizing enzyme complex or result in disassociation of the water-oxidizing reactions from the photochemical events of the reaction center complex of PS II. Under these conditions, it is possible to study partial reaction(s) of reaction I (Fig. 2) through the use of exogenous electron donors for PS II. KATOH and SAN PIETRO (1968) first demonstrated the photooxidation of ascorbate by PS II using heat-treated chloroplasts. Subsequently, YAMASHITA and BUTLER (1968a, 1968b, 1969) established that, under appropriate conditions, various compounds serve as electron donors for PS II. The list of effective artificial donors for PS II has since been expanded to include phenylenediamine, hydroquinone, benzidine, Mn^{2+}, NH_2OH, diphenylcarbazide, catechol and high concentrations of ferrocyanide or iodide (see this vol. Chap. II.15). As a result of these donor studies, evidence suggesting possible multiple sites for electron donation on the oxidizing side of PS II has appeared (KIMIMURA and KATOH, 1972; BABCOCK and SAUER, 1975b). However, at present it is not possible to state definitely the exact number of distinct sites capable of interaction with exogenous donors for PS II because neither chemical characterization nor identification of the endogenous component(s) of the proposed site(s) has yet been achieved. Furthermore, as pointed out in a recent review by TREBST (1974), the observed multiplicity of donation sites may simply reflect a differential degree of modification of the structural integrity of PS II in the preparations employed.

Another advance in our knowledge concerning the oxidizing side of PS II came from the detailed kinetic studies of oxygen evolution pioneered by the flash experiments of ALLEN and FRANCK (1955) and extended more recently by other investigators, in particular by B. KOK and by P. JOLIOT. These studies provide the following picture of PS II: with dark-adapted chloroplasts or algae, PS II is fully active when viewed from the reducing side; i.e. a single 200 μsec saturating flash converts most of the Q into Q^- (FORBUSH and KOK, 1968). However, with similarly dark-adapted samples, a single short saturating flash yields no detectable oxygen (ALLEN and FRANCK, 1955; JOLIOT and KOK, 1975); that is, PS II appears to be largely inactive when viewed from the oxidizing side. Furthermore, it was found that with appropriately spaced saturating short flashes, the yield of oxygen per flash fluctuates in a damped oscillation with a periodicity of four; maximal yields occur on flash numbers 3, 7, 11, etc. while minimal yields occur on flash numbers 1, 5, 9, etc. In the scheme proposed by KOK et al. (1970), the oxidation of two water molecules is accomplished by the accumulation of four positive charges within the oxygen-evolving enzyme complex; that is, four successive photochemical acts are involved as depicted below:

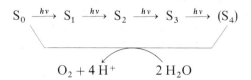

$$S_0 \xrightarrow{hv} S_1 \xrightarrow{hv} S_2 \xrightarrow{hv} S_3 \xrightarrow{hv} (S_4)$$

$$O_2 + 4 H^+ \qquad 2 H_2O$$

where S_0 to S_4 denote different oxidation states of the water-oxidizing enzyme complex. Based on kinetic data, it appears that S_0 and S_1 are stable in the dark whereas S_2 and S_3 decay to lower oxidation states within a few minutes. Once formed S_4 rapidly oxidizes two water molecules to yield molecular oxygen and thereby returns to S_0. The scheme explains nicely the oscillatory nature of the flash yields of oxygen with dark-adapted samples. It also accounts for the inhibitory effect (on water oxidation) of various treatments or inhibitors which act on the oxidizing side of PS II to prevent generation of state S_4 either by retardation of forward reactions or by acceleration of the decay of states S_2 and S_3 (Joliot and Kok, 1975).

II. The Reaction Center Complex of PS II

1. The Trap

It was postulated some years ago that T_{II}, the trap of PS II, is most likely a chlorophyll a molecule (or its dimeric complex) in some specialized environment. Indirect evidence derived from studies of fluorescence emission spectra at low temperature (Govindjee, 1963; Murata, 1968) are consistent with this postulation. A flash-induced and rapid absorbance change was described initially by Döring et al. (1967) and attributed to the reversible photo-bleaching of a pigment designated as P-680 or P-690. The complete flash-induced absorbance difference spectrum of P-680 shows a sharp red band at 680–690 nm and a blue band at 435 nm together with other spectral properties typical of chlorophyll a. The kinetic behavior of the P-680 absorbance change and its response to various treatments known to affect PS II activities are consistent with the proposed role of P-680 as the trap of PS II (Döring, 1975; Gläser et al., 1974). Furthermore, P-680 is associated with subchloroplast particle preparations enriched in PS II activity but deficient in PS I activity. More recently, a light-induced free radical EPR signal associated with PS II was detected at cryogenic temperatures. The characteristics of this PS II signal (g=2.002, line width ≃ 8 gauss), and its occurrence in medium buffered at a high redox potential, are the basis for the interpretation that it originated from $P-680^+$ (Bearden and Malkin, 1973; Malkin and Bearden, 1975).

2. The Primary Electron Acceptor and Donor

As in the case of the trap, no direct chemical characterization of either the primary acceptor or donor of PS II is available currently. In most studies concerning the primary reactants of PS II, changes of absorbance or fluorescence yield have been measured under conditions which inhibit or significantly retard secondary reactions (i.e. in the presence of inhibitors such as DCMU or at cryogenic temperatures). These studies indicate that the primary electron acceptor, symbolically designated as Q in Figure 2, behaves as a one-electron acceptor with a mid-point potential in the neighborhood of -100 mV (Cramer and Butler, 1969), and it is present in a ratio of approximately 1 equivalent per 500 Chl in normal chloroplasts from higher plants or algae (Kok et al., 1966). According to the hypothesis originally proposed by Duysens and Sweers (1963), the oxidized form of Q is a "quencher" of chlorophyll a fluorescence while the reduced form of Q is not. Thus, under appropriate conditions, the redox level of Q can be inferred from fluorescence yield measurements. Further, the light-induced absorbance change around 550 nm (associated with a blue-shift of a minor absorbance band of a component of PS II designated as C-550) has also been shown to be an excellent monitor for the redox state of the primary acceptor of PS II. However, these indirect parameters, fluorescence yield and C-550 absorbance change are not free from complications induced by factors not directly related with the primary photochemistry of PS II (for example, cation modification of fluorescence yield and membrane potential effect on the C-550 absorbance change). Furthermore, they provide no direct information concerning the chemical nature of the acceptor.

STIEHL and WITT (1969) and VAN GORKOM (1974) observed a very rapid light-induced increase in absorbance at 320 nm with both chloroplasts and PS II particles. The shape and magnitude of the absorbance difference spectrum were consistent with the formation of the semiquinone form of plastoquinone and they proposed, therefore, that a tightly bound plastoquinone serves as the primary acceptor of PS II. This proposal is in accord with the earlier reconstitution experiments where plastoquinone restored the PS II associated activities of chloroplast preparations which had been diminished by organic solvent extraction (BISHOP, 1957; HENNINGER and CRANE, 1967; COX and BENDALL, 1974).

Information concerning the chemical nature of the primary donor, Z, is also lacking at present. Tightly bound manganese has been implicated to be identical or closely associated with Z (HOMANN, 1968). On the other hand, cytochrome b-559 has been proposed as the primary donor of PS II based primarily on the efficient photooxidation of the cytochrome by PS II under cryogenic temperatures (BUTLER et al., 1972; ERIXON and BUTLER, 1971). The assignment of cytochrome b-559 for Z is complicated by the mismatch of their redox potentials. Even though this cytochrome can exist in high and low potential forms, the value of $+0.37$ V for the high potential form is still too low for Z. The role of b-559 in PS II reactions is treated in detail by CRAMER (cfl. see this vol. Chap. II. 12).

Recently the oxidized form of the primary donor, Z^+, has been correlated with a fast decaying free-radical species detected by EPR spectroscopy. The now classic EPR Signal II induced by light was shown to be associated with the oxidizing side of PS II. However, its slow kinetic behavior $t_{1/2}=1$ sec for formation and $t_{1/2}=4$ h for decay) precludes direct participation of the species responsible for the classic Signal II (designated as Signals IIs by BABCOCK and SAUER, 1975a) in the main path of the electron transport reactions of PS II. Upon inhibition by various treatments, a kinetically distinct component of Signal II (designated as Signal IIf) became detectable. Although the spectral shape of Signal IIf is identical to Signal IIs, it exhibits rapid rise kinetics ($t_{1/2}=500$ µsec) and decays with a half-time of several hundred milliseconds. Further, the decay kinetics of Signal IIf are affected by the redox potential of the medium and greatly accelerated by the addition of lipophilic electron donors for PS II (BABCOCK and SAUER, 1975b). BABCOCK and SAUER (1975a) concluded from electrochemical and kinetic studies that the species responsible for Signal IIf is identical with Z. However, the chemical nature of the Signal IIf species has not yet been identified.

III. The Reducing Side of PS II

The interaction between the secondary electron acceptor pool, A, of PS II with the primary electron acceptor, Q, constitutes the reactions on the reducing side of PS II and is designated as reaction II in Figure 2. The size of the secondary pool, A, in terms of redox equivalents is approximately 20 times that of Q based on analysis of the rise curve of chlorophyll a fluorenscence (Malkin and Kok, 1966; Forbush and Kok, 1968) and on measurement on the oxygen gush or spike (de Kauchkousky and Joliot, 1967). Pool A is not homogeneous and at least two kinetically distinct sub-pools, A_1 and A_2, can be differentiated. Based on the kinetics of photoreduction of DPIP by chloroplasts under flash illumination, Forbush and Kok (1968) concluded that sub-pool A_1, approximately one-third of the size of A, equilibrated rapidly with Q^- while the rest of the pool (A_2) reacted slowly with Q^-. Lien and Bannister (1971) studied the transient dark reduction of DPIP by chloroplasts and sub-chloroplast preparations immediately following the cessation of a strong, steady illumination and concluded that a sub-pool which amounts to approximately one-half of the A pool is located on the PS II side of a phosphorylation-coupled step. Kok et al. (1969) measured the relative number of electrons flowing through P-700 as induced by either a short or long flash and arrived at a ratio of 6 to 10 for [A]/[Q]. They concluded further from kinetic analysis that this PS I-linked sub-pool of A is probably identical with the sub-pool A_1 which interacts rapidly with Q^- as measured by the DPIP photoreduction technique. Thus it appears as if the slowly equilibrating sub-pool, A_2, is not located within the main linear chain connecting the two photosystems.

The chemical components of sub-pools A_1 and A_2 are not known. Reports from Witt's laboratory (Stiehl and Witt, 1968; Schmidt-Mende and Rumberg, 1968), concerning light-induced absorption changes around 254 nm ascribed to plastoquinone(s), revealed a close similarity between the plastoquinone pool and A_1 in terms of kinetics and pool size. Recently, a quinone antagonist, dibromothymoquinone (DBTQ, or DBMIB), has been shown to inhibit photosynthetic electron transport activities in a manner which is consistent with the interpretation that DBMIB interacts with certain component(s) of the plastoquinone pool linking PS II to PS I (Trebst et al., 1970; Böhme et al., 1971). Further, an unidentified component, designated M, was shown to be located in the electron transfer chain linking the two photosystems based on studies with mutant strain ac-21 of Chlamydomonas (Smillie and Levine, 1963; Levine and Gorman, 1966). Böhme and Cramer (1972) suggested that component M many be analogous to the plastoquinone component(s) affected by DBMIB. Based on data obtained with other mutant strains of Chlamydomonas cytochrome b-559 was included as a component of the A pool located very close to the primary acceptor, Q, and is in the main linear path linking the two photosystems (Levine and Gorman, 1966; Levine, 1969), although alternative assignments of the role of cytochrome b-559 have appeared (see this vol. Chap. II. 12). In addition to plastoquinone and cytochrome b-559, other known secondary electron acceptors of PS II are cytochrome f and plastocyanin which are more closely associated with PS I than PS II.

C. Photosystem I

We have witnessed within the past five years an impressive series of developments that represent the first concrete steps in the identification of the molecular components involved in the conversion by photosystem I of solar to chemical potential energy. Elucidation of the biochemistry of this process has been encumbered by the membrane-bound nature of the components involved and the recent advances are, in large part, due to increased application of sophisticated biophysical techniques to provide the clues necessary for their identification. (Indeed, if extraction and reconstitution studies are to biochemistry as analysis and synthesis are to organic chemistry, a certain amount of confusion in interpretation of data may well be the rule rather than the exception — at least until membrane biochemistry really comes of age.) These advances are presented primarily in terms of bioenergetics and component identity. Particular attention is afforded the methodologies employed and the assumptions involved so as to provide both an appreciation of the limitations as well as the potential inherent in these techniques. Neither the problems of artificial electron donors nor of the membrane localization of components associated with PS I are treated in detail. Therefore, the reader may wish to consult ANDERSON's (1975) recent review on the molecular organization of the chloroplast thylakoid and TREBST's (1974) excellent review on energy conversion in the photosynthetic electron transport pathway for further information on these topics.

As in the earlier section, it is convenient also to consider photosystem I from three vantage points: (1) The nature of the primary acceptor; (2) The reducing side; and (3) The oxidizing side.

I. The Primary Acceptor of PS I

The data available within the last few years appears to have inspired many more questions than answers. Has the primary acceptor of photosystem I actually been identified? Do the newly-discovered non-heme iron-sulfur centers have the proper qualifications, as some investigators suppose, or is the newest EPR component a better candidate? A more immediate problem involves the stabilized reducing power generated by illuminated chloroplasts. What is the midpoint potential of the primary acceptor of PS I? Or perhaps a more realistic question — what is actually being measured in the redox experiments, a transient primary acceptor or a more stable secondary acceptor? Finally we must consider the possibility of components existing as membrane-bound electron carriers after the primary acceptor. Are, for example, some of the components championed in the past as primary acceptors actually secondary acceptors? Is the current excitement over the non-heme iron-sulfur centers premature and may they, too, eventually fall into this category?

The chemical identity of the primary electron acceptor of PS I has been the topic of considerable debate and speculation for the last several decades. An abbreviated listing of some of the chemical candidates proposed is presented in Table 1. The chlorophyll ferredoxin interactions are, perhaps, the most interesting particularly in light of recent findings that non-heme iron proteins act near the site of initial charge separation in photosystem I.

In terms of experimental verification, any candidate for the role of primary electron acceptor should, ideally, exhibit these minimum properties:

1. The light-induced reduction kinetics should mirror the oxidation kinetics of the primary donor, P-700;

Table 1. Proposed primary acceptors of photosystem I

Molecule	Comments	Reference
Chlorophyll· heme	Absorption of a photon would bring about an excited singlet state in the chlorophyll. Subsequent electron transfer would result in a charge separation: chlorophyll$^+$·chlorophyll$^-$	Kamen (1961)
Chlorophyll· chlorophyll	Same as above except that the charge separation would occur within the dimer to produce: chlorophyll$^+$·chlorophyll$^-$	Kamen (1963)
Unconjugated pteridines	Naturally occurring and presumed low redox potential. Tetrahydropteridine reduces ferredoxin and interacts with bacterial and plant reaction centers resulting in a spectral shift similar to that produced by light	Fuller and Nugent (1969)
Pteridines	Redetermination of the midpoint potentials of tetrahydropteridine derivatives show the midpoint potential to be 0.15 V. 6,7-Dimethyl-tetra-hydropterin does not reduce spinach ferredoxin even at 150-fold excess in their experiments	Archer and Scrimgeour (1970)
Flavins	Formation of molecular complexes with chlorophyll was demonstrated with absorbance maxima near 700 nm. Can be reversibly bleached by light	Wang (1970)
Ferredoxin	Photoexcited and chemically reduced chlorophyll can transfer an electron to ferredoxin in a charge-transfer reaction	Vernon and Ke (1966)
Porphyrins	Photoexcited porphyrins can catalyze the reduction of ferredoxin with dithioerythritol serving as the reductant	Kassner and Kamen (1967)

2. The kinetics of the back reaction between the reduced acceptor and oxidized donor should be identical for both species; and

3. A kinetic correlation should exist between the reoxidation of the primary acceptor and the reduction of known secondary acceptors.

In 1971, a spectroscopic component, P-430, was discovered that exhibited two of these properties. Hiyama and Ke (1971a, 1971b) found that the dark reduction kinetics of P-700$^+$ to P-700 showed as fast and a slow component in the Soret region. The slower phase was accelerated with increasing amounts of tetramethylphenylenediamine (TMPD, an artificial PS I donor) thereby implicating this kinetic phase as the reduction of oxidized P-700. The fast component, however, showed an acceleration in its dark decay with several well-known PS I acceptors — methyl viologen (-446 mV), safranine T (-290 mV) and spinach ferredoxin (-430 mV) among others. It is associated with PS I based on its enrichment in PS I particles and because the quantum yield increases from one to two for both P-700 and P-430 when far red light is substituted by red light. The onset times (Ke, 1972) for both P-700 and P-430 are less than 0.1 μsec but resolution is limited to the rise time capabilities of the instrumentation. Since risetimes of 20 nsec have been measured for P-700 photooxidation (Witt and Wolff, 1970), P-430 reduction will have to show the same risetime before its role as the reaction partner to P-700 is confirmed. A backreaction between P-430$^-$ and P-700$^+$ has been demonstrated that showed special-case second order kinetics indicating the production of equimolar amounts of P-700$^+$ and P-430$^-$ during the photoact. Finally, the oxidation kinetics of P-430$^-$ correspond kinetically to the reduction kinetics of several artificial PS I acceptors. Notably, the oxidation kinetics of P-430$^-$ were pseudo-first order with respect to the acceptors methyl viologen and safranine T. Further, the midpoint potential of P-430, determined with a series of viologen dyes as redox buffers and measured as the magnitude of the absorbance change during illumination, is -470 mV and it appears to be a one-electron acceptor. This value, of course, is somewhat higher (more positive) than the midpoint potential determined by viologen reduction and redox poising experiments. The chemical identity of P-430 remains unknown (KE, 1973). Its light-minus-dark absorbance difference spectrum shows bands at 460 nm, 430 nm, and 405 nm. To label it as a ferredoxin, though, may be a bit premature, particularily since it has an extinction coefficient at 430 nm larger than that known for any plant type ferredoxin. It should be noted, however, that this extinction coefficient is ultimately based upon a determination of the amount of P-700. Since the extinction coefficient for P-700 is still in dispute, no definitive determination of the extinction coefficient for P-430 is possible at present.

Ever since MALKIN and BEARDEN (1971) showed the existence of a membrane-bound non-heme iron-sulfur center at 25 K, electron spin resonance techniques have become increasingly popular as sensitive probes of the reducing side of PS I. Any low temperature technique, though, has limitations as well as advantages. Since most kinetic motion has ceased at near-liquid temperatures, all solution chemistry is at a standstill and only the primary photochemistry is thought to occur. In fact, most investigators in this field believe that demonstration of an event at cryogenic temperatures is proof that the event is associated with the primary photoact. While this may be true of solution chemistry, closely spaced membrane-bound components may, indeed, demonstrate electron carrier abilities totally unexpected at these temperatures. Whether any temperature-sensitive changes have been induced in the membrane structures of these particules is also unknown. Non-heme iron-sulfur proteins must be observed at these temperatures, nevertheless, because there is a broadening of their EPR signals at temperatures above a few tens of degrees Kelvin.

There is now reasonably good evidence that two non-heme iron-sulfur centers exist on the reducing side of photosystem I in chloroplasts. Center A (g values 1.86, 1.94 and 2.05) becomes visible during illumination at 77 K after preparation in the dark or during illumination throughout the freezing process (EVANS et al., 1972). Dithionite alone in the dark causes a partial reduction of this center (KE et al., 1973) and subsequent illumination increases the magnitude of the signals even further. The signals can be elicited with equal effectiveness in both red and far-red light and there is an enrichment of the signals in various PS I preparations (BEARDEN and MALKIN, 1972a). Although there is a report to the contrary (YANG and BLUMBERG, 1972), equal spin concentrations of oxidized P-700 and reduced non-heme iron-sulfur center A have been demonstrated in chloroplasts and PS I particles (BEARDEN and MALKIN, 1972b). Potentiometric titrations (KE et al., 1973) reveal a midpoint potential of -530 mV for the g$=1.94$ and 2.05 lines. The g$=1.86$ line exhibits rather puzzling behavior; it appears at -502 mV, increases to a maximum around -562 mV and begins to decrease in intensity through -612 mV. Several explanations for this behavior have been proposed and will be dealt with later.

Center B (g values 1.89, 1.92 and 2.05) only becomes visible photochemically when chloroplasts are prepared under normal room illumination and frozen in the light (Evans et al., 1972). Dithionite alone in the dark causes no reduction of this center but its presence apparently increases the magnitude of this component during illumination. When dithionite and methyl viologen are both present, however, both centers A and B become reduced even in total darkness. Potentiometric titration of PS I subchloroplast fragments reveal a midpoint potential of less than or equal to -580 mV. Curiously, two authors (Ke et al., 1973; Evans et al., 1974) report evidence for two electron changes in both centers A and B. All plant type ferredoxins are one electron carriers.

The $g = 1.86$ component is thought to be associated with center A because it becomes visible after illumination at 77 K and after the first stage of electrochemical reduction. Conditions that allow for the reduction of center B, however, reduce its intensity. Ke et al. (1973) postulate that this behavior may be the result of an intermediate oxidation state, or alternatively, the structure giving rise to it may undergo a change as the potential is lowered. Evans et al. (1974) offer a different explanation. They feel that this line is a component of center A which broadens or undergoes a shift such that it is inseparable from the line at $g = 1.89$. Such a shift has been seen with eight-iron bacterial ferredoxins (Zubieta et al., 1973) where two redox centers are components of the same protein. They suggest that cooperation between subunits and reduction of one center may make the midpoint potentials less negative and also give an apparent value of n in the Nernst equation a value greater than one but less than two. This might explain the apparently high value of $n = 2$ reported by Ke et al. (1973).

Although there is some disagreement as to the extent (Ke et al., 1974; Visser et al., 1974), increasing amounts of P-700 become irreversibly photooxidized as the temperature is lowered from 200 K to 11 K. Several groups argue that the reaction partner for P-700 oxidation should also exhibit this behavior. However, since non-heme iron-sulfur proteins are not visible by EPR much above 25 K, direct observation of the irreversible extent of their photoreduction (at these temperatures) is not possible. Two groups have designed techniques aimed at overcoming this problem; both rely on the stability of the EPR signals in the dark at 11–13 K.

Visser et al. (1974) illuminated the samples at two different temperatures and measured the light-off extent of irreversible photooxidation and photoreduction at 11 K. They observed 20% of the total P-700 and non-heme iron-sulfur change to be reversible at this temperature. Assuming the signal height during illumination to be 100% of the total photooxidizable P-700 and photoreducible bound non-heme iron, illumination at 77 K and measurement three minutes later at 11 K showed 70% of both P-700 and bound non-heme iron remaining. They suggest that a different iron atom may be reduced in centers that are irreversibly photoconverted compared to those in which a back reaction is possible.

Ke et al. (1974) used a similar strategy only instead of illuminating the samples at higher temperatures, they illuminated them at liquid helium temperatures but exposed them to higher temperatures for a brief period of time before observing the remaining irreversible signal at 13 K. At temperatures of 13, 75, 125, 150,

175, and 225 K, the amount remaining of photooxidized P-700 and photoreduced non-heme iron (center A) agree.

There is no ready explanation for this irreversible phenomenon. In fact, one might expect charge recombination to occur with the same facility that charge separation takes place at these temperatures. If this were true, the data might suggest that another component is the primary acceptor, providing it has a redox potential more negative than the non-heme iron-sulfur centers. These centers might then function as secondary acceptors. Once separation of charge has been accomplished, the electron would begin a downhill flow from the primary acceptor to the non-heme iron-sulfur center(s). Recombination would be prevented by the energetic "barrier" it must overcome before it can react with P-700$^+$.

Also puzzling about the EPR data are the redox potentials of the two centers. Why is the higher potential non-heme iron center photoreducible at 25 K and not the lower potential center? Are both centers even on the same pathway?

That the non-heme iron-sulfur proteins may not be the reaction partner to P-700 has been suggested by the recent experiments by EVANS and CAMMACK (1975). P-700 photooxidation and non-heme iron-sulfur center A photoreduction are normally irreversible at 30 K. However, after a short exposure to dithionite, a slight reduction of the iron-sulfur protein occurs and a small amount of P-700 becomes reversibly photooxidizable. Under conditions where the non-heme iron-sulfur protein is almost fully reduced, the P-700 signal becomes almost fully and reversibly photooxidizable. McINTOSH et al. (1975) find that although 5–10% of the P-700 signal at 6 K is reversible, neither of the non-heme iron centers show any reversibility. They did, however, find a reversible component at g=1.75 and 2.07 that exhibited the same kinetics as P-700. In their experiments, particles frozen in the dark and illuminated at low temperature exhibited photoreduction of both non-heme iron-sulfur centers.

EVANS et al. (1975) subsequently showed that samples frozen while illuminated in the presence of dithionite and methyl viologen not only show the reduction of both non-heme iron-sulfur centers but reveal a new signal at g=1.76 and 1.86. This new component is not detectable in samples frozen without methyl viologen and illuminated or in dark samples frozen only with dithionite. Neither is it visible under conditions where the non-heme iron-sulfur proteins are oxidized before illumination; in this case, the irreversible oxidation of P-700 and the irreversible reduction of non-heme iron-sulfur centers occurs and no signal at g=1.76 is observed.

Under conditions where the iron-sulfur centers are reduced, this new component exhibits reversible behavior in parallel to that of P-700. The onset of both signals occurs immediately upon illumination at 9 K and both are fully reversible in the dark. Finally, if the g=1.76 component is frozen in the reduced state, the extent of P-700 photooxidation is greatly decreased. EVANS et al. (1975) believe the component has g values of 1.76, 1.86 and 2.06 and that it may be the primary acceptor of PS I. Although the g values are outside the range known for ferredoxins, it may be an iron-sulfur protein in a different environment or an altogether different type of iron-sulfur protein.

Still another potential iron-sulfur protein was uncovered (MALKIN and APARICIO, 1975) by electron spin resonance techniques in PS I particles. Its high redox potential ($E_{m, 7} = +0.29$ V) eliminates any function on the reducing side of PS I

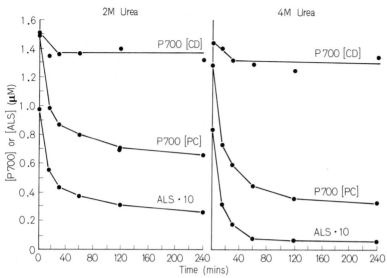

Fig. 3. Effect of urea and ferricyanide on the content of labile sulfide and P-700. The particles were incubated with 2 M or 4 M urea containing 5 mM ferricyanide the durations indicated on the graph. *P-700(CD)*: P-700 determined as an ascorbate-minus-ferricyanide difference spectrum; *P-700(PC)*: P-700 determined as light-induced bleaching at 697.5 nm; *ALS*: the amount of labile sulfide remaining in the particle

but it may nevertheless function in cyclic electron transfer associated with this photosystem. The g values (1.90 and 2.02) and midpoint potential are similar to the g = 1.90 component found in mitochondrial complex III (DUTTON and WILSON, 1974; WILSON and LEIGH, 1972).

Examination of the bound iron-sulfur proteins within our own laboratory has proceeded along more biochemical lines. Chemical analysis (MALKIN and BEARDEN, 1971) has not only previously revealed the existence of a very large pool of non-heme iron in whole spinach chloroplasts but it has shown that an enrichment of iron and labile sulfide occurs with respect to chlorophyll in PS I particles (BEARDEN and MALKIN, 1972a). Our own studies with highly enriched PS I particles (prepared by Triton X-100 treatment of washed thylakoid fragments followed by chromatography on Sephadex G-200 and DEAE Bio-Gel A) repeatedly show the presence of 8–10 mol of labile sulfide and non-heme iron per mol of P-700 (assuming $\varepsilon = 64,000 \ M^{-1} \ cm^{-1}$). This corresponds to a rather large pool of iron-sulfur protein; in fact, it is four to five times the in vivo molar concentration of P-700, plastocyanin, cytochrome f and soluble ferredoxin assuming 2 g atoms of iron and sulfide per mol of iron-sulfur protein.

We find that treatment of these photosystem I particles with either 2 or 4 M urea containing 5 mM ferricyanide produces a time-dependent loss of labile sulfide in the membrane-bound iron-sulfur proteins (Fig. 3). Because control levels of sulfide can be reestablished in all classes of depleted particles by overnight incubation with 1 mM dithiothrietol and because this enhancement can be prevented by preincubation of depleted particles with 5 mM cyanide, we believe that the labile sulfide is being oxidized to zero-valence sulfur in the bound iron-sulfur proteins in a manner analagous to that first described by PETERING et al. (1971).

There is little or no decline in the amount of P-700 when measured by a chemical oxidized-minus-reduced difference spectrum indicating that the primary electron donor to photosystem I has remained unaffected by the treatment. There is, however, a complex effect on the ability of P-700 to undergo light-induced oxidation. The extent of the normally fast photooxidation of P-700 correlates directly with the amount of labile sulfide remaining in the particle but a slow phase of P-700 photooxidation becomes increasingly evident in depleted particles that shows no relationship with labile sulfide. Although the meaning of the slow phase is still unclear, the strict relationship between the content of labile sulfide and the ability of P-700 to undergo photooxidation provides primary biochemical evidence that an iron-sulfur protein is a component of the primary acceptor complex and that it may participate directly in the primary photochemistry of PS I.

Progress in the isolation of membrane-bound iron-sulfur proteins has lagged noticeably behind biophysical and biochemical in situ investigations.

There has been one report, though, of isolation of an iron-sulfur protein from photosynthetic membranes. MALKIN et al. (1974) purified to homogeniety a 4-iron, 4-sulfur protein having a molecular weight of approximately 12,000 daltons. Its spectrum resembled more closely that of bacterial ferredoxin (in line with its iron and sulfur content) showing a broad absorbance in the 400 nm region, a shoulder at 310 nm and a peak at 280 nm. It is chemically reducible with dithionite, showing an absorbance decrease in the region of 430 nm and has an extinction coefficient similar to that of other ferredoxins — 4,000 per mole of iron or sulfur. Upon photochemical reduction by spinach chloroplasts, it can transfer electrons to mammalian cytochrome c but could not replace soluble ferredoxin in $NADP^+$ reduction. Curiously, the photoreduced protein underwent a rapid reoxidation after the cessation of illumination under anaerobic conditions where soluble ferredoxin remained reduced. Chemical and photochemical reduction resulted in a temperature-sensitive EPR absorbance centered at $g = 1.94$. No report was made of any ability to reduce soluble ferredoxin.

Our attempts at isolation and purification of membrane-bound non-heme iron-sulfur proteins have met with limited success due to facile formation of inorganic iron-sulfur-thiol (thiol = mercaptoethanol, cysteine, or dithioerythritol) complexes when utilizing "harsh" extraction procedures. We have found that, under the most promising extraction conditions, the iron and sulfur from the bound non-heme iron-sulfur proteins are released and form an inorganic complex with the thiol reductant in the extraction mixture. Under one set of conditions, the β-mercaptoethanol complex is retained on Whatman DE-32 and released with some protein at 0.15 M KCl. The reddish protein-containing fractions is "apparently" reducible with dithionite and "apparently" photoreducible by PS I particles. The fraction, however, was eventually purified free of protein and found equivalent to a synthetically produced iron-β-mercaptoethanol complex (Fig. 4).

YANG and HUENNEKENS (1969, 1970) have recently prepared iron-sulfur-β-mercaptoethanol complexes that show properties remarkably like those of iron-sulfur proteins (ferredoxins). This complex had an iron-to-sulfur ratio of one and an absorbance spectrum showing maxima at 325 nm and 412 nm; the extinction coefficient at 412 nm was calculated to be 4,200 per mole of iron. Low temperature EPR spectra show a large paramagnetic absorbance at $g = 4.1$ and weaker signals at $g = 2.01$ and 1.96 that are presumed to be associated with a transient species

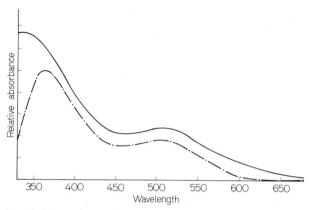

Fig. 4. Comparison of the absorbance spectra of a synthetically produced iron-β-mercaptoetha-nol complex (–·–·–) and the compound isolated by DE-32 chromatography (——). Synthetically produced complex contains 50 μM $FeCl_3$, 25 mM β-mercaptoethanol, 50 mM Tris·Cl, pH= 8.6. Iron analysis of both samples indicate $\simeq 40$ nmol $Fe^{2+}/0.1$ OD unit at 520 nm with a $S_2O_4^-$ vs untreated difference spectra. See text for further details

containing an unpaired electron. Oxygen is required to keep the iron in the trivalent state (β-mercaptoethanol slowly reduces it); the complex is therefore inherently unstable due to an eventual consumption of either the oxygen or β-mercaptoethanol. When the reaction mixture is allowed to become anaerobic, the spectrum of the complex slowly collapses but can be regenerated easily (as long as the β-mercapto-ethanol has not been totally consumed) by the simple reintroduction of oxygen.

In terms of our own experiments, we now know that the "apparent" reducibility with dithionite and "apparent" photoreducibility by PS I particles is actually a destabilization of the inorganic complex due to the removal of oxygen. Reintroduc-tion of oxygen allowed for regeneration of the spectrum. Due to the formation of these inorganic complexes, we might also wonder whether the isolated non-heme iron-sulfur protein preparation earlier discussed (Malkin et al., 1974) contains an inorganic iron-sulfur complex as a contaminant.

II. The Reducing Side of PS I

The reducing side of PS I is, undoubtedly, the best understood segment of the photosynthetic electron transport pathway. With the aid of hindsight, we can speculate why progress in this area was more rapid than progress on the reaction center itself. First, only two soluble proteins are involved in the direct transfer of electrons from the reduced membrane-bound components to $NADP^+$. Since each exhibit absorbance in the visible spectrum and undergo distinct changes upon reduction, the interaction can be studied by standard optical techniques. Secondly, the proteins intimately involved in $NADP^+$ reduction are either soluble in the chloroplast stroma or loosely bound (extrinsic) to the thylakoid membrane and are therefore readily studied by the methods of protein chemistry. The functionality of the chloroplast lamellae is, for the most part, unaffected by the extraction procedure thereby allowing facile extraction and reconstitution studies to be performed.

In terms of both bioenergetics and biochemistry, much progress has been made since WESSELS (1954) postulated that chloroplasts might be incapable of reducing compounds with $E_{m,7}$ values of less than zero volt. SAN PIETRO and LANG (1956) demonstrated in 1956 that NADP$^+$ ($E_{m,7} = -0.32$ V) could be quantitatively reduced by illuminated chloroplasts, thereby establishing a more reducing capacity for illuminated chloroplasts and the link between the product of the light reactions and the substrate for the dark assimilation of carbon dioxide. Its reduction requires the cooperation of both photosystems in vivo where water serves as an electron donor. GOVINDJEE et al. (1964) and BEN-HAYYIM and AVRON (1966) showed that the rate of NADP$^+$ reduction when chloroplasts are illuminated simultaneously with red and far-red light is higher than the sum of the rates when illuminated separately. This enhancement effect is the NADP$^+$ counterpart to the enhancement of oxygen evolution discovered by EMERSON and LEWIS (1943) which was a major basis for the concept of a two-light reaction scheme in plant photosynthesis. The quantum yield for the reduction of NADP$^+$ with far-red light was shown to approach one quantum per electron transferred when an ascorbate-DPIP couple was used (a specific photosystem I donor) but was much higher when water served as the electron donor (HOCH and MARTIN, 1963; SAUER and BIGGINS, 1965). Because far red light almost exclusively activates photosystem I, the electron transfer from this artificial donor system implies the intimate involvement of this photosystem in NADP$^+$ reduction.

The reducing power of illuminated chloroplasts was, once again, reevaluated with the demonstration that a reddish protein having an $E_{m,7}$ of -0.43 V (later known as ferredoxin) could be photoreduced by isolated chloroplasts (WHATELY et al., 1963; HORIO and SAN PIETRO, 1964; ARNON et al., 1965). Thereafter, kinetic studies (CHANCE and SAN PIETRO, 1963; CHANCE et al., 1965) demonstrated that ferredoxin was capable of being photoreduced at a rate similar to that at which cytochrome f was photooxidized. Since cytochrome f is closely coupled to the oxidizing side of PS I, the participation of ferredoxin in NADP$^+$ reduction seemed likely. Plant type ferredoxins are small ($\sim 10,000$ daltons) water-soluble proteins containing 2 mol of non-heme iron and acid-labile sulfur per mol of protein (FRY and SAN PIETRO, 1967). They function as one electron carrier as shown by the ability of 1 mol of NADP$^+$ to oxidize 2 mol of reduced ferredoxin (WHATELY et al., 1963; HORIO and SAN PIETRO, 1964).

Significant developments in the mid-1960s led to the realization that a second protein, ferredoxin-NADP$^+$ oxidoreductase, is necessary for the transfer of electrons from ferredoxin to NADP$^+$. The isolated protein was first found to stimulate the photoreduction of NADP$^+$ by illuminated chloroplasts (KEISTER et al., 1962; SHIN et al., 1963; DAVENPORT, 1963) and later shown to be involved in the in vitro oxidation of reduced ferredoxin by NADP$^+$ (SHIN and ARNON, 1965). (The reductase has additional electron transfer capabilities.) AVRON and CHANCE (1967) showed that in the absence of NADP$^+$, the rate of ferredoxin-NADP$^+$ oxidoreductase reduction is proportional to the amount of ferredoxin added to illuminated, ferredoxin-depleted chloroplasts.

Three events, then, are well characterized on the reducing side of PS I: (1) The reduction of ferredoxin by illuminated chloroplasts; (2) The dark transfer of electrons from reduced ferredoxin to the reductase enzyme; and (3) The dark transfer of electrons from the reduced reductase to NADP$^+$.

Because ferredoxin was the lowest potential, natural acceptor known as well as being the first stable reduced product of PS I, it was suggested to be the primary acceptor of PS I (Arnon, 1965). Kinetic data, however, showed that this was not the case. Chance et al. (1965) disclosed that cytochrome f photooxidation could be observed in well-washed chloroplast fragments devoid of ferredoxin and that addition of exogenous ferredoxin had little effect on the rate of cytochrome f photooxidation. It was now apparent that the immediate acceptor of electrons in PS I was endogenous to the membrane and different from soluble ferredoxin.

Three other groups, meanwhile, provided evidence from the reduction of low potential, γ,γ' dipyridyl bases (viologens) that PS I had an ultimate redox capability much more negative than that of ferredoxin. These viologens have properties which make good redox indicators: (1) They interact with the photosynthetic electron transport chain as shown by their ability to catalyze photophosphorylation in isolated chloroplasts; (2) They are available with $E_{m,7}$ values of -432 mV to -656 mV, the range expected for the primary reductant of PSI; and (3) They have $E_{m,7}$ values which are independent of pH, an important consideration when working with membrane-bound components. Parenthetically, derivatives of these dyes have found widespread use as herbicides and were used in 1933 by Michaelis and Hill in their early studies of enzyme kinetics.

Table 2. Reduction of viologens by chloroplasts

$E_{m,7}$ of dipyridyl (mV)	% photoreduction		
	Black (1966)	Kok et al. (1965)	Zweig and Avron (1965)
-342	100	100	—
-426	100	100	84
-521	50	100	30, 20
-636	7	26	—
-656	3	4	—

Zweig and Avron (1965), Kok et al. (1965) and Black (1966) monitored the reduction of several viologen dyes by illuminated chloroplasts under anaerobic conditions and observed some photoreduction even with the lowest potential dye (Table 2). Assuming that the naturally occurring reductant is one half reduced (and one half oxidized) at equilibrium, the midpoint potential of the reductant (that interacts with the viologens) is approximately -525 mV. Significantly, all reductions proceeded in the absence of ferredoxin. In addition, Kok et al. (1965) found that: (1) All viologens are capable ultimately of mediating oxygen evolution at rates equivalent to those observed with ferricyanide as substrate; (2) All are capable of catalyzing rapid P-700 turnover as monitored by the reversible photooxidation of P-700; and (3) All are efficient mediators of photophosphorylation.

Lozier and Butler (1974) used a slightly different approach in determining the midpoint potential of the reaction partner to P-700. The primary reductant was poised at various redox levels with *clostridial* hydrogenase and 1,1'-trimethylene-2,2'-dipyridilium dibromide and the extent of photooxidation of the primary electron donor, P-700, was measured. Under these conditions, the greater level of reduction of the primary acceptor, the smaller will be the amount of photooxidizable P-700, assuming a 1:1 stoichiometry between the two components. At $-196°C$, the potential was calculated to be -530 mV and shown to be pH independent.

All of these methods, of course, assume that the viologen dye interacts directly with the primary reductant and not indirectly via a secondary reductant. If, for example, the primary electron acceptor is inaccessable to the dye, or alternatively will not react with it on kinetic grounds, and if the primary donor is in rapid, reversible equilibrium with a secondary acceptor, then the redox level of a secondary acceptor(s) might actually have been measured.

III. The Oxidizing Side of PS I

It is now reasonably well established that P-700 is the primary donor of PS I (KOK, 1961, 1963). Its concentration in the chloroplast is about 1 per 400 chlorophylls and although attempts at purification (while retaining photoactivity) have not yet been successful, subchloroplast preparations have been obtained with chlorophyll to P-700 ratios as low as 25 (BOARDMAN, 1970). The light-induced bleaching of P-700 at 700 nm and 430 nm is identical to ferricyanide-induced oxidation. Redox titrations establish it as a single electron carrier with an $E_{m,7}$ of $+0.45$ V and it can be photooxidized to various degrees at liquid helium temperatures (VISSER et al., 1974; KE et al., 1974). The oxidized form of P-700 (P-700$^+$) exhibits an EPR signal at $g=2.0025$ observable either by chemical oxidation with ferricyanide and by photochemical oxidation (BEINERT et al., 1962). The absorbance band at slightly longer wavelength than the bulk light-harvesting chlorophylls assures that most of the excitation energy captured by these pigments will be funneled into P-700.

In spite of the wealth of data supporting P-700 as the primary donor of photosystem I, RURAINSKI et al. (1971) have suggested that P-700 photooxidation may not be coupled to NADP$^+$ reduction. They find that the electron flux through P-700 per unit time in fresh chloroplast preparations decreases as a function of Mg^{2+} concentration while there is a concurrent increase in the rate of NADP$^+$ reduction.

It is becoming quite evident that plastocyanin, a small copper-containing protein, and not cytochrome f, is the direct electron donor to oxidized P-700. Chloroplast preparations devoid of plastocyanin are impaired in cytochrome f photooxidation (AVRON and SHNEYOUR, 1971; NELSON and RACKER, 1972; SIEDOW et al., 1973) and incapable of catalyzing NADP$^+$ reduction when an ascorbate-DPIP donor system (AVRON and SHNEYOUR, 1971) is provided. Both activities are restored by the addition of plastocyanin. A mutant of *Chlamydomonas reinhardi* (GORMAN and LEVINE, 1966) lacking functional plastocyanin similarily shows impairment of cytochrome f oxidation that can be restored by addition of purified plastocyanin.

In conclusion, PS I appears to be much more complex than we might have imagined a few years ago. Some of the new components may well be involved in linear electron flow to NADP$^+$ while others will be concerned with cyclic flow around PS I. Indeed, until the later process is more fully characterized in vivo, the assignment of specific functional roles for the individual components may remain difficult.

Speculating even further, it appears that the "reducing end" of PS I is an excellent site for invoking regulatory or feedback control on both linear and cyclic electron flow in the photosynthetic electron transport chain. The ratio of ATP to NADPH could be carefully controlled by regulating the pathway for electrons through either cyclic or non-cyclic carriers. It may be that one of the recently discovered components functions in this role.

Acknowledgements. Some of the reported herein was supported by Grant BMS 75-03415 (to A.S.P.) from the National Science Foundation.

References

Allen, F.L., Franck, J.: Arch. Biochem. Biophys. **58**, 124–143 (1955)
Anderson, J.M.: Biochim. Biophys. Acta **416**, 191–235 (1975)
Archer, M.C., Scrimgeour, K.G.: Can. J. Biochem. **48** (4), 526–527 (1970)
Arnon, D.I.: Science **149**, 1460–1469 (1965)
Arnon, D.I., Tsujimoto, H.Y., McSwain, B.D.: Nature (London) **207**, 1367–1372 (1965)
Arnon, D.I., Whatley, F.R., Allen, M.B.: Science **127**, 1026–1034 (1958)
Avron, M. (ed.): Proc. 3rd. Int. Congr. Photosynth. (1975)
Avron, M., Chance, B.: Brookhaven Symp. Biol. **19**, 149–160 (1967)
Avron, M., Shneyour, A.: Biochim. Biophys. Acta **226**, 498–500 (1971)
Babcock, G.T., Sauer, K.: Biochim. Biophys. Acta **376**, 329–344 (1975a)
Babcock, G.T., Sauer, K.: Biochim. Biophys. Acta **396**, 48–62 (1975b)
Bearden, A.J., Malkin, R.: Biochem. Biophys. Res. Commun. **46**, 1299–1305 (1972a)
Bearden, A.J., Malkin, R.: Biochim. Biophys. Acta **283**, 456–468 (1972b)
Bearden, A.J., Malkin, R.: Biochim. Biophys. Acta **325**, 266–274 (1973)
Beinert, H., Kok, B., Hoch, G.: Biochem. Biophys. Res. Commun. **7**, 209–212 (1962)
Ben-Hayyim, G., Avron, M.: Israel J. Chem. **4**, 73 (1966)
Bennoun, P., Joliot, A.: Biochim. Biophys. Acta **189**, 85–94 (1969)
Bishop, N.: Proc. Nat. Acad. Aci. U.S. **45**, 1696–1702 (1957)
Black, C.C.: Biochim. Biophys. Acta **120**, 332–340 (1966)
Blankenship, R.E., Babcock, G.T., Sauer, K.: Biochim. Biophys. Acta **387**, 165–175 (1975)
Blinks, L.R.: In: Research in Photosynthesis. Gaffron, H. (eds.). New York: Interscience, 1957, pp. 444–449
Boardman, N.K.: Ann. Rev. Plant Physiol. **21**, 115–140 (1970)
Böhme, H., Cramer, W.A.: Biochim. Biophys. Acta **283**, 302–315 (1972)
Böhme, H., Reimer, S., Trebst, A.: Z. Naturforsch. **26B**, 341–352 (1971)
Butler, W.L., Erixon, K., Okayama, S.: In: Proc. 2nd. Intern. Congr. Photosynthesis. The Hague: Dr. Junk, 1972, Vol. I, 73–80
Butler, W.L., Visser, J.W., Simons, H.L.: Biochim. Biophys. Acta **325**, 539–545 (1973)
Chance, B., San Pietro, A.: Proc. Nat. Acad. Sci. U.S. **49**, 633–638 (1963)
Chance, B., San Pietro, A., Avron, M., Hildreth, W.W.: In: Non-heme Iron Proteins. San Pietro, A. (ed.). Ohio: Yellow Springs, 1965, pp. 225–236
Cheniae, G.M., Martin, I.F.: Brookhaven Symp. Biol. **19**, 406–417 (1966)
Cheniae, G.M., Martin, I.F.: Plant Physiol. **47**, 568–575 (1971a)
Cheniae, G.M., Martin, I.F.: Biochim. Biophys. Acta **253**, 167–181 (1971b)
Cheniae, G.M., Martin, I.F.: Plant Physiol. **50**, 87–94 (1972)
Cox, R.P., Bendall, D.S.: Biochim. Biophys. Acta **347**, 49–59 (1974)
Cramer, W.A., Butler, W.L.: Biochim. Biophys. Acta **172**, 503–510 (1969)
Davenport, H.E.: Nature (London) **184**, 524–526 (1959)
Davenport, H.E.: Proc. Roy. Soc., Ser. B **157**, 332–345 (1963)
Döring, G.: In: Proc. 3rd Intern. Congr. Photosynthesis. Avron, M. (ed.). Amsterdam: Elsevier, 1975, Vol. I, pp. 149–158

Döring, G., Stiehl, H.H., Witt, H.T.: Z. Naturforsch. **22 B**, 639–644 (1967)
Döring, G., Witt, H.T.: In: Proc. 2nd. Intern. Congr. Photosynthesis. The Hague: Dr. Junk, 1972, Vol. I, pp. 39–45
Dutton, L.P., Wilson, D.F.: Biochim. Biophys. Acta **346**, 165–212 (1974)
Duysens, L.N.M.: Progr. Biophys. **14**, 1–104 (1964)
Duysens, L.N.M., Sweers, H.E.: In: Studies on Microalgae and Photosynthetic Bacteria. Tokyo: Univ. Tokyo, 1963, pp. 353–372
Emerson, R., Lewis, C.M.: Am. J. Botany **30**, 165–178 (1943)
Erixon, K., Butler, W.L.: Photochem. Photobiol. **14**, 427–433 (1971)
Evans, M.C.W., Cammack, R.: Biochem. Biophys. Res. Commun. **63**, 187–193 (1975)
Evans, M.C.W., Reeves, S.G., Cammack, R.: FEBS Lett. **49**, 111–114 (1974)
Evans, M.C.W., Sihra, C.K., Bolton, J.R., Cammack, R.: Nature (London) **256**, 669–670 (1975)
Evans, M.C.W., Telfer, A., Lord, A.V.: Biochim. Biophys. Acta **267**, 530–537 (1972)
Forbush, B., Kok, B.: Biochim. Biophys. Acta **162**, 243–253 (1968)
Fry, K.T., San Pietro, A.: Biochem. Biophys. Res. Commun. **9**, 218–225 (1967)
Fuller, R.C., Nugent, N.A.: Proc. Nat. Acad. Sci. U.S. **63**, 1311–1318 (1969)
Gläser, M., Wolff, Ch., Buchwald, H.E., Witt, H.T.: FEBS Lett. **42**, 81–85 (1974)
Gorham, P.R., Clendenning, K.A.: Arch. Biochem. Biophys. **37**, 199–223 (1952)
Gorman, D.S., Levine, R.P.: Plant Physiol. **41**, 1648–1656 (1966)
Govindjee: In: Photosynthesis Mechanisms of Green Plants. NASNRC publication **1145**, 318–334 (1963)
Govindjee, R. (ed.): Bioenergetics of Photosynthesis New York: Academic Press, 1975
Govindjee, R., Govindjee, Hock, G.: Plant Physiol. **39**, 10–14 (1964)
Govindjee, R., Papageorgiou, G.: Photophysiology **6**, 1–46 (1971)
Heath, R.L., Hind, G.: Biochim. Biophys. Acta **172**, 290–299 (1969)
Henninger, M.D., Crane, F.L.: J. Biol. Chem. **242**, 1155–1159 (1967)
Hill, R.: In: Essays in Biochemistry, Campbell, P.N., Greville, G.D. (eds.). London: Academic Press, 1965, Vol. I, pp. 121–151
Hill, R., Bendall, F.: Nature (London) **186**, 136–137 (1960)
Hind, G., Nakatani, H.Y., Izawa, S.: Biochim. Biophys. Acta **172**, 277–289 (1969)
Hiyama, T., Ke, B.: Proc. Nat. Acad. Sci. U.S. **68**, 1010-1013 (1971 a)
Hiyama, T., Ke, B.: Arch. Biochem. Biophys. **147**, 99–108 (1971 b)
Hoch, G., Martin, I.: Arch. Biochem. Biophys. **102**, 430–438 (1963)
Hommann, P.H.: Biochem. Biophys. Res. Commun. **33**, 229–234 (1968)
Horio, T., San Pietro, A.: Proc. Nat. Acad. Sci. U.S. **51**, 1226–1231 (1964)
Izawa, S., Heath, R.L., Hind, G.: Biochim. Biophys. Acta **180**, 388–398 (1969)
Joliot, P., Kok, B.: In: Bioenergetics of Photosynthesis. Govindjee (ed.). New York: Academic Press, 1975, pp. 387–412
Kamen, M.D.: In: Light and Life. McElroy, W.D., Glass, B. (eds.). Baltimore, MD: Johns Hopkins, 1961, pp. 438–488
Kamen, M.D.: Primary Processes in Photosynthes in Advanced Biochemistry. San Pietro, A. (ed.). New York: 1963, 183 pp.
Kassner, R.J., Kamen, M.D.: Proc. Nat. Acad. Sci. U.S. **58**, 2445–2450 (1967)
Katoh, S., San Pietro, A.: Arch. Biochem. Biophys. **128**, 378–386 (1968)
Ke, B.: Arch. Biochem. Biophys. **152**, 70–77 (1972)
Ke, B.: Biochem. Biophys. Acta **301**, 1–33 (1973)
Ke, B., Hansen, R.E., Beinert, H.: Proc. Nat. Acad. Sci. U.S. **70**, 2941–2945 (1973)
Ke, B., Sugahara, K., Shaw, R.R., Hansen, R.E., Hamilton, W.D., Beinhet, H.: Biochim. Biophys. Acta **368**, 401–408 (1974)
Keister, D.L., San Pietro, A., Stolzenbach, F.E.: Arch. Biochem. Biophys. **98**, 235–244 (1962)
Kessler, E.: Planta (Berl.) **49**, 435–454 (1957)
Kessler, E., Arthur, W., Brugger, J.E.: Arch. Biochem. Biophys. **71**, 326–335 (1957)
Kimimura, M., Katoh, S.: Plant Cell Physiol. **13**, 287–296 (1972)
Knox, R.S.: J. Theor. Biol. **21**, 244–259 (1968)
Kok, B.: Biochim. Biophys. Acta **48**, 527–533 (1961)
Kok, B.: Proc. Intern. Congr. Biochem. **6**, 73–81 (1963)
Kok, B., Forbush, B., McGloin, M.: Photochem. Photobiol. **11**, 457–475 (1970)

Kok, B., Joliot, P., McGloin, M.P.: In: Progress in Photosynthesis Research. Metzner, H. (ed.). Tübingen: Laupp, 1969, Vol. II, pp. 1042–1056
Kok, B., Malkin, S., Owens, O., Forbush, B.: Brookhaven Symp. Biol. **19**, 446–459 (1966)
Kok, B., Rurainski, H.J., Owens, O.V.H.: Biochim. Biophys. Acta **109**, 347–356 (1965)
Kouchkovsky, Y., de, Joliot, P.: Photochem. Photobiol. **6**, 567–587 (1967)
Levine, R.P.: In: Progress Photosynthesis Research. Metzner, H. (ed.). Tübingen: Laupp, 1969, Vol. II, pp. 971–977
Levine, R.P., Gorman, D.S.: Plant Physiol. **41**, 1293–1300 (1966)
Lien, S., Bannister, T.T.: Biochim. Biophys. Acta **245**, 465–481 (1971)
Lozier, R., Baginsky, M., Butler, W.L.: Photochem. Photobiol. **14**, 323–328 (1971)
Lozier, R.H., Butler, W.L.: Biochim. Biophys. Acta **333**, 460–464 (1974)
Malkin, R., Aparicio, P.J.: Biochim. Biophys. Res. Commun. **63**, 1157–1160 (1975)
Malkin, R., Aparicio, P.J., Arnon, D.I.: Proc. Nat. Acad. Sci. U.S. **71**, 2362–2366 (1974)
Malkin, R., Bearden, A.J.: Proc. Nat. Acad. Sci. U.S. **68**, 16–19 (1971)
Malkin, R., Bearden, A.J.: Biochim. Biophys. Acta **396**, 250–259 (1975)
Malkin, S., Kok, B.: Biochim. Biophys. Acta **126**, 413–432 (1966)
McIntosh, A.R., Chu, M., Bolton, J.R.: Biochim. Biophys. Acta **376**, 308–314 (1975)
Michaelis, M., Hill, R.: J. Gen. Phys. **16**, 859–873 (1933)
Miles, D.: Stadler Symposium. Columbia, MO.: Univ. Missouri, 1975, Vol. VII, pp. 135–154
Murata, N.: Biochim. Biophys. Acta **162**, 106–121 (1968)
Nelson, N., Racker, E.: J. Biol. Chem. **247**, 3048–3055 (1972)
Petering, D., Fee, J.A., Palmer, G.: J. Biol. Chem. **246**, 643–653 (1971)
Pietro, A., San, Lang, H.M.: Science **124**, 118–119 (1956)
Radmer, R., Kok, B.: Ann. Rev. Biochem. **44**, 409–433 (1975)
Rurainski, H.J., Randles, J., Hoch, G.E.: FEBS Lett. **13**, 98–100 (1971)
Sauer, K., Biggins, J.: Biochim. Biophys. Acta **102**, 55–72 (1965)
Schmidt-Mende, P., Rumberg, B.: Z. Naturforsch. **23B**, 225–228 (1968)
Shin, M., Arnon, D.I.: J. Biol. Chem. **240**, 1405–1411 (1965)
Shin, M., Tagawa, K., Arnon, D.I.: Biochem. Z. **338**, 84–96 (1963)
Siedow, J.N., Curtis, V.A., San Pietro, A.: Arch. Biochem. Biophys. **158**, 889–897 (1973)
Smillie, R.M., Levine, R.P.: J. Biol. Chem. **238**, 4058–4062 (1963)
Stiehl, H.H., Witt, H.T.: Z. Naturforsch. **23B**, 220–224 (1968)
Stiehl, H.H., Witt, H.T.: Z. Naturforsch. **24B**, 1588–1598 (1969)
Tel-Or, E., Avron, M.: In: Proc. 3rd. Intern. Congr. Photosynthesis. Avron, M. (ed.). Amsterdam: Elsevier, 1975, Vol. I, pp. 569–578
Trebst, A.: Ann. Rev. Plant Physiol. **25**, 423–458 (1974)
Trebst, A., Harth, E., Draber, W.: Z. Naturforsch. **25B**, 1157–1159 (1970)
Van Gorkom, H.J.: Biochim. Biophys. Acta **347**, 439–442 (1974)
Van Gorkom, H.J., Donze, M.: Photochem. Photobiol. **17**, 333–342 (1973)
Vernon, L.P., Ke, B.: In: The Chlorophylls. Vernon, L.P., Seely, G.R. (eds.). New York: Academic Press, 1966, pp. 569–641
Visser, J.W.M., Rijgersberg, K.P., Amesz, J.: Biochim. Biophys. Acta **368**, 235–246 (1974)
Wang, J.H.: Science **167**, 25–30 (1970)
Warburg, O., Lüttgens, W.: Biokhimya **11**, 303–322 (1946)
Wessels, J.S.C.: Thesis, University of Leyden (1954)
Whatley, F.R., Tagawa, K., Arnon, D.I.: Proc. Nat. Acad. Sci. U.S. **49**, 266–270 (1963)
Wilson, D.F., Leigh, J.S.: Arch. Biochem. Biophys. **150**, 154–163 (1972)
Witt, K., Wolff, C.: Z. Naturforsch. **25B**, 387–388 (1970)
Yamashita, T., Butler, W.L.: Plant Physiol. **43**, 1978–1986 (1968a)
Yamashita, T., Butler, W.L.: Plant Physiol. **43**, 2037–2040 (1968b)
Yamashita, T., Butler, W.L.: Plant Physiol. **44**, 435–438 (1969)
Yamashita, T., Tomita, G.: Plant Cell Physiol. **15**, 69–82 (1974)
Yamashita, T., Tsuji, J., Tomita, G.: Plant Cell Physiol. **12**, 117–126 (1971)
Yang, C.S., Blumberg, W.E.: Biochem. Biophys. Res. Commun. **46**, 422–428 (1972)
Yang, C.S., Huennekens, F.M.: Biochem. Biophys. Res. Commun. **35**, 643–641 (1969)
Yang, C.S., Huennekens, F.M.: Biochemistry **9**, 2127–2133 (1970)
Zubieta, J.A., Mason, R., Postgate, J.R.: Biochem. J. **133**, 851–854 (1973)
Zweig, G., Avron, M.: Biochem. Biophys. Res. Commun. **19**, 397–400 (1965)

2. Porphyrins, Chlorophyll, and Photosynthesis

D. MAUZERALL

A. Introduction

The porphyrins have been selected by the fine comb of evolution as the functional pigment of biology. As heme they are part of the electron transport system of almost all cells. As chlorophyll they are at the heart of photosynthesis. The role of porphyrins in electron transport appears to be as handmaidens to the redox properties of iron ions, but their role in photosynthesis is far more central and unique. They are involved in the critical energy conversion step of photosynthesis. As far as is known, only two pigments are used for this crucial reaction: chlorophyll a and bacteriochlorophyll a. So the question arises: why these particular chemical structures?

This article will be concerned with the relation of the structure of chlorophyll to its function. The reader is referred to special compilations for quantitative data on absorption spectra, chemical properties, etc. Various authors have discussed the purification, spectral and chemical properties of these pigments in *The Chlorophylls* (VERNON and SEELY, 1966). The older work is very well summarized in Rabinowitch's opus (RABINOWITCH, 1945, 1951, 1956). An encyclopedic work on *The Porphyrins* will contain chapters on chlorophyll and its derivatives (DOLPHIN, 1975).

The best available spectral data for the chlorophylls a and b and their pheophytins are summarized in Table 1. Only primary sources where the absorbance index was determined are used. For analytical methods, and data on the less known chlorophylls see in addition to the above, SMITH and BENITEZ (1955), FRENCH (1960), VERNON (1960), WINTERMANS and DE MOTS (1965) and STRAIN et al. (1971).

B. Structure

The molecular structure of chlorophyll is shown in Figure 1. It is a large conjugated macrocycle containing four convoluted nitrogen-containing subcycles and having a variety of side chains attached to its periphery. We will discuss these characteristics in turn.

The conjugated double bonds or π-electron system result in strong absorption of light. The size of the molecule is determined by the necessity to absorb solar radiation, which is that of a black body peaking near 600 nm. To absorb these wavelengths, the conjugated system must be 15 Å long or 7 Å in diameter if cyclic, according to an exercise in quantum mechanics. It is better to be cyclic, since one then gains aromatic stability, a classical effect now also understood with

Table 1. Spectral data on chlorophylls and pheophytins. The units are nm for λ and mM cm^{-1} for ε. The latter are calculated on the basis of anhydrous chlorophyll a (MW 893) or b (MW 907) and pheo a (MW 870) or b (884). The values of ε for chlorophyll may be 2–4% low because of the presence of 1–2 mol of water

Solvent	λ	ε	λ	ε	λ	ε
Chlorophyll a						
Ether[a]	661	84	427	107	408	68
Ether[b]	662	87	430	111	410	69
Ether[c]	661	91	428	121	409	85
Ether[d]	662	90	430	118	410	85
Benzene[a]	664	80	432	104	412	70
Chlorophyll b						
Ether[b]	642	58	452	161		
Ether[c]	643	52	453	155	430	53
Ether[d]	644	56	455	159	430	57
Pheophytin a						
Benzene[a]	669	49	508	10.5	414	100
Ether[c]	667	52	506	11.7	409	118
Ether[d]	667	55	505	12.7	409	115
Pheophytin b						
Ether[c]	653	33	523	11.1	433	174
Ether[d]	655	37	526	12.6	434	191

[a] Bellamy et al. (1963)
[b] Sauer et al. (1966). The ether contained 5% CCl_4
[c] Zscheile and Comor (1941)
[d] Smith and Benitez (1955)

Chlorophyll a   Fig. 1. Structure of chlorophyll a

quantum mechanics. Far UV radiation has often been suggested as a source of photosynthetic energy in early evolution because the high energy of far UV photons is sufficient to break chemical bonds indiscriminately. This may be useful in "scrambling" molecules in the prebiotic soup, but tends to disfavor the formation of complex molecules because of their increased optical cross section. Thus the selectivity, and the greater number of visible photons would make them the preferable source of energy.

The convoluted ring structure and the inner nitrogen atoms produce a rich optical spectrum, and an even richer supply of redox levels. These are important for the photochemical function discussed below. These apparent complexities are actually the result of the very simple biosynthesis of this macrocycle (Vol. 8 this Encyclopedia). The basic ring structure is that of a porphyrin, which has a four-fold symmetry, rare in organic compounds. It can be assembled from four α-substituted pyrrole units (Fig. 2). The natural pyrrole is called porphobilinogen and it is itself made up of two molecules of δ-aminolevulinic acid. Thus, the apparently complex macrocycle is a disguised octamer of δ-aminolevulinic acid. This building block is itself made from glycine and succinate (SHEMIN, 1955) or possibly form glutamate (BEALE and CASTELFRANCO, 1974) in higher plants.

The assorted side chains on the chlorophyll molecule originate in a simple biosynthetic way. The condensation to the macrocycle forms uroporphyrin (actually the hexahydro form or porphyrinogen) which has one propionate and one acetate side chain on each of the four pyrrole units. The change of this highly ionic, water soluble macrocycle along the biosynthetic pathway is relentlessly uni-directional (Fig. 2). The ionic groups are systematically removed via decarboxylations and the molecule is made increasingly hydrophobic. This property is reinforced by the esterification of the remaining propionic acid group with phytol, a C-20 isoprenoid alcohol. Clearly chlorophyll is a lipophilic molecule and it, and the photosynthetic apparatus is always found associated with intra-cellular membranes.

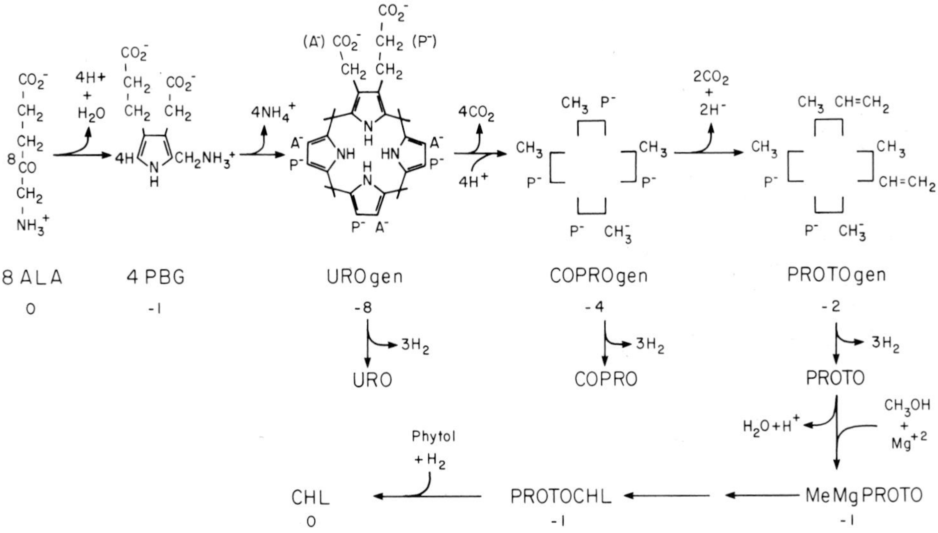

Fig. 2. Outline of biosynthetic pathway to chlorophyll. The numbers below the names are the ionic charges per molecule near pH 7

A hypothesis which rationalizes these changes to the lipophilic state is presented below.

The macrocycle of chlorophyll is a chlorin, not a porphyrin. The hydrogens on the outer edges of a single pyrrole unit convert the high four-fold symmetry of the π-electron system to a simple plane of symmetry. This seemingly minor change increases the strength of the absorption bands in the visible since they were forbidden in optical terms by the four-fold symmetry of the porphyrins. The absorption also moves to the red. By this means, the absorption of solar radiation is increased, at least for an organism on the surface of the land or ocean. This trick can be repeated on the opposite pyrrole unit to form bacteriochlorophyll which extends the absorption to the near-infrared. Reducing the adjacent pyrrole unit causes absorption at shorter wavelengths, and further reduction of the pyrrole units or of the meso or bridge positions ruins the aromatic and useful photochemical properties of the macrocycle. These steps have been presumably discarded by evolutionary selection.

The chelated metal in chlorophyll is magnesium ion, in place of the iron ion of the cytochromes. Since magnesium is held in the ring much more loosely and is difficult to chelate from aqueous solution, it seemed previously that evolution had chosen an unnecessarily difficult path. We now believe the magnesium ion is critical to the photochemistry. Transition metals quench this photochemistry, as they do the fluorescence, probably by rapid internal electron transfer. Our studies (MAUZERALL, 1962a, b, 1973, 1975; MAUZERALL and FEHER, 1964; HONG and MAUZERALL, 1974; CARAPELLUCCI and MAUZERALL, 1975) have shown that the presence of closed shell divalent ions in the porphyrin or chlorin, makes the excited state a powerful electron donor. This is what is required to reduce carbon dioxide to the level of sugars. However, the other half of modern green plant photosynthesis requires the oxidation of water to form oxygen. In connection with this function it is interesting that the ground state redox potential of chlorophyll is >0.4 V more oxidizing then simple magnesium chlorins (FUHRHOP and MAUZERALL, 1969; FUHRHOP, 1970). Thus, the detailed structure of chlorophyll, in particular the five-membered meso-condensed carbonyl ring and the vinyl group, makes the cation resulting from electron donation a good oxidant. The structure of chlorophyll is at a level of complexity that these interweaving, or synergistic, effects are quite noticeable. Relevant to this point, the two extra hydrogen atoms added to form the chlorin also favor the cyclization of the fifth ring by reducing steric crowding (WOODWARD, 1961).

The fifth ring of chlorophyll has mesmerized many investigators. It concentrates the polar residues: the carbonyl and the carbomethoxy groups. Even the ester group to phytol is close by. Several suggestions as to the possible role of this grouping have been made. WARBURG and KRIPPAHL (1956) argued that a bound carbon dioxide molecule transferred an oxygen atom to the carbonyl oxygen to form a peroxide and then molecular oxygen. FRANK (1957) thought that the C-10 hydrogen of triplet chlorophyll would reduce phosphoglyceric acid while the C-9 oxygen formed oxygen via "The Enzyme". VISHNIAC and ROSE (1958) tried to obtain evidence via tracer experiments that the C-10 hydrogen was involved in transfer from water. Our suggestion is that, together with manganese, the carbomethoxy cyclopentanone ring may provide a mechanism for oxygen formation through a dioxolium ion (MAUZERALL and CHIVVIS, 1973; MAUZERALL, 1976).

The similarity of the in vivo spectrum of chlorophyll to that of aggregates (RABINOWITCH, 1945, 1951, 1956; KRASNOVSKI, 1960) and of monolayers of the pigment (KE, 1966) has led to the identification of one with the other. KATZ et al. (1966) have greatly clarified the structure of the aggregated forms of chlorophyll by detailed studies of NMR and optical spectra. They have proposed a structure for these aggregates which contains one mol of water per mol of chlorophyll. The water serves as a donor to the magnesium of one chlorophyll and as the acceptor from the carbonyl of the cyclopentanone ring of a second overlapping chlorophyll. KATZ and NORRIS (1973) have suggested that this structure is present in most of the chlorophyll in green plants. FONG (1974) has proposed a more symmetrical variant of this dimeric structure for the trap or reaction center chlorophyll. The symmetry allows the possibility of a long-lived lowest energy state. These proposed structures give another function to the magnesium and the fifth ring of chlorophyll, but as for the dioxolium ion theory, there is as yet no direct experimental evidence for their existence in vivo.

C. Function

The function of the photosynthetic apparatus is to convert the energy of solar photons to chemical energy useful to the cell. It does this by capturing a solar photon in accessory or antenna pigments and rapidly transferring the excitation to the trap or reaction center pigments. Rapid electron transfer in the donor-acceptor complex of the trap converts the excitation to chemical energy followed by subsequent transfer to secondary donors and acceptors. The key point is that a barrier exists to electron return to the ground state of the trap following the first charge transfer, thus allowing sufficient time for secondary electron transfer to occur and thus biochemistry to take over. In photosynthetic bacteria, this barrier amounts to 30 msec (MCELROY et al., 1974) and is independent of temperature from 80 to 1 K. Consideration of the most probably mechanism, that of electron tunnelling, shows that the critical parameters are the energy levels of the donor and acceptor and the distance and orientation of the molecules involved (MAUZER-ALL, 1975). The energy levels are determined by the pigment chlorophyll, and the primary acceptor, with the requirement that a considerable fraction ($\geq 1/2$) of the excitation energy be conserved in the charge transfer state. Too little interaction and electron transfer does not occur during the lifetime of the excitation in the pigment complex. Too much interaction and, although the charge transfer state is formed rapidly and in high yield, its lifetime before degradation to the ground state by reverse electron transfer becomes shorter than that of the original excited state. A narrow optimum thus exists for the yield-lifetime product of the charge-transfer state (MAUZERALL and HONG, 1975; MAUZERALL, 1975). I believe that the distance and orientation of the molecules forming the primary charge transfer complex in photosynthesis have been selected by evolution to achieve this optimum situation.

There is evidence that the donor in bacterial photosynthesis is a dimer of bacteriochlorophyll (NORRIS et al., 1973; FEHER et al., 1975). This may also be true for system I of green plant photosynthesis (NORRIS et al., 1971, 1974). The function of the dimer may be many fold. As mentioned above, Katz and Norris (1973) suggest it may be part of the primary charge transfer function and Fong (1974) suggests it may provide a long-lived excited state. It is clear from the theory presented above that lengthening of the excited state lifetime will in general favor the electron transfer, and to longer distances (Mauzerall, 1975). The bacterial reaction center in fact contains four bacteriochlorophylls and two bacteriopheophytins. These molecules could, by their distance and mutual orientations, act as electronic way stations for the transfer of an electron to the "primary" acceptor: the iron-quinone complex. The near resonant levels of the bacteriochlorophylls, and the possible favoring of the cation of bacteriochlorophyll and anion of bacteriopheophytin (Mauzerall, 1975) would allow rapid tunnelling away from the primary donor. The slightly downhill energy tilt would favor the "forward" electron transfer and so could be a mechanism of enhancing the efficiency of the primary energy storage. A summary of the recent evidence concerning the primary donors and acceptors, and results of picosecond flash experiments is available in the excellent review by Parsons and Cogdell (1975) on bacterial photosynthesis. Sauer (1975) has presented a similar but more general review on primary events.

D. Evolution

The present function of photosynthesis is to reduce carbon dioxide to sugars and oxidize water to oxygen, thus forming the thermodynamic gradient of free energy necessary for all other forms of life. However, in the early stages of biogenesis, the earth's atmosphere is thought to have been chemically reducing. Thus, at least in the early stages of the evolution of life, a useful form of photosynthesis would have had to oxidize organic compounds and evolve hydrogen. Details of this theory have been presented elsewhere (Mauzerall, 1973, 1975, 1976) and only a brief outline will be given here. Since life originated in the sea, the first pigment should be water soluble and capable of oxidizing reduced organic compounds to reactive intermediates useful in prebiotic evolution. Such a pigment is uroporphyrin, the first macrocycle formed in the biogenesis of both chlorophyll and heme. The reduced porphyrin formed in the photo-oxidation of amines etc. can disproportionate (Mauzerall, 1962a) and so could possibly form hydrogen and porphyrin. An additional point in favor of porphyrins as the earliest photosynthetic agents in the sea is that water rapidly attenuates the red region of the solar spectrum. The intensity at 680 nm (Chl peak) is decreased 100 times in 10 m, and the peak quantal irradiance is near 550 nm even at a few meters depth (Jerlov, 1968). This is the region of the admittadly weak porphyrin absorption bands. This argument suggests that the stage of chlorophyll was reached when the early forms of life inhabited tidal flats and other such shallow water. The UV radiation and/or the low ozone layer (Ratner and Walker, 1972) of the early earth very likely favored the development of cellular life in deeper water.

In contrast to the behavior of porphyrin itself, when the porphyrin chelates a closed shell metal ion such a zinc or magnesium, the favored photochemical reaction is the reduction of an acceptor and the formation of the metalloporphyrin cation (Fuhrhop and Mauzerall, 1969; Carapellucci and Mauzerall, 1975).

Even nicotinamide adenine dinucleotide is reduced to the radical state ($E_0 \sim -0.7$ V) by the triplet state of zinc porphyrin. I believe the evolutionary occurrence of this kind of photochemical reaction may have been the turning point in photosynthetic evolution. When this reaction was coupled to the ultimate donor, water, oxygen was formed and modern photosynthesis was born. The rise of oxygen in the atmosphere in addition to forming the ozone shield for UV light and allowing the evolution of respiratory and more complex forms of life, possibly had a more direct effect on photosynthetic evolution. Previous to its formation photoreactions could proceed at a leisurely pace because of the long lifetimes (msec) of triplet states. These states are efficiently formed in porphyrins and chlorophylls. However, they are extremely efficiently quenched ($> 10^{10}$ M^{-1} sec^{-1}) by oxygen. Thus, the primary charge transfer reaction has to occur rapidly, in less than 100 nsec at the present level of oxygen. This requires a tight organization of the donor and acceptor pair. The tunnelling theory of electron transfer allows a maximum efficiency at a fixed distance and orientation (MAUZERALL, 1975) and so fits this requirement very well. Thus, the oxygen evolution and the more structured photosynthetic unit may have evolved synergistically. As mentioned previously there is a strong decrease in ionic charge on these molecules on passing from uroporphyrin to chlorophyll (Fig. 2). It is interesting that the charge has decreased to unity at the level of the first metalloporphyrin, magnesium protoporphyrin monomethyl ester. These and succeeding molecules along the pathway are highly lipophilic and will concentrate in such hydrophobic structures as lipid membranes. If this photosynthetic unit is arranged across a membrane-water interface, vectorial photoelectrochemistry becomes possible and a potential can be formed across the membrane (MAUZERALL and HONG, 1975). From this potential, ion transport and even ATP formation à la Mitchell hypothesis are possible (WITT, 1971).

Most of the above arguments are based on model systems. Whereas the photoreactions of porphyrins and metalloporphyrins in anaerobic solution occur through the triplet state (MAUZERALL, 1962b; CARAPELLUCCI and MAUZERALL, 1975) and at rates determined by the concentration of reactants, that of metalloporphyrins in lipid bilayers occurs in the presence of air (HONG and MAUZERALL, 1974) and with a rise time of less than 100 nsec. The bilayer reactions are in fact monitored the very potential discussed above.

E. Summary

A rationalization of the apparent complexity of the structure of chlorophyll has been given by relating this structure to its function, and to its evolutionary selection. At the moment the argument is largely an outline, but the advancing knowledge of both the photosynthetic apparatus and the detailed photochemical properties of this magnificent pigment of life promises the development of a more complete argument in the future.

Acknowledgements. This article owes much of its origins to probing discussions with Dr. S. GRANICK. Its preparation and the work mentioned were supported by grants from the NIH and NSF.

References

Beale, S.I., Castelfranco, P.A.: Plant. Physiol. **53**, 297–303 (1974)
Bellamy, W.D., Gaines Jr., G.L., Tweet, A.G.: J. Chem. Phys. **39**, 2528–2538 (1963)
Carapellucci, P., Mauzerall, D.: Ann. N. Y. Acad. Sci. **244**, 214–237 (1975)
Dolphin, D. (ed.): The Porphyrins. New York: Academic Press. In preparation (1975)
Feher, G., Hoff, A.J., Isaacson, R.A., Ackerson, L.C.: Ann. N. Y. Acad. Sci. **244**, 239–259 (1975)
Fong, F.K.: Proc. Nat. Acad. Sci. U.S. **71**, 3692–3695 (1974)
Frank, J.: In: Research on Photosynthesis. Gaffron, H. (ed.). New York: Interscience, 1957, pp. 142–144
French, C.S.: In: Handbook of Plant Physiology. Ruhland, W. (ed.). Berlin-Heidelberg-New York: Springer, 1960, Vol. V, Pt 1, pp 252–297
Fuhrhop, J.H.: Z. Naturforsch. **25 B**, 255–265 (1970)
Fuhrhop, J.H., Mauzerall, D.: J. Am. Chem. Soc. **91**, 4174–4181 (1969)
Hong, F., Mauzerall, D.: Proc. Nat. Acad. Sci. U.S. **71**, 1564–1568 (1974)
Jerlov, N.G.: Optical Oceanography. Amsterdam-London-New York: Elsevier, 1968, pp. 115–132
Katz, J.J., Dougherty, R.C., Boucher, L.J.: In: The Chlorophylls. Vernon, L.P., Seely, G.R. (eds.). New York: Academic Press, 1966, pp. 186–251
Katz, J.J., Norris, J.R.: Current Topics Bioenerg. **5**, 41–75 (1973)
Ke, B.: In: The Chlorophylls. Vernon, L.P., Seely, G.R. (eds.). New York: Academic Press, 1966, pp. 253–279
Krasnovski, A.A.: Ann. Rev. Plant. Physiol. **11**, 363–410 (1960)
Mauzerall, D.: J. Am. Chem. Soc. **84**, 2437–2445 (1962a)
Mauzerall, D.: J. Phys. Chem. **66**, 2531–2533 (1962b)
Mauzerall, D.: Ann. N. Y. Acad. Sci. **206**, 483–494 (1973)
Mauzerall, D.: Phil. Trans. Roy. Soc., Ser. B. **273**, 287–294 (1976)
Mauzerall, D.: In: The Porphyrins. Dolphin, D. (ed.). New York: Academic Press, in preparation, 1975
Mauzerall, D., Chivvis, A.: J. Theor. Biol. **42**, 387–395 (1973)
Mauzerall, D., Feher, G.: Biochim. Biophys. Acta **79**, 430–432 (1964)
Mauzerall, D., Hong, F.: In: Porphyrins and Metalloporphyrins. Smith, K.M. (ed.). Amsterdam: Elsevier, 1975, Chap. 17, pp. 701–725
McElroy, J.D., Mauzerall, D., Feher, G.: Biochim. Biophys. Acta **333**, 261–278 (1974)
Norris, J.R., Druyan, M.E., Katz, J.J.: J. Am. Chem. Soc. **95**, 1680–1682 (1973)
Norris, J.R., Scheer, H., Druyan, M.E., Katz, J.J.: Proc. Nat. Acad. Sci. U.S. **71**, 4897–4900 (1974)
Norris, J.R., Uphaus, R.A., Crespi, H.L., Katz, J.J.: Proc. Nat. Acad. Sci. U.S. **68**, 625–628 (1971)
Parson, W.W., Cogdell, R.J.: Biochim. Biophys. Acta **416**, 105–149 (1975)
Rabinowitch, E.I.: Photosynthesis. New York: Interscience, Vol. I, 1945; Vol. II, pt. 1, 1951; pt. 2, 1956
Ratner, M.I., Walker, J.C.G.: J. Atmos. Sci. **29**, 803–808 (1972)
Sauer, K.: In: Bioenergetics of Photosynthesis. Govindjee (ed.). New York: Academic Press, 1975, pp. 115–181
Sauer, K., Smith, J.R.L., Schultz, A.J.: J. Am. Chem. Soc. **88**, 2681–2688 (1966)
Shemin, D.: Harvey Lectures **50**, 258–284 (1955)
Smith, J.H.C., Benitez, A.: In: Modern Methods of Plant Analysis. Paech, K., Tracey, M.V. (eds.). Berlin-Heidelberg-New York: Springer, 1955, Vol. IV, pp. 142–196
Strain, H.H., Cope, B.T., Svec, W.A.: Methods Enzymol. **23 A**, 452–476 (1971)
Vernon, L.P.: Anal. Chem. **32**, 1144–1150 (1960)
Vernon, L.P., Seely, G.R. (eds.): The Chlorophylls. New York: Academic Press 1966
Vishniac, W., Rose, I.A.: Nature (London) **182**, 1089–1090 (1958)
Warburg, O., Krippahl, G.: Z. Naturforsch. **11 B**, 179–180 (1956)
Wintermans, J.F.G.M., DeMots, A.: Biochim. Biophys. Acta **109**, 448–453 (1965)
Witt, H.T.: Quart. Rev. Biophys. **4**, 365–477 (1971)
Woodward, R.B.: Pure Appl. Chem. **2**, 383–404 (1961)
Zscheile, F.P., Comor, C.L.: Botan. Gaz. **102**, 463–481 (1941)

3. Light Conversion Efficiency in Photosynthesis

R.J. RADMER and B. KOK

A. Basic Principles

The process of photosynthesis involves the conversion of light energy into chemical energy. Electromagnetic radiation is emitted and absorbed in discrete quantities (quanta) of radiant energy called photons possessing an energy $E=h\nu$, where h is Planck's constant and ν the frequency of the radiation.

$\nu=c/\lambda$, where c is the speed of light and λ the wavelength. Since $c=3\cdot10^8$ m sec^{-1} and $h=6.6\cdot10^{-34}$ Joule-sec, a quantum of red light (a red photon) of $\lambda=680$ nm ($6.8\cdot10^{-7}$ m) is equal to $2.9\cdot10^{-19}$ Joules or $6.9\cdot10^{-20}$ calories. Thus a mole of these quanta ($6\cdot10^{23}$ quanta, defined as an Einstein) is equivalent to about 40 kilocalories.

The question of photosynthetic efficiency can be approached in two ways: the more fundamental measurement is the determination of the fraction of the radiation absorbed by the photosynthetic pigments that is utilized. It is usually expressed as quantum yield ($O_2/h\nu$) or energy efficiency (calories fixed as chemical potential/calories absorbed), and is related to the efficiency of the photosynthetic machinery. In the more holistic "field" approach, one determines the fraction of incident radiant energy on a given area that is fixed as organic plant material. Such an approach reflects losses due to dissimilatory processes of the plant and incomplete light absorption (a field that is only partially covered with leaves). In the following sections, we will consider photosynthetic efficiency from each of these viewpoints.

Generally, it has been tacitly assumed that light energy is freely convertible into chemical energy, and that there are no thermodynamic restrictions on such a conversion. However, DUYSENS (1958) concluded that a significant fraction of light energy is lost when it is converted to chemical energy: the source-sink system is really a heat engine for which the maximum amount of work (W) is given by the expression $W=Q(T_2-T_1)/T_2$ (where Q is the amount of heat transferred and T_2 and T_1 the absolute temperatures of the source and reaction vessel respectively). A similar conclusion was obtained by MORTIMER and MAYO (1961) on the basis of a much more rigorous analysis of the problem (see also SHOCKLEY and QUEISSER, 1961). The calculated maximum efficiency is thus a function of the assumed source temperature (the sink temperature can be determined). DUYSENS assumed a low (1,100 K) source temperature and arrived at a maximum efficiency of about 70%. The other workers assume that the source is a 6,000 K black body (the radiation of which roughly matches the solar spectrum) and obtain maximum efficiencies of about 95%.

The law of photochemical equivalence states that the photochemical transformation of a molecule is induced by the absorption of a single photon. Consequently, the ratio of transformed molecules to absorbed photons, usually denoted as the quantum yield, ϕ, can be equal to one for primary photoconversions. In practice, most photochemical processes appear to deviate from this photochemical law, due to the less-than-perfect efficiency of the subsequent reaction steps that serve to stabilize the highly reactive primary photoproducts. In the process of photosynthesis, there are many reaction steps subsequent to the initial photoreaction and thus the relationship between absorbed quanta and chemical end products is rather indirect.

The energy efficiency of a photochemical process is generally less than the quantum efficiency. The longest wavelength of light able to initiate the two photoconversions in green plants is about 700 nm; a quantum of this light possesses an energy of about 1.8 eV. Since both photoacts result in a charge separation of about 1.0 eV, the maximum energy efficiency of each photoact is about 50%. Chlorophylls and accessory pigments absorb (with variable effectiveness) light of wavelengths between 400 and 700 nm. However, the photoexcitation of photosystems I or II results in the same photochemical transformation regardless of the energy of the absorbed photon; any energy in excess of the threshold (680–700 nm) is dissipated as heat. Since the photosynthetic quantum yields for 430 nm and 680 nm light are equal (see below), the energy efficiencies for these two wavelengths differ by about 60%. Because of the strong wavelength dependence of the computed energetic efficiency (compared to the quantum efficiency, which is almost wavelength independent), this parameter is of little theoretical significance in the study of the photosynthetic process.

The energy efficiency for a photoprocess of a given band gap (which corresponds to the maximum effective wavelength λ_e) is a compromise between the loss of energy for photons $<\lambda_e$ (proportional to $1/\lambda - 1/\lambda_e$) and the non-utilization of photons $>\lambda_e$. A detailed analysis of the upper limit for the energy efficiency of a solar process by SHOCKLEY and QUEISSER (1961) showed that the maximum efficiency is only about 40%. At this maximum the band gap (which corresponds to the voltage spanned by the photoact) is about 1.1 eV, so that λ_e is about 1,100 nm. This suggests that aerobic photosynthetic organisms are not optimized with respect to energy efficiency (photosynthetic bacteria approach this ideal more closely). One might speculate that this is related to the energy level of the charge separation, and that more than the usually accepted 1.0 eV is obtained from each photoact. For example, a charge separation taking place across the thylakoid membrane could create a potential utilized for the formation of ATP.

B. The Maximum Efficiency of Photosynthesis: Quantum Yields Under Optimum Conditions

The true maximum quantum yield of photosynthesis has been a hotly controversial issue. For a period of nearly four decades following their original measurements (WARBURG and NEGELEIN, 1922), WARBURG and his associates propounded and defended manometric experiments of O_2 and CO_2 exchange using whole algae that, according to their interpretation, indicated a quantum requirement ($1/\phi$) of about $4\,hv/O_2$. [For a critical review of these experiments see KOK (1960).] However, in whole cells this seemingly simple measurement is confounded by the anomalous transient behavior of CO_2 and O_2 at the onset of illumination (EMERSON and LEWIS, 1941). In addition, great caution is required when correcting the observed light-induced gas exchange for "dark respiration" since (a) dark respiration can be inhibited by light (KOK, 1949) and (b) a similar process (called photorespiration) can occur which is stimulated by light (see below).

A compilation of quantum yield measurements made with whole cells under steady-state conditions with minimal interference by respiratory phenomena reveals that the minimum quantum requirement observed by various investigators (excepting WARBURG and collaborators) is at least $8–10\ hv/O_2$ (see KOK, 1960) for a catalog of these experiments). Despite its somewhat tenuous relationship to the primary photoevents, the maximum photosynthetic quantum yield has strong mechanistic implications, and the general acceptance of a quantum requirement ≥ 8 aided in the development of the concept of two series-connected photoacts. According to this scheme each electron equivalent is photoactivated twice in its journey from H_2O to CO_2.

Green plants are able to produce molecular O_2 and, at the same time, generate a reductant at least as strong as molecular H_2. From the potential of the hydrogen electrode $(-0.42\,V)$ and the oxygen electrode $(+0.81\,V)$ it appears that a total chemical potential of about 1.2 V must be created through photosynthesis. In addition, the turnover of the photosynthetic apparatus in vivo results in the synthesis of at least two equivalents of ATP per O_2 evolved (see below). In both photosystems the absorption of a light quantum results in the movement of one electron equivalent against a gradient of about 1 eV. Therefore, two light reactions using two photons to move each equivalent through the electron transport chain are required. According to current concepts, these two light reactions operate "in series." Photoact I utilizes photons which are mainly absorbed by chlorophyll a, and produces a very strong reductant X^- and a weak oxidant P-700$^+$. Photoact II is sensitized by both chlorophyll a and accessory pigments, and produces a weak reductant Q^- and a very strong oxidant Z^+.

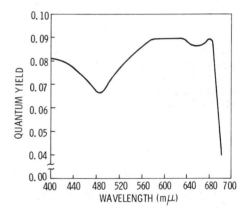

Fig. 1. Quantum yield of O_2 evolution as a function of wavelength in the green alga *Chlorella* (EMERSON and LEWIS, 1943)

Figure 1 shows the quantum yield of O_2 evolution as a function of wavelength measured in the green alga *Chlorella*. Note that the wavelength dependence is, if anything, the reverse of what one would predict from the energy of the incident photons. The 25% dip at ~ 480 nm and a smaller ($\sim 10\%$) dip at ~ 650 nm suggest that the energy absorbed in these spectral regions (by carotenoids and chlorophyll b respectively) is somewhat inefficiently utilized. In the spectral region of maximum efficiency the observed quantum requirement is about $11\ hv/O_2$.

There appear to be two different types of pigments associated with the photosynthetic apparatus. Pigments such as chlorophyll a and accessory pigments are primarily light gathering molecules and transfer excitation energy to the trap with a high efficiency. Other pigments such as carotenoids serve a dual role; in addition to harvesting quanta these molecules also serve to suppress undesirable chlorophyll triplet excitations. These pigments apparently have a lower efficiency of energy transfer to the trap.

Because of the problems inherent in the measurement of quantum yields in whole cells, the determination of ϕ in partial reactions (using isolated chloroplasts, chloroplast preparations, or so-called "reaction centers" from photosynthetic bacteria) has been an important facet of the study of the limiting photosynthetic quantum yield. In these experiments it is not necessary to cope with ancillary gas exchange reactions (CO_2 gushes, respiratory interactions, etc.) and other endogenous metabolic cycles. However, one is hampered by the necessarily brief measurements with inherently unstable material. In addition, the reaction pathways involved are not always unambiguous.

The primary photoprocesses can apparently proceed with almost perfect quantum efficiency; a crude quantum yield determination for the photooxidation of P-700 by Kok and Hoch (1961) gave a value of about one, and more precise determination by Wraight and Clayton (1973) for the photooxidation of P-870 (a pigment analogous to P-700) in photosynthetic bacteria reaction centers yielded a value of 1.02 ± 0.04. Close-to-perfect quantum yields have also been reported for other chloroplast reactions, for example: $\phi = 1.0 \pm 0.05$ has been reported for the photooxidation of ferrocytochrome c by a photosystem I chloroplast preparation in 710 nm light (Schwartz, 1967), the photoreduction of $NADP^+$ using ascorbate-$DCPIPH_2$ in 700 nm light, the photoreduction of DCPIP with water as the electron donor in the range 630–660 nm (Sun and Sauer, 1971), and the photoreduction of ferricyanide with water as the electron donor in 640 nm light (Avron and Ben-Hayyim, 1969). $\phi = 2.0 \pm 0.1$ has been reported for the photoreduction of $NADP^+$ with water as the electron donor in the range 640–678 nm (Sun and Sauer, 1971).

We should note that quantum yield measurements deviating significantly from integral values are often obtained. However, in analyzing quantum yield data, one tends to select the highest values and ascribes the others to suboptimal conditions. In addition, the statistical limits often given along with these determinations (e.g. 1.0 ± 0.05 above) only reflect the precision of the measurements, and cannot give any information regarding accuracy.

Table 1 illustrates experiments in which the quantum yields of some partial reactions of the chloroplast photosynthetic chain were compared. We note the following points:

1. When only photosystem I is operative (Exp. I) the quantum requirement is 1.2 in 710 nm (photosystem I) light, and 2.4 in 650 nm (photosystem II) light. This suggests that half of the absorbed 650 nm light cannot be utilized by photosystem I.

Table 1. Quantum requirement for some partial photosynthetic reactions in isolated chloroplasts using photosystem II (650 nm) light and photosystem I (710 nm) light (G. Cheniae, quoted in Kok, 1972)

Reaction	Photo-system(s)	$h\nu$/equivalent	
		650 nm	710 nm
1. $DCPIPH_2 \rightarrow MV$	I	2.4	1.2
2. $H_2O \rightarrow FeCN$	I+II	2.6	–
3. $H_2O \rightarrow MV$	I+II	2.6	14.5
4. Hydroxylamine $\rightarrow$ MV	I+II	2.2	9.3

MV = methylviologen; FeCN = ferricyanide

2. When both photosystems are operative (Exps. II and III) the quantum requirement for O_2 evolution is 2.6 $h\nu$/equivalent in 650 nm light in the presence of either a low potential or high potential acceptor (viologen or ferricyanide, respectively). However, the efficiency in 710 nm light is very poor due to the mismatch in the quantum absorption of the two photosystems.

3. When the O_2 evolution system is bypassed by using hydroxylamine as an electron donor the quantum requirement approaches 2 $h\nu$/equivalent. This suggests that the suboptimal quantum yields often observed when H_2O is the electron donor can be ascribed to the well-known fragility of the O_2 evolving enzyme.

One of the important ramifications of the presently accepted Z scheme is that the two photosystems must be balanced so that trapping centers in both systems are maintained in the correct state. This requires that the two photosystems interact such that both the transfer of electrons and the distribution of light energy are optimized. Since these interactions are not independent of each other, factors which affect one interaction (e.g. electron transfer) can also influence the other (e.g. distribution of light energy).

Under normal growth conditions both photosystems are excited at an approximately equal rate, despite the fact that the two photosystems can have quite different absorption spectra. In green algae and higher plants growing in sunlight only minor adjustments in energy distribution are necessary; the relative constancy of the solar spectrum between 400 and 700 nm, coupled with the close similarity of the action spectra of the two photosystems, results in approximately equal rates of excitation for the two photosystems. The problem is more critical, however, in red and blue-green algae, where the action spectra of the two photosystems are quite different. To a large extent, however, these organisms are self-correcting: the pigment composition of the two photosystems varies with the spectral composition of the light in which the algae are grown such that the balance in the turnover rate of the two photosystems is maintained.

Figure 2 shows the action spectra of the two photosystems of *Anacystis* grown under incandescent light. The relative number of quanta at each wavelength is shown by the thin full line. Note that the areas of photosystem II and photosystem I excesses are about equal when corrected for the light field. However, under normal sunlight, with its relatively flat emission spectrum, these particular organisms would have an excess of photosystem I activity.

There are currently two main hypotheses which attempt to explain the mechanism by which quanta are distributed between the two photosystems (see MYERS, 1963). The concept of "separate packages" assumes that the two photosystems have their own properly matched independent pigment beds. This hypothesis is consistent with many observations but does not readily explain the near-perfect constancy of the effectiveness spectrum in green cells below 690 nm. An amendment to this hypothesis was proposed by JOLIOT et al. (1968) to explain the relative flatness of the ϕ vs. λ plot. They proposed that the balance between the two photosystems is due to a low overall equilibrium constant (K $\sim$ 5) for the interaction between Q and P-700. However, this model predicts that the traps are not all open even when quanta are equally distributed between the two photosystems. This implies about 20% loss in quantum efficiency, probably more than can be tolerated in view of the quantum yield data described above.

The alternate "spill-over" concept of quantum distribution assumes that, except for the long wave photosystem I component, most pigments (and thus light quanta) are shared by the two systems such that a quantum preferentially flows to a photosystem II trap; if this trap is closed, it migrates to a photosystem I trap. This hypothesis predicts a flat quantum

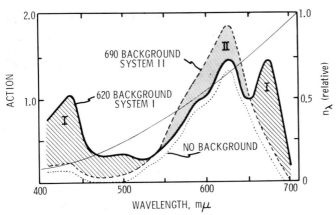

Fig. 2. Action spectra of O_2 evolution in the blue-green alga *Anacystis*. The action of each wavelength was measured either without a background light *(dotted line)* or on a strong background light of complementary wavelength, so that the effectiveness was limited by the variable wavelength. *Solid line:* photosystem I, measured with background light of 620 nm (photosystem II maximum); excess hatched. *Dashed line:* photosystem II, measured with background light of 690 nm (photosystem I maximum); excess shaded. *Thin solid line* (right hand ordinate): relative number of quanta at each wavelength (n_λ relative) for incandescent light. (Adapted from Jones and Myers, 1964)

yield spectrum of O_2 evolution for $\lambda < 690$, but fails to explain the poor quantum yield of photosystem I at wavelengths shorter than 690 nm observed in the presence of DCMU and reduced DCPIP (see above).

A compromise between these two extreme views has also been proposed (Thornber and Highkin, 1974). According to this picture (the so-called tripartite scheme) there are three different pigment complexes: in addition to the photosystem I and photosystem II complexes there exists a light-harvesting chlorophyll complex which can transfer excitation energy to either of the photosystem complexes.

The distribution of light between the two photosystems can also be affected in vivo by the spectral composition of the light; the light is distributed in such a way as to tend to balance the turnover of the two photosystems. This effect is probably due to the transport of ions across the chloroplast membranes, and as such, is probably related to the effect of Mg^{2+} on light distribution in isolated chloroplasts (see Myers, 1971).

C. ATP Production and Utilization

Up to this point we have only considered the quantum yields and efficiency of the photosynthetic redox reactions (e.g. O_2 evolution and CO_2 reduction). However, the efficiency for the photoproduction of ATP is equally important. Many of the early quantum yield determinations were obtained under steady state conditions in vivo; the products of photosynthesis under these conditions are O_2 and new organisms, and thus the observed quantum requirement of about 10 $h\nu/O_2$ pertains to growth. The many metabolic conversions subsequent to the reduction of CO_2 are driven by energy (ATP) generated as electrons are promoted from the redox level of H_2O to that of $NADP^+$ (i.e. without a requirement for additional photons).

It is still uncertain how the ATP demand for growth is correlated with the quantum efficiency of photosynthesis. Studies with heterotrophic and chemosynthetic organisms indicate that at least 5 mol of ATP are required for the conversion of 1 mol of CO_2 into cellstuff (BONGERS, 1970). In the context of the photosynthetic process this would be equivalent to an ATP/2e ratio of at least 2.5. However, the direct application of these data to photosynthetic organisms is questionable, and a more realistic ATP requirement might be obtained by summing the ATP requirements of the partial processes involved. The overall stoichiometry of the CO_2 fixation cycle in C_3 plants is such that 3 mol of ATP (and 2 of NADPH) are required to reduce 1 mol of CO_2 to the redox level of a hexose and regenerate the components of the cycle. In C_4 plants an additional 2 mol of ATP are required. Since growth experiments with heterotrophic organisms indicate that two to three moles of ATP are required to convert a mole of carbohydrate to cellstuff (BAUCHOP and ELSDEN, 1960) it appears that under growing conditions 5–8 mol of ATP are required per mole of O_2 evolved or CO_2 incorporated. This means that the ATP/2e ratio must be 2.5–4 in vivo.

The amount of ATP generated via photophosphorylation observed in isolated chloroplasts does not appear to be sufficient for cell growth. Observed ATP/2e ratios for photophosphorylation in high light are usually 1–1.5 (IZAWA and GOOD, 1968); at low light intensity ratios approaching two have been reported (HEATHCOTE and HALL, 1974). Somewhat higher ratios can be inferred indirectly; if, in the context of the chemiosmotic theory, the translation of three protons results in the production of one molecule of ATP (SCHRÖDER et al., 1972), then the reported H^+/e ratio of 3–4 (FOWLER and KOK, 1975) implies an ATP/2e ratio as high as 2.67.

For the case of non-cyclic phosphorylation associated with electron transport from water to $NADP^+$, the quantum requirement for ATP is half the ATP/2e ratio or one-fourth the ATP/O_2 ratio. Direct measurements of ATP/hv and ATP/2e ratios have been made in several laboratories; a detailed discussion of these measurements is beyond the scope of this article.

The source of any "extra ATP" in vivo is difficult to rationalize. For instance, in the case of a cyclic operation of photosystem I the ATP would be generated at the expense of additional photons that would yield no net O_2 and reducing power (resulting in a lower value for ϕ). The same holds true for an enhanced respiration in the light. One is hard pressed to account for the overall ATP budget and still remain within the bounds imposed by the quantum requirement of about 10 hv/O_2 observed with growing algae.

D. Quantum Yields of Growing Cells and Photosynthetic Productivity Under Natural Conditions

The most direct measure of photosynthetic efficiency in vivo is a comparison of the amount of absorbed light energy to the amount of energy obtained upon combustion of the plant material produced. Other more indirect methods have been used: if one assumes that the redox state of the fixed carbon is known, or CO_2 uptake is monitored along with O_2 evolution, it is possible to derive the quantity of energy fixed from the amount of O_2 evolved. However, the validity of this method is compromised by the participation of nitrogen-containing compounds in the energy transformation processes (see KOK, 1960) for a more thorough discussion of this topic).

If we assume that photosynthesis involves the production of compounds at the redox level of glucose, then the energy fixed per O_2 evolved will be about 115 kcal (one-sixth the heat of combustion of glucose). Since 1 mol of red (680 nm) quanta contains about 40 kcal, a requirement of 10 hv/O_2 corresponds to an energy efficiency of about 30%. This value drops to about 20% in (weak) sunlight due to the higher energy of blue quanta (see above). Only about one-half of the solar radiation is absorbed by green plants (< 700 nm), so that the maximum overall energy conversion is about 10% of the total incident solar radiation.

Algae and higher plants have actually been grown in weak light with efficiencies of about 20% of the absorbed ($\bar{\lambda} = 550$ nm) radiation (i.e. 10% of total radiation) which corresponds to a quantum requirement of about 10 hv/O_2 (KOK, 1952; WASSINK, 1959). However, as a result of the various factors that tend to degrade photosynthetic efficiency in the field, actual "net field production" seldom exceeds 2%, with typical values being in the neighborhood of 0.5–1.5%. In the following paragraphs we will discuss some of the reasons for this discrepancy.

Since natural illumination fluctuates widely during a diurnal cycle, the "dynamic range" of the photosynthetic apparatus with respect to light intensity is an important parameter affecting photosynthetic efficiency. A graphical representation of the "working range" and some of the limiting parameters is shown in Figure 3. Under conditions of very weak light the quantum yield is limited by the stability

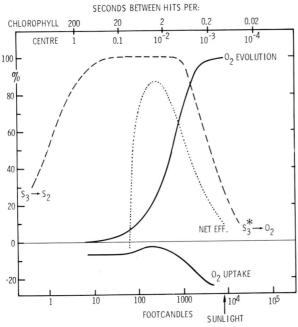

Fig. 3. Illustration of the working range of photosynthesis and some of the limiting parameters. The S_n's refer to the various oxidation states according to the nomenclature of KOK et al. (1970): $S_3 \rightarrow S_2$ is a so-called deactivation reaction, and $S_3^* \rightarrow O_2$ the concerted oxidation of H_2O (see Chap. 2.8). (Adapted from KOK, 1972)

of the O_2 precursors ($S_3 \rightarrow S_2$ in Fig. 3) (JOLIOT, 1966). In whole cells in weak light the half life of the least stable precursor is about 3 sec, and thus the efficiency of light conversion is half maximal at an intensity in which each photosystem II reaction center receives a quantum every 3 sec. For a unit of 200 chlorophyll per trap, this corresponds to one quantum absorption per 10 min for each chlorophyll, or about 10^{-4} times the intensity of full sunlight.

Under high light conditions the quantum yield is limited by the relatively slow turnover of the dark reactions (e.g. the transfer of electron from Q to P-700 and the turnover of the CO_2 fixation cycle). These slow limiting reactions cause electrons to accumulate within the electron transport chain with a resulting loss in photochemical efficiency. (A rate limitation between Q and P-700 results in Q being in the reduced state while P-700 is oxidized, while a limitation in the CO_2 fixation cycle results in the accumulation of all intermediates in the reduced state, including X. In both instances substrates which participate in the photoacts are in the wrong redox state and thus unable to photoreact.)

In full sunlight a single chlorophyll molecule absorbs about 10 photons per sec; thus a trapping center with its 200 light-harvesting chlorophylls receives two thousand quanta per sec. However, the endogenous rate-limiting step located in the electron transport chain between the photoacts has a maximum turnover rate of about 100 equivalents per sec. Consequently, even in the absence of other limiting factors (such as low CO_2) photosynthesis is half saturated at about one-tenth the intensity of full sunlight. This is reflected in the observed rate of O_2 evolution using artificial electron acceptors (solid curve in Fig. 3). It thus appears that the photosynthetic apparatus is designed to operate under low-light conditions; it can proceed relatively efficiently over a one-hundred to one-thousand fold intensity range, all well below the intensity of bright sunlight.

Under normal "field" conditions the concentration of CO_2 in the air (0.03%) is not high enough to saturate the CO_2-fixing enzyme. Consequently, except under low light conditions the CO_2 concentration limits the overall photosynthetic efficiency.

The relationship between [CO_2] and the rate of photosynthesis is complex, and the final chapter in this area of study is yet to be written. It is generally observed that the photosynthetic rate in high light can be accelerated roughly two-fold by increasing the CO_2 concentration; this does not, however, necessarily imply that the K_m of the carboxylating system itself is about 0.03% CO_2. There is evidence that the rate of electron transport rather than the rate of CO_2 fixation is limiting in high light and high CO_2, and that the saturation characteristics of the carboxylating enzyme do not contribute substantially to the overall rate limitation.

According to LILLEY and WALKER (1975) the true "apparent K_m" for intact chloroplasts and ribulose bisphosphate carboxylase (the carboxylating enzyme) are both 46 µM. Since 0.03% CO_2 corresponds to about 10 µM CO_2 in solution this implies that the enzyme is only about 10% saturated when the photosynthetic rate is half-maximal (at the normal 0.03% CO_2 concentration). According to this picture the ultimate capabilities of the carboxylating system are in excess of that required to sustain the observed maximum photosynthetic rates under high [CO_2] conditions.

One of the most popular areas of photosynthetic research during the past ten years has involved the ·interrelated topics of photorespiration, glycolic acid metabolism, and C_3 vs C_4 plants (see reviews by ZELITCH, 1971 and TOLBERT, 1974). It has been known for some time that certain plant species which tend to show rapid rates of CO_2 assimilation in air produce 4-carbon compounds (asparate or malate) as the first detectable product of CO_2 incorporation, and hence are known as C_4 plants. (The more common C_3 species produce 3-phosphoglyceric acid as the initial product of CO_2 fixation.) A relatively low efficiency in high light and atmospheric [CO_2] in several C_3 species has been attributed to the "photorespiratory" release of CO_2 from recently fixed photosynthetic compounds; this can amount to as much as 50% of net photosynthetic production.

Since there is no energy obtained from the process of photorespiration, it is apparently a waste process which can be ascribed to the high concentration of O_2 and the suboptimal concentration of CO_2 in the atmosphere. C_4 plants such as maize have evolved a mechanism to provide a high internal concentration of CO_2, and thus exhibit very little apparent photorespiration. However, since this CO_2 pump is driven by ATP, these plants must still pay the price of lowered efficiency exacted by the suboptimal CO_2 and O_2 concentrations.

At the present time, the importance of the "C_4 syndrome" as a necessary adjunct to high photosynthetic efficiency is far from proven. Several species, such as sunflower (*Helianthus*) and cattail (*Typha latifolia*), have high rates of net photosynthesis despite their C_3-metabolism. It may be that the physiological and anatomical characteristics of C_4 plants reflect a xerophytic adaptation rather than a step toward greater photosynthetic efficiency in high light (BERRY, 1975). (A CO_2 fixation pathway similar to that of C_4 plants is present in xerophytic plants utilizing the so-called Crassulacean acid metabolism.)

Acknowledgements. The authors would like to thank Drs. ROBERT G. LYE and CHARLES F. FOWLER for helpful discussions during the preparation of this manuscript.
This work was supported in part by a grant from the Energy Research and Development Administration Contract E(11-1)-3326. This report was also prepared with the support of the National Science Foundation Grant No. BMS74-20736 and Grant No. AER73-03291. Any opinions, findings, conclusions or recommendations expressed herein are those of the author(s) and do not necessarily reflect the views of NSF.

References

Avron, M., Ben-Hayyim, G.: In: Progress in Photosynthesis Research. Metzner, H. (ed.). Tübingen: Laupp, 1969, Vol. III, pp. 1185–1196
Bauchop, T., Elsden, R.: J. Gen. Microbiol. **23**, 457–469 (1960)
Berry, J.A.: Science **188**, 644–650 (1975)
Bongers, L.: J. Bacteriol. 145–151 (1970)
Duysens, L.N.M.: Brookhaven Symp. Biol. **11**, 10–25 (1958)
Emerson, R., Lewis, C.M.: Am. J. Botany **28**, 789–804 (1941)
Emerson, R., Lewis, C.M.: Am. J. Botany **30**, 165–178 (1943)
Fowler, C.F., Kok, B.: Biochim. Biophys. Acta **423**, 510–523 (1976)
Heathcote, P., Hall, D.O.: Biochem. Biophys. Res. Commun. **56** (3), 767–774 (1974)
Izawa, S., Good, N.E.: Biochim. Biophys. Acta **162**, 380–391 (1968)
Joliot, P.: Brookhaven Symp. Biol. **19**, 418–433 (1966)
Joliot, P., Joliot, A., Kok, B.: Biochim. Biophys. Acta **153**, 635–652 (1968)
Jones, L.W., Myers, J.: Plant Physiol. **39**, 738–746 (1964)
Kok, B.: Biochim. Biophys. Acta **3**, 625–638 (1949)
Kok, B.: Acta Botan. Neerl. **1**, 445–467 (1952)

Kok, B.: In: Encyclopedia of Plant Physiology. Ruhland, W. (ed.). Berlin-Göttingen-Heidel-
  berg: Springer, 1960, Vol. V, pp. 566–633
Kok, B.: In: Horizons of Bioenergetics. New York-London: Academic Press, 1972, pp. 153–170
Kok, B., Forbush, B., McGloin, M.: Photochem. Photobiol. **11**, 457–475 (1970)
Kok, B., Hoch, G.: In: Light and Life. McElroy, W.D., Glass, B. (eds.). Baltimore: Johns
  Hopkins, 1961, pp. 397–423
Lilley, R.McC., Walker, D.A.: Plant Physiol. **55**, 1087–1092 (1975)
Mortimer, R.G., Mayo, R.M.: J. Chem. Phys. **35**, 1013–1017 (1961)
Myers, J.: In: Photosynthetic Mechanisms of Green Plants. Nat. Acad. Sci., Nat. Res. Council
  1145, 301–317 (1963)
Myers, J.: Ann. Rev. Plant Physiol. **22**, 289–312 (1971)
Schröder, H., Muhle, H., Rumberg, B.: In: Proc. 2nd Intern. Congr. Photosynthesis. Forti,
  G., Avron, M., Melandri, A. (eds.). The Hague: Dr. Junk, 1972, pp. 919–930
Schwartz, M.: Nature (London) **213**, 1187–1188 (1967)
Shockley, W., Queisser, H.J.: J. Appl. Phys. **32**, 510–519 (1961)
Sun, A.S.K., Sauer, K.: Biochim. Biophys. Acta **234**, 399–414 (1971)
Thornber, J.P., Highkin, H.R.: Europ. J. Biochem. **41**, 109–116 (1974)
Tolbert, N.E.: In: Algal Physiology and Biochemistry. Stewart, W.D.P. (ed.). Univ. of Califor-
  nia Press, Berkeley and Los Angeles 1974, Vol. X, pp. 474–504
Warburg, O., Negelein, U.E.: Z. Physik. Chem. **102**, 235–266 (1922)
Wassink, E.C.: Plant Physiol. **34**, 356–361 (1959)
Wraight, C.A., Clayton, R.K.: Biochim. Biophys. Acta **333**, 246–260 (1973)
Zelitch, I.: Photosynthesis, Photorespiration, and Plant Productivity. New York-London: Aca-
  demic Press, 1971, pp. 1–347

4. P-700

G.E. HOCH

A. General

P-700 was discovered by KOK (1957) following DUYSENS' (1952) observation of a similar pigment in photosynthetic bacteria. Both pigments absorbed on the long wavelength side of the main red chlorophyll band, a suitable position to accept transfer of excitation energy from the bulk pigments and thereby capture energy collected over a wide range of wavelengths. Both pigments represented only a small fraction of the total absorber concentration (the ratio of P-700 to total chlorophyll is about 1:400) and were bleached during illumination. This bleaching was shown to be a result of oxidation, the pigments being relatively strong oxidants (midpoint potentials of +400 to 500 mV).

In contrast to the bacterial pigment, P-700 was efficiently bleached in weak light by only a narrow spectral region. This region is confined to far red wavelength ($\lambda > 690$ nm) in green algae, but somewhat wider in blue-green algae where both the main blue and red bands of chlorophyll, but not the accessory pigments, would sensitize effectively.

It was precisely these regions where the quantum yield of photosynthesis was low in green and blue-green algae (DUYSENS, 1952). In order to align P-700 with the main stream of electron flow, it was necessary to postulate that P-700 was reduced by one photoreaction and oxidized by another, the reduction being too rapid to measure (KOK and HOCH, 1961), a hypothesis which concurred with Emerson's "enhancement" effect (1958). This hypothesis was supported by the observation that P-700, brought into a partially oxidized state by far red light, could be (in the steady state) reduced by additional light absorbed by accessory pigment (KOK and GOTT, 1960). DUYSENS et al. (1961) obtained similar results for cytochrome f and both observations were consistent with a hypothesis of HILL and BENDALL (1960) for the interaction of cytochromes b and f, a hypothesis which was clearly stated and provides the basis for interpretation of many phenomena to this time, even though the place of the cytochromes in the scheme is presently disputed.

Continuous low intensity illumination of algae with light absorbed by photosystem I (far red in green algae or blue or red in blue-green algae) results in a steady state oxidation level of P-700 which is proportional (at low intensities) to the intensity used (KOK and GOTT, 1960). If the illumination is absorbed by accessory pigment (photosystem II, chlorophyll b in green algae and phycobilins in blue-green algae), no oxidation occurs until the intensity is raised sufficiently to nearly saturate photosynthesis (see FORK and AMESZ, 1970). Similar effects are observed with flashing illumination. Photosystem I yields an oxidation during the flash and photosystem II gives no signal if the intensity is moderate (KOK and HOCH, 1961). If a background light of photosystem I is used (sufficient to

bring P-700 into a partially oxidized state) a continuous beam of photosystem II
will cause a lessening of this oxidation level (KOK and GOTT, 1960) and a photosystem II flash will produce a reduction after the flash (see RUMBERG, 1964a).

P-700 can also undergo cyclic oxidation and reduction in blue-green algae
in which oxygen evolution is prevented by DCMU (KOK and HOCH, 1961). This
effect has been correlated with a inhibition of respiration (HOCH et al., 1962),
where the cyclic action of P-700 presumably results in a production of ATP.
The action spectrum of this inhibition of respiration indicates that it is sensitized
by photosystem I (RIED, 1971).

The generally accepted series formulation (in its most basic form) is

$$H_2O \xrightarrow{h\nu_{II}} Q \rightarrow A \rightarrow D \rightarrow P\text{-}700 \xrightarrow{h\nu_I} X \rightarrow NADP^+.$$

A great deal of evidence (not reviewed here) suggests that a scheme of this general
type accounts for most observations of photosynthetic mechanism (of course, some
cyclic reactions must be allowed to account for energy conserving processes occurring in the absence of O_2 evolution). Strong support for the role of P-700 in
this scheme has come from studies on algal mutants (see BISHOP, 1973). In particular
Scenedesmus mutant 8 lacks P-700 and can neither photoassimilate glucose (an
endogeneous cyclic reaction) nor can particles prepared from it reduce $NADP^+$
(PRATT and BISHOP, 1968; KOK and DATKO, 1965). The particles can reduce ferricyanide however.

B. Optical Properties

The absolute absorption spectrum of P-700 cannot be determined as even the
most enriched preparations still contain bulk chlorophylls. Consequently difference
spectra, light minus dark or oxidized minus reduced, must be used. The two
methods agree well, at least in the red region (KOK, 1961). The difference spectrum
obtained by light oxidation on two enriched preparations is shown in Figure 1a
and 1b (HIYAMA and KE, 1972). Two main bands, nominally located at 430 and
700 nm, and a satellite red band at about 682 nm decrease in absorption while
small and broad increases occur in the green and far red. The latter increases
might be ascribed to a π cation radical (BORG et al., 1970). All three of these
bands can have similar kinetics (RUMBERG and WITT, 1964; MURATA and TAKAMIYA,
1969). However, some spectra have been reported in which the 430 band was
not apparent despite a large and easily measurable change at 700 nm (HOCH and
RANDLES, 1971)

The 682 and 700 nm bands were first proposed to result from exciton splitting
of a dimer by DÖRING et al. (1968). This interpretation has been supported by
the difference circular dichroism measurements of PHILIPSON et al. (1972). PHILIPSON
et al. suggest the bands result from the disappearance of two exciton bands and
the appearance of a chlorophyll radical-chlorophyll species. The latter is presumed
to absorb at about 686 nm. Thus the observed spectrum would be complex. The
rather skewed shape of the 700 nm band (the extent of skew is variable) was
early commented upon (KOK, 1967). In any case the rather variable ratios of

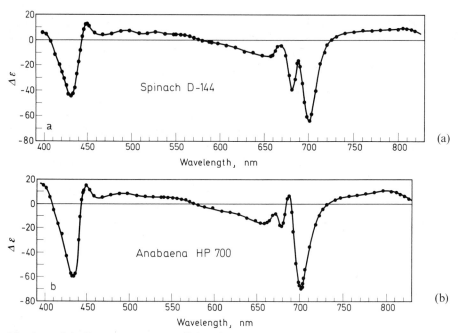

Fig. 1a and b. Room temperature light minus dark spectra of purified P-700 fractions; (a) from spinach; (b) from Anabaena (Hiyama and Ke, 1972)

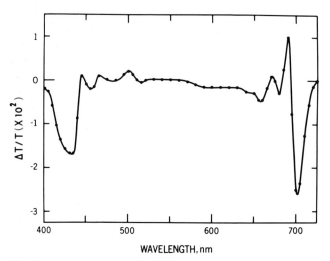

Fig. 2. Low temperature (86 K) light minus dark spectrum of purified P-700 fraction from Anabaena (Ke, personal communication)

the extinctions and peak positions would tend to support a somewhat loose structure of P-700.

The low temperature spectrum can be seen in Figure 2, again a light-induced spectrum from a purified preparation. The spectrum was recorded at 86 K (B. Ke, personal communication). Other low temperature spectra have been reported by FLOYD et al. (1971), WITT (1973), LOZIER and BUTLER (1974) and VISSER et al. (1974). It has been generally noted that the absorption increases at 690 nm, VISSER et al. (1974) have discussed various proposals to explain the band shape.

The extinction coefficient of the 700 nm band, which was originally assumed as that of a chlorophyll ($8 \cdot 10^4 \, M^{-1}$), has been a source of some dispute. An accurate knowledge of this value underlies all quantitative measurements of photochemical yields. All measurements have been made indirectly; P-700 has not been titrated with a known oxidant or reductant. The indirect measurements assume a reaction stoichiometry between photooxidized P-700 and an externally supplied and measurable reductant. HIYAMA and KE (1972) have done these experiments and discussed the problems involved. Their estimate for the 700 nm band is $6.4 \cdot 10^4 \, M^{-1} \, cm^{-1}$. This estimate has been supported by correlation of the optical bleaching at 700 nm and the number of unpaired electrons produced during light or chemical oxidation (BAKER and WEAVER, 1973).

C. Oxidation-Reduction

The bleached (light dependent) state of P-700 was shown to be an oxidized state by KOK (1961). Enriched preparations of P-700, obtained by acetone extraction, when titrated with mixtures of ferro-ferricyanide showed P-700 to have a midpoint potential of $+430 \, mV$ and behave as a one-electron transfer agent (KOK, 1961). The potential was independent of pH indicating protons were not involved in the reduction. Similar results were obtained by RUMBERG (1964b) with aged and broken chloroplasts. YAMAMOTO and VERNON (1969), using an enriched preparation made with the detergent triton X-100, measured a midpoint potential of $+480 \, mV$. KNAFF and MALKIN (1973) observed a midpoint potential of $+530 \, mV$ both optically and with EPR measurements on a digitonin preparation. This variation in measurements, a full 100 mV, is somewhat disquieting and may cause some of the problems associated with understanding the mechanism of P-700 reduction discussed below.

No information is available about oxidants which directly oxidize P-700 but the reduced forms of phenazine methosulfate (PMS) and $N_1 N_1 N_1' N'$ tetramethyl-phenylenediamine yield first order time courses for the reduction of P-700, in which the time constant is a linear function of their concentrations which suggests a direct interaction (WITT et al., 1963; KE, 1973).

The EPR data, which show that an unpaired electron results from P-700 oxidation, are discussed by WEAVER and KORKER (this vol., Chap. II.6). The EPR spectrum is lacking in hyperfine structure, indicating a delocalized electron.

Fluorescence yield changes do not result from a change in oxidation state of P-700 (KOK, 1963). This indicates that:

1. the sensitizing chlorophylls are inherently non-fluorescent, or

2. the trapping efficiency of P-700 is the same in its oxidized or reduced form, or

3. P-700 is sensitized from a metastable state from which fluorescence does not compete. None of these alternatives is completely satisfactory. A suitable

explanation remains to be found. In this respect P-700 differs from the corresponding pigment in bacteria where fluorescence yield changes accompany in oxidation
state (see DUYSENS, 1967). The kinetics indicate a "separate unit" model for P-700
and its associated bulk chlorophyll. That is, a photon entering the sensitizing
chlorophyll apparently does not have a choice of P-700 molecules to oxidize but
is restricted to one.

D. Models

KATZ and coworkers (see KATZ and NORRIS, 1973) have proposed a dimeric model
of P-700 in which one water molecule serves as ligand to two chlorophyll molecules.
Bonding occurs through the magnesium atom of one chlorophyll and two positions
of the other (carbonyl and carbomethoxy group of ring 5). This and the photoconverted dimer would have the following structure:

KATZ views the unpaired electron on the keto carbon as subsequently being delocalized into the π system of the porphyrin ring producing Chl$^-$ while the electron
deficient water oxygen would withdraw an electron from the right-hand chlorophyll
molecule producing an oxidized chlorophyll. The electron from Chl$^-$ must be
rapidly withdrawn leaving the oxidized chlorophyll. Much of the evidence for
this dimeric formulation comes from analysis of the EPR signal linewidth (of
the oxidized chlorophyll), which KATZ and NORRIS take as indicating the unpaired
electron is spread over two chlorophyll molecules. Just how this occurs is not
clear in this model.

FONG (1975) has modified the above model to include two water molecules
in the dimer. His model has both Mg atoms coordinated to a water molecule
and to only the carbomethoxy group of ring 5. The structure is:

which, after light excitation gives a charge transfer state having two tautomeric forms:

$$\text{[two tautomeric chemical structures shown, connected by a double-headed arrow]}$$

Left structure: OCH$_3$, O$-$C$^+$, Mg$-$O, H, H, R', R', H, O$\cdots$Mg, H, C$=$O

Right structure: OCH$_3$, O$=$C, Mg$\cdots$O, H, H, R', R', H, O$^-$$-Mg, H, C_+$ O

FONG suggests that this structure gives rise to observed ESR signal where the spin appears to be delocalized over two molecules. Note however that the dimer has not yet lost an electron. FONG (1974) has quite a different model of the photosynthetic mechanism from that which is generally accepted. He proposes that the dimeric structure provides a long triplet lifetime and accessible charge transfer state.

KOESTER et al. (1975) have recently shown that cooling a solution of chlorophyll a in peptone-methylcyclohexone to 151 K yields an absorption spectrum resembling P-700 and that this absorption can be reversibly bleached by iodine.

E. Localization of P-700

Numerous attempts to physically separate photosystems I and II (extensively reviewed by ARNTZEN and BRIANTAIS, 1975) have been made. Disruption of chloroplasts by detergents or physical stress, such as the French Press, yield "heavy" and "light" particles. Both particles contain P-700 but the "light" particle is incapable of oxygen evolution. Further fractionation of the "heavy" particle by detergents gives another "light" particle with characteristics similar to the first (WESSELS and VOORN, 1971). The generally accepted interpretation of these fractionations is that the "heavy" particle is derived from the grana lamellae and the "light" from the stroma. With this interpretation, P-700 appears to be distributed throughout the chloroplast lamellae system, but photosystem II is confined to grana. SANE and PARK (1971) have suggested that the stroma lamellae also reduce pyridine nucleotide and the resulting oxidized P-700 is reduced by a diffusable reductant produced in the grana by photosystem II. They also suggest that P-700 can function in a cyclic path, producing ATP, in both grana and stroma.

The light particles obtained by detergent or French Press treatment are enriched in both P-700 and chlorophyll a (WESSELS and VOORN, 1971; SANE and PARK, 1971) and will photoreduce triphosphopyridine nucleotide (NADP$^+$) if provided with plastocyanin, ferredoxin-NADP$^+$ reductase, ferredoxin and a suitable electron donor. Surprisingly, the quantum yield of this reaction increases in far red light similarly to intact chloroplasts. Short wavelength absorbing pigments inactive in this reaction occur therefore in both whole chloroplasts and in the photosystem I particles. Further purification of P-700 and the composition of the particles are reviewed by Thornber and Alberti (this vol., Chap. IV.5).

Certain plants contain normal chloroplasts in mesophyll cells and agranal chloroplasts in bundle sheath cells. Arntzen and Briantais (1975) conclude that P-700 is distributed throughout both of the chloroplast types.

F. Orientation of P-700

Geacintov et al. (1972) discovered that algae and chloroplasts align in a magnetic field. The alignment is such that the plane of the photosynthetic membranes is normal to the magnetic field. Breton et al. (1975) have used this effect to study the orientation of P-700. The flash induced absorbance change at 700 nm was 2.3-fold greater when the measuring beam was polarized parallel to the plane of the membranes than when polarized perpendicular to the plane. If the membranes were perfectly aligned in the magnetic field, this indicates the 700 nm transition oscillator (which is in the plane of the porphyrin ring for monomeric chlorophyll) is oriented 20° to the membrane plane. Since perfect membrane alignment seems hardly likely, a tilt of 20° is probably an upper limit. Junge and Eckhof (1973) earlier concluded that the angle between P-700 and the membrane plane was small (see also Junge and Eckhof, 1974). Junge (1974) has pointed out that the unknown nature of the porphyrins alignment in the proposed dimer, suggested to be at an angle by Döring et al. (1968) or parallel by Fong (1974), makes studies on transition moments less easily extrapolated to alignment of the porphyrins. Junge also argues for a placement of P-700 on the inner surface of the thylakoid membrane.

G. Oxidation of P-700

P-700 is rapidly oxidized by flash illumination; all measurements of the oxidation time are still limited by instrumentation. The fastest time reported is ≤ 20 nsec (Witt and Wolff, 1970). The oxidation occurs at temperatures as low as 92 K (Ke et al., 1974). The oxidation is sensitized by chlorophyll a, perhaps through a small amount of chlorophyll termed C-700 which is present in about 10 fold excess over P-700 (Butler, 1961; Kok, 1963).

H. Reduction of P-700

The relationship between P-700, cytochrome f and plastocyanin has been the subject of many types of investigations. Gorman and Levine (1966) obtained a mutant algae which was deficient in plastocyanin. Photosystem II light caused a reduction of cytochrome f in this mutant but photosystem I was unable to oxidize it. They concluded the electron transfer sequence was cytochrome f → plastocyanin → P-700. Strong support for this has come from extraction-reconstitution experiments by SIEDOW et al. (1973). Sonication of chloroplasts results in a loss of plastocyanin, indicating it is localized on the interior of the chloroplast membranes (see TREBST, 1974). The particles were unable to oxidize cytochrome f (which remains bound). Readdition of plastocyanin restored the photochemical oxidation of cytochrome f. The restoration was particularly effective if plastocyanin was added together with sonication, in this case one bound plastocyanin per cytochrome f gave full recovery of oxidation (HAUSKA et al., 1971).

The difficulty in measuring plastocyanin by spectrophotometric means has, unfortunately, precluded a great deal of useful information. Much, therefore, on the reduction of P-700 must be inferred from kinetic observations on P-700 and on more tractable components, such as the cytochromes and plastoquinone. One such approach is to measure the oxidation of P-700 by far red light under a variety of conditions. This light should oxidize all of the reduced components in the chain between the light reactions. For a given far red beam the time required to oxidize P-700 should be proportional to the number of equivalents reduced. This of course presumes that the intrinsic yield of P-700 oxidation remains constant. The results are complex.

After strong white or photosystem II illumination, P-700 is left in the oxidized form and Q in the reduced presumably because of a rate limiting step between A and D above. A reduction of P-700 follows in the subsequent dark period (KOK et al., 1969). The reduction follows a first order time course whose half-time is a complex function of pH, uncoupler and duration of preceding illumination and varies from a low value of 4 msec to ∼100 msec within the physiological pH range. The data are most simple in the presence of ammonia where increasing duration of preillumination causes a decreasing half time, from 15 to 4 msec. This can be reasonably accounted for by the filling up of the A pool. In the absence of uncoupler and with long (seconds) preillumination, higher pH favored faster reduction. The first-order nature suggested to the above authors that the reducing substances were in noninteracting chains. The reader is referred to the above article for a discussion of the reasons for the changing rate constant.

After this dark reduction P-700 can be oxidized by far red light. This oxidation may show a "lag" phase before oxidation starts and it is the interpretation of this "lag" which causes some controversy. The extent of this phase is commonly measured by taking the area under the time course of P-700 oxidation by photosystem I light. Divalent cations in the millimolar range or monovalent cations at 100 mM are required for this lag and their effect can be abolished by low concentrations of carbonyl cyanide m-chlorophenylhydrazone (CCCP) but not by ammonia (MARSHO and KOK, 1974). In any event the area follows an intensity times time law with regard to the photosystem I beam. Two interpretations are possible for this biphasic response. (1) A pool of reductants exist which have a sufficiently large equilibrium constant between them and P-700. This will cause P-700 to stay reduced until the greater part of this pool has been exhausted and only then will P-700 become oxidized, (2) The quantum yield of P-700 oxidation is initially low and remains low until some other competing reaction is finished. MALKIN (1969) has considered both of these models and was unable to distinguish between them. He points out that the time course of oxidation is not consistent with too large an equilibrium constant. KOK et al. (1969) measured the area after increasing times of illumination and found it to increase with increasing illumination time. A half time of 11 msec and an increase in area of 8-fold was found. Since the area is so sensitive to cations and to uncouplers, a universal ratio of it to P-700 is experimentally unrealistic.

If a dark period is interposed between the strong illumination and the oxidizing far red light, the area decreases with dark time (MARSHO and KOK, 1970).

An additional observation was made. If a Hill acceptor was omitted from the reaction mixture, the area increased by about 20%. MARSHO and KOK (1970) argue that this probably represents the equivalents between the rate limiting step in strong light and P-700 (presumably P-700 stays reduced in strong light without acceptor).

If the preillumination is photosystem I light, P-700 is left in the oxidized form and returns in the dark to the reduced form with a half time of ∼1 min (MARSHO and KOK, 1970; MALKIN, 1969). Presumably this does not occur via the ordinary reductant but by some endogenous donors. MALKIN (1969) and KOK et al. (1969) found apparently monophasic oxidation by photosystem I after this reduction but MARSHO and KOK (1970) observed biphasic kinetics. The latter authors suggest from these and other experiments that another component becomes reduced and the equilibrium constant between P-700 and this component is ∼20. KOK et al. (1969) also presented evidence that another component was reduced even though the oxidation time course was monophasic. This could be accommodated if the equilibrium constant was about unity between P-700 and the additional reductant.

Both MARSHO and KOK (1970) and MALKIN (1969) made observations on cytochrome f. MARSHO and KOK concluded that, if cytochrome f reduced P-700, the equilibrium constant must be variable (50–100 in dark, 3–5 in weak light). MALKIN concluded that all these components could be on different chains.

If the biphasic response of P-700 to photosystem I light is interpreted as a replacement of electrons by various reductants, then certain interpretations appear to be somewhat reasonable.

1. The pool of reductants can exceed P-700 by about 15 equivalents (assuming the extinction coefficient of P-700 is about $8 \cdot 10^4 \, M^{-1}$).

2. Twelve of these lie on the photosystem II side of the rate limiting step in strong light and three on the photosystem I side.

3. The equilibrium constant between these equivalents and P-700 is not too large.

4. The bulk of this pool is seen only with divalent cations or high monovalent cations.

5. The bulk of the pool is eliminated by the uncoupler CCP.

6. If cytochrome f is a reductant, one must postulate changes in the redox potential between light and dark or perhaps a complex network of interacting chains not amenable to simple kinetics.

Both MALKIN (1969) and MARSHO and KOK (1970) have considered alternative schemes. The crux of the matter is simply whether the quantum yield of P-700 oxidation in photosystem I light is always high. There does not appear to be an easy experimental answer to this question.

"Flash titration" experiments have recently been reported by HAEHNEL (1973, 1974). Preillumination of chloroplasts by photosystem I light was followed by short flashes of photosystem II. The first flash, which finds the carriers all in the oxidized state, caused a rapid reduction of one equivalent of plastoquinone which was subsequently oxidized ($t_{1/2} \sim 20$ msec) in the dark. At the same time P-700 became reduced ($t_{1/2} \sim 20$ msec). If several flashes followed the preillumination in quick succession (1.6 msec) and data recorded after the last flash, the number of equivalents leaving plastoquinone rose to 3 without change in kinetics while about the same amount of P-700 was reduced as after the first flash, but now with a half-time of 4 msec. HAEHNEL reasoned that the observations resulted from the rate limiting step between plastoquinone and a pool of two equivalents existing on the P-700 side of this step. HAEHNEL has experimentally shown that plastocyanin exhibited kinetics consistent with this two-equivalent pool. He concluded that plastocyanin reduces P-700, that only one half of the total copper protein sufficed to provide the requisite equivalents and that an equilibrium constant of 20 between them (consistent with midpoint potentials of $+0.37$ and $+0.45$ volts) explained the observations. Cytochrome f again proved anomalous, the kinetics did not support its placement on the P-700 side of the rate limiting reaction.

HAEHNEL et al. (1971) and HAEHNEL and WITT (1971) have reported on an absorbance change at 700 nm considerably faster than that usually observed (~ 20 msec). Measuring beam intensities of 50 ergs/cm^2-sec together with photosystem II flashes gave a polyphasic decay of the flash induced absorbance change. This was resolved into three components having decay times of 10 μsec, 200 μsec and 20 msec. The microsecond times were attributed to a flash finding the primary donor (plastocyanin, cytochrome f?) of P-700 in the reduced form while if the primary donor was oxidized when the flash was absorbed, the electrons would have to traverse the rate limiting step in the electron transport chain (~ 20 msec). An increasing intensity of measuring light, or photosystem I background light, reduced the amplitudes of the fast component and increased that of the 20 msec absorbance change. They argued that this far red light gave an increasing fraction of units with the primary donor oxidized. Similar results were obtained with increasing concentrations of DCMU. This was explained by assuming that low concentration of DCMU partly inhibited, the flash induced electron transport.

The data seem, in the main, consistent with a reduction of P-700 by plastocyanin and perhaps cytochrome f as well. The reduction of plastocyanin by plastoquinone appears likely, but cytochrome f does not share these kinetics (HAEHNEL, 1974). Mutant studies (reviewed by BISHOP, 1973) suggest that a loss of cytochrome f is equivalent to a loss of P-700 for many photoreactions, suggesting an obligatory role for cytochrome f in the reduction of P-700.

Other proposals for the role of P-700 have been made. ARNON et al. (1971) proposed a three light reaction scheme based, in part, upon a chloroplast preparation which reduced pyridine nucleotide and which did not show a 700 nm band in the oxidized minus reduced difference spectrum. This scheme attributed a cyclic role to P-700 and proposed the reduction of pyridine nucleotide by added reductants via P-700 was a nonphysiological reaction. ESSER (1972) found both the absorption change and an EPR signal corresponding to P-700 if plastocyanin was added.

He concluded plastocyanin was required for P-700 to react with external reductants in this preparation.

The absence of divalent cations causes, in addition to abolishing the lag in reduction of P-700 by photosystem I light after strong illumination (Marsho and Kok, 1974), weak intensities of photosystem II to sensitize P-700 turnover with good efficiency (Rurainski et al., 1971). Upon addition of cations, Rurainski et al. (1971) found a stoichiometry between the increase in yield of $NADP^+$ reduction and the decrease in yield of P-700 turnover. This, together with an increase in the saturation rate of $NADP^+$ reduction and an increase in the yield of variable fluorescence, led them to conclude that the observed (20 msec) P-700 was in competition with the reduction of $NADP^+$ and that divalent cations switched quanta from one photosystem to the other (Rurainski and Hoch, 1971). The instrumentation employed would not resolve relaxations faster than 10^{-3} sec and the experiments of Haehnel et al. (1971) give strong support to a fast P-700 being involved in the reduction of $NADP^+$. They do not rule out a competition between the fast and slow.

The rapid reduction of P-700 seems to be restricted to the electron transport pathway from photosystem II. Added reductants such as ascorbate-dichlorophenol indophenol still yield a 20 msec P-700 turnover even at low light intensities or with widely spaced flashes, conditions which eliminate problems with the rate of reduction of plastocyanin by the indophenol (Witt et al., 1961; Bose, 1974).

References

Arnon, D.I., Knaff, D.B., McSwain, B.D., Chain, R.K., Tsujimoto, H.Y.: Photochem. Photobiol. **14**, 397–425 (1971)

Arntzen, C.J., Briantais, J.M.: In: Bioenergetics of Photosynthesis. Govindjee (ed.). New York-San Francisco-London: Academic Press, 1975, pp. 52–182

Baker, R.A., Weaver, E.C.: Photochem. Photobiol. **18**, 237–241 (1973)

Bishop, N.I.: In: Photophysiology. Giese, A.C. (ed.). New York-London: Academic Press, 1973, Vol. VIII, pp. 65–96

Borg, D.C., Fajer, J., Felton, R.H., Dolphin, D.: Proc. Nat. Acad. Sci. U.S. **67**, 813–820 (1970)

Bose, S.: Thesis, Univ. Rochester, 1974

Breton, J., Roux, E., Whitmarsh, J.: Biochem. Biophys. Res. Commun. **64**, 1274–1277 (1975)

Butler, W.L.: Arch. Biochem. Biophys. **93**, 413–422 (1961)

Döring, G., Bailey, J.L., Kreutz, W., Weikard, J., Witt, H.T.: Naturwissenschaften **55**, 219–220 (1968)

Duysens, L.N.M.: Thesis, Univ. Utrecht, 1952

Duysens, L.N.M.: In: Energy Conversion by the Photosynthetic Apparatus, Brookhaven Symposium in Biology, **19**, pp. 71–80. Brookhaven National Laboratory, 1967

Duysens, L.N.M., Amesz, J., Kamp, B.M.: Nature (London) **190**, 510–511 (1961)

Emerson, R.: Science **127**, 1059–1060 (1958)

Esser, A.F.: Biochim. Biophys. Acta **275**, 199–207 (1972)

Floyd, R.A., Chance, B., De Vault, D.: Biochim. Biophys. Acta **226**, 103–112 (1971)

Fong, F.K.: J. Theor. Biol. **46**, 407–420 (1974)

Fong, F.K.: Appl. Phys. **6**, 151–166 (1975)

Fork, D.C., Amesz, J.: In: Photophysiology. Giese, A.C. (ed.). New York-London: Academic Press, 1970, Vol. V, pp. 97–126

Geacintov, N.E., Van Nostrand, F., Becker, J.F., Tinkel, J.B.: Biochim. Biophys. Acta **267**, 65–79 (1972)

Gorman, D.S., Levine, R.P.: Plant Physiol. **41**, 1648–1656 (1966)

Haehnel, W.: Biochim. Biophys. Acta **305**, 618–631 (1973)

Haehnel, W.: In: Proc. 3rd Intern. Congr. Photosynthesis. Avron, M. (ed.). Amsterdam: Elsevier, 1974, pp. 557–568

Haehnel, W., Döring, G., Witt, H.T.: Z. Naturforsch. **26B**, 1171–1174 (1971)

Haehnel, W., Witt, H.T.: In: Proc. 2nd Intern. Congr. Photosynthesis Research. Forti, G., Avron, M., Melandri, A. (eds.). The Hague: Dr. Junk, 1972, Vol. I, pp. 469–476

Hauska, G., McCarty, R.E., Berzborn, R., Racker, E.: J. Biol. Chem. **246**, 3524–3531 (1971)

Hill, R., Bendall, F.: Nature (London) **186**, 136–137 (1960)

Hiyama, T., Ke, B.: Biochim. Biophys. Acta **267**, 160–171 (1972)

Hoch, G.E., Owens, O.V.H., Kok, B.: In: La Photosynthèse. Editions du Centre National de la Reserche Scientific, pp. 261–270, Paris (1962)

Hoch, G.E., Randles, J.: Photochem. Photobiol. **14**, 435–449 (1971)

Junge, W.: In: Proc. 3rd Intern. Congr. Photosynthesis. Avron, M. (ed.). Amsterdam: Elsevier, 1975, pp. 273–286

Junge, W., Eckhof, A.: FEBS Lett. **36**, 207–212 (1973)

Junge, W., Eckhof, A.: Biochim. Biophys. Acta **357**, 103–117 (1974)

Katz, J.J., Norris, J.R.: In: Current Topics in Bioenergetics, Sanadi, Packer (eds.). New York and London: Academic Press 1973, Vol. V, pp. 41–76

Ke, B.: Biochim. Biophys. Acta **301**, 1–33 (1973)

Ke, B., Sugahara, K., Shaw, E.R., Hansen, R.E., Hamilton, W.D., Beinert, H.: Biochim. Biophys. Acta **368**, 401–408 (1974)

Knaff, D.B., Malkin, R.: Arch. Biochem. Biophys. **159**, 555–562 (1973)

Koester, V.J., Galloway, L., Fong, F.K.: Naturwissenschaften, in press (1975)

Kok, B.: Biochim. Biophys. Acta **22**, 399–400 (1956)

Kok, B.: Acta Botan. Neerl. **6**, 316–336 (1957)

Kok, B.: Biochim. Biophys. Acta **48**, 527–533 (1961)

Kok, B.: In: Photosynthetic Mechanisms of Green Plants. Kok, B., Jagendorf, A.T. (eds.). pp. 45–55. Publication 1145 National Academy of Sciences-National Research Council, Washington, D.C. 1963

Kok, B., Datko, E.A.: Plant Physiol. **40**, 1171–1177 (1965)

Kok, B., Gott, W.: Plant Physiol. **35**, 802–808 (1960)

Kok, B., Hoch, G.E.: In: Light and Life. McElroy, W.D., Glass, B. (eds.). Baltimore: Johns Hopkins, 1961, pp. 397–416

Kok, B., Joliot, P., McGloin, M.P.: In: Progress in Photosynthesis Research. Metzner, H. (ed.). Tübingen: Intern. Union Biol. Sci., 1969, Vol. II, pp. 1042–1056

Lozier, R.H., Butler, W.L.: Biochim. Biophys. Acta **333**, 465–480 (1974)

Malkin, S.: In: Progress in Photosynthesis Research. Metzner, H. (ed.). Tübingen: Intern. Union Biol. Sci., 1969, pp. 845–856

Marsho, T.V., Kok, B.: Biochim. Biophys. Acta **223**, 340–350 (1970)

Marsho, T.V., Kok, B.: Biochim. Biophys. Acta **333**, 353–365 (1974)

Murata, N., Takamiya, A.: Plant Cell Physiol. **10**, 193–202 (1969)

Philipson, K.D., Sato, V.L., Sauer, K.: Biochemistry **11**, 4591–4594 (1972)

Pratt, L.H., Bishop, N.I.: Biochim. Biophys. Acta **153**, 664–674 (1968)

Ried, A.: In: Proc. 2nd Intern. Congr. Photosynthesis Research. Forti, G., Avron, M., Melandri, A. (eds.). The Hague: Dr. Junk, 1972, Vol I, pp. 763–772

Rumberg, B.: Nature (London) **204**, 860–862 (1964a)

Rumberg, B.: Z. Naturforsch. **19B**, 707–716 (1964b)

Rumberg, B., Witt, H.T.: Z. Naturforsch. **19B**, 693–707 (1964)

Rurainski, H.J., Hoch, G.E.: In: Proc. 2nd Intern. Congr. Photosynthesis. Forti, G., Avron, M., Melandri, A. (eds.). The Hague: Dr. Junk, Vol. I, pp. 133–141

Rurainski, H.J., Randles, J., Hoch, G.E.: FEBS Lett. **13**, 98–100 (1971)

Sane, P.V., Park, R.B.: In: Proc. 2nd Intern. Congr.Photosynthesis. Forti, G., Avron, M., Melandri, A. (eds.). The Hague: Dr. Junk, 1972, Vol. I, pp. 825–832

Siedow, J.N., Curtis, V.A., San Pietro, A.: Arch. Biochem. Biophys. **158**, 889–897 (1973)

Trebst, A.: Ann. Rev. Plant Physiol. **25**, 423–458 (1974)

Visser, J.W.N., Rijgersberg, K.P., Amesz, J.: Biochem. Biophys. Acta **368**, 235–246 (1974)

Wessels, J.S.C., Voorn, G.: In: Proc. 2nd Intern. Congr. Photosynthesis. Forti, G., Avron,
    M., Melandri, A. (eds.). The Hague: Dr. Junk, 1972, Vol. I, pp. 833–846
Williams, W.P.: In: Proc. 2nd Intern. Congr. Photosynthesis. Forti, G., Avron, M., Melandri,
    A. (eds.). The Hague: Dr. Junk, 1972, Vol. I, pp. 745–752
Witt, H.T., Müller, A., Rumberg, B.: Nature (London) **192**, 967–969 (1961)
Witt, H.T., Müller, A., Rumberg, B.: Nature (London) **197**, 987–991 (1963)
Witt, K.: FEBS Lett. **38**, 112–115 (1973)
Witt, K., Wolff, C.: Z. Naturforsch. **25B**, 387–388 (1970)
Yamamoto, H.Y., Vernon, L.P.: Biochemistry **8**, 4131–4137 (1969)

5. Chlorophyll Fluorescence:
A Probe for Electron Transfer and Energy Transfer

W. L. Butler

A. Introduction

Chlorophyll fluorescence was recognized quite early as a potentially powerful probe for the study of photosynthesis. The fact that the intensity of fluorescence was a direct measure of the concentration of excited chlorophyll molecules indicated that changes of fluorescence yield should be related to changes in the efficiency of photosynthesis. Thus, it was expected that basic photochemical mechanisms of photosynthesis would be revealed through studies of chlorophyll fluorescence. A number of important discoveries were made in the early investigations. KAUTSKY and HIRSCH (1931) found the characteristic fluorescence yield changes (the Kautsky effect) that occur when plants are first illuminated. McALISTER and MYERS (1940) showed an inverse relationship between the yield of fluorescence and the rate of photosynthesis during the induction period of photosynthesis. KAUTSKY and ZEDLITZ (1941) found that fluorescence yield increased when photosynthesis was inhibited by certain poisons. Such experiments showing an inverse correlation between the yield of fluorescence and the efficiency of photosynthesis gave promise that fluorescence *should* be an effective tool with which to explore the photochemical mechanisms involved. However, the photochemical mechanisms of photosynthesis were more complex than was then realized and few of the early fluorescence studies (before 1960) could be interpreted in a meaningful context.

To date fluorescence has found its greatest use as a monitor of electron transport rather than as a meter for energy flow through the photosynthetic apparatus. In the early years of the 1960s the groundwork was laid which established fluorescence as a method to study photosynthetic electron transport reactions associated with primary photochemistry of photosystem II (PS II). At the end of the 1960s absorbance changes related to the primary electron acceptor of PS II were discovered and these measurements, in concert with fluorescence measurements, elucidated many aspects of the primary photochemical electron transfer reaction of PS II. In turn, these investigations led to a recent formulation of a model for the photosynthetic apparatus which fulfills the early expectations of defining the yields of fluorescence, energy transfer and photochemistry in terms of fundamental photochemical parameters. The purpose of this chapter on fluorescence is to outline and trace some of the highlights of these developments.

B. Fluorescence Yield and Electron Transport

I. A (Q)

The basic relationship between fluorescence yield and photosynthetic electron transport was first recognized by KAUTSKY et al. (1960) in a remarkable kinetic analysis of fluorescence yield changes in *Chlorella* cells. They deduced that the yield of fluorescence was controlled by a substance which quenched fluorescence in its oxidized state, A_0, but did not quench in its reduced state, A_1. They also deduced from the kinetics of the fluorescence yield changes that the electron transport system of photosynthesis was driven by two photochemical reactions operating in series. One photoreaction, which produced the oxidizing power for oxygen evolution, also reduced A_0 to A_1, thereby increasing the yield of fluorescence. The other photoreaction, which produced the reducing power for the reduction of CO_2, also produced a product, B_1, which oxidized A_1 back to A_0, thereby decreasing the yield of fluorescence. Inhibitors of photosynthesis such as phenylurethane or phenanthroline blocked electron transport between the two photoreactions so that B_1 could not oxidize A_1 but A_0 could still be photoreduced. They even calculated from the first-order rate constant for the light-induced fluorescence yield increase measured in the presence of phenylurethane that the ratio of chlorophyll to A_0 was about 400/1. They suggested that this ratio supported the concept of a photosynthetic unit in which A_0 was the acceptor molecule for energy absorbed within the unit and that chlorophyll fluorescence was quenched by energy transfer to A_0. Thus, most of the quenching properties that were to be ascribed to the primary electron acceptor of PS II were deduced by Kautsky et al. in a remarkable work which tends to be overlooked today.

The work of Kautsky et al. was ahead of its time but only by a couple of years. They apparently were unaware of the emerging work of Emerson and co-workers (1957, 1960) on enhancement phenomena showing the involvement of two spectrally distinguishable photochemical pigment systems. Although Kautsky et al. proposed the operation of two separate photochemical reactions in an electron transport system remarkably similar to that which we accept today, they did not attempt any dual wavelength studies to show the operation of different pigment systems.

The involvement of two photochemical pigment systems had to be recognized before any significant progress could be made to understand the photochemical basis of photosynthesis. In order to reconcile otherwise paradoxical results with red and blue-green algae, Duysens (1952) had proposed the existence of two types of chlorophyll a; one, a fluorescent, photochemically active form, the other, a non-fluorescent, inactive form. Emerson and Rabinowitch (1960) suggested that these two types of chlorophyll a were the bases of the two pigment systems indicated by the enhancement studies. A differential influence of these two pigment systems on fluorescence yield was soon recognized. Govindjee et al. (1960) showed that the excitation of fluorescence at 685 nm by a 670 nm beam and a 700 nm beam was less than additive when the two beams were used simultaneously. They suggested that the action of the 700 nm beam was to quench the fluorescence excited by the 670 nm. Butler (1962) confirmed the differential action of red and far-red light by showing that the fluorescence yield, increased by a previous irradiation with red light, could be decreased by a subsequent irradiation with far-red light with an action spectrum which was similar to the enhancement spectrum for Emerson's long wavelength pigment system. Although the differential action of Emerson's two pigment systems on fluorescence yield was clearly recognized, the precise nature of the control was not.

Duysens and Sweers (1963) were the first to present a unified theory showing the relationship between the two pigment systems of photosynthesis and the fluorescence yield. The scheme of Duysens and Sweers was very similar to that presented by Kautsky et al. but Duysens and Sweers proposed that the two photochemical reactions were analogous to the two photochemical pigment systems elucidated by EMERSON. KAUTSKY's quencher, A_0, was recognized as the primary electron acceptor of the photosystem operating nearest the oxygen evolving end of the electron transport chain, PS II, and labeled Q. The theory of DUYSENS and SWEERS readily explained the earlier data on the effects of red and far-red light on fluorescence; red light absorbed by PS II reduced Q to QH causing the fluorescence yield to increase while far-red light absorbed preferentially by PS I caused QH to be oxidized back to Q thereby reimposing the quenching action of Q. The formulation of DUYSENS and SWEERS was immediately accepted since it integrated the recently formalized concepts of a linear electron transport system driven photochemically by two pigment systems (HILL and BENDALL, 1960; KOK and HOCH, 1961; DUYSENS et al., 1961; WITT et al., 1961) with an explanation of how fluorescence yield was controlled by the electron transport system. Fluorescence became, almost instantly, a powerful tool with which to study photosynthetic electron transport activity.

II. C-550

The discovery of KNAFF and ARNON (1969b) of light-induced absorbance changes in the 550 nm region provided an additional tool with which to explore the primary photochemistry of PS II and the relationship of that photochemistry to fluorescence yield. They labeled the component responsible for these absorbance changes C-550 and suggested that it was the primary electron acceptor for one of the two PS II photoreactions, PS IIb, which they postulated at that time (KNAFF and ARNON, 1969c). The light-induced absorbance changes of C-550 were mediated by PS II but not by PS I; they persisted in the presence of DCMU as well as at temperatures as low as liquid nitrogen; and they appeared to represent a reduction in that they could be induced chemically with strong reductants such as dithionite but not by ascorbate (KNAFF and ARNON, 1969c; ERIXON and BUTLER, 1971b). At $-196°$ C the photo-reduction of C-550 was accompanied by the oxidation of cytochrome b-559 which KNAFF and ARNON (1969a, c) attributed to the other PS II photoreaction, PS IIa.

The redox behavior of C-550 to dithionite and ascorbate suggested a similarity between C-550 and the Q of DUYSENS and SWEERS in that the fluorescence yield showed a similar behavior. Redox titration experiments gave a more rigorous demonstration that the absorbance changes of C-550 and the fluorescence yield changes attributed to Q were equivalent manifestations of redox changes of the primary electron acceptor of PS II (ERIXON and BUTLER, 1971b). Light-induced absorbance changes of C-550 and of cytochrome b-559 and light-induced fluorescence yield changes were measured at $-196°$ C in chloroplast samples which had been equilibrated to specific redox potentials before freezing. The redox titration curves for the extent of the light-induced changes of C-550 and of fluorescence yield as a function of redox potential were identical, both being closely approximated by a 1 electron Nernst equation with a midpoint potential of $+25$ mV. Furthermore, the titration of the photooxidation of cyt b-559 followed that same curve; i.e. to the extent that C-550 was reduced chemically before freezing it was not available as an electron acceptor and to that same extent the photooxidation of cyt b-559 was prevented. It was apparent from these redox titration experiments that even if there had been no assay for the primary electron acceptor of PS II, the titration curve for the photooxidation of cyt b-559 could have been used to determine the midpoint potential of the primary acceptor. These techniques were applied later using the photooxidation of P-700 at $-196°$ C to determine the midpoint potential of the primary electron acceptor of PS I (LOZIER and BUTLER, 1974b). These data were taken to indicate a single PS II photoreaction in which the photooxidation of cyt. b-559 was stoichiometric with the photoreduction of C-550. According to the simple $D \cdot P \cdot A$ model (BUTLER, 1972b) where A is C-550, P is the reaction center chlorophyll P-680 and D is cyt b-559,

$$D \cdot P \cdot A \xrightarrow{hv} D \cdot P^+ \cdot A^- \rightarrow D^+ \cdot P \cdot A^-.$$

C-550 was shown in the titration experiments to be an obligatory electron acceptor consistent with its proposed role as the primary electron acceptor for the photoreaction. Cyt b-559 was not an obligatory electron donor since the photoreduction of C-550 at $-196°$ C proceeded equally well in the presence of ferricyanide added prior to freezing to chemically oxidize the cyt b-559. The photooxidation of cyt b-559 accompanies the photoreduction of C-550 only at very low temperatures (BUTLER et al., 1973a). As the temperature is raised above $-160°$ C another secondary electron donor becomes increasingly effective until at $-100°$ C no photooxidation of cyt. b-559 occurs while the photoreduction of C-550 is unimpeded.

The equivalence between Q, as measured by fluorescence, and C-550 as measured by absorbance, was shown in a number of experiments involving photosynthetic mutants which lacked C-550 (ERIXON and BUTLER, 1971a; EPEL and BUTLER, 1972) or treatments which destroyed (OKAYAMA et al., 1971) or removed (OKAYAMA and BUTLER, 1972a) the C-550. Much of the data on C-550 as well as a discussion of the nature of the absorbance change are covered in previous reviews (BUTLER, 1973; BEARDEN and MALKIN, 1975) more thoroughly than can be done here.

While the absorbance measurements of C-550 have provided a powerful tool to explore the primary photochemistry of PS II, it is important to recognize that these data have been most useful when measurements were made at low temperature. The magnitude of the C-550 absorbance change is much greater at $-196°$ C than at room temperature, because of the band-shift nature of the change (BUTLER and OKAYAMA, 1971), and much easier to interpret when secondary biochemical influences are frozen out. It was shown in a comparison of the C-550 changes and the 518 nm changes at room temperature (BUTLER, 1972a) that about 50% of the absorbance change attributed to C-550 reflected changes of membrane potential while the remainder was due to redox changes of the primary electron acceptor of PS II. Previous work (MURATA and SUGAHARA, 1969; WRAIGHT and CROFTS, 1970) showed that fluorescence yield at room temperature was controlled by the pH gradient across thylakoid membranes as well as by the primary electron acceptor of PS II. At room temperature membrane potentials, pH gradients and conformational changes generated by PS I can be sensed by PS II in ways that modify the absorbance attributed to C-550 and the fluorescence yield attributed to Q so that the analysis of room temperature measurements can be quite confusing if the various elements involved are not resolved. When measurements of absorbance and fluorescence are made at low temperature, however, only changes associated with the primary photochemical apparatus are manifest and under these conditions important aspects of the primary photochemistry of PS II have been elucidated through comparisons of C-550 and of fluorescence.

III. P-680

It was noted (OKAYAMA and BUTLER, 1972b) in fluorescence induction experiments at $-196°$ C that if ferricyanide was added to oxidize cyt b-559 prior to freezing, the initial level of fluorescence, F_0, was unaffected but the final level of fluorescence, F_M, (or the fluorescence of variable yield, $F_V = F_M - F_0$) was markedly decreased. Two explanations were suggested to account for the quenching of fluorescence by ferricyanide; one was that when the secondary donor, cyt b-559, was previously oxidized, the oxidized form of the primary electron donor, P-680$^+$, persisted and that P-680$^+$ quenched fluorescence; the other was that when cyt b-559 was previously oxidized another secondary donor was oxidized by P-680$^+$ and that the oxidized form of this alternative secondary donor quenched fluorescence. Subsequent work (LOZIER and BUTLER, 1974a) favored the second explanation since P-680$^+$ is not stable in the dark even in the presence of ferricyanide at $-196°$ C. Nevertheless, these experiments with ferricyanide raised the question of whether P-680$^+$ quenched fluorescence.

The experiments which established that a quenching action arose from the oxidizing side as well as the reducing side of PS II were comparisons of the rate of photoreduction of C-550, the rate of photooxidation of cyt b-559 and rate of the fluorescence yield increase at $-196°$ C (BUTLER et al., 1973a). It had been expected that the fluorescence yield increase should parallel the photoreduction of C-550; it was found that the fluorescence yield increase was appreciably slower than the reduction of C-550 (the half time was approximately three times longer) and that it paralleled the slower photooxidation of the cyt b-559. In the context of the $D \cdot P \cdot A$ model the fluorescence yield increase followed the oxidation of the secondary donor, D, not the reduction of the primary acceptor, A. The fluorescence yield is high only when the reaction center is in the $P \cdot A^-$ state. These results were interpreted to indicate that P$^+$ is a quencher as well as A. If an additional electron transfer component functions between cyt b-559 and P-680, the oxidized form of that component could be the quencher instead of P-680$^+$. However, there is no evidence for such a component.

MAUZERALL (1972) showed that a very brief flash (10 nsec) at room temperature did not cause an immediate fluorescence yield increase (i.e. within the period of the flash) as should be expected if the fluorescence yield were controlled solely by the redox state of the primary electron acceptor. Rather the fluorescence yield increased in the dark after the flash with a half time of about 3 μsec. BUTLER (1972b) suggested that this dark reaction reflected the dark reduction of P-680$^+$ and that the results of MAUZERALL were due to quenching by P-680$^+$. DEN HAAN et al. (1973, 1974), however, presented evidence for the action of still another quencher, T. They suggested that normally P-680$^+$ decayed more rapidly than T and that it was the decay of T that limited the fluorescence yield increase in the flash experiments. They also found that the fluorescence yield increase induced by a flash was much slower in the presence of hydroxylamine and they attributed this slower increase to a longer lifetime for P-680$^+$ under these conditions. However, the relationship between the proposed quenching by T and by P-680$^+$ is not entirely clear at this time and further comments should await additional experimental results. It is interesting, however, that three different quenchers, A, P$^+$ and T, are now associated with the PS II reaction centers.

IV. The Back-Reaction

A backreaction in the primary electron transfer couple was elucidated by measurements of fluorescence and absorbance induced by single saturating flashes at $-196°$ C. It was found that such flashes, even though supersaturating in intensity, caused the fluorescence yield (Butler, 1972b; Murata et al., 1973) to be increased by only a small amount and the C-550 (Butler et al., 1973b; Vermeglio and Mathis, 1973) to be reduced to only a small extent (in both cases approximately 20% of the maximum change inducible by continuous irradiation at $-196°$ C). The ineffectiveness of the flash, which is observed only at low temperatures, was interpreted to indicate a backreaction between P^+ and A^- (Butler et al., 1973b; Murata et al., 1973).

$$D \cdot P \cdot A \underset{k_{-1}}{\overset{h\nu}{\rightleftharpoons}} D \cdot P^+ \cdot A^- \xrightarrow{k_2} D^+ \cdot P \cdot A^-.$$

At room temperature k_2 is large compared to k_{-1} so that the flash is effective in reducing A and the quantum yield for the forward reaction [proportional to $k_2/(k_{-1}+k_2)$] is high. [The quantum yield for the photoreduction of C-550 at room temperature was estimated to be close to unity (Butler et al., 1973b).] As the temperature is lowered, however, k_2 slows down relative to k_{-1} until at $-196°$ C k_{-1} is approximately four-fold greater than k_2. Thus, most of the charge separation achieved by the flash at $-196°$ C is dissipated in the backreaction. The effect of temperature to lower the quantum yield for the forward reaction in continuous light at low temperatures has also been observed both for fluorescence (Kok, 1963) and for C-550 (Butler et al., 1973b). The existence of this backreaction has been confirmed recently in direct kinetic measurements of the decay of reduced C-550 (Mathis and Vermeglio, 1974) and of oxidized P-680 (Mathis and Vermeglio, 1975; Malkin and Bearden, 1975) following a flash at low temperatures.

The realization that the yield of fluorescence indicated the redox state of the primary electron acceptor of PS II provided a powerful tool for the exploration of the photosynthetic electron transport system. These measurements of fluorescence coupled to absorbance measurements of C-550 and cyt b-559 (generally at low temperatures) revealed many aspects of the electron transport reactions associated directly with the reaction centers of PS II. However, the more fundamental problem of relating the yield of fluorescence and the yield of photochemistry to basic photochemical parameters remained illusive. Recently, however, measurements of fluorescence and C-550 stimulated the development of models for the photosynthetic apparatus which expresses the yields of fluorescence, photochemistry and energy transfer in terms of fundamental molecular parameters.

C. The Photochemical Model

I. Photosystem II

Simple photochemical theory predicts (Kamen, 1963) that the maximum yield of photochemistry, φ_{P_0}, should be equal to the ratio of the total variable yield of fluorescence to the maximum yield of fluorescence, $\varphi_{F_V}/\varphi_{F_M}$ (or to just F_V/F_M); the ineffeciency in the system being represented by the minimal yield of fluorescence, φ_{F_0}, obtained with dark adapted cells or chloroplasts. However, the simple theory does not appear to hold since, in many cases, the ratio of F_V/F_M is in the range of 0.6 to 0.7 whereas the yield of photochemistry is thought to be higher. This dichotomy was resolved by assuming that part of the F_0 fluorescence was due to "dead fluorescence" or fluorescence from PS I or had some origin other than the active photochemical apparatus of PS II (CLAYTON, 1969; LAVOREL and JOLIOT, 1972) so that the true value of F_V/F_M would be higher than the measured value if a proper correction could be made for the inoperative part of F_0. In the absence of experimental means to determine what fraction of F_0 is inoperative, such an assumption, in effect, precludes any meaningful correlations between yields of fluorescence and yields of photochemistry.

The problem, however, was reopened when it was noted that a particular quinone, dibromothymoquinone (DBMIB), was a very effective quencher of fluorescence even at $-196°$ C (LOZIER and BUTLER, 1972). Furthermore, photochemistry measured as the photoreduction of C-550 at $-196°$ C persisted even though fluorescence was quenched to less than 1% of the normal yield (LOZIER and BUTLER, 1974a). Thus, the question of the interrelationship between the yield of photochemistry and the yield of fluorescence could be approached experimentally as a function of quenching by DBMIB. A study was made of the relative yield of photochemistry, φ_{P_0}, (assumed to be proportional to the initial rate of photoreduction of C-550 at $-196°$ C) and of the relative yields of fluorescence at the F_0 and F_M levels (measured at 692 nm at $-196°$ C) as a function of the concentration of DBMIB added prior to freezing (KITAJIMA and BUTLER, 1975a). The yields of photochemistry and the yields of fluorescence (of both F_0 and F_M) were quenched by DBMIB and in such a manner that the ratio of F_V/F_M decreased to the same relative extent that φ_{P_0} was quenched. These results indicated that fluorescence could be used to analyze the photochemical utilization of excitation energy in the photosynthetic apparatus and without any special assumptions as to the heterogeneity of F_0. It was concluded that F_0 had the same origin as F_V in the measurement of fluorescence at 692 nm at $-196°$ C.

The conclusions drawn from the quenching by DBMIB, however, were not consistent with other kinds of fluorescence quenching data. It was noted above, for instance, that the presence of ferricyanide at $-196°$ C quenched F_V specifically with no effect on F_0 (OKAYAMA and BUTLER, 1972b). It was further shown that the rate of photoreduction of C-550 at $-196°$ C was the same in the absence or presence of ferricyanide even though the ratio of F_V/F_M changed from 0.8 to 0.5. Fortunately, both kinds of fluorescence quenching data could be brought into harmony with a simple photochemical model for PS II units which included energy transfer from the antenna chlorophyll to a reaction center chlorophyll (BUTLER and KITAJIMA, 1975a).

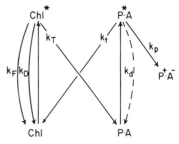

Excitation energy in the antenna chlorophyll is either transferred to the reaction center, k_T, emitted as fluorescence, k_F, or dissipated by nonradiative decay, k_D. Excitation energy in the reaction center chlorophyll is either used for photochemistry, k_p, dissipated by nonradiative decay, k_d, or transferred back to the antenna chlorophyll, k_t. There is no evidence for any fluorescence emission from the reaction center chlorophyll so none is assumed. Reaction centers were considered to exist either in the open $(P \cdot A)$ or closed $(P \cdot A^-)$ states. States of the reaction center involving P^+ were ignored because the fast turnover of P^+ should preclude any significant accumulation of these states under normal conditions.

The model was considered, assuming either no energy transfer between PS II units (the "separate package" model) or complete energy transfer so that all reaction centers were available to all of the antenna chlorophyll (the "matrix" model). Assuming the "separage package" model, photochemistry would be accomplished only by units with open reaction centers so that $\varphi_P = \varphi_T \varphi_p A$ where A is the fraction of the reaction centers that are open (or the fraction of the primary electron acceptors that are present in the oxidized state), $\varphi_T = k_T(k_F + k_D + k_T)^{-1}$ and $\varphi_p = k_p(k_p + k_t + k_d)^{-1}$. For those reaction centers that are closed, $1 - A$, photochemistry cannot be accomplished, $k_p = 0$, but the excitation energy trapped by the reaction center chlorophyll may be transferred back to the antenna chlorophyll by the k_t process. According to the "separate package" model excitation energy in a unit with a closed reaction center will continue to cycle into the trap and back out to the unit until it is dissipated by fluorescence from the antenna chlorophyll or by nonradiative decay. This cycling of excitation energy can be expressed as an infinite series which converges to a simple expression $(1 - \varphi_T \varphi_t)^{-1}$ where φ_t for the closed reaction centers is $k_t(k_t + k_d)^{-1}$. If we make a further simplifying assumption that k_p is much larger than k_t or k_d so that $\varphi_p \simeq 1$ for open reaction centers:

$$\varphi_P = \varphi_T A, \tag{1}$$

$$\varphi_F = \frac{k_F}{k_F + k_D + k_T} \left[A + \frac{1 - A}{1 - \varphi_T \varphi_t} \right] = \varphi_{F_0} \left[1 - \frac{(1 - A) \varphi_T \varphi_t}{1 - \varphi_T \varphi_t} \right]. \tag{2}$$

According to these equations,

$$\varphi_{P_0}(A = 1) = \varphi_T, \tag{3}$$

$$\frac{F_V}{F_M} = \frac{\varphi_{F_M}(A = 0) - \varphi_{F_0}(A = 1)}{\varphi_{F_M}} = \varphi_T \varphi_t. \tag{4}$$

Expressions for φ_P and φ_F were also derived for the "matrix model". Although the equations show different kinetic behavior for the two models, the expressions for φ_P and φ_F are identical at the initial $(A = 1)$ and final $(A = 0)$ states. Thus, Eqs. (3) and (4) are valid for any degree of energy transfer between PS II units.

Two kinds of fluorescence quenching processes are apparent in the model; nonradiative decay in the antenna chlorophyll analogous to a k_d process and nonradiative decay at the reaction center chlorophyll, a k_d process. Quenchers such as oxidized quinones or dinitrobenzene are of the first type. They create quenching centers in the antenna chlorophyll which compete with the reaction centers for the excitation energy. In the case of quenching by DBMIB, k_Q was shown to be directly proportional to the concentration of the quencher (KITAJIMA and BUTLER, 1975a). With this type of quenching, which quenches F_0 as well as F_V, Eqs. (3) and (4) predict that φ_{P_0} and the ratio F_V/F_M should be decreased to the same extent by DBMIB since both are proportional to $\varphi_T [\varphi_T = k_T (k_F + k_D + k_T + k_Q)^{-1}]$. With the other type of quenching exemplified by ferricyanide, quenching occurs at the reaction center chlorophyll by increasing k_d. Such quenching will not affect φ_p or φ_{P_0} so long as k_p at the open reaction centers remains large compared to k_d. For the closed reaction centers increasing k_d will decrease φ_t so that the ratio F_V/F_M will decrease. According to Eq. (2) increasing k_d should quench F_V but have no influence on F_0.

II. Photosystem I

Fluorescence at wavelengths longer than 730 nm at $-196°$ C is assumed to arise entirely from PS I. The emission from PS II appears to be negligible in this longer wavelength region (BUTLER and KITAJIMA, 1975c).

The photooxidation of P-700 at $-196°$ C does not result in any fluorescence yield changes even in the long wavelength region. Careful irradiation of chloroplasts at $-196°$ C with 720 nm light (so as to photooxidize P-700 but not photoreduce C-550) or irradiation of purified PS I particles at $-196°$ C oxidizes P-700 but causes no fluorescence yield changes. The observation that the fluorescence yield of PS I does not increase when P-700 is oxidized suggested that P-700$^+$ also functions as an energy trap (BUTLER and KITAJIMA, 1975b). It had been noted earlier (LOZIER and BUTLER, 1974a) that the bleaching of the 702 nm band of P-700 at $-196°$ C is accompanied by the increase of an absorption band at 690 nm which might represent P-700$^+$. Since both P-700$^+$ and P-680$^+$ are postulated to quench excitation energy, the same molecular mechanisms may be involved with both.

Although there are no fluorescence yield changes associated with the primary photochemistry of PS I, the fluorescence of PS I (e.g. at 730 nm at $-196°$ C) does show fluorescence yield changes which appear to be related to the redox state of the primary electron acceptors of PS II. The kinetics of the light-induced fluorescence yield changes at $-196°$ C are identical at 692 nm and at 730 nm (KITAJIMA and BUTLER, 1975b). F_V measured at 730 nm at $-196°$ C is attributed to energy transfer from PS II to PS I and as such is a very useful assay for "spillover".

III. The Photochemical Apparatus

The model was expanded to encompass both PS I and PS II units in a photosynthetic apparatus which included a light-harvesting chlorophyll complex, chlorophyll LH, which could transfer excitation energy to either of the two photosystems (BUTLER and KITAJIMA, 1975b, 1975c).

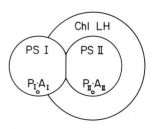

The three emission bands that are observed in fluorescence spectra at $-196°$ C at 685, 695 and 735 nm were ascribed to the antenna chlorophyll a present in chlorophyll LH, PS II and PS I, respectively. Measurements of the ratio of F_M/F_0 were made at various wavelengths of emission (Butler and Kitajima, 1975c). The maximum ratio (4.3 for that sample of chloroplasts) was found for fluorescence measured in a narrow passband around 692 nm. A ratio of 2.8 was measured at 679 nm and a constant ratio of 1.4 was found for wavelengths longer than 730 nm. The ratios measured at 679 and 730 nm were taken to be representative of the 685 and 735 nm emission bands, respectively. The ratio measured at 692 nm was assumed to be a minimum value for the 695 nm band since any overlap from the 685 or 735 nm bands would decrease the measured value. The only source of F_V at any wavelength is exciton transfer from closed PS II reaction centers out to the emitting species of chlorophyll a. Thus, chlorophyll a molecules closest to the reaction centers should contribute most to F_V. Since no attempt was made to describe or consider the spacial distribution of excitons within PS I, PS II or chlorophyll LH, the ratios of F_M/F_0 for a particular emission band are averaged values for all of the chlorophyll a molecules contributing to that band.

The initial distribution of quanta absorbed by the photochemical apparatus was specified in terms of α, the fraction absorbed directly by PS I or transferred to PS I from chlorophyll LH, and β, the fraction going into PS II or being dissipated in chlorophyll LH ($\alpha + \beta = 1$). A specific rate constant for energy transfer from the antenna chlorophyll of PS II to PS I, $k_{T(II\rightarrow1)}$, was also introduced. There is no evidence for energy transfer from PS I to chlorophyll LH or to PS II. There appears to be reasonably tight coupling between chlorophyll LH and PS II so that excitation energy can be transferred back and forth between these two types of chlorophyll complexes. Chlorophyll LH serves largely as antenna chlorophyll for PS II (if we assume that α is about 0.3 and β about 0.7, which will be justified vide infra, and that the amount of chlorophyll in PS I, PS II and chlorophyll LH is in the ratio of 1/1/2 we would estimate that the probability for energy transfer from chlorophyll LH to PS II would be about nine times greater than the probability for transfer to PS I). The rate constants for the antenna chlorophyll of PS II, k_{F692}, k_{DII}, $k_{T(II\rightarrow1)}$ and k_{TII} are averaged values which reflect the degree of energy coupling between PS II and chlorophyll LH. The yields of fluorescence from PS II and PS I and the yield of energy transfer from PS II to PS I were expressed as:

$$\varphi_{F692} = \frac{\beta k_{F692}}{k_{F692} + k_{DII} + k_{T(II\rightarrow1)} + k_{TII}} \left[A_{II} + \frac{1 - A_{II}}{1 - \varphi_{TII}\varphi_{tII}}\right], \tag{5}$$

$$\varphi_{F730} = \frac{k_{F730}}{k_{F730} + k_{DI} + k_{TI}} [\alpha + \beta \varphi_{T(II\rightarrow1)}], \tag{6}$$

$$\varphi_{T(II\rightarrow1)} = \frac{k_{T(II\rightarrow1)}}{k_{F692} + k_{DII} + k_{T(II\rightarrow1)} + k_{FII}} \left[A_{II} + \frac{1 - A_{II}}{1 - \varphi_{TII}\varphi_{tII}}\right]. \tag{7}$$

It is apparent that the yield of energy transfer, $\varphi_{T(II\rightarrow1)}$, has a constant part, $\varphi_{T(II\rightarrow1)(0)}$, and a variable part, $\varphi_{T(II\rightarrow1)(V)}$, that are in the same ratio as the constant and variable parts of φ_{F692}.

The previous schematic diagram of the photochemical apparatus (Butler and Kitajima, 1975b, 1975c) presented PS I and PS II units separated by chlorophyll LH.

Such a diagram suggests that chlorophyll LH might be necessary for energy transfer between PS II and PS I. That proposition was tested by measuring fluorescence induction curves at 750 nm at $-196°$ C on leaves which lacked chlorophyll LH. Leaves which contain active PS I and PS II units but no chlorophyll LH can be obtained from the chlorophyll b-less mutant of barley (THORNBER and HIGHKIN, 1974) or from etiolated bean leaves which have been partially greened by a series of brief flashes (HOFER et al., 1975). The induction curves of such leaves showed a normal degree of F_V at 750 nm (KITAJIMA and BUTLER, unpublished). Since chlorophyll LH does not appear to be required for energy transfer from PS II to PS I, in our current diagram we show the two photosystem units in contact with one another and we suggest that the chlorophyll LH accumulates primarily around PS II units but with some contact with PS I as well. Such a diagram is consistent with the observations that chlorophyll LH serves largely as antenna chlorophyll for PS II and that PS I units can be separated from chlorophyll LH much more readily than can PS II units. The diagram also suggests a morphological correlation with the large and small particles that are observed on the opposite fracture faces of freeze-fractured thylakoids. BOARDMAN et al. (1975) noted an absence of the large particles from the chlorophyll b-less mutant of barley. More work is needed, however, to establish the significance of such an observation.

IV. Energy Distribution Between PS I and PS II

The model was used to examine the effects of Mg^{2+} on the distribution of excitation energy between PS I and PS II (BUTLER and KITAJIMA, 1975c). MURATA (1969) originally proposed that energy distribution to PS I increased when chloroplasts were depleted of Mg^{2+}. Fluorescence induction curves were measured at 692 and 730 nm on chloroplasts which had been frozen to $-196°$ C in the absence and presence of Mg^{2+}. At 692 nm F_V was markedly quenched in the absence of Mg^{2+} while F_0 was affected very little: at 730 nm F_V was essentially unchanged while F_0 was markedly increased in the absence of Mg^{2+}. From such measurements of the relative values of F_0 and F_V at 692 and 730 made in the absence and presence of Mg^{2+}, relative values of α and $k_{T(II \to 1)}$ can be calculated within the context of the model. The absence of Mg^{2+} caused α to increase approximately 20% and $k_{T(II \to 1)}$ to increase approximately 100%. With the additional criterion that $\alpha + \beta = 1$, absolute values of α, β and $\varphi_{T(II \to 1)}$ were calculated (see Table 1). An example of these calculations is given in the appendix. It is seen from the analysis that both α and $\varphi_{T(II \to 1)}$ increase to direct more excitation energy into PS I when chloroplasts are depleted of divalent cations.

Table 1. Distribution of excitation energy between PS I and PS II in the absence and presence of 5 mM $MgCl_2$ assuming only a $\varphi_{T(II \to 1)}$ type of energy transfer from PS II to PS I

	α	β	$\varphi_{T(II \to 1)(0)}$	$\varphi_{T(II \to 1)(M)}$
$-Mg^{2+}$	0.32	0.68	0.12	0.28
$+Mg^{2+}$	0.27	0.72	0.065	0.23

The values of α in Table 1 should probably be considered as minimum values. It was noted (BUTLER and KITAJIMA, 1975c) that the ratio of F_M/F_0 measured at 692 nm might be somewhat less than the inherent value for the 695 nm emission band because of overlaps by the 685 and 735 nm emission bands. If that were the case the contribution of energy transfer to F_0 at 730 nm, $\beta \varphi_{T(II \to 1)(0)}$, would be overestimated and that due to α underestimated.

The wavelength dependence of α was determined by measuring the ratios of F_M/F_0 at 692 and 750 nm with various wavelengths of monochromatic exciting light in chloroplasts frozen to $-196°$ C in the presence of Mg^{2+} (KITAJIMA and BUTLER, 1975b). The value of α was found to be essentially independent of the wavelength of excitation from 400 to 680 nm. At wavelengths longer than 680 nm where PS I becomes the predominant absorber α increases reaching value near unity at 695 nm. The major variation in the spectrum of α at visible wavelengths shorter than 680 nm appears as a relatively broad maximum at 515 nm where α increases about 10%, suggesting a carotenoid pigment with some preferential energy transfer to PS I.

Pure excitation spectra of PS II and PS I were also determined (KITAJIMA and BUTLER, 1975b) within the context of the model. Once it is recognized that PS I contributes nothing to a fluorescence of variable yield and that F_V at any wavelength, even at 750 nm at $-196°$ C, is due to excitation of PS II, the determination of the excitation spectrum for PS II becomes relatively simple. Excitation spectra for 750 nm fluorescence at $-196°$ C were measured at the F_0 and F_M levels of fluorescence. The difference spectrum, $F_M - F_0$, is an excitation spectrum for PS II activity, and shows a maximum at 677 nm due to the antenna chlorophyll a in the PS II units, a shoulder at 670 nm due to the chlorophyll a in chlorophyll LH and a maximum at 650 nm due to the chlorophyll b in chlorophyll LH. Once the excitation spectrum of PS II is known and the fraction of F_0 at 750 nm which is due to energy transfer from PS II has been determined the excitation spectrum of PS I can be measured. Such a spectrum (at $-196°$ C) shows a sharp maximum at 681 nm due to the chlorophyll a in the PS I units and some indication of energy transfer from chlorophyll LH.

The model was refined further (BUTLER and KITAJIMA, 1975c) to consider two types of energy transfer processes; energy transfer from the antenna chlorophyll of PS II to PS I, $k_{T(II \to 1)}$, and energy transfer from the reaction center chlorophyll of PS II to PS I, $k_{t(II \to 1)}$. The latter type of energy transfer accounts for the specific quenching of F_V at 692 nm found in the absence of Mg^{2+}. Eq. (6) can be rewritten to include both types of energy transfer:

$$\varphi_{F\,730} = \frac{k_{F\,730}}{k_{F\,730} + k_{DI} + k_{TI}} \left[\alpha + \beta \varphi_{T(II \to 1)} + \beta \varphi_{T\,II} \varphi_{t(II \to 1)} (1 - A_{II}) \right] \tag{8}$$

where the expression for $\varphi_{T(II \to 1)}$, Eq. (7), is unchanged.

The influence of Mg^{2+} on both types of energy transfer can be estimated from the fluorescence data by procedures shown in the appendix. A set of values for α, β, $\varphi_{T(II \to 1)}$ and $\varphi_{t(II \to 1)}$ in the absence and presence of Mg^{2+} is presented in Table 2. Energy transfer from PS II to PS I when the PS II reaction centers are all closed ($A_{II} = 0$) is the sum of two parts, $\varphi_{T(II \to 1)(M)}$ plus $\varphi_{T\,II} \varphi_{t(II \to 1)}$, and is labeled $\Sigma_{(II \to 1)}$ in Table 2.

The values in Table 2 are based on the assumption that $\varphi_{T\,II}$ is 0.80 and that any $k_{d\,II}$ processes are negligible in comparison with $k_{t\,II}$ and $k_{t(II \to 1)}$. 0.80 should be regarded as a minimum value for $\varphi_{T_{II}}$; some data (KITAJIMA and BUTLER, unpublished) indicate that $\varphi_{T_{II}}$ can be as high as 0.90. Nor can we be certain that $k_{d_{II}}$ should be neglected, especially with chloroplast samples which may be subject to some disrepair by the preparation procedures. However, if we had assumed that $\varphi_{T_{II}}$ was 0.85 and that $\varphi_{d_{II}}$ was 0.05 the values in Table 2 would have been essentially the same. From a conceptional point of view the introduction of the $k_{t(II\to1)}$ energy transfer term is valuable in that it points out that conformational changes may facilitate energy transfer from the PS II reaction centers to PS I and thereby cause a specific quenching of F_v at 692 nm. On the basis of the calculations which include $k_{t(II\to1)}$ (Table 2) most of the control over the distribution of excitation energy is due to changes of $\alpha(\alpha^-/\alpha^+ = 1.3)$ and changes of $k_{t(II\to1)}(k_{t(II\to1)}^-/k_{t(II\to1)}^+ = 3.0)$ while changes of $k_{T(II\to1)}$ play a minor (or even insignificant) role. From an operational point of view, however, the simpler procedures based solely on $k_{T(II\to1)}$ may be adequate to estimate the influence of conditions, such as the absence of divalent cations, on the distribution of excitation energy.

Table 2. Distribution of excitation energy between PS I and PS II in the absence and presence of 5 mM $MgCl_2$ assuming both $k_{T(II\to1)}$ and $k_{t(II\to1)}$ types of energy transfer from PS II to PS I

	α	β	$\varphi_{T(II\to1)(0)}$	$\varphi_{T(II\to1)(M)}$	$\varphi_{T_{II}}\varphi_{t(II\to1)}$	$\Sigma_{(II\to1)}$
$-Mg^{2+}$	0.38	0.62	0.044	0.11	0.17	0.28
$+Mg^{2+}$	0.29	0.71	0.041	0.15	0.08	0.23

The distinction between $k_{t(II\to1)}$ and $k_{T(II\to1)}$ is an artificial one which arises from our inability to take into account the spacial distribution of excitons in the photochemical apparatus. Energy transfer from the PS II reaction centers to PS I could be considered as two sequential transfer processes, one from the reaction center to the antenna chlorophyll of PS II with a subsequent transfer to the antenna chlorophyll of PS I. In that context $\varphi_{t(II\to1)}$ could be expressed as the product $\varphi_{t_{II}}\varphi_{T(II\to1)}$. The problem is that the spacial distribution of excitons will depend on the state of the PS II reaction centers so that $k_{T(II\to1)}$, which depends on the average proximity of excitons to PS I, would be different when $A_{II} = 1$ than when $A_{II} = 0$. The procedures used in the appendix, which consider $k_{t(II\to1)}$ and $k_{T(II\to1)}$ as distinct transfer processes, are an attempt to treat the spacial distribution problem from at least a qualitative point of view.

The complete description of a model for the photosynthetic apparatus would include the geometrical relationship between the various elements and the spacial distribution of excitons throughout the apparatus. In the absence of such precise detail the model depends on approximations and operational definitions. The operational definition of α and β includes the spacial distribution of absorbed quanta and the probability that a quantum absorbed by chlorophyll LH will be transferred to PS I or to PS II. Likewise, as discussed in the previous paragraph, the operational use of $\varphi_{t(II\to1)}$ and $\varphi_{T(II\to1)}$ includes the spacial distribution of excitons transferred out of PS II reaction centers. Some uncertainty is introduced because the degree of energy coupling between chlorophyll LH and PS II cannot be defined rigorously. Even though the present description of the model includes some qualitative aspects, the overall framework provides a conceptual foundation which permits fluorescence, photochemistry and energy transfer to be considered in a more specific and rigorous manner than has been possible previously.

D. Appendix

I. $k_{T(II\rightarrow1)}$

The data from fluorescence induction curves at 692 and 730 nm measured on chloroplasts frozen to $-196°$ C in the absence and presence of 5 mM $MgCl_2$ (BUTLER and KITAJIMA, 1975c) will be used to demonstrate the calculation of energy distribution and energy transfer parameters within the context of the photochemical model presented in the text. At 692 nm in the absence of Mg^{2+} the fluorescence increased from a value of 28 (relative units on the chart paper) at F_0 to a value of 67 at F_M. In the presence of Mg^{2+} these values at 692 nm were 32 at F_0 and 114 at F_M. At 730 nm (on a different relative scale) F_0 and F_M were 76 and 97 respectively in the absence of Mg^{2+} and 60 and 83 in the presence of Mg^{2+}. Within the context of the model α, β and $\varphi_{T(II\rightarrow1)}$ can be calculated from these data.

The fluorescence induction curves measured at 730 nm in the absence and presence of Mg^{2+} are shown in Figure 1. The part of the F_0 level labeled A is the part α, [see Eq. (6)] due to the direct excitation of PS I fluorescence, the part labeled B is due to energy transfer from PS II: A is a relative measure of α, B is a relative measure of $\beta\varphi_{T(II\rightarrow1)(0)}$; B can be calculated from the relationship

$$\beta\varphi_{T(II\rightarrow1)(0)} = \frac{\varphi_{F\,692(0)}}{\varphi_{F\,692(V)}} \beta\varphi_{T(II\rightarrow1)(V)}$$

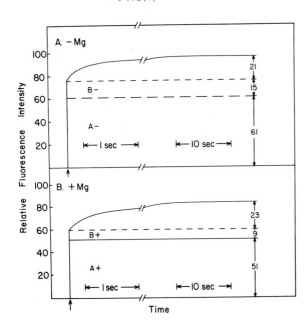

Fig. 1A and B. Fluorescence induction curves at 730 nm measured at $-196\,°C$ with broad band blue excitation, (A) in the absence of $MgCl_2$ and (B) in the presence of 5 mM $MgCl_2$. The F_0 level of fluorescence is analyzed in terms of A and B components as indicated in Appendix

where $\varphi_{F\,692\,(0)}/\varphi_{F\,692\,(V)}$ is obtained from the fluorescence induction curve at 692 nm and $\beta\,\varphi_{T(II\to1)(V)}$ is evaluated from the extent of F_V at 730 nm on the chart paper:

$$B^- = \frac{28}{39}\,21 = 15.0 \quad : \quad B^+ = \frac{32}{82}\,23 = 9.0$$

where $-$ and $+$ refer to the absence and presence of Mg^{2+}. Relative values of α are obtained by subtracting these values of B from the F_0 level of fluorescence:

$$A^- = 76 - 15 = 61 \quad : \quad A^+ = 60 - 9 = 51.$$

The model predicts that α should increase approximately 20 % in the absence of Mg^{2+}.

The influence of Mg^{2+} on the rate constant for energy transfer from PS II to PS I, $k_{T(II\to1)}$, can also be determined from these fluorescence data. It is assumed that a conformational change which brings PS I and PS II closer together will increase $k_{T(II\to1)}$ but that the other rate constants for the antenna chlorophyll of PS II, $k_{F\,692}$, k_{DII} and k_{TII}, will not be affected. Starting from the premise that F_V at 730 nm is due to energy transfer from PS II, it follows that the ratio of F_V at 730 nm to F_V at 692 nm should be a measure of the efficiency of that transfer. From Eqs. (2), (5), (6) and (7), it is apparent that:

$$\frac{\varphi_{F\,730\,(V)}}{\varphi_{F\,692\,(V)}} = \frac{k_{T(II\to1)}}{k_{F\,692}}.$$

If this ratio in the absence of Mg^{2+} is divided by the ratio in the presence of Mg^{2+}, we obtain

$$\frac{k_{T(II\to1)}^-}{k_{T(II\to1)}^+} = \frac{\varphi_{F\,730\,(V)}^-}{\varphi_{F\,692\,(V)}^-} \cdot \frac{\varphi_{F\,692\,(V)}^+}{\varphi_{F\,730\,(V)}^+} = \frac{F_{730\,(V)}^-}{F_{730\,(V)}^+} \cdot \frac{F_{692\,(V)}^+}{F_{692\,(V)}^-} = \frac{21}{23} \cdot \frac{82}{39} = 1.9.$$

According to these calculations $k_{T(II\to1)}$ should increase approximately 100 % in the absence of Mg^{2+}.

Absolute values of α, β, and $\varphi_{T(II\to1)(0)}$ in the presence and absence of Mg^{2+} can be determined from a set of six simultaneous equations, one being the ratio (R) of $k_{T(II\to1)}^-/k_{T(II\to1)}^+$, four being the relative values of α^-, α^+, $\beta^- \varphi_{T(II\to1)(0)}^-$ and $\beta^+ \varphi_{T(II\to1)(0)}^+$ and the sixth being the relationship that $\alpha^+ + \beta^+ = \alpha^- + \beta^- = 1$. The solution of these six simultaneous equations gives:

$$\varphi_{T(II\to1)(0)}^+ = \frac{RB^+ - B^-}{R(A^- - A^+) + (R-1)\,B^-}, \tag{9}$$

$$\alpha^+ = \frac{A^+}{A^+ + B^+/\varphi_{T(II\to1)(0)}^+} \quad : \quad \beta^+ = 1 - \alpha^+, \tag{10}$$

$$\alpha^- = \frac{A^-}{A^+ + B^+/\varphi_{T(II\to1)(0)}^+} \quad : \quad \beta^- = 1 - \alpha^-, \tag{11}$$

$$\varphi_{T(II\to1)(0)}^- = \frac{B^-}{A^+ + B^+/\varphi_{T(II\to1)(0)}^+ - A^-}. \tag{12}$$

The above data that $R = 1.9$, $A^- = 61$, $A^+ = 51$, $B^- = 15$ and $B^+ = 9$ gives the values in Table 1 with the additional observation that:

$$\varphi_{T(II \to 1)(M)} = \frac{\varphi_{F692(M)}}{\varphi_{F692(0)}} \varphi_{T(II \to 1)(0)}.$$

It was estimated previously (KITAJIMA and BUTLER, 1975a) (based on a natural fluorescence lifetime, τ_0, of 20 nsec and a fluorescence yield of 5% for the fluorescence at 692 nm at $-196°$ C) that $k_{F692} = 5 \cdot 10^7$ sec^{-1}, $k_{TII} = 80 \cdot 10^7$ sec^{-1} and $k_{DII} = 15 \cdot 10^7$ sec^{-1}. That value of k_{DII} included both nonradiative decay processes and energy transfer to PS I, $k_{T(II \to 1)}$. Based on those same assumptions we would now estimate that k_{DII} was about $8 \cdot 10^7$ sec^{-1} and that $k_{T(II \to 1)}$ was about $7 \cdot 10^7$ sec^{-1} in the presence of Mg^{2+} and $13 \cdot 10^7$ sec^{-1} in the absence of Mg^{2+}.

II. $k_{t(II \to 1)}$

The above treatment of the model does not account for the smaller extent of F_V at 692 nm found in the absence of Mg^{2+}. According to Eq. (4), F_V/F_M at 692 nm should be equal to $\varphi_{TII} \varphi_{tII}$. φ_{TII} will be slightly smaller (about 6%) in the absence of Mg^{2+} due to the larger value of $k_{T(II \to 1)}$ but not nearly enough to account for the observation that F_V/F_M at 692 nm is approximately 33% smaller in the absence of Mg^{2+}. Most of the quenching of F_V must be due to a decrease of φ_{tII} which was not considered in the above treatment. If there were a higher probability for energy transfer out of the closed PS II reaction centers to PS I in the absence of Mg^{2+}, such an energy transfer, $k_{t(II \to 1)}$, would be analogous to a nonradiative decay, k_{dII}, since the energy would be lost from PS II and would cause a specific quenching of F_V at 692 nm in the same manner as a k_{dII} process. This $k_{t(II \to 1)}$ type of transfer (as distinguished from $k_{T(II \to 1)}$ from the antenna chlorophyll of PS II to PS I) can be used to account for the quenching of F_V at 692 nm if certain assumptions are made.

We will assume that φ_{TII} is 0.80 and that any k_{dII} processes are negligible compared to k_{tII} and $k_{t(II \to 1)}$. Thus, at the closed reaction centers $\varphi_{tII} + \varphi_{t(II \to 1)} = 1$. From the relationship that $(F_V/F_M)_{692} = \varphi_{TII} \varphi_{tII}$, φ_{tII} and therefore $\varphi_{t(II \to 1)}$ can be calculated. From the fluorescence data in the presence of Mg^{2+}, $F_V/F_M = 0.72$ so that $\varphi_{tII}^+ = 0.9$ and $\varphi_{t(II \to 1)}^+ = 0.1$. In the absence of Mg^{2+}, $F_V/F_M = 0.58$, φ_{TII} decreased from 0.80 to 0.75, so that $\varphi_{tII}^- = 0.77$ and $\varphi_{t(II \to 1)}^- = 0.23$. From these values of $\varphi_{t(II \to 1)}$ it is apparent that $k_{t(II \to 1)}^-/k_{t(II \to 1)}^+ = 2.7$.

According to such a model, F_V at 730 nm can be divided into two components; one due to energy transfer from the closed PS II reaction centers, $\beta \varphi_{TII} \varphi_{t(II \to 1)} (1 - A_{II})$, and one due to the variable component of the energy transfer from the antenna chlorophyll of PS II, $\beta \varphi_{T(II \to 1)(V)}$. The latter component can be expressed as:

$$\beta \varphi_{T(II \to 1)(V)} = \beta \varphi_{T(II \to 1)(0)} \frac{\varphi_{TII} \varphi_{tII} (1 - A_{II})}{1 - \varphi_{TII} \varphi_{tII}}$$

where $\varphi_{TII} \varphi_{tII} = (F_V/F_M)_{692}$. These two types of energy transfer are indicated in Eq. (8) and diagrammatically in Figure 2. The relative extents of these two components of F_V at 730 nm can be estimated at $A_{II} = 0$ in the absence and presence of Mg^{2+}.

Fig. 2A and B. Same fluorescence induction curves as shown in Figure 1 but analyzed in terms of two sources of F_V and two sources of F_0 as indicated in Appendix

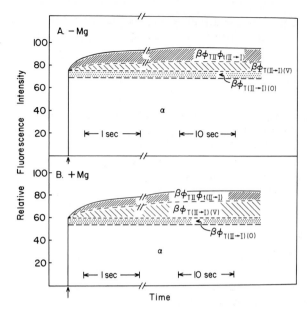

In the absence of Mg^{2+}:

$$\beta^-\,\varphi^-_{T\,II}\varphi^-_{t(II\to 1)}=\beta^-(0.75)\,(0.23)=0.17\beta^-$$

$$\beta^-\,\varphi^-_{T(II\to 1)(V)}=\beta^-(0.116)\,(0.58/0.42)=0.16\beta^-$$

where 0.116 is the value of $\varphi^-_{T(II\to 1)(0)}$ determined previously. The extent of F_V due to $\varphi^-_{T(II\to 1)(V)}$ is 0.16/0.33) 21 = 10.2 and relative value of $\beta^-\,\varphi^-_{T(II\to 1)(0)}$ will be:

$$B^-=\frac{28}{39}\cdot 10.2=7.3$$

and:

$$A^-=76-7.3=68.7.$$

In the presence of Mg^{2+}:

$$\beta^+\,\varphi^+_{T\,II}\varphi^+_{t(II\to 1)}=\beta^+(0.80)\,(0.10)=0.08\beta^+$$

$$\beta^+\,\varphi^+_{T(II\to 1)(V)}=\beta^+(0.065)\,(0.72/0.28)=0.17\beta^+$$

$$B^+=(32/82)\,(0.17/0.25)\,(23)=6.1$$

$$A^+=60-6.1=53.9$$

and

$$R=\frac{k^-_{T(II\to 1)}}{k^+_{T(II\to 1)}}=\frac{16\cdot 21\cdot 25\cdot 82}{33\cdot 39\cdot 17\cdot 23}=1.37.$$

From these new values of R, A^-, A^+, B^- and B^+

$$\varphi^+_{T(II\to1)(0)}=0.046 \quad : \quad \varphi^-_{T(II\to1)(0)}=0.062$$

$$\alpha^+=0.29 \qquad : \quad \alpha^-=0.37$$

$$\beta^+=0.71 \qquad : \quad \beta^-=0.63$$

The calculations of $\beta\varphi_{T(II\to1)(V)}$ used the previously calculated values of $\varphi_{T(II\to1)(0)}$ (0.116 in the absence of Mg^{2+} and 0.065 in the presence of Mg^{2+}). If we recalculate the values of α, β and $\varphi_{T(II\to1)(0)}$ using the newly calculated values of 0.045 and 0.057 for $\varphi_{T(II\to1)(0)}$ we obtain the values given in Table 2. For these latter calculations $R=1.08$, $B^-=5.1$, $B^+=5.4$, $A^-=70.9$, $A^+=54.6$.

References

Bearden, A.J., Malkin, R.: Quart. Rev. Biophys. **7**, 131–177 (1975)

Boardman, N.K., Björkman, O., Anderson, J.M., Goodchild, D.J., Thorne, S.W.: In: Proc. 3rd Intern. Congr. Photosynthesis. Avron, M. (ed.). Amsterdam: Elsevier, 1975, pp. 1809–1827

Butler, W.L.: Biochim. Biophys. Acta **64**, 309–317 (1962)

Butler, W.L.: FEBS Lett. **20**, 333–338 (1972a)

Butler, W.L.: Proc. Nat. Acad. Sci. U.S. **69**, 3420–3422 (1972b)

Butler, W.L.: Acc. Chem. Res. **6**, 177–184 (1973)

Butler, W.L., Kitajima, M.: Biochim. Biophys. Acta **376**, 116–125 (1975a)

Butler, W.L., Kitajima, M.: In: Proc. 3rd Intern. Congr. Photosynthesis. Avron, M. (ed.). Amsterdam: Elsevier, 1975b, pp. 13–24

Butler, W.L., Kitajima, M.: Biochim. Biophys. Acta **396**, 72–85 (1975c)

Butler, W.L., Okayama, S.: Biochim. Biophys. Acta **245**, 231–239 (1971)

Butler, W.L., Visser, J.W.M., Simons, H.L.: Biochim. Biophys. Acta **292**, 140–151 (1973a)

Butler, W.L., Visser, J.W.M., Simons, H.L.: Biochim. Biophys. Acta **325**, 539–545 (1973b)

Clayton, R.K.: Biophys. J. **9**, 60–76 (1969)

Den Haan, G.A., Duysens, L.N.M., Egberts, D.J.N.: Biochim. Biophys. Acta **368**, 409–421 (1974)

Den Haan, G.A., Warden, J.T., Duysens, L.N.M.: Biochim. Biophys. Acta **325**, 120–125 (1973)

Duysens, L.N.M.: Ph.D. Thesis, Univ. Utrecht, 1952

Duysens, L.N.M., Amesz, J., Kamp, B.M.: Nature (London) **190**, 510–511 (1961)

Duysens, L.N.M., Sweers, H.E.: In: Studies on Microalgae and Photosynthetic Bacteria. Tokyo: Univ. Tokyo, 1963, pp. 353–377

Emerson, R., Chalmers, R.V., Cederstrand, C.: Proc. Nat. Acad. Sci. U.S. **43**, 133–143 (1957)

Emerson, R., Rabinowitch, E.: Plant. Physiol. **35**, 477–485 (1960)

Epel, B.L., Butler, W.L.: Biophys. J. **12**, 922–929 (1972)

Erixon, K., Butler, W.L.: Photochem. Photobiol. **14**, 427–434 (1971a)

Erixon, K., Butler, W.L.: Biochim. Biophys. Acta **234**, 381–389 (1971b)

Govindjee, Ichimuri, S., Cederstrand, C., Rabinowitch, E.: Arch. Biochem. Biophys. **89**, 322–323 (1960)

Hill, R., Bendall, F.: Nature (London) **186**, 136–137 (1960)

Hofer, I., Strasser, R.J., Sironval, C.: In: Proc. 3rd Intern. Congr. Photosynthesis. Avron, M. (ed.). Amsterdam: Elsevier, 1975, pp. 1685–1690

Kamen, M.: Primary Processes in Photosynthesis. New York: Academic Press, 1963

Kautsky, H., Appel, W., Amann, N.: Biochem. Z. **332**, 277–292 (1960)

Kautsky, H., Hirsch, A.: Naturwissenschaften **19**, 964 (1931)

Kautsky, H., Zedlitz, W.: Naturwissenschaften **29**, 101–102 (1941)

Kitajima, M., Butler, W.L.: Biochim. Biophys. Acta **376**, 105–115 (1975a)

Kitajima, M., Butler, W.L.: Biochim. Biophys. Acta **408**, 297–305 (1975b)

Knaff, D.B., Arnon, D.I.: Proc. Nat. Acad. Sci. U.S. **63**, 956–962 (1969a)
Knaff, D.B., Arnon, D.I.: Proc. Nat. Acad. Sci. U.S. **63**, 963–969 (1969b)
Knaff, D.B., Arnon, D.I.: Proc. Nat. Acad. Sci. U.S. **64**, 715–722 (1969c)
Kok, B.: Nat. Acad. Sci. **1145**, 45–55 (1963)
Kok, B., Hoch, G.: In: Light and Life. McElroy, W.D., Glass, B. (eds.). Baltimore: Johns Hopkins, 1961, pp. 397–423
Lavorel, J., Joliot, P.: Biophys. J. **12**, 815–831 (1972)
Lozier, R.H., Butler, W.L.: FEBS Lett. **26**, 161–164 (1972)
Lozier, R.H., Butler, W.L.: Biochim. Biophys. Acta **333**, 465–480 (1974a)
Lozier, R.H., Butler, W.L.: Biochim. Biophys. Acta **333**, 460–464 (1974b)
Malkin, R., Bearden, A.J.: Biochim. Biophys. Acta **396**, 250–259 (1975)
Mathis, P., Vermeglio, A.: Biochim. Biophys. Acta **368**, 130–134 (1974)
Mathis, P., Vermeglio, A.: Biochim. Biophys. Acta **369**, 371–381 (1975)
Mauzerall, D.: Proc. Nat. Acad. Sci. U.S. **69**, 1358–1362 (1972)
McAlister, E.D., Myers, J.: Smithsonian Inst. Misc. Collections **99** (6), 1–37 (1940)
Murata, N.: Biochim. Biophys. Acta **189**, 171–181 (1969)
Murata, N., Itoh, S., Okada, M.: Biochim. Biophys. Acta **325**, 463–471 (1973)
Murata, N., Sugahara, K.: Biochim. Biophys. Acta **189**, 182–192 (1969)
Okayama, S., Butler, W.L.: Plant. Physiol. **49**, 769–774 (1972a)
Okayama, S., Butler, W.L.: Biochim. Biophys. Acta **267**, 523–529 (1972b)
Okayama, S., Epel, B.L., Erixon, K., Lozier, R., Butler, W.L.: Biochim. Biophys. Acta **253**, 476–482 (1971)
Thornber, J.P., Highkin, H.R.: Europ. J. Biochem. **41**, 109–116 (1974)
Vermeglio, A., Mathis, P.: Biochim. Biophys. Acta **292**, 763–771 (1973)
Witt, H.T., Müller, A., Rumberg, B.: Nature (London) **191**, 194–195 (1961)
Wraight, C.A., Crofts, A.R.: Europ. J. Biochem. **17**, 319–327 (1970)

6. Electron Paramagnetic Resonance Spectroscopy

E.C. Weaver and G.A. Corker

A. Introduction

Electron paramagnetic resonance (EPR, or ESR for electron spin resonance) spectroscopy is based on the fact that substances containing unpaired electrons are paramagnetic. This paramagnetism may be due to the presence of transition elements which have unfilled shells, or it may be due to the transitory presence of oxidized or reduced substances. In the case of photosynthetic materials, this oxidation and reduction (i.e., the transfer of an electron from one substance to another) is set in motion by light. The basic observation, made in the mid-1950s (Commoner et al., 1956), was that photosynthetic materials become paramagnetic when illuminated. Two prominent light-induced resonances in plants and a single one in bacteria were described in early papers, and numerous speculations on their origin and significance advanced. It is now generally accepted that the light-induced EPR signal at $g = 2.002$ (signal I) in plants is a direct measure of the oxidation state of the photosystem I reaction center, P-700. However, between the time when the signal was first described and a sense of certainty as to its significance, some 15 years elapsed. We now know that light-induced resonances in photosynthetic organisms or subcellular preparations of them are indeed probes into events essential to the overall process of photosynthesis.

There are several intrinsic advantages of the EPR method for the study of photosynthesis. The detecting radiation of the spectrometer is in the microwave region of the spectrum. Most commercial spectrometers employ a frequency of about 9.5 GHz, which is a wavelength of about 3 cm. This is less energetic than visible light by a factor of almost 10^5 and is probably far too weak to induce any transitions in living matter. If high microwave powers are used heating can result, but except for such thermal effects there is no known and specific effect on tissue. This means that the photosynthetic system is not disturbed by the means of observation.

A second advantage of the EPR method lies in the fact that only molecules undergoing oxidation-reduction reactions in response to light are detected. The great bulk of chlorophyll, more than 99%, serves merely to intercept photons. These molecules do not participate in chemical reactions under normal photosynthetic conditions, and hence are "invisible" by EPR spectroscopy. Again, the situation may be contrasted to that of spectroscopy in the visible region, where the absorption of bulk pigments represents an enormous background against which absorption changes by a small percentage of molecules must be detected.

Water itself is a strong absorber of microwave energy. However, the sensitivity of the modern spectrometer is such that this attenuation can be tolerated and whole cells, chloroplasts and subcellular preparations can be observed under physiologically favorable conditions.

This chapter will endeavor to give the reader a statement on the basic phenomenon, some explanation of instrumental design, an overview of current knowledge, and a glimpse of possible future developments. It will make no attempt to be exhaustive, but rather selective. Reviews have been published by KOHL (1972), WEAVER (1968a), and WEAVER and WEAVER (1972).

B. EPR Techniques

We will deal in this section only with the detection and analysis of very dilute paramagnetism in a diamagnetic matrix.

For a system to exhibit paramagnetism there must be present unpaired electrons whose recombination rates are slow enough to be observable. Reversible effects are the rule for experiments carried out in photosynthetic systems at physiological temperatures. In order to detect the presence of light-induced paramagnetism the sample is placed in a uniform magnetic field generated by an iron core electromagnet. The function of this polarizing magnetic field is to define a direction in space along which the spins will align themselves. This alignment results in an energy difference depending upon whether the ensemble of "spins" (electron magnetic moments) are, on the average, oriented parallel or antiparallel to this field. Starting with such a system in thermal equilibrium a change in the average polarization of this ensemble will require the absorption of energy from an externally applied perturbing force. The energy thus absorbed can then be detected and forms the basis of the EPR signal. If the perturbing force exceeds the ability of this spin system to give up its excess energy to the environment, saturation of the spin system will result.

Each of the unpaired electrons will have only two allowed quantum states in a magnetic field: one where a component of magnetization is parallel and one where a component is antiparallel to the field. However, a very small preference for the lower energy state will prevail and this small difference will yield a net magnetization for the sample. This net magnetization will then behave in a manner analogous to a magnetized spinning top whose angular momentum vector will precess about the vector of the magnetic field. This Zeeman precessional frequency can be determined from a measurement of the applied field by the following simple linear relationship:

$$h\nu_0 = g\beta H_0, \tag{1}$$

or

$$\nu_0 = 2.8 \cdot 10^6 H_0 (\text{Hz}) \tag{2}$$

where h = Planck's constant ($6.6 \cdot 10^{-27}$ erg·sec) and ν_0 = klystron frequency at resonance; H_0 is measured in gauss (G) and g is equal to 2.00229 for a free electron possessing no orbital angular momentum, β is the Bohr magneton ($0.9273 \cdot 10^{-20}$ ergs/G). A magnetic field of about 3,400 G results in a Zeeman precessional frequency of the order of $9 \cdot 10^9$ Hz. This frequency is in the microwave region and requires special generators (klystrons or solid state devices), conductors (wave guides), tuned circuits (cavities) and detectors (crystals) as spectrometer components. Various arrangements of these components form the basis for commercial spectrometers suitable for detection of the presence of light-induced paramagnetism in photosynthetic systems.

Apart from the electromagnet and its associated stabilized power supply, the microwave cavity, which provides the coupling to the sample, is a key element in a spectrometer of high sensitivity. It is in the cavity that good coupling of the microwave *magnetic* field to the sample is accomplished, a vital factor in achieving an adequate signal-to-noise (S/N) ratio, which is necessary for analysis of EPR spectra. The cavity defines a three-dimensional volume in which the microwaves are "standing", i.e., stationary in space but not in time. The sample must be placed with nearly optimal coupling in the region where the magnetic portion of the microwave field has a spatial maximum and the electric portion has a spatial minimum. At resonance, the magnetic rf field, H_1, can induce the unpaired electrons of the sample to change their orientation with respect to the dc magnetic field, H_0. In so doing, they absorb a small portion of the microwave energy stored in the cavity and this energy loss is detected, amplified and displayed for analysis.

Most biological samples are contained in water. The polar water molecules, in the liquid state, absorb very strongly over a broad range of frequencies in the microwave region of the electromagnetic spectrum via a coupling between the *electric* component of the microwave field and the electric dipole moment of the molecule. This absorption is orders of magnitude stronger than the magnetic interaction of the isolated electrons which we are trying to detect. Thus, the geometric shape of sample holders must be commensurate with the geometry of the cavities which are employed in order to reduce this water absorption to a manageable magnitude.

For persons familiar with optical spectroscopy, the appearance of the EPR absorption signal as displayed in a recorder with positive and negative excursions may at first seem contradictory (Fig. 1). This mode of presentation is a consequence of phase-sensitive detection of a sine wave modulation (usually at 100 kHz frequency) superimposed upon the dc polarizing field (H_0) as it is swept slowly through the signal region, and may be thought of as an approximation of a first derivative display, i.e., the slope of the absorption vs H_0 or the g value.

For a spectrometer equipped with field modulation and phase detection the signal amplitude measured by the net difference between plus and minus excursions on the recorder will vary according to the choice of field modulation amplitude. Furthermore, the separation in gauss between maximum and minimum points of the recorded signal will appear to separate with increasing values of modulation amplitude at values large compared to the natural line width of the resonance.

In order to perform electron paramagnetic resonance experiments on light-induced resonances in photosynthetic materials, one must use techniques which take into account the fact that these materials are aqueous suspensions of necessarily complex systems: algae, chloroplasts, subchloroplast fractions, bacteria, chromatophores and reaction center preparations.

By definition, the appearance of a light-induced signal tells us that the absorption of photons has resulted in the unpairing of previously paired electrons. How-

Fig. 1. Traces of EPR signals in wild type *Scenedesmus*, and in mutant Sc-8 (isolated by N.J. Bishop) which lacks P-700. The temperature is ambient, chlorophyll concentrations equal, and all instrumental settings the same. (From Giese, 1972.) (Fig. 7, from Giese, Photophysiology. New York-London: Academic Press, 1972, Vol. VII, Chap. 1)

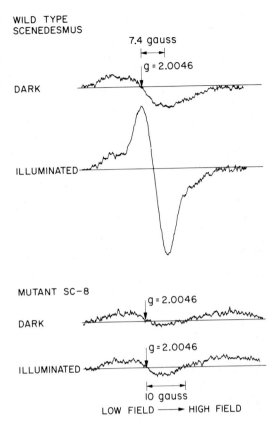

ever, there can be considerably more information gained from the position, or "*g*-value", of the resonance, its shape, intensity, temperature dependence, the kinetics of its formation and decay, and its response to exciting light of varying wavelength, duration and intensity. Computer techniques for signal-to-noise (S/N) enhancement enable the researcher to extract information otherwise lost in random noise.

The *g*-value corresponds in a qualitative way to the wavelength at which absorption peaks occur in optical spectroscopy. It is defined by the ratio:

$$g = \frac{h\nu}{\beta H_0}. \tag{3}$$

For the purposes of most experimenters, a comparison of the unknown resonance with a suitable known one can give a satisfactory estimate of the *g*-value. A given *g*-value is not necessarily unique to a single molecule, and by itself does not provide proof of the molecular identity of an unknown radical.

The light-induced resonances at physiological temperature in photosynthetic material are very nearly gaussian in form, which indicates that the unpaired electron is influenced by a statistically large number of weak magnetic interactions, stemming usually from neighboring protons. Hyperfine structure indicates that a relatively small number of magnetic nuclei are influencing the electrons.

The intensity of an EPR signal is proportional to the area under the absorption curve. The most reliable numerical method of intensity measurement on the typical signal is that of the first moment calculation on the "first derivative" signal as displayed. This method allows determination of the intensity of an unknown resonance by comparing with the intensity of a known standard, measured under precisely the same conditions even though the two may have different line shapes and structure. The only restriction is that the signals to be compared be well below being saturated with rf power. Saturation occurs at far lower powers when low temperatures are employed.

The time course of the formation and disappearance of a light-induced signal may be determined by setting the field at the point of maximum deflection; using a time-based recording system, the light may be pulsed.

It is possible to provide a controlled temperature enclosure for the sample while keeping the microwave cavity at room temperature by employing a quartz dewar insert, through which flows a stream of nitrogen or helium gas which is heated or cooled to the desired point.

C. EPR Studies in Photosynthesis

I. Bacterial Photosynthesis

When suspensions of whole cells, chromatophores or reaction centers of photosynthetic bacteria are illuminated within the cavity of an EPR spectrometer, a single resonance with a g-value of 2.0025 and a line width of 9.5 G appears. The action spectrum of the signal production in chromatophores of *Rhodospirillum rubrum* shows broad maxima coincident with the absorption peaks of the bacteriochlorophyll pigment at 800 and 880 nm (Androes et al., 1962). The fact that the signal could be produced at low temperature suggests that it is a product of the initial photochemistry.

Photo-induced optical density changes have been observed and characterized as arising from the oxidation of the photoactive bacteriochlorophyll in the reaction centers, P-870. Comparisons of the kinetics of the EPR and optical density changes at room temperature show that the decay behavior of both these physical changes is identical, suggesting that the g=2.0025 signal is due to the oxidized form of P-870. Results which provide support for this conclusion are: (1) the 1/1 correspondence between the amounts of EPR signal and the amount of oxidized P-870 (Bolton et al., 1969; Loach and Walsh, 1969); the redox potential of P-870 and the species giving rise to the EPR signal are the same (between +450 and +490 mV in most species) (Kuntz et al., 1964), (2) the quantum yield for both is one, (3) in mutants lacking reaction centers, the EPR signal is also lacking (Sistrom and Clayton, 1964).

Although the g-values of isolated bacteriochlorophyll cation radical and of oxidized P-870 radical agree, the former exhibits a line width of 12.8 G and the latter, 9.5 G. Norris et al. (1971) have postulated that the in vivo signal results from a spin which is delocalized over a pair of chlorophyll molecules. They envisage reaction centers of bacteria or plants as a (bacterio) chlorophyll-water-(bacterio) chlorophyll sandwich, not simply as a dimer of (bacterio) chlorophyll.

The experiments which address the question of the identity of the primary oxidant yield seemingly contradictory results. At present, there are two possible candidates: (1) an iron compound which upon reduction in the primary reaction exhibits EPR signals with g-values of 2.00, 1.86, and 1.68, or (2) ubiquinone which upon reduction to the semiquinone exhibits a signal with a g-value of 2.0047 and also influences the spin state of an associated ferrous ion.

The electron transport system involves thermal reactions which must be investigated at ambient temperatures and with the duration of the light exposure sufficient to cause a detectable electron flow through reactions which may be several steps removed from the primary reaction. The components of these reactions are metal-ion-containing proteins whose magnetic relaxation rates render the direct detection of their EPR spectra impossible at room temperatures. Thus, one must infer from the behavior of the P-870 signal what is occurring in the electron transport mechanism. The kinetics observed in the chromatophores are relatively simple, involving a monotonic rise to a steady state during illumination and a biphasic decay when the illumination terminated. The kinetics of the whole-cell signal are complex and show a pronounced dependence upon the light-dark history of the cells, their environment, and their structural integrity. The same kinds of differences are found in plants; signal I in intact algae and fresh chloroplasts exhibits complex changes in amplitude, reflecting electron flow in various parts of the photosynthetic apparatus. By subjecting suspensions of whole cell bacteria at room temperature to different periods of time in the dark or in the light followed by cooling, and then investigating the EPR spectra of these suspensions at temperatures between 6 and 77 K, CORKER and SHARPE (1975) showed that eleven of the EPR-detectable components were influenced by room temperature illumination of the organisms. The g-values of these signals were indicative of iron-heme proteins, iron-sulfur proteins and a free radical distinct from P-870$^+$.

II. Signals in Photosystem II (PS II)

1. Signal II

There were two light-induced signals described in plants in the earliest papers, and those descriptions remain valid. Their characteristics are summarized in Table 1 (see also Fig. 1). The association of signal II with photosystem (PS) II was demonstrated by the following observations: (1) Omitting manganese from the growth medium inhibited oxygen evolution and also eliminated the signal. (2) Mutants lacking PS II activity lacked signal II (BISHOP, 1964). (3) Photosynthetic bacteria, which do not evolve oxygen, also lack signal II.

The fact that signal II has a partially resolved hyperfine structure, indicating some localization of the unpaired electron, and a g-value of 2.0043, typical of quinones, strongly suggests a semiquinone of plastoquinone (or similar molecules) as the observed free radical. However, despite a large amount of work on the problem of the identity of the signal there has not been definitive proof. No optical signal with the kinetic characteristics of signal II has been described. The

Table 1

	Signal I	Signal II
g-value	2.0025	2.0046
Width (G)	7.2	20
Hyperfine structure	none	partially resolved
Microwave saturation (ambient temp.)	40 mW	20 mW[a]
Molecular species	P-700$^+$	Plastoquinone (?) (semiquinone)
Wavelength to excite	660–710 nm	<680 nm

[a] Babcock and Sauer (1975)

optical evidence linking C-550 (the PS II acceptor) to plastoquinone has not yet tied the absorbance changes to an EPR signal (van Gorkom, 1974). However, two groups have provided convincing evidence that signal II arises on the oxidizing side of PS II, and may be closely associated with the donor to P-680, the PS II reaction center (Babcock and Sauer, 1975; Lozier and Butler, 1973).

2. P-680 Signal

Recently, a signal has been described which has the same g value (2.0026) and width (ca. 8 G) as signal I, but which can be clearly attributed to the oxidized reaction center (P-680) of PS II. Its detection is based on a technique which involves oxidizing all the P-700 with $K_3Fe(CN)_6$, and then using red light to produce an additional signal at 77 K. The conclusion that this signal, which has a midpoint potential of ± 475 mV, originates in the reaction center chlorophyll of PS II (Bearden and Malkin, 1973; van Gorkom et al., 1974) is not acceptable to all (Lozier and Butler, 1974b). Light increases the ferricyanide-induced signal at both ambient (Weaver, 1968a) and low temperatures, and may indicate a heterogeneity of reaction centers.

III. Signals in Photosystem I (PS I)

1. Identification of Signal I

The first evidence that signal I might be identical to P-700$^+$ came in 1962 (Beinert et al., 1962) and was based on similar behavior of the light-induced EPR signal and absorbance changes at 700 nm. Additional evidence may be summarized as follows:

1. A mutant of *Scenedesmus*, with abundant chlorophylls a and b lacked both P-700 and signal I (Weaver and Bishop, 1963, and Fig. 1). Treatment which restored functioning P-700 to this mutant also restored the signal (Gee et al., 1969).

2. A concurrence of presence and behavior of both P-700 and signal I occurred in many studies involving physical separation of the photosynthetic apparatus; their midpoint potentials were the same ca. +0.520 V).

3. A large cross-section for spin production was demonstrated, together with a quantum efficiency of unity; this proved that the signal was intrinsic to photosynthesis, and was not an ancillary process (Weaver and Weaver, 1969).

4. The number of spins and concurrence of P-700$^+$ whether produced by an oxidizing agent or by light, was equal (BAKER and WEAVER, 1973).

5. Kinetic behavior of P-700 and signal was identical (WARDEN and BOLTON, 1972).

This body of evidence constitutes virtual proof that signal I is a manifestation of oxidized P-700. At all but the lowest temperatures, the signal intensity is due to a dynamic steady state balance of oxidation and reduction. Thus, the decay of the signal in the absence of light provides a direct and sensitive measure of the rate of reduction of the reaction centers as electrons flow to them via the electron transport chain, via cyclic electron flow, or through direct recombination of the primary acceptor and P-700$^+$.

2. Primary Acceptor of Photosystem I

In 1971, a group of light-induced EPR signals with g values of 2.05, 1.94, and 1.86, characteristic of a reduced iron-sulfur protein, was described in spinach chloroplasts which had been freed of soluble ferredoxin (MALKIN and BEARDEN, 1971; see also chapter II.7). This observation has been confirmed by others (EVANS et al., 1972; KE et al., 1974; VISSER et al., 1974b).

This bound ferredoxin appears to be identical with the iron-sulfur protein absorbing at 430 nm (P-430) which had first been described in 1971 (HIYAMA and KE, 1971) and subsequently confirmed in several laboratories; it had the attributes required to function as the primary acceptor of PS I, including a midpoint potential of -0.53 V (LOZIER and BUTLER, 1974a).

This reduced bound ferredoxin can only be observed at low temperatures (<25 K) for reasons stated above and since it recombines with P-700 in a back reaction at higher temperatures; nevertheless a satisfactory correlation of the decay kinetics of P-700$^+$ and the bound ferredoxin by spectrophotometric and EPR methods has been made (KE et al., 1974; VISSER et al., 1974b). In addition, there are equal numbers of spins due to P-700 and the new substance (BEARDEN and MALKIN, 1972).

It appears from the foregoing that the free radicals of four of the primary reactions in plant PS have been identified and characterized. There must be, in addition, several components of electron transport chains associated with the primary reactants which are also free radicals when the plant is functioning. However, none of these (except signal II) is normally seen at physiological temperatures.

There is a signal at $g=2.05$ that can be detected at 20 K in intact algae or chloroplasts, which corresponds to oxidized plastocyanin. It has been localized between the two photosystems (VISSER et al., 1974a). The membrane-bound plastocyanin midpoint potential is $+320$ mV, when measured in situ by EPR titration (KNAFF and MALKIN, 1973).

Another paramagnetic species involved in electron transport is Mn^{2+}. However, when manganese is bound, the peaks are so broadened as to be undetectable. Damaged chloroplasts release manganese to the medium, where it is easily detectable. Chloroplasts which are washed with the buffer Tris lose their ability to evolve oxygen; EPR studies indicate that manganese is released into the inner space of the thylakoid membrane (BLANKENSHIP and SAUER, 1974). This corresponds with the site of oxygen evolution.

3. Signal I as a Probe into Electron Transport

In whole cells at physiological temperatures the rate of rise of the signal I or the bacteriochlorophyll signal is the sum of the rates of oxidation and decay; it is frequently complex, especially in photosynthetically competent systems. In whole bacterial or algal cells, or in intact chloroplasts, the rate of signal formation is very slow following a prolonged dark period and can take minutes to reach a maximum. This delay is termed an "induction effect" and is characteristic of many photosynthetic phenomena. Fractions of systems have simpler rise and decay kinetics.

Given the fact that the light-induced signal indicates the loss of an electron from the PS I reaction center, one can predict that any factor which speeds the restoration of an electron to the reaction centers will make the signal smaller. In intact cells, the signal produced by long-wavelength light is reduced in amplitude when a beam of shorter-wavelength light, exciting PS II, is superimposed, because electrons are being driven from PS II to PS I. Kinetic analysis of electron flow through PS I is possible under varying light intensities, wavelengths, periodicity and at the range of temperatures in which the plant will function.

The behavior of signal I has also given independent evidence on the PS I side of the plant system for slow changes in the distribution of quanta which were inferred from the observations of others from fluorescence studies, fluorescence being a PS II parameter. The rate and extent of rise of the signal with PS I light and the rate of its decay are greater during or after an exposure of one minute or more to PS II light (Weaver, 1973).

IV. Spin Labels

Most of this chapter is concerned with light-induced paramagnetism of molecules that occur naturally in the photosynthetic organism. However, one can add a paramagnetic molecule called a "spin label" (Hamilton and McConnell, 1968) to the photosynthetic system. The molecules which have been used as spin labels are nitroxide radicals, stabilized by being surrounded by methyl groups. This fundamental radical can be extended to provide points of attachment to a great variety of groupings as needed.

Information can be gained from them because changes in their intimate environment can be reflected by changes in the EPR spectrum. The g-value and hyperfine interactions may be sensitive to the degree to which the environment is polar or ionic. If the EPR parameters are anisotropic one may determine to what extent the biomolecular structure is ordered, and the orientation of the spin label in respect to time. If the spin label is not stationary, the anisotropy allows one, under certain circumstances, to gain insight into the time dependence of the motion of the label. If the motion is sterically hindered by the structure of the molecule, a change in conformation of the biomolecule may be detectable. If the spin label is rigidly attached to the biomolecule, then the motion of that molecule may be studied.

Since light induces structural changes in the chloroplast, demonstrated by scattering and volume changes and illustrated with electron micrographs of chloroplasts fixed in the light vs those fixed in the dark, spin labels should provide a precise probe into events in the thylakoid. Study of the ionic flux accompanying photophosphorylation should be amenable to spin labeling techniques.

There are reports of change of the nitroxide radical with wavelength and with light-dark regimes, with time constants similar to those reported for scattering changes (WEAVER, 1973).

The fact that photosynthetic electron flow can reduce the nitroxide radical, thus abolishing its paramagnetism, makes it serve simultaneously as an internal indicator of electron flow.

D. Conclusion

Several EPR signals can safely be equated with optical ones, providing independent data and, in some instances, unique information. The power of EPR spectroscopy to measure directly free radicals produced by light without disturbing the organism has given those who study photosynthesis an important tool for exploration of the reactions. Several metal-containing enzymes can be observed in situ.

The details of the slower changes a functioning chloroplast undergoes in response to its light and temperature environment will also emerge by the application of EPR spectroscopy in combination with other techniques.

References

Androes, G.M., Singleton, M.F., Calvin, M.: Proc. Nat. Acad. Sci. U.S. **48**, 1022–1033 (1962)
Babcock, G.T., Sauer, K.: Biochim. Biophys. Acta **376**, 329–344 (1975)
Baker, R., Weaver, E.C.: Photochem. Photobiol. **18**, 237–241 (1973)
Bearden, A.J., Malkin, R.: Biochim. Biophys. Acta **283**, 456–468 (1972)
Bearden, A.J., Malkin, R.: Biochim. Biophys. Acta **325**, 266–274 (1973)
Beinert, H., Kok, B., Hoch, G.E.: Biochem. Biophys. Res. Commun. **7**, 209–212 (1962)
Bishop, N.F.: Record Chem. Prog. **25**, 181–195 (1964)
Blankenship, R.E., Sauer, K.: Biochim. Biophys. Acta **357**, 252–266 (1974)
Bolton, J.R., Clayton, R.K., Reed, D.W.: Photochem. Photobiol. **9**, 209–218 (1969)
Commoner, B., Heise, J.J., Townsend, J.: Proc. Nat. Acad. Sci. U.S. **42**, 710 (1956)
Corker, G.A., Sharpe, S.A.: Photochem. Photobiol. **21**, 49–61 (1975)
Evans, M.C.W., Telfer, A., Lord, A.V.: Biochim. Biophys. Acta **267**, 530 (1972)
Gee, R., Saltman, P., Weaver, E.C.: Biochim. Biophys. Acta **189**, 106–115 (1969)
Gorkom, H.J. van: Biochim. Biophys. Acta **347**, 439–442 (1974)
Gorkom, H.J. van, Tamminga, J.J., Haveman, J., van der Linder, E.K.: Biochim. Biophys. Acta **347**, 417–438 (1974)
Hamilton, C.L., McConnell, H.W.: In: Structural Chemistry and Molecular Biology. Rich, A., Davidson, N. (eds.). San Francisco: W.H. Freeman, 1968
Hiyama, T., Ke, B.: Proc. Nat. Acad. Sci. U.S. **68**, 1010–1013 (1971)
Ke, B., Sugahara, K., Shaw, E.R., Hansen, R.E., Hamilton, W.D., Beinert, H.: Biochim. Biophys. Acta **368**, 401–408 (1974)
Knaff, D.B., Malkin, R.: Arch. Biochem. Biophys. **159**, 556–562 (1973)
Kohl, D.: In: Biological Applications of Electron Spin Resonance. Swartz, H.M., Bolton, J.R., Borg, D.C. (eds.). New York-London-Sydney-Toronto: Wiley, 1972, pp. 213–264

Kuntz, I.D., Loach, P.A., Calvin, M.: Biophys. J. **4**, 227–249 (1964)
Loach, P.A., Walsh, K.: Biochemistry **8**, 1909–1913 (1969)
Lozier, R.H., Butler, W.: Photochem. Photobiol. **17**, 133–137 (1973)
Lozier, R.H., Butler, W.L.: Biochim. Biophys. Acta **325**, 460–464 (1974a)
Lozier, R.H., Butler, W.L.: Biochim. Biophys. Acta **333**, 465–480 (1974b)
Malkin, R., Bearden, A.J.: Proc. Nat. Acad. Sci. U.S. **68**, 16–19 (1971)
Norris, J.R., Uphaus, R.A., Crespi, H.L., Katz, J.J.: Proc. Nat. Acad. Sci. U.S. **68**, 625–628 (1971)
Sistrom, W.R., Clayton, R.K.: Biochim. Biophys. **88**, 61–73 (1964)
Visser, J.W.M., Amesz, J., Van Gelder, B.F.: Biochim. Biophys. Acta **333**, 279–287 (1974a)
Visser, J.W.M., Rijgersberg, K.P., Amesz, J.: Biochim. Biophys. Acta **368**, 235–246 (1974b)
Warden, J.T., Bolton, J.R.: J. Am. Chem. Soc. **94**, 4351 (1972)
Weaver, E.C.: Biochim. Biophys. Acta **162**, 286–289 (1968a)
Weaver, E.C.: In: Proc. 4th Intern. Biophys. Congr., Moscow (1972). Vol. II, Symposium V, "Structure and Function of Paramagnetic Centers and Free Radicals in Biological Systems", pp. 458–473 (1973)
Weaver, E.C., Bishop, N.J.: Science **140**, 1095–1097 (1963)
Weaver, E.C., Weaver, H.E.: Science **165**, 906–907 (1969)
Weaver, E.C., Weaver, H.E.: In: Electron Resonance Studies of Photosynthetic Systems in Photophysiology. Giese, A.C. (ed.). New York: Academic Press, 1972, Vol. VII, pp. 1–32

7. Primary Electron Acceptors

R. Malkin

A. Chloroplast Photosystem I

I. Background

The chemical identity of the primary electron acceptor of photosystem I has been an active area of investigation for several years. Early proposals were that a chlorophyll molecule might be involved as a primary acceptor for photosystem I (Kamen, 1961, 1963). The discovery and characterization of a soluble chloroplast iron-sulfur protein, ferredoxin, by Tagawa and Arnon (1962) led to the proposal that this protein functions as the primary acceptor. This view was based largely on the half-reduction potential (E_m) of -420 mV for soluble ferredoxin at pH 7.0, a value equivalent to that of the hydrogen electrode.

Subsequent studies indicated that a bound, not soluble, electron acceptor with a half-reduction potential of about -550 mV functions as the primary acceptor of photosystem I (Chance et al., 1965; Zweig and Avron, 1965; Zweig et al., 1965; Kok et al., 1965; Black, 1966). A variety of candidates that have been proposed to fill this role have been reviewed by Bishop (1971) and Siedow et al. (1973). Recent evidence, to be described in detail, now supports the role of a bound iron-sulfur protein as the photosystem I primary electron acceptor.

II. Electron Paramagnetic Resonance (EPR) Studies of Bound Iron-Sulfur Proteins

EPR studies by Malkin and Bearden (1971) first revealed the existence of bound iron-sulfur proteins in chloroplasts from higher plants. Of particular interest was the finding, shown in Figure 1, that a reduced iron-sulfur protein, characterized by g-values of 2.05, 1.94, and 1.86, could be observed after illumination of untreated spinach chloroplasts at cryogenic temperatures (~ 10 K). Because the light-induced change occurred at temperatures where diffusion-limited chemical reactions are precluded, it was suggested that this change could be associated with the photoreduction of a primary electron acceptor (Malkin and Bearden, 1971). Bound iron-sulfur proteins were later found in a variety of oxygen-evolving photosynthetic systems (Evans et al., 1973).

The association of the photoreducible bound iron-sulfur protein with chloroplast photosystem I, as opposed to photosystem II, has been demonstrated in a number of different photosystem I-enriched subchloroplast fragments (Bearden and Malkin, 1972a; Evans et al., 1972; Ke et al., 1974; Malkin, 1975). The photoreduction of this carrier proceeds as effectively in far-red light (715 nm), which activates primarily photosystem I, as in red light (645 nm) (Bearden and Malkin,

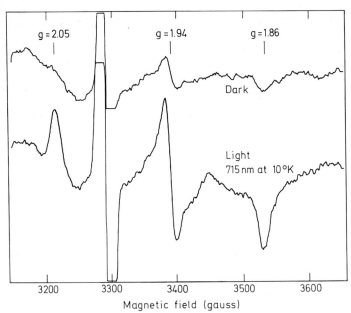

Fig. 1. Light-induced photoreduction of a bound chloroplast iron-sulfur protein detected by EPR spectroscopy after illumination at 10 K with 715-nm monochromatic light

1972a). It has therefore been proposed that the photoreducible bound iron-sulfur protein (also referred to as a bound ferredoxin) is the primary electron acceptor of photosystem I (Bearden and Malkin, 1972a; Evans et al., 1972).

Quantitative EPR studies of the primary reactants of photosystem I activated by far-red light at liquid-helium temperatures have shown that in both untreated chloroplasts and photosystem I subchloroplast fragments there are equal amounts of oxidized P-700 (the reaction-center chlorophyll of photosystem I; cf. Sect. 2.5) and reduced bound iron-sulfur protein (Bearden and Malkin, 1972b). In addition, investigations of the reversibility of the photosystem I primary reaction at cryogenic temperatures have shown that the decay kinetics of $P-700^+$ and those of the reduced bound iron-sulfur protein are identical in the temperature range of 10 to 200 K in both untreated chloroplasts and photosystem I subchloroplast fragments (Malkin and Bearden, 1974; Visser et al., 1974; Ke et al., 1974; see also McIntosh et al., 1975 and Evans and Cammack, 1975, for apparent kinetic discrepancies between $P-700^+$ and reduced iron-sulfur protein).

After chloroplasts or photosystem I subchloroplast fragments are illuminated at room temperature in the presence of an electron donor and frozen in the light, EPR examination at cryogenic temperature shows g-values of 2.05, 1.94, 1.92, 1.89, and 1.86 (Malkin and Bearden, 1971; Ke et al., 1973). This complex spectrum has been interpreted as arising from two different iron-sulfur centers, one of which is the center that is photoreducible at cryogenic temperatures (g-values of 2.05, 1.94, and 1.86). The second center, with g-values of 2.05, 1.92, and 1.89, does not undergo low-temperature photoreduction but can be photoreduced at room temperature or reduced by the addition of sodium dithionite (Malkin and Bearden, 1971; Evans et al., 1972; Ke et al., 1973). The iron-sulfur center that is photoreducible at cryogenic temperatures has a half-reduction potential of about -530 mV and the second center has a half-reduction potential of about -590 mV (KE et al., 1973; EVANS et al., 1974). Although it is possible that these two sets of EPR signals originate from two iron-sulfur centers in two different proteins, an observation by the above authors that the reduction of the most reducing center affects the EPR signal of the other center indicates a second alternative: interaction between two iron-sulfur centers, possibly because they are both located in the same protein. Such an interaction of multiple iron-sulfur centers located in one protein has been shown in bacterial type ferredoxins that contain two iron-sulfur clusters (MATHEWS et al., 1974).

A direct estimation of the half-reduction potential of the primary acceptor of photosystem I has come from studies using the extent of P-700 photooxidation at cryogenic temperatures as an index of the amount of primary acceptor in the oxidized state. The value of -530 mV in both untreated chloroplasts (LOZIER and BUTLER, 1974) and photosystem I subchloroplast fragments (KE, 1974) corresponds with the value obtained for one of the iron-sulfur centers detected by EPR spectroscopy in chloroplasts and photosystem I fragments and is consistent with the previous assignment to this center of the role of primary acceptor. In their titration studies, KE et al. (1973) could not detect any photoreduction of the low-potential iron-sulfur center at cryogenic temperatures, but EVANS and CAMMACK (1975) have reported a small photoreduction of the second center at 20 K. The role of the second center, with a half-reduction potential more electronegative than the center that appears to function as the primary electron acceptor, is not clear.

III. Flash Kinetic Spectroscopy of P-430

In studies with photosystem I subchloroplast fragments at room temperature, HIYAMA and KE (1971a) detected a biphasic absorption change at 430 nm after flash activation in the presence of an electron donor system and a secondary electron acceptor, such as methyl viologen. Analysis of the spectra of these two components indicated that the slowly decaying component was P-700 and that the second was a new component which showed a maximum absorption decrease at 430 nm and minor bands at 380, 325, 460, and 405 nm; this component was designated P-430 (HIYAMA and KE, 1971a; HIYAMA and KE, 1971b; KE, 1972). The reactions of P-430 have been described in detail in a recent review (KE, 1973) and will only be briefly summarized here.

P-430 was assigned as the primary electron acceptor of photosystem I because of its rapid rate of photoreduction (less than 0.1 µsec; KE, 1972) and the high rate of efficiency of its formation—a quantum efficiency of unity was determined for the formation of P-700$^+$ and P-430$^+$ at 710 nm (HIYAMA and KE, 1971b). Extensive kinetic studies have shown that the decay of P-430 could be accelerated by electron acceptors that are known to function as secondary photosystem I acceptors, including the physiological acceptor system, soluble ferredoxin $+ NADP^+$ (HIYAMA and KE, 1971a). That the bleaching of P-430 was a photoreduction was confirmed by the observed kinetic correspondence between the recovery of P-430 and the reduction of the dye safraine T (HIYAMA and KE, 1971a).

A preliminary value of $-475\,\text{mV}$ for the half-reduction potential of P-430 was obtained by monitoring the extent of P-700 photooxidation at room temperature as a function of reduction potential (KE, 1972).

IV. Relationship of P-430 to Bound Iron-Sulfur Protein

The independent assignment of both P-430 and a bound iron-sulfur protein as the primary electron acceptor of photosystem I has led to the consideration of a possible relationship between the two components. On the basis of spectral properties, it was suggested that P-430 might be an iron-sulfur protein (KE, 1973), although its spectrum showed differences from those of isolated iron-sulfur proteins, such as soluble chloroplast ferredoxin (HIYAMA and KE, 1971a; HIYAMA and KE, 1971b; KE, 1972).

KE and BEINERT (1973) obtained indirect evidence for the equivalence of P-430 and the bound iron-sulfur protein based on a comparison of their EPR and optical properties when P-700$^+$ accumulates and both P-430 and the bound iron-sulfur protein are in the reduced state. More direct evidence has been sought through oxidation-reduction titrations that have shown a half-reduction potential for the photoreducible bound iron-sulfur protein equal to that for the primary acceptor under comparable conditions (KE, 1974). Although there is some difference between the half-reduction potentials of P-430 and the bound iron-sulfur protein, it appears that the first estimation of the half-reduction potential of P-430 at room temperature (KE, 1972) may have been an underestimation. There is now good evidence based on half-reduction potentials for the relationship of a bound iron-sulfur protein, P-430, and the primary electron acceptor (KE, 1974).

Further evidence for the assignment of P-430 as a bound iron-sulfur protein has come from the isolation and characterization of one such chloroplast protein (MALKIN et al., 1974). As shown in Figure 2, this protein has an absorption band in the 400-nm region in the oxidized protein and the difference spectrum (dithionite reduced minus oxidized) shows a bleaching with a maximum at about 430 nm. Although characterization of the protein is not complete, the known properties are consistent with its identity as P-430.

To summarize experimental evidence to date, the primary electron acceptor of photosystem I appears to be a bound iron-sulfur protein with g-values of 2.05, 1.94, and 1.86 in the reduced state and an absorption change centered at 430 nm upon reduction.

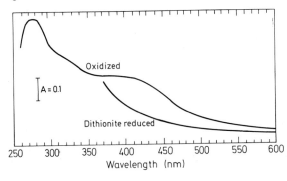

Fig. 2. Absorption spectrum of *oxidized* and *dithionite-reduced* chloroplast iron-sulfur protein in the isolated state. (From MALKIN et al., 1974)

B. Chloroplast Photosystem II

Knowledge of the chemical identity of the photosystem II primary acceptor is less complete than is that of the photosystem I acceptor. Two chloroplast components, X-320 and C-550, have been considered in relation to the primary electron acceptor. A third component, the fluorescence quencher, Q, has also been related to the acceptor (cf. Sect. 2.6).

I. X-320

STIEHL and WITT (1968) identified a component designated X-320 in flash kinetic studies of chloroplasts. The light minus dark difference spectrum of the component had an absorbance maximum near 320 nm, a smaller absorbance decrease at 270 nm, and negligible absorbance in the visible region of the spectrum.

A role for X-320 as the primary acceptor of photosystem II was initially assigned on the basis of the correlation of its decay time (0.6 msec) with the time of transfer of one electron from water to plastoquinone and the calculated rate of plastoquinone reduction (0.6 msec) (STIEHL and WITT, 1969; VATER et al., 1968). It has also been found that X-320 is formed rapidly (faster than 30 μsec) and that its kinetics are unaffected by photosystem I background illumination (STIEHL and WITT, 1968).

Additional evidence for the role of X-320 as the photosystem II primary acceptor was obtained in studies that have shown that X-320 can be photoreduced at low temperature and that its room-temperature reoxidation is inhibited by 3-(3',4'-dichlorophenyl)-1,1'-dimethylurea (DCMU), although the photoreduction is DCMU-insensitive (WITT, 1973). These properties are similar to those observed by others in measurements of the primary acceptor by fluorescence techniques (DUYSENS and SWEERS, 1963; MALKIN and KOK, 1966; ERIXON and BUTLER, 1971a).

The spectrum of the primary electron acceptor has also been reported by VAN GORKOM (1974) in deoxycholate-treated chloroplasts. In this preparation, the

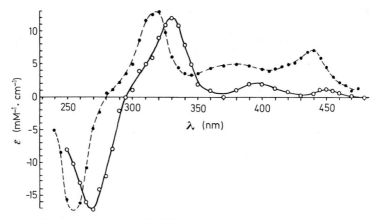

Fig. 3. Light minus dark difference spectrum of X-320 *(open circles)* in deoxycholate-treated chloroplasts and difference spectrum of plastosemiquinone anion minus plastoquinone *(closed circles)*. (From VAN GORKOM, 1974)

primary electron acceptor of photosystem II is reduced in the dark and conditions have been established to permit determination of its spectrum. The spectrum reported by van Gorkom is similar to that of X-320 previously described by Stiehl and Witt (1968) except for a much larger absorbance change in the 270-nm region. A spectrum of X-320 in chloroplasts has also been obtained at $-40°$C and it has been found that the reduction of X-320 at this temperature is dependent on the number of positive charges in the pathway to water (Pulles et al., 1974).

It has been suggested that the spectrum of X-320 arises from the reduction of plastoquinone to the plastosemiquinone anion (Stiehl and Witt, 1969; Witt, 1973). The spectrum obtained by van Gorkom (1974) in deoxycholate-treated chloroplasts is similar to that obtained by Bensasson and Land (1973) with the plastosemiquinone anion prepared by pulse radiolysis, as shown in Figure 3, except for a shift of about 15 nm to longer wavelengths in the chloroplasts.

II. C-550

A decrease in absorbance near 550 nm was observed after photosystem II illumination of chloroplasts at both room temperature and cryogenic temperature by Knaff and Arnon (1969). The component responsible for this change was designated C-550 (see also chapter II.5). This absorbance change was shown to result from a reduction because the addition of dithionite in the dark produced a similar change (Erixon and Butler, 1971 b). The spectrum of C-550 at low temperature and the effective activation by photosystem II illumination, as opposed to photosystem I illumination, are shown in Figure 4. Several properties, such as the low-temperature photoreduction (Knaff and Arnon, 1969; Erixon and Butler, 1971 b), the insensitivity of the room-temperature photoreduction to DCMU (Knaff and Arnon, 1969), and the oxidation-reduction potential dependence of the C-550 change (half-reduction potential about 0 mV at pH 7; Erixon and Butler, 1971 a) have related C-550 to the photosystem II primary electron acceptor.

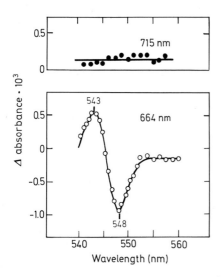

Fig. 4. Light minus dark difference spectrum of C-550 in chloroplasts after illumination at 77 K with photosystem I (715 nm) or photosystem II (664 nm) light

In further studies attempts have been made to identify the component responsible for the C-550 absorbance change. It was found by BUTLER and OKAYAMA (1971) that the absorbance change on reduction at 77 K corresponded to a shift of about 2 nm toward shorter wavelengths of a component which has an absorbance maximum at 546 nm. This finding suggested the absorbance change might arise from a band shift of a chromophoric molecule. This view was strengthened by the finding that in extraction and reconstitution experiments C-550 appears to be related to β-carotene (OKAYAMA and BUTLER, 1972; COX and BENDALL, 1974). Extraction of chloroplasts with hexane led to an almost complete loss of the low-temperature C-550 change and the change could be restored by the addition of β-carotene to the extracted material.

III. On the Chemical Identity of the Photosystem II Primary Electron Acceptor

Although X-320 and C-550 have both been related to the primary electron acceptor of photosystem II, the chemical identity of the acceptor is far from completely known. C-550 shares many of the properties of the primary electron acceptor but its association with β-carotene suggests that it is more likely a bound chromophoric molecule. It has also been possible to demonstrate photosystem II activity in the absence of a functional C-550, a finding that indicates C-550 itself is not the primary electron acceptor (MALKIN and KNAFF, 1973).

A role as the primary electron acceptor has also been assigned to plastoquinone; this electron carrier has been monitored by the X-320 absorbance change. The reactions of X-320 at room temperature are consistent with this role (STIEHL and WITT, 1968, 1969) but the recently reported low-temperature properties of X-320 (WITT, 1973) are not consistent with a widely accepted model for the photosystem II reaction at cryogenic temperatures. Several lines of evidence have indicated that the photosystem II reaction is reversible at low temperatures after flash activation and that charge recombination occurs with a half-time of about 5 msec (cf. Sect. 2.6 and BUTLER et al., 1973; MATHIS and VERMEGLIO, 1974; MALKIN and BEARDEN, 1975). The primary electron acceptor of this photosystem would be expected to show similar decay kinetics at low temperatures. Since X-320 is reported to be formed irreversibly after flash activation at low temperature (WITT, 1973), its assignment as the primary acceptor is still tentative and will await further experimental verification.

References

Bearden, A.J., Malkin, R.: Biochem. Biophys. Res. Commun. **46**, 1299–1305 (1972a)
Bearden, A.J., Malkin, R.: Biochim. Biophys. Acta **283**, 456–468 (1972b)
Bensasson, R., Land, E.J.: Biochim. Biophys. Acta **325**, 175–181 (1973)
Bishop, N.I.: Ann. Rev. Biochem. **40**, 197–226 (1971)
Black, C.C., Jr.: Biochim. Biophys. Acta **120**, 332–340 (1966)
Butler, W.L., Okayama, S.: Biochim. Biophys. Acta **245**, 237–239 (1971)

Butler, W.L., Visser, J.W.M., Simons, H.L.: Biochim. Biophys. Acta **325**, 539–545 (1973)
Chance, B.C., San Pietro, A., Avron, M., Hildreth, W.W.: In: Non-Heme Iron Proteins: Role in Energy Conversion. San Pietro, A. (ed.). Yellow Springs, Ohio: Antioch Press, 1965, pp. 224–236
Cox, R.P., Bendall, D.S.: Biochim. Biophys. Acta **347**, 49–59 (1974)
Duysens, L.N.M., Sweers, H.E.: In: Studies on Microalgae and Photosynthetic Bacteria. Tokyo: Univ. Tokyo, 1963, pp. 353–372
Erixon, K., Butler, W.L.: Biochim. Biophys. Acta **234**, 381–389 (1971a)
Erixon, K., Butler, W.L.: Photochem. Photobiol. **14**, 427–433 (1971b)
Evans, M.C.W., Cammack, R.: Biochem. Biophys. Res. Commun. **63**, 187–193 (1975)
Evans, M.C.W., Reeves, S.G., Cammack, R.: Federation Europ. Biochem. Soc. Lett. **49**, 111–114 (1974)
Evans, M.C.W., Reeves, S.G., Telfer, A.: Biochem. Biophys. Res. Commun. **51**, 593–596 (1973)
Evans, M.C.W., Telfer, A., Lord, A.V.: Biochim. Biophys. Acta **267**, 530–537 (1972)
Gorkom, H.J. van: Biochim. Biophys. Acta **347**, 439–442 (1974)
Hiyama, T., Ke, B.: Proc. Nat. Acad. Sci. U.S. **68**, 1010–1013 (1971a)
Hiyama, T., Ke, B.: Arch. Biochem. Biophys. **147**, 99–108 (1971b)
Kamen, M.D.: In: Light and Life. McElroy, W.D., Glass, B.D. (eds.). Baltimore: Johns Hopkins, 1961, pp. 483–488
Kamen, M.D.: Primary Processes in Photosynthesis. New York: Academic Press, 1963
Ke, B.: Arch. Biochem. Biophys. **152**, 70–77 (1972)
Ke, B.: Biochim. Biophys. Acta **301**, 1–33 (1973)
Ke, B.: In: Proc. 3rd Intern. Congr. Photosynthesis. Avron, M. (ed.). Amsterdam: Elsevier, 1975, pp. 373–382
Ke, B., Beinert, H.: Biochim. Biophys. Acta **305**, 689–693 (1973)
Ke, B., Hansen, R.E., Beinert, H.: Proc. Nat. Acad. Sci. U.S. **70**, 2941–2945 (1973)
Ke, B., Sugahara, K., Shaw, E.R., Hansen, R.E., Hamilton, W.D., Beinert, H.: Biochim. Biophys. Acta **368**, 401–408 (1974)
Knaff, D.B., Arnon, D.I.: Proc. Nat. Acad. Sci. U.S. **63**, 963–969 (1969)
Kok, B., Rurainski, H.J., Owens, O.V.H.: Biochim. Biophys. Acta **109**, 347–356 (1965)
Lozier, R.H., Butler, W.L.: Biochim. Biophys. Acta **333**, 460–464 (1974)
Malkin, R.: Arch. Biochem. Biophys. **174**, 414–419 (1976)
Malkin, R., Aparicio, P.J., Arnon, D.I.: Proc. Nat. Acad. Sci. U.S. **71**, 2362–2366 (1974)
Malkin, R., Bearden, A.J.: Proc. Nat. Acad. Sci. U.S. **68**, 16–19 (1971)
Malkin, R., Bearden, A.J.: Federation Proc. **33**, 378 (1974)
Malkin, R., Bearden, A.J.: Biochim. Biophys. Acta **396**, 250–259 (1975)
Malkin, R., Knaff, D.B.: Biochim. Biophys. Acta **325**, 336–340 (1973)
Malkin, S., Kok, B.: Biochim. Biophys. Acta **126**, 413–432 (1966)
Mathews, R., Charlton, S., Sands, R.H., Palmer, G.: J. Biol. Chem. **249**, 4326–4328 (1974)
Mathis, P., Vermeglio, A.: Biochim. Biophys. Acta **368**, 130–134 (1974)
McIntosh, A.R., Chu, M., Bolton, J.R.: Biochim. Biophys. Acta **376**, 308–314 (1975)
Okayama, S., Butler, W.L.: Plant Physiol. **49**, 769–774 (1972)
Pulles, M.P.J., Kerkhof, P.L.M., Amesz, J.: Federation Europ. Biochem. Soc. Lett. **47**, 143–145 (1974)
Siedow, J., Yocum, C.F., San Pietro, A.: Current Topics Bioenerget. **5**, 107–123 (1973)
Stiehl, H.H., Witt, H.T.: Z. Naturforsch. **23B**, 220–224 (1968)
Stiehl, H.H., Witt, H.T.: Z. Naturforsch. **24B**, 1588–1598 (1969)
Tagawa, K., Arnon, D.I.: Nature (London) **195**, 537–543 (1962)
Vater, J., Renger, G., Stiehl, H.H., Witt, H.T.: Naturwissenschaften **55**, 220–221 (1968)
Visser, J.W.M., Rijgersberg, K.P., Amesz, J.: Biochim. Biophys. Acta **368**, 235–246 (1974)
Witt, K.: Federation Europ. Biochem. Soc. Lett. **38**, 116–118 (1973)
Zweig, G., Avron, M.: Biochem. Biophys. Res. Commun. **19**, 397–400 (1965)
Zweig, G., Shavit, N., Avron, M.: Biochim. Biophys. Acta **109**, 332–346 (1965)

8. Oxygen Evolution and Manganese

B.A. DINER and P. JOLIOT

A. Introduction

The overall equation of photosynthesis does not permit a distinction between H_2O or CO_2 as the source of molecular oxygen. Both of these have been alternatively proposed as the oxygen source.

The establishment of water as the source of molecular oxygen became more firmly established with the thermodynamic arguments of WURMSER (1930) and the metabolic arguments of VAN NIEL (1931). Further experimental support was provided by EMERSON and LEWIS (1941) who showed that in the first few minutes of illumination, following a dark period, no parallelism existed between the uptake of CO_2 and the appearance of O_2. Furthermore the production of O_2, could be completely dissociated from the reduction of CO_2, as demonstrated by HILL and SCARISBRICK (1940), using ferrioxalate as an artificial electron acceptor. The most direct evidence for the formation of O_2 from water was provided by the mass spectrometric evidence of RUBEN et al. (1941) using isotopically labeled water.

The oxygen-evolving reaction in photosynthesis may be represented by the formal redox reaction:

$$2\,H_2O \rightleftharpoons 4e^- + 4\,H^+ + O_2$$

with a midpoint potential of 0.81 V at pH 7.0. The reducing equivalents are then subsequently used for the fixation of CO_2. The above reaction unfortunately gives little insight into the chemical mechanism involved in oxygen formation. Chemical studies have been hampered by the absence of suitable radio-isotopes of oxygen for isolating intermediates, the fragility of the oxygen-evolving site in isolated particles and the absence of evident spin resonance signals or absorbance changes linked to O_2 evolution. Nonetheless, both Mn and Cl^- have been shown to be required for the oxygen evolving reaction and probably participate at the oxygen evolving site. In addition, the study of proton release from this site has provided some preliminary conclusions as to the consecutive steps by which the two water molecules are oxidized. Various molecules compete with water for binding to the oxygen-evolving site, these include hydroxylamine, hydrazine, and ammonia. All of these chemical characterizations will be discussed below.

The kinetic characterization of oxygen production has provided considerable information as to how the primary photoreactions transfer oxidizing equivalents to the oxygen-evolving sites (see review by JOLIOT and KOK, 1975). EMERSON and ARNOLD (1932) studied oxygen production using repetitive flash techniques. These results combined with the quantum yield measurements of EMERSON and LEWIS (1941) (8–10 photon required per oxygen molecule) led to the concept of a unit of 200–250 chlorophyll molecules which together harvest the light energy necessary for the primary photoreactions of photosynthesis. In the primary photoreaction, occurring within a complex called a reaction center, one oxidizing and one reducing equivalent is generated per photon.

Two kinds of eight quanta schemes were proposed by RABINOVITCH (1945) to permit oxidation of water and reduction of CO_2. In one of these, two different kinds of photoreactions acted in series to sufficiently boost the reducing potential of four reducing equivalents to take those equivalents from water to CO_2 reduction. In the other, eight identical photoreactions occurred, with the energy of these eight primary reactions subsequently combined to drive four secondary reactions. The former model, of two series photoreactions was the precursor of the now generally accepted "Z scheme" of HILL and BENDALL (1960). Four photoacts then occurred in each of two photosystems: photosystems I and II (DUYSENS et al., 1961) differing in their action spectra.

This chapter will concentrate on the more oxidizing photosystem II, and the production of oxygen associated with it. We will see how an oxygen molecule is generated on a single site by the integration of four successive photoacts, and how photosystem II centers and oxygen evolving sites operate as a unit and independently of those located on other electron transport chains. Considerable emphasis has been placed on the newest results to minimize duplication with earlier reviews (KOK and CHENIAE, 1966; CHENIAE, 1970; TREBST, 1974; JOLIOT and KOK, 1975; RADMER and KOK, 1975).

B. Photosystem II

Photoexitation of a chlorophyll molecule contained within the reaction center of photosystem II results in an electron transfer generating a chlorophyll cation radical of at least 0.81 V and probably a plastosemiquinone (STIEHL and WITT, 1969) of about -30 mV potential at pH 7.0 (CRAMER and BUTLER, 1969). As a photon of 1.8 eV (680 nm) is sufficient to induce this electron transfer, approximately 50% of the light energy absorbed is immediately converted to electrical and chemical energy.

I. Acceptor Side

The fluorescence yield of photosystem II has been shown by DUYSENS and SWEERS (1963) to be a reflection of the redox state of the primary acceptor, Q, with the yield increasing upon reduction. The study of the decay of the fluorescence yield, following a flash, permitted FORBUSH and KOK (1968) to conclude that Q^- becomes reoxidized in about 0.6 msec. DELOSME et al. (1959) have shown that a linear relation exists between the photochemical rate and the fluorescence yield. A nonlinear relation, however, between the photochemical rate (or the fluorescence yield) and the fraction of Q reduced, has been interpreted by A. JOLIOT and P. JOLIOT (1964) as evidence for energy transfer between units of photosystem II.

STIEHL and WITT (1968) have demonstrated that a species absorbing at 335 nm (X-320) is formed in less than 10^{-5} sec and decays in 0.6 msec ($t_{1/2}$) following

a 10 μsec flash. The spectral properties of this species were also studied by VAN GOR-
KOM (1974), through a light minus dark difference spectrum of deoxycholate-treated
chloroplasts. The spectrum closely resembles that for plastosemiquinone anion
minus plastoquinone, thus providing a likely identity for Q^- and Q respectively.

Another component showing many characteristics of the primary acceptor is
C-550 discovered by KNAFF and ARNON (1969b). This species is also observed
as a band shift at 77° K and probably corresponds to the photoreduction in the
primary charge separation (ERIXON and BUTLER, 1971). The C-550 absorbance
change could be eliminated and then made to reappear upon extraction and reconstitution, respectively, with β carotene (OKAYAMA and BUTLER, 1972; COX and
BENDALL, 1974). In addition, C-550 could be eliminated by chemical oxidation
with a lesser effect upon photosystem II-mediated electron transport (MALKIN and
KNAFF, 1973). Thus, this absorbance change is probably only an indicator of
the redox state of the primary acceptor Q (BUTLER, 1973). RADMER and KOK
(1975) have summarized a number of correlations and contradictions between
the properties of this species and those measured by fluorescence yield and
X-320.

Plastoquinone also serves as a pool of about 8 to 14 equivalents (P. JOLIOT,
1961a; RUMBERG et al., 1963) (4–7 molecules) located between the two photosystems. P. JOLIOT (1961a) has shown that this pool is responsible for the oxygen
gush observed for intense illumination following dark adaptation.

Recently, BOUGES-BOCQUET (1973b) and VELTHUYS and AMESZ (1974) have demonstrated the existence of an intermediate acceptor, B, (R, in the terminology
of the latter authors) located between Q and the plastoquinone pool. This species
accepts electrons one at a time from Q and, having accumulated an electron
pair, transfers that pair to the pool. B is equipotential with the pool (DINER,
1975) and may be interpreted from the data of FOWLER and KOK (1974a) as
being a site of proton uptake. Thus B is probably also a plastoquinone. Nature
has apparently imposed three distinct functional roles on the same chemical entity.

II. Donor Side

Following the primary charge separation, the oxidized primary donor (Chl a_{II}^+)
becomes rereduced. DÖRING et al. (1967) and GLÄSER et al. (1974) have observed
a light-induced bleaching of chlorophyll a at 680 nm followed by a biphasic recovery
consisting of 35 μsec (80%) and 200 μsec (20%) components. According to the
latter authors, these components are probably sufficient to account for all the
photosystem II photoactive pigment. These results are difficult to reconcile with
certain turnover times of photosystem II which do not show either of these components. This problem will be discussed in more detail below and will lead to
the conclusion that there is an even more rapid reaction which reduces the primary
donor.

Recently, BUTLER et al. (1973) have shown that the reaction center chlorophyll
cation radical (Chl a_{II}^+) is a fluorescence quencher, as is Q. This property does
not invalidate the observation of the quenching properties of Q in that these
were generally made for reduced reaction center chlorophyll. This quenching should
permit a measure of the kinetics of reduction of the chlorophyll cation using

fluorescence techniques. Measurements of the rise in the fluorescence yield following a short flash (Zankel, 1973; A. Joliot, 1975) have revealed the existence of a 35 μsec component analogous to that observed by Gläser et al. (1974).

Another component associated with photosystem II is high-potential cytochrome b-559 (Bendall, 1968). This cytochrome is probably not associated with O_2 production (Cramer and Böhme, 1972; Cox and Bendall, 1972) and is oxidized by photosystem II only when the natural electron donor is prevented from functioning—Tris wash and low temperature ($-189°$ C) (Knaff and Arnon, 1969a) (see this vol., Chap. 2.12).

Electron spin resonance signal II (Commoner et al., 1957) (arising from the donor side of photosystem II) has been shown by Babcock and Sauer (1973a, 1975a) and Blankenship et al. (1975b) to consist of at least three kinetically distinct species—signals II_s (slow), II_f (fast) and II_{vf} (very fast). In addition signal II shows two interconvertible forms depending upon the ionic strength—both show identical light-dependent behavior (Babcock and Sauer, 1973b).

The species that gives rise to signal II_s in the light competes with water for oxidizing equivalents stored on the oxygen evolving site (Babcock and Sauer, 1973a). Signal II_s decays about 50–60% within 12 h in the dark in vivo, and in about 1 h in vitro. The overall decay is markedly accelerated by CCCP (carbonyl cyanide 3-chlorophenylhydrazone) or NH_2OH (Lozier and Butler, 1973) and by aging. In chloroplasts, isolated under conditions which minimize the preexistent signal II_s, a single turnover (10 μsec) saturating flash produces 80% of the potentially generated signal II_s within 1 sec (Babcock and Sauer, 1973a). Velthuys and Visser (1975) have shown that the reduced precursor to signal II_s cannot be oxidized by $K_3Fe(CN)_6$ but that signal II_s is reduced by DCPIP+ascorbate, thus implying an oxidizing potential of greater than 500 mV for signal II_s. The spin ratio of fully developed signal II_s to signal I (P-700 or Chl a_I^+) is equal to one.

Signal II_f is produced in the light following Tris wash, heating, or treatment with chaotropic reagents (Babcock and Sauer, 1975a). Following a 10μsec flash, it appears in less than 100 μsec (Blankenship et al., 1975b). The decay time, $t_{1/2} = 0.5$–1.0 sec, is accelerated by the addition of lipophylic reductants such as hydroquinone and phenylenediamine. The extent of formation of signal II_f is diminished by the addition of Mn^{2+} with no effect upon the decay kinetics. Babcock and Sauer (1975b) concluded that lipophylic donors reduce signal II_f directly or indirectly, while Mn^{2+} competes with the species, responsible for signal II_f, for the oxidizing equivalent of Chl a_{II}^+. Signal II_f is believed to have an oxidizing potential >575 mV (Babcock and Sauer, 1975a) and either is identical with or connected to a secondary or more distant photosystem II donor on the way to the oxygen evolving site. The spin ratio signal II_f/signal I is equal to 1.

Signal II_{vf} is observed in untreated chloroplasts, showing a rise time of <100 μsec and a $t_{1/2}$ decay time of 700 μsec (Blankenship et al., 1975b). This species is shown not to be located on the acceptor side of photosystem II as it is absent in Tris-washed chloroplasts, functioning with Mn^{2+} as the donor to Chl a_{II}^+. It, like signal II_f is thus located on the donor side and is present in a spin ratio signal II_{vf}/signal I equal to 1.1 ± 0.2. Signal II_f and II_{vf} possibly arise from the same species, while signal II_s with its much slower rise time, is probably a separate entity.

C. Kinetic Model of O_2 Production

ALLEN and FRANCK (1955) demonstrated for *Scenedesmus,* that in the presence of $5 \times 10^{-4} M$ benzoquinone, the oxygen yield of a 0.5 msec flash is increased 30% if that flash is preceded, one second earlier, by another. We now know that benzoquinone prevents the reduction of the plastoquinone pool under the anaerobic conditions of this experiment (DINER and MAUZERALL, 1973). ALLEN and FRANCK (1955) proposed that the first flash produces an O_2 precursor used by the second. P. JOLIOT (1961 b, 1965) extended these studies under aerobic conditions by demonstrating that, following a dark period, a lag was observed before O_2 was produced in continuous light. This lag exists despite the fact that oxidizing equivalents can be shown to be generated by photosystem II right from the start of illumination in the presence of $10^{-3} M$ NH_2OH (BENNOUN and A. JOLIOT, 1969). P. JOLIOT (1961 b) showed that, as the lag was inversely proportional to $I \times t$, a photoactivation reaction was required for the synthesis of oxygen. This reaction was shown to be driven by photosystem II.

Using a rapidly responding sensitive polarographic method (P. JOLIOT and A. JOLIOT, 1968; P. JOLIOT et al., 1969) studied O_2 emission from dark-adapted algae and chloroplasts resulting from a sequence of short saturating flashes given 300 msec apart. No oxygen was emitted on the first flash and only a small amount on the second. A maximal amount of O_2 was evolved on the third flash, followed, thereafter, by a damped oscillation of oxygen yields with period 4 (Fig. 1). No exception to this characteristic pattern has as yet been demonstrated.

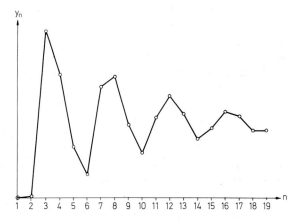

Fig. 1. Oxygen yields for each of a series of short saturating flashes (BOUGES-BOCQUET, 1974). *Spinach* chloroplasts were dark-adapted for 15 min. The duration of the xenon flashes were 2 μsec half amplitude. Y_n is the amount of O_2 evolved for flash number n

A number of kinetic models (P. JOLIOT et al., 1969; KOK et al., 1970; MAR and GOVINDJEE, 1972) have been proposed to explain this observation. The model which best explains the existing data is that proposed by KOK et al. (1970) in which oxidizing equivalents are successively accumulated at an oxygen evolving site attached to only one photosystem II reaction center. Oxidizing equivalents are neither exchanged between different photosystem II centers nor between O_2 producing sites. Each single turnover flash generates one oxidizing equivalent and the accumulation of four of these results in the liberation of an O_2 molecule. The cycle then begins again. This model may be formally expressed as follows, in which each state S of the oxygen evolving site has appended to it a subscript, indicating the number of oxidizing equivalents accumulated.

$$S_0 \xrightarrow{hv} S_0' \rightarrow S_1 \xrightarrow{hv} S_1' \rightarrow S_2 \xrightarrow{hv} S_2' \rightarrow S_3 \xrightarrow{hv} S_3' \rightarrow S_4$$

$$O_2$$

Each single turnover flash induces a separation of charge, transforming centers in redox state Chl Q (S_n) to redox state $Chl^+ Q^-$ (S_n'). The dark relaxation $S_n' \rightarrow S_{n+1}$ then corresponds to the transfer of the oxidizing equivalent to the oxygen evolving site (advancing the S state) and the reducing equivalent to secondary acceptor B.

A number of experimental facts and their relation to the model are discussed below:

1. Partial DCMU inhibition according to KOK et al. (1970) and P. JOLIOT and KOK (1975), has no effect on the oscillatory pattern of O_2 production. They concluded shat oxidizing equivalents generated by photosystem II centers are not shared with inactive centers — i.e. that photosystem II chains act independently with respect to accumulation of oxidizing equivalents. DINER (1974) confirmed the observation of KOK et al. (1970) in *Chlorella*. However, an increased double hitting was observed on the first or second flash of a sequence for chloroplasts inhibited > 90% by DCMU. DINER suggests that a cooperativity at the level of the photosystem II reaction center, on the donor or acceptor sides or an additional acceptor, might permit the generation of two oxidizing equivalents and their delivery to the same oxygen-evolving site. In the models proposed, this double hitting would occur primarily on the first flash of the sequence. The cooperativity apparent in chloroplasts may have arisen as a result of the chloroplast isolation procedure.

2. Weak continuous illumination, (at an intensity corresponding to the linear region of the light saturation curve) followed by a series of flashes showed essentially no oscillatory pattern (P. JOLIOT et al., 1969). This means that the optical cross sections and quantum yields are similar for each S state. A dissenting opinion has been published by DELRIEU (1973).

3. O_2 is first observed on the third flash, following dark adaptation. Either S_1 and not S_0 is the predominant state in the dark or one of the first three flashes generates two equivalents instead of one (DOSCHEK and KOK, 1972). Measurements in continuous light, following dark adaptation, indicate that apparently three photons per center are required to observe oxygen. Barring an unlikely doubling of the optical cross section or quantum yield at the start of the activation, the above observation is best explained by S_1 as the principal dark-adapted state.

4. The oscillatory pattern is damped and eventually attains a nonoscillatory stead-state. FORBUSH et al. (1971) postulated two damping phenomena: misses (α) in which not all centers sucessfully store oxidizing equivalents even in a saturating flash, and double hits (β) in which a small fraction of centers store two equivalents. The latter phenomenon, as evidenced by a small O_2 yield on the second flash, could be eliminated by using sufficiently short laser flashes (WEISS et al., 1971). Thus in algae and in uninhibited chloroplasts, the double hits are due solely to the fraction of centers able to turnover during a light flash and

to effectively use a second photon. The misses were not diminished by increasing the intensity of largely saturating flashes. This meant that all centers were excited by the flashes, but that for some reason, a fraction of these either (a) remained inactive or (b) underwent a back-reaction. If one accepts hypothesis (a) then one is obliged to admit that this period of inactivity exceeds the duration of a several microsecond flash. An analysis of the flash pattern indicated that the misses were randomly distributed over all the centers in a particular state. Thus the period of inactivity of a particular center is less than the time between flashes in the sequence. If one assumes for simplicity that the misses, α, are independent of S_n, then one can calculate an α of 0.20 for *Chlorella* and 0.12 for spinach chloroplasts. DELRIEU (1973) has argued that misses occur primarily in the S_2 state.

ETIENNE (1974) reported that the miss parameter α was increased by the addition of methylamine, $10^{-1} M$, and decreased by the addition of CCCP, $10^{-5} M$. These results were interpreted in terms of a modification of the equilibrium constant relating the primary and secondary donors of photosystem II, altering the probability of finding the center in an active or inactive state.

D. Interconversion of S-States in the Dark

I. Deactivation of States S_2 and S_3

States S_1 and S_0 are relatively stable; however, S_3 and S_2 are not and decay in the dark to S_1 (FORBUSH et al., 1971). This decay occurs primarily in one electron steps as a decrease in the concentration of S_3 transiently increases that of S_2 (P. JOLIOT et al., 1969). The deactivation reactions are studied by giving one or two flash preillumination to dark adapted algae or chloroplasts, followed, a variable time later, by two or one detecting flashes to determine, respectively, the concentrations of S_2 and S_3. Measurements obtained in this way are shown below:

State	Deactivation ($t_{1/2}$)
S_2 (*Chlorella*)	21 sec
S_3 (*Chlorella*)	4.9 sec
S_2 (*Chloroplasts*)	39 sec, 89 sec
S_3 (*Chloroplasts*)	49 sec, 92 sec

Data of FORBUSH et al. (1971) and P. JOLIOT et al. (1971).

The lifetime of these states may also be studied, following continuous illumination, by one or two-flash detection. LEMASSON and BARBIERI (1971) demonstrated a dependence of the lifetime of states S_2 and S_3 in state I algae (BONAVENTURA and MYERS, 1969), on whether they were generated in photosystem II light (480 or 650 nm) or photosystem I light (710 nm). Photosystem II illumination accelerated the decay of S_3 relative to photosystem I light, but little difference was observed for S_2. These authors suggest that different mechanisms may be involved in the deactivation of S_2 and S_3.

DINER (1975) showed that for *Chlorella* and chloroplasts (unpublished results), incubated under anaerobic conditions (reduced plastoquinone pool), S_2 deactivation is accelerated greater than 10-fold relative to the aerobic rate (oxidized pool), while S_3 is not at all affected by the pool redox state. RADMER and KOK (1973), on the other hand, have observed a 20-fold acceleration of S_3 deactivation in chloroplasts by reducing the pool with 650 nm light. These contradictory results for the deactivation of S_3 may be explained by a difference in preillumination preceding the two experiments. The rate of S_2 deactivation ($t_{1/2} = 0.8$ sec; DINER, 1975) when the pool is highly reduced, approaches the charge recombination time for Q reduced in the presence of saturating concentrations of DCMU ($t_{1/2} = 0.75$ sec; BENNOUN, 1970). These results suggest that the electron responsible for S_2 reduction, passes through Q. The correlation between delayed luminescence and the deactivation of states S_2 and S_3 suggests that the electron on Q^- recombines with Chl a_{II}^+ in equilibrium with the oxygen evolving site (BARBIERI et al., 1970; JOLIOT et al., 1971). BOUGES-BOCQUET et al. (1973) also present evidence for the reduction of S_3 via Q^- by demonstrating a stabilization of this state by DCMU. This inhibitor apparently prevents electrons in the plastoquinone pool from reducing Q yet allows the single charge separation necessary, for centers in state S_3, to produce oxygen. As we will see shortly another possible source of reductant for deactivation is the reduced precursor to signal II_s. As this species is normally present in its oxidized form (signal II_s), it is unlikely that it plays an important role in the above cited, deactivation measurements.

RENGER (1972) has demonstrated the deactivation accelerating effect of substituted thiophenes and CCCP (the so-called "ADRY" compounds). These compounds induce a cycle within the electron transport chain in which an electron donor other then Q reduces S_2 and S_3 (RENGER et al., 1973). LEMASSON and ETIENNE (1975) have shown that in the presence of CCCP, the oxygen yield per flash, in a flash sequence, decreases as the dark interval between flashes increases. Surprisingly, this decreased flash yield was not accompanied by an increase in the miss parameter α. It is difficult at present to interpret the mechanism by which states S_2 and S_3 disappear in these experiments.

II. Interconversion of States S_0 and S_1

An analysis of the oxygen flash sequence, particularly the ratio of the third to fourth flash yields (Y_3/Y_4) provides an indication of the initial concentrations of S_1 and S_0, respectively. Based on this kind of analysis, BOUGES-BOCQUET (1973a) has reported that the addition of 0.1 mM $K_3Fe(CN)_6$ to *Spinach* chloroplasts transforms all centers in state S_0 to S_1 such that all centers become synchronized in state S_1 in the dark, instead of the normal $S_1/S_0 = 3$. On the other hand, addition of DCPIP plus ascorbate results in a ratio of $S_1/S_0 = 1$. These results led BOUGES-BOCQUET to suggest that the reaction, $S_1 + e^- \rightleftharpoons S_0$, is characterized by a midpoint potential of close to 0.2 V. Depending upon the initial concentration of S_0 or S_1, a slow oxidation of S_0 to S_1 or reduction of S_1 to S_0 has been observed in the dark, in the absence of exogenous redox reagents, such that the final ratio of S_1/S_0 tends toward the dark equilibrium value of 3 (P. JOLIOT et al., 1971).

To produce an oxygen molecule from water, four electrons at an average potential of at least 0.81 V are predicted by the formal equation written in the Introduction. It is not obligatory, however, that each photoreaction generate this potential. In fact, the individual steps of water oxidation, in vitro, have potentials ranging from -0.45 to $+2.33$ V (GEORGE, 1965). A consequence of the proposal of BOUGES-BOCQUET (1973a) is that the remaining three photoreactions must collectively produce the 3.0 V potential required for the production of an O_2 molecule, or on the average 1.0 V apiece. ROSS and CALVIN (1967) have argued, on thermodynamic grounds, that the maximum possible available energy in the photosystem II photoreactions is 1.23 eV. Together, these proposals would predict an extremely high efficiency for the primary photosystem II reaction.

KOK et al. (1975) confirmed and gave an alternate interpretation to the results of BOUGES-BOCQUET (1973a). They proposed that the S_1/S_0 ratio is determined by a steady state oxidation of S_0 to S_1 by molecular oxygen offset by a reduction of S_1 by an endogenous or externally added reductant. The main advantage of this hypothesis is that it permits the $S_1 + e^- \rightleftharpoons S_0$ reaction to have a high midpoint potential, reducing the redox load of the other three photoreactions.

Another interpretation of the redox results of BOUGES-BOCQUET (1973a) has been offered by VELTHUYS and VISSER (1975) who proposed that all centers exist in state S_1 in the dark and that an oxidizing equivalent generated early in the sequence is lost to an endogenous reductant. These authors have identified such an endogenous reductant as the reduced precursor to signal II_s placed in the reduced state by DCPIP + ascorbate. This hypothesis does not, however, explain the slow conversion of S_0 to S_1 in the dark, which may still be a reaction of low potential, as proposed by BOUGES-BOCQUET or the steady-state redox poise proposed by KOK, RADMER and FOWLER. They also do not explain why the apparent ratio S_1/S_0 increased in the experiment of BOUGES-BOCQUET, upon addition of 0.1 M $K_3Fe(CN)_6$. In the interpretation of VELTHUYS and VISSER, such an observation would make sense only if the species, responsible for signal II_s could be oxidized by $K_3Fe(CN)_6$. They show that such an oxidation does not occur. In our opinion, the hypothesis of VELTHUYS and VISSER offers a convincing interpretation for the effect of reduced DCPIP on the oxygen flash sequence, but does not offer a satisfactory explanation for the apparent presence of S_0 in the absence of exogenous reductant.

From the results of BABCOCK and SAUER (1973a) and VELTHUYS and VISSER (1975), we can, however, conclude that the potential of reaction $S_1 + e^- \rightleftharpoons S_0$, is less oxidizing than that of reaction signal $II_s + e^- \rightleftharpoons$ signal II_s precursor, as reduced precursor and S_1 coexist. On the other, we will see shortly that $S_2 + e^- \rightleftharpoons S_1$ and $S_3 + e^- \rightleftharpoons S_2$ are more oxidizing than the signal II_s reaction in that S_2 and S_3 oxidize the reduced signal II_s precursor (BABCOCK and SAUER, 1973a).

E. Turnover Reactions of Photosystem II

The overall turnover kinetics of photosystem II have been measured by oxygen detecting methods. As mentioned earlier, a center in state S_n, is upon excitation and charge separation, converted to state S_n'. The dark relaxation of S_n' is represented

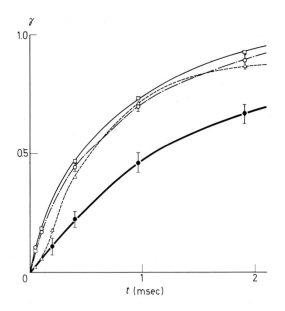

Fig. 2. Relaxation times for the transitions $S'_n \rightarrow S_{n+1}$ (Bouges-Bocquet, 1973a, 1974). □ $S'_0 \rightarrow S_1$; ○ $S'_1 \rightarrow S_2$; △ $S'_2 \rightarrow S_3$; ● $S'_3 \rightarrow S_4$. These experiments with the exception of $S'_0 \rightarrow S_1$, have been performed in the presence of 0.1 mM $K_3Fe(CN)_6$ which places all centers in state S_1 in the dark. γ represents the fraction of centers which have completed the relaxation

as $S'_n \rightarrow S_{n+1}$. These kinetics were first measured by Kok et al. (1970) who measured half times of 0.2 and 0.4 msec for $S'_1 \rightarrow S_2$ and $S'_2 \rightarrow S_3$, respectively.

Bouges-Bocquet (1973a) subsequently studied these relaxation times in more detail (Fig. 2).

The principal component observed in relaxations $S'_0 \rightarrow S_1$, $S'_1 \rightarrow S_2$ and $S'_2 \rightarrow S_3$ ($t_{1/2} \approx 0.45$ at 20°) is similar to that observed for the fast component of the fluorescence decay following a single flash (Forbush and Kok, 1968). These results are generally interpreted as indicating that the turnover is limited by the reoxidation of Q^-. The measurement of the turnover time under anaerobic conditions (Diner, 1975) provides added support for this hypothesis. Under these conditions, where the plastoquinone pool is greater than 90% reduced, the turnover $S'_1 \rightarrow S_2$ is slowed about 30-fold and is highly sigmoid. These results show that under anaerobic conditions, two consecutive reactions are required in the relaxation $S'_1 \rightarrow S_2$. These two reactions were interpreted as being a first oxidation of $B^=$ by the pool of plastoquinone ($t_{1/2} \approx 10$–15 sec) followed by an electron transfer from Q^- to B ($t_{1/2} \approx 0.5$ msec).

Under aerobic conditions (Bouges-Bocquet, 1973a), the absence of a sigmoidicity in relaxations $S'_0 \rightarrow S_1$, $S'_1 \rightarrow S_2$ indicates that only a single reaction is rate limiting, probably the reoxidation on Q^-. Thus it is difficult to understand why the components of 35 μsec (Gläser et al., 1974) and 200 μsec (Döring et al., 1967), attributed to the reduction of Chl a_{II}^+ are not observed. One hypothesis to explain these results is that both the primary and secondary donors are chlorophyll molecules, with an electron transfer between them occurring more rapidly than the time resolution of the turnover measurements (several μsec). In this case, the 35 μsec and 200 μsec components would correspond to the reduction of the secondary donor chlorophyll. This hypothesis is in accord with the proposal of Duysens et al. (1975) of a submicrosec reduction of the primary donor. A temperature dependent equilibrium should also exist between the primary and secondary donors

(A. JOLIOT, 1975). An alternate hypothesis is that of two chlorophyll molecules alternately acting as primary donors with a single primary acceptor.

Contrary to what was observed for $S'_0 \to S_1$ and $S'_1 \to S_2$, the aerobic $S'_2 \to S_3$ relaxation (Fig. 2) is sigmoidal (BOUGES-BOCQUET, 1973a) implying, in this case, the intervention of two rate limiting steps, it may be that one of these components arises on the donor side; however, the existing kinetic data is insufficient to arrive at a definitive conclusion. The $S'_3 \to S_0$ relaxation, under aerobic conditions, is much slower than the others (Fig. 2) and probably corresponds to the rate limiting step in the release of an oxygen molecule, as measured by P. JOLIOT et al. (1966), and recently confirmed by SINCLAIR and ARNASON (1974).

F. Phenomena Related to the S-States

A whole range of diverse phenomena have been shown to be correlated with the states of the photosystem II donor side. P. JOLIOT et al. (1971) have demonstrated, in dark-adapted *Chlorella*, that the fluorescence quantum yield, 2 sec after each of a series of saturating single turnover flashes, oscillates in parallel with the concentration of states S_2 plus S_3. This oscillation was tentatively linked to two photosystem II acceptors of different quenching ability; high-quenching Q_1, linked to states S_0 and S_1, low-quenching Q_2, linked to states S_2 and S_3 (P. JOLIOT and A. JOLIOT, 1972).

DELOSME (1972) showed that if the fluorescence yield were detected during a 2 µsec saturating flash, then the fluorescence is complementary to the S_2 plus S_3 oscillation, showing minima where the $S_2 + S_3$ concentration is maximal, immediately prior to the flash.

The intensity of luminescence emission, following each of a series of a saturating flashes, given to dark adapted algae, also oscillates with a periodicity 4 (BARBIERI et al., 1970; ZANKEL, 1971). These results have been interpreted in terms of the rate of charge recombination between the donor and acceptor sides of photosystem II which is a function of the S states.

VERMEGLIO and MATHIS (1973) showed that preillumination of dark-adapted chloroplasts at room temperature with a variable number of flashes, followed by continuous illumination at $-55°$, resulted in an oscillating ability to oxidize cytochrome b-559 HP (high potential; BENDALL, 1968) at this temperature. More of this cytochrome was oxidized in states S_2 and S_3 than in states S_0 and S_1. These results were interpreted as indicative of a competition between the natural donor, normally operating at ambient temperature, and the cytochrome, with the former operating less efficiently at low temperature in states S_2 and S_3.

BABCOCK and SAUER (1973a) observed that the formation of EPR signal II$_s$ is coincident with the presence of states S_2 and S_3. Signal II$_s$ forms with a rise time of about 1 sec at room temperature, a time long relative to the donor side equilibration times. Thus signal II$_s$, unlike cytochrome b-559 HP at low temperature, is not a competitive donor to Chl a$_{II}^+$ but rather a spoiling reaction that reduces already stabilized oxidizing equivalents on the donor side of photosystem II.

Most pertinent to the mechanism of oxygen production, is the demonstration by FOWLER and KOK (1974a), that the liberation of protons from the inside of the thylakoid membrane oscillates with period 4 in phase with oxygen release. This observation was interpreted as indicating that three or four photons are released in the reaction $S_4 \rightarrow S_0 + O_2$. Further calibration of the number of protons released should increase what one knows chemically of the S states and indicate the point at which water becomes fixed to the oxygen evolving site.

G. Chemical Treatments that Reversibly Affect the O_2 Evolving Site

I. Low Concentrations of NH_2OH

Algae or chloroplasts incubated in the dark in the presence of low concentrations of NH_2OH (50–100 µM) show an increased lag for the appearance of oxygen upon excitation by a series of saturating single turnover flashes (BOUGES, 1971) or by continuous illumination (BENNOUN and A. JOLIOT, 1969). The entire sequence of O_2 flash yields is displaced 2 flashes such that a maximum is obtained on the fifth or sixth instead of on the third flash. Very little oxygen is evolved on the preceding flashes. If the chloroplasts or algae are washed to remove the NH_2OH, following an incubation at 50 µM, the displaced flash pattern still appears, indicating a very tight binding by this molecule to the oxygen evolving site. All these properties are shown by NH_2NH_2, as well, at approximately the same concentration range as NH_2OH (DINER, unpublished results). These observations were interpreted by BOUGES as indicating that the fixation of NH_2OH results in the loss of two oxidizing equivalents. Two molecules of NH_2OH (or one molecule which could be doubly oxidized) were successively oxidized by photosystem II after which centers were in state S_1. The turnover time following the second flash did not correspond to $S_0' \rightarrow S_1$, indicating that the two successive oxidations occurred without passing through state S_0 (BOUGES-BOCQUET, 1973a).

II. Action of NH_3

VELTHUYS (1975) has shown that NH_3 (50 mM NH_4Cl, pH 7.8) binds to the oxygen-evolving site in states S_2 and S_3 but not in S_1 or S_0. NH_3 can be removed by lowering the pH to 6.8 in the absence of NH_4Cl. Excitation of a center in state S_3 (NH_3) produces S_4 (NH_3) which cannot produce O_2 and which undergoes a charge recombination producing an enhanced luminescence yield. On the other hand, the charge on S_2 is stabilized by the fixation of NH_3.

III. Artificial Donors

Photosystem II is capable of oxidizing in the steady state a considerable number of artificial donors (TREBST, 1974). These include benzidine, semicarbazide (YAMASHITA and BUTLER, 1968); ascorbate (CHIBA and OKAYAMA, 1962; TREBST et al.,

1963); phenylenediamine (YAMASHITA and BUTLER, 1968; IZAWA, 1970); NH_2OH (VAKLINOVA et al., 1966; BENNOUN and JOLIOT, 1969); hydroquinone (TREBST et al., 1963), Mn^{2+} (IZAWA, 1970) and diphenylcarbazide (VERNON and SHAW, 1969). CHENIAE and MARTIN (1970) have indicated that ascorbate and phenylenediamine are oxidizable with the oxygen evolving site intact. However, these donate electrons at lower rates after inactivation of the oxygen evolving site. Other donors including hydroquinone, benzidine, semicarbazide, Mn^{2+}, diphenylcarbazide and NH_2OH donate efficiently under these conditions. The oxidation of NH_2OH at high concentrations ($> 10^{-3}$ M) could be detected by an amperometric method (BENNOUN and A. JOLIOT, 1969) leading to the conclusion that no formation of precursors, as for O_2, was necessary for the oxidation of this donor.

IV. Mn and Inactivation and Reactivation of the Oxygen-Evolving Site

Among the treatments that inactivate the oxygen evolving site are heat treatment (KATOH and SAN PIETRO, 1967), NH_2OH ($> 10^{-3}$ M) (CHENIAE and MARTIN, 1970) and Tris wash (0.8 M, pH 8.0) (NAKAMOTO et al., 1959; YAMASHITA and HORIO, 1968). These methods were shown to result in the liberation of Mn bound to photosystem II centers (CHENIAE and MARTIN 1966, 1970; HOMANN, 1967).

Further support for the idea that Mn is linked to the oxygen evolving site arises from the study of Mn extraction using Tris or NH_2OH wash in spinach chloroplasts. CHENIAE and MARTIN (1970) established that two-thirds of the Mn bound to photosystem II centers is readily extracted by these treatments, resulting in a complete loss of oxygen evolving capacity. Oxygen activity was lost in parallel with the Mn comprising this labile pool. The remaining third stays tightly bound to the center following extraction. The most recent estimate of CHENIAE and MARTIN (1971 b) is that 4–6 Mn are bound to an active oxygen-evolving photosystem II center in *Spinach* chloroplasts.

The most convincing arguments in favor of the association of Mn with the oxygen-evolving site come from photoreactivation experiments in algae and more recently chemical and photoreactivation experiments in chloroplasts (YAMASHITA and TOMITA, 1975). It has been shown using algae grown in Mn-deficient medium (CHENIAE and MARTIN, 1966; GERHARDT and WIESSNER, 1967) or extracted with 2 mM NH_2OH (CHENIAE and MARTIN, 1971a) that oxygen evolving activity can be restored through a light induced fixation of Mn to the oxygen evolving site.

Reincorporation of Mn into the O_2 evolving site of Mn depleted algae requires first, an accumulation of Mn^{2+} within the thylakoids and second a light driven fixation (photoreactivation). The accumulation of Mn can occur in the dark, but is stimulated 5–10 fold, by an illumination of either or both photosystems (CHENIAE and MARTIN, 1969). This accumulation is linked to phosphorylation as it is inhibited by uncoupling agents DNP and CCCP (HOMANN, 1967). The photoreactivation which follows is characterized below (CHENIAE and MARTIN, 1971a; RADMER and CHENIAE, 1971):

The rate of photoactivation is a function of the Mn concentration and shows a first-order dependence on the number of inactive O_2-evolving sites.

The action spectra and optical cross section for the photoreactivation correspond to those of photosystem II.

Reactivation is inhibited by DCMU, at the same concentrations that block Hill activity in fully active cells.

No interaction between photosystem II centers is involved in the activation process.

Photoreactivation occurs with the low quantum yield of 0.01.

At least two dark reactions appear to be involved in the photoreactivation of Mn. A slow reaction of 0.2–1.0 sec halftime is followed by a first photoreaction and a 1–15 msec dark relaxation, possibly resulting in the oxidation of Mn^{2+} to Mn^{3+}. At this point, a second photoexcitation stabilizes the fixed Mn, possibly by a second oxidation to Mn^{4+}.

The series mechanism proposed by Radmer and Cheniae (1971) is shown below, where E is the fully active center.

$$A \xrightarrow{k_1} B \xrightarrow{\phi_1} C \xrightarrow{k_2} D \xrightarrow{\phi_2} E \qquad \begin{aligned} k_1 &= 1\text{--}3/\text{sec} \\ k_2 &= 50\text{--}100/\text{sec} \\ k_3 &= 0.5/\text{sec} \end{aligned}$$

Similar results have been observed by Inoue (1975) for photoactivation by flashes of intermittently illuminated greening wheat leaves. However, this author observes an additional photostep leading to the formation of A in the above scheme, as well as somewhat different rate constants.

Tris wash (0.8 M, pH 8.0) of chloroplasts inhibits O_2 activity by releasing 60% of the total chloroplast-bound Mn, which becomes trapped within the inner space of intact thylakoids (Blankenship and Sauer, 1974). This liberated Mn^{2+} diffuses across the thylakoid membrane with a $t_{1/2}$ of 2.5 h, but is shown to be rapidly released upon sonication. The Mn liberated by Tris wash was detected by EPR and was hexaquo Mn^{2+}. Yamashita et al. (1971) have shown that this loss in O_2 activity can be recovered through reactivation of the O_2 evolving site with DCPIP plus ascorbate. Blankenship et al. (1975a) show that this reactivation, which does not require light, results in the loss of EPR-detectable Mn^{2+}.

As mentioned earlier, EPR signal II_f, is formed in the light, following Tris wash (Babcock and Sauer, 1975a) but is absent in untreated or reactivated chloroplasts (Blankenship et al., 1975a). Reactivation occurs despite the presence of EDTA, indicating that the affinity of the O_2 evolving site for Mn^{2+} is greater than that of EDTA ($K_d = 10^{-14}$). Dark reactivated chloroplasts immediately show the oscillation of O_2 flash yields with periodicity 4 and a maximum yield on the third flash.

That the O_2 evolving site can be reactivated in the dark by the addition of reducing agents (reduced DCPIP) in the presence of Mn^{2+} strongly suggests that the reactivated O_2 evolving site contains Mn^{2+} rather than Mn^{4+} as proposed by Radmer and Cheniae (1971). It remains unclear, however, why Mn-deficient algae require light for reactivation.

V. Cl⁻ Depletion

The Hill reaction is stimulated from four to >10 fold by the addition of >5 mM Cl⁻ to chloroplasts depleted of Cl⁻ and uncoupled by EDTA (Izawa et al., 1969). The stimulation upon addition of Cl^-, for chloroplasts depleted of Cl^- in the absence of EDTA, was increased by the presence of uncouplers (HIND et al., 1969). The Cl^- effect is independent of the Hill oxidant used. Oxidation of NH_2OH with TCPI (2,3′,6-trichlorophenolindophenol) as acceptor, or ascorbate with FMN as acceptor, did not show a Cl^- effect, indicating that these donors act between the Cl^- dependent site and photosystem II (IZAWA et al., 1969). Photosystem I electron and cyclic phosphorylation were little effected by the chloride concentration. Cl^- can be replaced by Br^- and to a lesser extent by I^-, NO_3^- and ClO_4^-, all anions of strong acids with $pK < 2$ (HIND et al., 1969).

H. Localization of the Oxygen-Evolving Site

A major feature of the chemiosmotic theory of MITCHELL (1961) is that oxidizing equivalents are generated on the inner surface of the thylakoid membrane and reducing equivalents on the outer surface. WITT (1971) and co-workers have added considerable support to this theory in demonstrating that both photosystems generate, in their primary photoreactions, a separation of charges across the thylakoid membrane. FOWLER and KOK (1971 b) showed that the orientation of the resulting electric field was positive inside and negative outside. The orientation of the dipole in photosystem II centers, perpendicular to the plane of the membrane is consistent with the observed stimulation of the luminescence emission by salt injection (MILES and JAGENDORF, 1969; BARBER and KRAAN, 1970) and by the naturally generated proton motive force (WRAIGHT and CROFTS, 1971).

Further support for the location of the photosystem II donor side on the inner surface of the thylakoid membrane comes from the demonstration by HARTH et al. (1974a) of a photosystem II mediated phosphorylation linked to the oxidation of benzidine but not of tetramethyl benzidine. Using chloroplasts, inactive for oxygen production, these authors show that benzidine releases protons toward the inner thylakoid space upon oxidation whereas tetramethyl benzidine, oxidized at the same site, cannot release protons.

Additional arguments placing the oxygen evolving site on the inside are as follows:

1. Mn, which is clearly associated with the oxygen-evolving site, is selectively released toward the inside of the thylakoid membrane upon washing with Tris 0.8 M, pH 8.0 (BLANKENSHIP et al., 1975a).

2. The experiments of FOWLER and KOK (1974a) show that proton release, coupled to the production of oxygen and thus released from the oxygen evolving site, occurs toward the inside of the thylakoid membrane.

3. AUSLÄNDER et al. (1974) also found evidence for two sites of proton release toward the inside of the thylakoid. One of these, they proposed, was linked to oxygen production. AUSLÄNDER and JUNGE (1975) showed that, in the steady state, proton release from this site occurs with a $t_{1/2}$ of 0.3 msec, a rate four times faster than that measured by P. JOLIOT et al. (1966) for the liberation of an oxygen molecule. According to AUSLÄNDER and JUNGE (1975), this rapid proton release is not always observed and may indicate some variability in the site of liberation of the water protons.

4. Harth et al. (1974b) and Reimer and Trebst (1975) have reported a light-dependent inhibition of the oxygen evolving site, resulting from an alkaline *internal* pH (9.3) and an electron flow through photosystem II. Artificial electron donor oxidation is unaffected by this treatment, indicating a specific effect on the oxygen evolving site inactivated from the internal thylakoid space.

A model consisting of a fixed orientation and localization of photosystem II centers and the oxygen evolving site may, however, be oversimplified, as indicated by the following evidence.

5. Babcock and Sauer (1975b) show that Mn^{2+} competes with the (EPR) signal II_f presursors for the oxidizing equivalents of Chl a_{II}^+. As the thylakoid membrane is rather impermeable to Mn^{2+} (Blankenship and Sauer, 1974), these results imply an accessibility and a possible proximity of Chl a_{II} to the outside of the thylakoid.

6. P. Joliot and A. Joliot (1975) and Diner and P. Joliot (1976) studied the dependence of luminescence emission and the fluorescence quantum yield, respectively, on the transmembrane electric field. These authors found that photosystem II centers were not equally sensitive to the field and suggest that these centers may be variably oriented with respect to the plane of the membrane.

More germane to the oxygen evolving site are:

7. Giaquinta et al. (1974) have shown that increased labelling of chloroplast membranes with impermeant DABS [p-(diazonium)-benzenesulfonic acid] occurs in the light relative to a dark incubation for the same length of time. This additional labeling, unlike that in the dark, results in an inhibition of oxygen production. These authors propose that light-induced electron transport through photosystem II produces a conformational change in the photosystem which renders the oxygen evolving site accessible from the outside.

8. Using the DCMU-insensitive photosystem II electron acceptor, silicomolybdate, Giaquinta and Dilley (1975) argue that the oxygen-evolving site releases its protons toward the outside of the thylakoid membrane in the presence of DCMU and toward the inside in its absence.

9. Additional evidence (summarized by Trebst, 1974), based on other chemical probes, trypsin digest and antibody binding demonstrate the sensitivity of the photosystem II donor side to agents that act from the outer surface of the thylakoid. Some of the latter data plus that of paragraph (7) above are possible subject to the criticism that oxygen evolution may be inhibited indirectly from a site located at some distance from that of water oxidation.

There is, for the most part, considerable support for the location of the oxygen evolving site on the inside and for the orientation of the primary charge separation in photosystem II across and perpendicular to the plane of the membrane. It is likely, however, that there is some synamic variability in the structure of photosystem II which would account for some of the apparent contradictions observed and which might permit a regulation of photosystem II activity.

References

Allen, F.L., Franck, J.: Arch. Biochem. Biophys. **58**, 124–143 (1955)

Ausländer, W., Heathcote, P., Junge, W.: FEBS Lett. **47**, 229–235 (1974)

Ausländer, W., Junge, W.: FEBS Lett. **59**, 310–315 (1975)

Babcock, G.T., Sauer, K.: Biochim. Biophys. Acta **325**, 483–503 (1973a)

Babcock, G.T., Sauer, K.: Biochim. Biophys. Acta **325**, 504–519 (1973b)

Babcock, G.T., Sauer, K.: Biochim. Biophys. Acta **376**, 315–328 (1975a)

Babcock, G.T., Sauer, K.: Biochim. Biophys. Acta **396**, 48–62 (1975b)

Barber, J., Kraan, G.P.B.: Biochim. Biophys. Acta **197**, 49–59 (1970)

Barbieri, G., Delosme, R., Joliot, P.: Photochem. Photobiol. **12**, 197–206 (1970)

Bendall, D.: Biochem. J. **109**, 46P–47P (1968)

Bennoun, P.: Biochim. Biophys. Acta **216**, 357–363 (1970)

Bennoun, P., Joliot, A.: Biochim. Biophys. Acta **189**, 85–94 (1969)

Blankenship, R.E., Babcock, G.T., Sauer, K.: Biochim. Biophys. Acta **387**, 165–175 (1975a)

Blankenship, R.E., Babcock, G.T., Warden, J.T., Sauer, K.: FEBS Lett. **51**, 287–293 (1975b)

Blankenship, R.E., Sauer, K.: Biochim. Biophys. Acta **357**, 252–266 (1974)

Bonaventura, C., Myers, J.: Biochim. Biophys. Acta **189**, 366–383 (1969)

Bouges, B.: Biochim. Biophys. Acta **234**, 103–112 (1971)

Bouges-Bocquet, B.: Biochim. Biophys. Acta **292**, 772–785 (1973a)

Bouges-Bocquet, B.: Biochim. Biophys. Acta **314**, 250–256 (1973b)

Bouges-Bocquet, B.: Thesis, Paris (1974)

Bouges-Bocquet, B., Bennoun, P., Taboury, J.: Biochim. Biophys. Acta **325**, 247–254 (1973)

Butler, W.L.: Accounts Chem. Res. **6**, 177–184 (1973)

Butler, W.L., Visser, J.W.M., Simons, H.L.: Biochim. Biophys. Acta **292**, 140–151 (1973)

Cheniae, G.M.: Ann. Rev. Plant. Physiol. **21**, 467–498 (1970)

Cheniae, G.M., Martin, I.F.: Brookhaven Symp. Biol. **19**, 406–417 (1966)

Cheniae, G.M., Martin, I.F.: Plant Physiol. **44**, 351–360 (1969)

Cheniae, G.M., Martin, I.F.: Biochim. Biophys. Acta **197**, 219–239 (1970)

Cheniae, G.M., Martin, I.F.: Biochim. Biophys. Acta **253**, 167–181 (1971a)

Cheniae, G.M., Martin, I.F.: Plant Physiol. **47**, 568–575 (1971b)

Chiba, Y., Okayama, S.: Plant Cell. Physiol. **3**, 379–390 (1962)

Commoner, B., Heise, J.J., Lippincott, B.B., Norbert, R.E., Passonneau, J.V., Townsend, J.: Science **126**, 57–63 (1957)

Cox, R.P., Bendall, D.S.: Biochim. Biophys. Acta **283**, 124–135 (1972)

Cox, R.P., Bendall, D.S.: Biochim. Biophys. Acta **347**, 49–59 (1974)

Cramer, W.A., Böhme, H.: Biochim. Biophys. Acta **256**, 358–369 (1972)

Cramer, W.A., Butler, W.L.: Biochim. Biophys. Acta **172**, 503–510 (1969)

Delosme, R.: In: Proc. 2nd Intern. Congr. Photosynthesis. Forti, G., Avron, M., Melandri, A. (eds.). The Hague: Dr. Junk, 1972, Vol. I, pp. 187–195

Delosme, R., Joliot, P., Lavorel, J.: C.R. Acad. Sci. Paris **249**, 1409–1411 (1959)

Delrieu, M.J.: C.R. Acad. Sci. Paris **277**, 2809–2812 (1973)

Diner, B.: Biochim. Biophys. Acta **368**, 371–385 (1974)

Diner, B.: In: Proc. 3rd Intern. Congr. Photosynthesis. Avron, M. (ed.). Amsterdam: Elsevier, 1975, Vol. I, pp. 589–601

Diner, B., Joliot, P.: Biochim. Biophys. Acta **423**, 479–498 (1976)

Diner, B., Mauzerall, D.: Biochim. Biophys. Acta **305**, 329–352 (1973)

Döring, G., Stiehl, H.H., Witt, H.T.: Z. Naturforsch. **22B**, 639–644 (1967)

Doschek, W., Kok, B.: Biophys. J. **12**, 832–838 (1972)

Duysens, L.N.M., Amesz, J., Kamp, B.M.: Nature (London) **190**, 510–511 (1961)

Duysens, L.N.M., Den Haan, G.A., Van Best, J.A.: In: Proc. 3rd Intern. Congr. Photosynthesis. Avron, M. (ed.). Amsterdam: Elsevier, 1975, Vol. I, pp. 1–12

Duysens, L.N.M., Sweers, H.E.: In: Studies on Microalgae and Photosynthetic Bacteria. Tokyo: Univ. Tokyo, 1963, pp. 353–372

Emerson, R., Arnold, W.: J. Gen. Physiol. **16**, 191–205 (1932)

Emerson, R., Lewis, C.M.: Am. J. Botany **28**, 789–804 (1941)

Erixon, K., Butler, W.L.: Biochim. Biophys. Acta **234**, 381–389 (1971)

Etienne, A.L.: Thesis, Paris (1974)

Forbush, B., Kok, B.: Biochim. Biophys. Acta **162**, 243–253 (1968)
Forbush, B., Kok, B., McGloin, M.: Photochem. Photobiol. **14**, 307–321 (1971)
Fowler, C.F., Kok, B.: Biochim. Biophys. Acta **357**, 299–307 (1974a)
Fowler, C.F., Kok, B.: Biochim. Biophys. Acta **357**, 308–318 (1974b)
George, P.: The fitness of oxygen. In: Oxidases and Related Redox Systems. King, T.E.,
    Mason, H.S., Morrison, M. (eds.). New York: John Wiley, 1965, pp. 3–36
Gerhardt, B., Wiessner, W.: Biochem. Biophys. Res. Commun. **28**, 958–964 (1967)
Giaquinta, R.T., Dilley, R.A.: Biochim. Biophys. Acta **387**, 288–305 (1975)
Giaquinta, R.T., Dilley, R.A., Selman, B.R., Anderson, B.J.: Arch. Biochem. Biophys. **162**,
    200–209 (1974)
Gläser, M., Wolff, C., Buchwald, H., Witt, H.T.: FEBS Lett. **42**, 81–85 (1974)
Gorkom, H.J. van: Biochim. Biophys. Acta **347**, 439–442 (1974)
Harth, E., Oettmeier, W., Trebst, A.: FEBS Lett. **43**, 231–234 (1974a)
Harth, E., Reimer, S., Trebst, A.: FEBS Lett. **42**, 165–168 (1974b)
Hill, R., Bendall, F.: Nature (London) **186**, 136–137 (1960)
Hill, R., Scarisbrick, R.: Nature (London) **146**, 61–62 (1940)
Hind, G., Nakatani, H.Y., Izawa, S.: Biochim. Biophys. Acta **172**, 277–289 (1969)
Homann, P.H.: Plant Physiol. **42**, 997–1006 (1967)
Inoue, Y.: Biochim. Biophys. Acta **396**, 402–413 (1975)
Izawa, S.: Biochim. Biophys. Acta **197**, 328–331 (1970)
Izawa, S., Heath, R.L., Hind, G.: Biochim. Biophys. Acta **180**, 388–398 (1969)
Joliot, A.: In: Proc. 3rd Intern. Congr. Photosynthesis. Avron, M. (ed.). Amsterdam: Elsevier,
    1975, Vol. I, pp. 315–322
Joliot, A., Joliot, P.: C.R. Acad. Sci. Paris **258**, 4622–4625 (1964)
Joliot, P.: J. Chim. Phys. **58**, 570–583 (1961a)
Joliot, P.: J. Chim. Phys. **58**, 584–595 (1961b)
Joliot, P.: Biochim. Biophys. Acta **102**, 116–134 (1965)
Joliot, P., Barbieri, G., Chabaud, R.: Photochem. Photobiol. **10**, 309–329 (1969)
Joliot, P., Hofnung, M., Chabaud, R.: J. Chim. Phys. **63**, 1423–1441 (1966)
Joliot, P., Joliot, A.: Biochim. Biophys. Acta **153**, 625–634 (1968)
Joliot, P., Joliot, A.: In: Proc. 2nd Intern. Congr. Photosynthesis. Forti, G., Avron, M.,
    Melandri, A. (eds.). The Hague: Dr. Junk, 1972, Vol. I, pp. 26–38
Joliot, P., Joliot, A.: In: Proc. 3rd Intern. Congr. Photosynthesis. Avron, M. (ed.). Amsterdam:
    Elsevier, 1975, Vol. I, pp 25–39
Joliot, P., Joliot, A., Bouges, B., Barbieri, G.: Photochem. Photobiol. **14**, 287–305 (1971)
Joliot, P., Kok, B.: Oxygen evolution in photosynthesis. In: Bioenergetics of Photosynthesis.
    Govindjee (ed.). New York: Academic Press, 1975, pp. 387–412
Katoh, S., San Pietro, A.: Arch. Biochem. Biophys. **122**, 144–152 (1967)
Knaff, D., Arnon, D.: Proc. Nat. Acad. Sci. U.S. **63**, 956–962 (1969a)
Knaff, D., Arnon, D.: Proc. Nat. Acad. Sci. U.S. **63**, 963–969 (1969b)
Kok, B., Cheniae, G.M.: Current Topics Bioenerg. **1**, 1–47 (1966)
Kok, B., Forbush, B., McGloin, M.: Photochem. Photobiol. **11**, 457–475 (1970)
Kok, B., Radmer, R., Fowler, C.F.: In: Proc. 3rd Intern. Congr. Photosynthesis. Avron,
    M. (ed.). Amsterdam: Elsevier, 1975, Vol. I, pp. 485–496
Lemasson, C., Barbieri, G.: Biochim. Biophys. Acta **245**, 386–397 (1971)
Lemasson, C., Etienne, A.L.: Biochim. Biophys. Acta **408**, 135–142 (1975)
Lozier, R.H., Butler, W.L.: Photochem. Photobiol. **17**, 133–137 (1973)
Malkin, R., Knaff, D.: Biochim. Biophys. Acta **325**, 336–340 (1973)
Mar, T., Govindjee: J. Theor. Biol. **36**, 427–446 (1972)
Miles, C.D., Jagendorf, A.T.: Arch. Biochem. Biophys. **129**, 711–719 (1969)
Mitchell, P.: Nature (London) **191**, 144–148 (1961)
Nakamoto, T., Krogmann, D.W., Vennesland, B.J.: J. Biol. Chem. **234**, 2783–2788 (1959)
Niel, C.B. van: Arch. Mikrobiol. **3**, 1–112 (1931)
Okayama, S., Butler, W.L.: Plant Physiol. **49**, 769–774 (1972)
Rabinovitch, E.I.: Photosynthesis and Related Processes. New York: Interscience, 1945, p. 162
Radmer, R., Cheniae, C.M.: Biochim. Biophys. Acta **253**, 182–186 (1971)
Radmer, R., Kok, B.: Biochim. Biophys. Acta **314**, 28–41 (1973)
Radmer, R., Kok, B.: Ann. Rev. Biochem. **44**, 409–433 (1975)
Reimer, S., Trebst, A.: Biochem. Physiol. Pflanzn. **168**, 225–232 (1975)

Renger, G.: Biochim. Biophys. Acta **256**, 428–439 (1972)
Renger, G., Bouges-Bocquet, B., Delosme, R.: Biochim. Biophys. Acta **292**, 796–807 (1973)
Ross, R.T., Calvin, M.: Biophys. J. **7**, 595–614 (1967)
Ruben, S., Randall, M., Kamen, M., Hyde, J.: J. Am. Chem. Soc. **63**, 877–879 (1941)
Rumberg, B., Schmidt-Mende, P., Weikard, J., Witt, H.T.: In: Photosynthetic Mechanisms of Green Plants. Kok, B., Jagendorf, A.T. (eds.). Washington: Nat. Acad. Sci. Nat. Res. Council. **1145**, 18–34 (1963)
Sinclair, J., Arnason, T.: Biochim. Biophys. Acta **368**, 393–400 (1974)
Stiehl, H.H., Witt, H.T.: Z. Naturforsch. **23B**, 220–224 (1968)
Stiehl, H.H., Witt, H.T.: Z. Naturforsch. **24B**, 1588–1598 (1969)
Trebst, A.: Ann. Rev. Plant. Physiol. **25**, 423–458 (1974)
Trebst, A., Eck, H., Wagner, S.: In: Photosynthetic Mechanisms of Green Plants. Kok, B., Jagendorf, A.T. (eds.). Washington: Nat. Acad. Sci. Nat. Res. Council. **1145**, 174–194 (1963)
Vaklinova, S., Niklova-Tsenova, E., Anchelova, S.: C.R. Acad. Sci. Bulg. **19**, 1191–1194 (1966)
Velthuys, B.R.: Biochim. Biophys. Acta **396**, 392–401 (1975)
Velthuys, B.R., Amesz, J.: Biochim. Biophys. Acta **333**, 85–94 (1974)
Velthuys, B.R., Visser, J.W.M.: FEBS Lett. **55**, 109–112 (1975)
Vermeglio, A., Mathis, P.: Biochim. Biophys. Acta **314**, 57–65 (1973)
Vernon, L.P., Shaw, E.R.: Biochem. Biophys. Res. Commun. **36**, 878–884 (1969)
Weiss, C., Solnit, K.T., von Gutfeld, R.: Biochim. Biophys. Acta **253**, 298–301 (1971)
Witt, H.T.: Quart. Rev. Biophys. **4**, 365–477 (1971)
Wraight, C.A., Crofts, A.R.: Eur. J. Biochem. **19**, 386–397 (1971)
Wurmser, R.: Oxidations et Réductions. Paris: Presses Univ. France, 1930
Yamashita, T., Butler, W.L.: Plant Physiol. **43**, 1978–1986 (1968)
Yamashita, T., Butler, W.L.: Plant Physiol. **44**, 435–438 (1969)
Yamashita, T., Horio, T.: Plant Cell. Physiol. **9**, 267–284 (1968)
Yamashita, T., Tomita, G.: Plant Cell. Physiol. **16**, 283–296 (1975)
Yamashita, T., Tsuji, J., Tomita, G.: Plant Cell. Physiol. **12**, 117–126 (1971)
Zankel, K.: Biochim. Biophys. Acta **245**, 373–385 (1971)
Zankel, K.: Biochim. Biophys. Acta **325**, 138–148 (1973)

9. Ferredoxin

D.O. Hall and K.K. Rao

A. Introduction

Ferredoxins are a group of iron-containing proteins with relatively low molecular weights which function as electron carriers in a number of biological oxidation-reduction processes. The active centres of ferredoxins contain non-haem iron (i.e. iron atoms of a different type to that are found in haem proteins) and an equivalent amount of "labile sulphur" e.g. $2Fe-2S$, $4Fe-4S$, per molecule. Ferredoxins are members of that class of metalloproteins known as iron-sulphur proteins. Ferredoxins are ubiquitous; they are present in all types of bacteria, algae, plants and animals. We will limit this chapter to the soluble ferredoxins from algae and plants only, which contain $2Fe-2S$ active centres. A number of reviews on iron-sulphur proteins have appeared recently (Buchanan and Arnon, 1971; Orme-Johnson, 1973; Hall et al., 1973, 1974, 1975; Lovenberg, 1973) which may be consulted for details.

The name ferredoxin was aptly coined by Mortenson et al. (1962) for a protein they isolated from *Clostridium pasteurianum* which contained iron and which acted as an electron transfer factor in the pyruvate phosphoroclastic reaction. Tagawa and Arnon (1962) isolated an iron-containing protein from spinach leaves which resembled *C. pasteurianum* ferredoxin in composition and biological function and they extended the term ferredoxin to include all non-haem iron proteins with low molecular weights and redox potentials which are capable of photoreducing $NADP^+$. Earlier workers had reported ferredoxin-type activity in leaf extracts and had named the component as methaemoglobin reducing factor (Davenport et al., 1952), photosynthetic pyridine nucleotide reductase (San Pietro and Lang, 1958), etc.

B. Extraction and Purification

Ferredoxins are acidic, coloured proteins and they can be isolated by selective and repeated chromatography on the anion exchanger, DEAE cellulose, using a sodium chloride gradient for elution. The individual steps involved vary with the species from which the protein is extracted, but all procedures should be carried out in the cold ($<5°C$), and at a pH between 7 and 8. The extraction and elution media are usually buffered with 20 mM Tris-HCl, pH 7.5 (see also Buchanan and Arnon, 1971; Rao et al., 1971; Hall et al., 1972).

I. Preparation of Cell-Free Extracts

Algae or leaves are suspended in buffer and ground in a blender at full speed for 3 min. With some algal species and thick fibrous plant material, grinding with glass beads or grinding in 50% cold aqueous acetone ($-15°C$) is helpful to break open the cells. Addition of dithiothreitol, Tween 80, or polyvinyl pyrrolidone (PVP) to the grinding medium reduces the formation of brown and often troublesome oxidation products of phenols that are formed from certain plants. An acetone powder of the algae may also be prepared and then ground in buffer. The ground material is filtered through several layers of cheesecloth and spun in a centrifuge at low speed ($2,000 \cdot g$) to separate the cell debris. Ferredoxin is present in the supernatant.

II. Separation and Purification

The supernatant is treated with DEAE cellulose, suspended in 0.1 M Cl^-, either batchwise or in a column. The ferredoxin is absorbed on to the DEAE cellulose. Some of the contaminating proteins and nucleic acids are removed by washing the DEAE cellulose with 0.2 M Cl^- after which the ferredoxin, as a red band, is eluted with 0.8 M Cl^-. Further purification is achieved by ammonium sulphate fractionation (ferredoxins are precipitated at $>80\%$ saturated ammonium sulphate), absorption on DEAE cellulose, followed by elution with 0.2 M $\rightarrow$ 0.8 M chloride gradient (ferredoxins are eluted between 0.3 and 0.35 M Cl^- concentration), gel filtration using Sephadex G-25 or Biogel P-10, calcium phosphate gel chromatography, etc. Many ferredoxins have been crystallized from ammonium sulphate solutions. During the purification steps the ferredoxin can be monitored by its colour, optical absorption at 420 nm, or by the e.p.r. signal of a reduced sample (see E.III).

C. Assay

The biological assay is based on the principle that ferredoxin is required for the photoreduction of $NADP^+$ in chloroplasts by the electrons donated from water. The reaction is catalyzed by a flavoprotein (ferredoxin–$NADP^+$ oxidoreductase) which is usually found in the chloroplast membranes (SHIN et al., 1963):

$$NADP^+ + H_2O \xrightarrow[\text{Fd + Fd-NADP reductase}]{\text{illuminated chloroplasts}} NADPH_2 + \tfrac{1}{2}O_2.$$

Washed spinach chloroplasts are prepared according to WHATLEY and ARNON (1963). The assay is performed in a 3-ml quartz cuvette containing: chloroplasts (about 100 µg chlorophyll); $NADP^+$, 0.01 M; ferredoxin, 10 to 50 µg, and 50 mM Tris-HCl buffer, pH 7.8 to a final volume of 3 ml. The cuvettes are illuminated at room temperature and the NADPH formed is measured by the increase in absorbance at 340 nm (SAN PIETRO, 1963). The activity of a ferredoxin sample can also be determined by measuring the amount of oxygen evolved in the above reaction using an oxygen electrode (HASLETT et al., 1973).

D. Occurrence and Biosynthesis

As already mentioned ferredoxins are present in all types of algae and plants; so far it has been extracted from more than 30 different species (see YASUNOBU and TANAKA, 1973). The relative abundance of extractable ferredoxins varies very much from species to species—e.g. *Spirulina* yields more than 50 mg of ferredoxin per 100 g of dry cells, whereas less than 10 mg of Fd per kg of cells is obtained from *Cyanidium* (our unpublished data). Among plants it is convenient to prepare ferredoxin from spinach leaves in a fairly good yield. Molar ratios of ferredoxins to chlorophyll vary from 1/50 in *Nostoc* (MITSUI and ARNON, 1971) to 1/360 in green bean (*Phaseolus*) leaves (HASLETT et al., 1973). The ferredoxin to chlorophyll ratio is reported to be higher in the C_4 plants than in the C_3 plants (LEE et al., 1970). HASLETT et al. (1973) found that ferredoxins are synthesized on the cytoplasmic ribosomes of green bean leaves by a light-dependent reaction.

E. Properties

A summary of the properties of ferredoxins is listed in Table 1.

Table 1. Properties of ferredoxins

1. Molecular weight	10,500–11,000
2. Amino acid residues per molecule	95–100; 96–98 in the ferredoxins sequenced so far
3. No. of atoms of iron and labile sulphur	2 per molecule
4. Oxidation-reduction potential	$E_0' = -420$ mV at pH 7
5. No. of electrons transferred	One
6. Optical absorption spectra of oxidized ferredoxin	Peaks at 465, 420, 330, and 278 nm
7. Molar extinction coefficient	9,400–10,000 at 420 nm
8. EPR spectra of reduced ferredoxin measured at 77 K	$g_x = 1.88$, $g_y = 1.95$, $g_z = 2.04$

I. Oxidation-Reduction Potential

One of the characteristic properties of ferredoxins which distinguishes them from other proteins is their very negative oxidation-reduction potential, close to that of the hydrogen electrode, i.e. -420 mV. The ferredoxins transfer one electron during oxidation-reduction reactions (EVANS et al., 1968; TAGAWA and ARNON, 1968).

II. Optical Spectra

Ferredoxins are red in colour and they have typical optical absorption spectra with peaks around 465, 420, 390 and 278 nm (Fig. 1a). The visible absorption

Fig. 1 a. Optical absorption spectra of ferredoxins. Protein concentrations are ca. 1 mg ml^{-1}. (From HALL et al., 1973)

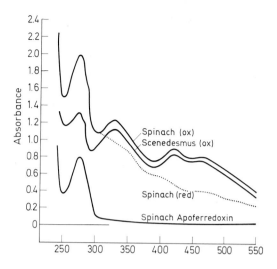

is mainly due to ligand to metal charge transfer bands of the (Fe−S) chromophore, whereas both the protein and the chromophore contribute to the absorbance in the UV region. The absorbance at 420 nm decrease by about 50% when ferredoxin is reduced anaerobically with sodium dithionite (or by light + chloroplasts) − the reaction is reversible and the original spectra can be restored by bubbling oxygen through the reduced protein. When the (Fe−S) chromophore is removed from ferredoxin (by precipitation with trichloroacetic acid) all the absorption in the visible region and about half the absorbance in the UV region is lost. Ferredoxins are optically active and show characteristic circular dichroism spectra in the visible and UV region (Fig. 1 b).

III. Electron Paramagnetic Resonance (EPR)

An important distinguishing property of ferredoxins is that they exhibit EPR spectra in the reduced state, which can be measured at liquid N_2 temperatures (77 K) (Fig. 1 c). The spectrum is broad, with rhombic symmetry, with a 'g'$_{ave}$ value of about 1.94. The lineshapes of the EPR spectra can be altered by treating the ferredoxins with chaotropic agents or salts (CAMMACK, 1975). Ferredoxins as prepared (in the oxidized state) do not show an EPR signal. The generation of an EPR signal with $g_{ave} = 1.94$ on reduction, is used as a qualitative test for the detection of ferredoxin (even in relatively impure extracts) and the intensity of the signal gives an estimate of the amount of ferredoxin in a sample.

IV. Mössbauer Spectroscopy

This is a very useful technique to probe into the valence and spin states of iron atoms liganded to proteins and has been extensively used to elucidate the nature of the active centre of ferredoxins (RAO et al., 1971; DUNHAM et al., 1971). Since Mössbauer spectra for iron are given only by the ^{57}Fe nucleus the iron atoms

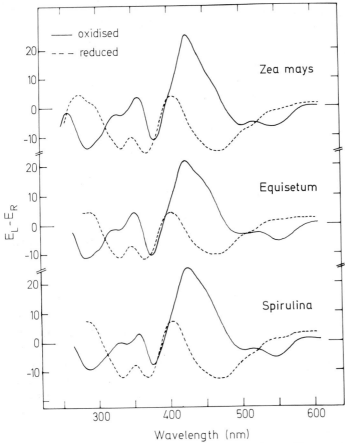

Fig. 1b. Circular dichroism spectra of ferredoxins. (From Hall et al., 1973)

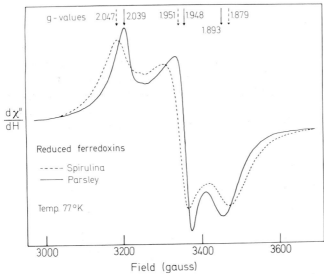

Fig. 1c. EPR spectra of reduced ferredoxins. (From Hall et al., 1973)

Fig. 1d. Mössbauer spectra of oxidized and reduced ferredoxins. (After Rao et al., 1971)

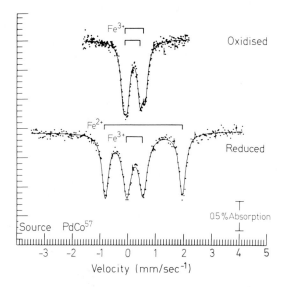

in native ferredoxin (along with the labile sulphur) are removed and the active
ferredoxin is then reconstituted from the apoprotein by adding ^{57}Fe in solution
with sodium sulphide. The Mössbauer spectra of oxidized and reduced ferredoxins
are shown in Figure 1d. The spectrum of oxidized ferredoxin consists of two
overlapping quadrupole-split doublets with splitting constants and chemical shifts
indicative of two high-spin Fe^{3+} atoms in almost similar chemical environment.
On reduction with a single electron the spectrum of reduced ferredoxin has two
sets of lines one due to Fe^{3+} and the other due to Fe^{2+} — thus one of the two
ferric atoms of ferredoxin is converted to ferrous on reduction.

F. Nature of the Active Center

The nature of the (Fe−S) chromophore in ferredoxins has been elucidated by
the concerted effort of various groups of investigators using a variety of physical
techniques like EPR, NmR, endor, and Mössbauer spectroscopy, magnetic suscepti-
bility measurements, and chemical methods. (For a discussion of various techniques
refer to Tsibris and Woody, 1970; Lovenberg, 1973; Hall et al., 1974.)

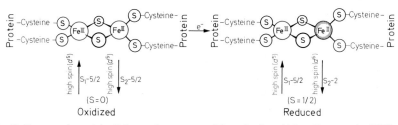

Fig. 2. Proposed model of the active center of ferredoxins. (After Rao et al., 1971)

Sequences of chloroplast ferredoxins

```
                1                                                              20
Aphanothece    Ala–Ser–Tyr–Lys–Val–Thr–Leu–Lys–Thr–Pro–Asp–Gly–Asp–Asn–Val–Ile–Thr–Val–Pro–Asp–
Spirulina M.   Ala–Thr–Tyr–Lys–Val–Thr–Leu–Ile–Ser–Glu–Ala–Gly  Asn–Glu–Thr–Ile–Asp–Cys–Asp–Asp–
                                                      Glu  Ile
Spirulina P.   Ala–Thr–Tyr–Lys–Val–Thr–Leu–Ile–Asp–Glu–Ala  Gly  Asn–Glu–Thr–Ile–Asp–Cys–Asp–Asp–
                                                      Glu  Ile
Scenedesmus    Ala–Thr–Tyr–Lys–Val–Thr–Leu–Lys–Thr–Pro–Ser–Gly–Asp–Gln–Thr–Ile–Glu–Cys–Pro–Asp–
Equisetum       –   –  Tyr–Lys–Thr–Val–Leu–Lys–Thr–Pro–Ser–Gly–Glu–Phe–Thr–Leu–Asp–Val–Pro–Glu–
Taro           Ala–Thr–Tyr–Lys–Val–Lys–Leu–Val–Thr–Pro–Ser–Gly–Gln–Gln–Glu–Phe–Gln–Cys–Pro–Asp–
Spinach        Ala–Ala–Tyr–Lys–Val–Thr–Leu–Val–Thr–Pro–Thr–Gly–Asn–Val–Glu–Phe–Gln–Cys–Pro–Asp–
Alfalfa        Ala–Ser–Tyr–Lys–Val–Lys–Leu–Val–Thr–Pro–Glu–Gly–Thr–Gln–Glu–Phe–Glu–Cys–Pro–Asp–
Leucaena        –  Ala–Phe–Lys–Val–Lys–Leu–Leu–Thr–Pro–Asp–Gly–Pro–Lys–Glu–Phe–Glu–Cys–Pro–Asp

                21                                                             40
Aphanothece    Asp–Glu–Tyr–Ile–Leu–Asp–Val–Ala–Glu–Glu  Glu–Gly–Leu–Asp–Leu–Pro–Tyr–Ser–│Cys│Arg–
Spirulina M.   Asp–Thr–Tyr–Ile–Leu–Asp–Ala–Ala–Glu–Glu–Glu–Glu–Leu–Asp–Leu–Pro–Tyr–Ser–│Cys│Arg–
Spirulina P.   Asp–Thr–Tyr–Ile–Leu–Asp–Ala–Ala–Glu–Glu–Glu–Glu–Leu–Asp–Leu–Pro–Tyr–Ser–│Cys│Arg–
Scenedesmus    Asp–Thr–Tyr–Ile–Leu–Asp–Ala–Ala–Glu–Glu–Ala–Gly–Leu–Asp–Leu–Pro–Tyr–Ser–│Cys│Arg–
Equisetum       –   –   –   –   –   –   –   –   –   –   –   –   –   –   –  Ser–│Cys│Arg–
Taro           Asp–Val–Tyr–Ile–Leu–Asp–Gln–Ala–Glu–Glu–Val–Gly–Ile–Asp–Leu–Pro–Tyr–Ser–│Cys│Arg–
Spinach        Asp–Val–Tyr–Ile–Leu–Asp–Ala–Ala–Glu–Glu–Gly–Ile–Asp–Leu–Pro–Tyr–Ser–│Cys│Arg–
Alfalfa        Asp–Val–Tyr–Ile–Leu–Asp–His–Ala–Glu–Glu–Glu–Gly–Ile–Asp–Val–Leu–Pro–Tyr–Ser–│Cys│Arg–
Leucaena       Asp–Val–Tyr–Ile–Leu–Asp–Gln–Ala–Glu–Glu–Leu–Gly–Ile–Asp–Leu–Pro–Tyr–Ser–│Cys│Arg

                41                                                             60
Aphanothece    Ala–Gly–Ala–│Cys│Ser–Thr–│Cys│Ala–Gly–Lys–Leu–Val–Ser–Gly–Pro–Ala–Pro–Asp–Glu–Asp–
Spirulina M.   Ala–Gly–Ala–│Cys│Ser–Thr–│Cys│Ala–Gly–Lys–Ile–Thr–Ser–Gly–Ser–Ile–Asp–Gln–Ser–Asp–
Spirulina P.   Ala–Gly–Ala–│Cys│Ser–Thr–│Cys│Ala–Gly–Thr–Ile–Thr–Ser–Gly–Thr–Ile–Asp–Gln–Ser–Asp–
Scenedesmus    Ala–Gly–Ala–│Cys│Ser–Ser–│Cys│Ala–Gly–Lys–Val–Glu–Ala–Gly–Thr–Val–Asp–Gln–Ser–Asp–
Equisetum      Ala–Gly–Ala–│Cys│Ser–Ser–│Cys│Leu–Gly–Lys  –   –   –   –   –   –   –   –   –
Taro           Ala–Gly–Ser–│Cys│Ser–Ser–│Cys│Ala–Gly–Lys–Val–Lys–Val–Gly–Asp–Val–Asp–Gln–Ser–Asp–
Spinach        Ala–Gly–Ser–│Cys│Ser–Ser–│Cys│Ala–Gly–Lys–Leu–Lys–Thr–Gly–Ser–Leu–Asn–Gln–Asp–Asp–
Alfalfa        Ala–Gly–Ser–│Cys│Ser–Ser–│Cys│Ala–Gly–Lys–Val–Ala–Ala–Gly–Glu–Val–Asn–Gln–Ser–Asp–
Leucaena       Ala–Gly–Ser–│Cys│Ser–Ser–│Cys│Ala–Gly–Lys–Leu–Val–Glu–Gly–Asp–Leu–Asp–Gln–Ser–Asp–

                61                                                             80
Aphanothece    Gln–Ser–Phe–Leu–Asp–Asp–Asp–Gln–Ile–Gln–Ala–Gly–Tyr–Ile–Leu–Thr–│Cys│Val–Ala–Tyr–
Spirulina M.   Gln–Ser–Phe–Leu–Asp–Asp–Asp–Gln–Ile–Glu–Ala–Gly–Tyr–Val–Leu–Thr–│Cys│Val–Ala–Tyr–
Spirulina P.   Gln–Ser–Phe–Leu–Asp–Asp–Asp–Gln–Ile–Glu–Ala–Gly–Tyr–Val–Leu–Thr–│Cys│Val–Ala–Tyr–
Scenedesmus    Gln–Ser–Phe–Leu–Asp–Asp–Ser–Gln–Met–Asp–Gly–Gly–Phe–Val–Leu–Thr–│Cys│Val–Ala–Tyr–
Taro           Gly–Ser–Phe–Leu–Asp–Asp–Glu–Gln–Ile–Gly–Glu–Gly–Trp–Val–Leu–Thr–│Cys│Val–Ala–Tyr–
Spinach        Gln–Ser–Phe–Leu–Asp–Asp–Asp–Gln–Ile–Asp–Glu–Gly–Trp–Val–Leu–Thr–│Cys│Ala–Ala–Tyr–
Alfalfa        Gly–Ser–Phe–Leu–Asp–Asp–Glu–Gln–Ile–Glu–Glu–Gly–Trp–Val–Leu–Thr–│Cys│Val–Ala–Tyr–
Leucaena       Gln–Ser–Phe–Leu–Asp–Asp–Glu–Gln–Ile–Glu–Glu–Gly–Trp–Val–Leu–Thr–│Cys│Ala–Ala–Tyr–

                81                                                      97
Aphanothece    Pro–Thr–Gly–Asp–Cys–Val–Ile–Glu–Thr–His–Lys–Glu–Glu–Ala–Leu–Tyr
Spirulina M.   Pro–Thr–Ser–Asp–Cys–Thr–Ile–Glu–Thr–His–Gln–Glu–Glu–Gly–Leu–Tyr
Spirulina P.   Pro–Thr–Ser–Asp–Cys–Thr–Ile–Lys–Thr–His–Gln–Glu–Glu–Gly–Leu–Tyr
Scenedesmus    Pro–Thr–Ser–Asp–Cys–Thr–Ile–Ala–Thr–His–Lys–Glu–Glu–Asp–Leu–Phe
Taro           Pro–Val–Ser–Asp–Val–Thr–Ile–Glu–Thr–His–Lys–Glu–Glu–Glu–Leu–Thr–Ala
Spinach        Pro–Val–Ser–Asp–Val–Thr–Ile–Glu–Thr–His–Lys–Glu–Glu–Glu–Leu–Thr–Ala
Alfalfa        Ala–Lys–Ser–Asp–Val–Thr–Ile–Glu–Thr–His–Lys–Glu–Glu–Glu–Leu–Thr–Ala
Leucaena       Pro–Arg–Ser–Asp–Val–Val–Ile–Glu–Thr–His–Lys–Glu–Glu–Glu–Leu–Thr–Ala
```

Fig. 3. Amino acid sequences of ferredoxins.
Reference: *Aphanothece* and *Spirulina platensis*: WADA et al. (1975); *S. maxima*: TANAKA et al. (1975); *Scenedesmus*: SUGENO and MATSUBARA (1969); *Taro*: RAO and MATSUBARA (1970); *Spinach*: MATSUBARA and SASAKI (1967); *Alfalfa*: KERESZTES-NAGY et al. (1969); *Leucaena*: BENSON and YASUNOBU (1969); *Equisetum*: KAGAMIYAMA et al. (1975)

Chemical analysis shows the presence of two atoms of ferric iron in a molecule of ferredoxin. When treated with dilute acids the non-cysteine sulphur of ferredoxin is liberated as hydrogen sulphide, each molecule liberating the equivalent of two sulphides—this sulphur is usually referred to as "labile sulphur". Amino acid analysis of various ferredoxins shows the presence of 95 to 100 amino acid residues per molecule, four to six of which are cysteines. None of these cysteines in native ferredoxin are reactive towards —SH reagents like iodoacetate. However, if the protein is denatured, or if the iron is removed, the cysteines react easily with iodoacetate suggesting the linkage of the cysteine residues with the iron atoms in the native protein. The primary structure of four algal and four plant ferredoxins are known (Fig. 3); they contain five to six cysteine residues per molecule. Four of the cysteines (at positions 39, 44, 47 and 77) are invarient in all the sequences. There are only four cysteines in a molecule of *Equisetum* ferredoxin; these cysteines occupy positions corresponding to the four invariant cysteines of ferredoxins whose complete sequences are known. Since there is no difference in the physical properties and biological activity between *Equisetum* ferredoxin and the other ferredoxins possessing more than four cysteines (KAGAMIYAMA et al., 1975), it is evident that only four cysteines are essential to maintain the structure and activity of ferredoxins. Though there are two high spin Fe^{3+} atoms in oxidized ferredoxin, the molecule as a whole is diamagnetic, since as proposed by GIBSON et al. (1966) the two iron atoms are antiferro magnetically coupled to produce a net spin $S=0$. When reduced, one of the Fe^{3+} atoms is converted to Fe^{2+} with a total spin for the molecule $S=\frac{1}{2}$. The available physicochemical and biological data suggest a structure as shown in Figure 2, for the active centre of ferredoxins. Confirmation of the structure must wait until the X-ray crystallographic data on the protein is available.

G. Stability

The physical and biological properties of ferredoxins depend on the intactness of the $(Fe-S)$ chromophore. The chromophore structure is altered considerably by exposure of ferredoxin to heat or oxygen, with concomitant loss in biological activity. A comparison of the relative biological activities of various ferredoxins after storage, at different temperatures in air, has shown that *Spirulina* ferredoxin is the most resistant to oxygen followed by ferredoxin from *Zea mays* (HALL et al., 1972). It is advisable to keep ferredoxins in small samples, in liquid N_2, for long-term storage.

H. Biological Function

The biological reactions in which chloroplast ferredoxins are known to participate are listed in Table 2. The most important biological role for ferredoxins is as a catalyst in cyclic and noncyclic photophosphorylation. In noncyclic photophosphorylation both $NADPH_2$ and ATP are formed.

$$NADP^+ + H_2O + 2\,ADP + 2\,Pi \xrightarrow[\text{chloroplasts}]{\text{Fd, light}} NADPH_2 + 2\,ATP + \tfrac{1}{2}O_2.$$

Cyclic photophosphorylation produces ATP only.

$$ADP + Pi \xrightarrow[\text{chloroplasts}]{\text{Fd, light}} ATP.$$

Table 2. Biological reactions catalyzed by ferredoxins, with illuminated chloroplasts

Reaction	Other essential components	Reference
1. NADP$^+$ photoreduction and noncyclic photophosphorylation	Ferredoxin-NADP$^+$ oxidoreductase	WHATLEY and ARNON (1963) SHIN et al. (1963)
2. Cyclic photophosphorylation	Chloroplast electron transport chain	WHATLEY and ARNON (1963)
3. Sulphite reduction $SO_3^= \to S^=$	Sulphite reductase	SCHMIDT and TREBST (1969)
4. Nitrite reduction $NO_2^- \to NH_3$	Nitrite reductase	LOSADA and PANEQUE (1971)
5. Fatty acid desaturation	Fatty acid desaturase, NADPH oxidase	NAGAI and BLOCH (1966)
6. Fructose 1-6 diphosphate $+ H_2O \to$ Fructose 6 phosphate $+ Pi$	Fructose diphosphatase	BUCHANAN et al. (1971)
7. Pyruvate decarboxylation Pyruvate $+ CoA \rightleftharpoons$ Acetyl CoA $+ CO_2$ (In blue-green algae only)	Pyruvate-ferredoxin oxidoreductase, ATP	LEACH and CARR (1971)

The algal and plant ferredoxins are interchangeable and function with the same efficiency in these reactions. Also there seems no difference between ferredoxins from C_4 and C_3 plants in the NADP photoreduction activity (CRAWFORD and JENSEN, 1971).

I. Immunological Studies

HIEDEMANN-VAN WYK and KANNANGARA (1971) prepared an antibody to ferredoxin for studying the localization of ferredoxin in the thylakoid membrane. The results indicated a binding of ferredoxin to the membrane via the ferredoxin–NADP$^+$ reductase. TEL-OR and AVRON (1974) prepared antibodies to Swiss chard ferredoxin and found that they interact with other plant and green algal ferredoxins, but not with clostridial ferredoxin. TEL-OR et al (1975) extended this work and showed that on the basis of binding affinity with anti-Swiss chard ferredoxin immunoglobulin the ferredoxins can be classed into three types: (1) Swiss chard, spinach and alfalfa which give a strong precipitin reaction; (2) parsley and maize ferredoxin which give a moderate interaction, and (3) *Equisetum* and algal ferredoxins which show only a moderate interaction. The studies show that there are subtle differences in the protein conformation between various ferredoxins.

J. Homology in the Primary Structures

The complete amino acid sequence of four algal and four higher plant ferredoxins are known (Fig. 3). There is a striking homology in their structure; 46 out of 98 residues occupy identical positions in the sequences, including the four invariant cysteines which are liganded to the iron atoms at the active centre. There are an additional 30 positions in the sequence where either all the algal ferredoxins or all the plant ferredoxins have identical amino acid residues. Noteworthy differences between algal ferredoxins and plant ferredoxins are: (1) all algal ferredoxins have a Cys at 85; (2) all plant ferredoxins have a Trp at position 73, and (3) the carboxy terminal of all algal ferredoxins is an aromatic amino acid, while those of all plant ferredoxins is alanine. These differences, however, do not change their biological function and activity though they may contribute to their relative stability and immunological cross reactivity.

Bacterial ferredoxins are believed to be one of the earliest proteins to be evolved (HALL et al., 1973). However, the evolutionary development of ferredoxin may have ceased with the formation of oxygen evolving photosynthetic organisms so that there is very little difference in the structure and function of plants and algal ferredoxins.

References

Benson, A.A., Yasunobu, K.T.: J. Biol. Chem. **244**, 955–963 (1969)
Buchanan, B.B., Arnon, D.I.: Methods Enzymol. **23A**, 413–440 (1971)
Buchanan, B.B., Schürmann, P., Kalberer, P.P.: J. Biol. Chem. **246**, 5952–5959 (1971)
Cammack, R.: Biochem. Soc. Trans. **3**, 482–488 (1975)
Crawford, C.G., Jensen, R.G.: Plant Physiol. **47**, 447–449 (1971)
Davenport, H.E., Hill, R., Whatley, F.R.: Proc. Roy. Soc., Ser. B **139**, 346–359 (1952)
Dunham, W.R., Bearden, A.J., Salmeen, I.T., Palmer, G., Sands, R.H., Orme-Johnson, W.H., Beinert, H.: Biochim. Biophys. Acta **253**, 134–152 (1971)
Evans, M.C.W., Hall, D.O., Bothe, H., Whatley, F.R.: Biochem. J. **110**, 485–489 (1968)
Gibson, J.F., Hall, D.O., Thornley, J.H.M., Whatley, F.R.: Proc. Nat. Acad. Sci. U.S. **56**, 987–990 (1966)
Hall, D.O., Cammack, R., Rao, K.K.: Pure Appl. Chem. **34**, 553–577 (1973)
Hall, D.O., Cammack, R., Rao, K.K.: In: Iron in Biochemistry and Medicine. Jacobs, A., Worwood, M. (eds.). London-New York: Academic Press, 1974, pp. 279–334
Hall, D.O., Rao, K.K., Cammack, R.: Biochem. Biophys. Res. Commun. **47**, 798–802 (1972)
Hall, D.O., Rao, K.K., Cammack, R.: Sci. Progr. Oxf. **62**, 285–317 (1975)
Haslett, B.G., Cammack, R., Whatley, F.R.: Biochem. J. **136**, 697–703 (1973)
Hiedemann-Van Wyk, D., Kannangara, C.G.: Z. Naturforsch. **26B**, 46–50 (1971)
Kagamiyama, H., Rao, K.K., Hall, D.O., Cammack, R., Matsubara, H.: Biochem. J. **145**, 121–123 (1975)
Keresztes-Nagy, S., Perini, F., Margoliash, E.: J. Biol. Chem. **244**, 981–995 (1969)
Leach, C.K., Carr, N.G.: Biochim. Biophys. Acta **245**, 165–174 (1971)
Lee, S.S., Travis, J., Black, C.C.: Arch. Biochem. Biophys. **141**, 676–689 (1970)
Losada, M., Paneque, A.: Methods Enzymol. **23A**, 487–491 (1971)
Lovenberg, W. (ed.): Iron sulphur proteins. New York-London: Academic Press, 1973, Vols. I and II
Matsubara, H., Sasaki, M.: J. Biol. Chem. **243**, 1732–1757 (1967)
Mitsui, A., Arnon, D.I.: Physiol. Plantarum **25**, 135–140 (1971)
Mortenson, L.E., Valentine, R.C., Carnahan, J.E.: Biochem. Biophys. Res. Commun. **7**, 448–452 (1962)
Nagai, J., Bloch, K.: J. Biol. Chem. **241**, 1925–1927 (1966)
Orme-Johnson, W.H.: Ann. Rev. Biochem. **42**, 159–204 (1973)
Rao, K.K., Cammack, R., Hall, D.O., Johnson, C.E.: Biochem. J. **122**, 257–265 (1971)
Rao, K.K., Matsubara, H.: Biochem. Biophys. Res. Commun. **38**, 500–506 (1970)
San Pietro, A.: Methods Enzymol. **6**, 439–445 (1963)
San Pietro, A., Lang, H.M.: J. Biol. Chem. **231**, 211–229 (1958)
Schmidt, A., Trebst, A.: Biochim. Biophys. Acta **180**, 529–535 (1969)
Shin, M., Tagawa, K., Arnon, D.I.: Biochem. Z. **338**, 84–96 (1963)

Sugeno, K., Matsubara, H.: J. Biol. Chem. **244**, 2979–2989 (1969)
Tagawa, K., Arnon, D.I.: Nature (London) **195**, 537–543 (1962)
Tagawa, K., Arnon, D.I.: Biochim. Biophys. Acta **153**, 602–613 (1968)
Tanaka, M., Haniu, M., Zeitlin, S., Yasunobu, K.T., Evans, M.C.W., Rao, K.K., Hall, D.O.: Biochem. Biophys. Res. Commun. **64**, 399–407 (1975)
Tel-Or, E., Avron, M.: Europ. J. Biochem. **47**, 417–421 (1974)
Tel-Or, E., Cammack, R., Hall, D.O.: FEBS Lett. **53**, 135–138 (1975)
Tsibris, J.C.M., Woody, R.W.: Coordin. Cem. Rev. **5**, 417–458 (1970)
Wada, K., Hase, T., Tokunaga, H., Matsubara, H.: FEBS Lett. **55**, 102–104 (1975)
Whatley, F.R., Arnon, D.I.: Methods Enzymol. **6**, 308–313 (1963)
Yasunobu, K.T., Tanaka, M.: In: Iron-Sulphur Proteins. Lovenberg, W. (ed.). London-New York: Academic Press, 1973, Vol. II, pp. 27–130

10. Flavodoxin

H. Bothe

A. Biological Properties

The first example of this type of flavoprotein was discovered in extracts from the blue-green algae *Anacystis nidulans* and *Anabaena cylindria* and named phyto-flavin (SMILLIE, 1965a, b). Phytoflavin was active in the photosynthetic NADP$^+$-reduction by isolated chloroplasts as a substitute for ferredoxin. Independently of SMILLIE's investigations, KNIGHT and HARDY (1966) found a flavoprotein related to phytoflavin and called flavodoxin in the obligate anaerobe bacterium *Clostridium pasteurianum* grown under iron-limiting culture conditions. Since both the biological and the chemical properties of flavodoxin and phytoflavin subsequently turned out to be very similar, it was proposed to abandon the name phytoflavin (BOTHE et al., 1971).

In the meantime, flavodoxins have been described from the photosynthetic microorganisms *Chlorella fusca* (ZUMFT and SPILLER, 1971), *Rhodospirillum rubrum* (CUSANOVICH and EDMONDSON, 1971), as well as from *Anacystis* and *Anabaena*, from the blue-green alga *Synechococcus lividus* (NORRIS et al., 1972), and from several other non-photosynthetic bacteria (cf. Table 1). In most cases, the formation of flavodoxins is induced by low iron strength in the media. However, flavodoxin and ferredoxin are constitutively formed in *Desulfovibrio, Escherichia coli, Azotobacter* and presumably in *Rhizobium*. A flavodoxin has not yet been found in higher plants.

In vitro, flavodoxins replace ferredoxin not only in the photosynthetic NADP$^+$-reduction by isolated chloroplasts, but also in a number of other ferredoxin-dependent reactions such as the pyruvate phosphoroclastic reaction and the hydrogen consumption or evolution in extracts from *Clostridium*, the nitrite reduction by spinach nitrite reductase, the ferredoxin-catalyzed cyclic photophosphorylation in preparations from spinach and *Anacystis*, the acetylene reduction by nitrogenases, and the reduction of sulfite in the presence of *Desulfovibrio* extracts. Since ferredoxin and flavodoxin are effective with more or less the same specific activity in all these assays, this type of experiment does not provide an indication of the natural function of flavodoxins, which is still not understood. Moreover, there is another interesting problem which has not been treated to date: the induction of the flavodoxin formation by the omission of the iron cation from the media.

The other flavoprotein of the photosynthetic electron transport, the ferredoxin-NADP$^+$ oxidoreductase, is also a NADPH diaphorase and a transhydrogenase. These activities are not observed with the flavodoxin from *Anacystis* (SMILLIE, 1965a). Both the ferredoxin- and the flavodoxin-dependent photoreduction of NADP$^+$ require the presence of the ferredoxin-NADP$^+$ oxidoreductase (SMILLIE, 1965a; TREBST and BOTHE, 1966). This enzyme forms a one-to-one complex with flavodoxin as well as with ferredoxin (FOUST et al., 1969). Iron deficiency presum-

Table 1. Properties of the flavodoxins so far described

Organism	1 Clostridium pasteurianum	2 Clostridium MP	3 Peptostreptococcus elsdenii	4 Desulfovibrio vulgaris	5 Escherichia coli	6 Azotobacter	7 Rhodospirillum	8 Anacystis	9 Chlorella	10 Klebsiella	11 Rhizobium
Long-wavelength absorbance max. (oxid.)	443	445	445	458	467	452	460	465	464	452	
Prosthetic group	FMN	FMN	FMN	FMN	FMN	FMN	FMN	FMN	FMN		
Molecular weight	14,600	13,800	15,000	16,000	14,500	23,000	23,000	21,000	22,000	21,000	occurrence communicated
E'_0 oxid./semiquinone (mV, pH 7.0)	−132	−92	−115	−102	−240	+50		−221			
E'_0 semiquinone/hydroquinone (mV, pH 7.0)	−419	−399	−371	−431	−410	−495		−447			
Active in the photosynthetic $NADP^+$-reduction	+	+	+	+	+	+	+	+	+	?	
Induced by iron deficiency	+	+	+	−	−	−	+	+	+	−	

References: 1. MAYHEW (1971); 2. MAYHEW (1971); 3. MAYHEW et al. (1969); 4. DUBOURDIEU et al. (1975); 5. VETTER and KNAPPE (1971) (redox potentials at pH 7.7); 6. SHETHNA et al. (1966), BARMAN and TOLLIN, (1972), VAN LIN and BOTHE (1972); 7. CUSANOVICH and EDMONDSON (1971); 8. BOTHE et al. (1971), ENTSCH and SMILLIE (1972); 9. ZUMFT and SPILLER (1971); 10. YOCH (1974); 11. WONG et al. (1971).

ably induces only one enzyme and not a second, flavodoxin-specific NADPH oxidoreductase, since antibody preparations against the spinach ferredoxin-$NADP^+$ oxidoreductase inhibit both the ferredoxin- and the flavodoxin-dependent photoreduction of $NADP^+$ by particles from *Anacystis* (BOTHE and BERZBORN, 1970).

B. Chemical Properties

All flavodoxins so far isolated have molecular weights of about 20,000, and their prosthetic group consists of one flavin mononucleotide (FMN) per molecule (see Table 1). The flavodoxins are the simplest flavoproteins and are suitable for studies on the chemistry of the flavin-dependent catalysis. Some flavodoxins are chemically well-characterized now (e.g. from *Clostridium*: MAYHEW, 1971; from *Azotobacter*: EDMONDSON and TOLLIN, 1971a, b; MACKNIGHT et al., 1973; and from *Desulfovibrio*: DUBOURDIEU et al., 1975). The amino acid sequence was completely determined for the protein from *Peptostreptococcus elsdenii* (TANAKA et al., 1973), from *Clostridium* MP (TANAKA et al., 1974) and from *Desulfovibrio vulgaris* (DUBOURDIEU et al., 1973) and partially from others (MACKNIGHT et al., 1974). X-ray crystallographic studies with the flavodoxin from *Desulfovibrio* (WATENPAUGH et al., 1973) and from *Clostridium* (ANDERSEN et al., 1972; BURNETT et al., 1974) revealed the structure of the protein and the location of the flavin in the molecule. The chemistry of the flavodoxins from photosynthetic microorganisms has been investigated in less detail.

Flavodoxins have the three redox states: oxidized, half-reduced (= semiquinone or radical), and fully reduced (= hydroquinone) form. The oxidized form has visible absorbance maxima in the vicinity of 370 nm and 450 nm and a shoulder at 490 nm which is not observed with free flavin. When anaerobically titrated with dithionite or photoreduced at the expense of EDTA, a solution of oxidized flavodoxin gradually turns blue with the generation of a broad absorbance band at 600 nm, which is due to the formation of the neutral semiquinone species. Depending on the pH value used, up to 90% of the oxidized flavodoxin is converted to the radical form upon half-reduction. In this respect flavodoxins are clearly distinguished from free FMN where less than 10% of the radical is generated when half-reduced.

Further addition of dithionite results in the formation of the fully reduced form in most of the flavodoxins. Some semiquinone forms, particularly that of the *Azotobacter* protein (= Shethna flavoprotein), show a marked resistance to further reduction at neutrality. The fully reduced form is very sensitive to reoxidation, whereas the radical can only slowly be reoxidized by oxygen. The generation of the semiquinone form in flavodoxins is still a matter of dispute (cf. discussion in BOTHE et al., 1971).

MAYHEW et al. (1969) gave the first accurate determination of the redox potentials of a flavodoxin, and their publication formed the basis for further measurements (Table 1). The E_0' value of the hydroquinone/semiquinone varies between -370 mV and -495 mV in the different flavodoxins and is hence as negative as that of ferredoxin. Contrary to this, the potential of the transition oxidized to semiquinone form is between -100 mV and -250 mV with the marked exception of the *Azotobacter* flavodoxin which gave a value of $+50$ mV (Barman and Tollin, 1972).

As expected from the redox potentials, flavodoxins act as one-electron carriers shuttling between the fully reduced and the semiquinone form when substituting for ferredoxin in the photosynthetic $NADP^+$-reduction. This statement was deduced from the observation that chloroplasts and light reduce the flavodoxin from *Anacystis* completely to the hydroquinone form, whereas it is converted only to the semiquinone by NADPH and ferredoxin-$NADP^+$ oxidoreductase, regardless of a stoichiometric amount or high excess of NADPH in the assay (Bothe et al., 1971). More direct evidence for this statement came from the experiments of Norris et al. (1972) who were able to detect the EPR signal of the semiquinone in intact cells of *Synechococcus* grown in D_2O and under iron deficiency. This signal is closely associated with the acceptor side of photosystem I. The signal intensity decreases upon transition from dark to light, indicating a reduction from the semiquinone to the hydroquinone.

Hall et al. (1971) proposed that a ferredoxin may have been among the earliest proteins formed in biological evolution. On the other hand, it many be speculated that pterin and flavin-like substances or even flavodoxin were developed in early earth history. In laboratory experiments, thermal polymerization of amino acids results in brownish products among which flavin-like substances can be isolated (Fox, 1972 and personal communication). When the earth atmosphere was enriched in oxygen due to photosynthesis, the hydroquinone form of flavodoxin may have turned out to be too sensitive to air in catalysis and therefore had to be replaced by the more oxygen-stable reduced ferredoxin. Therefore, the genetic information for the formation of a flavodoxin may only be retained in some microorganisms which, however, can be reactivated upon low iron strength.

References

Anderson, R.D., Apgar, P.A., Burnett, R.M., Darling, G.D., LeQuesne, M.E., Mayhew, S.G., Ludwig, M.L.: Proc. Nat. Acad. Sci. U.S. **69**, 3189–3191 (1971)

Barman, B.G., Tollin, G.: Biochemistry **11**, 4755–4759 (1972)

Bothe, H., Berzborn, R.: Z. Naturforsch. **25B**, 529–534 (1970)

Bothe, H., Hemmerich, P., Sund, H.: In: Flavins and Flavoproteins. 3rd Intern. Symp. Kamin, H. (ed.). Baltimore: Univ. Park; London: Butterworths, 1971, pp. 211–237

Burnett, R.M., Darling, G.D., Kendall, D.S., LeQuesne, M.E., Mayhew, S.G., Smith, W.W., Ludwig, M.L.: J. Biol. Chem. **249**, 4383–4392 (1974)

Cusanovich, M.A., Edmondson, D.E.: Biochem. Biophys. Res. Commun. **45**, 327–335 (1971)

Dubourdieu, M., LeGall, J., Favaudon, V.: Biochim. Biophys. Acta **376**, 519–532 (1975)

Dubourdieu, M., LeGall, J., Fox, J.L.: Biochem. Biophys. Res. Commun. **52**, 1418–1425 (1973)

Edmondson, D.E., Tollin, G.: Biochemistry **10**, 124–132 (1971a)

Edmondson, D.E., Tollin, G.: Biochemistry **10**, 133–145 (1971b)

Entsch, B., Smillie, R.M.: Arch. Biochem. Biophys. **151**, 378–386 (1972)

Foust, G.P., Mayhew, S.G., Massay, V.: J. Biol. Chem. **244**, 964–970 (1969)

Fox, J.L.: In: Molecular Evolution. Prebiological and Biological. Rohlfing, D.L., Oparin, A.J. (eds.). New York: Plenum, 1972, pp. 23–34

Hall, D.O., Cammack, R., Rao, K.K.: Nature (London) **233**, 136–138 (1971)
Knight, E., Jr., Hardy, R.W.F.: J. Biol. Chem. **241**, 2752–2756 (1966)
Lin, B. van, Bothe, H.: Arch. Mikrobiol. **82**, 155–172 (1972)
MacKnight, M.L., Gillard, J.M., Tollin, G.: Biochemistry **12**, 4200–4205 (1973)
MacKnight, M.L., Gray, W.R., Tollin, G.: Biochem. Biophys. Res. Commun. **59**, 630–637 (1974)
Mayhew, S.G.: Biochim. Biophys. Acta **235**, 276–288 (1971)
Mayhew, S.G., Foust, G.P., Massey, V.: J. Biol. Chem. **244**, 803–810 (1969)
Norris, J.R., Crespi, H.L., Katz, J.J.: Biochem. Biophys. Res. Commun. **49**, 139–146 (1972)
Shethna, Y.J., Wilson, P.W., Beinert, H.: Biochim. Biophys. Acta **113**, 225–234 (1966)
Smillie, R.M.: Biochem. Biophys. Res. Commun. **20**, 621–629 (1965a)
Smillie, R.M.: Plant Physiol. **40**, 1124–1128 (1965b)
Tanaka, M., Hanui, M., Yasunobu, K.T., Mayhew, S.G.: J. Biol. Chem. **249**, 4393–4396 (1974)
Tanaka, M., Hanui, M., Yasunobu, K.T., Mayhew, S.G., Massey, V.: J. Biol. Chem. **248**, 4354–4366 (1973)
Trebst, A., Bothe, H.: Ber. Deut. Botan. Ges. **79**, 44–47 (1966)
Vetter, H., Jr., Knappe, J.: Hoppe- Seyler's Z. Physiol. Chem. **352**, 433–466 (1971)
Watenpaugh, K.D., Sieker, L.C., Jensen, L.H.: Proc. Nat. Acad. Sci. U.S. **70**, 3857–3860 (1973)
Wong, P., Evans, H.J., Klucas, R., Russell, S.: Plant Soil, Special Vol. 525–543 (1971)
Yoch, D.C.: J. Gen. Microbiol. **83**, 153–164 (1974)
Zumft, W.G., Spiller, H.: Biochem. Biophys. Res. Commun. **45**, 112–118 (1971)

11. Flavoproteins

G. FORTI

A. Introduction

The presence of flavoproteins in the photosynthetic apparatus was first reported by AVRON and JAGENDORF (1956), who discovered and purified from chloroplasts an NADPH-specific diaphorase. These authors also identified FAD as the prosthetic group (AVRON and JAGENDORF, 1957), but were unable to assign a physiological role to the enzyme in photosynthetic electron transport.

MARRÈ and SERVETTAZ (1958) speculated that the enzyme may reduce cytochrome f in the chloroplasts. The role of the enzyme was later established as an essential catalyst of $NADP^+$ photoreduction (SAN PIETRO and KEISTER, 1962; SHIN et al., 1963), and participates in the phosphorylating cyclic electron flow of isolated chloroplasts when no exogenous cofactor of electron transport is added (FORTI and ZANETTI, 1969).

The isolated pure enzyme has been shown to catalyze several reactions: (a) the reduction of $NADP^+$ with ferredoxin as the electron donor (SHIN et al., 1963), (b) the reduction of cytochrome f by NADPH (FORTI et al., 1963), (c) the reduction of NAD^+ by NADPH, or of analogs of NADPH (transhydrogenase reaction; KEISTER et al., 1960), and (d) the oxidation of NADPH by a number of dyes and ferricyanide (diaphorase activity; AVRON and JAGENDORF, 1956).

All these activities are due to the same FAD-containing flavoprotein, as all of them are purified together with the flavoprotein (ZANETTI and FORTI, 1966; FORTI and STURANI, 1968), and they are all inhibited to the same extent by an antibody to the enzyme (FORTI and ZANETTI, 1967).

The chloroplast flavoprotein is ubiquitous in photosynthetic organisms, though only the enzyme from spinach (*Spinacia oleracea*) has been extensively studied.

B. Isolation and Physico-Chemical Properties of the Chloroplast Flavoprotein, Ferredoxin-NADP⁺ Reductase

I. Isolation of the Enzyme

Several procedures have been described for the purification to homogeneity of the chloroplast flavoprotein. After the pioneering work of AVRON and JAGENDORF (1956, 1957), based on ammonium sulfate fractionation and batch-wise calcium phosphate gel adsorption, the methods developed were based on acetone fractionation, ammonium sulfate fractionation and DEAE-cellulose chromatography (SHIN et al., 1963) with the introduction of a column chromatographic step on hydroxyapatite or calcium phosphate gel followed by Sephadex G-150 dextran gel (FORTI and STURANI, 1968). This last step separates from the enzyme trace amounts of a phosphodiesterase activity (FORTI and STURANI, 1968) which would make enzyme reconstitution studies from the apoprotein and FAD quite impossible, in that it hydrolyzes free FAD, though not the FAD bound to the native enzyme,

II. Properties of Ferredoxin-NADP$^+$ Reductase

The enzyme has a molecular weight of approximately 40,000, as determined by FAD content (AVRON and JAGENDORF, 1957; ZANETTI and FORTI, 1966) and amino acid analysis (FORTI and STURANI, 1968). It contains one $-S-S$ bridge and four $-SH$ groups, as determined by amperometric titration and reaction kinetics with mercurials (ZANETTI and FORTI, 1969), in agreement with six mol of cysteic acid on amino acid analysis (FORTI and STURANI, 1968). One of the sulfhydryl groups, not readily available to SH reagents in the native enzyme, is essential for catalytic activity (ZANETTI and FORTI, 1969).

The native enzyme shows little FAD fluorescence (FORTI, 1966). However, the fluorescence of FAD appears after a number of denaturing treatments, such as urea treatment (FORTI, 1966), and treatment with mercurials (FORTI and STURANI, 1968; ZANETTI and FORTI, 1969). Simultaneously with its increased fluorescence yield, FAD is made accessible to phosphodiesterase attack, and the fluorescence of the protein moiety mostly contributed by tryptophan residues increase (FORTI and STURANI, 1968). These observations indicate that the FAD prosthetic group must be hidden in the protein structure in such a way as to dissipate the light energy absorbed by non-radiative processes. The nature of the binding of FAD to the protein is unknown; however, covalent binding can be excluded, because FAD can be separated from the holoenzyme by reduction with NADPH in the presence of urea (FORTI and ZANETTI, 1967) or in the presence of mercurials (FORTI and STURANI, 1968). BÖGER et al. (1973) report that chloroplast flavoprotein shows high FAD fluorescence. However, this result could not be reproduced in a number of laboratories (B. CURTI, personal communication; G. Zanetti, personal communication).

Ferredoxin-NADP$^+$ reductase shows a typical flavoprotein spectrum; the extinction coefficients at the characteristic wavelenghts for the oxidized form, the semiquinone and the fully reduced enzyme are given in Table 1.

Table 1. Molar extinction coefficients of $F_p.FAD$, $F_p.FADH^.$ and $F_p.FADH_2$. (From FORTI et al., 1970)

Redox state	Wavelength (nm)			
	458	540	555	600
$F_p.FAD$	10,700	0.00	0.00	0.00
$F_p.FADH^.$	6,060	1,900	1,725	1,810
$F_p.FADH_2$	1,120	—	0.00	0.00
$F_p.FAD-F_p.FADH^.$	+4,640	−1,900	−1,725	−1,810
$F_p.FAD-F_p.FADH_2$	+9,580	—	—	—

As can be seen, the chloroplast flavoprotein forms a neutral semiquinone upon reduction with one electron per FAD. Complete reduction of the enzyme requires two electrons, indicating that FAD is the only electron acceptor group present (Massey et al., 1970). Upon reaction with one molecule of NADPH, the enzyme very rapidly forms an intermediate partially reduced form, probably resulting from one electron transfer from bound NADPH to FAD (Massey et al., 1970), to yield the neutral semiquinone (Massey et al., 1970; Forti et al., 1970). In the presence of excess NADPH and an NADPH-regenerating system in anaerobic conditions, the fully reduced form of the enzyme is obtained (Forti et al., 1970). The need for the NADPH-regenerating system to achieve full reduction of the enzyme is related to its very low potential of ca. -380 mV at pH 7 (Massey et al., 1970). Ferredoxin-NADP$^+$ reductase forms equimolecular complexes with ferredoxin, with the electron transport proteins rubredoxin and flavodoxin (from *Peptostreptococcus elsdenii*) and with NADP$^+$ (Foust et al., 1969). These complexes cause changes in the visible spectrum, which provide a convenient method to measure their formation and stoichiometry. Their formation is complete in less than 3 msec, and they are dissociated by ions (Foust et al., 1969). The formation of the ferredoxin-flavoprotein complex is important for the catalytic reduction of NADP$^+$ by chloroplasts (see below).

C. Kinetic Properties of Ferredoxin-NADP$^+$ Reductase

I. Reaction Kinetics

Ferredoxin-NADP$^+$ reductase reduces NADP$^+$ rapidly with substrate amounts of reduced ferredoxin as the electron donor or with catalytic amounts of ferredoxin and illuminated chloroplast fragments (Shin et al., 1963; Shin and Arnon, 1965). It is likely that the formation of the ferredoxin-flavoprotein complex (Foust et al., 1969) is essential for this reaction, because pyrophosphate inhibits NADP$^+$ photoreduction competitively with ferredoxin and also inhibits the formation of the complex (Forti and Meyer, 1969). Furthermore, it has been shown that the photoreduction of the flavoprotein by chloroplast fragments, but not the photoreduction of ferredoxin is inhibited by pyrophosphate (Forti et al., 1970).

The kinetics of the oxidation of NADPH by the chloroplast flavoprotein are quite different if cytochrome f from higher plants or ferricyanide or different dyes are the electron acceptors. Steady-state kinetic studies have demonstrated that cytochrome f reduction proceeds through the formation of a ternary complex of flavoprotein, NADPH and ferricytochrome f (Zanetti and Forti, 1966; Forti and Sturani, 1968). On the other hand, no evidence for a ternary complex could be found in the diaphorase activity with ferricyanide or indophenol dyes as acceptors (Zanetti and Forti, 1966; Forti and Sturani, 1968; Massey et al., 1970). Though the detailed mechanism of the diaphorase reaction is still unclear, the participation of the semiquinone FADH has been established (Massey et al., 1970).

Cytochrome 552 of *Euglena gracilis*, the low molecular weight cytochrome of this organism, functions as an electron acceptor for the spinach chloroplast flavoprotein, while mammalian cytochrome c does not (Forti and Sturani, 1968).

However, the NADPH-cytochrome 552 reductase activity resembles kinetically the diaphorase reaction, with no indication of a ternary complex (FORTI and STU-RANI, 1968). The K_m values for NADPH in the cytochrome f reductase, ferricyanide reductase and cytochrome 552 (*Euglena*) reductase reactions are 1.8 μmolar, 31 μmolar and 2.3 μmolar respectively (FORTI and STURANI, 1968). When plastocyanin is used as the electron acceptor, the apparent K_m for NADPH is much higher, of the order of 5 mmolar, and inhibition by NADPH is observed (FORTI and STURANI, 1968). This observation suggests that plastocyanin accepts electrons from the reduced enzyme at the NADPH site. The concept that the chloroplast flavoprotein has a binding site for oxidized and reduced pyridine nucleotides and a distinct site for ferredoxin, at which the acceptor dyes also interact, has been proposed by BÖGER (1971a). This concept is based on steady-state kinetic studies, which demonstrated that ferredoxin stimulates the transhydrogenase reaction (BÖGER, 1971a, b; FREDRICKS and GEHL, 1971), while NADH and NADPH inhibit it (BÖGER, 1971a). Confirming evidence that the pyridine nucleotide is distinct from the ferredoxin site has recently been provided by ZANETTI (1975, personal communication), who observed that dansylation of an ε-lysil residue of the enzyme causes loss of its activity without affecting its capacity to bind ferredoxin. The binding of $NADP^+$ or NADPH to the enzyme prevents the reaction with dansyl chloride, indicating that the modified lysil residue is the $NADP^+$ and NADPH binding site or is very close to it.

Ferredoxin-$NADP^+$ reductase has a low oxidase activity (ZANETTI and FORTI, 1966) of about 1% of the specific activity of the diaphorase reaction. If ferredoxin is added to the NADPH flavoprotein system, a fast oxygen uptake is observed, due to the autoxidation of ferredoxin. Such a system, if cytochrome c is added, will reduce cytochrome c at high rates through its interaction with ferredoxin (LAZZARINI and SAN PIETRO, 1962).

II. Inhibitors

All activities of the chloroplast flavoprotein, with the exception of cytochrome f reductase, are inhibited by a brief preincubation of the enzyme with NADPH (ZANETTI and FORTI, 1966; FORTI and STURANI, 1968).

This inhibition is prevented and reversed by $NADP^+$. It is noteworthy that the only activity which is not inhibited by NADPH preincubation, cytochrome f reductase, is the only one where a ternary complex is part of the catalytic route. Also, it should be noted that higher plant (parsley) cytochrome f used in the experiments referred to is present in solution as a dimer or a tetramer (FORTI et al., 1965), and the ternary complex containing flavoprotein, NADPH and cytochrome f could conceivably be a two-electron transfer system. Whether this could be associated with the lack of inhibition by NADPH preincubation of the enzyme is a matter of speculation.

The chloroplast flavoprotein has a low sensitivity to inhibition by mercurials, which is enormously enhanced if the enzyme is preincubated with such reagents as parachloromercuriphenylsulphonate in the presence of NADPH (FORTI and STURANI, 1968). Under these conditions, the inactivation is irreversible and FAD is split from the protein moiety.

The inactivation of the enzyme, in its oxidized form, by mercurials follows first-order kinetics, and only partial reactivation is obtained upon addition of excess —SH (Zanetti and Forti, 1969). One of the four —SH groups is involved in the catalytic site for NADP(H), on the basis of the inactivation kinetics and the protection afforded by NADP (Zanetti and Forti, 1969).

D. Multiple Forms of the Chloroplast Flavoprotein

The chloroplast flavoprotein seems to consist of two components, which can be separated by chromatography on Sephadex G-100 (Fredricks and Gehl, 1973), by isoelectric focusing (Zanetti, 1975, personal communication) or by phosphocellulose chromatography (Curti, 1975, personal communication). The understanding of the physiological significance of these isozymes await further studies.

References

Avron, M., Jagendorf, A.T.: Arch. Biochem. Biophys. **65**, 475–490 (1956)
Avron, M., Jagendorf, A.T.: Arch. Biochem. Biophys. **72**, 17–24 (1957)
Böger, P.: In: Proc. 2nd Intern. Congr. Photosynthesis. Forti, G., Avron, M., Melandri, B.A. (eds.). The Hague: Dr. Junk, 1972a, pp. 449–458
Böger, P.: Planta (Berl.) **99**, 319–338 (1971b)
Böger, P., Lien, S.S., San Pietro, A.: Z. Naturforsch. 28 C, Vol. 9/10 (1973)
Forti, G.: Brookhaven Symp. Biol. **19**, 195–201 (1966)
Forti, G., Bertholè, M.L., Parisi, B.: In: Photosynthetic Mechanisms of Green Plants. Nat. Acad. Sci., Nat. Res. Council **1145**, 284–290 (1963)
Forti, G., Bertholè, M.L., Zanetti, G.: Biochim. Biophys. Acta **109**, 33–40 (1965)
Forti, G., Melandri, B.A., San Pietro, A., Ke, B.: Arch. Biochem. Biophys. **140**, 107–112 (1970)
Forti, G., Meyer, E.M.: Plant Physiol. **44**, 1511–1514 (1969)
Forti, G., Sturani, E.: Europ. J. Biochem. **3**, 461–472 (1968)
Forti, G., Zanetti, G.: In: Biochemistry of Chloroplasts. Goodwin, T.W. (ed.). New York-London: Academic Press, 1967, Vol. II, pp. 523–529
Forti, G., Zanetti, G.: In: Progress in Photosynthesis Research. Metzner, H. (ed.). Tübingen: Laupp, 1969, Vol. III, pp. 1213–1216
Foust, G.P., Mayhew, S.G., Massey, V.: J. Biol. Chem. **244**, 964–970 (1969)
Fredricks, W.W., Gehl, J.M.: J. Biol. Chem. **246**, 1202–1205 (1971)
Fredricks, W.W., Gehl, J.M.: Federation Proc. **32**, 477 (1973)
Keister, D.L., San Pietro, A., Stolzenbach, F.E.: J. Biol. Chem. **235**, 2898–2908 (1960)
Lazzarini, R.A., San Pietro, A.: Biochim. Biophys. Acta **62**, 417 (1962)
Marrè, E., Servettaz, O.: Arch. Biochem. Biophys. **75**, 309–323 (1958)
Massey, V., Matthews, R.G., Foust, G.P., Howell, L.G., Williams, C.H., Jr., Zanetti, G., Ronchi, S.: In: Pyridine Nucleotide Dependent Dehydrogenases. Sund, H. (ed.). Berlin-Heidelberg-New York: Springer, 1970, pp. 393–409
San Pietro, A., Keister, D.L.: Arch. Biochem. Biophys. **98**, 235–243 (1962)
Shin, M., Arnon, D.I.: J. Biol. Chem. **240**, 1405–1411 (1965)
Shin, M., Tagawa, K., Arnon, D.I.: Biochem. Z. **338**S, 84–96 (1963)
Zanetti, G., Forti, G.: J. Biol. Chem. **241**, 279–285 (1966)
Zanetti, G., Forti, G.: J. Biol. Chem. **244**, 4757–4760 (1969)

12. Cytochromes

W.A. CRAMER

A. Introduction

The cytochromes are intracellular chromoproteins found by KEILIN to undergo reversible spectral changes in tissue from animals, plants and bacteria. The first observations on plants were made with a microspectroscope by KEILIN on dithionite-treated slices of non-chlorophyllous plant tissue. The early history of this work has been described by KEILIN and KEILIN (1966). Spectroscopic distinction between c- and b-type cytochrome in the green part of the leaf material was first made by HILL and SCARISBRICK (1951), and redox determinations in etiolated barley chloroplasts by HILL (1954). DUYSENS and AMESZ (1962) applied sensitive spectrophotometric techniques to infer the position of an f- or c-like cytochrome between two different pigment systems of the alga *Porphyridium cruentum*. At about the same time LUNDEGÅRDH (1961) studied the light-induced redox changes of cytochrome f as well as the two b cytochromes. Recently published sources of general information on plant and algal cytochromes are BENDALL et al. (1971), YAKUSHIJI (1971), MITSUI (1971), and LEMBERG and BARRETT (1973).

B. Isolated Higher Plant Cytochromes

Three different cytochromes have been isolated and purified from higher plant chloroplasts, and the properties of the purified components are shown in Table 1. Two of the three purified plant chloroplast cytochromes, b_6 and b-559, are defined to be b-type since the heme can be split from the protein by acid-acetone and the reduced α-band maximum of the pyridine hemochromagen is approximately 556 nm. The third cytochrome, which historically was the first to be purified, is c-type in terms of a thioether heme linkage to cysteines of the apoprotein through positions 2 and 4 of the porphyrin rings. This c-type cytochrome was originally denoted cytochrome f (Latin *folium*, leaf) because of its association with the green part of the plant. The cytochrome f differs from mammalian cytochrome c, however, in having an oxidation-reduction potential approximately 0.1 V more positive in vitro and in situ an acid isoelectric point, a larger reduced Soret/α ratio (~ 7), and in the case of the plant cytochrome a larger molecular weight. The total molecular weight of cytochrome f is uncertain, possibly because of aggregation problems. There is agreement in three of the five determinations shown in Table 1 on a molecular weight per heme of approximately 65 kilodaltons, and in the other two cases on a value half as large. The 35 kilodalton heme-containing preparation of NELSON and RACKER (1972) appears, in any case, to have the

Table 1. Properties of purified higher plant cytochromes

Cytochrome	Source	MW/Heme	MW/Protein	Polypeptides	pI	E_{m7} (V)	λ_m^α–LT (nm)
f (λ_m^α, 554 nm)	Parsley	68 (DAVENPORT and HILL, 1952)	110	–	4.7	0.365	
	Parsley	61 (FORTI et al., 1965)	245	–			
	Spinach	62 (SINGH and WASSERMAN, 1971)	–	2/heme			548, 552 (split)
	Spinach	35 (NELSON and RACKER, 1972)	–	1/heme			
	Radish	33 (TAKAHASHI and ASADA, 1975)	33	–			
b_6 (λ_m^α, 563 nm)	Spinach	40 (STUART and WASSERMAN, 1975)	40	4 (3 non-equiv.)	–	–0.08	557, 561 (split)
		60 (+ lipid, same authors)	60				
b-559	Spinach	46 (GAREWAL and WASSERMAN, 1974)	46				
		111 (+ lipid, same authors)	111	8 (3 non-equiv.)	–	low	556–557
b-560	Spinach	30 (MATSUZAKI and KAMIMURA, 1972)	30	–	–	0.13	–

Notation: λ_m^α, reduced α-band maximum; MW, molecular weight, in kilodaltons; pI, isoelectric point; E_{m7} (V), midpoint potential at pH 7, in volts; λ_m^α–LT, reduced α-band maximum at liquid nitrogen temperature (77° K).

necessary properties for photoreactivity, as it was photooxidized in depleted system I particles supplemented with very small amounts of plastocyanin.

The purification of the two b-type cytochromes from chloroplasts has only recently been achieved. Both cytochromes have been purified as multisubunit lipoproteins. Lipid-containing cytochrome b_6 has a molecular weight of 60 kilodaltons per heme, with 75–80% of the lipid being accounted for by seven chlorophyll a and six cardiolipin molecules per heme. The 40 kilodalton protein part of the molecule consists of subunits separated by sodium dodecylsulfate gel electrophoresis of 20, 9.6, and 6.6 (2) kilodaltons. Of importance in in situ spectroscopic experimentation on chloroplast cytochromes is the finding that the reduced α-band of this cytochrome is split at 77° K (STUART and WASSERMAN, 1973). Cytochromes

b_6 and f can be isolated together in a complex containing non-heme iron, carotene and phospholipid from a photosystem I-enriched chloroplast fraction, with a ratio of approximately two b_6 hemes per f (NELSON and NEUMAN, 1972). The spectral properties of cytochrome b_6 in the purified complex are labile, in that there is some CO binding to dithionite-reduced particles and the reduced α-band can shift from 563 nm to 561 nm. The midpoint oxidation-reduction potential of cytochrome b_6 in the particles was estimated to be approximately −100 mV on the basis of comparison with a titration of pyocyanine ($E_{m7} = −35$ mV). This value also appears to indicate some change in the properties of b_6 relation to the fresh membrane preparation. The values in the literature for the midpoint potential of cytochrome b_6 at pH 7 are: (1) −60 mV in etiolated barley chloroplasts obtained by comparison with ferri-ferrooxalate (HILL, 1954), recalculated (2) to be 0.0 V (HILL and BENDALL, 1967). (3) The technique of potentiometric redox titration using redox buffers and measurement of cuvette potential with a platinum electrode yielded a value of $−180 \pm 20$ mV (FAN and CRAMER, 1970). It was later concluded that this value was obtained in chloroplasts which had sustained structural damage during the course (2–3 h) of the redox titrations. The potential was remeasured over a shorter time interval and the midpoint was found in fresh preparations to be approximately +5 mV at pH 8 with the best fit of the Nernst equation occurring for a two-electron ($n = 2$) oxidation-reduction (BÖHME and CRAMER, 1973). Older preparations showed more negative midpoints. The midpoint could also be systematically lowered by illumination of the chloroplasts in the presence of 10 μM FCCP or 4 mM NH$_4$Cl. In the latter case approximately half the b_6 was rapidly (i.e., within minutes) shifted to a midpoint between −100 and −150 mV. Thus, it appears that the best value for the midpoint potential of cytochrome b_6 in situ is approximately 0.0 V with an uncertainty of perhaps +20 mV. The midpoint of at least half the b_6 is shifted in a negative direction by as much as 150 mV by chloroplast treatments which uncouple and/or alter membrane structure. (4) The midpoint potential of cytochrome b_6 in the b_6–f particles mentioned above is about −100 mV (NELSON and NEUMANN, 1972), and −80 mV in purified b_6 (STUART and WASSERMAN, 1975). The more negative midpoint potentials of isolated or purified b_6 preparations may be the inevitable consequence of removal of the cytochrome from its hydrophobic environment (KASSNER, 1972).

   A 111 kilodalton detergent-depleted cytochrome b-559 molecule has been isolated which contains 3 and 4 molecules of β-carotene and chlorophyll a, respectively, and two unknown polar lipids. The 8 subunits of the 46 kilodalton b-559 protein all have the same approximate molecular weight ($5,600 \pm 1,000$) but N-terminal amino acid analysis indicates at least three different kinds of polypeptide chain (GAREWAL and WASSERMAN, 1974). A 30-kilodalton soluble b cytochrome with a reduced α-band at 560 nm has also been isolated from spinach leaves, and this cytochrome can be photoreduced in a DCMU-sensitive manner (MATSUZAKI and KAMIMURA, 1972). The relationship of this soluble cytochrome to the membrane bound b-559 isolated by GAREWAL and WASSERMAN is not known. The midpoint of the soluble b-560 ($E_{m7} = +0.13$ V), and the midpoint of the isolated b-559 lipoprotein, which is not reducible by hydroquinone, are much more negative than that of the bulk of the b-559 in situ (see Sect. D). This again may be a result of increasing the average polarity or hydrophilicity of the heme environment (KASSNER, 1972).

C. Isolated Algal Cytochromes

Cytochromes of the f or c type of the photosynthetic electron transport chain have been purified from many algal sources, as documented by Mitsui (1971), Yakushiji (1971), and Lemberg and Barrett (1973). Additional comparative information can be found in Gorman and Levine (1966). These acidic cytochromes are of lower molecular weight ($\sim 12,000$) than those purified from higher plants, and are monomeric. An f-type cytochrome recently characterized from the alga *Bumilleriopsis filiformis* has the smallest molecular weight ($\sim 7,100$) yet reported for a cytochrome (Lach et al., 1973), and is of appreciable interest since it cannot contain many amino acids beyond the position of the sixth ligand to the heme. The purified *Euglena* cytochrome c-552 has a smaller Soret/α ratio and a slightly less acid isoelectric point (pI $= 5.5$). Like the f-type cytochromes, it has a relatively positive oxidation-reduction potential ($E_{m7} = +0.37$ V). The acidic algal cytochromes generally react rapidly with cytochrome oxidase from the bacterium *Pseudomonas aeruginosa* and very slowly with mammalian cytochrome oxidase from beef heart. *Euglena* cytochrome c-552 does not react as rapidly with *Pseudomonas* cytochrome oxidase as the more acidic algal cytochromes (Yamanaka and Oku-nuki, 1968).

The amino acid sequence of *Euglena* cytochrome c-552 has been determined (Pettigrew, 1974). Important features in the sequence are (a) the characteristic conserved heme-attachment site, Cys–X–Y–Cys–His, with this histidine being the fifth heme ligand, and a methionine around position 62 suggested to function as the sixth ligand; (b) aromatic residues at positions 10 and 78 which may be part of an electron conducting "right channel" from the heme to cytochrome reductase and/or oxidase. It is of interest that the *Euglena* sequence does not contain an array of conserved aromatic residues in a "left channel" as observed in cytochrome c and postulated to be involved in the pathway of cytochrome c reduction (Takano et al., 1973). The *Euglena* sequence has also been compared (Ambler and Bartsch, 1975) with those of a blue-green bacterium and different algae, including the crysophyte *Monochrysis lutheri* (Laycock, 1972). The blue-green algal cytochrome f is at least as similar to the eukaryotic cytochromes f as the latter are to each other, and all are somewhat different from the *Euglena* cytochrome.

The ability to employ algal mutants with specific lesions in the electron transport chain has provided a powerful approach to studies of the pathways and function of photosynthetic electron transport in situ (Levine, 1969a; Bishop, 1973). Algal mutants can be obtained, some of which are missing cytochromes f and b-559, while others contain reduced amounts of cytochrome b-559 and a different distribution between low and high potential forms of cytochrome b-559 (Epel and Butler, 1972; Epel et al., 1972).

D. Cytochrome Function in Electron Transport

The function of the cytochromes in electron transport may be in the coupled synthesis of ATP, and particularly in the case of cytochrome b-559 one should also consider a possible role in water splitting (LUNDEGÅRDH, 1965) and in the biochemical mechanism of intramembranal proton movement. Three important parameters which relate to these functions are: (1) position in the two-pigment system framework of the electron transport chain, (2) position in the membrane relative to the aqueous interface of the membrane, and (3) midpoint potential. Information on the latter two quantities should be obtained in illuminated membranes as well as those incubated in the dark, since the general properties of chloroplast membranes are well known to be light-dependent. Firstly, cytochromes f and b-559 are believed to be closely associated with photosystems I and II, respectively, on the basis of differential fractionation into fractions containing the appropriate photoactivities (BOARDMAN and ANDERSON, 1967; KE et al., 1972), and in the case of cytochrome b-559 because of preferential photooxidation of most of the cytochrome b-559 (most probably in a high potential form with $E_{m7} > +0.2$ V) by photosystem II at 77° K (KNAFF and ARNON, 1969; VERMEGLIO and MATHIS, 1974b). The question has been raised as to whether a low potential form of cytochrome b-559 might be associated with both photosystems I and II (ANDERSON and BOARDMAN, 1973; KNAFF and MALKIN, 1973). This question will be further discussed below in relation to the topic of midpoint potentials of b-559. On the basis of detergent fractionation (BOARDMAN and ANDERSON, 1967) and the close association of cytochromes b_6 and f, cytochrome b_6 is also linked structurally to photosystem I. The question of cytochrome position in the membrane has been approached in terms of the relative accessibility to the charged oxidant ferricyanide of cytochrome f and high potential b-559, both of which in the dark have approximately the same midpoint potential ($E_{m7} \simeq +0.35$ V). The rate of chemical oxidation of $^1/_2$–$^2/_3$ of the cytochrome f in the dark is no larger than 1–2% that of high potential b-559 (HORTON and CRAMER, 1974). The accessibility of cytochrome f to ferricyanide was increased after illumination of the chloroplasts. This leads to a schematic model for the relative position of high potential cytochrome b-559 and f in the membrane relative to the aqueous phase (Fig. 1), in which the chemical

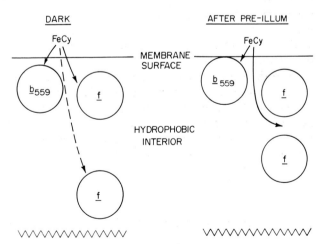

Fig. 1. Relative positions of cytochrome f in the thylakoid membrane relative to the aqueous phase as inferred by accessibility to charged oxidant ferricyanide in the dark and after exposure to actinic illumination

oxidation pathway of the b-559 proteins in the dark is closer to the outside surface of the membrane than is cytochrome f, although cytochrome f may move toward the outside upon illumination. *Euglena* cytochrome c-552 is also thought to be trapped inside the thylakoid (WILDNER and HAUSKA, 1974). Changes in the properties of the chloroplast cytochromes upon illumination and upon illumination in the presence of uncouplers complicate the analysis of cytochrome function and position in the electron transport chain, and may have physiological significance in membrane function. The existence of conformational changes linked to differential ion binding is documented for eukaryotic cytochromes c (TAKANO et al., 1973), though such extensive changes do not occur in cytochrome c_2 from the photosynthetic bacterium *Rhodospirillum rubrum* (SALEMME et al., 1973). It should not be surprising that the properties of the cytochromes may change upon illumination of the membrane, since these cytochromes are tightly bound membrane proteins and it is well known that the properties (and conformation) of chloroplast membranes and proteins such as coupling factor (RYRIE and JAGENDORF, 1972; McCARTY and FAGAN, 1973) change upon illumination. Besides the light-induced negatively directed shift in cytochrome b_6 midpoint potential mentioned above, and the change in cytochrome f accessibility to ferricyanide, another example of a light-induced change in cytochrome properties is that of the negatively directed shift in midpoint of cytochrome b-559 inferred from DBMIB inhibition of b-559 photooxidation (HORTON et al., 1975).

The midpoint potential of purified cytochrome f is $+0.365$ V (DAVENPORT and HILL, 1952). The midpoints of cytochrome b_6 and high potential b-559 determined under equilibrium conditions in the dark, using chloroplasts well coupled for phosphorylation, are approximately 0.0 V (HILL and BENDALL, 1967; BÖHME and CRAMER, 1973), as discussed above, and 0.35 V (BENDALL, 1968; IKEGAMI et al., 1968) respectively. Most of the cytochrome b-559 in well-coupled chloroplasts in the dark has a much more positive redox potential (E_{m7} $+0.35$ V) than that of purified b-559 (GAREWAL and WASSERMAN, 1974) or of energy-transducing b cytochromes in other membrane systems (CRAMER and HORTON, 1975). Because the high potential state is measured in the dark, it is neither caused by reversed electron transport nor by a transmembrane electrical potential. It may, however, be a consequence of a relatively positive ionic environment in the neighborhood of the heme conferred by an organized membrane structure. The high potential state can be irreversibly converted to a lower potential state(s), with much lower midpoint potential, through treatment with low concentrations of detergents, proteolytic enzymes, and other agents which may perturb membrane structure (CRAMER et al., 1971; WADA and ARNON, 1971; COX and BENDALL, 1972; OKAYAMA and BUTLER, 1972). A single lower potential species of b-559 with a midpoint potential of $+0.08$ V (FAN and CRAMER, 1970) or $+0.055$ V (HIND and NAKATANI, 1970) has been titrated. This midpoint is approximately at the ascorbate midpoint potential and this may be responsible for the finding that ascorbate will not completely reduce cytochrome b-559 (WADA and ARNON, 1971). The midpoint potential of b-559 can be reversibly shifted to a lower potential value(s) at which it is reduced by ascorbate, but not hydroquinone, through extraction of plastoquinone (COX and BENDALL, 1974). Addition of plastoquinone to the extracted preparation will restore the hydroquinone reduction. This suggests the possibility that the quinone pool may form the environment of the cytochrome b-559. A variable

(0–30%) amount of low potential ($E_{m7} < 0.25$ V) b-559 can be found in freshly prepared chloroplasts. There is disagreement as to whether these high and low potential forms are separate protein components or different potential states of the same protein with the heme in different ionic or polar environments. However, it is agreed that total irreversible conversion of b-559 to a low potential form not reducible by hydroquinone generally results in inactivation of system II. Such a correlation between the appearance of high potential b-559 (again measured by hydroquinone reduction which defines components with $E_{m7} \gtrsim 0.25$ V) and photosystem II activity in developing barley leaves (PLESNIČAR and BENDALL, 1973) again implies a structural or functional role for high potential b-559 in water splitting. This correlation in developing leaves involving cytochrome b-559 has not, however, been found by all workers (HENNINGSEN and BOARDMAN, 1973).

The high potential component of cytochrome b-559 does not easily fit in the main electron transport chain joining the two photosystems, as might be inferred from analogy with classical respiratory and photosynthetic electron transport pathways. The following pathways and functions near photosystem II have been most discussed for cytochrome b-559 in the electron transport chain:

1. It may function in the pathway of water oxidation (LUNDEGÅRDH, 1965; BENDALL and SOFROVA, 1971).

2. It may operate in a cycle around photosystem II (KOK et al., 1974).

3. It may function as an electron donor at the level of plastoquinone after a transient reversible light-induced negatively directed potential change (BÖHME and CRAMER, 1971).

These possibilities for cytochrome b-559 function are indicated in Figure 2.

The basic experimental problem with cytochrome b-559 in situ is that it is difficult to observe photooxidation of significant amplitude without perturbing the membrane and/or electron transport system in some non-physiological manner. In particular, much of the evidence for b-559 functioning in a cycle around photosys-

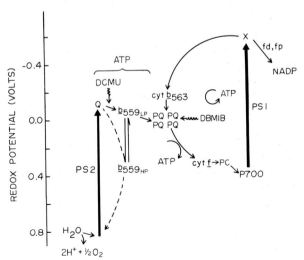

Fig. 2. Position of the cytochromes in two-pigment system framework of green plant photosynthetic electron transport chain. The framework as depicted here is most probably incomplete in at least one respect, absence of non-heme iron components in the electron transport chain

tem II or very close to the water splitting reaction is derived from experiments (at 77° K or after incubation in high concentrations of Tris buffer) in which the water splitting system is blocked and cytochrome b-559 becomes a donor to photosystem II. The hypothesis that cytochrome b-559 can function in the light at the level of plastoquinone is inferred from several experimental situations (e.g., the presence of FCCP, PMS, low pH, preillumination with strong actinic light) in which photosystem I light can be shown preferentially to oxidize the cytochrome, and this oxidation is blocked by the plastoquinone analog, DBMIB (Böhme and Cramer, 1971; Horton et al., 1975). Even though this pathway of oxidation to photosystem I occurs, it may be a relatively inefficient secondary pathway for b-559 oxidation, since the fastest rate of b-559 oxidation by photosystem I thus far detected is much slower than the rate-limiting step between the two photosystems of approximately 20 msec.

Recent data on cytochrome b-559 which may be relevant to its function are:

1. When chloroplasts are preilluminated by a number (n) of short flashes at room temperature, the amplitude of cytochrome b-559 oxidation by photosystem II at $-55°C$ is dependent upon n, with maxima at $n=2$ and 6 and minima at $n=0$ and 4, suggesting some correlation with the S-state dependence of oxygen evolution (Vermeglio and Mathis, 1974a).

2. In the dark, cytochrome b-559 becomes autooxidizable below pH 6 in a reversible manner so that it again becomes reduced when the pH is restored to 7–8. The autooxidizability appears to be a consequence of the cytochrome undergoing a negatively directed potential change at pH values below 6 (Horton and Cramer, 1975a). The effect of low pH in causing a potential change is inhibited by low ($< 1 \mu M$) concentrations of DCMU (Horton et al., 1976). These pH effects of DCMU on cytochrome b-559 as well as other effects (Ben-Hayyim and Avron, 1970; Satoh and Katoh, 1972) imply that the site of action of DCMU could be the cytochrome b-559 protein. Along with other data mentioned above on the correlation between irreversible disappearance of high potential cytochrome b-559 and loss of water splitting function, this might suggest that cytochrome b-559 has a proton-linked function in water splitting which is also linked to midpoint potential transitions of this cytochrome.

The function of cytochrome b_6 appears to be in photosystem I cyclic phosphorylation (Hind and Olson, 1966), although there is some evidence for a role in the main chain (Ikegami et al., 1970). The electron acceptor appears to be a component on the photosystem II side of cytochrome f in the chain, which may be plastoquinone (Levine, 1969b; Böhme and Cramer, 1972b; see Fig. 2). The electron donor to cytochrome b_6, however, is not known, nor is its role in cyclic electron transport in the presence of PMS as cyclic cofactor. The half-times for cytochrome b_6 photoreduction and dark reoxidation have recently been estimated to be (1.3 ± 0.1) msec, and (35 ± 4) msec (Dolan and Hind, 1974).

A role for cytochrome f in the electron transport chain of chloroplasts near a principal coupling site (Fig. 2) or closely linked to the coupling mechanism is implied by the antagonistic behavior of plastoquinone and cytochrome f in crossover experiments, where the rate of electron transport is controlled by addition of cofactors for phosphorylation or ATP (Avron and Chance, 1966; Böhme and Cramer, 1972a). On the other hand, the amplitude and kinetics of cytochrome f reduction in the dark and initial reoxidation by far-red light imply either that cytochrome *f* is not in the main chain (HAEHNEL, 1973) or that the position of cytochrome *f* in the chain and in the membrane is altered in a light-dark transition.

The question of whether plastocyanin is the obligatory oxidant of cytochrome *f* in the simple linear sequence from cytochrome *f* to P-700 as shown in Figure 2, as first concluded in the mutant studies of GORMAN and LEVINE (1965), has been discussed recently by HAEHNEL (1973, 1974) who found that the kinetic properties of cytochrome *f* are not consistent with such a pathway, and by WOOD (1974), who found that the rate of oxidation of cytochrome *f* by plastocyanin in solution is 30 times higher than that of other cytochromes used, with a rate coefficient of $3.6 \cdot 10^7 \text{ M}^{-1} \text{ sec}^{-1}$ which is also relatively large. It is very difficult in this case to establish the order of electron transport from in situ redox potential determinations, as both components have midpoints so close to $+0.35 \text{ V}$ that the potentiometric titration technique, which probably has an uncertainty of $\pm 20 \text{ mV}$, cannot separate them.

Recent measurements of the half-times for cytochrome *f* photooxidation and dark reduction in spinach chloroplasts are, respectively, 0.22 msec and 7.3 and 83 msec for a biphasic reduction (DOLAN and HIND, 1974). The technique of steady-state relaxation spectroscopy in which modulated actinic light is used to drive electron flow through the individual carriers (HOCH, 1972) allows the turnover of the individual carrier to be estimated from the amplitude and phase response to the driving force. This technique has emphasized differences in the cytochrome response of isolated chloroplasts and whole algal cells. The relaxation times of P-700 and cytochrome *f* in chloroplasts were found to be about 17 msec and 24 msec respectively, the faster relaxation time of P-700 consistent with its oxidizing for cytochrome *f* (RURAINSKI et al., 1970). The relaxation times of P-700 and cytochrome in whole cells of the red alga *Porphyridium cruentum* are similar to those in isolated chloroplasts, but the relaxation time for P-700 is equal to or slightly longer than for cytochrome *f*. These data do not support a sequential sequence of cytochrome f-P-700 shown in Figure 2. The ability of DCMU to inhibit the electron flux in chloroplasts, but not in *P. cruentum*, emphasizes the importance of a DCMU-insensitive cyclic pathway in the whole cells, or perhaps the loss of such function in isolated chloroplasts. The turnover rate of P-700 in the above experiments may have been limited by the modulation frequency, and the half-time for P-700 reduction with chloroplasts in the dark after a flash regime has been found to be 4 msec (HAEHNEL, 1973). More recent measurements of cytochrome *f* reduction times in *P. cruentum* yield half-times of 150 msec after photosystem I excitation in the presence or absence of DCMU, and a biphasic reduction after illumination with photosystem II light with half-times of 25 msec and 150 msec (BIGGINS, 1973). The faster phase, associated with electron flow from photosystem II to photosystem I, was accelerated by m-Cl-CCP and inhibited by DCMU. Both phases are inhibited by DBMIB (BIGGINS, 1974), supporting the role for plastoquinone in the pathway of cyclic phosphorylation shown in Figure 2.

References

Ambler, R.P., Bartsch, R.G.: Nature (London) **253**, 285–288 (1975)
Anderson, J.M., Boardman, N.K.: FEBS Lett. **32**, 157–160 (1973)
Avron, M., Chance, B.: Currents Photosynthesis (eds. Thomas, J.B., Goedheer, J.C.), 455–463 (1966)
Bendall, D.S.: Biochem. J. **109**, 46P (1968)
Bendall, D.S., Davenport, H.E., Hill, R.: Methods Enzymol. **23**, 327–344 (1971)
Bendall, D.S., Sofrova, D.: Biochim. Biophys. Acta **234**, 371–380 (1971)
Ben-Hayyim, G., Avron, M.: Europ. J. Biochem. **14**, 205–213 (1970)
Biggins, J.: Biochemistry **12**, 1165–1170 (1973)
Biggins, J.: FEBS Lett. **38**, 311–314 (1974)
Bishop, N.I.: Photophysiology (ed. Giese, A.C.) **8**, 65–96 (1973)
Boardman, N.K., Anderson, J.M.: Biochim. Biophys. Acta **143**, 187–203 (1967)
Böhme, H., Cramer, W.A.: FEBS Lett. **15**, 349–351 (1971)
Böhme, H., Cramer, W.A.: Biochemistry **11**, 1155–1160 (1972a)
Böhme, H., Cramer, W.A.: Biochim. Biophys. Acta **283**, 302–315 (1972b)
Böhme, H., Cramer, W.A.: Biochim. Biophys. Acta **325**, 275–283 (1973)
Cox, R.P., Bendall, D.S.: Biochim. Biophys. Acta **283**, 124–135 (1972)
Cox, R.P., Bendall, D.S.: Biochim. Biophys. Acta **347**, 49–59 (1974)
Cramer, W.A., Böhme, H.: Biochim. Biophys. Acta **256**, 358–369 (1972)
Cramer, W.A., Fan, H.N., Böhme, H.: J. Bioenerg. **2**, 289–303 (1971)
Cramer, W.A., Horton, P.: In: The Porphyrins. Dolphin, D. (ed.). New York: Academic Press, 1975
Davenport, H.E., Hill, R.: Proc. Roy. Soc., Ser. B **139**, 327–345 (1952)
Dolan, E., Hind, G.: Biochim. Biophys. Acta **357**, 380–385 (1974)
Duysens, L.N.M., Amesz, J.: Biochim. Biophys. Acta **64**, 243–260 (1962)
Epel, B.L., Butler, W.L.: Biophys. J. **12**, 922–929 (1972)
Epel, B.L., Butler, W.L., Levine, R.P.: Biochim. Biophys. Acta **275**, 395–400 (1972)
Fan, H.N., Cramer, W.A.: Biochim. Biophys. Acta **216**, 200–207 (1970)
Forti, G., Bertolè, M.L., Zanetti, G.: Biochim. Biophys. Acta **109**, 33–40 (1965)
Garewal, H.S., Wasserman, A.R.: Biochemistry **13**, 4072–4079 (1974)
Gorman, D.S., Levine, R.P.: Proc. Nat. Acad. Sci. U.S. **54**, 1665–1669 (1965)
Gorman, D.S., Levine, R.P.: Plant. Physiol. **41**, 1643–1647 (1966)
Haehnel, W.: Biochim. Biophys. Acta **305**, 618–631 (1973)
Haehnel, W.: In: Proc. 3rd Intern. Congr. Photosynthesis. Avron, M. (ed.). Amsterdam: Elsevier, 1975, pp. 557–567
Henningsen, K.W., Boardman, N.K.: Plant Physiol. **51**, 1117–1126 (1973)
Hill, R.: Nature (London) **174**, 501–503 (1954)
Hill, R., Bendall, D.S.: Oxidation-reduction potentials in relation to components of the chloroplast. In: Biochemistry of Chloroplasts. Goodwin, T.W. (ed.). London: Academic Press, 1967, Vol. II, pp. 559–564
Hill, R., Scarisbrick, R.: New Phytologist **50**, 98–111 (1951)
Hind, G., Nakatani, H.: Biochim. Biophys. Acta **216**, 223–225 (1970)
Hind, G., Olson, J.M.: Brookhaven Symp. Biol. **19**, 188–194 (1966)
Hoch, G.E.: In: Methods Enzymol. **238**, 297–303 (1972)
Horton, P., Böhme, H., Cramer, W.A.: In: Proc. 3rd Intern. Congr. Photosynthesis. Avron, M. (ed.). Amsterdam: Elsevier, 1975, Vol. I, pp. 535–545
Horton, P., Cramer, W.A.: Biochim. Biophys. Acta **368**, 348–360 (1974)
Horton, P., Cramer, W.A.: FEBS Lett. **56**, 244–247 (1975a)
Horton, P., Cramer, W.A.: Manuscript in preparation. (1975b)
Horton, P., Whitmarsh, T., Cramer, W.A.: Arch. Biochem. Biophys., in press (1976)
Ikegami, I., Katoh, S., Takamiya, A.: Biochim. Biophys. Acta **162**, 604–606 (1968)
Ikegami, I., Katoh, S., Takamiya, A.: Plant Cell Physiol. **11**, 777–791 (1970)
Kassner, R.: Proc. Nat. Acad. Sci. U.S. **69**, 2263–2267 (1972)
Ke, B., Vernon, L.P., Chaney, T.H.: Biochim. Biophys. Acta **256**, 345–357 (1972)
Keilin, D., Keilin, J.: The History of Cell Respiration and Cytochrome. Cambridge: Cambridge Univ., 1966

Knaff, D.B., Arnon, D.I.: Proc. Nat. Acad. Sci. U.S. **63**, 956–962 (1969)
Knaff, D.B., Malkin, R.: Arch. Biochem. Biophys. **159**, 555–562 (1973)
Kok, B., Radmer, R., Fowler, C.F.: In: Proc. 3rd Intern. Congr. Photosyn. Avron, M. (ed.). Amsterdam: Elsevier, 1975, pp. 485–496
Lach, H.J., Ruppel, H.G., Böger, P.: Z. Pflanzenphysiol. **70**, 432–451 (1973)
Laycock, M.V.: Can. J. Biochem. **50**, 1311–1325 (1972)
Lemberg, R., Barrett, J.: The Cytochromes. New York-London: Academic Press, 1973
Levine, R.P.: Ann. Rev. Plant Physiol. **20**, 523–540 (1969a)
Levine, R.P.: In: Progress in Photosynthesis Research. Metzner, H. (ed.). Tübingen: Laupp, 1969b, Vol. II, pp. 971–977
Lundegårdh, H.: Nature (London) **192**, 243–248 (1961)
Lundegårdh, H.: Proc. Nat. Acad. Sci. U.S. **53**, 703–710 (1965)
Matsuzaki, E., Kamimura, Y.: Plant Cell Physiol. **13**, 415–425 (1972)
McCarty, R.E., Fagan, J.: Biochemistry **12**, 1503–1507 (1973)
Mitsui, A.: Methods Enzymol. **23**, 368–371 (1971)
Nelson, N., Neumann, J.: J. Biol. Chem. **247**, 1817–1824 (1972)
Nelson, N., Racker, E.: J. Biol. Chem. **247**, 3848–3853 (1972)
Okayama, S., Butler, W.L.: Plant Physiol. **49**, 769–774 (1972)
Pettigrew, G.W.: Biochem. J. **139**, 449–459 (1974)
Plesničar, M., Bendall, D.S.: Biochem. J. **136**, 803–812 (1973)
Rurainski, H.J., Randles, J., Hoch, G.E.: Biochim. Biophys. Acta **205**, 254–262 (1970)
Ryrie, I.J., Jagendorf, A.T.: J. Biol. Chem. **247**, 4453–4459 (1972)
Salemme, F.R., Kraut, J., Kamen, M.D.: J. Biol. Chem. **248**, 7701–7716 (1973)
Satoh, K., Katoh, S.: Plant Cell Physiol. **13**, 807–820 (1972)
Singh, J., Wasserman, A.R.: J. Biol. Chem. **246**, 3532–3541 (1971)
Stuart, A.L., Wasserman, A.R.: Biochim. Biophys. Acta **314**, 284–297 (1973)
Stuart, A.L., Wasserman, A.R.: Biochim. Biophys. Acta **376**, 561–572 (1975)
Takahashi, M., Asada, K.: Plant Cell Physiol. **16**, 191–194 (1975)
Takano, T., Kallai, O.B., Swanson, R., Dickerson, R.: J. Biol. Chem. **248**, 5234–5255 (1973)
Vermeglio, A., Mathis, P.: Bioelectrochem. Bioenerg. **1**, 438–447 (1974a)
Vermeglio, A., Mathis, P.: In: Proc. 3rd Intern. Congr. Photosynthesis. Avron, M. (ed.). Amsterdam: Elsevier, 1974b, pp. 323–334
Wada, K., Arnon, D.I.: Proc. Nat. Acad. Sci. U.S. **68**, 3064–3068 (1971)
Wildner, G.F., Hauska, G.: Arch. Biochem. Biophys. **164**, 136–144 (1974)
Wood, P.M.: Biochim. Biophys. Acta **357**, 370–379 (1974)
Yakushiji, E.: Methods Enzymol. **23**, 364–368 (1971)
Yamanaka, T., Okunuki, K.: In: Structure and Function of Cytochromes. Okunuki, K., Kamen, M.D., Sekuzu, I. (eds.). Tokyo: Univ. Tokyo, 1968, pp. 390–403

13. Plastoquinone

J. AMESZ

A. Introduction and Properties

Plastoquinone was isolated nearly 30 years ago by KOFLER from dried lucerne, but interest in its physiological function arose only in the late fifties, when it became known that a related substance, ubiquinone, is involved in mitochondrial electron transport. Plastoquinone was found to occur almost exclusively in photosynthetic material: algae (including blue-green algae) and the green parts of higher plants (LESTER and CRANE, 1959; FULLER et al., 1961; BARR and CRANE, 1967; SUN et al., 1968). It does not occur in photosynthetic bacteria. It soon became apparent that plastoquinone is not a single substance, but that a variety of related substances can be isolated from green plant material. The most abundant (and probably most important) of these is plastoquinone A. Plastoquinone A, like ubiquinone, is a substituted benzoquinone; attached to the quinone ring are two methyl groups and a chain usually consisting of nine isoprenic groups (PQ-9, Fig. 1). Its structure was elucidated by TRENNER et al. (1959) and KOFLER et al. (1959). Plastoquinones with shorter side chains have also been isolated, from spinach chloroplasts by MISITI et al. (1965) (PQ-3) and by ECK and TREBST (1963) from chestnut leaves (PQ-4).

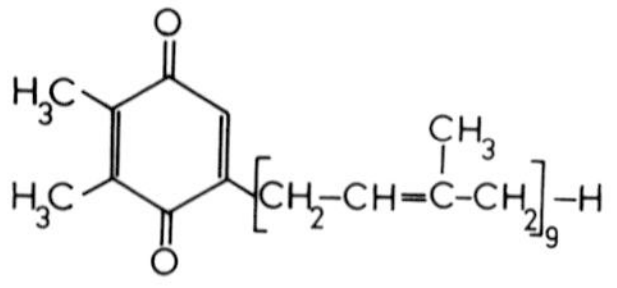

Fig. 1. Structure of plastoquinone A

Compared to other electron carriers, the amount of plastoquinone in chloroplasts of higher plants and in algae is high: a typical concentration in spinach chloroplasts is about $100 \, \mu mol \, mmol^{-1}$ of chlorophyll (CRANE, 1968), i.e. about 40 molecules per reaction center of photosystem I or II. In higher plants, only part of the plastoquinone is located in the photosynthetic lamellae, part of it is contained in the so-called osmiophilic granules or plastoglobuli (BAILEY and WHYBORN, 1963; LICHTENTHALER, 1969). The latter pool is probably not directly involved in photosynthesis (GRUMBACH and LICHTENTHALER, 1975). In addition to plastoquinone A smaller amounts of related compounds have also been isolated from chloroplasts and algae. These are the so-called plastoquinones B and C (KEGEL et al., 1962; HENNINGER et al., 1966; BUCKE et al., 1966), which according to more recent reports (GRIFFITHS et al., 1966; BARR and CRANE, 1967; SUN et al., 1968) again can be differentiated into several related substances. The detailed chemical structure of these compounds is not completely known yet (GRIFFITHS, 1966; DAS et al., 1967). The total amount of plastoquinones B and C present in algae and in chloroplasts of higher plants is often considerably less than that of plastoquinone A (see e.g. CRANE, 1968; SUN et al., 1968).

Analysis of so-called photosystems I and II particles (particles enriched in either of the two photosystems, prepared by detergent treatment) has not yielded clear-cut evidence about the association of plastoquinones with the two photosystems. Data obtained by various authors have been summarized by AMESZ (1973). Photosystem I particles prepared with Triton X-100 or digitonin showed a higher content of plastoquinones, especially of plastoquinone B than photosystem II particles, but even these contained 14–20 μmol plastoquinone A per nmol of chlorophyll, several times more than other electron carriers. According to TEVINI and LICHTEN-THALER (1970) these preparations may be contaminated with plastoglobuli; for a photosystem I preparation free of this contamination they reported a plastoquinone content of 15 μmol per mmol chlorophyll.

In the following we shall use the term plastoquinone in those cases where the identity of the plastoquinone involved is not known with certainty, or where mixtures of plastoquinones were used.

Plastoquinones are colorless substances, but they have a rather specific absorption spectrum in the ultraviolet with a characteristic double band near 260 nm. Figure 2 shows the absorption spectrum of plastoquinone A in ethanol; the spectra of plastoquinones B and C are similar. Upon reduction to the hydroquinone (plasto-quinol) the bands near 260 nm are replaced by a weaker one near 290 nm. This reaction involves the transfer of two electrons; the corresponding midpoint potential in an ethanol-light petroleum mixture, as determined by CARRIER (1967), is $+113$ mV. Plastoquinones, like other quinones, can also be reduced in a one-electron transfer reaction to the semiquinone radical; the absorption spectrum of the plastosemiquinone A anion, prepared by BENSASSON and LAND (1973) is also shown in Figure 2.

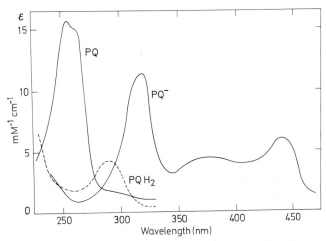

Fig. 2. Absorption spectrum of plastoquinone (PQ) and plastohydroquinone (PQH$_2$, *broken line*) in ethanol and of plastosemiquinone anion in alkaline methanol (PQ$^-$). Adapted from REDFEARN (1965) and from BENSASSON and LAND (1973) respectively. ε: millimolar extinction coefficient

B. Experiments with Extracted Chloroplasts

In 1957, Lynch and French reported that chloroplasts extracted with organic solvents lost the capacity to perform the Hill reaction, i.e. the light-induced reduction of added oxidants with concomitant oxygen evolution. Readdition of the extracted material gave partial reactivation. It soon became apparent (Bishop, 1959; Trebst, 1963) that reactivation could also be obtained by addition of plastoquinones or of synthetic quinones. These experiments suggested that plastoquinone might be an essential component of photosynthetic electron transport, probably functioning at a site near photoreaction 2.

More extensive experimentation with extracted chloroplast systems has yielded complicated results, some of which are perhaps mainly of historic interest now and will be mentioned here only in passing. Krogmann (1961), Krogmann and Olivero (1962), and Chang and Vedvick (1968) observed that photosynthetic phosphorylation could also be inactivated and restored, and it was found that different quinones showed different activity in restoring different functions (see also Henninger and Crane, 1963). After drastic extraction, mixtures of quinones, including tocopherol quinones, were found to be more active than plastoquinones A or B alone in restoring the Hill reaction with ferricyanide (Henninger et al., 1963). Crane (1968) has given a review of these and other results, which have been taken as evidence for a more or less specific function for various quinones in photosynthetic electron transport. However it must be borne in mind that readdition of quinones to extracted chloroplasts, which have lost many other lipids beside quinones, may be a rather crude way partially to restore the original structure. Moreover, recent evidence indicates that plastoquinone may also have a specific structural function. Cox and Bendall (1974) observed that extraction of plastoquinone lowered the midpoint potential of cytochrome b-559; readdition restored the potential to its original value. Okayama (1974) reported that mild extraction of chloroplasts with hexane did not affect the photoreduction of 2,6-dichlorophenol indophenol with the artificial electron donor for system 2, diphenyl-carbazide, but inhibited oxygen evolution. Addition of plastoquinone A partially restored the activity. This result was explained by a structural effect on a component of the electron pathway from water to system 2. Direct (e.g. spectrophotometric) evidence for a function of tocopherol quinones, which, like 2-methyl-3-phytyl-1,4-naphthoquinone (Vitamin K_1), are known to occur naturally in chloroplasts (see Crane, 1968), has not been obtained so far.

C. Reactions of Endogenous Plastoquinone as Secondary Electron Acceptor

Specific evidence about oxidation-reduction reactions of endogenous plastoquinone has come from measurements of changes in light absorption brought about by illumination, a technique that has been widely used for the study of other electron carriers in photosynthesis also (see e.g. this vol., Chaps. II. 1, II.4 and II.7). Absor-

bance changes in the ultraviolet that can be ascribed to photoreduction of plastoquinone were first observed by KLINGENBERG et al. (1962) in spinach chloroplasts.
Experiments with intact algae by AMESZ (1964), RUMBERG et al. (1964), and AMESZ
et al. (1972) showed that plastoquinone was reduced by photosystem II and oxidized
by photosystem I. The reduction was inhibited by DCMU, which indicated that
plastoquinone acted as a secondary electron acceptor, presumably between cytochrome f and the primary acceptor of photosystem II. The difference spectra showed
maxima near 260 nm and isosbestic points at 275–280 nm; the general shape indicated conversion of plastoquinone to plastohydroquinone and vice versa in a two-
electron transfer reaction (Fig. 3). As the sensitivity of measurement is lower than
in the visible part of the spectrum, and the absorbance changes are rather small,
averaging techniques and repetitive illumination have been used in more recent
investigations in order to improve the accuracy. From the size of the absorbance
changes and the specific extinction coefficients of oxidized and reduced plastoquinone, it could be calculated that the pool of plastoquinone that reacts in the
light is several times larger than that of other electron carriers. For spinach chloroplasts this was also demonstrated in a somewhat different way by STIEHL and
WITT (1968), who found that the average amount of plastoquinone reduced in
a short saturating flash (during which only one electron is transferred per reaction
center) was about one-tenth of that reduced by longer illumination, which permitted
a large number of turnovers of the reaction center of photosystem II (Fig. 4).
This large pool of electron acceptor was observed earlier by FORK (1963) and
others in an indirect way, by measuring the amount of oxygen evolved by chloroplasts to which no external electron acceptor had been added. The pool is also
reflected in the kinetics of chlorophyll fluorescence. For a discussion of this point
the reader is referred to AMESZ (1973).

In strong light plastoquinone accumulates mainly in the reduced, and cytochrome f in the oxidized form. This demonstrates that the rate-limiting step in
the electron transport chain between photosystem I and II is between plastoquinone

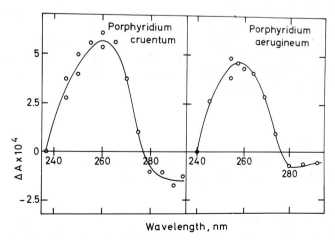

Fig. 3. Absorbance difference spectra of two species of unicellular red algae. Spectra obtained
upon alternating illumination with light mainly absorbed by photosystems I and II respectively.
(From AMESZ et al., 1972.) Reprinted by permission of Elsevier Publ. Co.

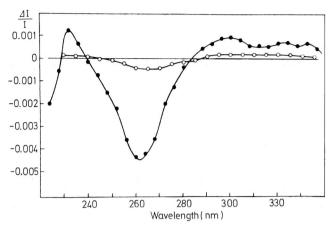

Fig. 4. Absorbance difference spectra of spinach chloroplasts in presence of benzylviologen as electron acceptor. *Open circles*: induced by short flashes (20μsec); *solid circles*: induced by long flashes (125 msec). Long flashes given on "background" of far-red light to reoxidize plastoquinone between flashes (STIEHL and WITT, 1968)

and cytochrome f. From kinetic analysis STIEHL and WITT (1969) calculated the total pool of plastoquinone to be 7 molecules (14 electron equivalents) per reaction center of photosystem II in spinach chloroplasts. Reaction times of about 0.5 and 20 msec were obtained for the reduction and oxidation of plastoquinone by photosystems II and I respectively. Experiments with *Chlorella pyrenoidosa* and spinach chloroplasts in which part of the reaction centers had been inhibited with DCMU indicated that the plastoquinone pool connects several electron transport chains (SIGGEL et al., 1972; DUYSENS, 1972).

D. Identity of the Primary Electron Acceptor of Photosystem II

Although it was shown several years ago (DUYSENS and SWEERS, 1963) that the redox level of the primary electron acceptor of photosystem II can be determined by measuring the fluorescence yield of chlorophyll a, until recently no clear-cut information was available about its chemical nature. STIEHL and WITT (1968), 1969) observed absorbance changes in the ultraviolet region, with a more rapid decay than those discussed in the previous section, in chloroplasts that were illuminated with repetitive, short flashes. The substance responsible for these absorbance changes was called X-320 after the location of the main peak in the difference spectrum. The relaxation time of the absorbance changes (0.6 msec) agreed with that of the overall system for oxygen production (VATER et al., 1969) and with that of the decay of the fluorescence yield of chlorophyll a (FORBUSH and KOK, 1968). It was proposed that X-320 is the primary electron acceptor of photosystem II. Although the absorbance changes below 300 nm were relatively small and the spectrum was not well defined below 265 nm, X-320 was tentatively defined as plastoquinone being reduced to the semiquinone anion upon illumination.

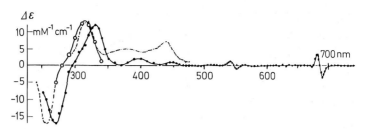

Fig. 5. Difference spectrum (*solid circles*) obtained on illumination of subchloroplast fragments prepared with deoxycholate. Spectrum can be ascribed to reduction of primary electron acceptor of photosystem II. (From van Gorkom, 1975.) In region below 450 nm, spectrum presumably caused by reduction of plastoquinone to semiquinone anion; *broken line* shows difference spectrum of this reduction in vitro (see also Fig. 2). *Open circles*: light-induced difference spectrum obtained immediately after addition of deoxycholate to more concentrated chloroplast suspension (not to scale). Reprinted by permission of Elsevier Publ. Co.

Clearer evidence that the primary acceptor of photosystem II is indeed plasto-quinone has come from recent experiments which were done under conditions in which reduction of the large, secondary pool of plastoquinone did not occur upon illumination. With chloroplasts treated with the detergent deoxycholate, van Gorkom (1974) obtained a difference spectrum which was similar to that obtained upon reduction of plastoquinone to its semiquinone anion in vitro by Bensasson and Land (1973). The spectrum of van Gorkom, however, appeared to be shifted toward the red by about 15 nm, apparently due to the detergent (Fig. 5). No such shift in the difference spectrum was observed upon illumination of chloroplasts at −40° C in the absence of deoxycholate (Pulles et al., 1974) and in chloroplasts treated only briefly with deoxycholate at room temperature (van Gorkom, 1975). Both at −40° C and in the presence of deoxycholate, the kinetics of the UV absorbance changes were the same as those of C-550, a substance that has been found to act as indicator for the reduction of the primary acceptor by displaying a blue shift of a band near 550 nm (see this vol., Chap. II.5, and also Butler, 1973). The UV changes were also correlated with changes in the yield of chlorophyll a fluorescence. Witt (1973) recently also observed at 140 K light-induced absor-bance changes with a spectrum similar to that of X-320 in the region 310–350 nm. This supports the hypothesis that the reduction to semiquinone is a primary reac-tion. However, Witt reported that the UV absorbance changes induced by a short flash were irreversible, whereas Mathis and Vermeglio (1974) observed that most of the absorbance change due to C-550 reversed with a half-time of about 4 msec under similar conditions, presumably by a back reaction with the oxidized electron donor (P-680$^+$) of photosystem II. The discrepancy between the results of Witt and those of Mathis and Vermeglio remains to be resolved by further experimentation. Another point that needs clarification is the observation by Vernon et al. (1971) that chloroplast fragments which are enriched in photosys-tem II reaction centers contain only traces of plastoquinone.

There are indications that the primary electron acceptor does not react directly with the large plastoquinone pool. Velthuys and Amesz (1974) have shown that the rate of increase of chlorophyll a fluorescence brought about by addition of the reductant dithionite is dependent on the number of light flashes given before dithionite addition. After an uneven number of flashes the fluorescence yield increased relatively fast; after an even number of flashes, or without preillumination, a relatively slow increase occurred. This indicated that "charge accumulation" with periodicity two occurs at the reducing side of photosystem II, comparable to the charge accumulation with periodicity four (the so-called S-states, see this vol. Chap. II. 6) occurring at the oxidizing side. The experiments were done with chloroplasts treated with a high concentration of Tris and in the presence of an artificial donor, i.e. under conditions where the S-state phenomena do not occur. On the basis of these experiments it was concluded that a secondary acceptor, R, functions between the primary acceptor and the large plastoquinone pool. R reacts with the primary acceptor by two successive electron transfers; when fully reduced, $R^=$ reacts in a two-electron reaction with the secondary plastoquinone pool. The same conclusion was obtained independently by Bouges-Bocquet (1973) from a periodicity two observed in methylviologen reduction by flashes. Measurements of the fluorescence kinetics of intact, anaerobic cells of *Chlorella pyrenoidosa* after flashes also indicated the existence of a secondary acceptor R (van Best and Duysens, to be published) and indicated the existence of various endogenous electron donors for R and plastoquinone. The identity of R is still unknown; it might be a plastoquinone molecule but there is no evidence on this point as yet.

E. Specific Inhibitors of Plastoquinone

Some years ago, Trebst and co-workers (Trebst et al., 1970; Böhme, et al., 1971) observed that 2,5-dibromo-3-methyl-6-isopropyl-p-benzoquinone (DBMIB) inhibits electron transport in the chain between photosystem I and II. Similar results were obtained by Arntzen et al. (1971) with 2,3-dimethyl-5-hydroxy-6-phytyl benzoquinone. DBMIB inhibited cyclic phosphorylation with menadione as cofactor, but not the photooxidation of diaminodurene by photosystem I or cyclic phosphorylation catalyzed by PMS. The Hill reaction with 2,6-dichlorophenol indophenol or ferricyanide as electron acceptor was partially inhibited; the photoreduction of endogenous cytochrome f was completely inhibited by DBMIB (Böhme and Cramer, 1971). Addition of plastoquinone (Böhme et al., 1971) and of phenylenediamines (Trebst and Reimer, 1973) relieved the inhibition. From the above data it was concluded that DBMIB (and the other quinone) inhibits electron transport between plastoquinone and cytochrome f. Dichlorophenol indophenol and ferricyanide were assumed to be able to react directly with reduced plastoquinone, a reaction not blocked by DBMIB. Electron transport to Hill acceptors via cytochrome f would of course be inhibited. The different sensitivities of phosphorylation with different cofactors were explained similarly: menadione-induced phosphorylation would involve electron transport from plastoquinone to cytochrome f, whereas PMS would short circuit this part of the chain. Experiments of Lozier and Butler (1972) indicate that DBMIB can also be reduced by reduced plastoquinone.

Recent, more extensive investigations on the effect of DBMIB on cyclic phosphorylation (Hauska et al., 1974) indicate that three types of cofactors can be distinguished on the basis of sensitivity to DBMIB and stimulation by tetramethylphenylenediamine. With some cofactors plastoquinone is needed as transmembrane shuttle for electrons and protons, needed to generate energy for the production of ATP, according to the well-known Mitchell concept; other cofactors may act as shuttle themselves (see also GOULD, 1975). Measurement of the effect of DBMIB on $NADP^+$ reduction by photosystem I with artificial electron donors indicates that with nearly all donors the plastoquinone shuttle is inoperative and is replaced by a donor shuttle capable of driving ATP synthesis (HAUSKA et al., 1975).

Results obtained with intact algae are in agreement with the proposed site of inhibition by DBMIB. DBMIB inhibited the reduction of cytochrome f by photosystem II in the red alga *Porphyridium cruentum*, while measurements of chlorophyll flourescence indicated that the oxidation, but not the reduction of plastoquinone by photosystem II was inhibited (GIMMLER and AVRON, 1972; BAUER and WIJNANDS, 1974). As in chloroplasts, phenylenediamines relieved this inhibition.

References

Amesz, J.: Biochim. Biophys. Acta **79**, 257–265 (1964)

Amesz, J.: Biochim. Biophys. Acta **301**, 35–51 (1973)

Amesz, J., Visser, J.W.M., Engh, G.J. van den, Dirks, M.P.: Biochim. Biophys. Acta **256**, 370–380 (1972)

Arntzen, C.J., Neumann, J., Dilley, R.A.: J. Bioenerg. **2**, 73–83 (1971)

Bailey, J.L., Whyborn, A.G.: Biochim. Biophys. Acta **78**, 163–174 (1963)

Barr, R., Crane, F.L.: Plant Physiol. **42**, 1255–1263 (1967)

Bauer, R., Wijnands, M.J.G.: Z. Naturforsch. **29C**, 725–732 (1974)

Bensasson, R., Land, E.J.: Biochim. Biophys. Acta **325**, 175–181 (1973)

Bishop, N.I.: Proc. Nat. Acad. Sci. U.S. **45**, 1696–1702 (1959)

Böhme, H., Cramer, W.A.: FEBS Lett. **15**, 349–351 (1971)

Böhme, H., Reimer, S., Trebst, A.: Z. Naturforsch. **26B**, 341–352 (1971)

Bouges-Bocquet, B.: Biochim. Biophys. Acta **314**, 250–256 (1973)

Bucke, C., Leech, R.M., Hallaway, M., Morton, R.A.: Biochim. Biophys. Acta **112**, 19–34 (1966)

Butler, W.L.: Acc. Chem. Res. **6**, 177–184 (1973)

Carrier, J.-M.: In: Biochemistry of Chloroplasts. Goodwin, T.W. (ed.). London-New York: Academic Press, 1967, Vol. II, pp. 551–557

Chang, S.B., Vedvick, T.S.: Plant Physiol. **43**, 1661–1665 (1968)

Cox, R.P., Bendall, D.S.: Biochim. Biophys. Acta **347**, 49–59 (1974)

Crane, F.L.: In: Biological Oxidations. Singer, T.P. (ed.). New York: Interscience, 1968, pp. 533–580

Das, B.C., Lounasmaa, M., Tendille, C., Lederer, E.: Biochem. Biophys. Res. Commun. **26**, 211–215 (1967)

Duysens, L.N.M.: In: Proc. 2nd Intern. Congr. Photosynthesis. Forti, G., Avron, M., Melandri, A. (eds.). The Hague: Dr. Junk, 1972, Vol. I, pp. 19–25

Duysens, L.N.M., Sweers, H.E.: Special Issue Plant Cell Physiol., pp. 353–372. Tokyo: University of Tokyo Press 1963.

Eck, H., Trebst, A.: Z. Naturforsch. **18B**, 446–451 (1963)

Forbush, B., Kok, B.: Biochim. Biophys. Acta **162**, 243–253 (1968)

Fork, D.C.: Plant Physiol. **38**, 323–332 (1963)

Fuller, R.C., Smillie, R.M., Rigopoulos, N., Yount, V.: Arch. Biochem. Biophys. **95**, 197–202 (1961)

Gimmler, H., Avron, M.: In: Proc. 2nd Intern. Congr. Photosynthesis. Forti, G., Avron, M., Melandri, A. (eds.). The Hague: Dr. Junk, 1972, Vol. I, pp. 789–800

Gorkom, H.J. van: Biochim. Biophys. Acta **347**, 439–442 (1974)

Gorkom, H.J. van: In: Proc. 3rd Intern. Congr. Photosynthesis. Avron, M. (ed.). Amsterdam: Elsevier, 1975, Vol. I, pp. 159–162

Gould, J.M.: Biochim. Biophys. Acta **387**, 135–148 (1975)

Griffiths, W.T.: Biochem. Biophys. Res. Commun. **25**, 596–602 (1966)

Griffiths, W.T., Wallwork, J.C., Pennock, J.F.: Nature (London) **211**, 1037–1039 (1966)

Grumbach, K.H., Lichtenthaler, H.K.: In: Proc. 3rd Intern. Congr. Photosynthesis. Avron, M. (ed.). Amsterdam: Elsevier, 1975, Vol. I, pp. 515–523

Hauska, G., Oettmeier, W., Reimer, S., Trebst, A.: Z. Naturforsch. **30C**, 37–45 (1975)

Hauska, G., Reimer, S., Trebst, A.: Biochim. Biophys. Acta **357**, 1–13 (1974)

Henninger, M.D., Barr, R., Crane, F.L.: Plant Physiol. **41**, 696–700 (1966)

Henninger, M.D., Crane, F.L.: Biochemistry **2**, 1168–1171 (1963)

Henninger, M.D., Dilley, R.A., Crane, F.L.: Biochem. Biophys. Res. Commun. **10**, 237–242 (1963)

Kegel, L.P., Henninger, M.D., Crane, F.L.: Biochem. Biophys. Res. Commun. **8**, 294–298 (1962)

Klingenberg, M., Müller, A., Schmidt-Mende, P., Witt, H.T.: Nature (London) **194**, 379–380 (1962)

Kofler, M., Langemann, A., Rüegg, R., Chopart-Dit-Jean, L.H., Rayroud, A., Isler, O.: Helv. Chim. Acta **42**, 1283–1292 (1959)

Krogmann, D.W.: Biochem. Biophys. Res. Commun. **4**, 275–277 (1961)

Krogmann, D.W., Olivero, E.: J. Biol. Chem. **237**, 3292–3295 (1962)

Lester, R.L., Crane, F.L.: J. Biol. Chem. **234**, 2169–2175 (1959)

Lichtenthaler, H.K.: Protoplasma **68**, 65–77 (1969)

Lozier, R.H., Butler, W.L.: FEBS Lett. **26**, 161–164 (1972)

Lynch, V.H., French, C.S.: Arch. Biochem. Biophys. **70**, 382–391 (1957)

Mathis, P., Vermeglio, A.: Biochim. Biophys. Acta **368**, 130–134 (1974)

Misiti, D., Moore, H.W., Folkers, K.: J. Am. Chem. Soc. **87**, 1402–1403 (1965)

Okayama, S.: Plant Cell Physiol. **15**, 95–101 (1974)

Pulles, M.P.J., Kerkhof, P.L.M., Amesz, J.: FEBS Lett. **47**, 143–145 (1974)

Redfearn, E.R.: In: Biochemistry of Quinones. Morton, R.A. (ed.). London-New York: Academic Press, 1965, pp. 149–181

Rumberg, B., Schmidt-Mende, P., Witt, H.T.: Nature (London) **201**, 466–468 (1964)

Siggel, U., Renger, G., Stiehl, H.H., Rumberg, B.: Biochim. Biophys. Acta **256**, 328–335 (1972)

Stiehl, H.H., Witt, H.T.: Z. Naturforsch. **23B**, 220–224 (1968)

Stiehl, H.H., Witt, H.T.: Z. Naturforsch. **24B**, 1588–1598 (1969)

Sun, E., Barr, R., Crane, F.L.: Plant Physiol. **43**, 1935–1940 (1968)

Tevini, M., Lichtenthaler, H.K.: Z. Pflanzenphysiol. **62**, 17–32 (1970)

Trebst, A.: Proc. Roy. Soc., Ser. B. **157**, 355–364 (1963)

Trebst, A., Harth, E., Draber, W.: Z. Naturforsch. **25B**, 1157–1159 (1970)

Trebst, A., Reimer, S.: Z. Naturforsch. **28C**, 710–716 (1973)

Trenner, N.R., Arison, B.H., Erickson, R.E., Shunk, C.H., Wolf, D.E., Folkers, K.: J. Am. Chem. Soc. **81**, 2026–2027 (1959)

Vater, J., Renger, G., Stiehl, H.H., Witt, H.T.: In: Progress in Photosynthesis Research. Metzner, H. (ed.). Tübingen: Laupp, 1969, Vol. II, pp. 1006–1008

Velthuys, B.R., Amesz, J.: Biochim. Biophys. Acta **333**, 85–94 (1974)

Vernon, L.P., Shaw, E.R., Ogawa, T., Raveed, D.: Photochem. Photobiol. **14**, 343–357 (1971)

Witt, K.: FEBS Lett. **38**, 116–118 (1973)

14. Plastocyanin

Plastocyanin is a copper protein which was first discovered in the green alga, *Chlorella ellipsoidea* (KATOH, 1960). It is now established that the protein functions as an electron carrier between photosystems I and II of chloroplasts. The protein was designated plastocyanin because of its localization in chloroplasts and its blue color in the oxidized form (KATOH and TAKAMIYA, 1961).

A. Distribution and Localization

Plastocyanin occurs in a wide variety of higher plants and algae (KATOH et al., 1961; GORMAN and LEVINE, 1966; PLESNIČAR and BENDALL, 1970). In prokaryote, the protein is present in blue-green algae (LIGHTBODY and KROGMANN, 1967), but not in photosynthetic bacteria. The occurrence of plastocyanin in the red alga, *Porphyridium aerugineum*, which had been suspected for some time, was recently confirmed by VISSER et al. (1974), who used a refined technique of EPR spectroscopy for the detection of plastocyanin in the cells.

In higher plants, plastocyanin is exclusively found in the green photosynthetic tissues (KATOH et al., 1961). Exceptionally, an occurrence of the protein in etiolated leaves of barley seedlings was reported by PLESNIČAR and BENDALL (1970). The protein is present in association with the photosynthetic membranes in the cells. Spinach chloroplasts contain plastocyanin at a ratio of about 300 chlorophyll molecules per atom of copper of the protein, which accounts for about half the total copper in the chloroplasts (KATOH et al., 1961). Recent estimates yielded somewhat higher values for the plastocyanin content of 200 chlorophyll molecules per atom of copper.

B. Extraction and Purification

Plastocyanin is loosely bound to the lamella membrane and is readily solubilized by acetone- or detergent-treatment, or sonic oscillation, of chloroplasts. The protein is purified by fractionation with ammonium sulfate followed by a column chromatography with diethylaminoethyl cellulose (KATOH et al., 1962). Column chromatography on Sephadex G-75 or G-100 can also be used at the final stage of the purification.

A convenient measure of the purity of plastocyanin is the absorbance index, $E_{278/597}$, which is defined as the ratio of absorbance at 278 nm to that at 597 nm, the maximum of the main visible absorption peak of the oxidized protein. Purified protein shows a value for the index of about 1.0. Plastocyanin from several plants has been crystallized (YAKUSHIJI, personal communication; see also BLUMBERG and PEISACH, 1966; SCAWEN and HEWITT, 1971).

C. Molecular Properties

Plastocyanin can readily be identified by its absorption spectrum (KATOH et al., 1962). The oxidized protein is blue in color and shows three absorption bands in the visible and far-red regions; the main band with absorption maximum at 597 nm and two minor bands at 770 nm and 460 nm (Fig. 1). On reduction of the protein, the visible absorption disappears completely.

The absorption in the ultraviolet region is characterized by the presence of vibrational fine structure bands of amino acid constituents. Besides the main absorption peak at 278 nm, four peaks at 253, 259, 265, and 269 nm, corresponding to the fine structure bands of phenylalanine, and a peak at 284 nm, corresponding to that of tyrosine, are clearly identified.

The blue color of the oxidized plastocyanin is due to Cu^{2+} bound to the protein (KATOH and TAKAMIYA, 1964). Spinach plastocyanin contains 0.58% copper. Extinction coefficient of the copper chromophore, $\varepsilon_{g\ atom\ copper}^{1\ cm}$, is $4.9 \cdot 10^3$ at 597 nm in the oxidized spinach plastocyanin (KATOH et al., 1962). Although the molecular weight of the proteins isolated from different plants varies between $1 \cdot 10^4$ and $2 \cdot 10^4$ (RAMSHAW et al., 1973), values for copper content and for the extinction coefficient are similar with most of the plastocyanin samples studied.

The amino acid compositions for plastocyanin from several plants are similar in that the protein contains two methionine and one half-cysteine per copper atom. Neither tryptophane nor arginine is present. Plastocyanin contains relatively large amounts of acidic amino acids and shows an isoelectric point of about

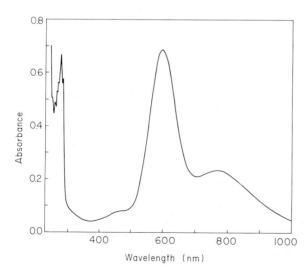

Fig. 1. Absorption spectrum of spinach plastocyanin (KATOH et al., 1961)

4.0 (KATOH et al., 1962; MILNE and WELLS, 1970; SCAWEN and HEWITT, 1971). One exception is plastocyanin isolated from a blue-green alga, *Anabaena variabilis*, which was a basic protein (LIGHTBODY and KROGMANN, 1967).

The oxidation-reduction potential of spinach plastocyanin is 370 mV between pH 5.4 and 9.9 (KATOH et al., 1962). The protein can be reduced or oxidized by various substances. The reduced protein, however, cannot react with molecular oxygen. Rate data for the oxidation and reduction of plastocyanin so far reported are summarized in Table 1.

The copper-protein linkage in plastocyanin is stable toward various treatments. Incubation of the protein for 5 min at temperature up to 60°C, or at pH between 5.0 and 9.5 causes no change in the absorption spectrum. Various chelating reagents and potassium cyanide (KCN) do not react with the copper of plastocyanin under the conditions where these reagents are usually employed as inhibitors against copper-containing oxidases (KATOH and SAN PIETRO, 1966a). However, the copper can be removed from the protein slowly by a prolonged dialysis against the solution of the chelating reagents, or rather rapidly by incubation with KCN of higher concentration at an alkaline pH (OUITRAKUL and IZAWA, 1973).

The copper-protein binding in plastocyanin involves the sulfhydryl (SH) group of the cysteine residue and is sensitive towards SH reagents. The copper atoms of plastocyanin can be replaced by Hg^{2+} rapidly in the reduced state but very slowly in the oxidized state of the protein even in the presence of 6 M urea (KATOH and TAKAMIYA, 1964; KIMIMURA and KATOH, 1972).

Table 1. Rate data for oxidation and reduction of plastocyanin by various substances

Electron donors or acceptors	k ($M^{-1}s^{-1}$)	Reaction conditions		Source of plastocyanin
		pH	Temp. (°C)	
Reduction				
Ascorbate[a]	280	7.0	25	cucumber
Ascorbate[b]	115	7.0	25	parsley
Hydroquinone[a]	88	7.0	25	cucumber
Hydroquinone[b]	80	7.0	25	parsley
p-Phenylenediamine[a]	$3.2 \cdot 10^4$	7.0	25	cucumber
Dimethyl p-phenylenediamine[a]	$1.7 \cdot 10^4$	7.0	25	cucumber
Dithionite[c] $S_2O_4^{2-}$	$1.4 \cdot 10^5$	8.0	25	spinach
SO_2^-	$2.9 \cdot 10^7$	8.0	25	spinach
Chromous ion (CrII)[d]	$1\text{--}3.3 \cdot 10^4$	4.2	4–32	spinach, bean
Cytochrome f (parsley)[b]	$3.6 \cdot 10^7$	7.0	25	parsley
Cytochrome f (red alga)[b]	$5.0 \cdot 10^5$	7.0	25	parsley
Cytochrome c (horse heart)[b]	10^6	7.0	25	parsley
Reduced P-430 (illuminated chloroplasts)[e]	$2.0 \cdot 10^6$	7.0	25	parsley
Oxidation				
Ferricyanide[b]	$7.0 \cdot 10^4$	7.0	25	parsley
O_2^- [e]	$2.6 \cdot 10^5$	7.0	25	parsley
$P\text{-}700^+$ [e]	$8.0 \cdot 10^7$	7.0	25	parsley

[a] NAKAMURA and OGURA (1968); [b] WOOD (1974); [c] LAMBETH and PALMER (1973); [d] DAWSON et al. (1972); [e] WOOD and BENDALL (1975).

D. Function in Photosynthetic Electron Transport System

No evidence for the presence of an enzymic activity of plastocyanin has been reported.

Earlier works on the sonically treated or detergent-treated chloroplasts established the role of plastocyanin as an electron carrier in photosynthetic electron transport, functioning at the electron-donating side of photosystem I (KATOH and TAKAMIYA, 1963; KATOH and SAN PIETRO, 1966b; WESSELS, 1966; ELSTNER et al., 1968). However, the location of plastocyanin in relation to cytochrome f and P-700 has long been disputed and several models for this part of the electron transport chain have been proposed:

Model 1. plastocyanin→cytochrome f→P-700

Model 2. plastocyanin
$\qquad\qquad\qquad\searrow$ P-700
$\qquad$ cytochrome f $\nearrow$

Model 3. cytochrome f → plastocyanin → P-700

Model 4. Photosystem IIb → plastocyanin → Photosystem IIa → NADP$^+$
$\qquad$ ┌─ cytochrome f → P-700 ─┐
$\qquad$ └──────────────◄──────────┘

Model 5. cytochrome f → P-700

Models 1 and 2, by FORK and URBACH (1965) and KOK and RURAINSKI (1965) respectively, were suggested mainly from the results of spectrophotometric investigations on the light-induced absorbance changes of cytochromes and plastocyanin in vivo and in vitro.

The sequence in model 3 was supported by a number of investigators, based on the observations that in subchloroplast particles which were depleted in plastocyanin, NADP$^+$ photoreduction was restored by the addition of plastocyanin but not of cytochrome f, and that cytochrome f photooxidation was markedly accelerated by the addition of plastocyanin (WESSELS, 1966; AVRON and SHNEYOUR, 1971). Further strong evidence for this model was presented by GORMAN and LEVINE (1966), who found that chloroplasts isolated from a *Chlamydomonas* mutant lacking plastocyanin were inactive in cytochrome f photooxidation, NADP$^+$ photoreduction and phenazinemethosulfate-supported cyclic photophosphorylation. The addition of plastocyanin was effective in restoring all the activities. The rate data presented in Table 1 show that reduction of plastocyanin by reduced cytochrome f and oxidation by oxidized P-700 are the fastest reactions among those listed (WOOD and BENDALL, 1975). This is also compatible with model 3.

Recently, the sequence in model 3 was challenged by two groups of investigators. ARNON and his co-workers (1971) and FORK and MURATA (1971) reported that photosystem I-enriched particles of spinach chloroplasts prepared with the French press showed high rates of NADP photoreduction and cytochrome f photooxidation. However, the particles were free of plastocyanin and the addition of the protein failed to affect significantly the rates of the two reactions. With this and other results, ARNON et al. proposed model 4, in which plastocyanin is located in a non-cyclic electron transport system connecting photosystems IIa and IIb, whereas cytochrome f is placed in a cyclic system involving only photosystem I. Similarly, FORK and MURATA concluded that plastocyanin is not involved in the electron transfer between cytochrome f and P-700 (model 5).

The above observations were, however, not confirmed by BASZYNSKI et al. (1971) and SANE and HAUSKA (1972) who found that the French press particles still retained substantial amounts of the protein.

MALKIN and BEARDEN (1973) demonstrated that redox reaction of plastocyanin bound to the chloroplast membrane could be studied with EPR spectroscopy. Their finding that plastocyanin is oxidized and reduced by photosystems I and II respectively is incompatible with model 4.

Spectrophotometric analysis of light-induced changes of plastocyanin in situ is difficult because of its broad absorption bands with low extinction coefficients. However, flash photometric studies of redox reactions of endogenous electron carriers suggested the location of plastocyanin in the main electron transport chain connecting the two photosystems. HAEHNEL et al. (1971) found that dark reduction of P-700 by endogenous electron carriers between plastoquinone and P-700 showed two kinetic components with half times of 20 and 200 μsec respectively. The 200 μsec component is due to electron transfer from cytochrome f to P-700 while the 20 μsec component is assumed to be due to reduction of P-700 by plastocyanin.

Comparison of the kinetics of known electron carriers revealed that no more than 15% of the electrons from plastoquinone passes cytochrome f (HAEHNEL, 1973). More recently, HAEHNEL (1975) reported that an absorbance change due to plastocyanin could be determined without superposition of other absorbance changes in spinach chloroplasts. Relative concentrations of cytochrome f, plastocyanin and P-700 which respond to illumination were estimated to be 0.4/1/1. HAEHNEL suggested, therefore, that most of the electrons from plastoquinone are transferred via plastocyanin to P-700, whereas cytochrome f is situated in a side path of the linear electron transport.

It is now generally accepted that plastocyanin is located inside the lamella membrane. HAUSKA et al. (1971) showed that an antibody against plastocyanin could not inhibit electron transfer associated with photosystem I unless chloroplasts were sonicated in the presence of the antibody. This implies that the antibody could not react with plastocyanin in the intact membrane. Recently, several substances which were able to inactivate plastocyanin in situ have been reported. Participation of plastocyanin in electron flow from cytochrome f to P-700 was confirmed by experiments in which a plastocyanin-specific inhibitor was employed. KIMIMURA and KATOH (1972) showed that, by incubating the chloroplasts at a chlorophyll to $HgCl_2$ ratio of unity, $HgCl_2$ preferentially inactivated plastocyanin bound to the lamella membrane, by replacing the copper of the protein with Hg^{2+}. In the treated chloroplasts, electron transport associated with photosystem I, such as methylviologen photoreduction with reduced indophenol dye as electron donor, was completely inhibited. Photooxidation of cytochrome f was similarly blocked, whereas photooxidation of P-700, as well as photoreduction of cytochrome f, was highly resistant to $HgCl_2$. These results conclusively show that plastocyanin participates in the electron flow between cytochrome f and the reaction center of photosystem I.

Polylysine (Brand and San Pietro, 1972) and KCN (Ouitrakul and Izawa, 1973) appear to act as inhibitors of the membrane-bound plastocyanin, since they block electron transfer from cytochrome f to P-700. They react with solubilized plastocyanin, and recent evidence (Selman et al., 1975) indicates that KCN does inactivate plastocyanin in situ.

References

Arnon, D.I., Knaff, D.B., McSwain, B.D., Chain, R.K., Tsujimoto, H.Y.: Photochem. Photobiol. **14**, 397–425 (1971)
Avron, M., Shneyour, A.: Biochim. Biophys. Acta **226**, 498–500 (1971)
Baszynski, T., Brand, J., Krogmann, D.W., Crane, F.L.: Biochim. Biophys. Acta **234**, 537–540 (1971)
Blumberg, W.E., Peisach, J.: Biochim. Biophys. Acta **126**, 269–273 (1966)
Brand, J., San Pietro, A.: Arch. Biochem. Biophys. **152**, 426–428 (1972)
Dawson, J.W., Gray, H.B., Holwerda, R.A., Westhead, E.W.: Proc. Nat. Acad. Sci. U.S. **69**, 30–33 (1972)
Elstner, E., Pistorius, E., Böger, P., Trebst, A.: Planta (Berl.) **79**, 146–161 (1968)
Fork, D.C., Murata, N.: Photochem. Photobiol. **13**, 33–44 (1971)
Fork, D.C., Urbach, W.: Proc. Nat. Acad. Sci. U.S. **53**, 1307–1315 (1965)
Gorman, D., Levine, R.P.: Plant Physiol. **247**, 1948–1956 (1966)
Haehnel, W.: Biochim. Biophys. Acta **305**, 618–631 (1973)
Haehnel, W.: Proc. Third Intern. Congr. Photosyn. Avron, M. (ed.). Amsterdam: Elsevier, 1975, pp. 557–566
Haehnel, W., Döring, G., Witt, H.T.: Z. Naturforsch. **26**, 1171–1174 (1971)
Hauska, G.A., McCarty, R.E., Berzborn, R.J., Racker, E.: J. Biol. Chem. **246**, 3524–3531 (1971)
Katoh, S.: Nature (London) **186**, 533–534 (1960)
Katoh, S., San Pietro, A.: Biochem. Biophys. Res. Commun. **24**, 903–908 (1966a)
Katoh, S., San Pietro, A.: Biochemistry of Copper. Peisach, J., Aisen, P., Blumberg, W.E. (eds.). New York: Academic Press, 1966b, pp. 407–422
Katoh, S., Shiratori, I., Takamiya, A.: J. Biochem. **51**, 32–40 (1962)
Katoh, S., Suga, I., Shiratori, I., Takamiya, A.: Arch. Biochem. Biophys. **94**, 136–141 (1961)
Katoh, S., Takamiya, A.: Nature (London) **189**, 665–666 (1961)
Katoh, S., Takamiya, A.: Plant Cell Physiol. **4**, 335–347 (1963)
Katoh, S., Takamiya, A.: J. Biochem. **55**, 378–387 (1964)
Kimimura, M., Katoh, S.: Biochim. Biophys. Acta **283**, 279–292 (1972)
Kok, B., Rurainski, H.J.: Biochim. Biophys. Acta **94**, 588–590 (1965)
Lambeth, D.O., Palmer, G.: J. Biol. Chem. **248**, 6095–6103 (1973)
Lightbody, J.J., Krogmann, D.W.: Biochim. Biophys. Acta **131**, 508–515 (1967)
Malkin, R., Bearden, A.J.: Biochim. Biophys. Acta **292**, 169–185 (1973)
Milne, P.R., Wells, R.E.: J. Biol. Chem. **245**, 1566–1574 (1970)
Nakamura, T., Ogura, Y.: J. Biochem. **64**, 267–270 (1968)
Plesničar, M., Bendall, D.S.: Biochim. Biophys. Acta **216**, 192–199 (1970)
Ouitrakul, R., Izawa, S.: Biochim. Biophys. Acta **305**, 105–118 (1973)
Ramshaw, J.A.M., Brown, R.H., Scawen, M.D., Boulter, D.: Biochim. Biophys. Acta **303**, 269–273 (1973)
Sane, P.V., Hauska, G.A.: Z. Naturforsch. **27 B**, 932–938 (1972)
Scawen, M.D., Hewitt, E.J.: Biochem. J. **124**, 32p (1971)
Selman, B.R., Johnson, G.L., Giaquinta, R.T., Dilley, R.A.: Bioenergetics **6**, 221–231 (1975)
Visser, J.W., Amesz, J., Van Gelder, B.F.: Biochim. Biophys. Acta **333**, 279–287 (1974)
Wessels, J.S.C.: Biochim. Biophys. Acta **126**, 581–583 (1966)
Wood, P.M.: Biochim. Biophys. Acta **375**, 370–379 (1974)
Wood, P.M., Bendall, D.S.: Biochim. Biophys. Acta **387**, 115–128 (1975)

15. Artificial Acceptors and Donors

G. Hauska

A. Introduction

The use of artificial electron donors and acceptors has been one of the biochemical approaches to the elucidation of complex electron transport chains in biological membranes. In photosynthesis it started with Hill's discovery that ferric salts could induce light-dependent oxygen evolution in cell-free leaf extracts (Hill, 1937). Since then the results of this approach have taught us about the sequence, about the energy-conserving steps, and more recently about the topography of the photosynthetic electron transport chain in the chloroplast membrane. It has also made possible the design of assays for parts of the chain in a physiologically non-functional state, i.e. for the reaction center complexes, either during biochemical isolation (see this vol., Chap. IV.5) or during biogenesis of the photosynthetic apparatus (see this vol. Chap. IV.6).

Several reviews pertinent to the subject exist in the literature (Avron, 1967; Avron and Neumann, 1968; Jagendorf, 1962; Kandler, 1960; Trebst, 1974). For a summary with more experimental background, see Trebst (1972b). For examples of the approach applied to electron transport of mitochondria or photosynthetic bacteriy, consult Hinkle (1973, Colowick and Kaplan (1967), Bose and Gest (1963), Del Valle Tascon and Ramirez (1975).

B. General Aspects

Figure 1 represents the photosynthetic electron transport chain of chloroplasts in a linear form (see this vol., Chap. II.1b), with five sites for inhibition (see this vol., Chap. II.16), and the emphasis on the exit and entry of reducing equivalents, to or from artificial redox compounds. These can be grouped into four classes: electron donors for photosystem I (D_1), electron donors for photosystem II (D_2), electron acceptors for photosystem I (A_1), and electron acceptors for photosystem II (A_2). For each group multiple possibilities to react with the electron transport chain exist; the more important ones are shown in the scheme. Several partial reactions can be induced by the combined use of redox compounds with the appropriate inhibitors. Most important among them is the functional isolation of photosystem II, by applying inhibitor 2 or 3 in combination with A_2 (Fig. 1), and of photosystem I with inhibitor 3 or 4 and D_1.

On a redox potential scale the chain spans a region from about 1.0 V positive for the oxidant in photosystem I to about 0.6 V negative for the reductant in

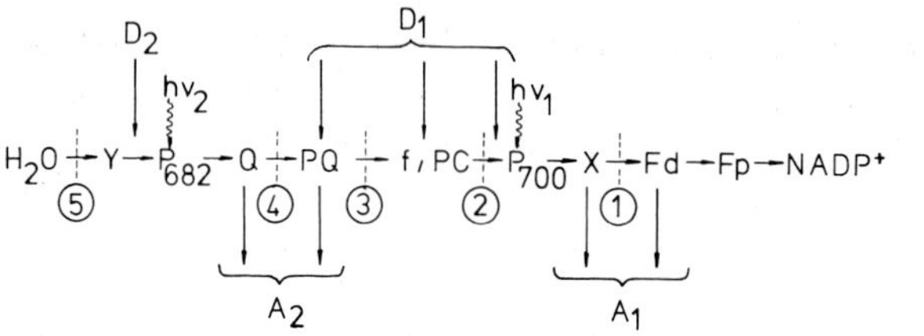

Fig. 1. Sites of inhibition, electron acceptance and electron donation in photosynthetic electron transport chain of chloroplasts. *Vertical, dashed lines*: Different sites of inhibition: ② block of ferredoxin; ① inactivation of plastocyanin; ④ inhibition of plastoquinone oxidation by DBMIB; ③ site of inhibition by DCMU; ⑥ inactivation of water splitting reaction. *Vertical arrows*: sites of artificial electron acceptance and donation. D_1 and D_2: electron donor for photosystem I and photosystem II respectively. A_1 and A_2: electron acceptor in photosystem I and II

Y: water splitting complex; *P-682*: reaction center chlorophyll in photosystem II; *Q*: primary electron acceptor in photosystem II; *PQ*: plastoquinone; *f*: cytochrome f; *PC*: plastocyanin; *P-700*: reaction center chlorophyll in photosystem I; *X*: primary electron acceptor in photosystem I; *Fd*: ferredoxin; *Fp*: ferredoxin-NADP$^+$-reductase

photosystem I. The reductant of photosystem II has a redox potential of about 100 mV negative, and the oxidant of photosystem I, P-700, has a redox potential around 400 mV positive (see this vol., Chap. II.1 b). These potentials set the limits for the reactions of artificial redox compounds on a thermodynamic basis, i.e. photosystem I should reduce any compound with a potential more positive than -600 mV, and oxidize any compound with a potential more negative than 400 mV. Photosystem II, on the other hand, should reduce compounds more positive than 100 mV, and oxidize compounds more negative than 1.0 V. Consequently, specificity of an acceptor for photosystem I (A_1) is to be expected if its redox potential is sufficiently more negative than -100 mV, as is the case for dipyridylium salts and anthraquinones; specificity of a donor for photosystem II is expected if its redox potential is sufficiently more positive than 400 mV, which applies to benzidines (see below). Specificity for A_2 and D_1 can only be achieved with the aid of inhibitors which block photosystem I or II respectively.

Besides the sensitivity to inhibitors other criteria distinguish reactions involving photosystem I, photosystem II, or both (Avron, 1967; Hoch and Martin, 1963). A photosystem I reaction will be active in light above 700 nm, while a photosystem II reaction will show the so-called "red drop". If both photosystems are involved the reaction will require the double number of quanta, and exhibit the phenomenon of enhancement.

Redox potentials, giving the thermodynamic frame, do not determine the rate of the reactions. Moreover, the relative rates of the reactions for artificial redox compounds with the respective reaction sites are not only determined by the activation energies of the reaction proper. In a polyphasic system, such as a suspension of chloroplasts, the relative accessibility of the reaction sites, and thus the relative solubilities of the redox compounds in aqueous and lipid phase, plays a decisive role. All the considerations about the biological activity of drugs apply here (Hansch et al., 1968). Accordingly, the intactness of the chloroplasts has to be considered; for example, one of the most convenient methods to estimate the percentage of chloroplasts which have lost the outer envelope is to measure the

Hill reaction with ferricyanide before and after osmotic shock (HEBER and SANTA-
RIUS, 1970). Fragmentation of chloroplasts by mechanical treatment as well as
by detergents liberates soluble redox carriers, like plastocyanin, and changes the
relative accessibility of the components in the chain, especially that of the reaction
center of photosystem I (SANE and HAUSKA, 1972). Furthermore, in detergent-
treated preparations reactions photosensitized by chlorophyll in micellar state
("Krasnovsky reactions"; see SEELY, 1966) might overrule the reactions under
investigation.

In summary: the criteria for an ideally suitable redox compound are its redox
potential, the accessibility of the reactive site, the specificity for only one site,
and the lack of side effects such as uncoupling or inhibition. In most cases exclusive
specificity is not given. Many compounds are reduced at one site and oxidized
at another. These can be used to mediate cyclic electron flow, or bypass parts
of the linear chain. If combinations of electron donors and acceptors are used,
a further trivial criterium must be met: they must not react with each other
in a dark reaction.

On the other hand, with experience and care, the complex possibilities for
the reaction of artificial redox compounds represent a powerful approach which
substantially contributes to present knowledge about electron transport in chloroplasts.

C. Electron Acceptors

Oxygen evolution in chloroplasts at the expense of an artificial electron acceptor
is called the Hill reaction. The first acceptors used were ferrioxalate (HILL, 1937),
ferricyanide (HOLT and FRENCH, 1946), and benzoquinone (WARBURG and
LÜTTGENS, 1944). Since then a great number of compounds have been introduced
as Hill reactants, their different chemical constitution showing the inspecificity
of the reaction. If the compound changes color upon reduction, this may be
used instead of oxygen evolution to measure the reaction. In many cases the
reduced form of the acceptor is autooxidizable, converting oxygen into H_2O_2
(MEHLER, 1951; WALKER and HILL, 1959). This oxygen uptake more than compen-
sates for the concomitant oxygen evolution, because it involves only 2 as against
4 electrons per oxygen molecule. It is called the Mehler reaction (MEHLER, 1951),
which is measured by oxygen uptake. If catalase is present no net change in
oxygen concentration occurs, but oxygen exchange can be measured (KRALL et al.,
1961) (pseudocyclic electron transport).

I. Reduction of Electron Acceptors by Photosystem I

Ferricyanide is the most frequently employed Hill reagent. Oxygen evolution, disappearance of ferricyanide at 420 nm (JAGENDORF and SMITH, 1962), or formation
of ferrocyanide by complexing with phenanthroline (AVRON and SHAVIT, 1963)
can be measured. The site of ferricyanide reduction has been debated on and

off in the literature (Trebst, 1974). Older reports, f.i. based on measurement of quantum-requirement (Avron and Ben-Hayyim, 1969), favored a direct reduction by photosystem II. In flashing light it also seems to be reduced by photosystem II (Rumberg et al., 1965). However, it now seems established that in conventionally isolated chloroplasts, under continuous illumination, ferricyanide is predominantly reduced by the reductant of photosystem I, as visualized by the sensitivity to the inhibitors DBMIB (Böhme et al., 1971), KCN (Ouitrakul and Izawa, 1973) and $HgCl_2$ (Kimura and Katoh, 1973) (2 and 3 in Fig. 1). In fragmented chloroplasts this sensitivity is lost (Böhme et al., 1971), suggesting that the reductant of photosystem II has become more accessible to ferricyanide. Chloroplasts from *Euglena* also do not show sensitivity to DBMIB in the Hill reaction with ferricyanide, probably reflecting fragmentation during isolation (Trebst, 1972a).

Both 2,6-dichlorophenolindophenol (DPIP) and 2,6,2'-trichlorophenolindophenol (Avron et al., 1958; Böhme et al., 1971; Holt and French, 1948; Lien and Bannister, 1971) behave very similarly to ferricyanide as Hill reactants. The reaction can be conveniently measured by the decrease in absorbance at 600 nm.

Methylred and tetrazolium blue are other examples of dyes which are bleached in Hill reactions (Ash et al., 1961).

Cytochrome c (Niemann and Vennesland, 1959; Oettmeier and Lockau, 1973) and metmyoglobin (Davenport, 1960) can also be reduced by photosystem I, but the reaction is dependent on ferredoxin.

Quinones have been systematically studied by Trebst and Eck (1961a). Quinones with relatively high redox potentials, and without hydrophilic substituents, especially the benzoquinones, are preferentially reduced by photosystem II (Saha et al., 1971). They are not rapidly autooxidizable. Quinones of low redox potential, like anthraquinones and alkylated naphthoquinones, are reduced by photosystem I only. They are autooxidizable and thus catalyze a Mehler reaction. They may also catalyze cyclic electron flow, as discussed below (Hauska et al., 1974; Trebst and Eck, 1961a). Anthraquinone-2-sulfonate is a particularly suitable catalyst for a clear-cut Mehler reaction. A Mehler reaction with endogenous o-benzoquinone derivatives is thought to play a physiological role (Elstner and Konze, 1974; Heber, 1973).

Dipyridylium salts, like the viologens, diquat and triquat, are also very often used to catalyze the Mehler reaction (Black, 1966; Kok et al., 1965; Zweig and Avron, 1965). They have very low redox potentials and are extremely autooxidizable. Their effective competition with ferredoxin for reducing equivalents is thought to explain their action as herbicides. Under anaerobic conditions the reduction of viologens can also be measured directly, by the formation of the reduced, blue radical form (Zweig and Avron, 1965).

Flavine derivates, such as FMN, have also been shown to catalyze a Mehler reaction (Arnon et al., 1961).

Hill reactions and Mehler reactions involving photosystem I are coupled to photophosphorylation with a stoichiometry higher than 1 (Reeves and Hall, 1973). It is now established that two sites of energy conservation participate (Trebst, 1974). The reactions are stimulated by uncouplers (see this vol. Chap. III.11). An exception is the Hill reactions with indophenol dyes, which are not coupled. An explanation for this was found by the observation that these compounds are uncouplers themselves (Gromet-Elhanan and Avron, 1964; Keister, 1963).

II. Reduction of Electron Acceptors by Photosystem II

Work mainly from two laboratories shows that rather lipophilic compounds with high redox potentials, such as benzoquinones and oxidized p-phenylenediamines, are preferentially reduced by photosystem II. The compounds were employed in catalytic amounts kept oxidized by excess ferricyanide (SAHA et al., 1971). These Hill reactions differ from those involving photosystem I. They are not sensitive to DBMIB (IZAWA et al., 1973; TREBST and REIMER, 1973a) or KCN (OUITRAKUL and IZAWA, 1973), they show inhibition by uncouplers (GOULD and ORT, 1973; SAHA et al., 1971; TREBST and REIMER, 1973b) at pH 8.0 and stimulation only at lower pH (TREBST and REIMER 1973b), and their pH optimum is shifted to lower values (GOULD and IZAWA, 1973a; SAHA et al., 1971; TREBST and REIMER, 1973b). This shift was previously observed for the Hill reaction with ferricyanide in chloroplasts fragmented by sonication (KATOH and SAN PIETRO, 1966).

The site of reduction seems to be plastoquinone (TREBST and REIMER, 1973b, c). A Mehler reaction with photosystem II alone has been described with DBMIB, which is rapidly autooxidizable (GOULD and IZAWA, 1973b). DBMIB at higher concentrations can obviously intercept electrons like other benzoquinones, besides acting as an inhibitor.

Like classic Hill reactions, the reactions with benzoquinones or phenylenediamines are sensitive to DCMU. They are coupled to photophosphorylation, although only with about half the stoichiometry (GOULD and ORT, 1973; TREBST, 1974; TREBST and REIMER, 1973b), which suggests that only one energy-conserving site is operative (TREBST, 1974).

DCMU-insensitive Hill reactions with $HgCl_2$ (MILES et al., 1973) and silicotungstic (GIRAULT and GALMICHE, 1974) or silicomolybdic acid (GIAQUINTA and DILLEY, 1975; GIAQUINTA et al., 1974) as acceptors have been reported. The reaction with this last compound has been investigated in detail. Most interesting is the fact that it is not coupled to photophosphorylation, although the compound is not an inhibitor of energy conservation (GIAQUINTA and DILLEY, 1975).

D. Electron Donors

Reactions in which the physiological electron donor, water, is replaced by artificial systems are often called photoreductions. The first photoreduction was introduced by VERNON and ZAUGG (1960), with reduced DPIP. This system was used to reconstitute electron transport to $NADP^+$ in DCMU-inhibited chloroplasts. It is now known that this reaction involves photosystem I only.

I. Electron Donors for Photosystem I

Besides DPIP and substituted p-phenylenediamines (TREBST, 1964; TREBST and PISTORIUS, 1965; VERNON and ZAUGG, 1960; WESSELS, 1964) related compounds (HAUSKA et al., 1975) such as indamines (OETTMEIER et al., 1974) and diaminobenzidine (Ben-Hayyim et al., 1975; Goffer and Neumann, 1973) are suitable electron donors for photosystem I, in the presence of DCMU. Ascorbate is not an efficient electron donor for photosystem I, and can be used to keep catalytic amounts of the donors reduced. Hydroquinones are poorly oxidized by photosystem I, if they have redox potentials higher than plastoquinone. Low potential quinones are not suitable for photoreductions with photosystem I, because ascorbate will not reduce them, and substrate amounts cannot be used, because of low solubility.

The reaction can be measured by $NADP^+$-reduction in the presence of ferredoxin, or more conveniently by oxygen uptake, employing a low-potential, autooxidizable acceptor, like methylviologen or anthraquinone-2-sulfonic acid, as described above. In the latter case, for the measurement of the true electron transport rate, care must be taken to avoid oxidative chain reactions started by the superoxide radical ion, which is primarily formed in the autooxidation process (Elstner and Kramer, 1973; Ort and Izawa, 1974). In fragmented chloroplasts (see this vol. Chap. IV.3) ascorbate can be oxidized in a plastocyanin-dependent reaction without an additional redox compound (Elstner et al., 1968; Sane and Hauska, 1972). Also cytochrome c can be photooxidized via plastocyanin in such a preparation (Niemann and Vennesland, 1959), a reaction which can be conveniently measured in a spectrophotometer. These reactions have been used as assays for plastocyanin (Plesničar and Bendall, 1970; Sane and Hauska, 1972). In intact chloroplasts photoreductions with p-phenylenediamines also seem to occur mainly via plastocyanin. This is suggested by the inhibition of the reactions by KCN (Ouitrakul and Izawa, 1973), and by the very rapid reaction of these compounds with isolated plastocyanin (Wood, 1974). Part of the reaction with reduced DPIP is insensitive to KCN (Ouitrakul and Izawa, 1973) in chloroplasts and independent of plastocyanin in deficient subchloroplast fragments (Hauska, 1975). Thus this compound seems to react partially with P-700 in direct way. An acridane-derivative related to DPIP (Hill, 1970; Hill et al., 1973: "Liebermanns dye stuff"), was shown to interact very efficiently with the reaction center, P-700, of photosystem I. Photoreductions with DPIP and p-phenylenediamines are partially sensitive to DBMIB, showing interaction also with the oxidation site of plastoquinone (Böhme et al., 1971).

The reactions are coupled to photophosphorylation, provided the artificial donor compounds lose protons upon oxidation (Hauska et al., 1973). The reaction with N,N,N′,N′-tetramethyl-p-phenylenediamine (TMPD) is not coupled, while the reaction with 2,3,5,6-tetramethyl-p-phenylenediamine is (Trebst and Pistorius, 1965, 1967). The stoichiometry of the coupled reaction is about half the stoichiometry of the complete chain from water to $NADP^+$ (Goffer and Neumann, 1973; Ort and Izawa, 1974; Trebst, 1974), suggesting that only one energy-conserving site is involved. An older argument that the observed phosphorylation might merely result from superimposed cyclic electron flow with a special site for energy conservation (Avron, 1964; Gromet-Elhanan, 1968), seems to have been ruled out (Hauska et al., 1970; Ort, 1975; Trebst, 1974). On the basis of a recent, more detailed study, it was suggested that only the KCN-insensitive part of the reaction with reduced DPIP was coupled, and not the direct reaction with the photocenter (Gould, 1975). The photoreduction with indophenol is stimulated by uncouplers, while the one with p-phenylene diamines is not, although both reactions are coupled. Recently an explanation for this differential effect was offered, based on the properties of the compounds as weak acids and bases, and the distribution within the chloroplasts in relation to the high energy state and uncoupling (HAUSKA, 1975; HAUSKA et al., 1975).

II. Electron Donors for Photosystem II

In order to study artificial electron donation to photosystem II, the water oxidation has to be inhibited on the oxidizing side of photosystem II. Several methods have been described (TREBST, 1974) (see this vol. Chap. II.16). Hydroquinones, phenylenediamines, benzidine and semicarbazide were introduced by YAMASHITA and BUTLER (1969) as donors for photosystem II. Cysteine had already previously been shown to be photooxidized in a DCMU-sensitive reaction (KATOH and SAN PIETRO, 1967). Other donors reported are diphenylcarbazide (VERNON and SHAW, 1969), Hydrazobenzene (HAVEMANN et al., 1972), hydrazine (HAVEMANN, et al., 1972; MANTAI and HIND, 1971), hydroxylamine (BENNOUN and JOLIOT, 1969; HAVEMANN et al., 1972), ascorbate (BÖHME and TREBST, 1969; BEN-HAYYIM and AVRON, 1970a), diketogulonic acid (HABERMANN et al., 1968), H_2O_2 (INOUE and NISHIMURA, 1971) and tetraphenylboron (HOMANN, 1972). This list demonstrates that the oxidizing end of photosystem II is as inspecific as the reducing end of photosystem I. Mn^{2+} ions can also donate electrons to photosystem II (BEN-HAYYIM and AVRON, 1970b; IZAWA, 1970). Since manganese is involved in the water splitting reaction (CHENIAE, 1970), this reaction is of special interest. Recently it was found that ferrocyanide and iodide, at rather high concentrations, can also be oxidized (IZAWA and ORT, 1974).

Different sites of interaction of the donor compounds between water oxidation and the reaction center of photosystem II have been discussed (KIMIMURA and KATOH, 1972), although varying activity might rather reflect varying accessibility of one and the same reaction site (TREBST, 1974).

Several complications have to be faced with photoreductions by photosystem II: some donors are also oxidized by photosystem I, especially the p-phenylenediamines, but also hydroquinones and ascorbate, the latter only in fragmented chloroplasts after addition of plastocyanin (see above); other compounds, like tetraphenylboron (HOMANN, 1972) and hydroxylamine (BENNOUN and JOLIOT, 1969), inactivate photosystem II (see Chap. II.16); some compounds, like hydrazine (MANTAI and HIND, 1971) and ascorbate (ELSTNER and KRAMER, 1973), after initial oxidation by photosystem II, can be further oxidized by oxygen in a chemical chain reaction, obscuring the photosynthetic reaction; oxidation products might have secondary effects (HAVEMANN et al., 1972; SHNEYOUR and AVRON, 1971); finally, restoration of the water splitting reaction by poising the redox state with a redox compound, after mild inactivation treatment, has been reported (YAMASHITA et al., 1971; BLANKENSHIP et al., 1975), which might interfere.

Photoreductions involving photosystem II are normally measured via photosystem I-dependent electron flow (see above). Two systems have been reported for photoreductions which involve photosystem II alone. One is diphenylcarbazide as D_2 with DPIP as A_2 (VERNON and SHAW, 1969), the other is semicarbazide as D_2 with ferricyanide as A_2 (YAMASHITA and BUTLER, 1969). In both cases the direct chemical reaction is too slow to interfere. The first is especially suitable for following biochemical isolation of the reaction center complex of photosystem II. Neither works well in intact chloroplasts.

Many of the donor systems for photosystem II do not allow the study of accompanying phosphorylation, because they themselves, or their reaction products, inactivate the energy-conserving system. Benzidine is a very appropriate electron donor, not only because its high redox potential of about 600 mV renders it specific for photosystem II, but also because its reaction with $NADP^+$ as electron acceptor is coupled to photophosphorylation with a stoichiometry similar to that of water oxidation (HARTH et al., 1974; YAMASHITA and BUTLER, 1969). It is interesting to note that N,N,N',N'-tetramethyl-benzidine, which in contrast to benzidine does not liberate protons upon oxidation, gives a reaction which is coupled with only half the stoichiometry (HARTH et al., 1974), reminiscent of the differential coupling of photooxidation with the tetramethyl-p-phenylenediamines in photosystem I. The relatively low stoichiometry observed in photooxidation of iodide and ferrocyanide (IZAWA and ORT, 1974) by photosystem II was also attributed to the lack of proton liberation. This will be discussed further in the last paragraph of this article.

E. Compounds Accepting and Donating Electrons — Cyclic Electron Transport and Bypasses

Many redox compounds can be reduced and oxidized by illuminated chloroplasts at the same time, mediating cyclic electron flow. Also bypasses of parts in the chain are feasible for such compounds. In theory three cyclic reactions are expected: a cycle around photosystem I combining the features of D_1 and A_1 (see Fig. 1), a cycle around photosystem II combining D_2 and A_2, and a cycle around both photosystems combining D_2 and A_1. Cyclic electron flow in the steady state cannot be measured directly, but only by the secondary reaction of coupled photophosphorylation. Therefore, only cycles including a site of energy conservation are measurable in continuous light.

Phenazine derivatives, PMS (JAGENDORF and AVRON, 1958) and pyocyanine (WALKER and HILL, 1959) give the highest rates of cyclic photophosphorylation in photosystem I. FMN (ARNON et al., 1955) and naphthoquinones, like menadione (WHATLEY et al., 1955), were introduced earlier as the first mediators of cyclic electron flow. The reaction with FMN was shown later to depend on oxygen (ARNON et al., 1961; JAGENDORF, 1962), and therefore might not catalyze a clear-cut reaction. Since then it has gradually emerged that many artificial donors D_1 and also acceptors A_1 in photosystem I can mediate a cycle under special conditions (HAUSKA et al., 1974, 1975 for recent detailed studies). Most prominent among them are DAD (TREBST and PISTORIUS, 1967), DPIP (TREBST and PISTORIUS, 1967) as examples for D_1, and diquat (ZWEIG et al., 1965) or methylviologen (HAUSKA et al., 1974) as examples for A_1. Optimal cyclic electron flow requires a subtle poising of the redox state depending on the mediator used (HAUSKA et al., 1970; HAUSKA et al., 1974; JAGENDORF and AVRON, 1959; TAGAWA et al., 1963; TREBST and ECK, 1961b; see also BOSE and GEST, 1963). Quinones have been systematically investigated for their ability to catalyze cyclic electron flow (HAUSKA et al., 1974; TREBST and ECK, 1961a). It was established that only those with a redox potential more negative than zero V catalyze the reaction, implying that plastoquinone is the site of interaction (HAUSKA et al., 1974; KANDLER, 1960; TREBST and ECK, 1961a). This was corroborated later by the finding that cyclic electron flow with quinones was sensitive to DBMIB (BÖHME et al., 1971). The systems with DAD and DPIP are only partially sensitive, and the one with PMS is insensitive to this inhibitor (BÖHME et al., 1971). The former systems are totally sensitive to KCN, while the latter is only partially inhibited (OITRAKUL and IZAWA, 1973). Obviously at least three different pathways for cyclic photophosphorylation in photosystem I are possible, one involving plastoquinone, one via plastocyanin, and one which seems to connect the reductant X of photosystem I (Fig. 1) directly with its oxidant P-700 (HAUSKA et al., 1974; TREBST, 1974). In consequence the question of the sites of energy conservation involved arises. It has been suggested that the pathway via plastoquinone uses the native site, which also operates in non-cyclic electron flow from photosystem II, while the shorter pathways with DAD, DPIP or, PMS as mediators might create artificial types of energy conservation (see below) (HAUSKA et al., 1974; TREBST, 1974).

Cyclic electron transport around photosystem II, or around both photosystems is less well documented. Mn^{2+} ions (BEN-HAYYIM and AVRON, 1970a) and diaminobenzidine (BEN-HAYYIM et al., 1975) have been reported to mediate such reactions.

Bridging reactions along the photosynthetic electron transport chain should be possible for any redox compound within its redox potential range, but may be difficult to observe. Recently, however, it has been demonstrated that TMPD and other phenylenediamines are able to bypass the site of inhibition by DBMIB, reconstituting coupled electron flow from water to $NADP^+$ (TREBST and REIMER, 1973c). In this reaction TMPD combines the action of an acceptor for photosystem II, A_2, with the action of a donor for photosystem I, D_1. In addition it was found that TMPD also overcomes the DBMIB-sensitivity of cyclic electron transport systems around photosystem I, e.g. with low potential quinones. Most probably it connects reduced plastoquinone with oxidized plastocyanin. It also enables high potential quinones to act as catalysts for cyclic phosphorylation. In this case it probably bridges electron flow from the artificial quinone to plastocyanin (HAUSKA et al., 1974).

F. The Topography of the Chloroplast Membrane and Artificial Redox Compounds

The asymmetric architecture of electron transporting membranes has been established in the last few years. Several approaches have been applied (see this vol., Chaps. II.1b, II.17, II.18 and 4.2), one of them being the use of artificial electron donors and acceptors (TREBST, 1974).

I. Lipophilicity and Reactivity

The first report classifying artificial redox compounds according to their lipid solubility came from Saha et al. (1971). They showed that in chloroplasts hydrophilic compounds, such as ferricyanide or benzoquinone-2-sulfonate, accept electrons almost exclusively from photosystem I, while more lipophilic compounds, such as benzoquinone or p-phenylenediamines, may function as A_2. Subsequently it was demonstrated that for D_1 activity compounds must be soluble in organic phase (Hauska, 1972; Hauska et al., 1973). For example, PMS-3-sulfonate and DPIP-3'-sulfonate, both insoluble in organic phase and unable to pass a lipid membrane (Hauska and Prince, 1974), were inactive in cyclic phosphorylation and as D_1, respectively, although both catalyzed a Hill reaction acting as A_1. On the other hand, too lipophilic compounds, like quinones, are also unable to react with the oxidizing end of photosystem I directly. They seem to react only via plastoquinone or via TMPD (see above). In conclusion, the oxidizing end of photosystem I seems to be located beyond a lipid barrier, while the reducing end is accessible from the aqueous medium suspending the chloroplasts, in agreement with results from other approaches (Trebst, 1974). A similar vectorial orientation of the reaction center of potosystem II is under discussion (Trebst, 1974), but could not be clearly established by the use of artificial redox compounds. It seems that both A_2 (Saha et al., 1971) and D_2 (Harth et al., 1974; Izawa and Ort, 1974; Katajima et al., 1973) should be lipophilic to be active at low concentrations. As pointed out above, in fragmentated chloroplasts A_2 may be hydrophilic (Böhme et al., 1971; Kimimura and Katoh, 1973; Junge and Ausländer, 1974).

II. The Concept of Artificial Energy Conservation

As described above, the oxidation of artificial donors D_1 is coupled to photophosphorylation only if protons are liberated from the compound during the reaction (Hauska et al., 1973; Trebst, 1974; Hauska et al., 1975). These protons would be literated on the inner side of the chloroplast membrane, according to the notion about the topography of photosystem I. Under these circumstances the chemiosmotic hypothesis of energy conservation (Mitchell, 1966) lends itself to the explanation of the results (see this vol., Chap. 3.1), with the specification that the proton-translocating role of plastoquinone might be fulfilled by the artificial donor employed. Therefore, with respect to proton translocation, we might be dealing with an artificial type of energy conservation (Hauska et al., 1973; Trebst, 1974; Hauska et al., 1975). Similar considerations apply for cyclic phosphorylation systems, especially for the system with PMS (Hauska, 1972; Hauska et al., 1974), and for energy conservation in photosystem II (Harth et al., 1974).

References

Arnon, D.I., Losada, M., Whatley, F.R., Tsujimoto, H.Y., Hall, D.O., Horton, A.A.: Proc. Nat. Acad. Sci. U.S. **47**, 1314–1340 (1961)

Arnon, D.I., Whatley, F.R., Allen, M.B.: Biochim. Biophys. Acta **16**, 607–608 (1955)

Ash, O.K., Zaugg, W.S., Vernon, L.P.: Acta Chem. Scand. **15**, 1629 (1961)

Avron, M.: Biochem. Biophys. Res. Commun. **17**, 430–432 (1964)

Avron, M.: Current Topics Bioenerg. **2**, 1–22 (1967)

Avron, M., Ben-Hayyim, G.: Interaction between two photochemical systems in photoreductions of isolated chloroplasts. In: Progress in Photosynthesis Research. Metzner, H. (ed.). Tübingen: Laupp, 1969, Vol. III, pp. 1185–1196

Avron, M., Krogmann, D.W., Jagendorf, A.T.: Biochim. Biophys. Acta **30**, 144–153 (1958)

Avron, M., Neumann, J.: Ann. Rev. Plant Physiol. **19**, 137–166 (1968)

Avron, M., Shavit, N.: Anal. Biochem. **6**, 549–554 (1963)

Ben-Hayyim, G., Avron, M.: Europ. J. Biochem. **15**, 155–160 (1970a)

Ben-Hayyim, G., Avron, M.: Biochim. Biophys. Acta **205**, 86–94 (1970b)

Ben-Hayyim, G., Drechsler, Z., Goffer, J., Neumann, J.: Europ. J. Biochem. **52**, 135–141 (1975)

Bennoun, P., Joliot, A.: Biochim. Biophys. Acta **189**, 85–94 (1969)

Black, C.C.: Biochim. Biophys. Acta **120**, 332–340 (1966)

Blankenship, R.E., Babcock, G.T., Sauer, K.: Biochim. Biophys. Acta **397**, 165–176 (1975)

Böhme, H., Reimer, S., Trebst, A.: Z. Naturforsch. **26B**, 341–352 (1971)

Böhme, H., Trebst, A.: Biochim. Biophys. Acta **180**, 137–148 (1969)

Bose, S.K., Gest, H.: Proc. Nat. Acad. Sci. U.S. **49**, 337–345 (1963)

Cheniae, G.M.: Ann. Rev. Plant Physiol. **21**, 467–498 (1970)

Colowick, S.P., Kaplan, N.O.: Methods Enzymol. **10**, 33–37, 38–40, 216–224 (1967)

Davenport, H.E.: Biochem. J. **77**, 471–477 (1960)

Del Valle Tascon, S., Ramirez, J.M.: Z. Naturforsch. **30C**, 46–52 (1975)

Elstner, E., Konze, J.R.: FEBS Lett. **45**, 18–21 (1974)

Elstner, E., Kramer, R.: Biochim. Biophys. Acta **314**, 340–353 (1973)

Elstner, E., Pistorius, E., Böger, P., Trebst, A.: Planta (Berl.) **79**, 146–161 (1968)

Giaquinta, R.T., Dilley, R.A.: Biochim. Biophys. Acta **387**, 288–305 (1975)

Giaquinta, R.T., Dilley, R.A., Crane, F.L., Barr, R.: Biochem. Biophys. Res. Commun. **59**, 985–991 (1974)

Girault, G., Galmiche, J.M.: Biochim. Biophys. Acta **333**, 314–319 (1974)

Goffer, J., Neumann, J.: FEBS-Lett. **36**, 61–64 (1973)

Gould, J.M.: Biochim. Biophys. Acta **387**, 135–147 (1975)

Gould, J.M., Izawa, S.: Biochim. Biophys. Acta **314**, 211–223 (1973a)

Gould, J.M., Izawa, S.: Europ. J. Biochem. **37**, 185–192 (1973b)

Gould, J.M., Ort, D.R.: Biochim. Biophys. Acta **325**, 157–166 (1973)

Gromet-Elhanan, Z.: Arch. Biochem. Biophys. **123**, 447–456 (1968)

Gromet-Elhanan, Z., Avron, M.: Biochemistry **3**, 365–373 (1964)

Habermann, N.M., Handel, M.A., McKellar, P.: Photochem. Photobiol. **7**, 211–224 (1968)

Hansch, C., Lien, E.J., Helmer, F.: Arch. Biochem. Biophys. **128**, 319–330 (1968)

Harth, E., Oettmeier, W., Trebst, A.: FEBS Lett. **43**, 231–234 (1974)

Hauska, G.: FEBS Lett. **28**, 217–220 (1972)

Hauska, G.: In: Proc. 3rd Intern. Congr. Photosynthesis. Avron, M. (ed.). Amsterdam: Elsevier, 1975, Vol. I, pp. 689–696

Hauska, G., McCarty, R.E., Racker, E.: Biochim. Biophys. Acta **197**, 206–218 (1970)

Hauska, G., Oettmeier, W., Reimer, S., Trebst, A.: Z. Naturforsch. **30C**, 37–45 (1975)

Hauska, G., Prince, R.C.: FEBS Lett. **41**, 35–39 (1974)

Hauska, G., Reimer, S., Trebst, A.: Biochim. Biophys. Acta **357**, 1–13 (1974)

Hauska, G., Trebst, A., Draber, W.: Biochim. Biophys. Acta **305**, 632–641 (1973)

Havemann, J., Daysens, L.N.M., T.C.M. van der Gest, Gorkom, H.J. van: Biochim. Biophys. Acta **283**, 316–327 (1972)

Heber, U.: Ber. Deut. Botan. Ges. **86**, 187–195 (1973)

Heber, U., Santarius, K.A.: Z. Naturforsch. **25B**, 718–728 (1970)

Hill, R.: Nature (London) **139**, 881–882 (1937)
Hill, R.: J. Bioenerg. **4**, 229–237 (1973)
Hill, R., Bedford, G.R., Webster, B.R.: J. Chem. Soc. Sect. C, 2462–2466 (1970)
Hinkle, P.C.: Federation Proc. **32**, 1988–1992 (1973)
Hoch, G., Martin, I.: Arch. Biochem. Biophys. **102**, 430–438 (1963)
Holt, A.S., French, C.S.: Arch. Biochem. **9**, 25–43 (1946)
Holt, A.S., French, C.S.: Arch. Biochem. **19**, 368 (1948)
Homann, P.H.: Biochim. Biophys. Acta **256**, 336–344 (1972)
Inoue, H., Nishimura, M.: Plant Cell Physiol. **12**, 739–741 (1971)
Izawa, S.: Biochim. Biophys. Acta **197**, 328–331 (1970)
Izawa, S., Gould, J.M., Ort, D.R., Felker, P., Good, N.E.: Biochim. Biophys. Acta **305**, 119–128 (1973)
Izawa, S., Ort, D.R.: Biochim. Biophys. Acta **357**, 127–143 (1974)
Jagendorf, A.T.: In: Survey of Biological Progress. Glass, B. (ed.). New York-London: Academic Press, 1962, Vol. IV, pp. 183–344
Jagendorf, A.T., Avron, M.: J. Biol. Chem. **231**, 277–290 (1958)
Jagendorf, A.T., Avron, M.: Arch. Biochem. Biophys. **80**, 246–257 (1959)
Jagendorf, A.T., Smith, M.: Plant Physiol. **37**, 135–141 (1962)
Junge, W., Ausländer, W.: Biochim. Biophys. Acta **333**, 59–70 (1974)
Kandler, O.: Ann. Rev. Plant. Physiol. **11**, 37–54 (1960)
Katajima, M., Ogawa, T., Inoue, Y., Shibata, K.: Plant Cell Physiol. **14**, 787–790 (1973)
Katoh, S., San Pietro, A.: The role of plastocyanine in NADP-photoreduction by chloroplasts. In: The Biochemistry of Copper. Peisach, J., Aisen, P., Blumbach, W.E. (eds.). New York-London: Academic Press, 1966, pp. 407–422
Katoh, S., San Pietro, A.: Arch. Biochem. Biophys. **122**, 144–152 (1967)
Keister, D.L.: J. Biol. Chem. **238**, 2590–2592 (1963)
Kimimura, M., Katoh, S.: Plant Cell Physiol. **13**, 287–296 (1972)
Kimimura, M., Katoh, S.: Biochim. Biophys. Acta **325**, 167–174 (1973)
Kok, B., Rurainski, H.J., Owens, O.V.H.: Biochim. Biophys. Acta **109**, 347–356 (1965)
Krall, A.B., Good, N.E., Mayne, B.C.: Plant Physiol. **36**, 44–47 (1961)
Lien, S., Bannister, T.T.: Biochim. Biophys. Acta **245**, 465–481 (1971)
Mantai, K.E., Hind, G.: Plant Physiol. **48**, 5–8 (1971)
Mehler, A.H.: Arch. Biochem. Biophys. **33**, 65–77 (1951)
Miles, D., Polen, P., Farag, S., Goodin, R., Lutz, J., Moustafa, A., Rodriquez, B., Well, C.: Biochem. Biophys. Res. Commun. **50**, 1113–1119 (1973)
Mitchell, P.: Bodmin: Glynn Research, 1966
Niemann, R.H., Vennesland, B.: Plant Physiol. **34**, 255–262 (1959)
Oettmeier, W., Lockau, W.: Z. Naturforsch. **28C**, 717–721 (1973)
Oettmeier, W., Reimer, S., Trebst, A.: Plant Sci. Lett. **2**, 37–45 (1974)
Ort, D.R.: Arch. Biochem. Biophys. **116**, 629–638 (1975)
Ort, D.R., Izawa, S.: Plant Physiol. **53**, 370–376 (1974)
Ouitrakul, R., Izawa, S.: Biochim. Biophys. Acta **305**, 105–118 (1973)
Plesničar, M., Bendall, D.S.: Biochim. Biophys. Acta **216**, 192–199 (1970)
Reeves, S.G., Hall, D.O.: Biochim. Biophys. Acta **314**, 66–78 (1973)
Rumberg, B., Schmidt-Mende, P., Skerra, B., Vater, J., Weikard, J., Witt, H.T.: Z. Naturforsch. **20B**, 1086–1101 (1965)
Saha, S., Ouitrakul, R., Izawa, S., Good, N.E.: J. Biol. Chem. **246**, 3204–3209 (1971)
Sane, P.V., Hauska, G.: Z. Naturforsch. **27B**, 932–938 (1972)
Seely, G.R.: Photochemistry of chlorophylls in vitro. In: The Chlorophylls. Vernon, L.P., Seely, G.R. (eds.). New York-London: Academic Press, 1966, pp. 523–568
Shneyour, A., Avron, M.: Biochim. Biophys. Acta **253**, 412–420 (1971)
Tagawa, K., Tsujimoto, H.Y., Arnon, D.I.: Nature (London) **199**, 1247–1251 (1963)
Trebst, A.: Z. Naturforsch. **19B**, 418–421 (1964)
Trebst, A.: Methods Enzymol. **24**, 146–164 (1972a)
Trebst, A.: In: Proc. 2nd Intern. Congr. Photosynthesis. Forti, G., Avron, M., Melandri, B.A. (eds.). The Hague: Dr. Junk, 1972b, Vol. I, pp. 399–417
Trebst, A.: Ann. Rev. Plant Physiol. **25**, 423–458 (1974)
Trebst, A., Eck, H.: Z. Naturforsch. **16B**, 44–49 (1961a)
Trebst, A., Eck, H.: Z. Naturforsch. **16B**, 455–461 (1961b)

Trebst, A., Pistorius, E.: Z. Naturforsch. **20B**, 143–147 (1965)
Trebst, A., Pistorius, E.: Biochim. Biophys. Acta **131**, 580–582 (1967)
Trebst, A., Reimer, S.: Biochim. Biophys. Acta **305**, 129–139 (1973a)
Trebst, A., Reimer, S.: Biochim. Biophys. Acta **325**, 546–557 (1973b)
Trebst, A., Reimer, S.: Z. Naturforsch. **28C**, 710–716 (1973c)
Vernon, L.P., Shaw, E.R.: Plant Physiol. **44**, 1645–1649 (1969)
Vernon, L.P., Zaugg, W.: J. Biol. Chem. **235**, 2728–2733 (1960)
Walker, D.A., Hill, R.: Plant Physiol. **34**, 240–245 (1959)
Warburg, O., Lüttgens, W.: Naturwissenschaften **32**, 161 (1944)
Wessels, J.S.C.: Biochim. Biophys. Acta **79**, 640–642 (1964)
Whatley, F.R., Allen, M.B., Arnon, D.I.: Biochim. Biophys. Acta **16**, 605–606 (1955)
Wood, P.M.: Biochim. Biophys. Acta **357**, 370–379 (1974)
Yamashita, T., Butler, W.L.: Plant Physiol. **44**, 435–438 (1969)
Yamashita, T., Tsuji, J., Tomita, G.: Plant Cell Physiol. **12**, 117–126 (1971)
Zweig, G., Avron, M.: Biochem. Biophys. Res. Commun. **19**, 397–400 (1965)
Zweig, G., Shavit, N., Avron, M.: Biochim. Biophys. Acta **109**, 332–346 (1965)

16. Inhibitors of Electron Transport

S. IZAWA

A. Introduction

The subject of inhibitors was covered not long ago by two review articles (IZAWA and GOOD, 1972; GOOD and IZAWA, 1973). However, large portions of these previous articles have becomes outdated, missing most of the important inhibitors introduced in the early 1970s, such as the plastoquinone antagonist dibromothymoquinone (DBMIB) and inhibitors of plastocyanin. Some of the indispensable roles which these new inhibitors played in the recent development of electron transport and photophosphorylation studies have been discussed by TREBST (1974).

Figure 1 represents a simplified version of the currently accepted "Z scheme" for photosynthetic electron transport, in which several inhibitor-sensitive regions and sites are indicated by the numbered arrows. (Following discussions are arranged according to these numberings.)

Since the number of inhibitors that are known to act in regions 1 and 2 (around photosystem II) is so great, only a selected few compounds or families of compounds are discussed. The remainder are simply listed with references. Inhibitory treatments, such as heat treatment and UV irradiation, are included but with only brief notes. The effects of some of these treatments have been discussed

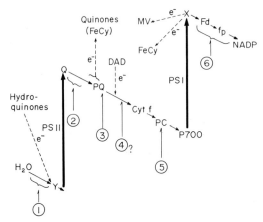

Fig. 1. Simplified "Z scheme" for photosynthetic electron transport showing sites of inhibition (*numbered arrows*) and main sites of electron donation and acceptance by artificial reductants and oxidants (*broken arrows*). Y: unidentified primary electron donor for PS II; Q: unidentified primary electron acceptor for PS II; PQ: plastoquinone (plastoquinone pool); $Cyt f$: cytochrome f; PC: plastocyanin; P-700: reaction center chlorophyll of PS I; X: unidentified primary electron acceptor for PS I; Fd: ferredoxin; fp: flavoprotein (ferredoxin-NADP$^+$ reductase); MV: methylviologen; $FeCy$: ferricyanide; DAD: diaminodurene

in some detail by CHENIAE (1970) in his review on O_2 evolution. A recent review on photosystem II by RADMER and KOK (1975) also includes comments on the effects of several inhibitors. Immunochemical inhibitors (antisera) are not included because they are dealt with separately by BERZBORN and LOCKAU (this vol. Chap. II.17).

B. Description of Inhibitors

I. Inhibitors that Act on Water-Oxidizing Side of Photosystem II

1. Hydroxylamine

This classical inhibitor of photosynthesis (SHIBATA and YAKUSHIJI, 1933) has been a subject of vigorous reinvestigations in recent years. JOLIOT (1966) found that hydroxylamine inhibition of algal photosynthesis did not change the steady-state fluorescence yield, in contrast to CMU inhibition which markedly increased it (even in algae which were already inhibited by hydroxylamine). From these and other findings, she deduced that hydroxylamine probably reacts with the primary electron donor of photosystem II. Although later studies did reveal various responses of the fluorescence yield to this reductant-inhibitor (e.g. BENNOUN, 1970; MOHANTY et al., 1971), it is now well-established that hydroxylamine inhibition indeed takes place at the level of water splitting (BENNOUN and JOLIOT, 1969; IZAWA et al., 1969; CHENIAE and MARTIN, 1970; ELSTNER et al., 1970).

Hydroxylamine inhibition, however, is an intricate phenomenon. At very low concentrations (50 μM) the amine causes a lag (by two flashes) in the O_2 flash-yield oscillation in algae as well as in chloroplasts without affecting the steady-state rate of O_2 production (BOUGES, 1971; BENNOUN and BOUGES, 1972). This was interpreted in terms of photo-reversible binding of two molecules of hydroxylamine to each O_2-evolving center (BOUGES-BOCQUET, 1973). Chloroplasts and algae exposed to higher concentrations of hydroxylamine (1–10 mM) in the dark lose their O_2-evolving ability in a process which shows biphasic kinetics: an "immediate" (< 30 sec) partial inhibition followed by a first-order decay (CHENIAE and MARTIN, 1971). Approximately two-thirds of chloroplast-bound Mn is released in the process, and this results in a total destruction of O_2-evolving centers (CHENIAE and MARTIN, 1970, 1971). However, hydroxylamine is quite ineffective below pH 6, which is the pK_a of this weak amine, suggesting that the unprotonated base is the inhibitory form of hydroxylamine (CHENIAE and MARTIN, 1971). This seems true of any amine-type inhibitors (see below).

In algae, but not in isolated chloroplasts, the hydroxylamine-inactivated O_2-evolving centers can be reactivated by exposing the cells to a weak photosystem II light for 20 min in a hydroxylamine-free medium. A rebinding of Mn to chloroplasts occurs within the cells during illumination (CHENIAE and MARTIN, 1972).

Hydroxylamine-poisoned chloroplasts are capable of oxidizing, via photosystem II, a variety of artificial reductants, including hydroxylamine itself (IZAWA, et al., 1969; BENNOUN and JOLIOT, 1969; CHENIAE and MARTIN, 1970), ascorbate, catechol, benzidine, ferrocyanide, iodide, etc. (e.g. IZAWA et al., 1975) at rates equivalent to 20–60% of the rate of water oxidation. However, exposure of chloro-plasts to strong light in the presence of hydroxylamine has been reported to destroy not only the ability of photosystem II to oxidize water but also its ability to oxidize exogenous reductants (KATOH et al., 1970; CHENIAE and MARTIN, 1971). The variable fluorescence was nearly irreversibly quenched (KATOH et al., 1970). Similar destructive effects of light had previously been reported by YAMASHITA and BUTLER (1969a) for Tris-washed chloroplasts. Noting the preventive action of exogenous reductants, YAMASHITA and BUTLER explain the photodamage as being due to the Tris block-facilitated accumulation of oxidants produced by photo-system II. Indeed, if a reductant level of hydroxylamine ($\geqq 50$ mM) had been used by the former authors, the results might have been different.

2. Ammonia and Methylamine

The stimulation of electron flow (uncoupling) by ammonia and methylamine is a well-known phenomenon (see this vol. Chap. III.11). However, excessive concen-trations of these bases inhibit photosystem II as first noted by VERNON and ZAUGG (1960) for ammonia (>100 mM). IZAWA et al. (1969) placed the site of inhibition by these uncoupler amines near the water-oxidizing step of photosystem II on the basis of their finding that the DCMU-sensitive photoreduction of indophenol dyes supported by hydroxylamine (50 mM) was totally insensitive to the concentra-tions of ammonia and methylamine which severely inhibited the Hill reaction with the same dyes as oxidants.

Ammonia and methylamine inhibit only at alkaline pHs. In the EDTA-treated chloroplasts used by IZAWA et al., 30 mM ammonia (pK_a 9.2) became inhibitory as the pH exceeded 7.5, while 150 mM methylamine (pK_a 10.6) did so only above pH 8. Thus, the unprotonated forms of these amines seem to be the acitve species. (Computations will show that, on the free amine basis, the inhibitor potencies of ammonia and methylamine are quite comparable to that of hydroxylamine—all effective at $\leqq 1$ mM).

Chloroplasts suspended in pH 8 medium containing 50 mM ammonia lose their membrane-bound Mn in much the same way as chloroplasts exposed to hydroxyl-amine (IZAWA, unpublished data). Thus, it would appear that all amine inhibitors share basically the same mechanism of action, namely some form of nucleophilic attack on the O_2-producing "Mn enzyme". The structure-activity relationships of inhibitory amines, briefly examined by CHENIAE and MARTIN (1971) for several hydroxylamine derivatives, seem well worth pursuing further.

3. Tris

The inhibitory effect of high concentrations of tris(hydroxymethyl)aminomethane (Tris) on the Hill reaction was first noted by NAKAMOTO et al. (1959). YAMASHITA and BUTLER (1968a, 1969b) reinvestigated the phenomenon of Tris inhibition and established that the Tris-block occurs at the level of water oxidation. The sharp pH dependence of concentration requirement (e.g. effective concentrations 0.8 M at pH 8 and 50 mM at pH 8.8) again suggests that unprotonated Tris base (pK_a 8.3) is the inhibitory species (YAMASHITA and BUTLER, 1969b).

A release of bound Mn (up to 70%) was reported to accompany the Tris inactivation of chloroplasts (ITOH et al., 1969; CHENIAE and MARTIN, 1970; YAMASHITA and TOMITA, 1974) but clearly Tris inhibition can be achieved without significant loss of Mn from the membrane, since YAMASHITA et al. (1971) were able to reverse the inhibition by simply washing the Tris-treated chloroplasts with a buffer containing ascorbate and reduced indophenol dyes. The same reactivation procedure proved ineffective for hydroxylamine inhibition (CHENIAE and MARTIN, 1972). Recently YAMASHITA and TOMITA (1974) have reported on a two-stage reactivation procedure for chloroplasts which have been inactivated by treatment with Tris plus acetone (which removes Mn). The procedure includes a dark incubation of inactivated chloroplasts with reduced DCIP followed by a weak illumination in the presence of Mn^{2+}, dithiothreitol and several additional substances.

One of the important findings made by YAMASHITA and BUTLER (1969b) is the disappearance of variable fluorescence upon Tris inhibition and its reappearance on addition of reductants which are known to be capable of donating electrons to photosystem II (e.g. hydroquinone). This phenomenon, which was explained in terms of the availability of electrons, through photosystem II, to the fluorescence-quenching primary electron acceptor (Q) of photosystem II, has since been accepted as being diagnostic of an inhibition on the water oxidizing side of photosystem II.

Chloroplasts treated with Tris and then washed with a regular buffer can be used for studies of phosphorylation associated with the transfer of electrons through photosystem II from exogenous reductants, such as p-phenylenediamine, benzidine, p-aminophenol and hydroquinone (YAMASHITA and BUTLER, 1968a, 1969b). ORT and IZAWA (1974), however, prefer the use of hydroxylamine-washed chloroplasts for this line of experiments.

4. Carbonylcyanide m-Chlorophenylhydrazone and Other Anomalous Inhibitors

High concentrations ($> 10 \mu M$) of carbonylcyanide m-chlorophenylhydrazone (CCCP), a typical phosphorylation uncoupler (see GOOD, this vol. Chap. III.11), inhibits the Hill reaction in isolated chloroplasts (DE KIEWIET et al., 1965; KATOH and SAN PIETRO, 1967). The inhibition progresses rapidly in the light (HIND et al., 1969; HOMANN, 1971), accompanied by a photobleaching of carotenoids (YAMASHITA et al., 1969; ITOH et al., 1969). The photobleacing can be prevented by reducing agents or by DCMU, as has been observed with Tris-inhibited chloroplasts (YAMASHITA and BUTLER, 1969a). Inhibitory concentrations of CCCP suppress the fluorescence yield (ITOH et al., 1969; HOMANN, 1971).

These are signs of an inhibition on the water side of photosystem II. However, CCCP inhibition differs from amine inhibition in that none of the known electron donors for photosystem II seems able to restore electron flow through CCCP-inhibited photosystem II (KATOH and SAN PIETRO, 1967; IZAWA et al., 1969; KIMIMURA et al., 1971) or to recover the CCCP-depressed fluorescence yield (HOMANN, 1971; KIMIMURA et al., 1971). Addition of DCMU or dithionite did increase the fluorescence yield, but the stimulatory effect rapidly disappeared as the CCCP concentration exceeded 0.1 mM. No release of Mn from the chloroplast membrane was detected (KIMIMURA et al., 1971). It was therefore suggested that CCCP acts directly on the primary electron donor of photosystem II (HOMANN, 1971) or on a site very close to it (KIMIMURA et al., 1971). In either case, the CCCP effect does not seem simple. HOMANN (1972) has found that tetraphenylboron behaves similarly, and has shown that its inhibitory action is related to its ability to reduce the oxidant produced by photosystem II.

RENGER (1969) noticed in his O_2 flash-yield experiments that low concentrations of CCCP markedly shorten the life-time of an intermediate of O_2 production. Anilinothiophenes (see also GREGORY, 1974) had a similar effect. RENGER (1972) designated these substances ADRY reagents, meaning reagents which accelerate the deactivation of the water-splitting enzyme system Y.

5. Inhibitory Treatments and Miscellaneous Inhibitors

During the last decade a number of treatments and substances have been reported to cause inhibitions on the water side of photosystem II. In all cases, restoration of DCMU-sensitive electron flow by adding reductants has been demonstrated. The brief listing here by no means implies that these treatments and inhibitors are unimportant.

The sensitivity of photosystem II to mild heating was recognized early (HINKSON and VERNON, 1959). KATOH and SAN PIETRO (1967) demonstrated that heating (40–50° C) can selectively destroy the water-splitting step of photosystem II by showing that their heated *Euglena* chloroplasts, which exhibited no Hill reaction activity, were still capable of ascorbate-supported, DCMU-sensitive $NADP^+$ photoreduction. This was the first demonstration of electron donation to photosystem II by an artificial reductant. YAMASHITA and BUTLER (1968b) showed that UV irradiation could exert a Tris-treatment-like effect (see above) on chloroplasts, except that the variable fluorescence, abolished by UV treatment, was not restored by adding photosystem II electron donors. Chloride ion can be removed from the chloroplast membrane with difficulty. This causes a reversible suppression of the ability of photosystem II to oxidize water (IZAWA et al., 1969). Other treatments and inhibitors reported are: exposure of chloroplasts to extreme pH (below 4 or above 9; KATOH and SAN PIETRO, 1968), chaotropic agents (LOZIER et al., 1971), high concentrations of KCl (ASADA and TAKAHASHI, 1971) and high concentrations of 8-hydroxyquinoline (INOUÉ and NISHIMURA, 1971). Heating, extreme pH and chaotropic agents have been shown to cause a liberation of Mn from the lamellar membrane (CHENIAE and MARTIN, 1967; LOZIER et al., 1971).

II. Inhibitors that Block Exit of Electrons from Photosystem II

1. Phenylureas

3-(3,4-Dichlorophenyl)-1,1-dimethylurea (DCMU or Diuron) and its 4-chlorophenyl analog (CMU or Monuron), introduced by WESSELS and VAN DER VEEN in 1956, are the best known of all inhibitors of photosynthetic electron transport. Today, however, DCMU is by far the more frequently used of the two, because of its superior inhibitory potency.

It was inferred early that phenylurea inhibition occurs close to the O_2 evolution step (photosystem II), from the lack of inhibitory effect on H_2-supported photosynthesis (BISHOP, 1958), on cyclic photophosphorylation (JAGENDORF and AVRON, 1959) and $NADP^+$ photoreduction supported by reduced DPIP (VERNON and

ZAUGG, 1960). Most workers now believe that DCMU blocks electron transfer between the primary electron acceptor and the secondary electron acceptor (plastoquinone pool) of photosystem II. This view originates from DUYSENS and SWEERS (1963) who, in proposing the now widely accepted concept of the fluorescence-quenching primary electron acceptor (Q) of photosystem II, explained the stimulatory effect of DCMU on the fluorescence yield as being due to a blocking of reoxidation of reduced Q.

Besides the fluorescence experiments of KOK et al. (1967) which indicated a DCMU block before a large electron pool, there is spectroscopic evidence that DCMU inhibition occurs before the plastoquinone pool (AMESZ, 1964; STIEHL and WITT, 1969; SIGGEL et al., 1972). As for new evidence of the DCMU site being after the photosystem II center, DUYSENS (1972) showed that a strong flash can induce an O_2 burst in DCMU-poisoned algae, and WITT (1973) found that DCMU inhibits the dark decay of X-320, a spectral change which may represent the reduction of the primary acceptor of photosystem II (DÖRING et al., 1967). An interesting piece of evidence came from the work of RIENITS et al. (1973) who discovered a phenomenon which seemed to represent a back flow of electrons to photosystem II: an ATP hydrolysis-induced enhancement of the fluorescence yield. The enhancement was abolished by DCMU, which suggests that the endogenous electrons (from plastohydroquinone?) did not reach the quencher when DCMU was present.

A rapid spectral change peaking at 682 nm, and its sensitivity to DCMU, have led a few groups of workers to believe that DCMU interacts directly with the reaction center chlorophyll of photosystem II (DÖRING et al., 1967; DÖRING et al., 1969), possibly in such a way that one molecule of DCMU inactivates two reaction centers (SIGGEL et al., 1972). WESSELS and VAN DER VEEN (1956) had suggested a similar mechanism earlier. However, some of the aforementioned results (e.g. DUYSENS, 1972) are very difficult to explain on the basis of these models, and even more so is the recently demonstrated DCMU-insensitive Hill reaction (GIRAULT and GALMICHE, 1974; GIAQUINTA et al., 1974), although this new type of the Hill reaction requires more thorough investigation before any definitive conclusions can be drawn from it.

Apart from the possibility of the reaction center chlorophyll, nothing is known about the chemical nature of the target site of DCMU inhibition. Results from the vigorous structure-activity studies of the 1960s on phenylureas and related inhibitors had little predictive power in this repect. However, these and other studies (e.g. MORELAND and HILL, 1959; GOOD, 1961; CAMPER and MORELAND, 1966; HANSCH and DEUTSCH, 1966) did clarify, to a considerable extent, the roles of various structural elements governing the inhibitor potency. As for the functional group, it was agreed that in practically all types of herbicidal inhibitors (but see the next section), the biologically acitve moiety is the group $\geqq C-NH-$ (or $-CO-NH-$ in DCMU). Suggestions have been made that this moiety interacts with an amide linkage of the target "protein" through a charge transfer mechanism (HANSCH, 1969) or through hydrogen bonding (MORELAND, 1969).

The small amount of DCMU required to inhibit chloroplast electron transport (DCMU/chlorophyll $\leqq 1/100$; mol/mol) has drawn the attention of many workers (e.g. WESSELS and VAN DER VEEN, 1956; GOOD, 1961; DUYSENS and SWEERS, 1963) because of its obvious relevance to the photosynthetic units (believed to consist of 400–500 chlorophyll molecules). Izawa and Good (1965) attempted to determine the exact ratio of the number of DCMU-sensitive sites to the number of chlorophyll molecules in the chloroplast, from the analysis of the partition behavior of DCMU and of the inhibition kinetics. The ratio they obtained was very small: 1 site per 2,500 chlorophyll molecules. Cheniae and Martin (1968), however, questioned the validity of some of the kinetic arguments used by these authors, and concluded that their value was an underestimation. In any case, it is clear that a truly incontrovertible value would have to be sought by some radically different approach.

2. Other Herbicidal Inhibitors

The symmetrical triazines Atrazine (2-chloro-4-(2-propylamino)-6-ethylamino-s-triazine) and Simazine (2-chloro-4,6-bis(ethylamino)-s-triazine) are the best known of non-phenylurea type herbicidal inhibitors (Moreland et al., 1959) and are still used occasionally by photosynthesis researchers. No important difference was found between DCMU and these compounds in terms of the effect on electron transport and phosphorylation reactions (Good, 1961; Izawa and Good, 1965), on fluorescence (Zweig et al., 1963) and on cytochrome changes in algae (Nishimura, 1967). Gabbot (1969) has discussed the structure-activity relation of some of s-triazine derivatives. Chlorophenylcarbamates (Moreland and Hill, 1959) and acylchloroanilides (Good, 1961) are structurally similar to phenylureas but are in general much less potent than DCMU. Other herbicidal inhibitors are uracils (Hilton et al., 1964), benzimidazoles (Büchel and Draber, 1969), triazinones (Draber et al., 1969), N-alkylated ureas and ring-closed N-acylamides (Trebst and Harth, 1974). The last two families of compounds are distinct from all the others in that they have no imino group (or amino as in triazinones) in their structure. The supposedly vital imino hydrogen is replaced by a methyl group or by a ring formation; i.e., in short, the functional nitrogen is in the form of a tertiary amine. Thus, Trebst and Harth (1974) suggest, with Büchel (1972), that it is the free electron pair at the nitrogen that is responsible for the binding (inhibitory action) of all herbicidal inhibitors.

3. 1,10-Phenanthroline and Related Inhibitors

1,10-Phenanthroline (o-phenanthroline) is well known for its ability to complex with heavy metals. However, it has never been demonstrated that metal complexing indeed underlies the inhibition by this nucleophilic base, nor has its site of inhibition been clearly distinguished from the site DCMU inhibition. The original finding of Warburg and Lüttgens (1946) that 1,10-phenanthroline inhibition of the Hill reaction in chloroplasts was prevented preferentially by Zn^{2+} (suggestive of a Zn enzyme involvement) was confirmed earlier by MacDowall (1949), but contradictory results have been reported recently by Satoh (1974a). Satoh found that metal ions such as Ni^{2+} and Co^{2+}, which form more stable complexes with 1,10-phenanthroline than does Zn^{2+}, were more effective than Zn^{2+} in preventing or reversing the inhibition, and those forming less stable complexes were less effective (Mn^{2+} and Hg^{2+}). The inhibition was also readily reversed by washing the chloroplasts. Thus, Satoh finds no evidence of 1,10-phenanthroline removing any essential trace metals from the membrane, or of the metal ion-prevention (or reversal) of inhibition involving anything more than a simple "deactivation" (complexing) of the inhibitor by the metal ions added. Nevertheless, SATOH's (1974a) comparative study of 1,10-phenanthroline derivatives did produce results which show a correlation between inhibitor potency and Fe^{2+}-complexing capability. As for the inhibition site, fluorescence experiments allowed him to conclude that 1,10-phenanthroline and its inhibitory derivatives (4,7-dimethyl and 5-methyl derivatives) all block the exit of electrons from photosystem II at the same site (SATOH, 1974a). Some interesting effects of 1,10-phenanthrolines on the fluorescence emission spectra have also been noted (SATOH, 1974b).

OETTMEIER and GREWE (1974), who studied the structure-activity relationship of various mono- and diazaphenanthrenes (including 1,10-phenanthrolines), have concluded that the Fe^{2+}-complexing capability of 1,10-phenanthroline has nothing to do with its inhibitory action. Their conclusion is based on the finding that some of the non-Fe^{2+}-complexing azaphenanthrenes (e.g. l-azaphenanthrene) had a rather strong inhibitory activity in sharp contrast to the feeble inhibitory activities of the strong Fe^{2+}-chelators 2,2-bipyridine and 8-hydroxyquinoline (for the latter two compounds, see also MacDOWALL, 1949; KROGMANN and JAGENDORF, 1959). Their attempts to relate the inhibitor potency to the basicity and lipophilicity of the molecule (Hansch approach), however, met with only limited success, although the approach did help them clarifiy some of the steric and other structural factors governing the inhibitory activity of azaphenanthrenes. 1,10-phenanthroline and some of its derivatives have been confirmed to be still the most inhibitory of all the azaphenanthrenes tested.

OETTMEIER and GREWE carefully confirmed that none of the azaphenanthrenes they used had an inhibitory effect on photosystem I activity, but apparently did not consider the possibility that some of them might act on the water side of photosystem II. This possibility seems real, since BARR et al. (1975) have recently reported that the DCMU-insensitive Hill reaction with silicomolybdate as the electron acceptor was insensitive to 1,10-phenanthroline but was sensitive to bathocuproin (4,7-diphenyl-2,9-dimethyl-1,10-phenanthroline). The observation of BARR et al. reemphasizes the need for caution in applying a standard structure-activity analysis to multi-enzyme systems such as the chloroplast electron transport system.

4. Miscellaneous

2-Heptyl-4-hydroxyquinoline oxide (HQNO) and its 2-nonyl analog (NQNO) inhibit chloroplast electron transport strongly (AVRON, 1961; AVRON and SHAVIT, 1965) in a manner very similar to DCMU (IZAWA et al., 1967), but the exact site of action may be different from that of DCMU (HIND and OLSON, 1967). Spectroscopy suggested that in *Chlamydomonas* cells, HQNO and salicylaldoxime both block between cytochrome b-564 and cytochrome f (HILDRETH, 1968). Salicylaldoxime is a copper chelator originally introduced by TREBST (1963) as a possible plastocyanin inhibitor. Antimycin was claimed to block specifically the electron transport pathway utilized by ferredoxin-cyclic photophosphorylation (TAGAWA et al., 1963; but see IZAWA et al., 1967). Antimycin is an inhibitory uncoupler (DRECHSLER et al., 1969). It stimulates electron flow at lower pH and inhibits at higher pH, as do many phenolic substances (see GOOD's article in this volume).

n-Butyl-3,5-diiodo-4-hydroxybenzoic acid (Avron and Shavit, 1965) also inhibits. The relation of these phenol inhibitions to DCMU inhibition seems worth exploring.

III. Plastoquinone Antagonists

1. Dibromothymoquinone (DBMIB)

2,5-Dibromo-3-methyl-6-isopropyl-p-benzoquinone (dibromothymoquinone or DBMIB), a powerful inhibitor of photosynthetic electron transport, introduced by Trebst et al. (1970), is believed to act as an antagonist to plastoquinone in situ. The belief is based on several lines of evidence: (1) inhibition of reactions that require both photosystems (e.g. $NADP^+$ Hill reaction), (2) lack of inhibition of photosystem I-mediated reactions (e.g. PMS-cyclic photophosphorylation), (3) lack of inhibition of reactions that are supported by photosystem II alone (e.g. DCIP Hill reaction in sonicated chloroplasts) (Trebst et al., 1970; Trebst, 1972; Böhme et al., 1971), and (4) spectroscopic evidence showing an inhibition of electron flow between cytochrome b-559 and cytochrome f (Böhme and Cramer, 1971). Besides the obvious structural analogy between DBMIB and plastoquinone, these experiments clearly point to an inhibition site in an area near photosystem II where plastoquinone is known to be functioning as an electron carrier. A relief of DBMIB inhibition by the addition of a great excess of exogenous plastoquinone (0.1 mM) to the reaction mixture has also been reported (Böhme et al., 1971), but the significance of this observation is not very clear. (Various other quinones and oxidized p-phenylenediamines have also been reported to restore $NADP^+$ photoreduction in DBMIB-inhibited chloroplasts; Trebst and Reimer, 1973a, b.)

The suggestion of Böhme et al. (1971) that the exit of electrons from plastoquinone may be the DBMIB-sensitive process acquired experimental basis when Gimmler and Avron (1972), using algae, and Lozier and Butler (1972), using chloroplasts, showed that DBMIB inhibition had little effect on the fluorescence induction kinetics. This suggests that DBMIB did not affect the process of the electrons from photosystem II filling up the plastoquinone pool. However, in chloroplasts such clear results were obtained only with the reduced form of DBMIB (with excess ascorbate); the regularly used quinone form ($>2\,\mu M$) strongly quenched not only the variable fluorescence from photosystem II (even in the presence of DCMU) but also the low-temperature fluorescence of photosystem I (Lozier and Butler, 1972), although the suppressed fluorescence slowly recovered as the DBMIB was reduced by photosystem II (Lozier and Butler, 1972; for the properties of DBMIB as an electron acceptor of photosystem II, see Gould and Izawa, 1973). Kouchkovsky and Kouchkovsky (1974) have concluded from their O_2 gush experiments that the effect of oxidized form of DBMIB extends gradually from the photosystem I side to the photosystem II side of the plastoquinone pool as the concentration of DBMIB is increased.

The long-suspected variability of the accessibility of photosystem II to ferricyanide was clearly demonstrated when the sensitivity of the ferricyanide Hill reaction to DBMIB was found to vary greatly depending on the chloroplast preparation — from zero inhibition in sonicated chloroplasts (Trebst, 1972; Böhme et al., 1971) to 90% inhibition in chloroplasts with high membrane integrity (Izawa et al.,

1973a). Similar results were obtained with DPIP Hill reaction (Böhme et al., 1971). In chloroplasts with high membrane integrity, the only electron acceptors that are able to intercept electrons efficiently from photosystem II appear to be those with lipophilicity and high oxidizing potentials, such as certain quinones and oxidized p-phenylenediamines (Class III acceptors of Saha et al., 1971). Izawa et al. (1973a) and Trebst and Reimer (1973a, b) have found that DBMIB-resistant electron flow from water to these quinoid compounds is invariably coupled to phosphorylation, thus providing strong evidence for the existence of an energy conservation mechanism associated with photosystem II.

2. Other Possible Plastoquinone Antagonists

Trebst (1972) has presented brief data which show that the naturally occurring isoprenylbenzoquinones, helveticone and bovinone, have inhibitory effects on chloroplasts which are similar to that of DBMIB. However, these are at least 100 times weaker inhibitors than DBMIB. 2,3-Dimethyl-5-hydroxy-6-phytylbenzoquinone was found by Arntzen et al. (1971) to inhibit chloroplast electron transport apparently at a point between the two sites of electron donation by reduced DCIP to photosystem I. PMS-cyclic photophosphorylation was inhibited. If this is a plastoquinone antagonist, as supposed, clearly the target plastoquinone must be located closer to photosystem I than is the target plastoquinone of DBMIB inhibition.

IV. Inhibitors of Electron Transfer Between Plastoquinone and Cytochrome f

1. 1-Ethyl-3-(3-dimethylaminopropyl)-carbodiimide (EDAC)

When used in usual ways, this water soluble carbodiimide is only a poor inhibitor of chloroplast electron transport (Uribe, 1972). However, McCarty (1974) has found that the electron transport inhibition is greatly enhanced when chloroplasts are preilluminated in the presence of EDAC (0.5 mM) for 5 min. The pattern of the light-potentiated inhibition McCarty found was characteristic of an inhibition between the two photosystems and was similar to that of KCN inhibition (at plastocyanin, see below) rather than that of DBMIB inhibition. Reactions involving both photosystems (e.g. methylviologen Hill reaction) were inhibited, and so were reactions which required photosystem I only and were insensitive to DBMIB (e.g. diaminodurene oxidation). Reactions supported largely by photosystem II alone (photoreduction of oxidized diaminodurene) was not inhibited. Despite such similarities between EDAC inhibition and KCN inhibition, there was no indication of plastocyanin being chemically affected by EDAC. Moreover, spectroscopy indicated a block before cytochrome f. Thus, McCarty logically concluded that EDAC attacked an unknown component between plastoquinone and cytochrome f.

It is perhaps fair to add here that the placing of this interesting inhibitor under the heading is tentative. Because of the important implication of the results reported by McCarty, confirmative reports and further exploration of characteristics of EDAC inhibition are awaited. As for other possible inhibitors of this category, it seems possible that the benzoquinone derivative of Arntzen et al. (1971) listed under plastoquinone antagonists (see above) actually belongs here.

V. Inhibitors of Plastocyanin

1. Polycations

Histone and high molecular poly-L-lysine are interesting inhibitors because of their unique membrane-probe-like properties (Brand et al., 1971, 1972a; Berg et al., 1973, 1974).

Low-salt conditions are requisite for effecting polycation inhibition (Brand et al., 1971). When chloroplasts are exposed to a low-salt medium, they swell and lose their granal structure as a result of loosening of membrane-membrane association (Izawa and Good, 1966). Addition of small amounts of polycation (final 10 μg/ml) to such a suspension causes the swollen chloroplasts to shrink (Berg et al., 1973) and the dissociated membranes to reassociate (Berg et al., 1974), much as salt addition does (Izawa and Good, 1966). However, these polycation-shrunken chloroplasts are unable to photoreduce weak oxidants via photosystem I (Brand et al., 1971, 1972a) because the transfer of electrons from cytochrome f to P-700 does not occur (Brand et al., 1972b). The inhibition is prevented, though not reversed, by adding salts (0.1 M KCl) or other agents (e.g. glutathione) which cause chloroplast shrinking (Brand et al., 1971; Berg et al., 1973) or by adding polyanions (Brand et al., 1972a). Purified spinach plastocyanin, also a polyanion, is not only exceedingly effective in preventing inhibition, but it even partially reverses the inhibition. Moreover, it readily forms aggregates with polycations in vitro (Brand et al., 1972a). Thus Brand et al. suggest that, in the unfolded lamellar membranes of swollen chloroplasts, endogenous plastocyanin might be accessible to polycation in the medium and thus a direct interaction similar to that observed in vitro might occur in situ. Berg et al. (1973) point out the interesting fact that polycation does not inhibit *Anabaena* lamellae which contain basic (cationic) plastocyanin (Lightbody and Krogmann, 1967).

Meanwhile, evidence is accumulating that the binding of polycation to the lamellar membrane is a surface binding (Berg et al., 1973) involving the negative charge of carboxyl groups which are exposed when the lamellar membranes are disarrayed by low-salt conditions (Berg et al., 1974). This raises many interesting questions regarding the structure-function relation between the membrane surface and the membrane-bound copper protein.

Poly-L-lysine acts as a potent uncoupler when the chloroplast suspension contains salts, as in phosphorylation experiments (Dilley, 1968). However, Ort et al. (1973) have devised a procedure which permits the use of poly-L-lysine as an electron transport inhibitor under phosphorylating conditions. The effects of polylysine on various electron transport reactions and the associated phosphorylation thus observed were quite similar to those of KCN.

2. Mercuric Chloride

Kimimura and Katoh (1972) reported that incubation of chloroplasts with $HgCl_2$ (Hg/chl ≃ 1) for 30 min at 0° C resulted in a complete inhibition of methylviologen photoreduction with water or reduced DPIP as the electron donor. Oxidation of cytochrome f by photosystem I was inhibited, but neither the photoreduction of the cytochrome nor the photooxidation of P-700 was affected, indicating a block at the plastocyanin site. The biochemical evidence KIMIMURA and KATOH presented for the copper protein being the target site is clear-cut and absolute. However, at somewhat higher concentrations (Hg/chl ≃ 10), $HgCl_2$ inhibited P-700 oxidation significantly, and partially suppressed the variable fluorescence (even in the presence of DCMU and dithionite), which is indicative of secondary effects on both photosystems. In line with the results from KCN and polylysine experiments (IZAWA et al., 1973 a; ORT et al., 1973), KIMIMURA and KATOH (1973) found that the photoreductions of high potential quinoid oxidants contained portions which were insensitive to $HgCl_2$ and were therefore considered to be mediated by photosystem II alone. The activities of these photosystem II reactions reported by KIMIMURA and KATOH were, however, surprisingly low.

Very low concentrations of mercuric acetate and $HgCl_2$ (Hg/chl < 1/10) had previously been shown to act as an energy transfer inhibitor of chloroplasts (IZAWA and GOOD, 1969). This inhibition is instantaneous. Such low concentrations of $HgCl_2$ have recently been used for analysis of the relation of phosphorylation reactions to electron transport pathways (BRADEEN et al., 1973).

3. Cyanide

High concentrations (> 10 mM) of this well-known inhibitor of photosynthetic CO_2 fixation (WARBURG, 1919) inhibit chloroplast electron transport (BISHOP and SPIKES, 1955; TREBST, 1963) primarily at plastocyanin (OUITRAKUL and IZAWA, 1973; IZAWA et al., 1973 b). A high degree of selective inhibition at plastocyanin can be achieved by incubating chloroplasts with 20–30 mM KCN at pH 7.8 at 0° C for 60 to 90 min. The development of inhibition can be quickened by raising the temperature or by raising the pH, but these modifications were found to aggravate the weak tendency of KCN to inhibit photosystem II.

The Hill reaction involving both photosystems (methylviologen or ferricyanide as acceptor) is abolished, but the much faster Hill reaction with lipophilic oxidants, and associated phosphorylation, are only partially inhibited (20–50% depending on the oxidant), indicating an electron interception from photosystem II. Photosystem I-mediated reactions tested were all inhibited, with the exception of cyclic photophosphorylation mediated by high concentration of PMS. Spectroscopy clearly located a block between cytochrome f and P-700, and EPR spectrometry confirmed the ability of reduced PMS to interact directly with P-700 (IZAWA et al., 1973 b). BERG and KROGMANN (1975), extending the preliminary work of SELMAN et al. (1975), have investigated in detail the interaction of KCN with plastocyanin both in situ and in vitro. Their findings include a stoichiometric release of plastocyanin Cu from the lamellar membrane exposed to KCN.

KCN inhibition has been found useful in determining energy output from various segments of the electron transport chain (KRAAYENHOF et al., 1972; OUITRAKUL and IZAWA, 1973). At the present time, cyanide seems to be the only plastocyanin inhibitor that can be applied to chloroplasts without inflicting any visible side effect on the mechanism of phosphorylation.

4. Amphotericin B

Nolan and Bishop (1975) have reported that incubation of chloroplasts in a medium containing 0.5 mM amphotericin B (a steroid complexing agent) for 60 min at 25° C causes a release of plastocyanin from the chloroplast membrane and thereby destroys the ability of the chloroplasts to carry out methylviologen reduction. Not surprisingly, the inhibition of ferricyanide reduction was not complete (70%). Despite the harsh treatment, the chloroplast membrane was reported to be structurally undamaged. Yet the addition of exogenous plastocyanin completely reversed the inhibition. Of the various observations described by Nolan and Bishop, this free reversibility of inhibition is the most interesting.

VI. Inhibitors of Reactions in Ferredoxin-NADP$^+$ Region

1. Disalicylidendiamines

Disalicylidens are known as metal-chelating agents. Disalicylidenpropanediamine (1,3) (DSPD) and its water soluble 1,2-analog (sulfo-DSPD) have been demonstrated by Trebst and Burba (1967) to be useful inhibitors of ferredoxin-dependent reactions in algae and in chloroplasts respectively. Thus, sulfo-DSPD (1 mM) did not inhibit the Hill reaction in chloroplasts with non-physiological oxidants (ferricyanide or anthraquinone) but it did inhibit the reaction when the ferredoxin/NADP$^+$ couple served as the acceptor system. The inhibition was partially reversed by increasing the concentration of ferredoxin. All the other ferredoxin-dependent reactions tested (photoreduction of nitrate or nitrite, and the enzymic reduction of nitrate) were also inhibited, with the exception of the enzymic reduction of nitrite (dithionite $\rightarrow$ ferredoxin $\rightarrow$ nitrite reductase $\rightarrow$ nitrite). This exception led Trebst and Burba to conclude that ferredoxin itself is not the target enzyme of sulfo-DSPD inhibition. Consequently the site of sulfo-DSPD inhibition in the electron transport pathway was placed between photosystem I and ferredoxin. However, Ben-Amotz and Avron (1972) suggest that ferredoxin itself is probably the target enzyme, based on their observation that the ferredoxin-dependent enzymic reduction of cytochrome c was inhibited by sulfo-DSPD.

The membrane-permeating DSPD (1–2 mM) inhibited photosynthetic CO$_2$ fixation in algae (Trebst and Burba, 1967; Gimmler et al., 1968; Ben-Amotz and Avron, 1972). However, in algal cells, ferredoxin is clearly not the only enzyme that is sensitive to DSPD, since in *Ankistrodesmus*, dark respiration and associated ^{32}P uptake were more sensitive to DSPD than was photosynthesis (Gimmler et al., 1968). On several grounds Ben-Amotz and Avron (1972) even question the involvement of ferredoxin in photosynthetic CO$_2$ fixation in the alga *Dunaliela* even though the reaction was inhibited (60%) by 1.5 mM DSPD (see below).

2. Pyrophosphate

Forti and Meyer (1969) reported that pyrophosphate ($>$10 mM) had no effect on the methylviologen Hill reaction in chloroplasts (spinach) but strongly inhibited the NADP$^+$ Hill reaction and other reactions involving both NADP$^+$ and ferredoxin, such as the enzymic reduction of cytochrome c by the process: NADPH $\rightarrow$ ferredoxin-$NADP^+$ reductase→ferredoxin→cytochrome c. The inhibition was competitive with respect to ferredoxin. However, neither the photoreduction of ferredoxin in chloroplasts nor its aerobic reoxidation was affected (FORTI et al., 1970). From these findings and spectroscopic evidence, the authors concluded that pyrophosphate interfered with the interaction (complexing) between ferredoxin and ferredoxin-$NADP^+$ reductase. BEN-AMOTZ and AVRON (1972), however, reported that pyrophosphate did inhibit the photoreduction of ferredoxin in lettuce chloroplasts as strongly as it inhibited the ferredoxin-mediated photoreduction on $NADP^+$. They concluded that pyrophosphate probably interacted with ferredoxin itself. In either case, the inhibitory effect of pyrophosphate on $NADP^+$ photoreduction is in sharp contrast with the well-known stimulatory effect of pyrophosphate on photosynthetic CO_2 fixation in isolated chloroplasts (JENSEN and BASSHAM, 1968), and this led FORTI et al. (1970) to stress the possibility of reduced ferredoxin directly serving as a reductant for CO_2 fixation.

3. Phosphoadenine Diphosphate Ribose

This degradation product of $NADP^+$ is a highly specific inhibitor of $NADP^+$-dependent dehydrogenases including ferredoxin-$NADP^+$ reductase (BEN-HAYYIM et al., 1967). BEN-AMOTZ and AVRON (1972) showed that this $NADP^+$ analog inhibited the activity of ferredoxin-$NADP^+$ reductase in cell-free preparations from *Dunaliela* without inhibiting photosynthetic CO_2 fixation by the same preparations. These authors therefore suggest that in this halophilic alga some unknown reductant preceding ferredoxin might be involved in photosynthetic CO_2 reduction.

Acknowledgement. The author wishes to thank Dr. STEVEN BERG for reading through the manuscript.

References

Amesz, J.: Biochim. Biophys. Acta **79**, 257–265 (1964)

Arntzen, C.J., Neumann, J., Dilley, R.A.: J. Bioenergetics **2**, 73–83 (1971)

Asada, K., Takahashi, M.: In: Photosynthesis and Photorespiration. Hatch, M.D., Osmond, C.B., Slatyer, R.O. (eds.). New York: Wiley-Interscience, 1971, pp. 387–393

Avron, M.: Biochem. J. **78**, 735–739 (1961)

Avron, M., Shavit, N.: Biochim. Biophys. Acta **109**, 317–331 (1965)

Barr, R., Crane, F.L., Giaquinta, R.T.: Plant Physiol. **55**, 460–462 (1975)

Ben-Amotz, A., Avron, M.: Plant Physiol. **49**, 244–248 (1972)

Ben-Hayyim, G., Hochman, A., Avron, M.: J. Biol. Chem. **242**, 2837–2839 (1967)

Bennoun, P.: Biochim. Biophys. Acta **216**, 357–363 (1970)

Bennoun, P., Bouges, B.: Effect of hydroxylamine and DCMU on Photosystem II. Proc. 2nd Intern Congr. Photosynthesis. Forti, G., Avron, M., Melandri, A. (eds.). The Hague: Dr. Junk, 1972, pp. 569–576

Bennoun, P., Joliot, A.: Biochim. Biophys. Acta **189**, 85–94 (1969)

Berg, S., Cipollo, D., Armstrong, B., Krogmann, D.W.: Biochim. Biophys. Acta **305**, 372–383 (1973)

Berg, S., Dodge, S., Krogmann, D.W., Dilley, R.A.: Plant Physiol. **58**, 619–627 (1974)

Berg, S., Krogmann, D.W.: J. Biol. Chem. (1975) (in press)

Bishop, N.I.: Biochim. Biophys. Acta **27**, 205–206 (1958)

Bishop, N.I., Spikes, J.D.: Nature (London) **176**, 307–308 (1955)

Böhme, H., Cramer, W.A.: FEBS Lett. **15**, 349–351 (1971)
Böhme, H., Reimer, S., Trebst, A.: Z. Naturforsch. **26B**, 341–352 (1971)
Bouges, B.: Biochim. Biophys. Acta **234**, 103–112 (1971)
Bouges-Bocquet, B.: Biochim. Biophys. Acta **292**, 772–785 (1973)
Bradeen, D.A., Winget, G.D., Gould, J.M., Ort, D.R.: Plant Physiol. **52**, 680–682 (1973)
Brand, J., Baszynski, T., Crane, F.L., Krogmann, D.W.: Biochem. Biophys. Res. Commun. **45**, 538–543 (1971)
Brand, J., Baszynski, T., Crane, F.L., Krogmann, D.W.: J. Biol. Chem. **247**, 2814–2819 (1972a)
Brand, J., San Pietro, A., Mayne, B.C.: Arch. Biochem. Biophys. **152**, 426–428 (1972b)
Büchel, K.H.: Pesticide Sci. **3**, 89–100 (1972)
Büchel, K.H., Draber, W.: In: Progress in Photosynthesis Research. Metzner, H. (ed.). Tübingen: Laupp, 1969, Vol. III, pp. 1777–1788
Camper, N.D., Moreland, D.E.: Biochim. Biophys. Acta **94**, 383–393 (1966)
Cheniae, G.M.: Ann. Rev. Plant Physiol. **21**, 467–498 (1970)
Cheniae, G.M., Martin, I.F.: Brookhaven Symp. Biol. **19**, 406–417 (1967)
Cheniae, G.M., Martin, I.F.: Biochim. Biophys. Acta **153**, 819–837 (1968)
Cheniae, G.M., Martin, I.F.: Biochim. Biophys. Acta **197**, 219–239 (1970)
Cheniae, G.M., Martin, I.F.: Plant Physiol. **47**, 568–575 (1971)
Cheniae, G.M., Martin, I.F.: Plant Physiol. **50**, 87–94 (1972)
Dilley, R.A.: Biochemistry **7**, 338–346 (1968)
Döring, G., Renger, G., Vater, J., Witt, H.T.: Z. Naturforsch. **24B**, 1139–1143 (1969)
Döring, G., Stiehl, H.H., Witt, H.T.: Z. Naturforsch. **22B**, 639–644 (1967)
Draber, W., Büchel, K.H., Dickoré, K., Trebst, A., Pistorius, E.: In: Progress in Photosynthesis Research. Metzner, H. (ed.). Tübingen: Laupp, 1969, Vol. III, pp. 1789–1795
Drechsler, Z., Nelson, N., Neumann, J.: Biochim. Biophys. Acta **189**, 65–73 (1969)
Duysens, L.N.M.: In: Proc. 2nd Intern. Congr. Photosynthesis. Forti, G., Avron, M., Melandri, A. (eds.). The Hague: Dr. Junk, 1972, pp. 19–25
Duysens, L.N.M., Sweers, H.E.: In: Studies on Microalgae and Photosynthetic Bacteria. Tokyo: Univ. Tokyo, 1963, pp. 353–372
Elstner, E.F., Heupel, A., Vaklinova, S.: Z. Pflanzenphysiol. **62**, 173–183 (1970)
Forti, G., Melandri, B.A., San Pietro, A., Ke, B.: Arch. Biochem. Biophys. **140**, 107–112 (1970)
Forti, G., Meyer, M.: Plant Physiol. **44**, 1511–1514 (1969)
Gabbot, P.A.: In: Progress in Photosynthesis Research. Metzner, H. (ed.). Tübingen: Laupp, 1969, Vol. III, pp. 1712–1727
Giaquinta, R.T., Dilley, R.A., Crane, F.L., Barr, R.: Biochem. Biophys. Res. Commun. **59**, 985–991 (1974)
Gimmler, H., Avron, M.: In: Proc. 2nd Intern. Congr. Photosynthesis. Forti, G., Avron, M., Melandri, A. (eds.). The Hague: Dr. Junk, 1972, pp. 789–800
Gimmler, H., Urbach, W., Jeschke, W.D., Simonis, W.: Z. Pflanzenphysiol. **58**, 353–364 (1968)
Girault, G., Galmiche, J.M.: Biochim. Biophys. Acta **333**, 314–319 (1974)
Good, N.E.: Plant Physiol. **36**, 788–803 (1961)
Good, N.E., Izawa, S.: In: Metabolic Inhibitors. Hochster, R.M., Kates, M., Quastel, J.H. (eds.). New York-London: Academic Press, 1973, Vol. IV, pp. 179–214
Gould, J.M., Izawa, S.: Europ. J. Biochem. **37**, 185–192 (1973)
Gregory, R.P.F.: Biochim. Biophys. Acta **368**, 228–234 (1974)
Hansch, C.: In: Progress in Photosynthesis Research. Metzner, H. (ed.). Tübingen: Laupp, 1969, Vol. III, pp. 1685–1692
Hansch, C., Deutsch, E.W.: Biochim. Biophys. Acta **112**, 381–391 (1966)
Hildreth, W.W.: Plant Physiol. **43**, 303–312 (1968)
Hilton, J.L., Monaco, T.J., Moreland, D.E., Gentner, W.A.: Weeds **12**, 129–131 (1964)
Hind, G., Nakatani, H.Y., Izawa, S.: Biochim Biophys. Acta **172**, 277–289 (1969)
Hind, G., Olson, J.M.: Brookhaven Symp. Biol. **19**, 188–194 (1967)
Hinkson, J.W., Vernon, L.P.: Plant Physiol. **34**, 268–272 (1959)
Homann, P.H.: Biochim. Biophys. Acta **245**, 129–143 (1971)
Homann, P.H.: Biochim. Biophys. Acta **256**, 336–344 (1972)

Inoué, H., Nishimura, M.: Plant Cell Physiol. **12**, 739–747 (1971)

Itoh, M., Yamashita, K., Nishi, T., Konishi, K., Shibata, K.: Biochim. Biophys. Acta **180**, 509–519 (1969)

Izawa, S., Connolly, T.N., Winget, G.D., Good, N.E.: Brookhaven Symp. Biol. **19**, 169–187 (1967)

Izawa, S., Good, N.E.: Biochim. Biophys. Acta **102**, 20–38 (1965)

Izawa, S., Good, N.E.: Plant Physiol. **41**, 544–552 (1966)

Izawa, S., Good, N.E.: In: Progress in Photosynthesis Research. Metzner, H. (ed.). Tübingen: Laupp, 1969, Vol. II, pp. 1288–1298

Izawa, S., Good, N.E.: Methods Enzymol. **24B**, 335–377 (1972)

Izawa, S., Gould, J.M., Ort, D.R., Felker, P., Good, N.E.: Biochim. Biophys. Acta **305**, 119–128 (1973a)

Izawa, S., Heath, R.L., Hind, G.: Biochim. Biophys. Acta **180**, 388–398 (1969)

Izawa, S., Kraayenhof, R., Ruuge, E.K., De Vault, D.: Biochim. Biophys. Acta **314**, 328–339 (1973b)

Izawa, S., Ort, D.R., Gould, J.M., Good, N.E.: In: Proc. 3rd Intern. Congr. Photosynthesis. Avron, M. (ed.). Amsterdam: Elsvier, 1975, pp. 449–461

Jagendorf, A.T., Avron, M.: Arch. Biochem. Biophys. **80**, 246–257 (1959)

Jensen, R.G., Bassham, J.A.: Biochim. Biophys. Acta **153**, 219–226 (1968)

Joliot, A.: Biochim. Biophys. Acta **126**, 587–590 (1966)

Katoh, S., Ikegami, I., Takamiya, A.: Arch. Biochem. Biophys. **141**, 207–218 (1970)

Katoh, S., San Pietro, A.: Arch. Biochem. Biophys. **122**, 144–152 (1967)

Katoh, S., San Pietro, A.: Arch. Biochem. Biophys. **128**, 378–386 (1968)

Kiewiet, D.Y. de, Hall, D.O., Jenner, F.L.: Biochim. Biophys. Acta **109**, 284–292 (1965)

Kimimura, M., Katoh, S.: Biochim. Biophys. Acta **283**, 279–292 (1972)

Kimimura, M., Katoh, S.: Biochim. Biophys. Acta **325**, 167–174 (1973)

Kimimura, M., Katoh, S., Ikegami, I., Takamiya, A.: Biochim. Biophys. Acta **234**, 92–102 (1971)

Kok, B., Malkin, S., Owens, O.V.H., Forbush, B.: Brookhaven Symp. Biol. **19**, 446–459 (1967)

Kouchkovsky, Y. de, Kouchkovsky, F. de: Biochim. Biophys. Acta **368**, 113–124 (1974)

Kraayenhof, R., Izawa, S., Chance, B.: Plant Physiol. **50**, 713–718 (1972)

Krogmann, D.W., Jagendorf, A.T.: Arch. Biochem. Biophys. **80**, 421–430 (1959)

Lightbody, J.J., Krogmann, D.W.: Biochim. Biophys. Acta **131**, 508–515 (1967)

Lozier, R.H., Baginsky, M., Butler, W.L.: Photochem. Photobiol. **14**, 323–328 (1971)

Lozier, R.H., Butler, W.L.: FEBS Lett. **26**, 161–164 (1972)

MacDowall, F.D.H.: Plant Physiol. **24**, 462–480 (1949)

McCarty, R.E.: Arch. Biochem. Biophys. **161**, 93–99 (1974)

Mohanty, P., Mar, T., Govinjee: Biochim. Biophys. Acta **253**, 213–221 (1971)

Moreland, D.E.: In: Progress in Photosynthesis Research. Metzner, H. (ed.). Tübingen: Laupp, 1969, Vol. III, pp. 1693–1711

Moreland, D.E., Gentner, W.A., Hilton, J.L., Hill, K.L.: Plant Physiol. **34**, 432–435 (1959)

Moreland, D.E., Hill, K.L.: Agr. Food Chem. **7**, 832–837 (1959)

Nakamoto, T., Krogmann, D.W., Vennesland, B.: J. Biol. Chem. **234**, 2783–2788 (1959)

Nishimura, M.: Brookhaven Symp. Biol. **19**, 132–142 (1967)

Nolan, W.G., Bishop, D.G.: Arch. Biochem. Biophys. **166**, 323–329 (1975)

Oettmeier, W., Grewe, R.: Z. Naturforsch. **29C**, 545–551 (1974)

Ort, D.R., Izawa, S.: Plant Physiol. **53**, 370–376 (1974)

Ort, D.R., Izawa, S., Good, N.E., Krogmann, D.W.: FEBS Lett. **31**, 119–122 (1973)

Ouitrakul, R., Izawa, S.: Biochim. Biophys. Acta **305**, 105–118 (1973)

Radmer, R., Kok, B.: Ann. Rev. Biochem. **44**, 409–433 (1975)

Renger, G.: Naturwissenschaften **56**, 370 (1969)

Renger, G.: Biochim. Biophys. Acta **256**, 428–439 (1972)

Rienits, K.G., Hardt, H., Avron, M.: FEBS Lett. **33**, 28–32 (1973)

Saha, S., Ouitrakul, R., Izawa, S., Good, N.E.: J. Biol. Chem. **246**, 3204–3209 (1971)

Satoh, K.: Biochim. Biophys. Acta **333**, 127–135 (1974a)

Satoh, K.: Biochim. Biophys. Acta **333**, 107–126 (1974b)

Selman, B.R., Johnson, G.L., Giaquinta, R.T., Dilley, R.A.: Bioenergetics **6**, 221–231 (1975)

Shibata, K., Yakushiji, E.: Naturwissenschaften **21**, 267–268 (1933)

Siggel, U., Renger, G., Stiehl, H.H., Rumberg, B.: Biochim. Biophys. Acta **256**, 328–335 (1972)
Stiehl, H.H., Witt, H.T.: Z. Naturforsch. **24B**, 1588–1598 (1969)
Tagawa, K., Tsujimoto, H.Y., Arnon, D.I.: Proc. Nat. Acad. Sci. U.S. **49**, 567–572 (1963)
Trebst, A.: Z. Naturforsch. **13B**, 817–821 (1963)
Trebst, A.: In: Proc. 2nd Intern. Congr. Photosynthesis. Forti, G., Avron, M., Melandri, B.A. (eds.). The Hague: Dr. Junk, 1972, pp. 399–417
Trebst, A.: Ann. Rev. Plant Physiol. **25**, 423–458 (1974)
Trebst, A., Burba, M.: Z. Pflanzenphysiol. **57**, 419–433 (1967)
Trebst, A., Harth, E.: Z. Naturforsch. **29C**, 232–235 (1974)
Trebst, A., Harth, E., Draber, W.: Z. Naturforsch. **25B**, 1157–1159 (1970)
Trebst, A., Reimer, S.: Biochim. Biophys. Acta **305**, 129–139 (1973a)
Trebst, A., Reimer, S.: Z. Naturforsch. **28C**, 710–716 (1973b)
Uribe, E.G.: Biochemistry **11**, 4228–4235 (1972)
Vernon, L.P., Zaugg, W.S.: J. Biol. Chem. **235**, 2728–2738 (1960)
Warburg, O.: Biochem. Z. **100**, 230–270 (1919)
Warburg, O., Lüttgens, W.: Biokhimiya **11**, 303–321 (1946) (in Russian)
Wessels, J.S.C., Van der Veen, R.: Biochim. Biophys. Acta **19**, 548–549 (1956)
Witt, K.: FEBS Lett. **38**, 116–118 (1973)
Yamashita, T., Butler, W.L.: Plant Physiol. **43**, 1978–1968 (1968a)
Yamashita, T., Butler, W.L.: Plant Physiol. **43**, 2037–2040 (1968b)
Yamashita, T., Butler, W.L.: Plant Physiol. **44**, 1342–1346 (1969a)
Yamashita, T., Butler, W.L.: Plant Physiol. **44**, 435–438 (1969b)
Yamashita, T., Konishi, K., Itoh, M., Shibata, K.: Biochim. Biophys. Acta **172**, 511–524 (1969)
Yamashita, T., Tomita, G.: Plant Cell Physiol. **15**, 69–82 (1974)
Yamashita, T., Tsuji, J., Tomita, G.: Plant Cell Physiol. **12**, 117–126 (1971)
Zweig, G., Tamas, I., Greenberg, E.: Biochim. Biophys. Acta **66**, 196–205 (1963)

17. Antibodies

R.J. Berzborn and W. Lockau

A. Introduction

Antibodies have supported the study of soluble enzymes for a long time (Cinader, 1963; Arnon, 1971), but only a few research groups have used them for studies on membrane-associated reactions. We have tried to collect all publications up to now on investigations of photosynthetic electron transport in chloroplasts with the aid of antibodies. Before describing the results and conclusions we will recall the possibilities and limitations of the combination of methods which to many readers still might seem odd.

B. General Considerations on the Application of Antibodies to Studies of Membrane Function

I. Properties of Antibodies

The usefulness of antibodies is founded on their unique properties. First, antibodies can be elicited; if a macromolecule, particle or membrane preparation with a rigid structure or at least some rigid determinant groups is injected into animals according to certain immunization schedules, the animal will either tolerate the material as identical to one of its own substances, or recognize the antigen (immunogen) as foreign and respond by synthesizing specific antibodies. Second, the antibodies used in our context are soluble, hydrophilic proteins found in the class of gammaglobulins in the blood serum. Third, they contain in the case of IgG two identical binding sites (Edelman, 1973) (in the case of IgM there are ten) located at the tips of the Fab parts (about 35 Å across and 60 Å long) of the Y-shaped molecule (Valentine and Green, 1967). The third part of the molecule, removable as Fc by papain (Porter, 1973), about 45 Å across and 40 Å long, does not contribute to the specific binding, but can react as antigen with antibodies from other animals or bind to membrane surfaces (Lauf, 1975). Under proper conditions Fc can bind and start the complement system (Kabat and Mayer, 1967). Fourth, the specificity (Landsteiner, 1945) of the antibody binding sites towards the antigen determinant groups is comparable to or even surpasses the specificity of enzymes. The binding is usually very strong (Kabat and Mayer, 1967). Measurements of binding of haptens with antibodies by equilibrium dialysis or stopped flow gave association constants between 10^5 and 10^{10} M^{-1} (cf. Smith and Skubitz, 1975).

II. Usefulness of Antibodies

1. Inhibition of the Antigen Function

From the above it thus results that antibodies may be used as specific crosslinking (precipitating or agglutinating) reagents or as specific inhibitors blocking a reaction sequence at a predictable site. The action of an inhibiting antibody can only be seen if it interferes with the limiting step of a reaction sequence, or if it makes a step in the reaction sequence limiting. From research on soluble enzymes the effects of antibodies on membrane-associated reactions can be extrapolated: an inhibition can be complete, partial or not measurable, even if plenty of precipitating antibodies are reacting. Competitive inhibition of soluble enzymes is rarely found, probably due to the absence of antibodies against active sites, due to tolerance caused by identical active sites on enzymes of the animal which produced the antibodies (CINADER, 1963). But after prolonged immunization antibodies inhibiting competitively can also be expected. Even a competitive inhibition need not indicate binding of the antibody to the catalytic site, since the inhibition may be caused by steric hindrance, in particular if the substrate is large (CINADER, 1963; ARNON, 1971). Since the strength of the binding of antibodies is pH-dependent, the same antibody might inhibit competitively or non-compeitively at different pH values. In non-competitive inhibition maximal inhibition by one antibody/enzyme is the rule (CINADER, 1963).

Allosteric action of antibodies is also conceivable, either by fixing the enzyme in a non-active conformation or by distorting it into a non-active conformation (CINADER, 1963). In the latter case the inhibition can be less, if enzyme and antibody are incubated in the presence of substrate. Also the activation of an enzyme by pulling it into an active conformation has been observed (MELCHERS and MESSER, 1973), and has to be considered as a possibility when working with membrane-bound or membrane-embedded enzymes. An allosteric fixation or distortion by an antibody becomes less probable if monovalent Fab fragments are used.

Even the simple formation of the insoluble antigen antibody matrix or of large agglutination aggregates of particles might cause inhibition (ARNON, 1971). But with soluble enzymes the contribution of aggregate formation to inhibition by antibodies is believed to be small (CINADER, 1963), and agglutination of chloroplast thylakoid systems alone does not seem to cause inhibition. Agglutinating antibodies against the coupling factor of photophosphorylation (CF_1) (McCARTY and RACKER, 1966) or even against entire thylakoid systems (TS) (SPIKES, 1959; BERZBORN et al., 1966) do not inhibit electron transport; also agglutinating antibodies against the β-subunit of CF_1 do not inhibit photophosphorylation (BERZBORN, 1972; NELSON et al., 1973). On the other hand Fab fragments of an antibody against CF_1 did not inhibit photophosphorylation any longer (SCHMID et al., 1976); we suggest that antibodies against CF_1 inhibit due to interference with the conformational change of CF_1 during photophosphorylation (RYRIE and JAGENDORF, 1972).

2. Accessibility of Antigens

A reaction of an antibody with a membrane component is only possible if the determinant group of the membrane component is accessible. Since antibodies are hydrophilic macromolecules, it seems a safe assumption that antibodies in vitro do not penetrate membranes (CINADER, 1963; JERNE, personal communication), although exceptions seem to exist; antibodies against receptors at the surface of lymphocytes are taken up after attachment and aggregation of the antigen antibody complexes in the plane of the membrane ("capping"); after synthesis antibodies are secreted from lymphocytes, and they can pass the placenta. All these phenomena may not require the penetration of antibodies through membranes, however, since the transport of antibodies could proceed via pinocytosis or exocytosis vesicles respectively. The above assumption is corroborated by the agreement of the results on the localization by antibodies (RACKER et al., 1970) of components in the inner mitochondrial membrane with the results of other techniques (reconstitution and labeling with impermeable probes), and by studies on black lipid membranes (DEL CASTILLO et al., 1966), using albumines as antigens. For the discussion of a conflicting result, which was achieved by using insulin as antigen, and which was dependent on the addition of serum complement (BARFORT et al., 1968), see LAUF (1975). Recently WOBSCHALL and MCKEON (1975) confirmed that antibodies do not penetrate black lipid membranes; insulin, however, and at least one component of bovine serum, which they used as antigen, are able to penetrate and to react with antibodies added to the compartment at the other side of the bilayer.

On the basis of the above assumption a reaction of an antibody with a membrane, manifested as agglutination, adsorption or inhibition, allows the localization of the reacting component at the surface of the membrane (BERZBORN, 1967; MENKE, 1967). In the special case of the chloroplast thylakoid system, however, even a part of the outer surface may be inaccessible: the stacked regions in the grana (partitions) (BERZBORN, 1969a; SCHMID and RADUNZ, 1974), as is discussed in this vol. Chap. IV.2 by SANE.

3. Experimental Approaches

Information about the function of a component in a reaction sequence can be obtained by using an inhibiting antibody against the isolated component. Even after immunization with hydrophobic proteins, isolated in the presence of sodium-dodecylsulfate, antibodies may be found which crossreact with the protein in the membrane in its native conformation (MENKE et al., 1975). Sometimes it would be of interest to test whether partial inhibitions by different antisera are additive.

Antisera against a complex structure (e.g. the thylakoid system) or against unpurified antigen mixtures can also be useful; if they contain inhibiting antibodies, components may be isolated which neutralize the inhibition (REGITZ, 1970; REGITZ et al., 1970; REGITZ and OETTMEIER, 1972). The neutralizing component can release the inhibition by "bypassing" the site of inhibition, i.e. by introducing a new reaction sequence, or by blocking the binding site of the antibody.

4. Conclusions and Limitations

The following conclusions can be drawn from studying the effects of antibodies on membrane-associated reactions:

a) If specific direct inhibition occurs, the antigen or an extremely similar determinant group must have been accessible to the antibody (for the possibility of indirect inhibition see below).

b) If an antibody against a component isolated from a membrane inhibits a reaction sequence carried out by this membrane, the component probably is involved in the reaction.

c) If an antiserum against a complex particle or a membrane contains inhibiting antibodies against a reaction sequence carried out by the membrane, and if an isolated component neutralizes the inhibition *and* reacts with the antibody, the inhibition is probably due to the reaction of the antibody with the neutralizing component as a part of the membrane (surface).

d) If a stimulation of a reaction occurs by a mixture of antigens, and if this stimulation is absent after precipitation of the mixture with a monospecific antiserum, the precipitated antigen caused the stimulation.

e) If no inhibition occurs, it does not follow that the antigen is inaccessible, because accessibility is necessary but not sufficient for inhibition (Cinader, 1963; Berzborn et al., 1966; Koenig et al., 1972). No inhibition might also mean that the particular step in the reaction sequence was not made rate-limiting, because not enough antibody was added, or the time of preincubation was not long enough, or the salt concentration or the pH were not suitable for an antibody reaction; the presence of detergent might have inactivated the antibody, or in the special case of chloroplasts, the osmotic state of the thylakoids might have hindered access of antibodies, even to the stroma regions.

f) If no inhibition occurs of a membrane-associated reaction by an antiserum against a component isolated from the membrane, it does not follow that the component is not involved in the reaction sequence, even if a reaction of antibodies with the membrane surface can be documented, because not *all* precipitating antibodies are inhibiting.

At first glance 100% inhibition should only occur if all single reaction chains are saturated (Radunz, 1975) by antibodies, whereas agglutination could occur after only a few determinant groups at the surface of the thylakoid systems were connected. Inhibition, however, can be conceived to be caused by indirect action of the antibodies, e.g. by uncoupling, or changes in conformation of many membrane components by a cooperative type of action. In this case the reacting antigen and the inhibited enzyme need not be identical.

A biphasic titration curve indicates that the antigen was not rate-limiting before addition of antibodies.

The specificity of an inhibition (or stimulation) can be tested by comparing the effect of control sera or control gammaglobulin. For critical experiments the control serum from the same animal before immunization is used, since rabbits sometimes have antibodies reacting with the chloroplast membrane without being injected with chloroplast preparations (Berzborn, unpublished), indicating that these rabbits contain antibodies against an unknown substance (microorganism) crossreacting with chloroplast antigens. In particular this was observed, when envelope containing chloroplasts were used (Larsson, Berzborn and Hauska, unpublished).

In neutralization tests the order of addition is important: added ferredoxin-$NADP^+$ reductase will release the inhibition of an antiserum against the reductase on $NADP^+$-reduction; added CF_1 will not release the inhibition of antibodies against CF_1 on photophosphorylation, once they have reacted with membrane-bound CF_1. In this way a "bypass" type of inhibition neutralization may be uncovered.

When drawing conclusions concerning the site of enzyme reaction one has to keep in mind that the antigen might be an integral protein (SINGER and NICOLSON, 1972; ANDERSON, 1975) and span the membrane, and therefore could combine with an antibody at the outer surface, whereas the site of the inhibited reaction could be located on the other side of the membrane.

Finally, the complement system can interfere, both in quantitative precipitation and quantitative adsorption measurements, and due to uncoupling in measurements of photophosphorylation. To avoid uncertainties the serum can be decomplementized, or only the gammaglobulin fraction may be used.

5. Reproducibility

In addition to the problems mentioned above, the immunization of each animal leads to an individual antiserum, which is often unique. To grant as much reproducibility as possible, knowledge is needed of the immunization schedule used, of the immunogenicity of the antigen (i.e. the amount of antigen to be injected, and the number of responding animals/treated animals), and of the quantitative amount of antibody/ml of antiserum. The purity of the immunogen can be tested by any sensitive method, in particular by an immunoelectrophoretic analysis with an antiserum against the crude antigen, and the specificity of an antiserum is best tested with a crude antigen mixture.

Since not all precipitating antibodies are inhibiting, the amount (titer) of precipitating and inhibiting antibodies may vary independently. Using the same immunogen and immunization schedule, one animal produces more precipitating antibodies, the other more inhibiting antibodies (CINADER, 1963). If therefore a precipitin test shows only one arc if equivalent amounts concerning one antigen are allowed to react, other precipitating or inhibiting antibodies could be not at all equivalent to the corresponding antigens; because under these conditions no precipitation occurs, these antibodies would not be detectable. Thus precipitation, inhibition and neutralization should be titrated over a wide range. Agglutination reactions also exhibit a zone phenomenon, i.e. a negative reaction could be due to a large excess of either antigen or antibody.

C. Results and Conclusions from Experiments with Antisera Against Individual Chloroplast Antigens

When J.D. SPIKES in a discussion in 1959 stated that antibodies to chloroplasts would agglutinate the chloroplasts but had "no effect on the Hill reaction", he asked, "... if you add this antibody to your particular material, what effect would this have on surface reactions?". The groups of MENKE and RACKER started independently to use antibodies for investigations on chloroplast structure and function. The basis of the method was surveyed in Sect. B, the results and conclusions

Table 1. Effects of antisera or antibodies on photosynthetic reactions. Antisera listed according to antigens used for immunization. Photosynthetic reactions listed according to electron donors and acceptors used (ET, photosynthetic electron transport; ATP, photosynthetic ATP formation). Numbers given are maximal

Antigen	Ref.	H2O / NADP+ ET	H2O / NADP+ ATP	H2O / Fd/Cytc ET	H2O / Cytc ET	H2O / AQ/O2 ET	H2O / MV/O2 ET	H2O / MV/O2 ATP	H2O / FeCy ET	H2O / FeCy ATP	H2O / DCPIP ET	DCPIP/Asc NADP+ ET
TS	SPIKES (1959)	·	·	·	·	·	·	·	0	·	·	·
TS	BERZBORN et al. (1966) slow IgG	90	·	·	·	·	·	·	24[a]	67	·	·
TS	fast IgG	70	·	·	·	·	·	·	0	4	·	·
TS	REGITZ et al. (1970) type 11 C_{10}	71[c]	·	50	39	68	·	·	0	·	·	56[d]
TS	type 6 C_{14}	85	·	81	0	2	·	·	4	·	·	71[d]
Red	KEISTER et al. (1962)	85	90	0	·	·	·	·	·	·	5[h]	·
Red	BERZBORN (1967, 68, 69a, b)	97[b]	95[b]	·	·	·	·	·	0[b]	0[b]	·	·
Red	BOTHE and BERZBORN (1970)	97	·	·	·	·	·	·	·	·	·	·
Red	SCHMID and RADUNZ (1974)	·	·	·	·	·	·	st	·	·	·	100[i]
Red	FORTI and ZANETTI (1969)	·	·	·	·	·	·	·	·	·	·	·
Fd	TEL-OR et al. (1973)	90	·	62	5	·	·	·	·	·	·	·
Fd	TEL-OR and AVRON (1974)	·	·	·	·	·	·	·	·	·	·	·
PS I	BENGIS and NELSON (1975)	78	*	·	·	·	0	·	·	·	·	·
Fr.1	KOENIG et al. (1972)	·	·	·	·	·	·	st	·	0	·	·
Fr.2	KOENIG et al. (1972)	·	·	·	·	40	50[i]	57	·	50	·	·
Fr.3	KOENIG et al. (1972)	·	·	·	·	·	·	st	17	st	·	·
1 ex	KOENIG et al. (1972)	·	·	·	·	·	·	st	·	13	·	·
66 K	MENKE et al. (1975)	·	·	·	·	·	·	22	·	0	·	·
62 K	MENKE et al. (1975)	65[a]	·	·	·	·	·	65	·	66	·	·
33 K	MENKE et al. (1975)	·	·	·	·	·	·	0	·	0	·	·
24 K	MENKE et al. (1975)	16	·	·	·	·	16	0	·	3	·	·
PC	HAUSKA et al. (1971)	0(77[m])	0(62[m])	·	·	DCPIP/ Asc Aq/O2 ET	·	·	0[m]	50[m]	·	·
PC	SCHMID et al. (1975)	45	·	·	·		·	st	·	·	·	60
552	WILDNER and HAUSKA (1974)	0	·	·	·	ET	·	·	·	·	·	·
Cyt f	RACKER et al. (1972)	·	·	·	·		·	70[m]	·	60[m]	·	·
PQ	RADUNZ and SCHMID (1973)	·	·	·	·	st	·	·	*	0	·	0
Lut	RADUNZ and SCHMID (1973)	·	·	·	·	0	·	·	65[p]	st	·	·
Chl	RADUNZ et al. (1971)	*	·	·	·	0[l], 0[k]	·	·	22	0	·	0
TS	RADUNZ et al. (1971)	·	·	·	·	40[l, r]	·	·	·	·	·	·
PS II	BRAUN and GOVINDJEE (1972)	12	·	·	·	·	·	·	·	·	15	0
Ft ex	BRAUN and GOVINDJEE (1974)	14	·	·	·	·	·	·	·	·	17, 27[s]	1
CF$_1$	McCARTY and RACKER (1966)	·	·	·	·	·	·	·	0	60	·	·
CF$_1$	GOULD (1975)	·	·	·	·	·	30[a]	75	·	·	·	·

ublished values of inhibition (in % of control), taken partly from authors' tables and partly calculated ·om figures. (Explanation and Footnotes see p. 290)

AD/ sc ADP+	DCPIP/Asc MV/O_2		DAD+ TMPD/Asc MV/O_2		NH_2OH FeCy	DPC FeCy	DCPIP/Asc O_2	Endog. aerobic	H_2O Fd/O_2	Cyclic Fd	Cyclic PMS	Agglutination	Adsorption
T	ET	ATP	ET	ATP	ET	ET	ET	ATP	ATP	ATP	ATP		
·	·	·	·	·	·	·	·	·	·	·	·	+	
·[b]	·	·	·	·	·	·	·	·	·	·	80	+[b]	
7[b]	·	·	·	·	·	·	·	·	·	·	·	·	·
5[c,e]	·	·	3[c,e]	·	·	·	50[c], 74[g]	·	·	·	·	+[b]	
7[f]	·	·	4[c,e]	·	·	·	63[g]	·	·	·	·	+[b]	
·	·	·	·	·	·	·	·	·	·	·	15	·	·
·	·	·	·	·	·	·	·	·	·	·	·	−	+
·	·	·	·	·	·	·	·	·	·	3	·	·	·
·	·	·	·	·	·	·	·	·	·	st	st	+[q]	++[q]
·	·	·	·	·	·	·	·	58	·	·	·	·	·
·	·	·	·	·	·	·	·	·	·	·	·	·	·
·	·	·	·	·	·	·	·	60	40	70	8	·	·
·	·	·	·	·	·	·	·	·	·	·	·	+	
·	·	·	·	·	·	·	·	·	·	·	0	−	+
·	0[k], 0[l]	·	·	·	·	·	·	·	·	·	0	+	
·	·	·	·	·	21	0	·	·	·	·	st	+	
·	100[k], 25[l]	·	·	·	·	·	·	·	·	·	27	−	+
·	25	·	·	·	·	·	·	·	·	·	7	+	
·	·	·	·	·	·	·	·	·	·	·	87	+	
·	40	·	·	·	·	·	·	·	·	·	30	+	
·	·	·	·	·	·	·	·	·	·	Cyclic K_3	0	+	
·	40(0[n])	st	20	21	·	·	·	·	·	54	0(54[m])	−	−
·	·	·	·	·	·	·	·	·	·	·	0(st[i])	+[q]	+[q]
·	·	·	·	·	·	·	·	·	·	·	0	−	−
·	·	·	·	·	·	·	·	·	·	·	0(80[m])	−	−
·	·	·	·	·	·	60	·	·	·	·	st[q], 19[q]	−	+
·	·	·	·	·	st	st	·	·	·	·	st[q], 0[q]	−	+
[k], 0[l]	·	·	·	·	·	·	·	·	·	·	st	+[q]	
3[k], st[l]	·	·	·	·	$MnCl_2$ DCPIP ET	DPC DCPIP ET	·	·	H_2O DAD ox ATP	·	·	·	·
·	·	·	·	·	13	0	·	·	·	·	·	·	·
·	·	·	·	·	17	3	·	·	·	·	·	·	·
·	·	·	·	·	·	·	·	·	·	·	70	·	·
·	·	·	·	85[t]	·	·	·	·	75	·	·	·	·

have been reviewed (Trebst, 1974; Anderson, 1975; Arntzen and Briantais, 1975). Conclusions relevant to thylakoid structure and topography are dealt with in this vol. Chaps. IV.1 and IV.2. Therefore our discussion of the effects of antisera on photosynthetic reactions can be concise (Table 1).

Fd-NADP⁺-Reductase. Work with antibodies to a transhydrogenase led to the recognition of the function of this flavin enzyme as $NADP^+$ reductase (Keister et al., 1962). Then, using the same antibody, it was found that the reductase would oxidize ferredoxin (Avron and Chance, 1966), i.e. that it functions as a ferredoxin-$NADP^+$ oxidoreductase. With the help of antibodies the chloroplast-specific diaphorase activity could be distinguished from other diaphorases (Nelson and Neumann, 1969). San Pietro (1963) discussed whether besides the soluble reductase there were also a bound reductase in the chloroplast membrane; this was proved to be the case by studying the agglutination of thylakoid systems by antibodies against the reductase (Berzborn, 1968, 1969a, b). On the other

---

Explanation of Table 1

Antigens:
TS: washed thylakoid systems of chloroplasts
Red: ferredoxin-$NADP^+$ reductase
Fd: ferredoxin
PS I: photosystem I reaction center preparation
Fr. 1⎫
Fr. 2⎬   proteid fractions after cholate solubilization of thylakoids and separation on columns
Fr. 3⎭
1 ex: fraction 1 as above, lipids extracted
66 K⎫
62 K⎪   polypeptids after solubilization of thylakoids with sodiumdodecylsulfate
33 K⎪   and separation in the presence of this detergent
24 K⎭
PC: plastocyanin
552:       cytochrome 552 from *Euglena*
Cyt f: cytochrome f from *Spinach*
PQ: plastoquinone
Lut: lutein
Chl: chlorophyll a
PS II: photosystem II particles according to Huzisige
FT ex: extract related to photosystem II according to Black, obtained by freezing and thawing thylakoids
CF_1:   chloroplast coupling factor 1 (ATP-synthetase of photophosphorylation)

Footnotes to Table 1

· no results published
* positive inhibition mentioned in text only
st rate stimulated
[a] due to photosynthetic control
[b] data from Berzborn (1967)
[c] data from Regitz (1970)
[d] data from Yocum and San Pietro (1970)
[e] donor: TMPD/Asc
[f] value from Regitz and Oettmeier (1972)
[g] data from Honeycutt and Krogmann (1970)
[h] acceptor: TCPIP

[i] chloroplast from aurea mutant
[k] tested in sonication supernatant
[l] tested in sonication sediment
[m] sonicated in the presence of antiserum
[n] after sonication
[p] if thylakoid systems are tightly coupled
[q] dependent on chloroplasts used
[r] inhibition of initial rate only
[s] tested with PS II particles
[t] donor: DAD/Asc

hand, it was concluded from the absence of inhibition by antisera that the reductase is not involved in electron transport from H_2O to pyridine analogs (BÖGER et al., 1966) or in cyclic, ferredoxin-dependent photophosphorylation (BOTHE and BERZBORN, 1970; FREDERICKS, 1968). ATP formation without added mediator seems to proceed via reductase (FORTI and ZANETTI, 1969; FORTI and ROSA, 1972). Whether thylakoid systems are agglutinated by antibodies against the reductase directly or only indirectly (BERZBORN, 1969b) — after addition of anti-gammaglobulin (Coombs test; COOMBS et al., 1951) or soluble reductase (mixed antigen agglutination; UHLENBRUCK and PROKOP, 1967) — depends on the chloroplast preparation used (BERZBORN, 1969a; SCHMID and RADUNZ, 1974; HIEDEMANN-VAN WYK and KANNANGARA, 1971); also a weak agglutination could have been interpreted as negative (BERZBORN, 1969a; HIEDEMANN-VAN WYK and KANNANGARA, 1971) or positive (SCHMID and RADUNZ, 1974). However, since maximal inhibition of $NADP^+$ reduction in spinach chloroplasts (KEISTER et al., 1962; BERZBORN, 1967) was more than 90%, in other systems (SCHMID and RADUNZ, 1974) only about 60%, the question (HEBER, 1967; BERZBORN, 1969a), of whether in general all sites of $NADP^+$ reduction are accessible to antibodies or not, and whether $NADP^+$ reduction occurs also in the partition (SCHMID and RADUNZ, 1974) or at the stroma membranes only (cf. this vol. Chap. IV.2) has to be investigated.

Ferredoxin. A bound ferredoxin is reported (TEL-OR and AVRON, 1974) to be involved in ATP formation by isolated thylakoids without mediator added, and a complex formation of ferredoxin with membrane-bound reductase was deduced from agglutination experiments (HIEDEMANN-VAN WYK and KANNANGARA, 1971). Unfortunately ferredoxin is a very bad immunogen (TEL-OR et al., 1973; HIEDEMANN-VAN WYK and KANNANGARA, 1971).

Thylakoid Systems. If gammaglobulin from an antiserum against thylakoid systems was separated into a lower charged (slow) and a higher charged (fast) fraction, inhibiting antibodies against CF_1 were found in the slow IgG and against reductase in the fast fraction (BERZBORN, 1967; BERZBORN et al., 1966). In the slow fraction also antibodies were found which seem to block the access of ferredoxin to the reducing site of photosystem I. It was possible to isolate (REGITZ, 1970; REGITZ et al., 1970) components from the thylakoid which neutralized the inhibiting effect of similar antisera on photoreductions by photosystem I. The mixture of neutralizing antigens (S_{L-eth}) could be separated into a low molecular weight, heat-stable fraction with redox properties, and a protein (REGITZ and OETTMEIER, 1972) without detectable activity but still neutralizing certain inhibiting antibodies. The interchangeability of inhibiting antisera and stimulating or neutralizing substances within several systems (REGITZ et al., 1970; HONEYCUTT and KROGMANN, 1970; YOCUM and SAN PIETRO, 1970) led to the suggestion that all these substances are somehow related to the primary electron acceptor of photosystem I (for a detailed discussion see TREBST, 1974).

Photosystem I. Antisera against an enriched photosystem I reaction center preparation agglutinated thylakoid systems (RACKER et al., 1972; BRIANTAIS and PICAUD, 1972), as did antisera against a highly purified photosystem I reaction center preparation (BENGIS and NELSON, 1975). The antisera against the enriched preparation inhibited (RACKER et al., 1972) plastocyanin-dependent photosynthetic electron transport from ascorbate to methylviologen/O_2; the serum against the purified reaction center did not interfere with methylviologen reduction, but inhibited NADP reduction (Bengis and Nelson, 1975). This type of inhibition is usually seen with sera against the thylakoid system (Regitz, 1970; Regitz et al., 1970; Radunz et al., 1971). Some of these antisera (e.g. type $11C_{10}$) also inhibit the photosynthetic reduction of anthraquinone sulfonate (Regitz et al., 1970; Radunz et al., 1971). It follows, according to the possibilities for conclusions deduced above, that the reducing site of photosystem I is susceptible to antibody inhibition, i.e. very probably directed against the stroma side of the thylakoid membrane.

Antibodies which also inhibit methylviologen reduction could be obtained if a lipid extracted fraction 1 after cholate solubilization of thylakoids was injected (Koenig et al., 1972) or a 66 K or a 33 K peptide after separation in the presence of sodiumdodecylsulfate (Menke et al., 1975). The antiserum against the extracted fraction 1 only inhibited after the thylakoids were opened by sonication (Koenig et al., 1972). On the other hand, the cholate treatment seems to have altered some properties of the membranes' antigens: an antiserum (Kannangara et al., 1970) against CF_1 precipitated (Koenig et al., 1972) the cholate fraction 1; this means that this fraction contained crossreacting determinant groups; but these groups were no longer immunogenic, because the antiserum against fraction 1 neither agglutinated chloroplasts nor inhibited PMS-mediated cyclic photophosphorylation (Koenig et al., 1972).

Plastocyanin and Cytochromes. The results and in particular the conclusions concerning the accessibility of plastocyanin are contradictory (Hauska et al., 1971; Schmid et al., 1975). Again, agglutination is very sensitive and only a semi-quantitative measurement, and a few broken thykaloids or adsorbed plastocyanin molecules may have caused a positive reaction in one case which was weak and interpreted as negative in the other group. But it cannot be explained why Schmid et al. (1975) find inhibition in $NADP^+$ reduction without sonication in contrast to Hauska et al. (1971), and why cyclic photophosphorylation with menadion was inhibited, whereas PMS mediated was not.

The soluble cytochrome 552, a substitute for plastocyanin in *Euglena,* seems to be inaccessible to antibodies (Wildner and Hauska, 1974); inhibition could only be achieved in the presence of cholate, and no agglutination or adsorption occurred. The agglutination and inhibition experiments with an antiserum against cytochrome f (Racker et al., 1972) corroborated the hypothesis of the inaccessibility of the oxidizing site of photosystem I. Inhibition occurred only if the thylakoids were sonicated in the presence of the antiserum. The inhibition of photophosphorylation mediated by PMS could be bypassed by plastocyanin added before sonication, which shows that internal plastocyanin can be oxidized by P-700 without the need for a functional cytochrome f (cf. Hauska, this vol. Chap. II.15) and Haehnel, 1973, for a discussion of the sequence of electron carriers on the oxidizing site of photosystem I).

Plastoquinone and Carotenoids. Inhibition of plastoquinone (Radunz and Schmid, 1973) showed that at least some plastoquinone is accessible. Antisera to lutein (Radunz and Schmid, 1973) and neoxanthin (Radunz and Schmid, 1975) inhibit photosynthetic H_2O oxidation but not DPC oxidation; the authors suggested that lutein and neoxanthin might feed electrons into photosystem II before the DPC site. Since the inhibition was often much less in the presence of methylamine, which besides uncoupling causes increased stacking (IZAWA and GOOD, 1966), the inhibition might either depend on the conformational state of the thylakoid under coupled conditions (RADUNZ and SCHMID, 1973; TREBST, 1974; BRAUN and GOVINDJEE, 1974) or on the degree of accessibility of partition surface (RADUNZ and SCHMID, 1973; TREBST, 1974).

Photosystem II. Agglutination experiments with antisera against photosystem II-enriched particles (BRIANTAIS and PICAUD, 1972) suggested that photosystem II is not well accessible, and is either buried in the membrane or hidden in the partition.

Inhibitions of photosystem II are consistently low, but several antisera with this property are described: against chlorophyll (RADUNZ, 1972a; RADUNZ et al., 1971; SCHMID, 1972), against a proteid fraction (fraction 3 after cholate treatment of thylakoids), (KOENIG et al., 1972); an antiserum (BRAUN and GOVINDJEE, 1972) against a photosystem II particle prepared according to HUZISIGE et al. (1969), and against an extract obtained by freezing and thawing thylakoids (BRAUN and GOVINDJEE, 1974). This extract was described as stimulating electron transport in Tris-treated chloroplasts (BLACK, 1968). These inhibitions are significant, but could be due to an effect on the conformation of the membrane (BRAUN and GOVINDJEE, 1974).

"Structural Protein" and Lipids. Antisera (BERZBORN, 1967; HIEDEMANN-VAN WYK, 1971) against thylakoid proteids, solubilized by formic acid (WEBER, 1963; MENKE and RUPPEL, 1971), did not inhibit Hill reactions or photophosphorylation (TREBST and BERZBORN, unpublished). No data on inhibition have been published in the case of antisera against the chloroplast sulfolipid (RADUNZ and BERZBORN, 1970), phosphatidyl glycerol (RADUNZ, 1971), monogalactosyldiglycerid (RADUNZ, 1972b) and a glycolipopeptid (HIEDEMANN-VAN WYK, 1971); all these antisera contain antibodies which agglutinate thylakoid systems directly or, in the case of sulfolipid and phosphatidyl glycerol, are adsorbed to determinant groups located towards the surface.

Coupling Factor. Finally the possibility of an inhibition of electron transport by antibodies against CF_1 due to photosynthetic control has to be considered. The inhibition of photophosphorylation by an antibody could be caused by uncoupling, i.e. no ATP is made because H^+ efflux is accelerated, ΔpH decreased and thus electron transport accelerated; or the antibody could act as an energy transfer inhibitor, i.e. no ATP is made due to a fixation of the CF_1 in a conformation with the "H^+ hole" closed, so that H^+ efflux is slowed down, ΔpH at a maximum and thus electron transport inhibited (PORTIS et al., 1975). The antibodies could also leave H^+ fluxes and electron transport unchanged. All possibilities have been observed: rabbit antisera against CF_1 (MCCARTY et al., 1971) or against the α-sub-unit of CF_1 (NELSON et al., 1973) did inhibit the increase of ΔpH by ATP, which is probably due to a closing of an H^+ hole in the ATP synthetase by ATP; the serum against CF_1 even accelerated H^+ efflux in the absence of ATP (MCCARTY et al., 1971). All other antisera tested did not uncouple, however, including an antiserum against mitochondrial coupling factor F_1 (SCHATZ et al., 1967). A mouse antiserum against CF_1 had no effect on the extent of ΔpH or H^+ efflux and did not inhibit or stimulate electron transport (McCarty and Racker, 1966). Some antisera, however, do inhibit photosynthetic electron transport (Berzborn et al., 1966; Menke et al., 1975; Gould, 1975) and this inhibition is released by preincubation with a CF_1 preparation (Berzborn et al., 1966) or by uncouplers (Menke et al., 1975; Gould, 1975). [If maximal rates of electron transport are desired, and an influence of antibodies against CF_1 in antisera against thylakoid systems on electron transport, which is due to photosynthetic control, is to be ruled out, uncoupler can be added (Regitz and Oettmeier, 1972).] In accordance with the finding that the inhibition of electron transport by antibodies against CF_1 was due to an indirect influence via photosynthetic control, an antiserum against CF_1 inhibited the acceleration of the decay of membrane potential by ADP + phosphate, as measured as 520 nm absorption change (Schmid et al., 1976). Fab fragments of the antibody did not give this effect, indicating that the antibody probably inhibited by preventing the conformational change in the ATP synthetase. No studies on inhibition of electron transport by antisera against CF_1-subunits have been published. (For a discussion of antibody influence on ion fluxes in the case of $Na^+ K^+$ ATPase see Lauf, 1975.)

D. Summary and Outlook

The studies on effects of antibodies upon photosynthetic electron transport, combined with precipitation, agglutination, adsorption, and neutralization experiments, show clearly that the reducing site of photosystem I is exposed at the stroma-facing side of the thylakoid; the oxidizing site is probably inaccessible, but the immunological data supporting this are questioned. The effects of antisera on photosystem II might be explained by an influence of the antibodies on integral proteins or more indirectly on the conformation of the membrane and its constituents; if this can be proved not to be the case, it would follow that some of the oxidizing sites of photosystem II are accessible, in contrast to the localization of this site by other methods (cf. this vol. Chap. II.1.b). On the other hand, it follows from the antibody experiments that most of photosystem II is inaccessible, buried either in the membrane or in the partition.

The specificity and the macromolecular nature of antibodies, and the possibility to obtain them against nearly all substances of interest to an investigator of photosynthetic membranes, justify the hope that despite all the problems mentioned, their use will yield still more information about the correlation of structure with function in these multicomponent, biological energy transducers.

References

Anderson, J.M.: Biochim. Biophys. Acta **416**, 191–235 (1975)
Arnon, R.: Current Topics Microbiol. Immunol. **54**, 47–93 (1971)
Arntzen, C.J., Briantais, J.M.: In: Bioenergetics of Photosynthesis. Govindjee (ed.). New York-San Francisco-London: Academic Press, 1975, pp. 51–113

Avron, M., Chance, B.: Brookhaven Symp. Biol. **19**, 149–160 (1966)
Barfort, P., Arquilla, E.R., Vogelhut, P.O.: Science **160**, 1119–1121 (1968)
Bengis, C., Nelson, N.: J. Biol. Chem. **250**, 2783–2788 (1975)
Berzborn, R.J.: Thesis, Cologne, 1967
Berzborn, R.J.: Z. Naturforsch. **23B**, 1096–1104 (1968)
Berzborn, R.J.: Z. Naturforsch. **24B**, 436–446 (1969a)
Berzborn, R.J.: In: Progress in Photosynthesis Research. Metzner, H. (ed.). Tübingen: Laupp, 1969, 1969b, Vol. I, 106–114
Berzborn, R.J.: Hoppe-Seyler's Z. Physiol. Chem. **353**, 693 (1972)
Berzborn, R.J., Menke, W., Trebst, A., Pistorius, E.: Z. Naturforsch. **21B**, 1057–1059 (1966)
Black, C.C.: Plant Physiol. **43**, S-13 (1968)
Böger, P., Black, C.C., San Pietro, A.: Arch. Biochem. Biophys. **115**, 35–43 (1966)
Bothe, H., Berzborn, R.J.: Z. Naturforsch. **25B**, 529–534 (1970)
Braun, B.Z., Govindjee: FEBS Lett. **25**, 143–146 (1972)
Braun, B.Z., Govindjee: Plant Sci. Lett. **3**, 219–227 (1974)
Briantais, J.-M., Picaud, M.: FEBS Lett. **20**, 100–104 (1972)
Cinader, B.: Ann. N.Y. Acad. Sci. **103**, 495–548 (1963)
Coombs, R.R.A., Gleeson-White, M.H., Hall, J.C.: Brit. J. Exp. Path. **32**, 195–202 (1951)
Del Castillo, J., Rogriguez, A., Romero, C.A., Sanchez, V.: Science **153**, 185–188 (1966)
Edelman, G.M.: Angew. Chem. **85**, 1083–1096 (1973)
Forti, G., Rosa, L.: In: Proc. 2nd Intern. Congr. Photosynthesis. Forti, G., Avron, M., Melandri, A. (eds.). The Hague: Dr. Junk, 1972, Vol. II, pp. 1261–1270
Forti, G., Zanetti, G.: In: Progress in Photosynthesis Research. Metzner, H. (ed.). Tübingen: Laupp, 1969, Vol. III, 1213–1216
Fredericks, W.W.: Biochem. Biophys. Res. Commun. **31**, 582–587 (1968)
Gould, J.M.: Biochem. Biophys. Res. Commun. **64**, 673–680 (1975)
Haehnel, W.: Biochim. Biophys. Acta **305**, 618–631 (1973)
Hauska, G.A., McCarty, R.E., Berzborn, R.J., Racker, E.: J. Biol. Chem. **246**, 3524–3531 (1971)
Heber, U.: Arbeitsgemeinsch. Forsch. Landes NRW **171**, 47 (1967)
Hiedemann-Van Wyk, D.: Z. Naturforsch. **26B**, 1052–1054 (1971)
Hiedemann-Van Wyk, D., Kannangara, C.G.: Z. Naturforsch. **26B**, 46–50 (1971)
Honeycutt, R.C., Krogmann, D.W.: Biochim. Biophys. Acta **197**, 267–275 (1970)
Huzisige, H., Usiyama, H., Kikuti, T., Azi, T.: Plant Cell Physiol. **10**, 441–455 (1969)
Izawa, S., Good, N.E.: Plant Physiol. **41**, 544–552 (1966)
Kabat, E.A., Mayer, M.M.: Experimental Immunochemistry, 2nd Ed. Springfield, Illinois: Charles C. Thomas, 1967
Kannangara, C.G., Van Wyk, D., Menke, W.: Z. Naturforsch. **25B**, 613–618 (1970)
Keister, D.L., San Pietro, A., Stolzenbach, F.E.: Arch. Biochem. Biophys. **98**, 235–244 (1962)
Koenig, F., Menke, W., Craubner, H., Schmid, G.H., Radunz, A.: Z. Naturforsch. **27B**, 1225–1238 (1972)
Landsteiner, K.: The Specificity of Serological Reactions, 2nd Ed. Cambridge, Mass.: Harvard Univ. 1945. Reprinted: New York: Dover, 1962
Lauf, P.K.: Biochim. Biophys. Acta **415**, 173–229 (1975)
McCarty, R.E., Fuhrman, J.S., Tsuchiya, Y.: Proc. Nat. Acad. Sci. U.S. **68**, 2522–2526 (1971)
McCarty, R.E., Racker, E.: Brookhaven Symp. Biol. **19**, 202–214 (1966)
Melchers, F., Messer, W.: Europ. J. Biochem. **35**, 380–385 (1973)
Menke, W.: Arbeitsgemeinsch. Forsch. Landes NRW **171**, 1–26 (1967)
Menke, W., Koenig, F., Radunz, A., Schmid, G.H.: FEBS Lett. **49**, 372–375 (1975)
Menke, W., Ruppel, H.-G.: Z. Naturforsch. **26B**, 825–831 (1971)
Nelson, N., Deters, D.W., Nelson, H., Racker, E.: J. Biol. Chem. **248**, 2049–2055 (1973)
Nelson, N., Neumann, J.: J. Biol. Chem. **244**, 1926–1931 (1969)
Porter, R.R.: Angew. Chem. **85**, 1097–1101 (1973)
Portis, A.R., Magnusson, R.P., McCarty, R.E.: Biochem. Biophys. Res. Commun. **64**, 877–884 (1975)
Racker, E., Burstein, C., Loyter, A., Christiansen, R.O.: In: Electron Transport and Energy Conservation. Tager, J.M., Papa, S., Quagliariello, E., Slater, E.C. (eds.). Bari: Adriatica Editrice, 1970, pp. 235–252

Racker, E., Hauska, G.A., Lien, S., Berzborn, R.J., Nelson, N.: In: Proc. 2nd Intern. Congr. Photosynthesis. Forti, G., Avron, M., Melandri, A. (eds.). The Hague: Dr. Junk, 1972, Vol. II, pp. 1097–1113

Radunz, A.: In: Proc. 2nd Intern. Congr. Photosynthesis. Forti, G., Avron, M., Melandri, A. (eds.). The Hague: Dr. Junk, 1972a, Vol. II, pp. 1613–1618

Radunz, A.: Z. Naturforsch. **26B**, 916–919 (1971)

Radunz, A.: Z. Naturforsch. **27B**, 822–826 (1972b)

Radunz, A.: Z. Naturforsch. **30C**, 484–488 (1975)

Radunz, A., Berzborn, R.J.: Z. Naturforsch. **25B**, 412–419 (1970)

Radunz, A., Schmid, G.H.: Z. Naturforsch. **28C**, 36–44 (1973)

Radunz, A., Schmid, G.H.: Z. Naturforsch. **30C**, 622–627 (1975)

Radunz, A., Schmid, G.H., Menke, W.: Z. Naturforsch. **26B**, 435–446 (1971)

Regitz, G.: Thesis, Göttingen, 1970

Regitz, G., Berzborn, R.J., Trebst, A.: Planta (Berl.) **91**, 8–17 (1970)

Regitz, G., Oettmeier, W.: In: Proc. 2nd Intern. Congr. Photosynthesis. Forti, G., Avron, M., Melandri, A. (eds.). The Hague: Dr. Junk, 1972, Vol. I, pp. 499–506

Ryrie, I.J., Jagendorf, A.T.: J. Biol. Chem. **247**, 4453–4459 (1972)

San Pietro, A.: Ann. N.Y. Acad. Sci. **103**, 1093–1105 (1963)

Schatz, G., Penefsky, H.S., Racker, E.: J. Biol. Chem. **242**, 2552–2560 (1967)

Schmid, G.H.: In: Proc. 2nd Intern. Congr. Photosynthesis. Forti, G., Avron, M., Melandri, A. (eds.). The Hague: Dr. Junk, 1972, Vol. II, pp. 1603–1611

Schmid, G.H., Radunz, A.: Z. Naturforsch. **29C**, 384–391 (1974)

Schmid, G.H., Radunz, A., Menke, W.: Z. Naturforsch. **30C**, 201–212 (1975)

Schmid, R., Shavit, N., Junge, W.: Biochim. Biophys. Acta **430**, 145–153 (1976)

Singer, S.J., Nicolson, G.L.: Science **175**, 720–731 (1972)

Smith, T.W., Skubitz, K.M.: Biochemistry **14**, 1496–1502 (1975)

Spikes, J.D.: Brookhaven Symp. Biol. **11**, 294 (1959)

Tel-Or, E., Avron, M.: Europ. J. Biochem. **47**, 417–421 (1974)

Tel-Or, E., Fuchs, S., Avron, M.: FEBS Lett. **29**, 156–158 (1973)

Trebst, A.: Ann. Rev. Plant Physiol. **25**, 423–458 (1974)

Uhlenbruck, G., Prokop, O.: Deut. Med. Wochenschr. **92**, 940–945 (1967)

Valentine, R.C., Green, N.M.: J. Mol. Biol. **27**, 615–617 (1967)

Weber, P.: Z. Naturforsch. **18B**, 1105–1110 (1963)

Wildner, G.F., Hauska, G.: Arch. Biochem. Biophys. **164**, 136–144 (1974)

Wobschall, D., McKeon, C.: Biochim. Biophys. Acta **413**, 317–321 (1975)

Yocum, C.S., San Pietro, A.: Arch. Biochem. Biophys. **140**, 152–157 (1970)

18. Chemical Modification of Chloroplast Membranes

R. Giaquinta and R.A. Dilley

A. Introduction

Elucidation of the spatial arrangement of photochemical and biochemical functions in photosynthetic membranes is central to understanding the processes of electron transport, ion movements, and energy transduction. Chemical modification of specific membrane groups provides a new approach to the study of the molecular architecture of chloroplast membranes, finding application as (1) a covalently bound marker for the membrane functions, (2) a means of probing the role of specific chemical groups in relation to function, and (3) as a probe to study conformational changes associated with energy transduction.

The use of chemical probes in membrane studies is well documented for animal cells, particularly for the erythrocyte (cf. WALLACH, 1972), but has only recently been employed to study photosynthetic membrane topography. Discussed below are several recent studies using selective chemical group reagents as probes for structure–function relationships in chloroplast membranes.

B. N-ethylmaleimide (NEM)

McCARTY and co-workers (1972, 1973) have recently characterized a light-activated inhibition of photophosphorylation by NEM, a sulfhydryl specific alkylating reagent. The inhibition site was at the coupling factor, since the soluble calcium-dependent ATPase derived from NEM-treated membranes (+light) showed marked inhibition. Subsequent studies (McCARTY and FAGAN, 1973), using ^{3}H-NEM, revealed that light markedly enhanced ^{3}H-NEM incorporation into the γ-subunit of the coupling factor. The inhibition of ATP formation with concomitant incorporation of NEM into CF_1 protein seemed directly related to a high energy state (proton accumulation), since uncouplers prevented both the inhibition and NEM incorporation. These studies indicated that the coupling factor (CF_1) undergoes a light-dependent conformational change which exposes NEM-reactive groups in the γ-subunit, allowing reaction with NEM and subsequent inhibition of phosphorylation. Similar results were also noted for another lipophilic sulfhydryl group reagent, DNTB, 5,5'-dithio(bis-2-nitrobenzoic acid).

The energy transfer-type inhibition caused by the mercurials, p-chloromercuribenzoate and mercury (IZAWA and GOOD, 1969), are also in accord with the essentiality of sulfhydryl groups in the terminal stages of energy transduction. The observation that the uncoupled rate of electron transfer was insensitive to

NEM treatment in light (McCarty et al., 1972; Giaquinta et al., 1973) [even though there is increased exposure of sulfhydryl groups during illumination (Hirose et al., 1971)] suggests that either sulfhydryl groups are not involved in electron transport, or that the various sulfhydryl reagents tested were not accessible to sulfhydryl groups involved in electron transfer.

C. Carbodiimides

EDAC [1-ethyl-3-(3-dimethylaminopropyl)-carbodiimide];
CDIS [1-cyclohexyl-3-(2-morpholino-4-ethyl)-carbodiimide metho-p-toluene sulfonate];
and DCCD [dicyclohexyl carbodiimide]

I. Divalent Cation Binding

Carboxyl group modification has recently been employed to study the relationship between calcium binding, membrane structural changes, and chlorophyll a fluorescence. Prochaska and Gross (1975), using three different carbodiimides, demonstrated that carboxyl group modification decreased calcium binding, chlorophyll a fluorescence, and structural changes monitored by light scattering changes ($\Delta A540$). The reaction involved activating the carboxyl groups with either the water soluble (EDAC or CDIS) or lipid soluble (DCCD) carbodiimides followed by nucleophilic attack by the amine group of glycine ethyl ester. Upon amidation the negatively charged carboxyl group is modified and the potential calcium binding site masked. Carbodiimide-detected conformational changes have also been proposed by McCarty (1974) and Giaquinta et al. (1973) (see below).

II. Membrane Association and Polycation Inhibition

Chemical modification of membrane surface localized carboxyl groups by the water soluble carbodiimide, CDIS, was used by Berg et al. (1974) to study the role of carboxyl groups in grana stacking and polycation inhibition of photosystem I. Negatively charged carboxyl groups were activated by CDIS followed by nucleophilic substitution by glycine methyl ester, leaving the derivatized group uncharged. The charge neutralization caused restacking of low salt treated membranes similar to that caused by divalent cation additions. Additionally, neutralizing the negatively charged carboxyl groups prevented both polycation inhibition of photosystem I electron transfer and ^{14}C-polycation binding to the membranes. These data indicate that polycation inhibition of low salt treated membranes is associated with electrostatic interactions at the membrane surface. Derivatization of carboxyl groups also altered the ability of chloroplast membranes to undergo conformational changes in response to acidification, presumably by preventing protonation of membrane negative fixed-charge groups.

III. Electron Transfer Inhibition

The water soluble carbodiimide, EDAC, was shown to inhibit electron transfer and phosphorylation when the chloroplasts were treated with the reagent in the light (McCarty, 1974). The inhibition was characterized at the level of electron transport rather than at phosphorylation, and the inhibition site was localized between plastoquinone and cytochrome f. Plastocyanin function was not inhibited by EDAC (but see Selman et al., 1974b). The light requirement of EDAC inhibition could be explained by a conformational change exposing an EDAC-sensitive component to the water soluble modifier. Another possibility proposed was that EDAC, an amine, could be accumulated in response to proton accumulation, thus effectively increasing its internal concentration and inhibiting component(s) on the internal membrane surface. McCarty (1974) noted that another water soluble carbodiimide, CDIS, was ineffective in causing the electron transport inhibition under light or dark conditions. This lack of sensitivity could be due to steric factors, (CDIS, unlike EDAC, having both cyclohexyl and morpholino groups attached to the molecule) or because CDIS is a quaternary amine and would not be accumulated in response to a proton gradient.

The insensitivity of photosystem I electron transport to CDIS in the light or dark was shown by Giaquinta et al. (1973). However, photosystem II electron transport was inhibited when the CDIS reaction was performed in the light. It should be noted, however, that the protocol for CDIS treatment used by Giaquinta et al. (1973) was different from that used by McCarty (1974) for EDAC. The carboxyl modification by CDIS in the former study was accomplished in the presence of the nucleophile, glycine methyl ester, and a relatively high CDIS concentration (20 mM compared to 0.2 mM EDAC). At CDIS concentrations less than 5 mM no inhibition of photosystem II was noted in either light or dark (unpublished results).

IV. Location of Plastocyanin

The carbodiimide-dependent incorporation of the nucleophile, ^{14}C-glycine ethyl ester into plastocyanin was used to study the localization of plastocyanin within the membrane. Selman et al. (1974b) showed a three- to four-fold increase in the amount of nucleophile covalently bound to plastocyanin derived from sonicated or triton disrupted membranes compared to intact membranes. Similar results were noted for the water soluble diazonium compound, DABS. Using the lipophilic, penetrating DCCD as the activating carbodiimide, this ratio approached unity.

In contrast to the plastocyanin insensitivity to EDAC treatment reported by McCarty (1974), Selman et al. (1974b) observed that plastocyanin activity (measured by the Plesnicar-Bendall assay) was markedly inhibited by CDIS treatment independent of the nucleophile. In addition, CDIS treatment of purified plastocyanin caused the protein to migrate as two bands in SDS gel electrophoresis. DABS treatment did not inhibit plastocyanin activity nor alter its migration profile, even though covalently bonded to the protein. These results are consistent with plastocyanin, a hydrophilic protein, being partially accessible to the external phase since larger ratios would be predicated for a protein completely buried (Schneider et al., 1972). Recent evidence by Schmid et al. (1975), using antibody to plastocya-nin, also supports its membrane surface location.

D. Lactoperoxidase-Catalyzed Iodination

Arntzen and co-workers (1974a, b) have recently used enzyme-catalyzed iodina-tion of chloroplast membranes to study the topographical distribution of photo-chemical functions. The basis of this technique is that the enzyme lactoperoxidase, in the presence of H_2O_2 and iodide, catalyzes the iododination of tyrosine and histidine residues of proteins. Under proper conditions the procedure can be specific for surface-localized membrane components, since the adduct is based on enzyme-substrate recognition and the enzyme is large (M.W. 78,000) (Arntzen et al., 1974a).

Iodination of membranes caused marked inhibition of ATP formation and of electron transport on the reducing side of photosystem I ($NADP^+$ reduction), consistent with the recognized surface localization of those functions. Photosys-tem II electron transport was inhibited and only partially restored by the alternate donor, diphenyl-carbohydrazide (DPC), indicating two sites of inhibition, one asso-ciated with water oxidation, the other after the DPC donation site. Further studies (Arntzen et al., 1974b) demonstrated that lower levels of iodination preferentially inhibited the photosystem II reaction center (quenching of photosystem II variable fluorescence), possibly by introducing iodide into the reaction center complex and thus adding a competing quencher molecule. Higher levels of iodination inhibited water oxidation activity, presumably due to lactoperoxidase-induced structural changes in the membrane since glutaraldehyde prefixation prevented water oxida-tion inhibition. These results led Arntzen et al. (1974b) to propose that photosys-tem II reaction centers are located on the external membrane surface, while the proteins associated with H_2O oxidation are not exposed to iodination, at least under non-electron transport conditions. Interestingly, the inhibition of the photo-system II reaction center only occurred when the lactoperoxidase reaction was performed on unstacked lamellae (low salt), suggesting that the reaction center may be located in the grana partition regions and thus not accessible to the large enzyme molecule. A similar protection of photosystem II activity by grana stacking conditions (high salt or sucrose) has been reported previously for trypsin (see below).

E. Trypsin

Trypsin treatment of chloroplast membranes has been used as a probe for mem-brane structure/function relationships (Selman et al., 1971, 1973, 1974). Because of its molecular weight (24,000) and water solubility, the hydrolysis of peptide bonds is apparently limited to the external surface. Detailed studies on the effect of trypsin on chloroplast membranes revealed that trypsin inhibits photosystem II electron transport at two sites on the oxidizing side of photosystem II. The more sensitive site appeared to be at the level of water oxidation (SELMAN and BANNISTER, 1971). The inhibition of photosystem II was not caused by a reduction in the manganese content nor by gross disruption of membranes as visualized by electron microscopy. When membranes were in the stacked condition, trypsin was not effective in inhibiting photosystem II, suggesting that the trypsin reaction sites are not accessible when the thylakoids are appressed.

Photosystem I non-cyclic electron transport to methyl viologen and P-700 turnover rate during cyclic electron transport were not affected by trypsin treatment (SELMAN et al., 1973). Trypsin did, however, markedly inhibit electron transport on the reducing side of photosystem I, possibly by binding to the membrane and inhibiting electron transfer between the primary acceptor of photosystem I and ferredoxin (SELMAN and BANNISTER, 1974).

F. Diazoniumbenzenesulfonic Acid (DABS)

Based on freeze-fracture electron microscopy, ARNTZEN et al. (1969) proposed the binary model of chloroplast membranes which visualizes photosystem I as localized in the external portion of the membrane, while the majority of photosystem II being located on the internal surface. Experiments by DILLEY et al. (1972), using the non-penetrating water soluble probe, DABS, which covalently bonds ε-amino, histidine, tyrosine, and sulfhydryl groups of proteins, provided a test for this hypothesis. By treating intact chloroplast membranes with DABS, followed by digitonin fractionation of the membrane into photosystem I and photosystem II fractions, they showed that the photosystem I fraction contained ten- to twenty-fold more DABS than the photosystem II fraction. It was concluded that the proteins associated with photosystem I are localized at the external membrane surface, while the majority of photosystem II is buried and not accessible to the hydrophilic diazonium reagent. This 10/1 distribution of DABS between photosystem II and photosystem I was observed even when the coupling factor (surface located) was removed prior to DABS labeling (GIAQUINTA et al., 1974b).

Subsequent studies demonstrated that, when chloroplasts were treated with DABS under dark conditions, electron transport on the reducing side of the photosystem I primary acceptor was inhibited at two sites, one site being between the photosystem I primary acceptor and ferredoxin, the other site at the ferredoxin-NADP reductase (SELMAN et al., 1974a). Studies by MARUYAMA et al. (1974), using diazonium-1,2,4-triazole, also show marked inhibition of electron transport on the reducing side of photosystem I. Similar to DABS, the triazole derivative did not inhibit electron transfer from H_2O to methyl viologen.

DABS treatment in the dark also inhibited phosphorylation. The binding of DABS to coupling factor subunits (resolved by SDS gel electrophoresis), along with inhibition of the calcium dependent ATPase activity, suggested that the inhibition of ATP formation was due to chemical modification of the coupling factor (CF_1) complex (GIAQUINTA et al., 1974b). There was not a light requirement for the DABS inhibition of ATP synthesis as noted for NEM treatment, but this is probably because the diazonium has a broader spectrum of chemical group reactivity compared to sulfhydryl-group specific NEM, and, thus, would modify the CF_1 to a greater extent. There was, however, a light dependent increase in DABS labeling of the CF_1 subunits, supporting the concept of an electron transport dependent conformational change in the CF_1, as proposed by McCARTY et al. (1972). Studies by GIAQUINTA et al. (1973, 1974a) demonstrated that the DABS treatment of chloroplasts in the light inhibited photosystem II electron transport at the level of water oxidation. Concomitant with the inhibition, the light-treated membranes covalently bound three- to four-fold more DABS than membranes treated in the dark. The presence of DCMU during light treatment prevented both the inhibition of photosystem II and the incremental DABS binding, indicating that electron transport was necessary to elicit these effects. The light-potentiated binding of DABS to the membranes was interpreted as resulting from an electron transport-dependent membrane conformational change that exposes otherwise inaccessible diazo-reactive groups, the reaction of some of which leads to inhibition of water oxidation (GIAQUINTA et al., 1974a). The data indicate the components associated with photosystem II undergo a change in conformation during electron transport.

Activation of cyclic electron transport using PMS or menadione in the presence of DCMU did not result in the DABS effects, indicating that the redox reaction(s) necessary for DABS binding are located prior to plastoquinone. Electron transfer from water to conventional Class III acceptors (accepting at or prior to plastoquinone) was sufficient to drive the conformation change (GIAQUINTA et al., 1974c; GIAQUINTA and DILLEY, 1974d).

Recent studies have examined the relationship of membrane conformational changes and energy transduction associated with photosystem II (GIAQUINTA et al., 1975). Those studies suggest that the DABS-detected conformational change is a direct result of protons, released from water oxidation, being directed into the membrane where they probably are bound to fixed-charge groups of proteins. Protonation of membrane polypeptides would be expected to cause rearrangements in the peptides, thus exposing additional diazo-reactive groups. In brief, the data show that both electron flow between the photosystem II quencher, Q, and plastoquinone, as well as proton-liberation from water oxidation (or the oxidation of an alternate proton-liberating photosystem II donor) are required to give the DABS-detected conformational change. If photosystem II electron flow is from water to silicomolybdate ($+$DCMU) (GIAQUINTA and DILLEY, 1975) there is neither internal proton accumulation, ATP formation, nor the extra DABS binding. Using hydroxylamine-treated chloroplasts that are unable to oxidize water, and providing iodide as an electron-only photosystem II donor (IZAWA and ORT, 1974), it was also found that the DABS-detected conformational change did not occur (GIAQUINTA et al., 1975).

The mechanistic significance of the conformational change observed with the diazo compound is not fully understood. However, for the purpose of this report we wish simply to stress the potential usefulness of this technique for monitoring membrane conformational changes.

References

Arntzen, C.J., Armond, P.A., Zettinger, C.S., Vernotte, C., Briantais, J.M.: Biochim. Biophys. Acta **347**, 329–339 (1974a)
Arntzen, C.J., Dilley, R.A., Crane, F.L.: J. Cell Biol. **43**, 16–31 (1969)
Arntzen, C.J., Vernotte, C., Briantais, J.M., Armond, P.: Biochim. Biophys. Acta **368**, 39–53 (1974b)
Berg, S., Dodge, S., Krogmann, D.W., Dilley, R.A.: Plant Physiol. **53**, 619–627 (1974)
Dilley, R.A., Peters, G.A., Shaw, E.R.: J. Membrane Biol. **8**, 163–180 (1972)
Giaquinta, R.T., Dilley, R.A.: In: Proc. 3rd Intern. Congr. Photosynthesis. Avron, M. (ed.). Amsterdam: Elsevier, 1975, pp. 883–895
Giaquinta, R.T., Dilley, R.A.: Biochim. Biophys. Acta **387**, 288–305 (1975)
Giaquinta, R.T., Dilley, R.A., Anderson, B.J.: Biochem. Biophys. Res. Commun. **52**, 1410–1417 (1973)
Giaquinta, R.T., Dilley, R.A., Anderson, B.J., Horton, P.: J. Bioenergetics **6**, 167–177 (1974c)
Giaquinta, R.T., Dilley, R.A., Selman, B.R., Anderson, B.J.: Arch. Biochem. Biophys. **162**, 200–209 (1974a)
Giaquinta, R.T., Ort, D.R., Dilley, R.A.: Biochemistry **14**, 4392–4396 (1975)
Giaquinta, R.T., Selman, B.R., Bering, C.L., Dilley, R.A.: J. Biol. Chem. **249**, 2873–2878 (1974b)
Hirose, S., Yamashita, K., Shibata, K.: Plant Cell Physiol. **12**, 775–778 (1971)
Izawa, S., Good, N.E.: In: Progress in Photosynthesis Research. Tübingen: Laupp, 1969, Vol. III, pp. 1288–1298
Izawa, S., Ort, D.R.: Biochem. Biophys. Acta **357**, 127–143 (1974)
Maruyama, I., Nakaya, K., Ariga, K., Obata, F., Nakamura, Y.: FEBS Lett. **47**, 26–28 (1974)
McCarty, R.E.: Arch. Biochem. Biophys. **161**, 93–99 (1974)
McCarty, R.E., Fagan, J.: Biochemistry **12**, 1503–1507 (1973)
McCarty, R.E., Pittman, P.R., Tsuchiya, Y.: J. Biol. Chem. **247**, 3048–3052 (1972)
Prochaska, L.J., Gross, E.L.: Biochim. Biophys. Acta **376**, 126–135 (1975)
Schmid, G.H., Radunz, A., Menke, W.: Z. Naturforsch. **30**, 201–212 (1975)
Schneider, D.L., Kagawa, Y., Racker, E.: J. Biol. Chem. **247**, 4074–4079 (1972)
Selman, B.R., Bannister, T.T.: Biochim. Biophys. Acta **253**, 428–436 (1971)
Selman, B.R., Bannister, T.T.: Biochim. Biophys. Acta **347**, 113–125 (1974)
Selman, B.R., Bannister, T.T., Dilley, R.A.: Biochim. Biophys. Acta **292**, 566–581 (1973)
Selman, B.R., Giaquinta, R.T., Dilley, R.A.: Arch. Biochem. Biophys. **162**, 210–214 (1974a)
Selman, B.R., Johnson, G.L., Dilley, R.A., Voegeli, K.K.: In: Proc. 3rd Intern. Congr. Photosynthesis. Avron, M. (ed.). Amsterdam: Elsevier, 1974b, pp. 897–909
Wallach, D.F.H.: Biochim. Biophys. Acta **265**, 61–83 (1972)

III. Energy Conservation

1. Photophosphorylation

A.T. JAGENDORF

A. Relation of Electron Transport to Phosphorylation

I. Electron Transport Patterns

The discovery of photophosphorylation, starting with the major insights of VAN NIEL (1941, 1949), through the first biochemical discoveries with isolated chloroplasts (HILL, 1937), indications of phosphorylation in whole cell photosynthesis and then the discovery of ATP synthesis by isolated chloroplasts (ARNON et al., 1954) and by bacterial chromatophores (FRENKEL, 1954) is described in Chapter I of this volume. The latter discoveries led to the direct measurement of simultaneous, coupled electron transport and ATP synthesis in the cell-free milieu (ARNON et al., 1958, 1959; AVRON et al., 1958).

As in mitochondria, electron flow proceeds through a series of contiguous electron-carrying enzymes or (fat-soluble) quinones, bound in this case in the thylakoid membrane. Unlike mitochondria the electron transport system includes light-driven uphill electron transport steps which are largely irreversible; and whose existence makes a cyclic pattern of electron flow feasible. The distinction between non-cyclic electron flow, with net amounts of an electron donor (for instance, H_2O) oxidized and of an electron acceptor (for instance, ferricyanide) reduced; and cyclic electron flow where there is no net change in redox substrates, will be found Chapter I in this volume. (It should be noted, however, that when ATP synthesis accompanies cyclic electron flow it tends to be given the unfortunate term, "cyclic photophosphorylation". The electron flow pattern, not the phosphorylation, is cyclic.)

With isolated chloroplasts only non-cyclic electron flow is possible until either some synthetic electron-carrying dye (JAGENDORF, 1962) or soluble ferredoxin at high concentrations (TAGAWA et al., 1963) is added. This raised some questions about cyclic electron flow occurrence in vivo; however the internal ATP level of leaf discs and of algae is increased by light in a way insensitive to inhibitors of non-cyclic electron flow (FORTI and PARISI, 1963; BEDELL and GOVINDJEE, 1973). The phosphorylation-associated cyclic electron flow cycles around photosystem I, so an indication of its existence is ATP synthesis that is not inhibited by compounds (DCMU, etc.) that block off photosystem II electron transport.

A cyclic electron flow pattern requires both an existing electron (on one of the carriers) and oxidized carriers for it to hop around. That means if all the carriers are completely reduced at the start no electron can find an empty spot to which to move and so no electron flow occurs; or if the system is entirely oxidized at the start there will be no electron to move around. ATP synthesis driven by cyclic systems, therefore, is inhibited by either over-reduction (high concentrations of ascorbate or thiols, etc.) or by over-oxidation (JAGENDORF and FORTI, 1961). Evidence of this sort has led to speculation about the direct effect of redox potential on the coupling mechanism (HORIO and KAMEN, 1962); however since ATP synthesis is the only reaction measured it is not possible to distinguish inhibition of phosphorylation from inhibition of the driving electron transport reactions.

II. Coupling Between Electron Transport and Phosphorylation

1. Definition: "Photosynthetic Control"

Electron transport and phosphorylation are said to be "coupled". The term has two aspects. No phosphorylation will occur unless electron transport is proceeding (but this is violated when phosphorylation is driven by ion gradients; see below, and this vol. Chaps. III.6 and III.9). Conversely there should be no electron transport unless $ADP + P_i$ are present to permit simultaneous phosphorylation. Actually isolated chloroplasts always have a small amount of electron transport (a "basal rate") in the absence of added ADP or P_i presumably due to leaks in the system. The stimulation of this basal rate by added ADP and P_i, expressed as a ratio, is often given the name "photosynthetic control" by analogy with the term "respiratory control" by ADP in the case of mitochondrial oxidative phosphorylation. Another measure of the tightness of coupling is the ratio of ATP made per two electrons moving, or P/e_2 ratio.

2. Physical State of the Chloroplasts

The degree of coupling varies depending on the conditions for measuring phosphorylation and on the state of the isolated chloroplasts.

[The various morphological conditions of isolated chloroplasts are probably best described by the classification scheme of HALL (1972). Chloroplasts with outer envelope and full content of stroma are called Type A; those with torn but re-healed outer envelope having lost some but not all of the stroma are Type B; naked thylakoid matrix without stroma or outer envelope are Type C; osmotically disrupted thylakoids are Type D, etc. In this chapter the reactions of Type C and Type D chloroplasts will be described, for the most part.]

Using long illumination times (30 min or more) in a hypotonic medium, low light intensities and probably battered chloroplasts, WARBURG et al. (1959) observed no stimulation of electron transport by $ADP + P_i$ and declared reports of its occurrence to be founded on artifacts. However with better chloroplasts (Type B or C), shorter illumination times and brighter light, photosynthetic control ratios range from 2.5–3.5 (WEST and WISKICH, 1968; AVRON et al., 1958). When Type A or B chloroplasts are prepared carefully and given a hypotonic shock only at the last moment, control ratios of 4–6 or even higher are observed (REEVES et al., 1971). This procedure is obviously preferred for obtaining chloroplasts in vivo with the tightest possible coupling characteristics.

3. Uncouplers

Some synthetic or natural compounds will permit fast electron flow in spite of the absence of $ADP + P_i$. These "uncouplers" should be thought of as releasing the restraint imposed on electron flow by an inactive but coupled energy conservation machine. In theory any compound that destroyed or consumed an energy-conserving intermediate between electron transport and phosphorylation; or substituted for phosphate to form an unstable adenylate (as arsenate is supposed to do, see AVRON and JAGENDORF, 1959); or caused hydrolysis of ATP as soon as it was formed, ought to permit rapid electron flow without net ATP formation. These acts would lead to turnover of the machinery for energy conservation.

4. Energy Transfer Inhibitors

Perhaps eight to ten compounds have been found that block ATP synthesis by inhibiting the terminal reactions, rather than by causing rapid turnover of intermediate high energy states. These "energy transfer inhibitors" prevent ATP synthesis and the extra electron flow induced by and dependent on phosphorylation, but do not touch either basal or uncoupled electron flow (Fig. 1 illustrates these relationships for phlorizin, WINGET et al., 1969). Of the other energy-transfer inhibitors (see this vol. Chap. III.12 for details) dicyclohexylcarbodimide (DCCD) (McCARTY and RACKER, 1967) and the antibiotic Dio-9 (McCARTY et al., 1965) have been studied in detail, and one of especial interest is ATP itself when used at low ionic strength (SHAVIT, 1971).

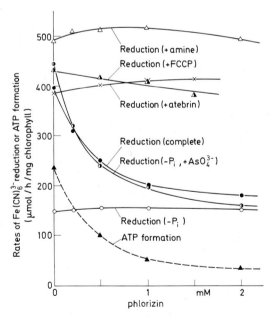

Fig. 1. Effect of energy transfer inhibitor on electron transport. Rates of ferricyanide reduction were measured as a function of increasing concentrations of phlorizin. (From WINGET et al., 1969)

III. Energy Conservation Sites

1. P/e_2 Ratios

As the elaborate nature of the electron transport chain became known, an important question was raised as to with which, and how many, of the electron transport steps is energy conservation and ATP synthesis associated? The first attempts at determining at least the number of sites were based on measurements of the P/e_2 ratio—the number of ATP molecules formed per pair of electrons moving from water to the reducing site of photosystem II. Assuming, as in mitochondrial oxidative phosphorylation, that the movement of two electrons past a given "site" in the chain permitted formation of one ATP, then a P/e_2 ratio of 1.0 would indicate one site, and of 2.0 would indicate two sites.

Observed P/e_2 ratios (ARNON et al., 1958) started out at 1.0. Lower ratios (down to 0) are easily obtained with maltreated or uncoupled chloroplasts, so the finding of higher ratios is more significant than the lower numbers. As preparative procedures improved, and as buffers other than Tris came into use so that the true pH optimum of 8.5 or higher was discovered, observed ratios rose to 1.2 or 1.3 (WINGET et al., 1965). Using intact (Type A or B) chloroplasts subject to hypotonic shock at the last moment only, REEVES and HALL (1973) obtained P/e_2 ratios above 1.5 and approaching 2.0. The minimal conclusion is that at least two phosphorylation sites exist between water and photosystem I and indeed at least two have been isolated by means of specific inhibitors, electron donors and acceptors.

The exact P/e_2 remains both unresolved at present, and of major interest when summing up the energy balance of chloroplast electron transport. It is not yet clear, for instance, whether the movement of two or of three electrons past one site is needed to drive ATP synthesis. The major difficulty is in trying to decide if the "basal" electron flow proceeds as usual during ongoing phosphorylation, transporting electrons but not making ATP. If so, it ought to be subtracted from the total electron transport before calculating the P/e_2 ratio. On doing this in an extensive variety of reactions, IZAWA and GOOD (1968) calculate a corrected ratio of 2.0. However the basal rate is inhibited by ATP (i.e. "State 4" electron flow) which is always present during phosphorylation; hence it seems more reasonable to subtract only the slower ATP-inhibited electron flow rate. A more subtle line of reasoning leads to an even lower subtraction: the basal rate is permitted by proton leakage out of the thylakoids; the leakage is driven by the total pH gradient (ΔpH) across the membranes; the height of the ΔpH is diminished by ongoing ATP synthesis. By measuring the pH gradient *during* phosphorylation (PORTIS and McCARTY, 1974) and using an earlier calibration curve for basal electron transport rate vs ΔpH, corrected values reach a maximum of 1.3 ATP formed per electron pair moving in Type C chloroplasts (PORTIS, 1975).

2. Isolation of Two Phosphorylation Sites

Recent advances, especially including the use of the inhibitor DBMIB to block the oxidation of plastoquinone, lipophilic electron donors or acceptors to interact with components in the middle of the chain between the two light reactions,

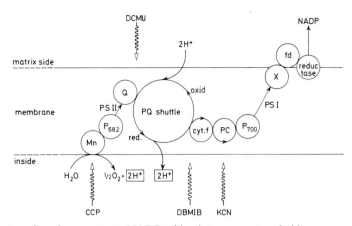

Fig. 2. Photosynthetic electron flow from water to NADP taking into account probable geometry of electron carriers in membrane and consequent vectorial proton translocation. Two
protons are released on the inside as water is oxidized, and two more are released as plastoquinone is oxidized by cytochrome f. Plastoquinone picks up two protons from the exterior
to accompany the two electrons coming from component Q (Quencher); and more protons
are consumed on the outside as $NADP^+$ is reduced to NADPH. (From TREBST, 1974)

and hydrophilic electron acceptors to react only after photosystem I, have permitted
the isolation of two separate phosphorylating sites in the span between water
and photosystem I (see TREBST, 1974, for an extensive review; also this vol.
Chaps. II.15 and II.16). The first site occurs between water and photosystem II;
the second is between the component Q (Quencher) and cytochrome f (Fig. 2).
Experimental results leading to this scheme constituted much better evidence for
two phosphorylations in non-cyclic electron transport than any number of P/e_2
determinations could have provided.

With the ordinary biochemists' chloroplasts, P/e_2 ratios at each site are measured between 0.3 and 0.6, adding to a total of almost 1.0 for the complete
span from water to ferricyanide (GOULD and IZAWA, 1973, for instance). When
the most tightly coupled thylakoids are used, P/e_2 ratios are found close to 1.0
at each site, and close to 2.0 for the complete span (HEATHCOTE and HALL, 1975).

B. Chemiosmotic Principles of Coupled Electron Flow and ATP Synthesis

I. The Chemiosmotic Hypothesis; and Others

Three major hypotheses, as well as variants and intermediate versions, have been
proposed to account for the mechanism of coupled electron transport and ATP
synthesis. The chemical hypothesis (SLATER, 1953; CHANCE and WILLIAMS, 1956)
proposes the formation of direct chemical bonds between electron carriers and
enzymes or even substrates of the phosphorylation reaction. These bonds are raised

to a high energy level by the oxidation of the complex, and this energy drives the anhydro bond formation of ATP. The conformational hypothesis proposed and more recently revised by BOYER (1965, 1974) suggests that transfer of energy from redox reactions to ATP synthesis is mediated by means of forced conformational changes on the part of the enzymes of phosphorylation, or the membrane, or both. The chemiosmotic hypothesis, originating in suggestions made by membrane physiologists (ROBERTSON, 1960) and developed and expounded in detail by P. MITCHELL (1963, 1966, 1967, 1968, 1969, 1972, 1973) depends on vectorial translocation of protons across the membrane as a necessary consequence of electron transport, and the subsequent proton electrochemical activity gradient playing a direct role in the dehydration of $ADP + P_i$. Somewhat related are hypotheses in which electron transport creates charge redistributions and electrochemical stress within the membrane rather than entirely across it; and these stresses represent the coupling points (WILLIAMS, 1969; CHANCE, 1972).

Of these, the hypothesis of a direct chemical coupling is not supported by any convincing evidence. The chemiosmotic hypothesis has provided the most coherent framework to understand the facts noted above, and has generated a large number of predictions (many of them irrelevant to the other hypotheses) most of which have been verified over the past decade. These relate especially to the sequence from electron transport to the proton activity gradient as a high energy intermediate. The chemical nature of the phosphorylation reaction is still largely unresolved and protein conformational changes may turn out to have a major role to play at this point. While a few discrepancies remain, the bulk of the evidence is sufficiently consistent that the chemiosmotic hypothesis will be taken here as the primary conceptual framework with which energy conservation in chloroplasts, between electron transport and ATP synthesis, can be understood.

There are several component parts to the chemiosmotic hypothesis:

1. Nature of the Membrane and Electron Transport

Phosphorylation must occur in an intact vesicle — a bubble with an aqueous phase both inside and out — whose membrane is relatively impermeable to protons or hydronium ions.

Electron transport carriers embedded in the membrane must have one or more alternations, in sequence, between electron carriers (i.e. with metal atoms as prosthetic group) and hydrogen atom carriers (plastoquinone, etc.). The hydrogen carriers must be oriented in the membrane so that wherever their electron may come from, the necessary proton to complement it comes from outside the vesicle. In turn, when giving up the electron it should be to an electron acceptor on the inside, with the complementary proton released on the inside. In this way, due to geometry, electron transport through the chain is coupled obligatorily to vectorial proton translocation across the membrane. These concepts provided a strong impetus to attempts to discover specific details of the sidedness of chloroplast membranes (TREBST, 1974).

The specific way in which H^+ translocation occurs in chloroplast membranes is outlined in Figure 2, taken from TREBST (1974). The first internal protons are those released from water (FOWLER and KOK, 1974); this constitutes Site II for energy conservation. The second set moves in during the sequential reduction then reoxidation of plastoquinone and represents Site I. It is possible that the "plastoquinone shuttle" depicted in the figure involves hydrogen atom transfer from one member of the pool of plastoquinone molecules to another, which are known to be present in 5- to 10-fold functional excess over the numbers of other components of the electron transport chain (STIEHL and WITT, 1968).

2. Proton Translocation, Charge Separation and the Protonmotive Force

As protons move to the inside of the vesicle they carry positive charge with them. Almost immediately, when electron flow starts a membrane potential will arise, positive on the inside, i.e. an "electrogenic" pump.

   In thylakoids, an earlier source of electric potential is the charge separation occurring in the primary photoact itself (Fig. 3a). Evidence for this comes mainly from the work of WITT and associates (WITT, 1971, 1975; WITT et al., 1968),

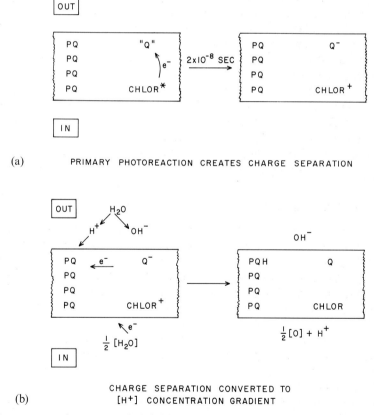

(a)   PRIMARY PHOTOREACTION CREATES CHARGE SEPARATION

(b)   CHARGE SEPARATION CONVERTED TO
             [H$^+$] CONCENTRATION GRADIENT

Fig. 3. (a) Initial charge separation across thylakoid membrane due to electron transport in primary photoact. Excited state chlorophyll transfers an electron to primary acceptor, perhaps a plastoquinone derivative. Result is oxidized, chlorophyll$^+$ localized on or near interior of membrane and reduced quinone localized on outside. (b) Proton addition and translocation. Water is oxidized, reducing and neutralizing chlorophyll cation, but leaving behind H$^+$ on interior. Plastoquinone anion on outside is neutralized by addition of proton from external aqueous phase. Both figures adapted from WITT (1971) who have used band shifts of carotenoids and other natural pigments of the membrane as indicators for the membrane potential. These shifts occur extremely rapidly, with rise times on the order of 10^{-8} to 10^{-7} sec. Protons move more slowly (in 10^{-3} to 10^{-2} sec), in a pattern which preserves the membrane potential on the basis of unequal proton distribution (Fig. 3b).

Existence of charge separation in the primary photoact is unique, of course, to photosynthetic systems. Correspondingly its role within the chemiosmotic framework represents a unique contribution of WITT and associates to the operating rationale of the chemiosmotic hypothesis.

Once a membrane potential has been created by proton flux, other ions will diffuse passively across the thylakoid membranes, perhaps facilitated by as yet undiscovered ion carriers. In particular Cl^- ions are observed to enter with protons (thylakoids pump HCl into themselves in the light!) and Mg^{2+} ions leave in varying proportions depending on the medium the plastids have equilibrated with (HIND et al., 1974). The counter-ion flux at least in part collapses the membrane potential but leaves a hydrogen ion concentration gradient undisturbed, with the interior more acid than the exterior.

In this way electron transport creates an electrochemical gradient of protons across the membrane. Both the higher proton concentration and any net positive charge on the inside tend to drive protons out again. The two parameters are additive, contributing to a "protonmotive force" differential across the membrane:

$$\Delta \mathrm{pmf} = \Delta \mathrm{pH} + \Delta \psi. \tag{1}$$

Where pmf stands for protonmotive force, and ψ is the symbol for membrane potential. This pmf is the high energy state driving ATP synthesis. The energy available from *one* mol equivalent of protons trapped inside, once the arithmetic details are worked out (MITCHELL, 1966; MCCARTY, 1976; JAGENDORF, 1975a) is given by:

$$\Delta G' = 1.36(\Delta \mathrm{pH}) + 1.36 \left(\frac{\Delta \psi}{59} \right) \tag{2}$$

where the units of $\Delta G'$ are in kcal/mol, and the numerical modifications of pH and ψ take into account the constants needed to convert their units to kcal/mol.

3. The Reversible ATPase and ATP Synthesis

The protonmotive forces drives ATP synthesis via a membrane-localized, reversible, vectorial ATPase. It is vectorial most importantly with respect to the protons and hydroxyl ions involved in the dehydration of $ADP + P_i$ leading to ATP synthesis. Vectorial discharge of the hydroxyl ion from ADP plus P_i (Fig. 4), into the proton-rich interior could make use of the pmf differential. Recombination of the hydroxyl group taken from the substrates with interior protons to form water is the reaction that pulls the enzymatic dehydration. In the reverse direction all arrows change direction and the chemiosmotic hypothesis predicts proton pumping inwards as a part of the vectorial hydrolysis of ATP.

Fig. 4. Thermodynamic relations in ATP syn-
thesis by membrane-bound, reversible, vectorial
ATPase according to chemiosmotic hypothesis.
(From JAGENDORF, 1967)

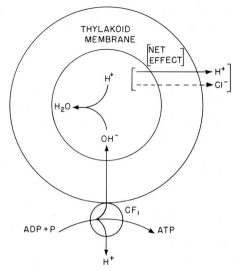

Note that this is a global argument, rather than a specific chemical mechanism describing the working of the enzyme functional groups and topography. It may be that the terminal enzyme functions to accomplish the reactions as drawn. However, there are other possibilities for the chemical mechanism of phosphorylation; using the pmf differential for instance to drive conformational and functional changes in the enzyme (see Chaps. III.7 and III.10, and Sect. D.IV below).

4. Controls Over Electron Transport: Uncouplers

Photosynthetic control is readily interpreted by the chemiosmotic theory. If vectorial translocation of a proton is part of a given electron transport step, then an existing pmf differential will put back-pressures on the reaction that has to put yet another proton into the acidic, positively charged interior. Given no release of internal protons or charge, electron transport might stop altogether. It is also possible that either the high internal acidity, or the stressed condition of a membrane with a high potential across it, might inhibit electron transport at steps other than those translocating protons.

Since electron transport rates are diminished by a high internal pmf, reactions that degrade the internal pmf will speed up further electron transport. Physiologically this would be the synthesis of ATP (Fig. 4); hence more rapid electron flow as phosphorylation is allowed to proceed ("photosynthetic control"). Also, leakage of protons or other ions through the membrane will permit non-phosphorylating electron transport. Indeed, known uncouplers have turned out to be compounds that carry protons, or other ions, through biological membranes. There are two major functional types—one (nigericin, NH_4Cl, etc.) facilitates an electrically neutral loss of a pH gradient; the other (valinomycin+K^+; dinitrophenol) permits loss of a membrane potential without necessarily eliminating a pH gradient (for details see Chap. III.11; GOOD et al., 1966; HIND and MCCARTY, 1973; JAGENDORF, 1975a). One or the other, or a combination of the two, may be needed for most effective uncoupling depending on whether the given system depends most on Δ pH or on $\Delta\psi$ under the conditions used.

5. Nature of Energy-Conserving Sites

It is apparent that an energy-conserving site is best defined as an electron transport step that results in vectorial translocation of protons across the membrane. There are at least 2 interesting consequences.

In the first place, the requirement may be filled by an entirely artificial dye acting as electron transport mediator, as long as it (1) dissolves in the membrane, and (2) can carry hydrogen atoms with it in the reduced form. Extensive work from the laboratory of TREBST (TREBST, 1974; HAUSKA et al., 1974) has demonstrated the reality of just such "artificial sites". For instance methyl phenazinium sulfate (PMS) completes a cyclic electron flow around photosystem I. The reduced form of this dye can give electrons directly to P-700 (RUMBERG and WITT, 1964) and accept electrons from the primary electron acceptor from photosystem I. It is thus able to complete an extremely short cycle, bypassing plastoquinone, cytochrome f, and even plastocyanine (Fig. 5). Protons are translocated by reason of associating with reduced PMS, which thus becomes the entirely artificial but quite effective energy conserving "site".

The second consequence is that the mechanism permits the electron carriers to be in one area of the membrane, and the phosphorylating enzyme in an entirely different area, as long as they are both on or in the continuous membrane defining a topological sphere with a common inside space. This is not conceivable in hypotheses requiring either a chemical intermediate, or direct contact for purposes of conformational interactions between electron carriers and the phosphorylating enzymes. This distinction provides significance to those observations in which the coupling factor seems localized on the outer membranes of a given granal system (MILLER and STAEHELIN, 1976; OLESZKO and MOUDRIANAKIS, 1974). It also means that "phosphorylation site", used initially when the energy-conserving electron carriers and the phosphorylation enzymes were thought to exist in immediate contact, has become an ambiguous term.

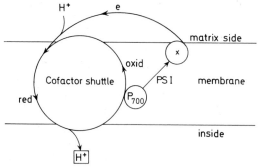

Fig. 5. Minimum electron flow pathway to support photophosphorylation: short cycle around photosystem I using proton carrying dye that is reduced by primary acceptor and oxidized by P-700 (e.g. PMS). (From TREBST, 1974)

C. Evidence Relating to Operation of the Chemiosmotic Principles in Chloroplasts

I. Light-Driven Proton Uptake

1. Occurrence and Extent: Counter-Ions and Internal Buffers

A basic piece of information is the rise in pH demonstrated by a suspension of illuminated chloroplasts during active electron transport (JAGENDORF and HIND, 1963; NEUMANN and JAGENDORF, 1964). This reaction was actually discovered while looking for the nature of stored high energy intermediates in ATP synthesis (see Sect. IV). Similar observations although with quantitative differences were observed thereafter with chromatophores from the bacterium *Rhodospirillum rubrum* (VON STEDINGK and BALTSCHEFFSKY, 1966) and since then with virtually every photosynthetic organelle studied.

The number of protons taken up from the medium with spinach thylakoids is very large, at times amounting to 0.7 to 1.0 equivalent H^+ per mg (approx. µmol) of chlorophyll, or up to 400 H^+ per cytochrome f. The extent is much less with other organelles and other conditions, however. The amount is limited, for a given system, by three factors. In the first place the observations relate to a steady state condition. Protons are pumped in by electron transport, but diffuse out again through various leaks or ports. Adding uncouplers can increase the rate of leakage and thereby diminish the observable steady state amount of proton uptake.

The other two limitations on H^+ uptake represent self-limiting features. If protons enter without any counter-ion flux the membrane potential would rise almost immediately to a point that would halt further functioning of the membrane. It is doubtful that any biological membrane system functions at potentials much above 350 mV. The number of protons that can enter, therefore, is strictly limited by the number of counter-ions that can be redistributed to neutralize the entry of positive charge. (Also the rate of proton flux can be limited by the rate of counter-ion flux.)

Extent of internal buffering is the other limitation. The internal volume of spinach chloroplasts is estimated at much less than 100 µl mg^{-1} of chlorophyll (GAENNSLEN and MCCARTY, 1971). To bring this volume to the unphysiological pH of 4.0, less than 0.01 equivalents of H^+ would be needed, or perhaps 1% as much as actually enter whole chloroplasts. Obviously 99% of the entering protons are neutralized by internal buffers; conversely it is obvious that proton uptake cannot occur beyond the extent of available internal buffering groups.

The reality of the $\Delta\psi$ limitation for small particles is demonstrated by the increases in proton uptake that occur on providing mobile counterions K^+ with valinomycin (VON STEDINGK and BALTSCHEFFSKY, 1966; MCCARTY, 1970; etc.) or permeant anions (ISAEV et al., 1970; GROMET-ELHANAN and LEISER, 1973, 1975).

The reality of the limitation by internal buffer capacity is seen when proton uptake by whole chloroplasts, often at pH 7 or above, is increased by providing effective internal buffers. These are weak bases having a pK of about 5 or 6; compounds such as tetramethylenediamine, imidazole, pyridine or aniline (CROFTS, 1968; LYNN, 1968; NELSON et al., 1971; AVRON, 1971).

2. Estimating Internal pH and the pH Gradient

It is obvious that the small size of chloroplast thylakoids precludes the use of ordinary electrode techniques to measure internal pH. Some recent advances in miniaturization offer the chance to obtain at least preliminary estimates, however, and in the monstrous chloroplast of a hornwort the internal pH was recently measured as moving from 6.2 in the dark to 4.6 in the light (Davis, 1974). However more definite results, and all that is known of isolated thylakoids, comes from the use of indirect methods.

The speed of membrane electron transfer reactions has been used as a pH indicator (Rumberg and Siggel, 1969). The relationship was calibrated in the presence of saturating amounts of gramicidin to maintain similar internal and external pH; then without gramicidin the decrease in rate of electron transport ($P\text{-}700^+$ reduction after a flash) was ascribed to a more acid pH on the interior than the outside. This estimate called for an internal pH of 5.1 when the outside pH was 8.0. However other evidence suggests that the rate of electron transport is sensitive not to the internal pH per se, but to the magnitude of the pH gradient across the membranes (Bamberger et al., 1973; Portis, et al., 1975). If so, the estimate is too low.

The most widely used methods depend on finding the relative concentrations of a weak base inside and outside illuminated thylakoid membranes. The procedure usually assumes rapid penetration by the free (non-protonated) base, and complete impermeability of the protonated form (if the latter is not the case internal acidity will be underestimated). When protons on the inside trap the base as the protonated form the total internal concentration will be higher than that measured outside (see McCarty, 1976, for a critical discussion of methodology and rationale). Bases (9-amino-acridine, atebrine) whose fluorescence is quenched upon membrane energization have been much used; however the quenching reflects binding rather than actual uptake into the interior (Fiolet et al., 1974) and the resulting pH estimates are not reliable. The distribution of tracer amounts of radioactive bases into chloroplasts which are then separated by centrifugation has been used by the laboratories of Avron (Rottenberg et al., 1972) and of McCarty; the procedures in which centrifugation is accomplished in an illuminated apparatus, in a matter of seconds, through inert silicone oil layers gives the most reliable measurements (Gaennslen and McCarty, 1971; Portis and McCarty, 1974; Portis, 1975).

These procedures have documented the remarkable fact that illuminated thylakoids, under optimum conditions, can maintain a pH gradient from 3 to 3.5 units more acid in the inside than in the outside medium. As might be expected the size of ΔpH increases with increasing light intensity and electron flow rates, and is decreased by uncouplers. At low concentrations either ATP or ADP, shown previously to decrease rates of H^+ leakage (McCarty et al., 1971; Telfer and Evans, 1972), enhance ΔpH to its maximum value of 3.5 units (Portis and McCarty, 1974). Addition of $ADP + P_i$ (or arsenate) which permits phosphorylation to occur decreases the ΔpH to 2.8 units, and these effects are prevented by an energy transfer inhibitor such as phlorizin. Interestingly, a clear straight line relation is found between the log of the phosphorylation rate (from 1 to almost 1,000 μmol mg chlor^{-1} h^{-1}) and the ΔpH (Portis and McCarty, 1974).

These significant correlations provide a strong indication that the pmf can be the driving force for phosphorylation in the steady state in the light as well as in model experiments.

II. The Membrane Potential

1. Micro-Electrode Measurements

Recent advances on the miniaturization of electrodes have finally made it possible to impale chloroplasts in vivo and obtain signals relating to membrane potential and even pH (BULYCHEV et al., 1972; DAVIS, 1974; VREDENBERG and TONK, 1975). Results consistently show a more positive potential, presumably at the interior of the thylakoids, in the light than in the dark. The values reported are variable however, between 5 and 40 mV. VREDENBERG and TONK (1975) point out that quantitation depends on knowing the salt concentration in the thylakoids better than is now the case; they found a sharp spike on turning the light on (and an "undershoot" after turning the light off) with an estimate of a 10 mV potential in the steady state in the light. Some problems do remain: the exact position of the electrode tip is hard to be certain of, and subtle even if not major damage to the membranes might occur when inserting a foreign object such as an electrode (DAVIS, 1974). Nevertheless the results are enough to show that this is a potentially very interesting direct approach, likely to provide valuable information especially when combined with other sorts of measurements.

2. Electrochromic Effects

Large band shifts in the absorption spectra of illuminated thylakoids have been traced to the effect of an electric field (membrane potential) on properties of embedded carotenoids and other pigments (see this vol. Chap. III.8; WITT, 1971, 1975; HIND and McCARTY, 1973, etc.). Direct evidence for this relationship came from classical experiments of JACKSON and CROFTS (1969) with bacterial chromatophores; a salt-induced diffusion potential in the dark duplicated the band shifts induced by light, and the consequent ability to relate height of the band shift to size of the membrane potential permitted the absorbancy change to be used as a membrane potential voltmeter. Similar experiments with chloroplasts are more difficult due largely to light-scattering changes; but STRICHARTZ and CHANCE (1972) were able to estimate a membrane potential of 30 mV in the steady state in the light, using this approach. On the other hand, measurements made with "single turnover" flashes of light (WITT, 1971, 1975; JOLIOT and DELOSME, 1974) and calibrated by means of assumptions concerning dielectric constant, thickness, and area of the membranes, led to estimates of membrane potentials in chloroplasts on the order of 100 mV more positive on the inside. It is difficult to rule out the possibility that the extent of the electrochromic shift might be altered by secondary considerations of the state of the membrane or its interactions with the electric field (LARKUM and BONNER, 1972). This variability in the calibration could explain the differences between estimates obtained by this method and those from electrode studies (VREDENBERG and TONK, 1975).

3. Mobile Ion Distributions

Radioactive chloride distributions, interpreted by means of the Nernst equation, should also indicate membrane potentials. Using centrifugation techniques that may not have been adequate (McCarty, 1976) Rottenberg et al. (1972) calculated a membrane potential of 10 mV in the light. Chloride electrode measurements also produced a calculated membrane potential on the order of 10 mV (Schröder et al., 1971).

4. Delayed Light Emission

The delayed light emitted by chloroplasts in the millisecond time range after actinic light is turned off, is enhanced by a salt-induced diffusion potential positive on the inside (see Chap. III.14). This has been interpreted as showing that the (+) charges which recombine with stored electrons are on the inside of the thylakoids. The relation between the height of the enhanced light emission and the salt-induced membrane potential in mV provided a calibration curve. When applied to the portion of the initial light emission which is sensitive to membrane potential collapse by ionophores, it permitted an estimate of a membrane potential of 75 to 105 mV present right after the light was turned off (Barber, 1972; Barber and Varley, 1972).

Although the quantitative estimates are not yet uniform, the existence of some degree of membrane potential in the light seems assured. As the estimate for the steady state condition becomes firm we may add the contribution of these potentials to the ΔpH in estimating the total pmf available in the light.

5. Alternative Use of ΔpH or of $\Delta\psi$

In the chemiosmotic mechanism, a functional protonmotive force may be composed either of a pH gradient, or of an equivalent membrane potential, or of any combination of the two. Since a minute and undetectable number of protons will create a large membrane potential if no counter-ions are moving, the failure to see a pH rise in the medium is not a critical indication of the absence of a protonmotive force. Subchloroplast particles or bacterial chromatophores have rather little net proton uptake in the steady state to begin with; this can be abolished by the sorts of uncouplers that collapse the pH gradient but do not affect an existing membrane potential (see this vol. Chap. III.11 or Jagendorf, 1975a), and phosphorylation continues. In all cases the need for maintaining a sufficient $\Delta\psi$ was discovered later; electrogenic uncouplers still abolish phosphorylation by small particles.

It is a matter of comparative interest that Type C (whole) chloroplasts have most of their pmf in the form of a pH gradient. They are uncoupled best by agents that collapse the ΔpH in a neutral way, and are not affected by ionophores that collapse only a membrane potential.

The difference between large and small chloroplasts is predictable (Mitchell, 1968) because of the vast difference in surface/volume ratio. Since chlorophyll and electron carriers are in the membrane, the amount of internal volume relative to these will be very much less in small particles. The ability to neutralize the electric charge of moving H^+ ions depends either on prior internal storage of cations, or the ability to accommodate entering Cl^- ions; and either of these needs internal space which is much diminished in small particles. In addition the smaller size of the particle probably means a higher surface tension and decreased ability to swell as Cl^- ions try to come in; the osmotic forces preventing ion entry will also be much greater for small particles (WALKER, 1975). On both grounds the $\Delta\psi$ will build up quickly. However the ordinary internal buffering groups are most likely components of the inner face of the membrane, hence as abundant for small as for large particles. The very small number of protons permitted to enter by the above considerations will therefore make only the smallest changes in internal pH, and a pH gradient will not be a major part of the Δpmf.

Even with whole chloroplasts, the function of a membrane potential can sometimes be demonstrated. For instance, phosphorylation starts almost immediately when a bright light is turned on, but it is obvious that much more time (probably one to several seconds) is needed for the proton gradient to reach its full value (HIND and JAGENDORF, 1963b; JAGENDORF and HIND, 1965). Indeed, when using repeated short, bright flashes of light to drive ATP synthesis, proton accumulation has no time to occur, valinomycin $+ K^+$ for the first time inhibit phosphorylation, and it is clear from changes in 515 nm absorbancy kinetics that $\Delta\psi$ drives phosphorylation (WITT, 1975).

A membrane potential can also contribute to phosphorylation by chloroplasts in the acid-base experiment. Both URIBE (1972) and SCHULDINER et al. (1972) showed that an insufficient pH gradient could be supplemented by a K^+ diffusion potential (with valinomycin) to increase the yield of ATP.

III. ATP-Driven Proton Uptake

The phosphorylating enzyme of chloroplasts contains an enzymatic center for hydrolyzing ATP (see Sect. D.I). Once activated by the high energy state plus $-SH$ compounds, it remains active in the dark as long as ATP is supplied. If this activity represents a true reversal of ATP synthesis (see Fig. 2), the chemiosmotic theory predicts that ATP hydrolysis should pump protons into thylakoid membranes, building up the internal pmf just as light does. In striking confirmation of this prediction, proton uptake was measured by uptake of ammonium salts (CROFTS, 1966), of amines (GAENNSLEN and McCARTY, 1971) and later more directly by a pH rise in the outer medium (CARMELI et al., 1975). This phenomenon is an integral part of the chemiosmotic hypothesis, but a surprising occurrence requiring ad hoc assumptions in most alternative formulations of the mechanism.

The sensitivity of this vectorial ATPase to uncouplers provides corroborating evidence for chemiosmotic principles at work. The high energy state (light, or an acid-base transition) is needed for activation; hence uncouplers which collapse the pH gradient inhibit activation. Once activated, ATP hydrolysis is coupled to vectorial transport of protons inwards. As in electron flow, an excessive internal pmf will oppose further operation of the total vectorial reaction. Relieving the internal pmf with moderate concentrations of uncouplers does indeed stimulate ATP hydrolysis rates (HOCH and MARTIN, 1963; PETRACK et al., 1965; CARMELI et al., 1975). However some Δpmf is probably needed to keep the enzyme active, just as it was needed for activation in the first place; thus excessive concentrations of uncouplers inhibit again. Alternatively once the ATP is used up and H^+ is no longer being pumped inwards, activity of the ATPase decays away.

IV. Post-Illumination ATP Synthesis ("X_E")

Our first efforts were directed towards searching for intermediates by a light to dark transition, adding other components of the phosphorylation mixture early in post-illumination darkness. On finding a residual ability of pre-illuminated chloroplasts to do phosphorylation (SHEN and SHEN, 1962; HIND and JAGENDORF, 1963a) we decided the stored intermediate was both unknown (and therefore "X") and energetic (therefore "E"). However it is clear by now that the intermediate is a stored proton concentration gradient, so the term "X_E" is vestigial at best.

Correlations between the ability to make ATP in the dark and the stored ΔpH were established in several ways (see Chap. III.6; also JAGENDORF, 1975a). The total amount of ATP was greater than stoichiometric with electron carriers amounting to 40 to 50 mol of ATP per mol of cytochrome f, for instance. Kinetics of proton entry and efflux corresponded to kinetics of onset or loss of the ability to make ATP later. Varying the amount of internal protons by means of: changing the pH of the light stage, use of inhibitors, neutralizing an excessive membrane potential via valinomycin plus K^+ or adding penetrating weak organic bases which act as penetrating buffers, all led to corresponding changes in the amount of post-illumination ATP that could be formed. The last two cases were especially striking. Subchloroplast particles did not demonstrate either H^+ uptake in the light or post-illumination ATP synthesis; but when supplemented with valinomycin both appeared (McCARTY, 1970). With whole chloroplasts at pH 7 addition of penetrating buffers permitted amounts of proton uptake to increase 10- to 20-fold; concurrently post-illumination ATP synthesis increased 5- to 10-fold (NELSON et al., 1971; AVRON, 1971). Similar results were obtained using either valinomycin plus K^+ or permanent anions to neutralize the unfavorable membrane potential of bacterial chromatophores (LEISER and GROMET-ELHANAN, 1975; GROMET-ELHANAN and LEISER, 1975).

Inhibitors can be added to each stage separately. Electron transport inhibitors were shown thereby to affect only the light stage; energy transfer inhibitors affect the dark stage (HIND and JAGENDORF, 1965; GROMET-ELHANAN and AVRON, 1965). These results reinforced the concept that the light-dark protocol truly dissociated ATP synthesis from electron transport.

V. Acid to Base Transition

Further evidence for thylakoids' ability to use a pH gradient to drive ATP synthesis comes from the experiments in which chloroplasts are placed first at pH 4, then rapidly into a pH 8 (or higher) phosphorylating medium (JAGENDORF and URIBE, 1966a; JAGENDORF, 1967 or 1975a for more complete documentation). ATP formation in complete darkness this way is entirely insensitive to electron transport inhibitors, nevertheless is sensitive to the same energy transfer inhibitors, and inhibitory antibody to CF_1 (McCarty and Racker, 1966) as is light and electron transport-driven phosphorylation. Even the intimate chemistry of oxygen exchanges between substrates and water induced by this "artificial" high energy state duplicates those found in the light (Shavit and Boyer, 1966). Again, mimicking the effects of light, the acid-base transition drives such things as activation of ATPase (Kaplan et al., 1967) and of a weak P_i-ATP exchange reaction (Bachofen and Specht-Jurgensen, 1967); conformational changes in CF_1 (Ryrie and Jagendorf, 1972) and increased sensitivity to some inhibitors such as $KMnO_4$ (Datta et al., 1974) or trinitrobenzenesulfonate (Oliver and Jagendorf, 1975).

Considerable evidence was accumulated to show that the actual pH gradient is the driving force. For instance, a more alkaline than usual phosphorylating pH permitted ATP formation in spite of a less acid and ordinarily ineffective acid pH (Jagendorf and Uribe, 1966a). Jumps in pH from 4 to 8 or from 6 to 10, both a net 4.0 pH units, were equally effective in activating the ATPase (Kaplan and Jagendorf, 1968), although from 6 to 8 was insufficient.

The amount of ATP made could be as much as 100/mol of cytochrome f present. This amount of ATP was strongly correlated with the amount of organic acid buffer entering at pH 4. The relationship held in comparing numerous non-physiological or physiological acids, or in looking at the kinetics of acid entry compared to the kinetics of ability to make ATP if transferred rapidly to pH 8, or in examining the osmotic squeezing effects of sucrose concentrations which have no effect on photophosphorylation proper but which prevent massive amounts of acid from entering.

VI. Stoichiometrics and Thermodynamics

One of the most important but difficult tasks in testing the operation of the chemiosmotic hypothesis is to find the precise stoichiometrics between electrons moving past each "site", protons translocated and ATP formed. The chemiosmotic mechanism predicts one H^+ per e^- moving past one site. As originally formulated (Mitchell, 1966) it envisaged $2H^+$ translocated out per ATP synthesized, or translocated in per ATP hydrolyzed. The overall ratios predicted from these intermediate reactions should come to 2 e^-/ATP, or a P/e_2 ratio of 1.0 at each site. Given two sites in the electron transport chain, the overall P/e_2 ratio should be 2.0. As long as these are rigid ratios, they also set the thermodynamic requirements in terms of the extent of the pmf to form ATP at a given phosphate potential. (More extended discussion of stoichiometrics and thermodynamics will be found in Jagendorf, 1975a; McCarty, 1976.)

1. H^+/e^- and $H^+/h\nu$

The best way to measure the ratio of protons translocated to electrons is to use a pulse of light, move a lump sum of each and count the numbers moved. Using optical indicators for pH and for electron transport down the chain, fast optical methods and flash spectroscopy, Witt and colleagues have done essentially that (Witt, 1975). They report definitive ratios of 2.0 H^+ for each electron moving all the way from water to photosystem I, indicating two proton-translocating sites.

As soon as one turns to ordinary glass electrode pH measurements and much longer time scales, the lump sum approach does not work. This is because the rate of leakage of protons from chloroplast thylakoids is high relative to the rate of entry. A 2- to 5-sec flash of light could be given and the increase in pH measured, but one cannot be sure all the protons moving in are measured because they are leaking out again at the same time. The rate of leakage is much lower for mitochondria, and there this approach is feasible (MITCHELL and MOYLE, 1967). It might work, and certainly should be tried again, with either much more tightly coupled chloroplasts (REEVES and HALL, 1973) or perhaps with chloroplasts whose proton leakage rate has been artificially slowed down (PACKER et al., 1974).

Instead, efforts have been made to compare the flux rates of protons either going in or leaking out, with rates of electron transport. Accurate kinetic measurements are difficult for numerous reasons with these methods (JAGENDORF, 1975a), however the most reliable measurements, using either an especially rapidly responding pH electrode (SCHWARTZ, 1968) or a flowing system (IZAWA and HIND, 1967) show an overall H^+/e^- ratio of 2.0, or 1.0 per site. TELFER and EVANS (1972) found this ratio was the same whether or not phosphorylation was proceeding. If the proton gradient were on a side pathway, using energy for no reason related to ATP synthesis, one would expect severe competition to show up and a decrease in the ratio to occur. That it did not, seems to be critical evidence supporting the concept that the ΔpH is an obligate intermediate between electron flow and ATP formation.

On the other hand disturbing evidence comes from carefully documented reports of ratios between 3.5 and 6.7 for protons moved to quanta absorbed in the far red region of the spectrum, during cyclic electron flow (see especially HEATH, 1972). The possibility that a light-induced re-equilibration involving protons, rather than net proton pumping, had been measured is yet to be evaluated. However until any such possible artifact is uncovered these data are a reminder that evidence concerning the chemiosmotic mechanism in chloroplasts is not yet unanimously favorable.

2. H^+/ATP

This ratio can be found in both the forward and the backward direction. Careful measurements of proton uptake during ATP hydrolysis indicate a H^+/ATP ratio approaching 2.0 at pH 8 (CARMELI et al., 1975). Measurements of the initial rates of proton uptake were made with a glass pH electrode and compared to the hydrolysis of radioactive ATP. No special precautions were taken to insure the most rapid response on the part of the pH meter or electrodes, hence the recorded rates of proton uptake could be underestimated.

For ATP synthesis, reported H^+/ATP ratios vary from 2 to 4. For instance, comparing the steady state rate of phosphorylation to the initial rate of H^+ loss when the light is turned off, SCHWARTZ (1968) calculated a ratio of 2. However SCHRÖDER et al. (1971) did the same experiment but added valinomycin and K^+, to overcome the delaying effects on proton leakage of a reversed membrane potential when the light goes out, and calculated a H^+/ATP ratio of 3. Calculations based on the faster kinetics of 515 nm shift decay (calibrated for number of ions leaving the electric field), caused by synthesis of ATP as measured by optical changes indicating a pH drift, also lead to a H^+/ATP ratio of 3.0 (WITT, 1971). More recent estimates of the Berlin group run to H^+/ATP ratios of 4 (WITT, 1975). Finally the slope of the linear relation between log of phosphorylation rate and Δ pH across the thylakoid membranes can be shown, mathematically, to be equal to the ratio of H^+ consumed to ATP formed (PORTIS and MCCARTY, 1974). This ratio was found in all cases to be close to 3. Thus for chloroplasts, most of the evidence now available suggests that three protons are needed to somehow cooperate and make one ATP.

The combination of these two ratios determines the expected overall ATP/e_2 ratio. As the two electrons go past two phosphorylating sites $4\,H^+$ will be translocated; these should be capable of the synthesis of 4/3 or 1.33 ATP molecules. This is consistent with calculated P/e_2 ratios of 1.2 or 1.3 measured with naked thylakoid systems (see Sect. A.III), but not with the observed values of 1.5 or above and calculated ratios of 2.0 found with very carefully prepared, recently opened plastids. However, so far none of the measurements of H^+/ATP have been made with these more physiological plastids. It is conceivable that their H^+/ATP ratio might be 2.0 and so internally consistent with the higher P/e_2 ratio.

3. Thermodynamic Considerations

A crucial test of the chemiosmotic theory is whether the pmf built up in the light has sufficient driving power to permit ATP synthesis under actual reaction conditions and with the established stoichiometrics. The proton driving force times the number of protons used to make one ATP, will have to equal the energy for dehydration of $ADP+P_i$.

The actual energy in a phosphoryl bond of ATP varies, of course, with several parameters of the reaction – concentrations of all substrates and products, and of Mg^{2+}, pH, temperature, etc. The chloroplast system should be able to form ATP up to the point of equilibrium in an actual phosphorylation mixture, i.e. when almost all of the ADP is used up, and a great deal of ATP is present, so that the phosphate potential is as high as possible. KRAAYENHOF (1969) made such measurements for chloroplasts; when corrected for the new and slightly lower $\Delta G_0'$ value of ATP (ROSING and SLATER, 1972) one can estimate about 14 kcal/mol as the energy to aim for (MCCARTY, 1976).

The energy for this must come from either 2 or 3 protons. The required pmf expressed either as $\Delta \psi$ in mV or Δ pH in pH units, can be calculated from equation 2:

$$\Delta G' = 1.36(\Delta\,pH) + 1.36 \left(\frac{\Delta \psi}{59}\right).$$

The required values for a pmf composed either of Δ pH alone, or of $\Delta \psi$ alone, are shown in Table 1. Since pH gradients of 3.5 units are found, adequate energy is available if three protons cooperate to make one ATP. If only two protons function then one would have to sum up a membrane potential of 100 mV and a 3.3 pH unit gradient, to have the required driving force.

Table 1. Energetic requirements for ATP synthesis

$\Delta G'$ for ATP	pmf required if $H^+/ATP=2$		pmf required if $H^+/ATP=3$	
	$\Delta\psi$	or ΔpH	$\Delta\psi$	or ΔpH
kcal/mol	mV	units	mV	units
−14	342	5.0	208	3.3

For the basis of these calculations, see text.

It is well to bear in mind that these calculations are based on the highest possible phosphate potential, that found when phosphorylation in the light comes to equilibrium. This will rarely if ever be the case for chloroplasts in vivo. Based on rather indirect measurements with intact chloroplasts, HEBER and KIRK (1975) make the interesting suggestion that the stoichiometry might be flexible with a higher P/e_2 ratio, when ATP consumption is rapid (and the phosphate potential low) than when the phosphate potential is high. Measurements of the phosphate potential, ΔpH and $\Delta\psi$ all from the same reaction mixture are needed, rather than numbers from different experiments, for a full evaluation of the energetic aspects of the chemiosmotic model.

The presently available comparisons do not make it appear inevitable that thermodynamics will require us to discard the chemiosmotic framework. Also notice that the units (mV) for $\Delta\psi$ are identical to those used in comparing redox potentials of sequential electron carriers. If the chemiosmotic hypothesis is stressed by the demand for a 312 mV potential, it is not much better for any mechanism that has to turn a two-electron transfer from plastoquinone (about 0 V) to cytochrome f ($+0.365$ V) into the force driving ATP synthesis at equilibrium.

D. Role of the Coupling Factor in Phosphorylation

I. CF₁ Enzymatic Activities

As yet only one enzyme — "coupling factor number one" or CF_1 — is known to be directly involved in phosphorylation by chloroplasts. Discovery came from two independent sources. AVRON successfully uncoupled chloroplasts by dilute EDTA, then accomplished partial recoupling with the aid of a protein that came off in the EDTA solution (1963). VAMBUTAS and RACKER (1965) were able to extract from the acetone powder of chloroplasts a protein with hidden ATPase activity, evoked by incubation with trypsin. After purification the coupling factor of AVRON was found to have cryptic ATPase activity, and vice-versa.

A third reaction in which CF_1 participates is the vectorial hydrolysis of ATP by thylakoid membranes. This also is a cryptic activity, aroused by either light (PETRACK and LIPMANN, 1961) or an artificial proton gradient (KAPLAN et al., 1967) in the presence of high concentrations of −SH containing compounds (cysteine, lipoic acid, DTT, etc.). Identity of the proteins with these three activities

Fig. 6. Relationships between CF_1 on and
off chloroplast membranes, active or inactive in ATP hydrolysis. *Top left diagrams:*
CF_1 on coupled chloroplast thylakoids; *Top
right:* they have been uncoupled by removing some of the CF_1 by dilute EDTA at
otherwise low ionic strength. Activation of
CF_1 for ATP hydrolysis is accomplished by
DTT and high energy state while on chloroplast; by trypsin or by heating in ATP or
by DTT when off membranes. Reconstitution of membranes with CF_1 is accomplished by incubation together with
Mg^{2+} ions. (From JAGENDORF, 1975a)

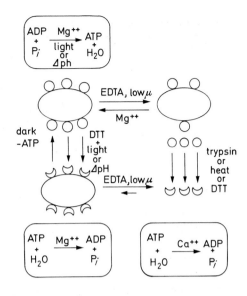

is indicated by similarities in sensitivity to inhibitors, most especially to the antibodies against purified coupling factor (MCCARTY and RACKER, 1966). Some of
the relationships between these functions of CF_1 are diagrammed in Figure 6.
A more detailed review of the biochemistry of the coupling factor will be found
in this volume, Chapter III.7 and JAGENDORF (1975a). The following section will
concentrate on the functioning of the coupling factor while on chloroplast membranes.

II. Nature, Visualization, and Location of the Protein

Purified CF_1 has a molecular weight of about 325,000 and is composed of five
different polypeptide subunits whose molecular weights range from 59,000 to
13,000. The cryptic nature of ATPase seems to be due to inhibition by the smallest
of the bound subunits ("epsilon"); activation apparently requires its digestion
by trypsin, or being temporarily pushed aside in conformational change as yet
not adequately characterized.

Both on and off the membrane, CF_1 can be visualized in the electron microscope
by negative staining, or by a newer positive staining technique (OLESZKO and
MOUDRIANAKIS, 1974) as a 90 Å knob. It sits on the surface of thylakoid membranes,
and in view of its release by dilute EDTA should be thought of as a peripheral,
rather than an integral, membrane protein (SINGER, 1974). From the more recent
studies (OLESZKO and MOUDRIANAKIS, 1974; MILLER and STAEHELIN, 1976) it appears to be localized on the outer, non-appressed membranes of grana stacks
and on stroma lamellae. As with other proteins and artificial probes, it appears
to be able to move laterally in the membranes since antibodies make the 90 Å
knobs clump together (BERZBORN et al., 1974). Based on counting knobs in the
electron microscope pictures (OLESZKO and MOUDRIANAKIS, 1974) or on measurements of total chloroplast ATPase levels compared to specific activity of pure
CF_1 (STROTMANN et al., 1973) there appears to be 1 mol per 500 to 850 chlorophylls.

III. Uncoupling, Recoupling, and Function in Proton Translocation

CF_1 is removed from chloroplasts when they are incubated at low ionic strength together with divalent cation chelators; the resulting chloroplasts are uncoupled. Recoupling is accomplished by mixing these stripped thylakoids with the extract, or with purified CF_1, then adding Mg^{2+} or other divalent cations. Phosphorylation rate is restored to a variable extent; usually better restoration occurs if it was not completely inhibited in the first place.

Actually CF_1 may have two functions in phosphorylation. In interpreting these experiments it is important to note that the EDTA treatment (MCCARTY and RACKER, 1966) or pyrophosphate pre-treatment (STROTMANN et al., 1973) usually removes only 50% of the CF_1 as judged from the Ca^{2+}-dependent, trypsin-activated ATPase left on uncoupled chloroplasts. Nevertheless uncoupling is complete; phosphorylation can disappear, and proton leakage rates are very high. Recoupling restores both of these; but so, to a certain extent, do some entirely irrelevant chemicals such as DCCD (MCCARTY and RACKER, 1967; URIBE and LI, 1973), chlorotributyltin (KAHN, 1968) and even silicotungstate anions in moderation (GIRAULT and GALMICHE, 1972). In these cases the restored phosphorylation is still inhibited by antibody to CF_1 so presumably the protein that was not removed by EDTA is performing its catalytic function. From these results one may infer that CF_1 does two things — fill up holes in the membrane through which protons can leak, and catalyze photophosphorylation. "Hole-plugging", but not catalytic activity, is replaceable by DCCD. Indeed, with chromatophores of *Rhodopseudomonas capsulata* (MELANDRI et al., 1970) removal of the coupling factor increases rather than diminishes net proton uptake in the light.

(It may be an oversimplification to think of CF_1 removal as opening up a specific hole for protons. The conditions for CF_1 removal include incubation at low ionic strength, known to cause numerous changes in thylakoid structural arrangements and reactivity, including unstacking. The proton leakage could be a function of the deranged structures primarily, and the presence of CF_1 or of an energy transfer inhibitor binding to underlying proteins might be critical for the membranes to revert to a proton impermeable state.)

The catalytic function of CF_1 in phosphorylation cannot be proven from recoupling experiments in which the stripped thylakoids retain significant amounts of their original CF_1. However 0.8% silico-tungstate was shown to remove CF_1 completely (LIEN and RACKER, 1971) and a weak but perceptible recoupling by fresh CF_1 demonstrated both the hole-plugging and catalytic functions.

The coupling factor has an intriguing relation to proton transport when on thylakoid membranes which have never been uncoupled. There are two aspects. In the first place, it should be recalled that during ATP hydrolysis protons are translocated inwards. This implies that CF_1 is a proton translocator as well as an ATPase. This seems anomalous in that there is no sign of CF_1 penetrating all the way through the membrane; whereas the protons do move all the way through. However in mitochondria and in non-photosynthetic bacteria, firm evidence has been obtained that highly hydrophobic proteins, intrinsic components of the membrane, are both the point of attachment of the coupling factor analogous to CF_1 and a part of the functioning ATPase system of the intact membrane (see, for instance, PEDERSEN, 1975). One or more of these proteins, for instance, is the site of binding of oligomycin and of DCCD when these inhibit the membrane ATPase in mitochondria. Similar proteins have not yet been found in photosynthetic systems, but their existence seems likely and their discovery, therefore, a challenge for the future. Assuming that they do occur the question can be re-phrased to what is the molecular mechanism for pumping protons through the entire ATPase system?

Finally, CF_1 seems to function in allowing protons to leak out from a pmf arising during basal electron transport. This concept comes from the fact that ATP, or reagents such as Dio-9 that interact with CF_1 make the thylakoids less permeable to protons, increase the pH in the light and thereby retard basal electron transport rates (McCarty et al., 1971; Telfer and Evans, 1972). A critical experiment showed that the ATP effect was prevented by the CF_1 antibody (McCarty et al., 1971). The proton leakage through CF_1, and its ATP reversal, are most pronounced at pH 8 and at high values of ΔpH (Portis et al., 1975); conditions which are correlated with a pronounced conformational change in CF_1 (see below). Thus in membranes the CF_1, probably together with as yet undiscovered membrane proteins, appears to act as a gated translocator for protons.

IV. Function in Phosphorylation: Conformational Changes and Ligand Binding

1. The Phosphorylation Reaction

At the moment of phosphorylation, the reaction must be exergonic. The function of the system must be to permit substrate binding, create the local chemical environment in which the reaction is exergonic, then permit release of the products into the medium. All of this may occur within the outlines of the chemiosmotic hypothesis [see Mitchell (1974) for more recent suggestions] in which components of the protonmotive force play a direct chemical role in the dehydration. An interesting alternative has been suggested by Boyer (1965, 1974) and by Slater (1974) in which the pmf acts indirectly via the coupling factor, altering its relation to tightly bound substrates. These concepts started with the suggestion by Boyer (1965) that conformational energies of the relevant enzymes might be the way in which energy of redox reactions is transmitted to the phosphorylation reaction. This was followed by the surprising evidence that pyrophosphate bond formation from substrates bound stoichiometrically at the active center of pyrophosphatase, or myosin, etc., may be entirely exergonic (Boyer et al., 1975). The energy requirements in this developing concept occur to some extent in getting $ADP + P_i$ tightly bound, and to much larger extent in getting ATP released to the medium. The pattern of oxygen exchange reactions in mitochondria and chloroplasts, and their response to uncouplers, is consistent with (and helped suggest) this pathway. The current speculation is that initial binding, the reaction itself and the release process may be facilitated by differing conformations of the coupling factor. The protonmotive force, in turn, might act to drive the particular cycle of conformational changes required. The mechanism is quite hypothetical at present, and the necessary experiments involving catalytic amounts of the substrates present a difficult challenge. However the concept adds significance to observations of conformational changes in CF_1 and to the variety of interactions with ligands that seem possible.

2. Evidence for Conformational Changes

Clear evidence for conformational changes of CF_1 attached to illuminated thyla-koids was obtained by hydrogen exchange techniques (RYRIE and JAGENDORF, 1971b, 1972) whose rationale is shown in Figure 7. Exchange between chloroplast protein hydrogen atoms and tritiated water was permitted in the high energy state; subsequent washings in the dark and purification of CF_1 allowed all exposed hydrogen atoms to exchange back with unlabeled water. The occurrence of radioac-tivity in the purified protein was evidence for the existence of groups ordinarily hidden from solvent water, and exposed only when the membrane pmf opened up interior parts of CF_1. Up to 100 hydrogen atoms per mol of CF_1 exchanged with water within a matter of seconds; hence extensive parts of the molecule must be involved in the conformational movements.

On the other hand the method suffers from a drawback — there is no way to find which parts of the molecule participate. Any attempt to separate the protein into component parts results in the tritium atoms exchanging out rapidly with unlabeled water.

A chemical approach is capable of yielding information about specific individual parts of the molecule exposed or hidden under different conditions. Covalent modification is preferable, applied either in the light or in the dark. This requires finding chemical modifiers which (a) will work effectively at a physiological pH, preferably at low concentrations, and act rapidly (within seconds); and (b) can both be present and bind covalently, without causing immediate destruction of the high energy state. Since there are many CF_1 molecules on one thylakoid, if one portion of the membrane is damaged it could ruin the high energy state and therefore the chances of any other CF_1 staying in the open configuration to be modified.

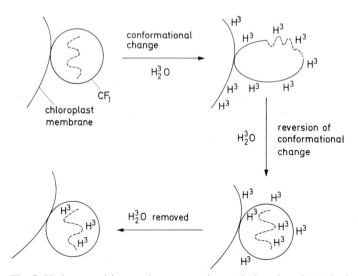

Fig. 7. Hydrogen-tritium exchange experiment designed to detect hydrogen-exchanging groups on internal portions of the CF_1 molecule, hidden from solvent water in the dark but exposed when the membrane goes into a high energy configuration. (From JAGENDORF, 1975b)

Chemical modification has been applied most successfully so far with N-ethyl maleimide (NEM) (McCarty et al., 1972; McCarty and Fagan, 1973). One mol of NEM per mol of CF_1 binds to a specific inhibitory site only if the membranes are in the high energy state. The NEM binding site has been found to be on the gamma (middle) subunit polypeptide of CF_1. Current work with trinitrobenzenesulfonate (TNBS) which binds to amino groups shows a light-dependent modification (and inhibition of ATPase) if the chloroplasts have first had their dark-available amino groups modified with methylacetimidate (Oliver and Jagendorf, 1975). In this case at least four amino groups are modified; one each on the alpha and beta subunit, and two on the gamma subunit. Work of this sort may map out the parts of CF_1 most strongly affected by the energy-dependent conformational change.

Other indications of the light-dependent conformational changes come from changing interactions with ligands—nucleotides, and phosphate analogs—in the light vs the dark. Perhaps most relevant are observations by Roy and Moudrianakis (1971 b) that added AMP is bound stoichiometrically to CF_1 on illuminated chloroplasts, and by Harris and Slater (1975) that illumination permits a rapid exchange of bound with added ADP or ATP. Neither of these events takes place in darkness, and the difference is ascribed as probably due to differing conformations of CF_1 before and during illumination of the chloroplasts.

Other indicators of the conformational change include a light-accelerated inhibition of the CF_1 ATPase function by permanganate (Datta et al., 1974) and an irreversible uncoupling of chloroplasts by pre-illumination with sulfate (Ryrie and Jagendorf, 1971 a; Grebanier and Jagendorf, in preparation). Both of these anions are in some sense analog of P_i which does prevent their respective inhibitions. In both of these cases there is a specific requirement for low concentrations of ADP as well as for the high energy state. Accordingly the proper conformation for recognition of these phosphate analogs has been inferred to require the interaction between energized CF_1 and ADP at an allosteric site.

3. Ligand Binding: Physiological Interactions

Interaction between substrate or effector molecules and the phosphorylation mechanism can be inferred from the characteristic roles these play in the reactions of chloroplasts. Specificity and effective concentrations of nucleotides, anions and divalent cations that function in phosphorylation provide the background for more esoteric analytical experiments. Among nucleotide acceptors ADP is the most effective, but GDP is also a primary acceptor (Bennun and Avron, 1965) allowing about half the maximal rate that ADP can support. K_M values for these nucleotides are variously reported between 50 or 60 and 100 μM. Evidence for direct phosphorylation of IDP, UDP and CDP (at lower rates) is less firm. Phosphate is, of course, the anion used; however arsenate is clearly a competitive inhibitor that allows rapid turnover of the high energy state (Avron and Jagendorf, 1959), and sulfate is competitive with phosphate (Asada et al., 1968) apparently acting as an energy transfer inhibitor (Pick and Avron, 1973). The Mg^{2+} requirement is replaced by Co^{2+} but not by Mn^{2+} (Arnon et al., 1958).

In the hydrolysis of nucleotides by chloroplasts, specificity is very similar to that for nucleotide phosphorylation (Bennun and Avron, 1965) which is consistent with the same active site being used in hydration and dehydration. These correlations tend to fall down when synthetic analogs of ADP or ATP are studied, however (Shahak et al., 1974).

Solubilized CF_1 hydrolyzes ATP after activation, using Ca^{2+} in most reaction media, but replaceable with Mg^{2n} as long as dicarboxylic acids or bicarbonate are present in the medium (Nelson et al., 1972).

In a surprising number of experiments ADP or ATP effects have been found that seem to relate to a controlling or allosteric role, often interacting with energy-dependent conformations of CF_1. These are all characterized by effective action of the nucleotides below 10 µM concentration, and (with one exception) no effects of other than adenyl nucleotides. However the nature of interactions of adenylates with anions, metals and each other are surprisingly diverse in the various test systems. Highlights of this diversity are summarized in the following list:

a) Inhibition of proton leakage in illuminated membranes (McCarty et al., 1971; Telfer and Evans, 1972). ATP in the µM range inhibits electron transport by decreasing the leakage of protons from thylakoid membranes. It is not clear if ADP shares this ability.

b) Inhibiting dark reversion of the membrane-bound ATPase (Carmeli and Lifshitz, 1972). When ATP runs out, chloroplast ATPase reverts to its inactive configuration. The reversion is accelerated by ADP at concentrations of less than 2 µM, and P_i decreases its effectiveness. GDP is only effective at 100 µM, P_i alone tends to prevent loss of activity.

c) NEM inhibition (Magnusson and McCarty, 1975). Protection against NEM binding and inhibition is provided by either ADP or ATP, with enhancement of either of their effects by the presence of P_i. Mg^{2+} is required for the protection by ATP but *not* for the protection by ADP.

d) Hydrogen exchange (Ryrie and Jagendorf, 1972). Half of the exchange of tritium atoms into cryptic parts of CF_1 was prevented by the combination of ADP plus P_i or arsenate, and neither GDP nor ATP would substitute for ADP. (However the necessary concentrations of ADP have not yet been determined.) The allosteric nature of this effect was indicated by the failure to prevent it by phlorizin or Dio-9 that prevent phosphorylation. The ADP function in this respect is unique by virtue of the absolute requirement for P_i (neither ADP nor P_i, alone, had any effect at all).

e) Irreversible uncoupling by sulfate (Ryrie and Jagendorf, 1971a; Grebanier and Jagendorf, in preparation). Illumination with sulfate (10 mM) present, and especially at low ionic strength, brings on a still poorly characterized uncoupling of chloroplasts. ADP is specifically required (ATP and other nucleotides do not work) at concentrations below 10 µM. Since sulfate is an analog of phosphate, this is reminiscent of the combined requirement for ADP and P_i in inhibiting 50% of the hydrogen exchange, or the synergistic action in preventing NEM inhibition. However in this case evidence was obtained that a different anion binding site is probably involved. Selenate and sulfate are both competitive against phosphate in their function as energy transfer inhibitors (Pick and Avron, 1973) and so presumably act at the same site. In the irreversible uncoupling caused by pre-illumination with sulfate, selenate has absolutely no effect and is not even competitive with sulfate. Thus this site is one that does not respond to selenate.

f) Light-facilitated inhibition by permanganate (DATTA et al., 1974). The permanganate effect, as in the inhibition by sulfate, requires both ADP and Mg^{2+} ions. Other nucleotides or ADP are ineffective.

g) Inhibition by TNBS (OLIVER and JAGENDORF, 1975). The light-dependent TNBS inhibition of methyl imidoacetic-blocked chloroplasts is prevented in part by ADP or ATP, and no other nucleotides are effective. It is also prevented in part by anions — phosphate, arsenate, molybdate or chromate but not sulfate or selenate. The anion and nucleotide protections appear to be at different sites, since saturating levels of each are additive to give complete protection. This system is further unique in that there is no requirement for added Mg^{2+} (and EDTA does not prevent the protection by adenylates or by anions).

h) Protection of the solubilized enzyme against inactivation by cold temperature and high salt concentrations (POSORSKE and JAGNDORF, in preparation). ATP prevents cold inactivation at concentrations below 10 µM. Purified ADP has no effect until Mg^{2+} is added. With Mg^{2+} also GDP, GTP, UTP and UDP give protection at low concentrations. This is the only instance so far where non-adenyl nucleotides are effective. Note also that Mg^{2+} ions are required here for ADP to work and not ATP, whereas in the NEM inhibition they are required for ATP to be effective, but not for ADP.

4. Ligand Binding: Direct Studies, and the Phosphorylation Reaction

Given results as diverse as those above, direct studies of adenylate binding by soluble CF_1, although capable of revealing more precise information, can only produce results relevant to a fraction of the differing experimentally elicited states of CF_1 on the thylakoid membranes. Studies of adenylate binding to solubilized CF_1 have included changes in the Circular Dichroism spectrum of ADP (GIRAULT et al., 1973), changes in fluorescence of added etheno-derivatives (CANTLEY and HAMMES, 1975), and binding of radioactive compounds (ROY and MOUDRIANAKIS, 1971a,; GIRAULT et al., 1973; CANTLEY and HAMMES, 1975). Even before activation of the ATPase there appear to be two rather tight binding sites for ADP or derivatives, with dissociation constants below 10 µM. CANTLEY and HAMMES consider these to be likely allosteric conformational switches, probably representing the binding sites for the ADP inhibition of ATP hydrolysis. ATP binding at the catalytic site after activation is much faster, with a similar dissociation constant. ROY and MOUDRIANAKIS (1971a) discovered that the two bound ADP moieties can undergo a transphosphorylation, leaving the enzyme with bound ATP+AMP; this has been confirmed in part by CANTLEY and HAMMES.

On the chloroplasts, added AMP was the favored nucleotide for binding (or exchange) in the experiments of ROY and MOUDRIANAKIS (1971b). Their evidence showed it was phosphorylated to ADP, then served as the phosphate donor for a more loosely bound (substrate) ADP which would then be released as the product ATP. While HARRIS and SLATER (1975) were not able to find binding or exchange of added AMP, they did observe that tightly bound nucleotides (ADP and ATP) received labeled ^{32}P in the beta position very early during phosphorylation; a result consistent with a bound AMP as first phosphate acceptor in photophosphorylation. Also BOYER et al. (1975), in studying the millisecond time course for phosphorylation of bound nucleotides, report that in chloroplasts, unlike mitochondria, beta-labeled ADP is formed in appreciable amounts (a little less than the CF_1 present) and earlier than bound or free ATP. Again, an AMP as primary acceptor in photophosphorylation seems probable.

Other measurements have been made of tightly bound nucleotides in chloroplasts or bacterial chromatophores (BACHOFEN, 1971; HORIO et al., 1966), and early phosphorylation is apparent in most of these experiments. In the case of bacterial chromatophores the evidence appears to be fairly good that phosphate from bound ATP can transphosphorylate added substrate amounts of ADP (YAMAMOTO et al., 1972; LUTZ et al., 1974). Because of the very small amounts of products considered and the superb sensitivity of the methods, the chance of detecting artifacts or side reactions is high and the results will have to be viewed with caution for some time. Nevertheless it is obviously a very important direction for further efforts.

5. A Concluding Remark

Whether the conformational changes in CF_1 and other proteins are intermediates in the chain of energy transmission leading from light to the formation of a dehydro bond in ATP, or whether they play non-energetic but controlling roles in the catalytic enzyme mechanism, elucidation of their nature and significance must be one of the important priorities for further understanding of the way in which chloroplasts work. One can think of no better statement of the challenge than the following quotation from BOYER (1974): "At some future time it may be possible to describe at the molecular level conformational events that participate in energy transductions. The magnitude of the task and the potential beauty of the result may be visualized by considering the developments leading to the present understanding of conformational transitions accompanying hemoglobin function. When and if a comparable degree of understanding is reached for energy-transducing processes in membranes, many will share in the achievement."

Acknowledgement. This paper was supported in part by grant GM 14479 from the National Institute of Health.

References

Arnon, D.I., Whatley, F.R., Allen, M.B.: J. Am. Chem. Soc. **76**, 6324–6329 (1954)
Arnon, D.I., Whatley, F.R., Allen, M.B.: Science **127**, 1026–1034 (1958)
Arnon, D.I., Whatley, F.R., Allen, M.B.: Biochim. Biophys. Acta **32**, 47–57 (1959)
Asada, K., Deura, R., Kasai, Z.: Plant Cell Physiol. **9**, 143–146 (1968)
Avron, M.: Biochim. Biophys. Acta **77**, 699–702 (1963)
Avron, M.: In: Proc. 2nd Intern. Congr. Photosynthesis. Forti, G., Avron, M., Melandri, A. (eds.). The Hague: Dr. Junk, 1972, Vol. II, pp. 861–871
Avron, M., Jagendorf, A.T.: J. Biol. Chem. **234**, 967–972 (1959)
Avron, M., Krogmann, D.W., Jagendorf, A.T.: Biochim. Biophys. Acta **30**, 144–153 (1958)
Bachofen, R.: In: Proc. 2nd Intern. Congr. Photosynthesis. Forti, G., Avron, M., Melandri, A. (eds.). The Hague: Dr. Junk, 1972, pp. 1151–1168
Bachofen, R., Specht-Jurgensen, I.: Z. Naturforsch. **22B**, 1051–1054 (1967)
Bamberger, E.S., Rottenberg, H., Avron, M.: Europ. J. Biochem. **34**, 557–563 (1973)
Barber, J.: FEBS Lett. **20**, 251–254 (1972)

Barber, J., Varley, W.J.: J. Exp. Botany **23**, 216–228 (1972)

Bedell, G.W., Govindjee: Plant Cell Physiol. **14**, 1081–1097 (1973)

Bennun, A., Avron, M.: Biochim. Biophys. Acta **109**, 117–127 (1965)

Berzborn, R.J., Kopp, F., Mühlethaler, K.: In: Proc. 3rd Intern. Congr. Photosynthesis. Avron, M. (ed.). Amsterdam: Elsevier, 1975, Vol. II, pp. 809–820

Boyer, P.D.: In: Oxidases and Related Redox Systems. King, T.H., Mason, H.S., Morrison, M. (eds.). New York: Wiley, 1965, pp. 994–1008

Boyer, P.D.: In: Dynamics of Energy-Transducing Membranes. Ernster, L., Estabrook, R., Slater, E.C. (eds.). Amsterdam: Elsevier, 1974, pp. 289–301

Boyer, P.D., Stokes, B.O., Wolcott, R.G., Degani, C.: Federation Proc. **34**, 1711–1717 (1975)

Bulychev, A.A., Andrianov, V.K., Kurella, G.A., Litvin, F.F.: Nature (London) **236**, 175–177 (1972)

Cantley, L.C., Hammes, G.G.: Biochemistry **14**, 2968–2975 (1975)

Carmeli, C., Lifshitz, Y.: Biochim. Biophys. Acta **267**, 86–95 (1972)

Carmeli, C., Lifshitz, Y., Gepshtein, A.: Biochim. Biophys. Acta **376**, 249–258 (1975)

Chance, B.: FEBS Lett. **23**, 3–20 (1972)

Chance, B., Williams, G.R.: Advan. Enzymol. **17**, 65–134 (1956)

Crofts, A.R.: Biochem. Biophys. Res. Commun. **24**, 725–731 (1966)

Crofts, A.R.: In: Regulatory Functions of Biological Membranes. Jarnfelt, J. (ed.). Amsterdam: Elsevier, 1968, pp. 247–263

Datta, D.B., Ryrie, I., Jagendorf, A.T.: J. Biol. Chem. **249**, 4404–4411 (1974)

Davis, R.F.: In: Membrane Transport in Plants. Zimmerman, U., Dainty, J. (eds.). Berlin-Heidelberg-Oxford: Springer, 1974, pp. 197–201

Fiolet, J.T.W., Bakker, E.P., Van Dam, K.: Biochim. Biophys. Acta **368**, 432–445 (1974)

Forti, G., Parisi, B.: Biochim. Biophys. Acta **71**, 1–6 (1963)

Fowler, C.F., Kok, B.: Biochim. Biophys. Acta **357**, 299–307 (1974)

Frenkel, A.: J. Am. Chem. Soc. **76**, 5568–5569 (1954)

Gaennslen, R.E., McCarty, R.E.: Arch. Biochem. Biophys. **147**, 55–65 (1971)

Galmiche, J.M., Girault, G., Tyszkiewicz, E., Fiat, R.: C. R. Acad. Sci. Paris **265**, 374–377 (1967)

Girault, G., Galmiche, J.M.: FEBS Lett. **19**, 315–318 (1972)

Girault, G., Galmiche, J.M., Michel-Villaz, M., Thiery, J.: Europ. J. Biochem. **38**, 473–478 (1973)

Good, N., Izawa, S., Hind, G.: Current Topics Bioenerg. **1**, 75–112 (1966)

Gould, J.M., Izawa, S.: Biochim. Biophys. Acta **314**, 211–223 (1973)

Gromet-Elhanan, Z., Avron, M.: Plant Physiol. **40**, 1053–1059 (1965)

Gromet-Elhanan, Z., Leiser, M.: Arch. Biochem. Biophys. **159**, 583–589 (1973)

Gromet-Elhanan, Z., Leiser, M.: J. Biol. Chem. **250**, 90–93 (1975)

Hall, D.O.: Nature New Biol. **235**, 125–126 (1972)

Harris, D.A., Slater, E.C.: Biochim. Biophys. Acta **387**, 335–348 (1975)

Hauska, G.A., Reimer, S., Trebst, A.: Biochim. Biophys. Acta **357**, 1–13 (1974)

Heath, R.L.: Biochim. Biophys. Acta **256**, 645–655 (1972)

Heathcote, P., Hall, D.O.: In: Proc. 3rd Intern. Congr. Photosynthesis. Avron, M. (ed.). Amsterdam: Elsevier, 1975

Heber, U., Kirk, M.R.: Biochim. Biophys. Acta **376**, 136–150 (1975)

Hill, R.: Nature (London) **139**, 881–882 (1937)

Hind, G., Jagendorf, A.T.: Proc. Nat. Acad. Sci. U.S. **49**, 715–722 (1963a)

Hind, G., Jagendorf, A.T.: Z. Naturforsch. **18B**, 689–694 (1963b)

Hind, G., Jagendorf, A.T.: J. Biol. Chem. **240**, 3202–3209 (1965)

Hind, G., McCarty, R.E.: The role of cation fluxes in chloroplast activity. In: Photophysiology. Giese, A.C. (ed.). New York-London: Academic Press, 1973, Vol. VIII, pp. 113–156

Hind, G., Nakatani, H.Y., Izawa, S.: Proc. Nat. Acad. Sci. U.S. **71**, 1484–1488 (1974)

Hoch, G., Martin, I.: Biochem. Biophys. Res. Commun. **12**, 223–228 (1963)

Horio, T., Kamen, M.D.: Biochemistry **1**, 144–153 (1962)

Horio, T., Nishikawa, K., Yamashita, J.: Biochem. J. **98**, 321–329 (1966)

Isaev, P.I., Liberman, E.A., Samuilov, V.D., Skulachev, V.P., Tsofina, L.M.: Biochim. Biophys. Acta **216**, 22–29 (1970)

Izawa, S.: Biochim. Biophys. Acta **223**, 165–173 (1970)

Izawa, S., Good, N.E.: Biochim. Biophys. Acta **162**, 380–391 (1968)

Izawa, S., Hind, G.: Biochim. Biophys. Acta **143**, 377–390 (1967)
Jackson, J.B., Crofts, A.R.: FEBS Lett. **4**, 185–192 (1969)
Jagendorf, A.T.: In: Survey of Biological Progress. Glass, B. (ed.). New York: Academic Press, 1962, Vol. IV, pp. 181–344
Jagendorf, A.T.: Federation Proc. **26**, 1361–1369 (1967)
Jagendorf, A.T.: In: Bioenergetics of Photosynthesis. Govindjee (ed.). New York: Academic Press, 1975a, pp. 414–492
Jagendorf, A.T.: Federation Proc. **34**, 1718–1722 (1975b)
Jagendorf, A.T., Forti, G.: In: Light and Life. McElroy, W., Glass, H.B. (eds.). Baltimore: Johns Hopkins, 1961, pp. 576–586
Jagendorf, A.T., Hind, G.: In: Photosynthetic Mechanisms in Green Plants. Nat. Acad. Sci. Nat. Res. Council **1145**, 559–610 (1963)
Jagendorf, A.T., Hind, G.: Biochem. Biophys. Res. Commun. **18**, 702–709 (1965)
Jagendorf, A.T., Uribe, E.: Proc. Nat. Acad. Sci. U.S. **55**, 170–177 (1966a)
Jagendorf, A.T., Uribe, E.: Brookhaven Symp. Biol. **19**, 215–245 (1966b)
Joliot, P., Delosme, R.: Biochim. Biophys. Acta **357**, 267–284 (1974)
Kahn, J.S.: Biochim. Biophys. Acta **153**, 203–210 (1968)
Kaplan, J., Jagendorf, A.T.: J. Biol. Chem. **243**, 972–979 (1968)
Kaplan, J., Uribe, E., Jagendorf, A.T.: Arch. Biochem. Biophys. **20**, 365–375 (1967)
Kraayenhof, R.: Biochim. Biophys. Acta **180**, 213–215 (1969)
Larkum, A.W.D., Bonner, W.D.: Biochim. Biophys. Acta **256**, 396–408 (1972)
Leiser, M., Gromet-Elhanan, Z.: J. Biol. Chem. **250**, 84–89 (1975)
Lien, S., Racker, E.: J. Biol. Chem. **246**, 4298–4307 (1971)
Lutz, H., Dahl, J.S., Bachofen, R.: Biochim. Biophys. Acta **347**, 359–370 (1974)
Lynn, W.S.: Biochemistry **7**, 3811–3820 (1968)
Magnusson, R.P., McCarty, R.E.: J. Biol. Chem. **250**, 2593–2598 (1975)
McCarty, R.E.: J. Biol. Chem. **244**, 4292–4298 (1969)
McCarty, R.E.: FEBS Lett. **9**, 313–316 (1970)
McCarty, R.E., Fagan, J.: Biochemistry **12**, 1503–1507 (1973)
McCarty, R.E., Fuhrman, J.S., Tsuchiya, Y.: Proc. Nat. Acad. Sci. U.S. **68**, 2522–2526 (1971)
McCarty, R.E., Guillory, R.J., Racker, E.: J. Biol. Chem. **240**, PC4282–PC4283 (1965)
McCarty, R.E., Pittman, P.R., Tsuchiya, Y.: J. Biol. Chem. **247**, 3048–3051 (1972)
McCarty, R.E., Racker, E.: Brookhaven Symp. Biol. **19**, 202–214 (1966)
McCarty, R.E., Racker, E.: J. Biol. Chem. **242**, 3435–3439 (1967)
Melandri, B.A., Baccarini-Melandri, A., San Pietro, A., Gest, H.: Proc. Nat. Acad. Sci. U.S. **67**, 477–484 (1970)
Miller, K.R., Staehelin, L.A.: J. Cell. Biol. **68**, 30–47 (1976)
Mitchell, P.: Biochem. Soc. Symp., Cambridge, Engl. **22**, 142–169 (1963)
Mitchell, P.: Biol. Rev. Cambridge Phil. Soc. **41**, 445–502 (1966)
Mitchell, P.: Federation Proc. **26**, 1370–1379 (1967)
Mitchell, P.: Chemiosmotic Coupling and Energy Transfer. Bodmin: Glynn Research, 1968
Mitchell, P.: In: Membranes and Ion Transport. Bittar, E.E. (ed.). New York: Interscience, 1969, pp. 192–256
Mitchell, P.: Bioenergetics **3**, 5–24 (1972)
Mitchell, P.: In: Mechanisms in Bioenergetics. New York: Academic Press, 1973, pp. 177–201
Mitchell, P.: FEBS Lett. **43**, 189–194 (1974)
Mitchell, P., Moyle, J.: Biochem. J. **105**, 1147–1162 (1967)
Nelson, N., Nelson, H., Naim, V., Neumann, J.: Arch. Biochem. Biophys. **145**, 263–267 (1971)
Nelson, N., Nelson, H., Racker, E.: J. Biol. Chem. **247**, 6506–6510 (1972)
Neumann, J., Jagendorf, A.T.: Arch. Biochem. Biophys. **107**, 109–119 (1964)
Niel, C.B. van: Advan. Enzymol. **1**, 263–328 (1941)
Niel, C.B. van: In: Photosynthesis in Plants. Franck, J., Loomis, W. (eds.). Ames, Iowa: Iowa State, 1949, pp. 437–495
Oleszko, S., Moudrianakis, E.N.: J. Cell. Biol. **63**, 936–948 (1974)
Oliver, D.J., Jagendorf, A.T.: Federation Proc. **34**, 596 (1975)
Packer, L., Torres-Pereira, J., Chang, P., Hansen, S.: In: Proc. 3rd Intern. Congr. Photosynthesis. Avron, M. (ed.). Amsterdam: Elsevier, 1975, pp. 867–872
Pedersen, P.L.: Bioenergetics **6**, 243–275 (1975)

Petrack, B., Craston, A., Sheppy, F., Farron, F.: J. Biol. Chem. **240**, 906–914 (1965)
Petrack, B., Lipmann, F.: In: Light and Life. McElroy, W.D., Glass, H.B. (eds.). Baltimore: Johns Hopkins, 1961, pp. 621–630
Pick, U., Avron, M.: Biochim. Biophys. Acta **325**, 297–303 (1973)
Portis, A.R.: PhD. Thesis, Cornell Univ., 1975
Portis, A.R., Jr., Magnusson, R.P., McCarty, R.E.: Biochem. Biophys. Res. Commun. **64**, 877–884 (1975)
Portis, A.R., Jr., McCarty, R.E.: J. Biol. Chem. **249**, 6250–6254 (1974)
Reeves, S.G., Hall, D.O.: Biochim. Biophys. Acta **314**, 66–78 (1973)
Reeves, S.G., Hall, D.O., Baltscheffsky, H.: Biochem. Biophys. Res. Commun. **43**, 359–366 (1971)
Robertson, R.N.: Biol. Rev. Cambridge Phil. Soc. **35**, 231–264 (1960)
Rosing, J., Slater, E.C.: Biochim. Biophys. Acta **267**, 275–290 (1972)
Rottenberg, H., Grunwald, T., Avron, M.: Europ. J. Biochem. **25**, 54–63 (1972)
Roy, H., Moudrianakis, E.N.: Proc. Nat. Acad. Sci. U.S. **68**, 464–468 (1971a)
Roy, H., Moudrianakis, E.N.: Proc. Nat. Acad. Sci. U.S. **68**, 2720–2724 (1971b)
Rumberg, B., Siggel, U.: Naturwissenschaften **56**, 130–138 (1969)
Rumberg, B., Witt, H.T.: Naturforsch. **19B**, 693–707 (1964)
Ryrie, I., Jagendorf, A.T.: J. Biol. Chem. **246**, 582–588 (1971a)
Ryrie, I., Jagendorf, A.T.: J. Biol. Chem. **246**, 3771–3774 (1971b)
Ryrie, I., Jagendorf, A.T.: J. Biol. Chem. **247**, 4453–4459 (1972)
Schröder, H., Muhle, H., Rumberg, B.: In: Proc. 2nd Intern. Congr. Photosynthesis. Forti, G., Avron, M., Melandri, A. (eds.). The Hague: Dr. Junk, 1972, Vol. III, pp. 919–930
Schuldiner, S., Rottenberg, H., Avron, M.: FEBS Lett. **28**, 173–176 (1972)
Schwartz, M.: Nature (London) **219**, 915–919 (1968)
Shahak, Y., Chipman, D.M., Shavit, N.: In: Proc. 3rd Intern. Congr. Photosynthesis. Avron, M. (ed.). Amsterdam: Elsevier, 1975, Vol. II, pp. 859–866
Shavit, N.: In: Proc. 2nd Intern. Congr. Photosynthesis. Forti, G., Avron, M., Melandri, A. (eds.). The Hague: Dr. Junk, 1972, Vol. II, pp. 1221–1231
Shavit, N., Boyer, P.D.: J. Biol. Chem. **241**, 5738–5740 (1966)
Shen, Y.K., Shen, G.M.: Sci. Sinica (Peking) **11**, 1097–1106 (1962)
Singer, S.J.: Ann. Rev. Biochem. **43**, 805–834 (1974)
Slater, E.C.: Nature (London) **172**, 975–978 (1953)
Slater, E.C.: In: Dynamics of Energy-transducing. Ernster, L., Estabrook, R., Slater, E.C. (eds.). Amsterdam: Elsevier, 1974, pp. 289–301
Stedingk, L.V. von, Baltscheffsky, H.: Arch. Biochem. Biophys. **117**, 400–404 (1966)
Stiehl, H.H., Witt, H.T.: Z. Naturforsch. **23B**, 220–224 (1968)
Strichartz, G.R., Chance, B.: Biochim. Biophys. Acta **256**, 71–84 (1972)
Strotmann, H., Hesse, H., Edelmann, K.: Biochim. Biophys. Acta **314**, 202–210 (1973)
Tagawa, K., Tsujimoto, H.Y., Arnon, D.I.: Proc. Nat. Acad. Sci. U.S. **49**, 567–572 (1963)
Telfer, A., Evans, M.C.W.: Biochim. Biophys. Acta **256**, 625–637 (1972)
Trebst, A.: Ann. Rev. Plant Physiol. **25**, 423–458 (1974)
Uribe, E.G.: Biochemistry **11**, 4228–4235 (1972)
Uribe, E.G., Li, B.C.Y.: J. Bioenerg. **4**, 435–444 (1973)
Vambutas, V.K., Racker, E.: J. Biol. Chem. **240**, 2660–2667 (1965)
Vredenberg, W.J., Tonk, W.J.M.: Biochim. Biophys. Acta **387**, 580–587 (1975)
Walker, N.A.: FEBS Lett. **50**, 98–101 (1975)
Warburg, O., Krippahl, G., Gewitz, H.S., Volker, W.: Z. Naturforsch. **14B**, 712–724 (1959)
West, K.R., Wiskich, J.T.: Biochem. J. **109**, 527–532 (1968)
Williams, R.J.P.: Current Topics Bioenerg. **3**, 79–156 (1969)
Winget, G.D., Izawa, S., Good, N.E.: Biochem. Biophys. Res. Commun. **21**, 438–443 (1965)
Winget, G.D., Izawa, S., Good, N.E.: Biochemistry **8**, 2067–2074 (1969)
Witt, H.T.: Quart. Rev. Biophys. **4**, 365–477 (1971)
Witt, H.T.: In: Bioenergetics of Photosynthesis. Govindjee (ed.). New York-London: Academic Press, 1975, pp. 493–554
Witt, H.T., Rumberg, B., Junge, W.: In: 19th Colloq. Gesell. Biol. Chem. Mosbach/Baden. Berlin-Heidelberg-New York: Springer, 1968, pp. 262–306
Yamamoto, N., Yoshimura, S., Higuti, T., Nishikawa, K., Horio, T.: J. Biochem. **72**, 1397–1406 (1972)

2. Proton and Ion Transport Across the Thylakoid Membranes

H. ROTTENBERG

A. Introduction

Upon illumination of a weakly buffered chloroplast suspension a prompt increase in the suspension pH is observed reaching a steady state after a short time (between $^1/_2$ and 2 min). When the light is turned off the pH gradually decreases to its initial value (JAGENDORF and HIND, 1963). This observation was interpreted as light-induced proton transport across the thylakoids membranes and opened the way for a vast number of studies. It was found that factors that stimulate electron transport greatly enhance both the *rate* and the *extent* of proton uptake, while uncouplers inhibit it (NEUMANN and JAGENDORF, 1964). MITCHELL (1961) has suggested earlier that proton transport is the process that couples electron transport to phosphorylation and the discovery of acid-base-induced ATP formation (JAGENDORF and URIBE, 1966; Chap. III.9), confirmed the existence of a link between light-induced electron transport, proton transport, and phosphorylation. Earlier HIND and JAGENDORF (1963) discovered that when chloroplasts are illuminated at low pH (6.5) and then transferred in the dark to a medium of high pH (8.5) phosphorylation takes place. The close relationships between postillumination ATP synthesis (see Chap. III.6) and proton uptake was established much later, as it was found that various buffers that penetrate the thylakoids membranes and increase the internal buffer capacity and thus the extent of proton uptake also increase several-fold the yield of postillumination ATP synthesis (NELSON et al., 1971; AVRON, 1972). Indeed the pH optimum for the extent of proton transport is 6.5 (NEUMANN and JAGENDORF, 1964) identical with the pH optimum of postillumination ATP synthesis (AVRON, 1972). However, the pH optimum for the *rate* of photophosphorylation was about 8.5 (WINGET et al., 1965), an external pH in which the *extent* of proton uptake is very low. The studies that followed these initial observations were aimed at an understanding of the mechanism of the light-induced proton pump, and the nature of the coupling of this active proton transport to electron transport, the transport of other ions and phosphorylation. A recent review of ion transport in chloroplasts can be found in HIND and MCCARTY (1973) and a comprehensive discussion of the measurement of proton electrochemical gradient in ROTTENBERG (1975).

B. The Mechanism of Light-Induced Proton Transport

It was early recognized that proton uptake is tightly coupled to electron transport (NEUMANN and JAGENDORF, 1964). However, despite enormous effort during the last decade to establish the mechanism of this coupling no satisfactory model for this process exists. This is partially due to the fact that our knowledge and understanding of the mechanism of electron transport per se is still very limited (TREBST, 1974; and see part II of this volume). The most basic piece of information which is required for establishing the mechanism of the coupling is the ratio between the number of transported protons and electrons H^+/e^-. However, the values reported from such measurements vary from 6 to 1 depending on the conditions, methods of measurements, and assumptions used in calculation (DILLEY, 1971). Some of the discrepancies are due to the fact that on illumination, both the rate of electron transport and proton transport are continuously changing and a true simultaneous determination of initial rate is not possible, once a pH gradient and a potential gradient is established, secondary proton movements through leaks, cotransport, or exchange carriers can mask the true rate of proton pumping. A calculation which is based on the steady-state values of electron transport, and on equating the initial rate of the pH dark decay with the steady rate of proton uptake generally yields lower values of H^+/e^- between 1 and 2. This ratio also depends very strongly on the external pH, being higher at low pH, a fact which most models cannot cope with. Moreover, the quantum efficiency of proton transport is invariably found to be high (DILLEY and VERNON, 1967; HEATH, 1972), being for cyclic electron transport 4–5 $H^+/h\nu$ (at 710 nm). Considering that for this system $e^-/h\nu = 1$, the H^+/e^- ratio is calculated as 4–5. These high values are difficult to reconcile with most of the models suggested for the coupling mechanism. Following the suggestion of MITCHELL (1968) it is generally assumed that the proton transport is electrically linked to electron transport (WITT, 1971; TREBST, 1974). This requires a shuttling of electrons from the external face of the membrane to the internal face (Fig. 1A). In each "loop" each electron can carry a single proton so that the ratio of H^+/e^- should equal the number of "loops". It is generally agreed that plastoquinone can act as a shuttle for electrons and protons and constitute a "loop", another loop is suggested to exist for noncyclic electron transport by separating the water-splitting reaction (inside) from the acceptor (outside). This can then account for H^+/e^- of two which is currently believed by many to be the "true" ratio (TREBST, 1974). However, for cyclic electron transport H^+/e^- should not exceed one and the high quantum efficiency cannot be reconciled with this mechanism. Of course, it is possible to suggest that the coupling between electron transport and proton transport is not electric but energetic. In this model (Fig. 1B) the free energy of the redox reaction is used by a proton carrier to pump protons. Such a coupling allows for a much higher and variable stoichiometry which is only limited by energetic factors (DILLEY, 1971). The kinetics of the light-induced pH rise and its dark decay are both exponential indicating a pseudo-first-order rate (KARLISH and AVRON, 1968). The dark decay can be fitted with a single-rate constant and can be considered as a leak through the membrane. The rise in pH is biphasic although the first phase is too fast for a study with a glass electrode (IZAWA and HIND, 1967). This phase is probably related to the binding of the protons to their carrier prior to their transport across the membrane, which is presumably the slower phase. A great deal of study on the first kinetics of proton binding and its relation to electron transport was carried out with chromatophores, utilizing pH indicators (JACKSON and CROFTS, 1969).

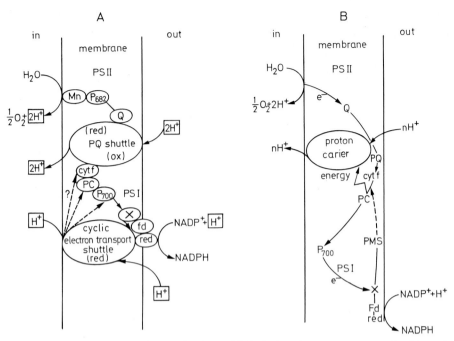

Fig. 1A and B. Models for the mechanism of light-induced proton transport in chloroplasts. (A) A chemiosmotic redox proton pump in which H⁺ are electrically coupled to electron shuttles. (After TREBST, 1974.) (B) A proton carrier pump in which H⁺ transport is not electrically linked to electron transport

C. Secondary Ion Transport

The large extent of proton uptake requires that other ions move simultaneously since the electroneutrality of the suspending medium must be maintained. Indeed, it was found (DILLEY and VERNON, 1965) that the uptake of H^+ is accompanied by the extrusion of Mg^{2+} and K^+. This indicates that at least part of the uptake of positive charges (H^+) is compensated by efflux of cations. Later, it was observed that Cl^- uptake can also be associated with proton uptake (DEAMER and PACKER, 1969; GAENSSLEN and McCARTY, 1971; ROTTENBERG et al., 1972). The problem of the charge balance during H^+ uptake was recently reinvestigated by HIND et al. (1974) with the aid of specific ion electrodes. They found that both Mg^{2+} efflux and Cl^- uptake take place simultaneously with H^+ uptake, carrying about equal amounts of charges (Fig. 2A). K^+ efflux normally constitutes only a small fraction of the H^+ influx but can be increased by increasing K^+/Mg^{2+} ratio, and Ca^{2+} can substitute for Mg^{2+} when sufficient amounts of Ca^{2+} are added to the incubation medium. Thus it appears that the ion fluxes that compensate the charges carried by the proton flux are determined by the membrane permeability, and the presence of these ions in the medium. Chloroplasts were shown to be quite permeable to Cl^- (SCHULDINER and AVRON, 1971) and the abundance of this anion in vivo and in most reaction medium suggests that normally Cl^- uptake is associated with H^+ uptake. Indeed, in the absence of permeant anions,

Fig. 2. Light-induced ion movement in chloroplasts. Top: the uptake of H^+ and Cl^- and the efflux of Mg^+ as traced by ion electrodes. (After HIND et al., 1974.) Bottom: a model of the movement of ion linked to the formation of ΔpH and $\Delta \psi$ by the proton pump. Left: ion movements driven by $\Delta \psi$; and right: driven by ΔpH

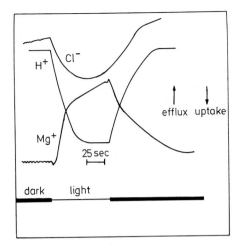

in a medium of sodium polygalacturonate (COHEN and JAGENDORF, 1972) the proton pump is inhibited. Other permeable anions such as NO_3^- and $SO_4^=$ can substitute for Cl^-. Mg^{2+} is found in large quantities in chloroplasts in vivo and is also routinely added to the reaction medium, so that normally Mg^{2+} efflux and Cl^- uptake are linked to H^+ uptake. These findings can be considered also as an additional evidence (see below) that the proton pump is electrogenic in nature. If the pump were an exchange pump (as the $Na^+ + K^+$ pump) or a cotransport pump a greater specificity for the accompanying ions is expected. On the other hand, if the H^+ pump is electrogenic, membrane potential is formed and the formation of membrane potential (positive inside) would induce permeable anion uptake and permeable cation efflux (Fig. 2B) leading to accumulation of proton and a formation of ΔpH. A completely reverse direction of movement is observed for organic anion of weak acids and for amines (weak bases). CROFTS et al. (1967) found that when chloroplasts are suspended in a salt of weak acid the rise of pH upon illumination is followed by an acidification and shrinkage. This they interpreted as efflux of the undissociated acid which is induced by the acidification of the internal volume (Fig. 2B). This phenomena was observed with a variety

of acids (PACKER and CROFTS, 1967), including the buffer tricine in high pH (ROT-
TENBERG, unpublished). Even more remarkable is the accumulation of ammonium
ion upon illumination (CROFTS, 1967). The mechanism suggested for this uptake
is the diffusion across the membrane of the free base which associated with proton
internally and leads to accumulation of NH_4^+ due to the lower pH inside (Fig. 2B).
The association of protons with the amines increases the extent of H^+ uptake
and the extent of Cl^- uptake. However, the increased extent of H^+ uptake is
not observed with amine of high pK because the dissociation of the amine externally
masks the H^+ uptake. The net result is accumulation of NH_4Cl and if the external
amine concentration is high enough, swelling is induced (SCHULDINER and AVRON,
1971).

D. Electrochemical Potential of Protons Across the Thylakoid Membranes

The observations described so far indicate that as the result of the light-induced H^+
uptake the internal volume of the chloroplast should become more acidic in relation
to the dark value and similarly that the membrane potential should become more
positive (inside). However, the magnitude of these changes has to be determined by
special techniques. The total energetic potential of proton transport is the proton
electrochemical potential, $\Delta\tilde{\mu}_H$. $\Delta\tilde{\mu}_H$ is composed of a concentration term, ΔpH,
and of an electrical potential term, $\Delta\psi$. When expressed in electrical units it is termed
"protonmotive force" PMF (MITCHELL, 1968).

$$ PMF = \Delta\tilde{\mu}_H/F = \Delta\psi - \frac{2.3\,RT}{F}\Delta pH \quad (in\ mV)[1]. \tag{1} $$

The method most commonly used for internal pH determinations in chloroplasts is
based on the observation that amines equilibrate across the membrane in their
neutral form (CROFTS, 1967). Thus, when the neutral species is at equilibrium the ratio
of the total internal amine to the external is given by (SCHULDINER et al., 1972)

$$ \frac{A_{in}^T}{A_{out}^T} = \frac{K_a + H_{in}}{K_a + H_{out}} \tag{2} $$

for amine of high pK (small K_a), where most of the amine is in the charged form

$$ \frac{A_{in}^T}{A_{out}^T} = \frac{H_{in}}{H_{out}}. \tag{3} $$

Thus a determination of the amine in/out ratio allows the calculation of ΔpH.
Various techniques were developed for this purpose. Radioactive labeled amines
such as ^{14}C-methylamine were used and the chloroplasts were separated in the
light by rapid centrifugation. The pellet/superantant count ratio is corrected for
the externally trapped water which was estimated from ^{14}C-sorbitol distribution

---

[1]   Editor's comment: the sign of the pH term depends on the direction, in which the H^+
gradient across the membrane is calculated.

(ROTTENBERG et al., 1971, 1972). Other labeled amines such as ^{14}C-ethylamine (GAENSSLEN and McCARTY, 1971) can also be used for that purpose. Once the internal volume is known, ΔpH can also be estimated from the uptake of NH_4^+ as measured by a specific electrode (ROTTENBERG and GRUNWALD, 1972). The most convenient method is the use of fluorescent amines such as 9-aminoacridine which are quenched upon uptake by the chloroplast (SCHULDINER et al., 1972). Although Atebrin quenching is also related to the formation of pH gradient (SCHUL-DINER and AVRON, 1971; GROMET-ELHANAN, 1971) its use for quantitative estimate of ΔpH is less satisfactory. Figure 3A shows the quenching of 9AA upon illumination. The values of ΔpH calculated from such measurements at different external pH are shown in Figure 3B. It is observed that the pH optimum of ΔpH is at a high pH above pH 8.5 as compared to the optimum for the extent of proton uptake (pH 6.5). The optimum for ΔpH nearly coincides with the optimum for the phosphorylation rate, indicating a correlation between the magnitude of ΔpH and the rate of phosphorylation. PICK et al. (1974), have shown that the optimum for photophosphorylation is determined by the external pH optimum of the ATP synthetase (7.5) and ΔpH (9.0). From the pH dependence of the extent of proton

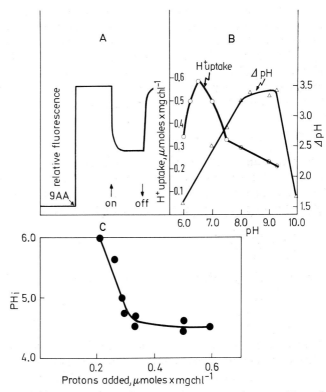

Fig. 3A–C. ΔpH and the extent of proton uptake in chloroplasts. (A) The quenching of 9AA fluorescence by illuminated chloroplasts. (After SCHULDINER et al., 1972.) (B) The ΔpH calculated from the extent of 9AA quenching and the extent of proton uptake as functions of pH. (After SCHULDINER et al., 1972; ROTTENBERG et al., 1972.) (C) The internal pH as function of the extent of uptake (data from B)

pump and ΔpH it is possible to calculate the internal pH as a function of the extent of proton uptake (Fig. 3C). It appears that the interior of the chloroplast has a strong buffering capacity below pH 5.0 which is mostly responsible for the low pH optimum for the extent of proton uptake. An identical conclusion was reached by comparing the total buffering power of chloroplasts suspensions in the dark and the light (WALZ et al., 1974). Thus the extent of proton uptake is mostly a measure of the internal buffer capacity and under most conditions does not reflect the magnitude of the pH gradient. ΔpH was found to be very sensitive to uncouplers, amines and proton-transporting ionophores such as nigericin or gramicidin (ROTTENBERG et al., 1972). It is lowered during phosphorylation and is enhanced by the energy transfer inhibitor Dio-9 (PICK et al., 1973). RUMBERG and SIGGEL (1969) have suggested that ΔpH can be estimated from the rate of electron transport assuming it is controlled by the internal pH. Indeed the assumption that internal pH controls the rate of electron transport explains fairly well (ROTTENBERG et al., 1971, 1972) the observations that in uncoupled chloroplasts the optimum for electron transport is shifted to lower pH and that at high pH, uncouplers inhibit electron transport instead of stimulating it. It also explains the multiphase kinetic of electron transport which is observed at low and high pH (BAMBERGER et al., 1973). However, it was found that in addition to the strong dependence on internal pH, the rate of electron transport is dependent on ΔpH and on external pH and thus cannot be used for accurate estimation of the internal pH (BAMBERGER et al., 1973).

Membrane potential in chloroplasts could be estimated from the distribution of permeable anions since when equilibrated ($\Delta\tilde{\mu}_n = 0$):

$$\Delta\psi = \frac{RT}{ZF} \ln \frac{(A^-)_{in}}{(A^-)_{out}}. \tag{4}$$

Thus if the potential is indeed positive inside, anions should be concentrated in the chloroplast. The observations of proton-linked ion transport (HIND et al., 1974) are best explained by the light-dependent formation of membrane potential (positive inside). However, the magnitude of this potential is found to be very low, -10 mV in the dark and about 0 in the light, giving a light dependence difference of 10 mV (ROTTENBERG et al., 1972). This low value cannot be explained by fast relaxation during centrifugation (HIND and McCARTY, 1973) since methylamine relaxes faster than either ^{36}Cl or ^{86}Rb (ROTTENBERG et al., 1972; and also see electrode traces in HIND et al., 1974). WITT (1971) and his collaborators have suggested that the light-induced 515 shift indicates formation of a membrane potential. There is a vast array of evidence to indicate that this is the case. However, the problem of calibrating the shift for quantitative estimation is unsolved as yet. Since the shift is caused by local fields within the membrane ("charge separation") its relation to transmembrane potential must be separately established. WITT (1971) has estimated a potential of 100 mV in steady-state light, based on estimation of the membrane capacity and charges transported. However, whether his simple model indeed describes the electrical events and properties of chloroplast membranes is not clear as yet. Another calibration procedure is based on the estimation of diffusion potential which was generated artificially (JACKSON and CROFTS, 1969). Its application in chloroplasts also gave relatively low potential values of about 30 mV (STRICHARTZ and CHANCE, 1972). However, this procedure is also questionable, first with regard to the diffusion potentials which are estimated but not measured, and secondly since the profile of the potential during light-dependent processes is certainly completely different from the profile of diffusion-generated potential. Since the probe indicates *local* potentials, calibration of one process by another does not seem to be justified. Considering the relative permeability of the chloroplast to both anions and cations, it is indeed not to be expected that a large steady-state potential would be established. Thus, although initially a membrane potential precedes the formation of a pH gradient (WITT, 1971), the induced secondary movements of ions reduce this potential, and ΔpH is built up. In steady-state illumination, $\Delta\tilde{\mu}_H$ in chloroplasts is probably entirely composed of ΔpH, with only a negligeable contribution by $\Delta\psi$. These conclusions are reinforced by experiments with ionophores and uncouplers (see below).

E. Ion Transport and the Mechanism of Uncoupling in Chloroplasts

MITCHELL (1968) has suggested that uncoupling in chloroplasts and mitochondria is due to collapsing of the proton motive force, and thus always caused by ion transport. Amines are known to be effective uncouplers of photophosphorylation (KROGMANN et al., 1959; HIND and WHITTINGHAM, 1963). The strong pH dependence of amine uncoupling and the good correlation between their pK and the concentration of the free base required for uncoupling were taken as evidence that the effectivness of the uncoupling is determined by the concentration of the free base (HIND and WHITTINGHAM, 1963). Since CROFTS (1967) has shown that NH_4^+ is taken up by chloroplasts and concluded that the transported species is NH_3, it was suggested that the uncoupling is the result of amine uptake. However, the uptake of the amine per se should not lead to uncoupling. During the uptake of an amine with a high pK, a proton is left in the medium for each NH_4^+ accumulated in the thylakoid. This is formally equivalent to an electroneutral exchange of $NH_4^+ - H^+$. This should decrease the internal proton concentration and allow increased proton pumping and increase $\Delta\psi$ but upon reaching a steady state, when the amine uptake has vanished, $\Delta\tilde{\mu}_H$ should not be reduced. Uncoupling should occur in steady state only when ammonium transport continues. Once the internal concentration of the charged amine has reached a high value, a leak of the *charged* amine would cause a circulation of amine, entering as a neutral amine and leaving as a charged amine. On each cycling H^+ is left in the medium with a net effect of H^+ leak. This cycling reduces $\Delta\tilde{\mu}_H$ even in a steady state, but both ΔpH and $\Delta\psi$ should be reduced. In fact this mechanism is completely analogous to uncoupling by classical uncouplers, except that in the latter case the uncoupler is present mostly in the membrane and the H^+ is carried by the neutral acid. The rate-determining step for amine cycling is the leak of the charged species and therefore, it is the concentration of the charged amine *internally* that is the determinant for uncoupling and not the free base externally, as was proposed previously. This explains why amines with high pK are better uncouplers (HIND and WHITTINGHAM, 1963) and also why amines in which the charged species can dissolve in the membrane, such as long chain alkylamines, are very effective uncouplers even in low concentration.

In chromatophores and subchloroplast particles amines are less effective uncouplers than in chloroplast. This fact was explained by assuming that amines only collapse ΔpH, where as $\Delta\psi$, which is high in these systems (see below), is not affected. However, in the mechanism which was outlined above, amines should collapse both ΔpH and $\Delta\psi$ and their relative ineffectivness is explained as the result of their lower *internal* concentration in these systems, which is due to the lower magnitude of ΔpH. Another event, often associated with amine uptake is swelling. Swelling can increase the membrane leak to charged amines or even to protons. ΔpH was found to be reduced in hypotonic medium (ROTTENBERG et al., 1972), and it was found that in high salt medium, amines are less effective uncouplers than in low salt (NEUMANN, 1972). Nevertheless, in comparing the effectivness of various ammonium salts, no correlation between swelling and uncoupling was found (SCHULDINER and AVRON, 1971). There are various other ion movements which are associated with uncoupling in chloroplasts. Nigericin which catalyzes an electroneutral $K^+ - H^+$ exchange was found to be a very effective uncoupler in chloroplasts (SHAVIT et al., 1968). If our argument for the mechanism of amine uncoupling is correct, then in the case of nigericin also, there should be no uncoupling when potassium transport has come to a steady state, unless circulation of potassium is possible by leak of the charged potassium. In chloroplasts, which are relatively permeable to potassium, the high accumulation of K^+ caused by the nigericin exchange is probably sufficient for inducing a leak and thus uncoupling. In chromatophores, which are apparently less permeable to potassium, valinomycin has to be added in addition to nigericin to allow circulation of K^+ and cause uncoupling (JACKSON et al., 1968).

F. ATP-Induced Proton Transport

Energy for proton uptake can be provided also by ATP, when the ATPase system is properly activated (CARMELI, 1970). This can be interpreted either as evidence for the existence of a separate proton pump or by assuming that ATP and electron transport are both capable of generating an intermediate which can be utilized by the same proton pump (as in Fig. 1B, for instance). ATP was also found to induce NH_4^+ (CROFTS, 1967) and other amine uptake (GAENSSLEN and MCCARTY, 1971) and the ΔpH which was calculated for the steady state from 9-aminoacridine uptake is about 3.0 (BAKKER-GRUNWALD and VAN-DAM, 1973). The stoichiometry of the ATPase-dependent H^+ transport was first estimated at $2H^+/ATP$ (CARMELI, 1970), but later higher values were observed in low pH (CARMELI et al., 1975). In view of the difficulties in evaluating the true stoichiometry of this pump, it is instructive to estimate this ratio from energetic considerations. From the relation between the phosphate potential and ΔpH both during ATP hydrolysis or during phosphorylation, a most plausible value of $3H^+/ATP$ was arrived at (ROTTENBERG, in press). It was shown that light-induced H^+ uptake is stimulated by low concentration of ATP, whereas the efflux is inhibited (McCARTY et al., 1971). Thus, during illumination both pumps affect proton transport. The interaction between the ATPase proton pump and the light-dependent pump is the essence of MITCHELL's chemiosmotic hypothesis and is discussed at length in several other sections of this volume (e.g. III. 1; III. 4; III. 8).

G. Proton Transport in Subchloroplast Particles and Chromatophores

The extent of proton transport in these systems is small compared with chloroplasts (NELSON et al., 1970; BALTSCHEFFSKY et al., 1971). This fact and their relative resistance to amine and nigericin uncoupling was generally interpreted as evidence that these systems are less permeable to anions and cations and, therefore, ΔpH is reduced and $\Delta\psi$ is larger than in chloroplasts. Indeed estimates of these parameters in chromatophores (SCHULDNER et al., 1974; CASADIO et al., 1974) shows that under normal conditions $\Delta\tilde{\mu}_H$ is about equally distributed between ΔpH and $\Delta\psi$. The large value of $\Delta\psi$ is assumed to be due to relative impermeability to ions, thus addition of truly permeating ions (i.e. those that penetrate in their charged form) decreases $\Delta\psi$, increases ΔpH and also the extent of proton uptake (GROMET-ELHANAN, 1972).

Table 1 summarizes measurement of $\Delta\tilde{\mu}_H$ in chloroplasts, chromatophores and subchloroplast particles. Although there is considerable disagreement between the

Table 1. Estimates of $\Delta\tilde{\mu}_H$ in steady-state light conditions

System	Method	ΔpH	$\Delta\psi$	$\Delta\tilde{\mu}_H$	References
Chloro-plasts	9AA quenching ^{86}Rb, ^{36}Cl distributions	3.5–4.0	0	210–240	SCHULDINER et al. (1972) ROTTENBERG et al. (1972) PICK et al. (1974)
	Rate of e.T 515 shift	3.0	30–100	210–280	RUMBERG and SIGGEL (1969) WITT (1971) STRICHARTZ and CHANCE (1972)
Chromato-phores	^{14}C-methylamine ^{14}C-SCN$^-$ distribution	1.8	90	210	SCHULDINER et al. (1974)
	9AA quenching 515 shift	2.5	190–280	300–400	CASADIO et al. (1974)
Subchloro-plasts	9AA quenching ^{131}I distribution	2.6	60	210	ROTTENBERG and GRUNWALD (1972) ROTTENBERG (unpublished)

various estimates, it indicates that in chloroplasts Δ pH is the major contribution to $\Delta\tilde{\mu}_H$, whereas in chromatophores and subchloroplast particles $\Delta\psi$ contributes significantly to $\Delta\tilde{\mu}_H$. Mitchell has stressed the role of $\Delta\tilde{\mu}_H$ in energy transduction. However, $\Delta\tilde{\mu}_H$ might also be important for energy storage. Because both the thylakoid and the lamellar space in the chloroplast have a high buffer capacity, a generation of ΔpH across the thylakoid membrane in the chloroplast can store large amounts of energy. Since chloroplasts can take up to 1 μmol H^+/mg chlorophill (which is equivalent to about 100 mM internal buffer) and their buffering capacity is larger at low pH (Fig. 3), and since there is in the light a gradient of up to 4 pH units, the total energy content that can be stored in this gradient is about 0.5 kcal/l. ATP 10 mM, with a phosphate potential of 15 kcal/mol can store only 0.15 kcal/l. Thus the energy stored in a proton gradient in the chloroplast is larger than the energy stored in ATP.

References

Avron, M.: In: Proc. 2nd Intern. Congr. Photosynthesis. Forti, G., Avron, M., Melandri, A. (eds.). The Hague: Dr. Junk, 1972, Vol. II, p. 861
Bakker-Grunwald, T., Van Dam, K.: Biochim. Biophys. Acta **292**, 808–814 (1973)
Baltscheffsky, H., Baltscheffsky, M., Thore, A.: Current Topics Bioenerg. **4**, 273–325 (1971)
Bamberger, E.S., Rottenberg, H., Avron, M.: Europ. J. Biochem. **34**, 557 (1973)
Carmeli, C.: FEBS lett. **7**, 297 (1970)
Carmeli, C., Lifshitz, Y., Gepshtein, A.: Biochim. Biophys. Acta **376**, 249–258 (1975)
Casadio, R., Baccarini-Melandri, A., Melandri, B.A.: Europ. J. Biochem. **47**, 121–128 (1974)
Cohen, W.S., Jagendorf, A.T.: Arch. Biochem. Biophys. **150**, 235 (1972)
Crofts, A.R.: J. Biol. Chem. **242**, 3352 (1967)
Crofts, A.R., Deamer, D.W., Packer, L.: Biochim. Biophys. Acta **131**, 97 (1967)
Deamer, D.W., Packer, L.: Biochim. Biophys. Acta **172**, 539 (1969)
Dilley, R.A.: Current Topics Bioenerg. **4**, 237 (1971)
Dilley, R.A., Vernon, L.P.: Arch. Biochem. Biophys. **111**, 365 (1965)
Dilley, R.A., Vernon, L.P.: Proc. Nat. Acad. Sci. U.S. **58**, 395 (1967)
Gaensslen, R.E., McCarty, R.F.: Arch. Biochem. Biophys. **147**, 55–65 (1971)
Gromet-Elhanan, Z.: FEBS Lett. **13**, 124 (1971)
Gromet-Elhanan, Z.: Biochim. Biophys. Acta **275**, 125 (1972)
Heath, R.L.: Biochim. Biophys. Acta **256**, 645–655 (1972)
Hind, G., Jagendorf, A.: Proc. Nat. Acad. Sci. U.S. **49**, 715 (1963)
Hind, G., McCarty, R.E.: In: Photophysiology. Giese, A.C. (ed.). New York: Academic Press, 1973, Vol. VIII, pp. 113–156
Hind, G., Nakatani, H.Y., Izawa, S.: Proc. Nat. Acad. Sci. U.S. **71**, 1484 (1974)
Hind, G., Whittingham, C.P.: Biochim. Biophys. Acta **75**, 194 (1963)
Izawa, S., Hind, G.: Biochim. Biophys. Acta **143**, 377 (1967)
Jackson, J.B., Crofts, A.R.: Europ. J. Biochem. **10**, 226 (1969)
Jackson, J.B., Crofts, A.R., von Steding, L.V.: Europ. J. Biochem. **6**, 41 (1968)
Jagendorf, A.T., Hind, G.: Nat. Acad. Sci. Nat. Council **1145**, 599–610 (1963)
Jagendorf, A.T., Uribe, E.: Proc. Nat. Acad. Sci. U.S. **55**, 170–177 (1966)
Karlish, S.J.D., Avron, M.: Biochim. Biophys. Acta **153**, 873 (1968)
Krogmann, D.W., Jagendorf, A.T., Avron, M.: Plant Physiol. **34**, 272 (1959)
McCarty, R.E., Fuhrman, J.S., Tsuchiya, Y.: Proc. Nat. Acad. Sci. U.S. **68**, 2522 (1971)
Mitchell, P.: Nature (London) **191**, 144–148 (1961)
Mitchell, P.: Chemiosmotic Coupling and Energy Transduction. Bodmin: Glynn, 1968
Nelson, N., Dechsler, Z., Neumann, J.: J. Biol. Chem. **245**, 143 (1970)
Nelson, N., Nelson, H., Naim, V., Neumann, J.: Arch. Biochem. Biophys. **145**, 263 (1971)

Neumann, J.: Congress of Photobiology, Bochum (abstract), 1972
Neumann, J., Jagendorf, A.T.: Arch. Biochem. Biophys. **107**, 109–119 (1964)
Packer, L., Crofts, A.R.: In: Current Topics in Bioenergetics. Sanadi, D.R. (ed.). New York: Academic Press, 1967, Vol. II, p. 23
Pick, U., Rottenberg, H., Avron, M.: FEBS Lett. **32**, 91 (1973)
Pick, U., Rottenberg, H., Avron, M.: FEBS Lett. **48**, 32–36 (1974)
Rottenberg, H.: J. Bioenergetics, **7**, 61–74 (1975)
Rottenberg, H.: In: Progress in Surface and Membrane Science. Danielli, J.F., Cadenhead, A., Rosenberg, M.D. (eds.). New York: Academic Press, in press
Rottenberg, H., Grunwald, T.: Europ. J. Biochem. **25**, 71–74 (1972)
Rottenburg, H., Grunwald, T., Avron, M.: FEBS Lett. **13**, 41 (1971)
Rottenberg, H., Grunwald, T., Avron, M.: Europ. J. Biochem. **25**, 54–63 (1972)
Rumberg, B., Siggel, U.: Naturwissenschaften **58**, 130–132 (1969)
Schuldiner, S., Avron, M.: Europ. J. Biochem. **19**, 227–231 (1971a)
Schuldiner, S., Avron, M.: FEBS Lett. **14**, 233 (1971b)
Schuldiner, S., Padan, E., Rottenberg, H., Gromet-Elhanan, Z., Avron, M.: FEBS Lett. **49**, 174 (1974)
Schuldiner, S., Rottenberg, H., Avron, M.: Europ. J. Biochem. **25**, 64–70 (1972)
Shavit, N., Dilley, R.A., San Pietro, A.: Biochemistry **7**, 2356 (1968)
Strichartz, G.R., Chance, B.: Biochim. Biophys. Acta **256**, 71 (1972)
Trebst, A.: Ann. Rev. Plant Physiol. **25**, 423–458 (1974)
Walz, D., Goldstein, L., Avron, M.: Europ. J. Biochem. **47**, 403 (1974)
Winget, S.P., Izawa, S., Good, N.E.: Biochem. Biophys. Res. Commun. **21**, 438 (1965)
Witt, H.T.: Quart. Rev. Biophys. **4**, 365–477 (1971)

3. Bound Nucleotides and Conformational Changes in Photophosphorylation

N. Shavit

A. Introduction

The mechanism by which energy-transducing membranes utilize energy from oxidation-reduction reactions to promote the synthesis of ATP remains a major unsolved problem in molecular biology. Hypotheses of energy coupling assume the existence of either a common high-energy intermediate or an energized state of the membrane linked, on the one hand, to electron transport reactions and, on the other, to the hydration–dehydration reactions of ATP formation. Investigations during the last decade have provided ample evidence that electron transport is coupled to the vectorial translocation of ions (particularly protons) across the membrane. The electrochemical gradient resulting from these ion activities, the so-called "protonmotive force" of the chemiosmotic hypothesis (MITCHELL, 1966) may be indeed directly linked to the dehydration of ADP and P_i to form ATP (for a comprehensive review see JAGENDORF, 1975). However, based on current knowledge, we are still unable to propose a molecular mechanism for energy transfer from an electrochemical gradient of protons across the membrane to ATP formation.

In chloroplast membranes the only enzyme known to participate in the catalysis of ATP synthesis is the "chloroplast-coupling factor 1" or CF_1. In analogy to the mitochondrial coupling enzyme, CF_1 is a protein of 325,000 molecular weight containing five different subunit peptide chains (RACKER et al., 1971). The participation of this protein in ATP hydrolysis and in the electron transport-coupled ATP formation has been widely demonstrated. However, it is not known how the reversible ATPase functions in the transfer and use of energy in ATP formation.

More recently firmly bound nucleotides have been found in beef heart and rat liver submitochondrial particles, chloroplast thylakoid membranes, *Rhodospirillum rubrum* chromatophores, *Escherichia coli* and *Streptococcus faecalis* membranes (see SLATER, 1974). The wide distribution of such nucleotides in a variety of energy-transducing membranes and the fact that these nucleotides undergo exchange or phosphorylation, upon energization of the membrane, support the idea of their involvement in energy tranductions. Indeed, the concept that energy coupling involving conformational changes of mitochondrial membrane-bound proteins was suggested more than ten years ago (BOYER, 1965; HACKENBROOCK, 1966; GREEN, 1970). More recently, a modified and highly attractive proposal of conformational coupling has been suggested by BOYER (1974) and SLATER (1974). Important in this proposal is the concept that energy input is required to effect the release of a performed ATP molecule firmly held at the catalytic site by means of a conformational change in the coupling ATPase. This chapter will deal with the binding of nucleotides and nucleotide analogs to thylakoid membranes and to the isolated CF_1 protein, and with their proposed role in photophosphorylation.

B. Tightly Bound Nucleotides on Isolated and Membrane-Bound CF_1

The presence of nonreactive ATP and ADP tightly bound to submitochondrial particles was first recognized by KLINGENBERG (1967). HARRIS et al. (1973) have shown that the isolated coupling factor (F_1) from beef-heart mitochondria contains 3 mol of ATP and 2 mol of ADP/mol of enzyme. These nucleotides remain noncovalently bound to the enzyme even after repeated washings and are only partially removed by treating the enzyme with activated charcoal or by filtration through Sephadex. The incorporation of $^{32}P_i$ into mitochondrial protein fractions observed by KEMP (1966) and CROSS and BOYER (1973) and into chromatophores of *R. rubrum* by YAMAMOTO et al. (1972) and LUTZ et al. (1974) probably results from phosphorylation or exchange reactions involving tightly bound nucleotides.

Binding of nucleotides to chloroplast-coupling factor and the presence of firmly attached nucleotides in the isolated and membrane-bound CF_1 have also been reported. Equilibrium dialysis experiments indicate the presence of two binding sites for ATP per CF_1 molecule (LIVNE and RACKER, 1969). The binding of ADP to two nonidentical sites on CF_1 was reported by ROY and MOUDRIANAKIS (1971a); this binding was not affected by the presence of other nucleoside diphosphates. Extending these studies and using circular dichroism GIRAULT et al. (1973) have shown that the binding of ADP was inhibited by P_i, pyrophosphate and polyphosphate. More recently, CANTLEY and HAMMES (1975a) report the binding of ADP to two identical "tight-binding" sites on the isolated CF_1 which are probably not the catalytic sites for ATP hydrolysis. An additional third nucleotide tight-binding site appears in the heat-activated enzyme.

HARRIS and SLATER (1975) found that spinach chloroplast membranes contain 2 mol of ADP and 1 mol of ATP/mol of CF_1. The AMP level was much lower and quite variable. The loss of membrane-bound nucleotides was correlated with the removal of the CF_1 from the membranes, indicating their binding to CF_1. The isolated CF_1 contained less than one mol of each nucleotide/mol of coupling factor ATPase. No AMP was detected. Formation of $[\gamma^{-32}P]$ ATP by short-term illumination (3 sec) of washed lettuce chloroplast membranes, in the absence of added nucleotides, show that these membranes contain between 1 and 2 mol of membrane-bound ADP (AFLALO, 1975). These tightly bound nucleotides in the isolated and the membrane-bound CF_1 were shown to participate in exchange reactions believed to represent steps in the overall mechanism of ATP synthesis. ROY and MOUDRIANAKIS (1971a) observed a slow incorporation of $[^{14}C]$ADP into isolated CF_1 with the formation of a relatively stable complex between ADP and CF_1. The ADP was then transphosphorylated in a myokinase type of reaction to ATP, which remained tightly bound, and to AMP, which was easily lost to the medium. The levels of ATP found in chloroplast membranes (HARRIS and SLATER, 1975) did not vary considerably if membranes, energized or nonenergized, were subjected to several washes with either P_i, pyrophosphate, $MgCl_2$, ATP or ADP. Loss of bound ADP occurred upon energization of the membrane under conditions known to induce the CF_1-ATPase activity. Exchange between the bound nucleotides or with other nucleotides in solution required energization of the membrane. Exchange between free P_i and the bound nucleotides however, required energization of the membrane and the presence of pyocyanine and free Mg^{2+}. The label was introduced into both the γ and β positions, indicating that

$P_{i\,free} \leftrightarrow ATP_{bound}$ and $P_{i\,free} \leftrightarrow ADP_{bound}$ exchange reactions occur. Roy and Moudrianakis (1971 b) reported the formation of a complex between ADP and mem-
brane-bound CF_1, when chloroplast membranes were energized in the presence
of AMP and P_i. The formation of this complex was faster than that with isolated
CF_1 and did not occur if AMP was replaced by ADP. Arsenate or sulfate inhibited
its formation. Based on these results they suggested AMP as the initial acceptor
of P_i in photophosphorylation. Results compatible with this role for AMP were
reported by Boyer et al. (1975). Investigations of the labeling pattern of ADP
and ATP by energized chloroplasts in the presence of $^{32}P_i$, using a rapid mixing
technique, showed a small lag in [^{32}P]ATP formation and a rapid labeling of
ADP at reaction times up to 50 msec. Although the amount of $^{32}P_i$ incorporated
was very small, less than the amount of CF_1 present, the detectable amount of
[^{32}P]ADP was equal or greater than the initial amounts of [^{32}P]ATP formed.
However by comparing the labeling of the bound nucleotide with that of the
free nucleotide formed in photophosphorylation (see Avron, 1961) and by taking
into consideration the fact that no CF_1-bound AMP was detected, Harris and
Slater (1975) ruled out the participation of AMP as an initial acceptor of phosphate in the major route of photophosphorylation. They suggest that the appearance
of label in both the γ and β positions of the bound nucleotides represent exchange
reactions rather then de novo synthesis. Indeed, recent experiments (Aflalo, 1975)
with washed lettuce chloroplasts which were illuminated for 2–15 sec in the presence
of $^{32}P_i$ and without added nucleotides show that 99% of the $^{32}P_i$ initially incorporated is in [γ-^{32}P]ATP. The phosphorylation of the bound ADP was dependent
upon illumination and was stimulated by PMS and Mg^{2+}. Upon prolonged illumination (up to a maximum of 15 min), [^{32}P] was found in both the γ and β positions.
The relative distribution of label in the β positions of ATP and ADP did not
exceed 10–15% of the total label incorporated into the bound nucleotides. Horak
and Zalik (1974), furthermore, found a slow but definite phosphorylation of an
α, β methylene ADP analog to ATP catalyzed by chloroplast membranes. This
analog cannot participate in the transfer of its β phosphate in a myokinase type
of reaction but is nevertheless an acceptor of P_i in ATP formation. Based on
the above-mentioned experiments, it is safe to conclude that the initial acceptor
in photophosphorylation is ADP and not AMP.

C. Nucleotide and Nucleotide Analogs: Binding and Activity

Substrate analogs resembling the natural substrate and able to interact with the
active site of an enzyme are useful tools in the elucidation of the mechanism
of enzymic reactions. Fluorescent substrate analogs can provide further information
on mechanism and on changes in the conformation of an enzyme active site.
Equilibrium binding, fluorescence polarization, fluorescence energy transfer and
steady-state activity measurements, using nucleotides and fluorescent adenine nucleotide analogs, have provided valuable data on the interactions of isolated or
membrane-bound CF_1 with ADP and ATP. Cantley and Hammes (1975a) have
shown that the isolated spinach CF_1 binds ADP, 1,N^6-ethenoadenosine diphos-

phate (εADP), adenylyl imidodiphosphate (AMP-PNP) and $1,N^6$-ethenoadenylyl imidodiphosphate (εAMP-PNP) to two apparently identical sites with dissociation constants less than $10\,\mu M$. The binding strength is similar in the presence of Mg^{2+} and Ca^{2+}. A third AMP-PNP binding site was found with the heat-activated enzyme. AMP-PNP was also shown to inhibit competitively the Ca^{2+}-ATPase activity of the heat-activated CF_1. From the observed equilibrium dissociation constant for this site and the competitive inhibition constant it was suggested that this is the active site for ATP hydrolysis. Fluorescence energy transfer measurements indicate that the tight-binding sites are quite far from the active site (CANTLEY and HAMMES, 1975b). These results led to the suggestion that the tight sites serve as an allosteric conformational switch for the ATPase activity. VANDERMEULEN and GOVINDJEE (1975) have shown that isolated CF_1 binds εADP and εATP but not εAMP. The binding of εADP was affected by the divalent cation present and by P_i. No effect of P_i was found on the Mg^{2+}-dependent binding of εATP. The binding of εADP and εATP to the membrane-bound CF_1 is indicated from the inhibition of the light-triggered ATPase activity by εADP and from the fact that εATP replaces ATP in effects believed to involve high-affinity binding sites on CF_1 (SHAHAK et al., 1973).

In view of the complexity in structure and function of the coupling proteins and the important role that the membrane plays in energy transduction, considerations of the in vivo function of CF_1 must take into account differences between the solubilized and the membrane-bound enzyme. Photophosphorylation and ATPase or P-ATP exchange reactions with the membrane-bound enzyme show a dissimilar substrate specificity. While GDP and IDP are good substrates in photophosphorylation (AVRON, 1960), GTP and ITP are rather poor substrates in the P-NTP exchange reaction (CARMELI and AVRON, 1967). Similarly, εADP is a good substitute for ADP in photophosphorylation while εATP is a very poor substitute for ATP in the ATPase and P-ATP exchange reactions (SHAHAK et al., 1975). Synthesis and hydrolysis or exchange reactions are carried out under somewhat different assay conditions and this may be the cause for the observed differences in specificity. However, εATP is not only an ineffective substrate for exchange or hydrolysis, but also inhibits ATP utilizing reactions competitively with ATP. This dissimilar specificity of the forward and back reactions of photophosphorylation led to the suggestion (SHAHAK et al., 1973) of different sites for ATP synthesis and hydrolysis. The lack of inhibition of ATP synthesis in mitochondria by AMP-PNP (PENEFSKY, 1974) was also interpreted as involving different sites for synthesis and hydrolysis. SHAHAK et al. (1975) suggested that different conformational states of membrane-bound CF_1 may exhibit altered nucleotide binding and turnover rates. εATP may have to bind to a site on the energized membrane-bound CF_1 very specific for adenosine. On the other hand, εADP could bind to the less specific alternate state of the protein, become phosphorylated and be released upon input of energy. However, the binding constants and turnover rates of nucleotides and nucleotide analogs to membrane-bound CF_1 are not known. With the soluble enzyme isolated from lettuce chloroplasts, BANAI (1975) confirmed the presence of two binding sites for ADP with a dissociation constant of about $1\,\mu M$. At $5\,mM\ Mg^{2+}$, the affinity of the protein for εADP, GDP and CDP is very similar to that for ADP. However strikingly different affinities for nucleotides were observed at lower Mg^{2+} concentrations. With $50\,\mu M\ Mg^{2+}$, only one mol of ADP is bound per mol of CF_1 with a dissociation constant less than 10 μM. Under these conditions the affinity of CF_1 for εADP is about 30 times lower than for ADP, and its affinity for GDP and CDP is lower still. Banai (1975) suggests that the concentration of a divalent cation affects the conformational state of soluble CF_1 and that these results add support to the proposal of Shahak et al. (1975) that different states of membrane-bound CF_1 may be involved in photophosphorylation (low specificity) and the partial reactions of photophosphorylation (very specific for adenosine).

D. Antisera to CF_1

Antibodies to spinach CF_1 obtained from mouse serum inhibited CF_1-ATPase acitivity as well as other chloroplast reactions involving adenine nucleotides (McCarty and Racker, 1966). No inhibition of proton movements or of coupled electron transport was observed. More recently, antisera to CF_1 elicited in rabbits inhibited ATP synthesis, H^+ uptake and coupled electron transport (McCarty et al., 1971). An immunoglobulin fraction also elicited in rabbits against lettuce CF_1 behaved similarly (Shoshan et al., 1975). However, Fab fragments of modified multivalent immunoglobulin fractions (obtained by chemical or enzymic treatment of the original anti-CF_1) strongly inhibited the P-ATP exchange and the soluble or membrane-bound ATPase activities, but had little effect on ATP formation (Shoshan et al., 1975; Shoshan and Shavit, unpublished). Moreover Fab fragments did not prevent the acceleration of the decay rate of the 520 nm change induced by phosphorylating conditions while multivalent anti-CF_1 did (Schmid et al., 1975).

This differential inhibition of ATP utilizing reactions by the modified anti-CF_1 molecules might reflect the structure and function relationships of the membrane-bound CF_1. The preferential inhibition of ATP hydrolysis and P-ATP exchange reactions by the "crippled" anti-CF_1 molecules could be explained by assuming that the inhibition is due to the interaction between antibody molecules and regulatory sites for ATPase activity on the α subunit (see McCarty et al., 1971; Cantley and Hammes, 1975a). However a reduced catalytic activity at the active site for ATP formation would only be achieved efficiently by interaction with an intact antibody molecule.

E. Conformational Coupling in Thylakoid Membranes

Energy derived from oxidation-reduction reactions must be converted to some form to be transmitted and used by the phosphorylation complex. Recent proposals assume that energy transduction occurs by conformationally linked changes in affinities of substrates for the coupling factor ATPase (Boyer, 1974; Slater, 1974). This concept is based on the existence of tightly bound nucleotides on membrane-bound F_1 and the well-documented uncoupler insensitive P_i-HOH exchange in mitochondrial particles (BOYER et al., 1973). An important step in the conformational coupling hypothesis, not considered in other suggested coupling mechanisms, is the formation of ATP at the catalytic site with limited or no energy input (Fig. 1). The ATP formed binds to the coupling protein with the formation of a tightly bound nucleotide-protein complex (reaction I). This conformation is stable and can be isolated as such from energy-transducing membranes. Upon energy input from the electron transport chain, the tightly bound ATP is released via an energy-linked conformational change (reaction II + III). It is possible that an ATP loosely bound at the catalytic site does exist (the transition from a tight to a loose or nonexistent binding site for ATP accompanies the conformational change during ATP release). The coupling protein regains its original conformation with little or no energy requirement (reaction IV). The release of ATP is accomplished in such a manner that hydrolysis at the catalytic site is avoided.

Several important observations point to the possibility that conformational changes in CF_1 may participate in the mechanism of photophosphorylation. In isolated chloroplasts, which usually do not catalyze ATP hydrolysis, energization of the membrane is required in order to activate the latent ATPase and nucleotide exchange activities. These results suggest that, as a result of a conformational change in membrane-bound CF_1, the epsilon subunit (proposed to be the natural inhibitor of ATPase activity) is moved aside (NELSON et al., 1972). RYRIE and JAGENDORF (1971) observed a hydrogen exchange between groups on membrane-bound CF_1 and medium water. This exchange depended upon energization of the membrane and was inhibited by uncouplers. Up to 100 hydrogen atoms/mol of CF_1 participated in the exchange and the incorporated tritium was not lost upon removal of CF_1 and partial purification by precipitation and column chromatography on Sephadex. Anti-CF_1 removed tritiated CF_1 from solution; urea treatment liberated the label from the enzyme. KRAAYENHOF and SLATER (1975) observed

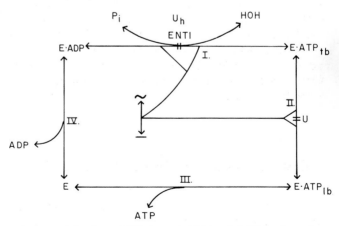

Fig. 1. A reaction cycle for catalytic participation of CF_1-bound ADP and ATP in photophosphorylation. E: coupling factor one; $ENTI$: energy transfer inhibitor; U and U_h: uncoupler at low and high concentrations, respectively; tb and lb: tightly and loosely bound nucleotides, respectively; $\sim$ energized state

a light-dependent conformational change in CF_1 covalently labeled with a fluorescent probe after reincorporation into the chloroplast membranes. These studies suggest that upon energization of the membrane, the membrane-bound CF_1 undergoes a conformational change involving a considerable part of the enzyme molecule.

The relationship of the bound nucleotides to electron transport-coupled ATP formation merits further investigation. Harris and Slater (1975) have shown that the exchangeability of bound nucleotides with either free nucleotides or free P_i and the net photophosphorylation reaction are inhibited to a similar degree by the uncoupler S_{13}. An energy-transfer inhibitor, Dio-9, strongly inhibited net photophosphorylation but had no effect on any of the exchange reactions. Similarly, the incorporation of $^{32}P_i$ into bound nucleotides in *R. rubrum* chromatophores was not affected by oligomycin while photophosphorylation was (Yamamoto et al., 1972). Aflalo (1975), however, has shown that the rapid incorporation of $^{32}P_i$ into bound ATP is much less sensitive to nigericin than is the capacity for net phosphorylation. The phosphorylation of both bound and free nucleotides was equally inhibited by phlorizin, an energy-transfer inhibitor. Thus, even when net photophosphorylation is blocked by uncouplers, the condensation of P_i and ADP to form ATP bound at the catalytic site is only partially inhibited. Higher concentrations of an uncoupler strongly inhibited the $^{32}P_i$ incorporation into the bound ATP. These results are consistent with the proposal that energy input is required to promote the release of tightly bound ATP. Moreover, from the inhibitory effects of high concentrations of uncouplers it seems that energy is also used to aid ATP formation at the catalytic site. These results are included in Figure 1, showing the steps in energy coupling as proposed by Boyer (1975). However, the interpretation of the experiments with uncouplers is complicated by the fact that the absolute rates of phosphorylation of the bound and free ADP are rather different and the relative effect of an uncoupler is being compared.

It is most probable that the chemiosmotic and conformational concepts may together provide a satisfactory explanation as to how the energy is transmitted and ATP synthesized. Further exploration of the role of CF_1-bound nucleotides and other aspects of the chemistry of CF_1 should provide direct insight into the mechanism of ATP formation.

Acknowledgment. The support during the preparation of this manuscript of the Israeli National Science Foundation is gratefully acknowledged.

References

Aflalo, C.: M. Sc. Thesis. Ben Gurion Univ. of the Negev, 1975
Avron, M.: Biochim. Biophys. Acta **40**, 257–272 (1960)
Avron, M.: Anal. Biochem. **2**, 535–543 (1961)
Banai, M.: M. Sc. Thesis. Ben Gurion Univ. of the Negev, 1975
Boyer, P.D.: In: Oxidases and Related Redox Systems. King, T.E., Mason, H.S., Morrison, M. (eds.). New York: Wiley, 1965, pp. 994–1000
Boyer, P.D.: In: Dynamics of Energy-Transducing Membranes. Ernster, L., Estabrook, R.W., Slater, E.C. (eds.). Amsterdam: Elsevier, 1974, pp. 289–301
Boyer, P.D., Cross, R.L., Momsen, W.: Proc. Nat. Acad. Sci. U.S. **70**, 2837–2839 (1973)

Boyer, P.D., Stokes, B.O., Wolcott, R.G., Degani, C.: Federation Proc. **34**, 1711–1717 (1975)
Cantley, L.C., Jr., Hammes, G.G.: Biochemistry **14**, 2968–2975 (1975a)
Cantley, L.C., Jr., Hammes, G.G.: Biochemistry **14**, 2976–2981 (1975b)
Carmeli, C., Avron, M.: Europ. J. Biochem. **2**, 318–326 (1967)
Carmeli, C., Lifshitz, Y.: Biochim. Biophys. Acta **267**, 86–95 (1972)
Cross, R.L., Boyer, P.D.: Biochem. Biophys. Res. Commun. **51**, 59–66 (1973)
Girault, G., Galmiche, J.M., Michel-Villaz, M., Thiery, J.: Europ. J. Biochem. **38**, 473–478 (1973)
Green, D.E.: Proc. Nat. Acad. Sci. U.S. **67**, 544–549 (1970)
Hackenbroock, G.R.: J. Cell Biol. **30**, 269–297 (1966)
Harris, D.A., Rosing, J., van de Stadt, R.J., Slater, E.C.: Biochim. Biophys. Acta **314**, 149–153 (1973)
Harris, D.A., Slater, E.C.: Biochim. Biophys. Acta **387**, 335–348 (1975)
Horak, A., Zalik, S.: Nature (London) **249**, 858–860 (1974)
Jagendorf, A.T.: In: Bioenergetics of Photosynthesis. Govindjee (ed.). New York-San Francisco-London: Academic Press, 1975, pp. 414–492
Kemp, A., Jr.: In: Regulation of Metabolic Processes in Mitochondria. Tager, J.M., Papa, S., Quagliariello, E., Slater, E.C. (eds.). Amsterdam: Elsevier, 1966, pp. 264–274
Klingenberg, M.: Adenine Nucleotides in Submitochondrial Particles: Mitochondrial Structure and Compartmentation. Quagliariello, E., Papa, S., Slater, E.C., Tager, J.M. (eds.). Amsterdam: Elsevier, 1967, pp. 320–324
Kraayenhof, R., Slater, E.C.: In: Proc. 3rd Intern. Congr. Photosynthesis. Avron, M. (ed.). Amsterdam: Elsevier, 1975, Vol. II, pp. 985–996
Livne, A., Racker, E.: J. Biol. Chem. **244**, 1332–1338 (1969)
Lutz, H.V., Dahl, J.S., Bachofen, R.: Biochim. Biophys. Acta **347**, 359–370 (1974)
McCarty, R.E., Fuhrman, J.S., Tsuchiya, Y.: Proc. Nat. Acad. Sci. U.S. **68**, 2522–2526 (1971)
McCarty, R.E., Racker, E.: Brookhaven Symp. Biol. **19**, 202–214 (1966)
Mitchell, P.: Chemiosmotic Coupling in Oxidative and Photosynthetic Phosphorylation. Bodmin: Glynn, 1966
Nelson, N., Nelson, H., Racker, E.: J. Biol. Chem. **247**, 7657–7662 (1972)
Penefsky, H.S.: J. Biol. Chem. **249**, 3579–3585 (1974)
Racker, E., Hauska, G.A., Lien, S., Berzborn, R.J., Nelson, N.: In: Proc. 2nd Intern. Congr. Photosynthesis. Forti, G., Avron, M., Melandri, A. (eds.). The Hague: Dr. Junk, 1971, Vol. II, pp. 1097–1113
Roy, H., Moudrianakis, E.N.: Proc. Nat. Acad. Sci. U.S. **68**, 464–468 (1971a)
Roy, H., Moudrianakis, E.N.: Proc. Nat. Acad. Sci. U.S. **68**, 2720–2724 (1971b)
Ryrie, I.J., Jagendorf, A.T.: J. Biol. Chem. **246**, 3771–3774 (1971)
Schmid, R., Shavit, N., Junge, W.: Biochim. Biophys. Acta **430**, 145–153 (1976)
Shahak, Y., Chipman, D.M., Shavit, N.: FEBS Lett. **33**, 293–296 (1973)
Shahak, Y., Chipman, D.M., Shavit, N.: In: Proc. 3rd Intern. Congr. Photosynthesis. Avron, M. (ed.). Amsterdam: Elsevier, 1975, Vol. II, pp. 859–866
Shoshan, V., Tel-Or, E., Shavit, N.: In: Proc. 3rd Intern. Congr. Photosynthesis. Avron, M. (ed.). Amsterdam: Elsevier, 1975, Vol. II, pp. 831–837
Slater, E.C.: In: Dynamics of Energy-Transducing Membranes. Ernster, L., Estabrook, R.W., Slater, E.C. (eds.). Amsterdam: Elsevier, 1974, pp. 1–20
Vandermeulen, D.L., Govindjee: FEBS Lett. **57**, 272–275 (1975)
Yamamoto, N., Yoshimura, S., Higuti, T., Nishikama, K., Horio, T.: J. Biochem. (Tokyo) **72**, 1397–1406 (1972)

4. The High Energy State

B.A. Melandri

In photosynthetic phosphorylation ATP synthesis is coupled to the redox reactions driven by light: this mechanism of coupling implies thermodynamically the existence of some common intermediate, generated by photosynthetic electron transport and able to store the energy to be utilized during the synthesis of ATP. The concept of high energy state (HES) is concerned with the nature and the properties of this "energized state" of phosphorylating membranes. Since the literature in this field, covering common aspects of respiration, photosynthesis and chemosynthesis, as well as aspects of the mechanism of active transport of metabolites, is very large, and since special topics in this area will be dealt with in other chapters of this book, the discussion will be limited here only to some important points:

1. current and historically important theories on the nature of the HES.

2. experimental evidence of the existence of an HES distinct from the immediate precursor of ATP.

3. experimental evaluation of the energy level and the energy capacity of the HES.

In this discussion it will be necessary to refer often to experiments, which cannot be described extensively in this short chapter; for details the reader should refer to Chapters III.2–III.10 of this volume, and to excellent reviews which have appeared on this topic (e.g. Walker and Crofts, 1970; Baltscheffsky and Baltscheffsky, 1974; Jagendorf, 1975; Witt, 1975).

A. Coupling Mechanism Hypotheses

Common features of all phosphorylating systems are the presence of redox enzymes of different nature and of an ATP synthetase, endowed generally with ATP hydrolyzing activity; these proteins are always contained within, or arranged on a membrane enclosing a vesicle, which appears to be essential for the mechanism of coupling. Any coupling hypothesis, in line with the experimental evidence, should, therefore, offer a proper role for all three of these components, which appear to be present in most if not all phosphorylating systems.

This condition is not met by the coupling hypothesis proposed first, the so-called chemical hypothesis (Slater, 1953; Chance and Williams, 1956), which postulates the existence of an HES in the form of a (covalent) intermediate between redox carriers and the phosphate donor to ADP. According to this hypothesis, a reduced electron carrier (AH) will interact (covalently) with another component (X) of the membrane (ATP synthetase?), yielding a low energy intermediate AH$-$

X [Eq. (1)]; this will be converted to a high energy intermediate $A \sim X$ upon oxidation of AH by the subsequent electron carrier B [Eq. (2)]. In turn $A \sim X$ will generate a phosphorylated intermediate $(X \sim P)$ [Eq. (3)], capable of donating phosphate to ADP [Eq. (4)]. Phenomena like photosynthetic control (stimulation of electron transport by ADP plus phosphate or by uncouplers) are interpreted as evidence for the need of forming an $AH - X$ complex as an obligate precursor of Eq. (2) and therefore of generating free A and X by Eqs. (3) and (4), or by dissipative breakdown of $A \sim X$ or $X \sim P$ by uncouplers. Such a scheme cannot be sustained anymore, at least in its simplest formulation, since it does not account for any role of the membrane in the process and does not explain other phenomena, like ion and proton transport and formation of membrane potentials, which are now accepted as being relevant for the mechanism of coupling.

$$AH + X \qquad \rightleftharpoons AH - X, \tag{1}$$

$$AH - X + B \quad \rightleftharpoons BH + A \sim X, \tag{2}$$

$$A \sim X + P_i \quad \rightleftharpoons A + X \sim P, \tag{3}$$

$$X \sim P + ADP \rightleftharpoons ATP + X. \tag{4}$$

The chemiosmotic hypothesis, proposed and developed principally by MITCHELL (1966, 1968, 1970), is based on the assumption that energy can be stored and utilized in the form of an electrochemical gradient of protons across the membrane, which should be rather impermeable to protons and form a closed compartment. The protonic gradient can be generated either by redox reactions or by ATP hydrolysis, which are both visualized as systems for primary active transport of protons, parallel-oriented in the membrane. Since both proton translocating systems are assumed to be electrogenic, separation of charges across the membrane and formation of a membrane potential are viewed as a consequence of proton translocation. Electrophoretic movements of other ions, passive or mediated by specific carriers, will tend to compensate this potential difference, but will not generally keep pace with the rate of proton translocation, and some membrane potential during steady-state electron transport will be established; this phenomenon will be particularly important in membrane fragments, like subchloroplast particles or bacterial chromatophores, characterized by a small internal volume and low electrical capacity (JACKSON et al., 1968; McCARTY, 1969; MONTAL et al., 1970). As a consequence the total difference in the electrochemical potential of protons inside and outside the vesicles, must take into account both the electrical contribution of the membrane potential and the chemical contribution of the concentration difference of protons. Expressed in electrical terms, the electrochemical potential difference of protons (proton motive force, PMF, in Mitchellian terms) will take the form:

$$\Delta \tilde{\mu}_{H^+} = PMF = \Delta \psi - \frac{2.3 RT}{F} \Delta pH^1 \tag{5}$$

where $\Delta \tilde{\mu}_{H^+}$ is the difference in electrochemical potential of protons on the two sides of the membrane, R and F the gas constant and the Faraday constant

---

[1]   Editor's comment: the sign of the pH term depends on the direction, in which the H^+ gradient across the membrane is calculated.

respectively and $\Delta\psi$ and ΔpH the transmembrane difference of electrical potential and pH.

Coupling between electron transport and ATP synthesis will ensue when the same ionic species (protons) are translocated with the same polarity by two active translocators placed on a membrane enclosing a single compartment. Thus when the $\Delta\tilde{\mu}_{H+}$ produced by the redox reactions exceeds the affinity of the reaction of ATP hydrolysis, this latter reaction is reversed and net ATP synthesis takes place, coupled to reversed translocation of protons, i.e. dissipation of the proton gradient.

Mechanistic models for proton translocation coupled to the function of the redox carriers have been proposed by MITCHELL (1966, 1968); he postulated specific orientation of the redox enzymes in "loops", formed by an alternation of proton-ated and nonprotonated carriers, arranged within the membrane so that during redox reactions protons are released at one face of the membrane and consumed on the other. Analogously, models for a reversible ATPase, coupling ATP hydro-lysis to the translocation of one or two protons per ATP, have been proposed (MITCHELL, 1966, 1968, 1974). Some aspects of these models have been confirmed by experimental evidence, in particular those postulating specific arrangements of redox proteins on one or the other side of photosynthetic membranes. The details of the mechanisms of protons translocation are, however, unimportant for the essence of the hypothesis, whose basic aspect is *indeed* the coupling of electron flow and ATP synthesis through an *obligatory* formation of a difference in the electrochemical potential of protons across the two sides of the membrane. This hypothesis identifies, therefore, the HES with the protonic electrochemical gradient and emphasizes the role of closed membranous systems for energy conser-vation and transduction.

In partial disagreement with the chemiosmotic view, WILLIAMS (1969) proposed that separation of charges and proton gradient formation take place *within* the membrane rather than *across* the membrane; in this author's opinion such a mecha-nism offers considerable advantages, such as a smaller dissipation of energy through passive proton leaks and a smaller number of sites to be protonated in order to energize the membrane fully. This proposal suggests also that phenomena like proton uptake and acid base phosphorylation represent events related only indi-rectly to the true mechanism of coupling, through slower and "rather artifactual" nonphysiological diffusion of protons from the large external bulk phases to the postulated protonable phase within the membrane.

Several theories propose that conformational changes of redox carriers (CHANCE, 1972) and of ATP synthetase (BENNUN, 1971; BOYER et al., 1973; SLATER et al., 1974) are the basis of energy coupling in energy transducing membranes. More specifically BOYER et al. (1973) has suggested that conformational changes of the ATP synthetase, induced by energy coupling with electron transport reactions, cause active release of preformed ATP, synthesized in a bound form in the absence of energy input. In general, in these theories emphasis is placed on the cooperative interactions of proteins (electron carriers *and* ATP synthezising enzymes), an inter-action induced by local fields and proton concentration differences. Undoubtedly these concepts are consistent with some recent findings, such as energy-induced changes of conformation of the coupling ATPase in chloroplasts (RYRIE and JAGEN-DORF, 1971; McCARTY and FAGAN, 1973), bacterial chromatophores (BACCARINI

MELANDRI et al., 1975) and mitochondria (VAN DE STADT et al., 1973). Analogously shifts in the apparent midpoint potential of some mitochondrial cytochromes (WILSON et al., 1973), dependent on membrane energization and indicating some functional modulation of specific carrier proteins, have also been reported; however, no clear cut evidence for such energy-dependent shifts has been presented so far for photosynthetic systems, either in higher plants (cf. BOARDMAN et al., 1971) or bacteria (DUTTON and BALTSCHEFFSKY, 1972).

Observations of this type fail to demonstrate a primary direct link between the HES and conformational changes: e.g. any conformational change is also compatible with an essentially indirect mechanism of coupling through active translocation of protons as chemiosmosis assumes. As WILLIAMS (1969) observed, any conformational change is essentially trivial unless the specific reason for such change is known, since "any change in the distribution of charges within proteins will always give rise to some conformational changes, but whether they are incidental or mandatory is quite unknown and might be unknowable".

B. Experimental Evidence for the Existence of a High Energy State

At different times several lines of evidence have been utilized in order to establish the existence of a HES of the membrane preexistent to, and independent from, the actual synthesis of ATP, which is catalyzed by the final enzyme of phosphorylation. Phenomena like photosynthetic control, which have been demonstrated in all photophosphorylating systems, are generally attributed to the build up of an "high concentration" of the high energy state, which will inhibit electron flow, exerting a back pressure on the energy conserving reactions of the electron transport chain. At variance with this view, however, important effects in the control of the rate of electron transport in chloroplasts by the internal pH of thylakoids has been also suggested (BAMBERGER et al., 1973). This accumulation of the HES will be inhibited by ADP and P_i or by uncouplers, but will be favored by energy transfer inhibitors, i.e. by substances blocking directly the activity of the coupling ATPase.

Besides photophosphorylation, photosynthetic membranes can carry on several other reactions, whose endergonic character implies the involvment of the HES: thus uptake of protons and amines (JAGENDORF and HIND, 1963; CROFTS, 1967), changes in osmotic volume, reverse electron transport between cytochrome f and Q and enhanced delayed fluorescence of chlorophyll (WRAIGHT and CROFTS, 1970) have been observed in chloroplasts. In bacterial chromatophores phenomena of this type are more numerous since other energy transducing enzymic systems, like reactions of the respiratory chain and energy linked transhydrogenase, can be demonstrated in the membrane. In chromatophores from facultative photosynthetic bacteria the impressive list of energy-dependent reactions includes: ATP synthesis, pyrophosphate synthesis (BALTSCHEFFSKY and VON STEDINGK, 1966), reduction of NAD^+ by succinate or by ascorbate-DCIP (KEISTER and YIKE, 1967a), reduction of $NADP^+$ by NADH (KEISTER and YIKE, 1967b), reduction of cyto-

chrome b by cyt c-2 (BALTSCHEFFSKY, 1969), proton (VON STEDINGK and BALT-
SCHEFFSKY, 1966) and amine uptake, extrusion of lipophylic cations or uptake
of lipophylic anions (ISAEV et al., 1970). To this list, active uptake of several
amino acids has been very recently added (HELLINGWERF et al., 1975). A number
of these reactions were shown to be fully reversible and to provide energy input
for driving other endoergonic processes. In general these reactions are sensitive
to inhibition by uncouplers, but are completely insensitive (and often stimulated)
by energy transfer inhibitors or antibodies against coupling ATPase (a very special
type of energy transfer inhibitor) (MCCARTY and RACKER, 1966; JOHANSSON, 1975),
unless ATP hydrolysis is utilized for energy input or ATP synthesis is directly
involved.

The concept of HES must be visualized, therefore, as a pool of some intermediate or a condition of the membrane, capable of interacting reversibly with a large
number of membrane enzymes; any hypothesis on the nature of the HES should
be compatible with these observations and must offer an adequate explanation,
on *mechanistic and steric* grounds, of all the interactions which have been observed.
In the author's opinion these considerations are strongly in favor of a largely
delocalized type of HES, as is proposed by the chemiosmotic hypothesis or by
its variations in which an intramembrane proton gradient is assumed.

Independent evidence for the existence of a condition of the membrane, precursor for the synthesis of ATP, has come from the classic observations by HIND
and JAGENDORF (1963), that a considerable amount of ATP can be synthesized
by chloroplasts in the dark after a brief irradiation with actinic light in the absence
of ADP and P_i. This type of observation has been recently extended to bacterial
chromatophores (LEISER and GROMET-ELHANAN, 1975). Although a detailed description of these experiments will be given in Chapter III.6 of this volume, it must
be noted in this context, that the experimental characteristics of this system, e.g.
the specific effects of inhibitors added during the light vs. the dark-stage of the
experiment, clearly indicate that a temporal separation of the formation and utilization of the HES can be achieved. In fact the stability in time of the ATP precursor
(several sec) exceeds by orders of magnitude the relaxation time of the redox
reactions driven by light in the postillumination period.

Physical separation of the enzymes generating and utilizing the HES in photosynthetic phosphorylation can be achieved by detaching the coupling ATPase from
the membrane; these procedures do not generally impair electron transport and
its potentiality in generating an HES. This fact is particularly evident in chromatophores of Rhodospirillaceae (MELANDRI et al., 1972), in which light driven energy
requiring reactions (e.g. proton uptake) are generally observed also in membranes
deprived of the coupling ATPase. In higher plant chloroplasts proton uptake is
not generally observed in preparations extracted with 1 mM EDTA, a procedure
which solubilizes large amounts of ATPase. Recent experiments have indicated
that the loss of a measurable proton uptake in decoupled chloroplasts is associated
with a large increase in conductivity of the membrane for ions (probably protons),
as indicated by the fast decay of the transmembrane electric field generated by
a flash of actinic light (SCHMID and JUNGE, 1975). The normal low conductivity
of the membrane, and proton uptake as well, can be restored by addition of
purified ATPase or by dicyclohexyl carbodiimide (MCCARTY and RACKER, 1967),
an energy transfer inhibitor interacting probably with hydrophobic components

of the ATPase still buried in the decoupled membrane. These experiments indicate that removal of the ATPase or inhibition of the activity of some of its hydrophobic components might have a profound effect on the membrane conductivity to ions in general and to protons in particular.

Along with the assiduous search for elucidating the nature of the HES many experimental parameters whose behavior is consistent with their relations with the HES have been described: uptake of ions and protons, uptake of amines, light scattering changes of chloroplast suspensions, interaction of photosynthetic membranes with fluorescent or colored probes, changes in properties of electron transport carriers (e.g. changes in E'^0 of cytochromes), spectral changes of endogenous pigments (cf. this vol. Chaps. III.1–III.14 for details). The observation that the response of these phenomena to energy transfer inhibitors and uncouplers is consistent with their association with the high energy state cannot however be considered as proof of their direct relation with the HES, since some of these phenomena are indeed quite clearly only side effects of the energization of the membrane. In order to make a decision on this point, great emphasis should be given to the kinetics of their generation by light, since the production of the HES should be as fast as the energy yielding redox reactions. Unfortunately some of the techniques used, expecially those involving potentiometric measurements, are often too slow for such studies. In this context, however, it should be recalled that spectral changes of photosynthetic pigments [518 nm signal of chloroplasts (WITT, 1972) or carotenoid band shift of bacterial chromatophores (JACKSON and CROFTS, 1971)] have rise-times in the nanosecond range and are, therefore, fast enough to be associated with the primary photochemical reactions of photosynthesis. Since these spectral signals have been amply demonstrated to be due to an electrochromic effect of a transmembrane electric field on the membrane pigments (WITT, 1972), these observations indicate, in line with the chemiosmotic hypothesis, that separation of charges across the membrane coincides with the primary act of photosynthesis. Equally consistent with the chemiosmotic concept is the fast proton uptake (or binding), observable in illuminated photosynthetic membranes with the aid of pH indicator dyes (COGDELL et al., 1973; JUNGE and AUSLÄNDER, 1973); these signals are, in general, fast enough to be associated with the redox reactions of the photosynthetic electron transport chain.

C. The Energy Level and the Energy Capacity of the High Energy State

The maximal energy level of the HES has been evaluated by measuring the affinity of the photophosphorylation reaction at which a transition from a high rate of electron transport to a low rate is observed ("state 3 to state 4 transition", in mitochondrial terminology). Under these conditions it was assumed that the ATP synthetase reaction is in quasi equilibrium with the HES and that, therefore, the free energy change associated with the dissipation of the HES can be calculated from the concentrations of ADP, P_i and ATP at the transition point [i.e. measuring the so-called phosphate potential (COCKRELL et al., 1966; KRAAYENHOF, 1969)].

It should be noted, however, that the values obtained by this procedure are only minimal values, since the assumption of a quasiequilibrium condition is equivalent to postulating a degree of coupling between ATP synthesis and the dissipation of the HES equal, or very close to unity; this condition is quite unlikely to be true in the in vitro systems in which phosphate potentials have been measured. If these considerations are taken into account, the general relation can be written:

$$\Delta G = \Delta G^0 + R T \ln \frac{[ATP]}{[ADP][P_i]} \leq n F \Delta E \qquad (6)$$

where ΔG is the free energy change for ATP synthesis at the concentrations of ATP, ADP and P_i characteristic of the transition point, and $n F \Delta E$ represents the work available from the dissipation of the HES. The meaning to be ascribed to n and ΔE differ according to the coupling hypothesis adopted; thus, generally n will represent the number of electrons per mol of ATP exchanged during the redox reactions at a coupling site characterized by an electromotive force ΔE. For the chemiosmotic hypothesis, n will represent the H^+/ATP stoichiometry of the coupling ATPase and ΔE will be equivalent to the protonmotive force.

The values obtained for the phosphate potential in "class 1" chloroplasts by KRAAYENHOF (1969) were in the range of 13.3–13.5 kcal/mol; analogous values (13.3 kcal/mol) were subsequently reported for *Rhodopseudomonas sphaeroides* chromatophores (CROFTS and JACKSON, 1970). In chromatophores of another photosynthetic bacterium, *Rhodopseudomonas capsulata,* characterized by high rates of photophosphorylation, values of 15.0 to 15.6 kcal/mol can be obtained (CASADIO et al., 1974b). Most importantly such values were reached both through net ATP synthesis, i.e. setting initial conditions with lower phosphate potentials, and through ATP hydrolysis, i.e. starting from higher phosphate potentials. These observations indicate that ATP synthetase in photophosphorylation is a perfectly reversible enzyme and that the maximal phosphate potential reached is in relation to the energy input of the HES. The figures here reported are all normalized to the most recent values for ΔG^0 for ATP hydrolysis (ROSING and SLATER, 1972); considerable variability for this parameter is, however, reported in the literature. Addition of increasing amounts of uncouplers (FCCP), detergents (Triton X-100) or ionophorous antibiotics plus K^+, decreases steadily the measured phosphate potential, in agreement with the concept that not only classical uncouplers, but also perturbation of the membrane conductivity to ions, impair the capability of storing energy of the phosphorylating apparatus (MELANDRI et al., 1975).

The relationships between the maximal phosphate potential and the redox characteristics of the electron transport chain have been discussed in detail by CROFTS et al. (1971). For such calculations, as Eq. (6) shows, the exact stoichiometry between electron transport and ATP synthesis must be known; unfortunately considerable uncertainty on this point still exists. If a stoichiometry of $2\,e^-$/ATP is assumed it can be calculated that 15.5 kcal/mol will correspond to 336 mV. This value is considerably higher than the potential gap estimated, e.g., between plastoquinone and cytochrome f [a segment of the chloroplast electron transport chain demonstrated to include a "coupling site" (BÖHME and CRAMER, 1972)]. Analogously in photosynthetic bacteria these values exceed the redox span between cytochrome b and cytochrome c-2 by at least 50 mV. Thus some other sort of energy input must be present in photosynthetic systems other than that provided by the redox reactions of the electron transport carriers. According to the suggestions of CROFTS et al. (1971) (cf. also WITT, 1972), and in line with MITCHELL's concepts, this energy could be provided by the electric field generated by the separation of charges taking place during the primary photochemical acts of photosynthesis.

Comparisons have been reported between phosphate potentials and the values of $\Delta \tilde{\mu}_{H^+}$. In spinach chloroplasts values of ΔpH ranging between 2.5 and 3.5 units (ROTTENBERG et al., 1972; PORTIS and MCCARTY, 1973) have been estimated; these values will correspond to 150–200 mV of proton motive force and are, therefore, insufficient to account alone for the high phosphate potentials observed. The contribution of a membrane potential to the energization of chloroplasts is still debated, since very low $\Delta \psi$ could be measured with tracer techniques (ROTTENBERG et al., 1972) or using spectroscopic methods (STRICHARTZ and CHANCE, 1972). Considerable higher contribution of membrane potential in steady state (around 100 mV) were however evaluated by other authors (cf. WITT, 1972; BARBER, 1972 and this vol. Chap. III.8). In any case the overall electrochemical proton gradient estimated in chloroplasts never exceeds 300 mV, a value hardly consistent with that of the energy input of photophosphorylation as estimated by the phosphate potential; this inconsistency does not hold, however, if a H^+/ATP ratio of 3, as determined independently by some authors (JUNGE et al., 1970; PORTIS and MCCARTY, 1974), is used. In bacterial chromatophores $\Delta \tilde{\mu}_{H^+}$ has been estimated by measuring the membrane potential on the basis of the electrochromic effect of carotenoids (cf. this vol. Chap. III.8) and by evaluating ΔpH either from the distribution of ^{14}C-organic amines (SCHULDINER et al., 1974) or of K^+ in the presence of nigericin (JACKSON et al., 1968) or by the quenching of fluorescence of 9-amino acridine (CASADIO et al., 1974a). While the first two techniques gave values around 1.5 to 2 pH units, 9-amino acridine indicated pH difference as high as 2.8–3.1 units. Altogether the PMF measured ranged between 280 mV to 400–420 mV. Only these last values are compatible with the high phosphate potentials determined in these preparations, if a stoichiometry of 2 H^+/ATP is assumed. When the total PMF is drastically reduced by additions of uncoupling agents or ionophores, the phosphate potential measured is decreased accordingly (CASADIO et al., 1974b; MELANDRI et al., 1975), and this, most significantly, also when only one of the two components of the PMF ($\Delta \psi$ or ΔpH) is specifically inhibited by proper experimental conditions (e.g. by addition of NH_4^+ to diminish ΔpH or additions of K^+ and valinomycin to decrease $\Delta \psi$).

The discrepancy in the values of the PMF measured in different laboratories is mainly due to the different response given by the use of fluorescent amines vs. radioactive amines for the measurement of ΔpH. Both these techniques are open to criticisms: no convincing explanation has been put forward so far in order to explain the quenching of fluorescence of amines upon entering into the inside of a membrane vesicle. On the other hand measuring the distribution of radioactive amines necessitates centrifugation of illuminated membrane suspensions, which involves the difficult maintenance of a saturating light intensity during the formation of the pellet; in addition the relatively high concentrations of amine used (10–40 μM) could perturb the maximal ΔpH reached in saturating light. In conclusion, the maximal potentiality of photosynthetic membranes to drive

ATP synthesis seems generally quantitatively related to the extent of the electro-chemical potential difference of protons existing across the membrane; the quantita-tive agreement between these two phenomena is still debated since uncertainty still exists on the true stoichiometry of the H^+/ATP ratio and on the validity of the methods used for the estimation of the Δ pH.

The extent of energy available for ATP synthesis should not be confused with the energy storing capacity of the photosynthetic machinery; this latter property can be evaluated only if the stage of formation of the high energy state is separated from that of ATP synthesis, i.e. in experiments of postillumination phosphorylation. The amount of ATP formed in the dark by preilluminated chloroplasts exceeds by about 50 times the number of electron transport chains in the membrane (HIND and JAGENDORF, 1963). This large storage capacity of chloroplasts for energy is incompatible with the concept of an HES only of a chemical type, which would be present in the membrane at concentrations comparable with the number of photosynthetic units. A large body of evidence indicates that most of the energy utilized in postillumination ATP formation is stored in the form of a pool of protons taken up during the light stage (cf. this vol. Chap. III.6 for details). The size of this pool depends on the internal volume of the membranes used and on the internal buffering capacity of the vesicles; the amount of ATP synthezised is, therefore, rather small in subchloroplast particles (McCARTY, 1970) or in bacte-rial chromatophores (LEISER and GROMET-ELHANAN, 1975). In all systems, postil-lumination phosphorylation can be greatly increased by addition during the light stage of permeant protonable substances (weak acids or amines), whose pK is such as to confer a high buffering power around pH 5; by addition of these substances stimulation up to 20-fold of both ATP synthesis in the dark stage and of proton uptake in the light stage have been obtained (compare CROFTS, 1968; AVRON, 1972). Most significantly postillumination ATP synthesis in dim light (SCHULDINER et al., 1973) or in subchloroplast particles (McCARTY, 1970) and bacterial chromatophores (GROMET-ELHANAN and LEISER, 1975) can be also stimu-lated by imposing a diffusion potential, positive inside, during the dark stage (by addition of K^+ and valinomycin); this observation indicates that a membrane potential artificially induced can greatly enhance the capacity of ATP synthesis in conditions in which Δ pH is small or is rapidly dissipated due to the small volume of the particles; these results also imply an electrogenic mechanism for the coupling between proton efflux and ATP synthesis. All these observations are again most satisfactorily explained by the chemiosmotic hypothesis.

References

Avron, M.: In: Proc. 2nd Intern. Congr. Photosynthesis. Forti, G., Avron, M., Melandri, B.A. (eds.). The Hague: Dr. Junk, 1972, pp. 861–871

Baccarini Melandri, A., Fabbri, E., Firstater, E., Melandri, B.A.: Biochim. Biophys. Acta **376**, 72–81 (1975)

Baltscheffsky, H., Baltscheffsky, M.: Ann. Rev. Biochem. **43**, 871–897 (1974)

Baltscheffsky, H., von Stedingk, L.-V.: Biochem. Biophys. Res. Commun. **22**, 722–728 (1966)

Baltscheffsky, M.: Arch. Biochem. Biophys. **133**, 46–53 (1969)

Bamberger, E.S., Rottenberg, A., Avron, M.: Europ. J. Biochem. **34**, 557–563 (1973)
Barber, J.: FEBS Lett. **20**, 251–254 (1972)
Bennun, A.: Nature New Biol. **233**, 5–8 (1971)
Boardman, N.K., Anderson, J.M., Hiller, R.G.: Biochim. Biophys. Acta **234**, 126–136 (1971)
Böhme, H., Cramer, W.A.: Biochemistry **11**, 1155–1160 (1972)
Boyer, P.D., Cross, R.L., Momsen, W.: Proc. Nat. Acad. Sci. U.S. **70**, 2837–2839 (1973)
Casadio, R., Baccarini Melandri, A., Melandri, B.A.: Europ. J. Biochem. **47**, 121–128 (1974a)
Casadio, R., Baccarini Melandri, A., Zannoni, D., Melandri, B.A.: FEBS Lett. **49**, 203–207
    (1974b)
Chance, B.: FEBS Lett. **23**, 3–20 (1972)
Chance, B., Williams, G.R.: Advan. Enzymol. **17**, 65–134 (1956)
Cockrell, R.S., Harris, E.J., Pressman, B.C.: Biochemistry **5**, 2326–2335 (1966)
Cogdell, R.J., Jackson, J.B., Crofts, A.R.: Bioenergetics **4**, 211–227 (1973)
Crofts, A.R.: J. Biol. Chem. **242**, 3352–3359 (1967)
Crofts, A.R.: In: Regulatory Functions of Biological Membranes. Jarnefelt, J. (ed.). Amster-
    dam: Elsevier, 1968, Vol. II, pp. 247–263
Crofts, A.R., Jackson, J.B.: In: Electron Transport and Energy Conservation. Tager, J.H.,
    Papa, S., Quagliariello, E., Slater, E.C. (eds.). Bari: Adriatica Editrice, 1970, pp. 383–398
Crofts, A.R., Wraight, C.A., Fleischmann, D.E.: FEBS Lett. **15**, 89–100 (1971)
Dutton, P.L., Baltscheffsky, M.: Biochim. Biophys. Acta **267**, 172–178 (1972)
Gromet-Elhanan, Z., Leiser, M.: J. Biol. Chem. **250**, 90–93 (1975)
Hellingwerf, K.J., Michels, P.A.M., Dorpema, J.W., Konings, W.N.: Europ. J. Biochem.
    55, 397–406 (1975)
Hind, G., Jagendorf, A.T.: Proc. Nat. Acad. Sci. U.S. **49**, 715–722 (1963)
Isaev, P.I., Liberman, E.A., Samuilov, V., Skulachev, V.P., Tsofina, L.M.: Biochim. Biophys.
    Acta **216**, 22–29 (1970)
Jackson, J.B., Crofts, A.R.: Europ. J. Biochem. **18**, 120–130 (1971)
Jackson, J.B., Crofts, A.R., von Stedingk, L.-V.: Europ. J. Biochem. **6**, 41–54 (1968)
Jagendorf, A.T.: In: Bioenergetics of Photosynthesis. Govindjee (ed.). New York-San Fran-
    cisco-London: Academic Press, 1975, pp. 413–492
Jagendorf, A.T., Hind, G.: In: Photosynthetic Mechanisms of Green Plants. Nat. Acad.
    Sci. Nat. Res. Council **1145**, 509 (1963)
Johansson, B.C.: Ph.D. Thesis, Univ. Stockholm 1975
Junge, W., Ausländer, W.: Biochim. Biophys. Acta **333**, 59–70 (1973)
Junge, W., Rumberg, B., Schröder, H.: Europ. J. Biochem. **14**, 575–581 (1970)
Keister, D.L., Yike, N.J.: Arch. Biochem. Biophys. **121**, 415–422 (1967a)
Keister, D.L., Yike, N.J.: Biochemistry **6**, 3847–3857 (1967b)
Kraayenhof, R.: Biochim. Biophys. Acta **180**, 213–215 (1969)
Leiser, M., Gromet-Elhanan, Z.: J. Biol. Chem. **250**, 84–89 (1975)
McCarty, R.E.: J. Biol. Chem. **244**, 4292–4298 (1969)
McCarty, R.E.: FEBS Lett. **9**, 313–316 (1970)
McCarty, R.E., Fagan, J.: Biochemistry **12**, 1503–1507 (1973)
McCarty, R.E., Racker, E.: Brookhaven Symp. Biol. **19**, 202–214 (1966)
McCarty, R.E., Racker, E.: J. Biol. Chem. **242**, 3435–3439 (1967)
Melandri, B.A., Baccarini Melandri, A., Crofts, A.R., Cogdell, R.J.: FEBS Lett. **24**, 141–145
    (1972)
Melandri, B.A., Zannoni, D., Casadio, R., Baccarini Melandri, A.: In: Proc. 3rd Congr.
    Photosynthesis. Avron, M. (ed.). Amsterdam: Elsevier, 1975, pp. 1147–1162
Mitchell, P.: Chemiosmotic Coupling in Oxidative and Photosynthetic Phosphorylation. Bod-
    min: Glynn, 1966
Mitchell, P.: Chemiosmotic Coupling and Energy Transduction. Bodmin: Glynn, 1968
Mitchell, P.: In: Membrane and Ion Transport. Bittar, E.E. (ed.). New York: Wiley, 1970,
    Vol. I, pp. 192–256
Mitchell, P.: FEBS Lett. **43**, 189–194 (1974)
Montal, M., Nishimura, M., Chance, B.: Biochem. Biophys. Acta **223**, 183–188 (1970)
Portis, A.R., McCarty, R.E.: Arch. Biochem. Biophys. **156**, 621–625 (1973)
Portis, A.R., McCarty, R.E.: J. Biol. Chem. **249**, 6250–6254 (1974)
Rosing, J., Slater, E.C.: Biochim. Biophys. Acta **267**, 275–290 (1972)

Rottenberg, H., Grunwald, T., Avron, M.: Europ. J. Biochem. **25**, 54–63 (1972)
Ryrie, I.J., Jagendorf, A.T.: J. Biol. Chem. **246**, 3771–3774 (1971)
Schmid, R., Junge, W.: In: Proc. 3rd Intern. Congr. Photosynthesis. Avron, M. (ed.). Amsterdam: Elsevier, 1975, pp. 821–830
Schuldiner, S., Padan, E., Rottenberg, H., Gromet-Elhanan, Z., Avron, M.: FEBS Lett. **49**, 174–177 (1974)
Schuldiner, S., Rottenberg, H., Avron, M.: Europ. J. Biochem. **39**, 455–462 (1973)
Slater, E.C.: Nature (London) **172**, 975–978 (1953)
Slater, E.C., Rosing, J., Harris, D.A., van de Stadt, R.J., Kemp, A.: In: Membrane Proteins in Transport and Phosphorylation. Azzone, G.F., Klingenberg, M.E., Quagliariello, F., Siliprandi, N. (eds.). Amsterdam: North Holland, American Elsevier, 1974, pp. 137–147
Stadt, R.J. van de, De Boer, B.L., Van Dam, K.: Biochim. Biophys. Acta **292**, 338–349 (1973)
Stedingk, L.-V. von, Baltscheffsky, H.: Arch. Biochem. Biophys. **117**, 400–404 (1966)
Strichartz, G.R., Chance, B.: Biochim. Biophys. Acta **256**, 71–84 (1972)
Walker, D.A., Crofts, A.R.: Ann. Rev. Biochem. **39**, 389–428 (1970)
Williams, R.J.P.: Current Topics Bioenerg. **3**, 79–156 (1969)
Wilson, D.F., Dutton, P.L., Wagner, M.: Current Topics Bioenerg. **5**, 233–265 (1973)
Witt, H.T.: Quart. Rev. Biophys. **4**, 365–477 (1972)
Witt, H.T.: In: Bioenergetics of Photosynthesis. Govindjee (ed.). New York-San Francisco-London: Academic Press, 1975, pp. 493–554
Wraight, C.A., Crofts, A.R.: Europ. J. Biochem. **19**, 386–397 (1970)

5. ATPase

T. Bakker-Grunwald

A. Introduction

In this section, the term ATPase will be used in a rather restricted sense: it will denote the breakdown of ATP into ADP and P_i under formation of a high-energy state, i.e., the reversal of photophosphorylation. Likewise, the term ATPase enzyme will be used for the membrane-bound equivalent of CF_1 (chloroplast coupling factor 1; see this vol. Chap. III.7).

After a short historical survey, a rather detailed description will be given of the mechanism of activation of the light-and-dithioerythritol — triggered Mg^{2+}-dependent ATPase in spinach chloroplasts; this description will serve as a framework for some generalizing remarks concerning (1) the relevance of the research on ATPase for other bioenergetical problems, and (2) the practical applicability of ATPase as an alternative means of energizing chloroplasts.

B. History of ATPase

It has long been thought (see for instance Avron and Jagendorf, 1959) that chloroplasts did not possess ATPase activity. In 1961, however, Petrack and Lipmann showed the existence of a Mg^{2+}-dependent ATPase reaction under special conditions: illumination in the presence of a high concentration of a thiol compound [later it was found (Kaplan et al., 1967) that dithiol compounds were effective at much lower concentrations]. Hoch and Martin (1963) and Marchant and Packer (1963) subsequently resolved this ATPase into a light and a dark stage: it appeared that light promoted some condition that favored the subsequent hydrolysis of ATP in the dark. The assumption that this ATPase represented a reversal of the later stages of photophosphorylation was supported by the demonstration of a simultaneous ATP-P_i exchange reaction (Rienits, 1967; Carmeli and Avron, 1967; McCarty and Racker, 1968); both ATPase and exchange were suppressed by energy-transfer inhibitors such as Dio-9 and DCCD.

The influence of uncouplers of photophosphorylation on the ATPase was rather ambiguous, until Rienits (1967) and Carmeli and Avron (1969) observed that after addition of uncoupler during the dark stage, ATPase rate was not constant in time: an initial stimulation of ATPase rate was followed by inhibition. They inferred that the high-energy intermediate built up in the light stage is maintained by dark ATPase activity; uncouplers dissipate the energy derived from ATP breakdown, and thus cause a reversal of the chloroplasts towards their ATPase-inactive state.

C. Feedback in ATPase

The gross features of the system after induction of ATPase activity by illumination in the presence of dithioerythritol, then, can be illustrated as in Figure 1 (BAKKER-GRUNWALD and VAN DAM, 1973).

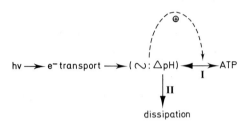

Fig. 1. Feedback in light-triggered ATPase. I: turnover capacity; II: dissipative capacity; dashed line: feedback control

As is a characteristic of all coupled ATPase systems, net ATPase rate is determined by the interplay between *two* capacities: the ATPase turnover capacity *and* the dissipative capacity. Thus, the net ATPase rate measured in the absence of uncoupler is a result of some natural degree of uncoupling of the chloroplasts as isolated: in ideally coupled chloroplasts, no ATP breakdown would be observed. Generally, it is important to realize that ATPase rate per se does not give any indication about the actual rate of turnover of the ATPase enzyme; thus, an effect on ATPase rate may either be caused by a change in ATPase turnover or by a change in dissipation rate, or both, and without further information about the parameters involved (degree of coupling and energy level) no conclusions whatsoever about the activity of the ATPase enzyme can be drawn.

As a special feature, indicated already, the chloroplast ATPase system exhibits feedback characteristics; even after the light induction step, ATPase activity remains dependent upon a high-energy state, that during the dark ATPase stage should be maintained by ATPase activity itself. This implies that, although in practice often a linear time course of ATPase has been found (PACKER and MARCHANT, 1964; PETRACK et al., 1965), one should not expect it a priori; in principle, a steady-state ATPase rate will only be reached after some degree of relaxation towards a situation in which the energy level kept up by ATPase activity is self-maintaining with respect to that activity. Especially after addition of uncoupler, but also after insufficient triggering of ATPase, it may take minutes before the ATPase rate is constant (BAKKER-GRUNWALD and VAN DAM, 1973).

ATPase activity has been shown to drive energy-requiring processes like proton uptake (CARMELI, 1970), swelling in ammonium salts (PACKER and CROFTS, 1967; GAENSSLEN and McCARTY, 1971), uptake of calcium salts (PACKER and CROFTS, 1967) and reversal of electron transport towards lower redox potential (RIENITS et al., 1974); it is clear that the built-in feedback device will put its own limitations here.

D. Conformational Changes Relevant for ATPase

As will be described in more detail in this vol. Chap. III.10, the ATPase enzyme changes conformation upon energization. The sequence of conformational events relevant for the light-and-dithioerythritol-triggered ATPase activity can be described as depicted in Figure 2 (BAKKER-GRUNWALD and VAN DAM, 1974).

First (step I) a light-induced conformational change takes place, that exposes both the active site for ATPase and the ATPase-regulating site sensitive towards dithioerythritol action. Then, in step II, dithioerythritol exerts its action on this regulating site; presumably, it interferes with the action of the so-called "inhibitor", an inhibitory subunit of the ATPase enzyme (see Chap. III.7). This step is essentially *irreversible,* while energization process I is an equilibrium reaction.

The energy-rich conformation (C), generated in step II, is the only one exhibiting ATPase activity: the active site is exposed to the substrate ATP *and* unmasked. Conformation (C) is in energy-dependent equilibrium III (comparable to equilibrium I) with a low-energy conformation (D); the latter, although still unmasked, does not have ATPase activity because the active site is not accessible towards ATP. (D) in its turn is subjected to an irreversible, Mg^{2+}-dependent degradation (IV) into another inactive conformation (E), the nature of which is yet unknown. This degradation is stimulated by ADP and retarded by P_i.

Addition of ATP to the dark system may, by mediation of the small fraction of ATPase molecules remaining in active form (C), start an autocatalytic energization process; if and to what degree this happens, depends on both the extent of unmasking by dithioerythritol, the progression of degradation IV and the degree of coupling of the chloroplasts. Further, both (D) and (E) can be converted back in active form (C) by light energy.

Thus, according to Figure 2, the decline of ATPase activity either after addition of uncoupler or in the dark, in the absence of ATP, is not due to a reversal of dithioerythritol action, as used to be more or less tacitly assumed (see for instance PETRACK et al., 1965; RIENITS, 1967; McCARTY and RACKER, 1968): the

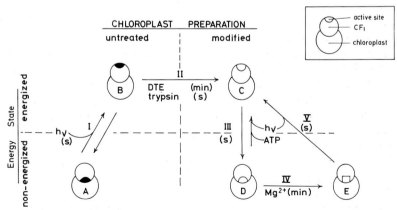

Fig. 2. Conformational events relevant to ATPase. *DTE:* dithioerythritol (min) and (s) indicate the time scale of the respective processes (BAKER-GRUNWALD and VAN DAM, 1974"); *black dote:* untreated active site; *empty dote:* DTE activated site

active site remains unmasked, but because of the prevailing low-energy conditions loses its accessibility towards ATP. Thus, the "dark decay" is a complex phenomenon, given by the decrease in time in the fraction of active conformation (C) finally regained by the ATP-induced autocatalytic activation process. In practice, this means that it is possible to omit dithioerythritol from the dark ATPase stage; this is important when ATP-driven processes are studied that are influenced by this compound, like the reversal of electron transport (RIENITS et al., 1974; BAKKER-GRUNWALD and VAN DAM, 1974). Furthermore, Figure 2 implies that the "dark decay" in the absence of ATP can be retarded by omission of Mg^{2+} and ADP and addition of P_i during the dark stage before addition of ATP (BAKKER-GRUNWALD and VAN DAM, 1974; for dark decay see also CARMELI and LIFSHITZ, 1972).

E. Component Requirements of Membrane-Bound ATPase in General

In order to elicit ATPase activity, the ATPase enzyme has to be unmasked (step II in Fig. 2). For this purpose, a high-energy conformation is required. Except by light, this can be accomplished by an acid-base jump (KAPLAN and JAGENDORF, 1968); for practical purposes, illumination is far more effective, however (BAKKER-GRUNWALD and VAN DAM, 1973). Because conformations (A) and (B) in Figure 2 are in energy-dependent equilibrium, reaction (II) takes also place in the dark, during prolonged incubation times (McCARTY and RACKER, 1968).

Except by dithioerythritol or by other (di) thiol compounds, unmasking can be accomplished by trypsin (LYNN and STRAUB, 1969); the latter compound has uncoupling side effects, however, and therefore for routine experiments the use of dithioerythritol may be preferable (BAKKER-GRUNWALD and VAN DAM, 1974).

As shown by NISHIZAKI and JAGENDORF (1969), there is no specific ion requirement during the light-induction period. The dark ATPase reaction itself is Mg^{2+}-dependent; Mg^{2+} fulfills a dual function here, both as a cofactor in the turnover of the enzyme and in coupling ATP breakdown to the formation of the high-energy state required for continuation of ATPase activity (BAKKER-GRUNWALD, 1974a).

F. Relations of ATPase with Other Topics in Bioenergetics

Energy-dependent conformation changes in the ATPase enzyme have been demonstrated in a number of ways (see Chap. III.10). Studies on ATPase activity have inspired to a quantitative approach of those conformation changes relevant for the ATPase system; from them, an unequivocal relation between the position of equilibrium III (Fig. 2) and the energy level has been shown (BAKKER-GRUNWALD, 1974b). Effectively, this relation results in an energy threshold for ATPase activity, that is independent of the degree of unmasking (step II in Fig. 2) and the presence of ADP and/or P_i, but increases with the pH of the medium. This strongly resembles the findings of PICK et al. (1974) for photophosphorylation (see also Chap. III.2), suggesting that in photophosphorylation also, a conformation type of control might be operative.

References

Avron, M., Jagendorf, A.T.: J. Biol. Chem. **234**, 967–972 (1959)
Bakker-Grunwald, T.: Biochim. Biophys. Acta **347**, 141–143 (1974a)
Bakker-Grunwald, T.: Biochim. Biophys. Acta **368**, 386–393 (1974b)
Bakker-Grunwald, T., Van Dam, K.: Biochim. Biophys. Acta **292**, 808–814 (1973)
Bakker-Grunwald, T., Van Dam, K.: Biochim. Biophys. Acta **347**, 290–298 (1974)
Carmeli, C.: Biochim. Biophys. Acta **189**, 256–266 (1969)
Carmeli, C.: FEBS Lett. **7**, 297–300 (1970)
Carmeli, C., Avron, M.: Europ. J. Biochem. **2**, 318–326 (1967)
Carmeli, C., Avron, M.: In: Progress in Photosynthesis Research. Metzner, H. (ed.). Tübingen: Laupp, 1969, Vol. III, pp. 1169–1175
Carmeli, C., Lifshitz, Y.: Biochim. Biophys. Acta **267**, 86–95 (1972)
Gaensslen, R.E., McCarty, R.E.: Arch. Biochem. Biophys. **147**, 55–65 (1971)
Hoch, G., Martin, I.: Biochem. Biophys. Res. Commun. **12**, 223–228 (1963)
Kaplan, J.H., Jagendorf, A.T.: J. Biol. Chem. **243**, 972–979 (1968)
Kaplan, J.H., Uribe, E., Jagendorf, A.T.: Arch. Biochem. Biophys. **120**, 365–375 (1967)
Lynn, W.S., Straub, K.D.: Proc. Nat. Acad. Sci. U.S. **63**, 540–547 (1969)
Marchant, R.H., Packer, L.: Biochim. Biophys. Acta **75**, 458–460 (1963)
McCarty, R.E., Racker, E.: J. Biol. Chem. **243**, 129–137 (1968)
Nishizaki, Y., Jagendorf, A.T.: Arch. Biochem. Biophys. **133**, 255–262 (1969)
Packer, L., Crofts, A.R.: Current Topics Bioenerg. **2**, 23–64 (1967)
Packer, L., Marchant, R.H.: J. Biol. Chem. **239**, 2061–2069 (1964)
Petrack, B., Craston, A., Sheppy, F., Farron, F.: J. Biol. Chem. **240**, 906–914 (1965)
Petrack, B., Lipmann, F.: In: Light and Life. McElroy, W.D., Glass, B. (eds.). Baltimore: Johns Hopkins, 1961, pp. 621–630
Pick, U., Rottenberg, H., Avron, M.: FEBS Lett. **48**, 32–36 (1974)
Rienits, K.G.: Biochim. Biophys. Acta **143**, 595–605 (1967)
Rienits, K.G., Hardt, H., Avron, M.: Europ. J. Biochem. **43**, 291–298 (1974)

6. Post-Illumination ATP Formation

J.M. Galmiche

A. Introduction

Numerous studies have dealt with the coupling mechanisms between electron transfer, induced by absorbed light, and ATP synthesis in chloroplasts and chromatophores. There should be some delay between the primary reactions, charge separation induced by absorbed photons in 10^{-9} sec, and the chemical reactions leading to ATP synthesis. The problem is to determine the length of the delay and the steps involved between the two events. These steps include electron transfer and formation of all other high energy states which precede ATP synthesis.

Research in this field has been carried out by separating light processes (electron transfer induction) and dark processes (postillumination ATP synthesis). Two lines of investigation have been followed.

The first is two-stage ATP synthesis initiated by KAHN and JAGENDORF (1960), who illuminated chloroplasts in the absence of a complete medium necessary for phosphorylation and added the missing compound(s) in the subsequent dark phase. The yield of the ATP synthesized gives a quantitative measurement of the high energy state intermediate. The half-life of the light-induced high energy state can be determined by adding the missing compound(s) at various times after the light is switched off. Using the same method, HIND and JAGENDORF (1963a) and SHEN and SHEN (1962) discovered independently the presence of a high energy state called X_e, with a half-life of a few seconds, during two-stage ATP synthesis by chloroplast suspensions from spinach and wheat seedlings respectively.

The second method is ATP synthesis by flashing light. NISHIMURA (1961) first accomplished this by illuminating chromatophores with short flashes (≤ 1.5 msec) separated by dark periods from 0 to 35 msec. The yield of ATP synthesis per flash increases with the duration of the dark period between two consecutive flashes until an optimum value is reached. The rate of this increase gives each time the ATP synthesis rate after one flash during the subsequent dark period and allows the half life of the high energy state produced by one flash to be computed. NISHIMURA (1962a, b), using this technique, found a very transient high energy state in chromatophores of *Rhodospirillum rubrum*. Its half-life was 1 to 30 msec.

All the studies in this postillumination ATP synthesis are aimed at finding out the nature of the high energy state intermediate. Three main hypotheses are currently proposed concerning the nature of the energy-conserving processes. According to the so-called chemical hypothesis (SLATER, 1953) the intermediate must be a chemical compound, and many attempts have been made to isolate this chemical intermediate especially a phosphorylated compound. The other two hypotheses, the chemiosmotic hypothesis (MITCHELL, 1961) and that of conformational changes (BOYER, 1965), could explain the lack of evidence for the existence of a true chemical high energy intermediate. Hence the high energy state must be studied in parallel with the three reversible processes implicated in the two last hypotheses: namely, directed proton and ion transport, field changes and conformational changes.

We do not intend to support any of the three hypotheses concerning the energy-conserving process. The general properties of postillumination ATP synthesis, the possible presence of a chemical intermediate, the relationships between the high energy state intermediate and the membrane properties revealed by ion fluxes, potential differences and conformational changes will be discussed. Before these subjects are reviewed a clear description of the material and methods used is necessary.

B. Materials and Methods

Research has been carried out with two types of material: chloroplasts from higher plants and chromatophores from photosynthetic bacteria. The chloroplasts used by most workers in this field have lost their outer membranes and are called C, D, E type in HALL's (1972) nomenclature. From chloroplasts are prepared different categories of subchloroplast particles which behave quite differently from the original material. The constant feature of all these preparations is a membranous structure delimiting an osmotic space or internal medium which responds to addition of permeant compounds in the external medium.

ATP synthesis is measured directly or deduced from the alkalinization of the external medium during its formation.

The latter type of measurement (NISHIMURA et al., 1962) is nondestructive and other events can be followed simultaneously with the pH variation of the external medium where the global reaction at pH around 8 is: $ADP^{3a-} + P_i^{2b-} + c\,H^+ \rightarrow ATP^{4d-}$, with a, b, c and $d \leq 1$. The pH of the external medium is measured with a glass electrode (NISHIMURA et al., 1962) or by recording absorbtion or fluorescence changes in pH-sensitive indicator dyes e.g. bromocresol purple (JACKSON and CROFTS, 1969a), umbelliferone (GRUNHAGEN and WITT, 1970). In this case the problem is to determine whether the optical changes reflect a pH shift in the external medium or changes in state of the membranes with which the dyes interact (CHANCE and MUKAI, 1971; FROMHERZ, 1973). Moreover the response of the glass electrodes is very slow if care is not taken in the choice of the electrodes, especially for temperatures near 0° C, and the amplifier. HIND and McCARTY (1974) give a good survey of all the pitfalls to be avoided. Then it is necessary to know the correspondence between the pH changes and the amount of NaOH equivalent. Since the buffer capacity of a chloroplast suspension is not the same in darkness and in light (POLYA and JAGENDORF, 1969) it is important to buffer the external medium with a sufficient concentration of chemicals to render the influence of light negligible.

On the other hand direct ATP measurements are performed after the reactions have been stopped by lowering the pH. ATP is determined enzymatically (the firefly assay is the most sensitive method, JAGENDORF and URIBE, 1966b). Otherwise by incorporation of ^{32}P into organic compounds (Avron, 1960) or into isolated ATP (Tyszkiewicz et al., 1966; Zaugg and Vernon, 1966) it is possible to compute the amount of phosphate esterified, knowing the specific radioactivity of the inorganic phosphate. Though the ^{32}P method is destructive it is the most widely used and also the most useful when phosphorylated intermediates other than ATP are investigated.

C. General Conditions for Post-Illumination ATP Formation

Postillumination ATP synthesis depends on the amount of high energy state intermediate and the yield of its recovery as ATP. These two factors have been studied during two-stage ATP synthesis experiments.

I. Two-Stage ATP Synthesis

The efficiency of this capture as ATP is optimum at pH 8 (Hind and Jagendorf, 1963a) and increases with the inorganic phosphate concentration (Jagendorf and Hind, 1963). Izawa (1970) estimates the capture efficiency to be 56%, assuming that the dark decay processes are the same with and without concurrent ATP synthesis. Even without knowing this efficiency accurately the amount of ATP synthesized allows a relevant determination of the intermediate X_e.

The level of this intermediate X_e depends on the rate of its light-dependent formation and on the rate of its decay, which is assumed to be independent of the illumination and is studied in the subsequent dark phase. The effect of various factors such as pH, cofactors, light intensity and temperature on these two phenomena have been investigated during two-stage photophosphorylation in chloroplast suspensions. The results must extend to other materials carefully.

1. pH Effects

Hind and Jagendorf (1963a) have shown that postillumination ATP synthesis in chloroplasts is maximum when the external pH is 5.6 during the light phase and 8 during the dark phase. However these values may be affected by very specific conditions such as imposing a diffusion potential across the chloroplast membrane (Schuldiner et al., 1973). Hind and Jagendorf (1963a) have called X_e "the substance which is responsible for ATP generation in the dark and X its ground or inactivated precursor" and have supposed an equilibrium $X \underset{k_f}{\overset{k_d}{\rightleftharpoons}} X_e$. k_d is the rate constant of the dark decay and k_f the rate constant of X_e formation. The rate constant k_d of the dark process increases with the external pH (half-life 2 sec at pH 8 and 32 sec at pH 6), the calculated X value decreases when the pH is raised and the rate constant k_f of the light process remains constant whatever the pH. This explains the drastic reduction in delayed ATP synthesis when the pH during the light phase is shifted from 6 to 8 (60 nmol at pH 6 to 10 nmol/mg of chlorophyll at pH 8).

Rhodospirillum rubrum chromatophores exhibit contrastive pH requirements. Postillumination ATP formation is higher at pH 7.5 or 8 during the light phase (LEISER and GROMET-ELHANAN, 1975), even if the X_e amount decays faster at pH 8 than pH 6 as with the chloroplasts.

2. Cofactor Effects

These effects have been extensively studied by HIND and JAGENDORF (1963b). Maximum postillumination ATP formation in chloroplasts is observed after 15 to 30 sec lighting in the presence of a cyclic photophosphorylation cofactor (phenazine-methosulfate, pyocyanine). With ferricyanide the ATP synthesis level is the same but reached after a longer light phase. Without cofactor the rate constants k_f and k_d are lower and the delayed ATP synthesis yield smaller but variable from one preparation to another. In the presence of a high concentration of phenazine-methosulfate (PMS ≥ 0.1 mM) the postillumination ATP synthesis yield is very large (330 nmol/mg of chlorophyll at pH 8), even at alkaline pH during the light phase (IZAWA, 1970). This PMS effect needs the presence of a high salt concentration and explains the significant ATP formation observed by SHEN and SHEN (1962) in darkness (30 nmol/mg of chlorophyll) after a light phase at pH 7.8, the system containing a high phenazine-methosulfate concentration (50 μM) and 0.3 M NaCl.

3. Light Effect

At the beginning of the light phase there is a linear relationship between the intermediate X_e formation and the duration of the illumination (GROMET-ELHANAN and AVRON, 1965). It appears that the amount of light energy absorbed is the determining factor.

4. Temperature Effect

Light-dependent reactions are less sensitive in chloroplasts to temperature than dark reactions, the respective Q_{10} being 1.5 and 2 (HIND and JAGENDORF, 1963a). Hence by lowering the temperature the dark decay is slowed down to a greater extent than the rate of formation of the intermediate X_e in light and the postillumination ATP synthesis is increased (GROMET-ELHANAN and AVRON, 1965). Light-dependent reactions are ineffective only at a temperature below $-13°$ C, and the high energy state intermediate X_e is stabilized by reducing the temperature to below $-30°$ C (TYSZKIEWICZ and ROUX, 1975).

Surprisingly in *Rhodospirillum rubrum* chromatophores postillumination ATP synthesis is higher at 25° C than at 5° C (LEISER and GROMET-ELHANAN, 1975).

5. Material Effect

All this research was carried out mostly with chloroplasts for which the optimum X_e formation conditions are: addition of pyocyanine or phenazine-methosulfate, pH around 6 during the light phase and low temperature close to 5° C. Without taking into account the X_e capture efficiency postillumination ATP formation has been reported to amount to 120–180 nmol/mg of chlorophyll at infinite inorganic phosphate concentration with pyocyanine as cofactor (Jagendorf and Hind, 1963; Girault et al., 1969), though a maximum yield as high as 370 nmol/mg of chlorophyll is found by Izawa (1970) in the presence of 0.3 mM phenazine-methosulfate. No such extensive studies have been reported with other materials using the two-stage ATP synthesis technique.

In subchloroplasts, postillumination ATP synthesis ranges from 25 nmol to even less 1–3 nmol/mg of chlorophyll in the smallest particles (McCarty, 1968, 1969, 1970).

In *Rhodospirillum rubrum* chromatophores Leiser and Gromet-Elhanan (1975) report that the optimum conditions are phenazine-methosulfate addition, pH around 7.5 to 8 during the light stage and temperature close to 25° C. The yield of postillumination ATP formation, in various chromatophores, is currently given as around a few nmol/mg of bacteriochlorophyll (McCarty, 1970; Schuldiner et al., 1973). In the same organisms Zaugg and Vernon (1966) observed a light-activated ATP inorganic phosphate exchange which makes it difficult to detect a true ATP synthesis with ^{32}P. But in special conditions (Leiser and Gromet-Elhanan, 1975) the ATP synthesis yield can amount to 100 nmol/mg of bacteriochlorophyll.

II. Delayed ATP Synthesis in Flashing Light

It is difficult to compare these above results with those obtained in flashing light. In the latter case the light and dark phases cannot be studied separately.

Basically the conditions of the experiments carried out in chromatophores and chloroplasts are quite different. Flash frequencies varied above 30 cycles/sec in chromatophores (Nishimura, 1962a) and from 1 to 30 cycles/sec in chloroplasts (Witt, 1971). It is, therefore, not surprising that the amount of delayed ATP synthesis after one flash is dependent in chromatophores and independent in chloroplasts of the flash frequency.

Very long dark phases between two subsequent flashes have also been used: e.g. 10 sec by Junge et al. (1970) working with chloroplasts and 60 sec by Nishimura (1962c) with chromatophores. Under these comparable conditions the delayed ATP synthesis yield after one flash amounts to 0.15 to 1 nmol/mg of chlorophyll in chloroplasts (Neumann et al., 1970; Junge et al., 1970) and 50 nmol/mg of bacteriochlorophyll in chromatophores (Nishimura, 1962c).

D. High Energy State Intermediate

The nature of the high energy state intermediate(s) is unknown. Gromet-Elhanan and Avron (1965) have shown that it is not diffusible but stuck on the chloroplast membrane. The identification of the intermediate as a chemical is the easiest understandable approach.

I. Chemical High Energy State Intermediate

1. Non-Phosphorylated Intermediate

The first idea is to assume that the intermediate is formed by accumulation of oxidized and reduced substances which on reacting together lead to ATP synthesis. This seems to be the case in NISHIMURA's experiments (1962c) where the amount of delayed ATP synthesized during the dark phase (60 sec) after one saturating flash (0.5 msec) is stoichiometrically linked, mol by mol, with that of cytochrome c. The delay in photophosphorylation reactions seems to be due to accumulation of oxidized cytochrome c and an unknown reduced substance, but in chloroplasts the high energy state X_e displays properties which forbid such an explanation. First the decay time is much higher, a few seconds, instead of a few milliseconds as in NISHIMURA's experiments, and it hardly seems that accumulation of oxidized and reduced substances back-reacting very quickly (a few msec) could be the cause of the delay in chloroplast photophosphorylation (a few sec). Furthermore in chloroplasts the plastoquinone A is the only electron carrier present in the same quantity as X_e and it is impossible to detect any significant change in the plastoquinone absorption spectrum under X_e forming conditions.

2. Phosphorylated Intermediate

The second idea is to search for a phosphorylated intermediate as proposed by SLATER's hypothesis. In support of this it is found that ATP synthesis is a reversible process in chloroplasts and chromatophores linked to the presence of a protein, the coupling factor. In intact A or B type chloroplasts (HALL, 1972) ATPase activity is found without induction (KRAAYENHOF et al., 1969), but in C, D or E types ATPase needs activation. Recently RIENITS et al. (1974) reported that a reverse electron flow in chloroplasts is observed during ATP hydrolysis by light-triggered ATPase. In chromatophores a transhydrogenase reaction (BALTSCHEFFSKY et al., 1966) is also observed during hydrolysis of ATP and pyrophosphate. It was already well known that the presence of ATP during the light phase enhances postillumination ATP synthesis and therefore high energy state intermediate X_e formation (JAGENDORF and HIND, 1965; McCARTY et al., 1971). In fact a phosphorylated intermediate has been detected during the functioning of most membraneous ATPases which are also able to drive ATP synthesis [in sarcoplasmic reticulum by MARTONOSI et al. (1974), in electric organs by FAHN et al. (1968) and in erythrocytes membranes by AVRUCH and FAIRBANKS (1972)], and this could be the case in photosynthetic organisms.

However, it turns out that the phosphorylated intermediate reported by KAHN and JAGENDORF (1960) in a two-stage photophorphorylation with chloroplasts is only intrachloroplastic ^{32}P-ATP which is split by the light-induced ATPase unless a sufficiently inhibitory concentration of ADP is added (JAGENDORF and HIND, 1963). The phosphorylated intermediate shown by SHEN and SHEN (1962) is also more probably internal or bound ATP, as reported in chromatophores by HORIO et al. (1966).

TYSZKIEWICZ et al. (1966) have shown that postillumination ATP synthesis, in the presence of unlabeled inorganic phosphate and magnesium ions during the two stages, is higher when carrier-free ^{32}P is added in the light phase than in the dark phase. This could be a matter of accessibility of ^{32}P to phosphorylating sites because no phosphorylated intermediate can be isolated. More convincing is TYSZKIEWICZ's observation (1972) that the X_e dark decay is faster without than with inorganic phosphate ions present during light and dark phases before ADP addition, at least in the presence of a high $MgCl_2$ concentration. Under these conditions a pyrophosphate synthesis occurs during the light phase at pH 6. HIND and JAGENDORF (1963a) fail to find any action of this sort, but use very low concentrations of $MgCl_2$ (around 5 µM) instead of 100 µM as in TYSZKIEWICZ's work. BACHOFEN et al. (1968) find a dark ATP synthesis in chloroplasts from pyrophosphate and ADP but do not report the phosphate transfer rate. Similarly GUILLORY and FISHER (1972) find pyrophosphate synthesis in chromatophores of *Rhodospirillum rubrum* at a rate which reaches about one fourth that of ATP synthesis at light saturation. In this organism hydrolysis of pyrophosphate is linked to an ATP synthesis (KEISTER and MINTON, 1971) but inorganic phosphate is not transferred from pyrophosphate to ATP, or at least exchanges very quickly with the inorganic phosphate present in the medium during transfer.

3. Conclusion

In conclusion, the only phosphorylated intermediates isolated are pyrophosphate and internal or bound ATP. It seems that energy-linked reactions leading to synthesis and conversely to hydrolysis of pyrophosphate and ATP involve a common intermediate or interactions between different intermediates, which are not necessarily phosphorylated. In chromatophores it is suggested that a common phosphorylated intermediate is not formed (KEISTER and MINTON, 1971). It remains that at high $MgCl_2$ concentration the presence of inorganic phosphate ions stabilizes the intermediate X_e in chloroplasts, but this effect could be indirect and not require a phosphorylating step.

Even if unknown phosphorylated intermediate(s) cannot be ruled out the high energy state intermediate X_e does not require the addition of inorganic phosphate or of $MgCl_2$ for its formation. Moreover the phosphorylated intermediate, if it exists, could be either an obligatory intermediate necessary for ATP synthesis or a compound synthesized in a side reaction not in the direct path of ATP synthesis.

II. High Energy State and Membrane Property Changes

Since it is not possible to specify what is the chemical nature of X_e and even less to decide whether X_e is a true chemical or only an "energized state", the next point is to establish the relationships between postillumination ATP synthesis and phenomena which have been said to be representative of this "energized state": e.g. proton and ion transport, field changes and conformation changes.

1. Proton Uptake and X_e Formation

JAGENDORF and HIND (1963) report a light-induced pH increase of an unbuffered chloroplast suspension. The kinetics of this pH change closely resemble those of the high energy state X_e (NEUMANN and JAGENDORF, 1964): e.g. the same dependency on cofactors acting catalytically, external pH, temperature and uncouplers (JAGENDORF and NEUMANN, 1965). However some discrepancies are reported by JAGENDORF and URIBE (1966a):

1. High salt concentrations or oleic acid addition inhibits postillumination ATP synthesis more than the light-dependent pH increase.

2. The kinetics of both phenomena are different in short-run experiments.

3. The light-dependent pH increase is more sensitive to low external pH.

"These discrepancies cannot be considered crucial until the pH changes and X_e are both measured in the same solution". These experiments have been carried out by GALMICHE et al. (1967) and the postillumination ATP formation is shown to be linearly related to the amount of proton uptake (equivalent to the pH increase) at least for an external pH between 4.5 and 6.5. This relation is verified during the light and subsequent dark phases. The full extent of the net proton uptake is titrated at constant pH in the light and is thus known with better accuracy than if it were calculated from pH changes and buffer capacity, which changes when the light is turned on (POLYA and JAGENDORF, 1969). At optimum concentration of inorganic phosphate GIRAULT et al. (1969) report a ratio H^+/ATP of 5 in chloroplasts during two-stage ATP synthesis. IZAWA (1970), taking into account the fact that only 50% of the high-energy state intermediate X_e is trapped as ATP, finds after correction a ratio H^+/ATP close to 2. IZAWA and HIND (1967) have resolved the fast kinetics of the pH changes by using a flow method. In this way they avoid the slow response of the glass electrodes and discrepancies between the X_e and pH change kinetics disappear. The pH change has been analyzed kinetically by KARLISH and AVRON (1968) in the same way as X_e and the rate constants of the dark and light processes of the two phenomena, X_e and pH change, show roughly the same sensitivity to external factors.

All these results were obtained with higher plant chloroplasts where the light-induced proton uptake amounts to around 1 μmol/mg of chlorophyll. In subchloroplasts the proton uptake and postillumination ATP synthesis are respectively, 0.3 μequivalents and 24 nmol/mg of chlorophyll for sonicated particles (McCARTY, 1968) and only 0.012 μequivalents and 1–4 nmol/mg of chlorophyll in digitonin treated particles (NELSON et al., 1970). Light-induced reversible pH changes have also been observed in chromatophores from *Rhodospirillum rubrum* by VON STEDINGK and BALTSCHEFFSKY (1966). Proton uptake is around 1 μequivalent and postillumination ATP synthesis only 10–20 nmol/mg of bacteriochlorophyll (McCARTY, 1970; SCHULDINER et al., 1973).

Two questions arise: the first concerns the meaning of this light-induced pH change, the second the relationship between pH changes and phosphorylating activities.

a) Proton Uptake and Proton Gradient

It seems that the pH rise in the external medium corresponds to a proton uptake by the chloroplasts, and postillumination ATP formation is driven by the reverse process, a proton efflux through the chloroplast membrane.

If the internal medium of the chloroplasts is buffered by penetrating weak acid the light-induced proton uptake is increased like the postillumination ATP formation. NELSON et al. (1971) use pyridine (pK = 6.2) and AVRON (1972) imidazole (pK = 6.9), aniline (pK = 4.5) and p-phenylenediamine (pK = 6.2). The report by IZAWA (1970) of a very high proton uptake and postillumination ATP synthesis in the presence of 100 µM phenazine methosulfate and high salt concentration is probably the result of internal buffering by the cofactor. Both proton uptake and postillumination ATP synthesis are increased (9 µequivalents and 0.5 µmol/mg of chlorophyll) and the pH optimum during the light phase is shifted to the alkaline range following the pK of the buffering compound. Such buffering seems ineffective in subchloroplasts (LYNN, 1968), but has been reported by LEISER and GROMET-ELHANAN (1975) to enhance postillumination ATP formation in *Rhodospirillum rubrum* chromatophores.

The osmotic space should respond to the proton uptake and its pH should decrease. This decrease has been found and measured by following the accumulation in the osmotic space of radioactive amines (ROTTENBERG et al., 1972), fluorescent amines (SCHULDINER et al., 1972b), ammonium ions (ROTTENBERG and GRUNWALD, 1972) or a weak radioactive electrolyte (DILLEY and ROTHSTEIN, 1967). RUMBERG and SIGGEL (1969) independently observed a similar pH drop in the "inner phase of the thylakoid membrane" by following the rise time of the P-700 reduction which is pH-dependent. The difference between external and internal pH is greatest for an external pH of 9 where the extent of proton uptake is minimum (ROTTENBERG et al., 1972). This difference amounts to around 3 in chloroplasts (SCHULDINER et al., 1972b), around 2.5 in subchloroplasts (ROTTENBERG and GRUNWALD, 1972) and from 1.8 (SCHULDINER et al., 1974) to 3.5 (CASADIO et al., 1974) in chromatophores at external pH between 7.5 and 8.

It is assumed that postillumation ATP synthesis is not only dependent on the amount of protons accumulated inside the membrane but also on the pH difference between the internal osmotic space and the external medium during dark phosphorylation reactions.

The first indication in that respect is the shift of the optimum pH to the alkaline range for the dark reaction when the pH of the chloroplast suspensions during the light stage is raised from 5.6 to 7.3 (HIND and JAGENDORF, 1963a). SCHULDINER et al. (1973) show that, in a two-stage ATP synthesis in chloroplasts, ATP formation is observed only if there is a minimum difference between internal and external pH during the dark phase. This critical proton gradient is dependent on the external pH during the dark stage: e.g. ΔpH is 2.4 at dark pH 7.5 and close to 3 at pH 8.5. The pH difference threshold could be related to the difference between the chemical potentials of the proton in the two media; $\Delta \tilde{\mu}_{H^+} = -2.3\, RT\, \Delta$pH (NOBEL, 1970) which must exceed a minimum value for the proton efflux to be linked to ATP formation.

In *Rhodospirillum rubrum* chromatophores (LEISER and GROMET-ELHANAN, 1975) the difference between external and internal pH during the light phase must exceed a ΔpH threshold close to 2.1. The pH shift in the external medium during the transition, light to dark phase, does not intervene in the phosphorylation reactions. Light-induced proton gradient is higher at pH 8 than at pH 6, as the postillumination ATP synthesis yield.

It must be explained why during two-stage photophosphorylation experiments the amount of ATP synthesized is found to be linearly connected only to the extent of the proton uptake in chloroplasts (GALMICHE et al., 1967). This is the result of internal pH buffering around 4.5. During the light phase at external pH below 6.5 the first few protons transported into the inner phase shift the internal pH to its final value (around the pK of the ionizable groups in the inner phase) while most of the other protons (around 70% at external pH 6.3) undergo buffering (SCHULDINER et al., 1972b). So conversely the pH gradient during the proton efflux in the dark phase is very stable at least until 30% of the protons are left inside the osmotic space and the most important factor for the yield of ATP formation is the extent of proton efflux, equivalent to the proton uptake.

b) Proton Gradient and Phosphorylation

The next question is the significance of this proton transfer for X_e formation and postillumination ATP synthesis, and tentatively for continuous photophosphorylations if it is assumed that X_e is an obligatory intermediate for ATP synthesis. If the proton gradient is necessary for ATP synthesis it should exist wherever this phenomenon is observed; furthermore, the proton gradient must be reduced during concomitant phosphorylation of added ADP and possibly should be created during the reverse reaction, ATP hydrolysis.

NELSON et al. (1970) report that digitonin treatment of chloroplasts suppresses both proton uptake and postillumination ATP formation whereas a high residual activity of phosphorylation is retained during a continuous illumination. Indeed ROTTENBERG and GRUNWALD (1972) show that a proton gradient still exists even if the proton uptake is so small that a rise in external pH is undetectable. MCCARTY (1968) reports that in subchloroplasts, which are prepared by sonication, NH_4Cl abolishes both proton uptake and postillumination ATP formation without inhibiting continuous photophosphorylation. Here also ROTTENBERG and GRUNWALD (1972) find that, in subchloroplast particles, the proton gradient and continuous photophosphorylation are inhibited together to the same extent by increasing the amine concentration. Thus in subchloroplasts as in chloroplasts the extent of proton uptake determines the yield of postillumination ATP synthesis and the proton gradient controls continuous photophosphorylation.

In *Rhodospirillum rubrum* chromatophores the light-induced proton uptake and the proton gradient, measured by the quenching of atebrine fluorescence, are inhibited by NH_4Cl or nigericine in the presence of KCl (GROMET-ELHANAN, 1972b). Under these conditions postillumination ATP synthesis is inhibited but not continuous photophosphorylations, assumed to be bound up with the presence of a membrane potential (GROMET-ELHANAN, 1972b; GROMET-ELHANAN and LEISER, 1973) which collapses in the presence of the permeant anions thiocyanate (SCN⁻), perchlorate (ClO$_4^-$) and tetraphenyl-boron (MONTAL et al., 1970). Thus in the presence of NH_4SCN or NH_4ClO_4 both proton gradient and membrane potential are inhibited as are continuous photophosphorylations. It has been suggested that the difference between the electrochemical potential of the proton in the two media, $\Delta \tilde{\mu}_{H^+} = F \Delta E$ (membrane potential) $-2.3 RT \Delta$pH, is implicated in ATP synthesis (SCHULDINER et al., 1974; CASADIO et al., 1974), the extent of proton uptake determining the yield of postillumination ATP synthesis.

In chloroplasts there are conflicting reports concerning the effect of photophosphorylation rates on the building of the proton gradient. MCCARTY et al. (1971) reconcile all the results by suggesting that the coupling factor (CF_1) alters its conformation by binding ATP. A very low concentration of ATP is active and it is not excluded that ADP acts in the same way by giving some traces of ATP. By changing its conformation the coupling factor would alter the proton permeability of the membrane and increase the proton gradient. This explains why suboptimal addition of ADP leads to an enhancement of the proton uptake instead of the expected inhibition. In the presence of saturating ADP, PORTIS and MCCARTY (1974) and PICK et al. (1973) report a significant decrease in the proton gradient under phosphorylation or arsenolysis conditions. These results are in agreement with the inhibition of NH_4^+ ion uptake (which replaces H^+ uptake) by phosphorylation conditions observed by CROFTS (1966). The ATPase activity of the chloroplasts, triggered by previous illumination in the presence of dithiothreitol, is linked with a proton uptake (CARMELI, 1970) and a quenching of the atebrine fluorescence (KRAAYENHOF, 1970) which is interpreted as the formation of a proton gradient through the membranes (BAKKER-GRUNWALD and VAN DAM, 1973; BAKKER-GRUN-WALD, 1974). In the same chloroplast suspensions hydrolysis of ATP induces uptake of amine (GAENSSLEN and MCCARTY, 1971) or NH_4^+ ion (CROFTS, 1966) instead of H^+.

In chromatophores, active photophosphorylation accelerates the light-dependent cyclic electron flow and the expected slowdown of the proton uptake is concealed (MELANDRI et al., 1970). Nevertheless GROMET-ELHANAN (1972b) reports a slight decrease in the proton gradient under phosphorylation conditions. MELANDRI et al. (1972) show that ATP addition in darkness increases the proton gradient, measured as fluorescence quenching of atebrine, in *Rhodospirillum capsulata*; this action is linked to the oligomycin-sensitive ATPase activity, dependent on the presence of the coupling factor.

These results are as expected if ATP synthesis is assumed to be a reversible reaction driven by the proton gradient, but other explanations can be put forward.

2. Membrane Potential and X_e Formation

In chloroplasts and subchloroplasts continuous photophosphorylation is observed only if the proton gradient is present. In chromatophores however continuous photophosphorylation takes place in the presence of either proton gradient or membrane potential. Nevertheless, X_e is related to the degree of proton uptake, at least in two-stage photophosphorylation experiments. Since, as already observed, the proton gradient controls the yield of postillumination ATP synthesis we are led to determine whether the so-called membrane potential also plays a role in this yield.

By adding valinomycin in the presence of K^+ ion during the dark phase of a two-stage photophosphorylation experiment a membrane potential (which is positive with respect to the inside of the membrane) is built up and is linked with the absorbance changes at 520 nm in chromatophores (JACKSON and CROFTS, 1969b) and in chloroplasts (STRICHARTZ and CHANCE, 1972). SCHULDINER et al. (1972a, 1973) show that in chloroplast suspensions such a membrane potential (positive inside), imposed during the dark phase, increases postillumination ATP formation by as much as 40–60 nmol/mg of chlorophyll. Under suboptimal conditions (low light intensity, external pH low during the dark phase or high during the light phase), the yield of postillumination ATP formation is increased by applying this potential and the threshold of the difference between external and internal pH at the beginning of the dark phase is lowered. The same yield increase by addition of valinomycin plus K^+ ion in the dark phase is observed in subchloroplast particles (McCarty, 1970) and in chromatophores (Schuldiner et al., 1973; Gromet-Elhanan and Leiser, 1975). In *Rhodospirillum rubrum* chromatophores addition of these compounds during the dark phase of a two-stage ATP synthesis changes the pH requirement and the yield of ATP synthesis increases when the pH of the light stage decreases from 8 to 6 (Gromet-Elhanan and Leiser, 1975).

Addition of valinomycin during the light phase of a two-stage photophosphorylation experiment in chloroplast or chromatophore suspensions containing K^+ ions drives a K^+ ion efflux down the potential membrane (Grunhagen and Witt, 1970; Junge and Schmid, 1971) and accelerates the dark decay of the membrane potential. In chloroplasts addition of valinomycin in the light phase speeds up the rate of formation and decay of the proton uptake without changing its extent nor the postillumination ATP formation (Karlish et al., 1969; McCarty, 1970). However in subchloroplasts (McCarty, 1970) and in chromatophores (Jackson et al., 1968; Schuldiner et al., 1973; Leiser and Gromet-Elhanan, 1975) the presence of valinomycin plus K^+ ions during the light phase enhances proton uptake and postillumination ATP formation whereas continuous photophosphorylations remain identical under such conditions. These differences could be due to the low permeability of the membrane to Cl^- ions in subchloroplasts (McCarty, 1969) and in chromatophores (Gromet-Elhanan, 1972b; Gromet-Elhanan and Leiser, 1973) with respect to that observed in chloroplasts (Hind et al., 1974). In chromatophores and subchloroplasts H^+ transport is not accompanied by a Cl^- transport in the same direction. The proton transfer is, therefore, electrogenic and induces formation of a membrane potential which becomes the factor limiting proton accumulation. In the presence of valinomycin and K^+ ions the membrane potential collapses and proton accumulation, no longer limited, thus increases. This limitation should not be observed in chloroplasts where the charges of the protons accumulated are compensated by a transport of Cl^- ions in the same direction. In *Rhodospirillum rubrum* chromatophores addition of permeant anions (thiocyanate, perchlorate) during the light stage has the same action as valinomycin plus K^+ (Leiser and Gromet-Elhanan, 1975). These additions do not change the pH requirement during postillumination ATP formation. The yield of ATP synthesis is maximum when pH of the light stage is 7.5–8 and amounts to 100 nmol/ mg of bacteriochlorophyll.

In chromatophores the proton gradient is not greatly reduced during active ATP synthesis but the membrane potential, measured as 8-anilino-1-naphthalene sulfonate (ANS) fluorescence enhancement, is strongly inhibited (Gromet-Elhanan, 1972b). Conversely, during ATP or pyrophosphate hydrolysis the membrane potential, measured as carotenoid absorbance changes at 515 nm (Baltscheffsky, 1969) or ANS fluorescence changes (Vainio et al., 1972), is enhanced. In tightly coupled chloroplasts Baltscheffsky and Hall (1974) report that the membrane potential, measured as the 515 nm absorbance change, also decreased under phosphorylating conditions. These results are consistent with the reversibility of photophosphorylation reactions and the involvement of the membrane potential in such reactions.

3. Conformational Changes and X_e Formation

The first conformational change which has been correlated with formation of X_e is a modification of the chloroplast structure measured by 90° light scattering changes (Dilley and Vernon, 1964). Hind and Jagendorf (1965a, b) however find very large discrepancies between X_e formation and absorbancy increase due to light scattering changes. It is also possible to find conditions where X_e is still present but light scattering changes absent: in the presence of sucrose (Hind and Jagendorf, 1965b) or with polygalacturonate-treated chloroplasts in the presence of permeant anions such as Cl^- (Cohen and Jagendorf, 1972). These light scattering changes are more likely to be related to the light-dependent redistribution of ions (Hind et al., 1974).

The conformational changes could be related to the structural changes of the proteins buried or attached to the membranes. These changes have been reported using fluorescent probes (Brand and Gohlke, 1972). Kraayenhof (1970) has described such changes in the fluorescence of reporter groups and claims that quenching of the 9-amino-acridine fluorescence reflects a binding of the probe to certain structures related to the energy conservation mechanism, or a change of the environment in which the probe is situated. Unfortunately the same quenching is interpreted by Schuldiner et al. (1972b) as a measurement of the proton gradient. It therefore seems that the calculation of the proton gradient from the quenching of fluorescent amine fluorescence is questionable (Fiolet et al., 1974). The same difficulties arise when the fluorescence of 8-anilino-1-naphthalene-sulfonate (ANS) is studied. Results of Gromet-Elhanan and Leiser (1973) are explained by a direct relation between enhancement of ANS fluorescence and formation of a membrane potential, but in electroplax membranes Patrick et al. (1971) interpret these phenomena as a change in the membrane conformation. In addition Kraayenhof and Katan (1972) report ANS and atebrine fluorescence change kinetics too slow to be consistent with those of membrane potential (measured as carotenoid absorption changes) and proton gradient (estimated from the proton uptake in the external medium) formation.

The result is that while it is easy to determine the light scattering changes which roughly show size and stacking modifications of the chloroplasts these changes are not directly related to X_e formation. In addition although it is difficult to refute a parallelism between the property changes of reporter groups attached to the membrane and the high energy state formation, interpretation of these changes is always a matter of speculation and cannot with certainty be based on a conformational change of the membrane proteins.

E. Hypotheses on the Nature of the High Energy State Intermediate X_e

Two interpretations of this high energy state intermediate X_e can be retained: one is related to the electrochemical potential of the proton in the two media separated by the membrane and the other involves a conformational high energy state of the membrane proteins. Indeed in both these interpretations protons are pumped from the external medium. In the first scheme the protons penetrate through the membrane and are accumulated inside the osmotic space and so far we have favored this hypothesis. In the second scheme the protons are fixed to the membrane as in the Bohr effect (CHANCE et al., 1970), a hypothesis supported by KRAAYENHOF's and SLATER's experiments (1974). Nevertheless, if proton fixation on the membrane does occur, the amount of dissociable groups is the same in light and in darkness and only the pH dependence of the proton binding changes drastically (WALZ et al., 1974). Hence there is an apparent pK shift of ionizable groups during illumination which could be:

1. According to the first hypothesis, the consequence of the differences between external and internal pH (WALZ et al., 1974).

2. According to the second hypothesis, the result of changes at the membrane interface (FROMHERZ, 1973).

In the discussion we shall refer to the first hypothesis which is easier to handle and allows, unlike the second, the introduction of two phenomena: the proton gradient and the membrane potential which are challenged by certain authors but are the expression of two different components of the membrane properties. Each of these two components can be cancelled separately and in the first scheme the inhibition of each enhances the extent of the other (HENDERSON, 1971). Table 1 lists some characteristics of these two components.

Table 1

Measurement	Inhibition
Proton gradient	
^{14}C amine uptake (ROTTENBERG et al., 1972)	Amines (HENDERSON, 1971)
Fluorescent amines uptake atebrine, 9-amino-acridine (SCHULDINER et al., 1972 b)	
NH_4^+ uptake (ROTTENBERG and GRUNWALD, 1972)	NH_4^+ (SCHULDINER and AVRON, 1971) Nigericin plus K^+ (SHAVIT et al., 1968)
Membrane potential	
515 nm absorption change (WITT, 1971)	
ANS fluorescence (GROMET-ELHANAN, 1972a)	
$^{14}C-SCN^-$ uptake (SCHULDINER et al., 1974)	SCN^-, ClO_4^- (GROMET-ELHANAN, 1972b)
Phenyldicarboundecaborane (PCB^-) uptake (GRINIUS et al., 1970)	PCB^- tetraphenylboron (MONTAL et al., 1970) Picrate (SCHULDINER et al., 1972a)

The high-energy state intermediate X_e is characterized not only by the extent of proton fixation but also by the amount of these two components which controls the yield of ATP formation during the release of protons in the external medium.

In chloroplasts the membrane potential seems not to contribute significantly to phosphorylation reactions. Addition of valinomycin in the presence of K^+ ions during the light phase destroys the membrane potential or at least the 515 nm absorption changes (LARKUM and BOARDMAN, 1974) but does not alter the extent of proton uptake, X_e formation and the rate of photophosphorylation in continuous illumination. However, nigericin in the presence of K^+ ions destroys the proton gradient but either does not change or slightly increases the membrane potential (LARKUM and BOARDMAN, 1974); in this case X_e formation is inhibited as are continuous photophosphorylations, and the proton gradient seems to be the limiting factor in these two reactions. Nevertheless, even if ROTTENBERG et al. (1972) fail to observe any membrane potential in continuously illuminated chloroplasts this potential, or at least the 515 nm absorption change which depends on it following WITT's interpretation (1971), is present to a debatable extent. The increased postillumination ATP formation reported by SCHULDINER et al. (1972a, b, 1973) when a membrane potential is imposed during the dark phase seems contradictory with the apparent ineffectiveness of this potential in phosphorylation above mentioned. But it should be recalled that this positive effect is observed with suboptimal proton gradients. It must be noted that in flashing light, if the frequency of the successive flashes is very low (0.1 per sec), the membrane potential becomes the determining factor and valinomycin in the presence of K^+ ions inhibits ATP synthesis (JUNGE et al., 1970). This could be explained by the generation of a suboptimal proton gradient under such conditions.

In chromatophores on the other hand the two components, proton gradient and membrane potential are interchangeable and both contribute significantly to photophosphorylation. NH_4^+, amines and nigericin plus K^+ ions destroy the proton uptake and proton gradient, inhibit X_e formation but increase the membrane potential and do not change the photophosphorylation rate in continuous illumination. The continuous photophosphorylation disappear only if the proton gradient and membrane potential are both inhibited e.g. by NH_4SCN or NH_4ClO_4. In the presence of SCN^-, ClO_4^- or valinomycin plus K^+ ions the membrane potential collapses but proton uptake and proton gradient are enhanced, as is X_e formation, while the continuous photophosphorylation rate remains the same.

Apparently subchloroplasts exhibit the same properties as chromatophores; inhibition of the proton gradient raises the membrane potential and vice versa. However while amines inhibit proton uptake and X_e formation, the proton gradient should not be inhibited by amines at the usual concentration active in chloroplasts. By increasing the amine concentration the proton gradient collapses progressively and the continuous photophosphorylation rate decreases proportionally. Thus in subchloroplasts as in chloroplasts the proton gradient is the essential factor in photophosphorylation.

The differences observed in the sensitivity of these different organisms to inhibition of the proton gradient or the membrane potential is still subject to controversial interpretations. This does not seem to be only the result of the different membrane permeabilities. The separate size of the vesicles are implicated by HAUSKA and SANE (1972) in the stability of the membrane potential, or by WALKER (1975) in the extent of the allowed internal osmotic pressure which governs the proton gradient. Specific properties of the different ATPases, especially the need for a proton gradient threshold to release them from inhibition, have also been proposed.

It is well known that the coupling factors from chloroplasts and chromatophores have different properties. Removal of the coupling factor (CF_1) abolishes the proton uptake and proton gradient in chloroplasts (McCarty and Racker, 1966) but not in chromatophores (Melandri et al., 1972; Gromet-Elhanan, 1974). Nevertheless, light-induced ANS fluorescence changes (interpreted as measures of the membrane potentials) are decreased by removal of the coupling factor from chromatophores and this decrease is reversed by oligomycin treatment or readdition of the coupling factor (Gromet-Elhanan, 1974).

More fundamental are the differences observed in the sensitivity of postillumination ATP synthesis to the proton gradient in chloroplasts and chromatophores. The determinative factor is the pH difference between external medium and internal osmotic space either during the dark stage (ATP formation) in chloroplasts or during the light stage in *Rhodospirillum rubrum* chromatophores. In these latter organisms the pH difference between the external medium in the light and dark stages is ineffective for ATP synthesis. In order to explain these discrepancies the membrane potential should have been determined in the same time.

F. Conclusions

These results are consistent with the presence of regulating and accumulating energy processes between the photoacts and the phosphorylation reactions. Postillumination ATP synthesis is the expression of these processes and is quantitatively linked to the quantity of protons "pumped" from the external medium.

Two questions remain unanswered and will finally remain unanswerable until the mechanism of ATP synthesis is explained.

First, what changes in the properties of the membrane are linked to proton "pumping" processes? Ion and charge concentrations in the different phases seems to be implicated and their changes have been described as the formation of a "proton gradient" and a "membrane potential". However, while the specificity of the proton in all these ion transports is unequivocal the measurement and the nature of the "proton gradient" and the "membrane potential" are controversial. These two properties characterize the membrane interface phenomena and should be referred to the known characteristics of such phenomena (Davies and Rideal, 1963). Indeed the proposed model, two media delimited by a membrane, has to be refined as it does not take into account the specific states of the membrane.

Second, what are the relationships between those "proton pumping" processes and the phosphorylation reactions? The proton influx may be either an obligatory prerequisit to ATP synthesis which is thus quantitatively linked to a proton efflux, or a reversible process, competing with the phosphorylation reactions for the dissipation of a common energized state intermediate. The first explanation seems more probable since the ratio H^+ to ATP (Schwartz, 1968; Izawa, 1970; Junge et al., 1970; Witt, 1971; Schröder et al., 1972; McCarty and Magnusson, 1974; McCarty et al., 1975; Heathcote and Hall, 1974; Trebst, 1974) is the same, around 2 or 3, during postillumination ATP formation as well as continuous photophosphorylation.

References

Avron, M.: Biochim. Biophys. Acta **40**, 257–272 (1960)

Avron, M.: In: Proc. 2nd Intern. Congr. Photosynthesis. Forti, G., Avron, M., Melandri, A. (eds.). The Hague: Dr. Junk, 1972, pp. 861–871

Avruch, J., Fairbanks, G.: Proc. Nat. Acad. Sci. U.S. **69**, 1216–1220 (1972)

Bachofen, R., Lutz, H., Specht-Jurgensen, I.: FEBS Lett. **1**, 249–251 (1968)

Bakker-Grunwald, T.: Biochim. Biophys. Acta **368**, 386–392 (1974)

Bakker-Grunwald, T., Van Dam, K.: Biochim. Biophys. Acta **292**, 808–814 (1973)

Baltscheffsky, M.: Arch. Biochem. Biophys. **130**, 646–652 (1969)

Baltscheffsky, M., Baltscheffsky, H., von Stedingk, L.-V.: Brookhaven Symp. biol. **19**, 246–257 (1966)

Baltscheffsky, M., Hall, D.O.: FEBS Lett. **39**, 345–348 (1974)

Boyer, P.D.: In: Oxidases and Related Redox Systems. King, T.E., Mason, H.S., Morrison, M. (eds.). New York: Wiley, 1965, Vol. II, pp. 994–1008

Brand, L., Gohlke, J.R.: Ann. Rev. Biochem. **41**, 843–868 (1972)

Carmeli, C.: FEBS Lett. **7**, 297–300 (1970)

Casadio, R., Baccarini-Melandri, A., Zannoni, D., Melandri, B.A.: FEBS Lett. **49**, 203–207 (1974)

Chance, B., McCray, J.A., Bunkenburg, J.: Nature (London) **225**, 705–708 (1970)

Chance, B., Mukai, Y.: In: Probes of Structure and Function of Macromolecules and Membranes. Chance, B., Chuan-Pu Lee, Blasie, J.K. (eds.). New York-London: Academic Press, 1971, Vol. I, pp. 239–243

Cohen, W.S., Jagendorf, A.T.: Arch. Biochem. Biophys. **150**, 235–243 (1972)

Crofts, A.R.: Biochem. Biophys. Res. Commun. **24**, 725–731 (1966)

Davies, J.T., Rideal, E.K.: Interfacial Phenomena. New York-London: Academic Press, 1963

Dilley, R.A., Rothstein, A.: Biochim. Biophys. Acta **135**, 427–443 (1967)

Dilley, R.A., Vernon, L.P.: Biochemistry **3**, 817–824 (1964)

Fahn, S., Koval, G.J., Albers, R.W.: J. Biol. Chem. **243**, 1993–2002 (1968)

Fiolet, J.W.T., Bakker, E.P., Van Dam, K.: Biochim. Biophys. Acta **368**, 432–445 (1974)

Fromherz, P.: Biochim. Biophys. Acta **323**, 326–334 (1973)

Gaensslen, R.E., McCarty, R.E.: Arch. Biochem. Biophys. **147**, 55–65 (1971)

Galmiche, J.M., Girault, G., Tyszkiewicz, E., Fiat, R.: C.R. Acad. Sci., Paris **265 D**, 374–377 (1967)

Girault, G., Tyszkiewicz, E., Galmiche, J.M.: In: Progress in Photosynthesis Research. Metzner, H. (ed.). Tübingen: Laupp, 1969, Vol. III, pp. 1347–1353

Grinius, L.L., Jasairis, A.A., Kadziauskas, Yu.P., Liberman, E.A., Skulachev, V.P., Topali, V.P., Tsofina, L.M., Vladimirova, M.A.: Biochim. Biophys. Acta **216**, 1–12 (1970)

Gromet-Elhanan, Z.: In: Proc. 2nd Intern. Congr. Photosynthesis. Forti, G., Avron, M., Melandri, A., (eds.). The Hague: Dr. Junk, 1972a, Vol. II, pp. 985–994

Gromet-Elhanan, Z.: Biochim. Biophys. Acta **275**, 124–129 (1972b)

Gromet-Elhanan, Z.: J. Biol. Chem. **249**, 2522–2527 (1974)

Gromet-Elhanan, Z., Avron, M.: Plant Physiol. **40**, 1053–1059 (1965)

Gromet-Elhanan, Z., Leiser, M.: Arch. Biochem. Biophys. **159**, 583–589 (1973)

Gromet-Elhanan, Z., Leiser, M.: J. Biol. Chem. **250**, 90–93 (1975)

Grunhagen, H.H., Witt, H.T.: Z. Naturforsch. **25 B**, 373–386 (1970)

Guillory, R.J., Fisher, R.R.: Biochem. J. **129**, 471–481 (1972)

Hall, D.O.: Nature New Biology **235**, 125–126 (1972)

Hauska, G.A., Sane, P.V.: Z. Naturforsch. **27 B**, 938–942 (1972)

Heathcote, P., Hall, D.O.: In: Proc. 3rd Congr. Photosynthesis. Avron, M. (ed.). Amsterdam: Elsevier, 1975, Vol. I, pp. 463–471

Henderson, P.J.F.: Rev. Microbiol. **25**, 393–428 (1971)

Hind, G., Jagendorf, A.T.: Proc. Nat. Acad. Sci. U.S. **49**, 715–722 (1963a)

Hind, G., Jagendorf, A.T.: Z. Naturforsch. **18 B**, 689–694 (1963b)

Hind, G., Jagendorf, A.T.: J. Biol. Chem. **240**, 3195–3201 (1965a)

Hind, G., Jagendorf, A.T.: J. Biol. Chem. **240**, 3202–3209 (1965b)

Hind, G., McCarty, R.E.: Current Topics Photobiol. Photochem. **8**, 114–156 (1974)

Hind, G., Nakatani, H.Y., Izawa, S.: Proc. Nat. Acad. Sci. U.S. **71**, 1484–1488 (1974)
Horio, T., von Stedingk, L.V., Baltscheffsky, H.: Acta Chem. Scand. **20**, 1–10 (1966)
Izawa, S.: Biochim. Biophys. Acta **223**, 165–173 (1970)
Izawa, S., Hind, G.: Biochim. Biophys. Acta **143**, 377–390 (1967)
Jackson, J.B., Crofts, A.R.: Europ. J. Biochem. **10**, 226–237 (1969a)
Jackson, J.B., Crofts, A.R.: FEBS Lett. **4**, 185–189 (1969b)
Jackson, J.B., Crofts, A.R., von Stedingk, L.V.: Europ. J. Biochem. **6**, 41–54 (1968)
Jagendorf, A.T., Hind, G.: Nat. Acad. Sci. **1145**, 599–610 (1963)
Jagendorf, A.T., Hind, G.: Biochem. Biophys. Res. Commun. **18**, 702–709 (1965)
Jagendorf, A.T., Neumann, J.: J. Biol. Chem. **240**, 3210–3214 (1965)
Jagendorf, A.T., Uribe, E.: Brookhaven Symp. Biol. **19**, 215–241 (1966a)
Jagendorf, A.T., Uribe, E.: Proc. Nat. Acad. Sci. U.S. **55**, 170–177 (1966b)
Junge, W., Rumberg, B., Schröder, H.: Europ. J. Biochem. **14**, 575–581 (1970)
Junge, W., Schmid, R.: J. Membrane Biol. **4**, 179–192 (1971)
Kahn, J.S., Jagendorf, A.T.: Biochim. Biophys. Res. Commun. **2**, 259–263 (1960)
Karlish, S.J.D., Avron, M.: Biochim. Biophys. Acta **153**, 878–888 (1968)
Karlish, S.J.D., Shavit, N., Avron, M.: Europ. J. Biochem. **9**, 291–298 (1969)
Keister, D.L., Minton, N.J.: Arch. Biochem. Biophys. **147**, 330–338 (1971)
Kraayenhof, R.: FEBS Lett. **6**, 161–165 (1970)
Kraayenhof, R., Groot, G.S.P., Van Dam, K.: FEBS Lett. **4**, 125–128 (1969)
Kraayenhof, R., Katan, M.B.: In: Proc. 2nd Intern. Congr. Photosynthesis. Forti, G., Avron, M., Melandri, A. (eds.). The Hague: Dr. Junk, 1972, Vol. II, pp. 937–949
Kraayenhof, R., Slater, E.C.: In: Proc. 3rd Intern. Congr. Photosynthesis. Avron, M. (ed.). Amsterdam: Elsevier, 1975, Vol. II, pp. 985–996
Larkum, A.W.D., Boardman, N.K.: FEBS Lett. **40**, 229–232 (1974)
Leiser, M., Gromet-Elhanan, Z.: J. Biol. Chem. **250**, 84–89 (1975)
Lynn, W.S.: Biochemistry **7**, 3811–3820 (1968)
Martonosi, A., Lagwinska, E., Oliver, M.: Ann. N.Y. Acad. Sci. **227**, 549–567 (1974)
McCarty, R.E.: Biochem. Biophys. Res. Commun. **32**, 37–43 (1968)
McCarty, R.E.: J. Biol. Chem. **244**, 4292–4298 (1969)
McCarty, R.E.: FEBS Lett. **9**, 313–316 (1970)
McCarty, R.E., Fuhrman, J.S., Tsuchiya, Y.: Proc. Nat. Acad. Sci. U.S. **68**, 2522–2526 (1971)
McCarty, R.E., Magnusson, R.P.: Interaction of chloroplastsbound coupling factor 1 with adenine nucleotiden. In: Proc. of the 3rd Intern. Congr. Photosynthesis. Israel Avron, M. (ed.). Amsterdam: Elsevier, 1974, Vol. II pp. 975–984
McCarty, R.E., Portis, A.R., Jr., Magnusson, R.P.: In: Proc. 3rd Intern. Congr. Photosynthesis. Avron, M. (ed.). Amsterdam: Elsevier, 1975, Vol. II, pp. 957–984
McCarty, R.E., Racker, E.: Brookhaven Symp. Biol. **19**, 202–214 (1966)
Melandri, B.A., Baccarini-Melandri, A., Crofts, A., Cogdell, R.J.: FEBS Lett. **24**, 141–145 (1972)
Melandri, B.A., Baccarini-Melandri, A., San Pietro, A., Gest, H.: Proc. Nat. Acad. Sci. U.S. **67**, 477–484 (1970)
Mitchell, P.: Nature (London) **191**, 144–148 (1961)
Montal, M., Nishimura, M., Chance, B.: Biochim. Biophys. Acta **223**, 183–188 (1970)
Nelson, N., Drechsler, Z., Neumann, J.: J. Biol. Chem. **245**, 143–151 (1970)
Nelson, N., Nelson, H., Naim, V., Neumann, J.: Arch. Biochem. Biophys. **145**, 263–267 (1971)
Neumann, J., Jagendorf, A.T.: Arch. Biochem. Biophys. **107**, 109–119 (1964)
Neumann, J., Ke, B., Dilley, R.A.: Plant Physiol. **46**, 86–92 (1970)
Nishimura, M.: Federation Proc. **20**, 374 (1961)
Nishimura, M.: Biochim. Biophys. Acta **57**, 88–95 (1962a)
Nishimura, M.: Biochim. Biophys. Acta **57**, 96–103 (1962b)
Nishimura, M.: Biochim. Biophys. Acta **59**, 183–188 (1962c)
Nishimura, M., Ito, T., Chance, B.: Biochim. Biophys. Acta **59**, 177–182 (1962)
Nobel, P.S.: In: Plant Cell Physiology. A Physiological Approach. Kennedy, D., Park, R.B. (eds.). San Francisco: W.H. Freeman, 1970, pp. 75–128
Patrick, J., Valeur, B., Monnerie, L., Changeux, J.P.: J. Membrane Biol. **5**, 102–120 (1971)

Pick, U., Rottenberg, H., Avron, M.: FEBS Lett. **32**, 91–94 (1973)
Polya, G.M., Jagendorf, A.T.: Biochem. Biophys. Res. Commun. **36**, 696–703 (1969)
Portis, A.R., McCarty, R.E.: J. Biol. Chem. **249**, 6250–6254 (1974)
Rienits, K.G., Hardt, H., Avron, M.: Europ. J. Biochem. **43**, 291–298 (1974)
Rottenberg, H., Grunwald, T.G.: Europ. J. Biochem. **25**, 71–74 (1972)
Rottenberg, H., Grunwald, T.G., Avron, M.: Europ. J. Biochem. **25**, 54–63 (1972)
Rumberg, B., Siggel, U.: Naturwissenschaften **56**, 130–132 (1969)
Schröder, H., Muhle, H., Rumberg, B.: In: Proc. 2nd Intern. Congr. Photosynthesis. Forti,
    G., Avron, M., Melandri, A. (eds.). The Hague: Dr. Junk, 1972, Vol. II, pp. 919–930
Schuldiner, S., Avron, M.: Europ. J. Biochem. **19**, 227–231 (1971)
Schuldiner, S., Padan, E., Rottenberg, H., Gromet-Elhanan, Z., Avron, M.: FEBS Lett. **49**,
    174–177 (1974)
Schuldiner, S., Rottenberg, H., Avron, M.: FEBS Lett. **28**, 173–175 (1972a)
Schuldiner, S., Rottenberg, H., Avron, M.: Europ. J. Biochem. **25**, 64–70 (1972b)
Schuldiner, S., Rottenberg, H., Avron, M.: Europ. J. Biochem. **39**, 455–462 (1973)
Schwartz, M.: Nature (London) **219**, 915–919 (1968)
Shavit, N., Dilley, R.A., San Pietro, A.: Biochemistry **7**, 2356–2363 (1968)
Shen, Y.K., Shen, G.M.: Sci. Sinica **11**, 1097–1106 (1962)
Slater, E.C.: Nature (London) **172**, 975–978 (1953)
Stedingk, L.V. von, Baltscheffsky, H.: Arch. Biochem. Biophys. **117**, 400–404 (1966)
Strichartz, G.R., Chance, B.: Biochim. Biophys. Acta **256**, 71–84 (1972)
Trebst, A.: In: Proc. 3rd Intern. Congr. Photosynthesis. Avron, M. (ed.). Amsterdam: Elsevier,
    1975, Vol. I, pp. 439–448
Tyszkiewicz, E.: In: Proc. 2nd Intern. Congr. Photosynthesis. Forti, G., Avron, M., Melandri,
    A. (eds.). The Hague: Dr. Junk, 1972, Vol. II, pp. 1303–1309
Tyszkiewicz, E., Girault, G., Galmiche, J.M., Roux, E.: In: Currents in Photo-Synthesis.
    Thomas, J.B., Goedheer, J.C. (eds.). Rotterdam: A.D. Donker, 1966, pp. 227–234
Tyszkiewicz, E., Roux, E.: Biochem. Biophys. Res. Commun. **65**, 1400–1408 (1975)
Vainio, H., Baltscheffsky, M., Baltscheffsky, H., Azzi, A.: Europ. J. Biochem. **30**, 301–306
    (1972)
Walker, N.A.: FEBS Lett. **50**, 98–101 (1975)
Walz, D., Goldstein, L., Avron, M.: Europ. J. Biochem. **47**, 403–407 (1974)
Witt, H.T.: Q. Rev. Biophys. **4**, 365–477 (1971)
Zaugg, W.S., Vernon, L.P.: Biochemistry **5**, 34–40 (1966)

7. Chloroplast Coupling Factor

N. NELSON

A. Introduction

The transfer of energy from the initial high energy state, that is produced by electron transport in chloroplasts, involves the activity of coupling factors. Recent studies suggest that an intact vesicular membrane is required for the energy conserving action to proceed (for a comprehensive review, see JAGENDORF, 1975, and this vol. Chap. III.1). Thus, it might be argued that all membrane components are in essence coupling factors. However, coupling factors are those components which directly play a role in energy transfer from the high energy state to ATP.

The elucidation of the mechanism of energy transfer involves various lines of investigation and includes work with isolated chloroplasts and resolved particles as well as purified components and reconstitution experiments. The purification and characterization of the individual components are essential for the study of their mode of action. Among the more powerful approaches for studying energy transfer are resolution and reconstitution experiments, in which various components are removed, thereby inactivating the overall process, and subsequently added back to reactivate the system.

Although it seems that there exist coupling factors other than coupling factor 1 (CF_1), almost nothing is known about them at the present time. Consequently, this chapter will deal almost exclusively with the properties and mode of action of CF_1. The rest of the coupling device will be referred to as the membrane counterpart (MC) of CF_1.

B. Reconstitution of CF_1 Depleted Chloroplasts

The study of coupling factors makes use of resolved particles and their ability to be reconstituted. An ideal system would consist of membrane particles that are entirely free of the relevant coupling factor and the coupling factor in pure form, upon addition of which the system can undergo reconstitution and regain a large degree of its original activity. By employing this approach, multiple coupling factors were found to exist in the mitochondrial energy transfer system (see RACKER, 1970). Coupling factor activity in chloroplasts was first described by AVRON (1963), who showed that the photophosphorylation activity of EDTA treated chloroplasts could be restored by the addition of their supernatant in the presence of Mg^{2+}. Subsequently this coupling factor (CF_1) was partially purified and shown to have a latent Ca^{2+}-dependent ATPase activity (VAMBUTAS and RACKER, 1965).

The removal of CF_1 from chloroplasts with EDTA treatment leads to several changes in the chloroplast particles. Their high energy state is dissipated and electron transport is accelerated (see Avron and Neumann, 1968). These EDTA particles have been used for reconstitution studies; however, they have been reported to be unstable (McCarty, 1971). The inclusion of 1% bovine serum albumin stabilized the particles and they can be stored at $-70°$ C for several months (Nelson et al., 1972a). It is now known that considerable amounts of CF_1 remain in EDTA treated particles (McCarty and Racker, 1967). In order to obtain an improved resolution between particles and CF_1, Lien and Racker (1971b) introduced a silicotungstate treatment. The treated subchloroplast particles obtained had lost all of their ATPase activity and it was concluded that the preparation was free of CF_1. Partial restoration of cyclic photophosphorylation could be achieved by addition of CF_1, however, the electron transport properties of the treated particles were badly damaged. Recently, Kamienietzky and Nelson (1975) have reported that sodium bromide treatment results in particles free of CF_1 as judged by the complete loss of Ca^{2+}-ATPase activity. Although the electron transport properties of such particles remained intact, no reconstitution of photophosphorylation activity could be obtained.

Depleted subchloroplast particles obtained by EDTA treatment usually possess a limited ability to undergo reconstitution by CF_1. Shoshan and Shavit (1973) have reported nearly 100% reconstitution by adding the purified CF_1 within a few minutes to their rapidly-prepared EDTA particles. Girault et al. (1975) conducted detailed studies on the degree of reconstitution obtainable in EDTA particles and found that it depends on the original specific activity of the chloroplasts, so that a preparation originally with low activity can be reconstituted to 100%, while very active chloroplasts can only be reconstituted to about 20%.

In view of the fact that over 30% of the original CF_1 remains attached to the membrane in most reconstitutable EDTA particles (McCarty and Racker, 1967), one should be careful in interpreting experiments involving reconstitution by purified CF_1. In most cases the obtained reconstitution might be due to the retained CF_1, while the added CF_1 plays only a structural role. Chemically modified CF_1 might be useful for the elucidation of the photophosphorylation mechanism. For this kind of experiment, a preparation totally lacking CF_1 and exhibiting a high reconstitution degree is badly needed.

C. Preparation of CF_1

The preparation of CF_1 is based on two main solubilization techniques: treatment with EDTA and acetone precipitation (Avron, 1963; Vambutas and Racker, 1965). Using acetone-treated chloroplasts Farron (1970) has purified CF_1 to homogeneity as judged by disc gel electrophoresis and analytical ultracentrifugation. An extra step of hydroxylapatite column chromatography was needed in order to obtain a homogeneous preparation although there was no significant increase in specific activity. Karu and Moudrianakis (1969) started with aqueous extracts of chloroplasts purified CF_1 (13S enzyme) by mainly using sucrose density gradient centrifugation. LIEN and RACKER (1971a) reported an improved purification procedure involving liberation of CF_1 by means of EDTA treatment, DEAE-Sephadex column chromatography and sucrose density gradient centrifugation. Their preparation was homogenous on acrylamide gel electrophoresis and had Ca^{2+}-ATPase specific activities as high as 34 μmol ATP hydrolyzed per mg of protein per min. They also introduced a convenient method for determination of CF_1 purity which is based on the fact that, unlike most proteins, CF_1 lacks tryptophan and therefore its fluorescence intensity ratio between 300/350 nm rises as purification progresses. A ratio exceeding 1.85 is considered as an indication of pure CF_1.

A complex similar to the soluble oligomycin-sensitive ATPase of mitochondria (TZAGOLOFF and MEAGHER, 1971) has never been reported in chloroplasts. The only work performed in this direction is that of CARMELI and RACKER (1973). They separated the energy transfer system including CF_1 from the chloroplast membrane by cholate treatment. Their preparation was almost free of chlorophyll, depleted of electron carriers, and active in N,N′-dicyclohexylcarbodiimide-sensitive ATPase and ATP-P_i exchange. More effort along these lines certainly seems called for.

D. Physical Properties of CF_1

Subchloroplast particles appeared in electron micrographs as vesicular structures with 90 Å spheres on their surface. These spheres could be removed by either EDTA (HOWELL and MOUDRIANAKIS, 1967) or silicotungstate treatment (LIEN and RACKER, 1971b) and could be shown to reappear after reconstitution with pure CF_1. Therefore, it was concluded that CF_1 is a spherical protein with a 90 Å diameter that is located on the surface of the inner chloroplast membranes.

By using equilibrium ultracentrifugation, FARRON (1970) has determined the molecular weight of CF_1 to be $325,000 \pm 600$. A partial specific volume of 0.73 and a sedimentation coefficient of $S_{20,w} = 13.8$ were obtained. Three classes of ionizable tyrosine groups were detected in CF_1 (LIEN et al., 1972). Half of the total tyrosine was found to be exposed to the aqueous medium, a quarter was located in intersubunit regions and the remaining residues were buried within hydrophobic regions. Like the mitochondrial coupling factor 1, isolated CF_1 was shown to be cold labile and lost its ATPase activity at 0° C in the presence of salts (McCARTY and RACKER, 1966). Cold inactivation led to the dissociation of CF_1. However, if the cold treatment was performed in the presence of ATP, this dissociation and inactivation were prevented (LIEN et al., 1972). Protection against cold-inactivation occurred also when CF_1 was recombined with chloroplast membranes or incubated with chloroplast lipids (LIVNE and RACKER, 1969). Reassembly and partial restoration of activity of cold-inactivated CF_1 was obtained by incubating the inactive enzyme at 25° C in the presence of glycerol and ATP (LIEN et al., 1972).

In the presence of high concentrations of sulfhydryl reagents, a light-induced Mg^{2+}-ATPase activity was observed in isolated chloroplasts (PETRACK and LIPMANN, 1961). The light-dependent Ca^{2+}-ATPase, demonstrated by AVRON (1962) in chloroplasts, was at least one order of magnitude slower than the light-triggered Mg^{2+}-dependent activity. However, isolated CF_1, catalyzes a rapid Ca^{2+}-dependent ATPase reaction (Vambutas and Racker, 1965) but a comparable Mg^{2+}-ATPase activity occurs only in the presence of carboxylic acids (Nelson et al., 1972a). Since all of these activities were shown to be catalyzed by the same enzyme (CF_1), it appears that the chloroplast membrane confers special properties on CF_1 (McCarty and Racker, 1968). Similarly, in addition to its catalytic activity in ATP formation, CF_1 fulfills a structural role and in so doing imposes upon the membrane the ability to accumulate high energy intermediates. The phenomenon of the conferral of properties to a soluble protein by a membrane and alteration of the membrane properties by a soluble enzyme was called "allotopy" (Racker, 1967). Ryrie and Jagendorf (1971b, 1972) studied most elegantly the relationship between the energized chloroplast membrane and the conformation of CF_1 by using tritiated water. They observed that, upon illumination of chloroplasts, tritiation of CF_1 took place and this tritiation could be prevented by uncouplers. Tritiated CF_1 retained its radioactivity after extraction by EDTA treatment and subsequent purification by ammonium sulfate precipitation and column chromatography on sephadex or precipitation with specific antibody. Denaturation of the tritiated CF_1 by urea treatment liberated all of the radioactivity from the enzyme. The tritium uptake was dependent on the presence of a pH gradient, which can be generated by electron transport, acid-base transition, or light-triggered ATPase in chloroplasts. Phlorizin and Dio-9, in concentrations that prevent photophosphorylation, had no effect. However, the inclusion of ADP and P_i decreased the tritium uptake into CF_1 independently of ATP formation. The data provided evidence for energy-linked conformational changes in membrane-bound CF_1, and it was suggested that these changes may be an obligatory prerequisite for ATP synthesis. Based on the studies of tritium uptake and chemical modification (see below) an attractive model for the various conformational forms of CF_1 has been put forward by Datta et al. (1974).

E. Catalytic Properties of Activated CF_1

In order to study the catalytic properties of CF_1, its latent ATPase activity has to be activated by treatment with heat, trypsin or dithiothreitol (Vambutas and Racker, 1965; McCarty and Racker, 1968; Farron, 1970). The activation by heat treatment was studied in great detail by Farron and Racker (1970). Heat treatment of the enzyme did not modify its amino acid composition, sedimentation velocity or electrophoretic mobility on acrylamide gels. The extent of activation increased with temperature elevation up to 60° C and it was found that during these activities the presence of ATP was necessary to prevent denaturation. Inclusion of 5 mM dithiothreitol doubled the final specific activity. The trypsin-induced ATPase activity was dependent on the presence of Ca^{2+}; other divalent ions such as Ni^{2+}, Mg^{2+}, Mn^{2+}, Co^{2+} and Sr^{2+} at 10 mM were less than 3% as effective as Ca^{2+}. Moreover, the Ca^{2+}-ATPase activity was inhibited approximately 50% by the addition of 0.3 mM Mg^{2+} to the assay medium (Vambutas and

RACKER, 1965). Later it was shown that in the presence of carboxylic acids and a proper Mg^{2+} concentration, heat or trypsin-activated CF_1 also has a considerable Mg^{2+}-ATPase activity (NELSON et al., 1972a). At pH 6 and 8 mM Mg^{2+}, sodium maleate accelerated the Mg^{2+}-ATPase activity by as much as 30-fold, approaching rates of 30 to 50% of the Ca^{2+}-ATPase activity under optimal conditions. At pH 8, the optimal Mg^{2+} ion concentration was 2 mM and about a three-fold stimulation by sodium maleate or sodium bicarbonate was observed. Other carboxylic acids were less effective. The specificity of Mg^{2+}-ATPase for substrates was similar to that reported for Ca^{2+}-ATPase. With ATP as the substrate, the activity was about six times higher than that with GTP or ITP, while the activities with UTP, CTP, ADP and pyrophosphate were virtually nil. The K_m for ATP, in the presence of 8 mM Ca^{2+} and 30 mM Tricinemaleate buffer at pH 8, was found to be 0.8 mM with heat or dithiothreitol activated CF_1 (NELSON and NELSON, unpublished observations). With 2-(N-morpholino) ethane sulfonic acid buffer, a K_m of 0.1 mM ATP and a V_{max} of 0.45 mol ATP hydrolyzed per milligram of protein per minute were observed for Mg^{2+}-ATPase (NELSON et al., 1972a). Inclusion of 60 mM maleate in the assay increased the K_m to 1.1 mM ATP and the V_{max} to 14. It was tentatively concluded that carboxylic acids increase the V_{max} by removing inhibitory reaction products from the active site.

The product of ATPase reaction, ADP, inhibited the Ca^{2+}-ATPase (VAMBUTAS and RACKER, 1965) and Mg^{2+}-ATPase (NELSON et al., 1972a) changing the affinity for ATP and the V_{max}. The plot of enzyme activity versus ATP concentration was hyperbolic but in the presence of 0.6 mM ADP it became sigmoidal. The apparent reaction order determined by Hill plots was 1 without ADP and 2.3 with 0.06 mM ADP. The existence of at least two active sites acting cooperatively was proposed.

F. Subunit Structure of CF_1

HOWELL and MOUDRIANAKIS (1967) were the first to detect a subunit structure for purified CF_1 in electron micrographs. RACKER et al. (1971) reported that treatment of CF_1 with SDS followed by electrophoresis in SDS gels revealed the presence of five different polypeptides. The polypeptides were designated as α, β, γ, δ, and ε subunits, in the order of decreasing molecular weights of 59,000, 56,000, 37,000, 17,500 and 13,000, respectively. The pattern and the molecular weights of the peptide chains are quite similar to those of ATPase coupling factors from various electron transport phosphorylating systems (see SENIOR, 1973; PENEFSKY, 1974; ABRAMS and SMITH, 1974). The five individual subunits have been isolated and their amino acid composition determined (NELSON et al., 1972b, 1973). Table 1 summarizes the amino acid composition, molecular weights and the number of the individual subunits in the holoenzyme. Whereas a ratio of 3/3/1/1/1 was proposed for the subunit distribution of mitochondrial ATPase (SENIOR, 1973), we propose that a molecule of CF_1 is composed of 2α, 2β, 1γ, 1δ, and 2ε subunits. This is deduced from the relative staining intensity of SDS gels and substantiated by the amino acid composition analysis. By growing pea plants in an atmosphere

Table 1. Amino acid analysis, molecular weights and number of CF_1 peptide components

	Residues per unit					
	α[a]	β[a]	γ[a]	δ[a]	ε[b]	CF_1[c]
Lysine	20	20	23		5	132
Histidine	3	5	1		1	24
Ammonia	51	36	43		23	—
Arginine	31	30	19		11	192
Half-cystine	2	3	6		1	12
Aspartic acid	38	44	36		15	228
Threonine	36	37	22		9	216
Serine	33	30	18		6	180
Glutamic acid	82	58	40		15	420
Proline	17	26	15		4	120
Glycine	43	49	20		8	252
Alanine	58	43	35		8	276
Valine	38	41	24		6	216
Methionine	11	14	7		1	84
Isoleucine	40	27	17		11	204
Leucine	53	50	34		13	288
Tyrosine	18	12	7		0	84
Phenylalanine	13	16	9		1	84
Number of subunits in CF_1	2	2	1	1	2	
Molecular weight of unit	59,000[a] 62,000[e]	56,000[a] 57,000[e]	37,000[a] 38,000[e]	17,500[a] 21,000[e]	13,000[a] 14,000[e]	310,500[d] 325,000[f, g]

[a] NELSON et al. (1973). [b] NELSON et al. (1972b). [c] Calculated from the data of FARRON (1970). [d] Calculated from the number of subunits in CF_1 and the molecular weights of "a". [e] McEVOY and LYNN (1973). [f] Calculated from the number of subunits in CF_1 and molecular weights of "e". [g] FARRON (1970)

of $C^{14}O_2$ and subsequent radioactive determination of the amount of label per subunit band in SDS gels, the above suggested subunit ratio was further supported (NELSON, unpublished observations).

An immunological approach was used to study the function of the individual subunits in the various reactions of CF_1 (NELSON et al., 1973). The antigens were prepared by DEAE cellulose column chromatography and subsequent SDS gel electrophoresis. Antibodies to each subunit were prepared. None of them inhibited, by itself, the ATPase activity of the isolated CF_1. Antibodies to subunits α or γ inhibited cyclic photophosphorylation, photophosphorylation coupled to Hill reaction and the light-triggered Mg^{2+}-ATPase. These data suggest that the α and γ subunits are likely to be involved in photophosphorylation. When antibody against α was checked for its effect on the reaction discovered by McCARTY et al. (1971), in which H^+ uptake is stimulated by ATP, inhibition was observed. This suggests that the high affinity binding site for ATP is probably located on the α subunit. Only antibodies against α or β subunits agglutinated the chloroplasts. Since the antibody against γ was found not to be monovalent and fails to cause agglutination, it was concluded that the α and β subunits protrude and that the γ subunit is located on the membrane side of CF_1. By employing the tritiated water technique developed by RYRIE and JAGENDORF (1971b), GREGORY and

RACKER (1973) showed that a monovalent antibody against γ subunit prevented CF_1 from undergoing those conformational changes necessary for photophosphorylation.

Among the five isolated subunits of CF_1, only the activity of ε subunit has as yet been clearly defined (NELSON et al., 1972b). Purified ε subunit was shown to inhibit the ATPase activity of heat activated CF_1 and, in analogy to F_1 inhibitor discovered by PULLMAN and MONROY (1963), the ε subunit was termed as CF_1 inhibitor. The CF_1 inhibitor was heat-stabile but sensitive to trypsin. It was specific for CF_1 and did not inhibit ATPase activity of F_1; similarly F_1 inhibitor did not affect the ATPase activity of CF_1. It was concluded that the ε subunit takes part in both the latency of CF_1 in vitro and the regulation of CF_1-ATPase activity in vivo. Antibodies against either γ or ε subunits prevented the inhibition of heat-activated CF_1 by the ε subunit. However, anti-γ did not interact directly with the ε subunit in immunological tests. The conclusion drawn was that the γ subunit possesses the binding site for CF_1 inhibitor. Hence, activation of latent ATPase of CF_1 probably occurs via the removal of ε subunit from the active site. Trypsin simply digests the ε and by this removes the inhibition. Heat and dithiothreitol treatments might remove the ε via conformational changes in which probably the γ subunits is involved. Recently, DETERS et al. (1975) were able to remove the three small subunits of CF_1 by trypsin treatment. The enzyme obtained was composed of only α and β subunits and exhibited normal levels of ATPase activity. This preparation failed to bind to depleted chloroplasts membranes; hence it lost completely its coupling activity. This indicates that the three small subunits are not required for the ATPase activity of CF_1, but might be necessary for its binding ability. The CF_1 inhibitor no longer inhibited the α and β ATPase which is in line with the finding that antibody against γ subunits also prevented this inhibition. Both sets of data favor the possibility that the binding site for CF_1 inhibitors is on γ subunit, while its inhibition is on α and β. The α and β ATPase preparation was still cold labile, showing that the three smaller subunits are not required for this property of CF_1. With a great degree of certainty it can be concluded that the active site for ATPase activity of CF_1 is located on the α or β subunit.

G. Chemical Modification of CF_1 and the Nature of its Active Site

Chemical modification of CF_1 can be performed on isolated or membrane-bound CF_1. With a purified preparation, the active site for ATPase activity and the membrane-binding site of the enzyme can be studied. With membrane-bound enzyme it is possible to examine the conformational changes which take place during electron transport or high energy intermediate production. The forward reaction of ATP synthesis can be studied mainly with membrane-bound CF_1. The involvement of SH groups in the catalytic activity of the purified CF_1 has been tested by reacting the enzyme with sulfhydryl reagents. Treatment of CF_1 with N-ethylmaleimide or iodoacetamide did not inhibit its ATPase activity, even though two of the SH groups of CF_1 were modified (FARRON and RACKER, 1970). After heat activation of CF_1, two additional SH groups became exposed to these reagents. Iodoacetamide did not inhibit the activated enzyme, however, pretreatment with this reagent prevented the heat activation of the enzyme. The same reagent did not affect activation of CF_1 by trypsin, so it seems clear that although it interacts with SH groups involved in the process of heat activation, trypsin activation by-passes the necessity for these SH groups. Deters and Nelson (1973) reported that the incorporation of either iodoacetate or iodoacetamide into the γ subunit was increased after heat activation of CF_1. Upon denaturation of the enzyme, further alkylation took place mainly on the β subunit. This suggests that the least accessible sulfhydryl groups reside in the β subunits. The nature of the active site of CF_1 has been investigated by using 7-chloro-4-nitrobenzo-2-oxa-1,3-diazole (NBD-Cl) (Deters et al., 1975). Overnight incubation of two equivalents of NBD-Cl per one equivalent of CF_1 caused about 80% inhibition of the ATPase activity of CF_1. This inhibition could be reversed by the inclusion of dithiothreitol either during the assay or by pretreatment of the inhibited enzyme. NBD-Cl was shown to be covalently bound to the enzyme and by the use of radioactively labeled NBD-Cl and subsequent SDS-gel electrophoresis it was found to be attached mainly to β subunits. The data suggest that NBD-Cl modifies a tyrosine residue on each of the two β subunits in a CF_1 molecule. The best guess at the present time is that the active site for ATPase is situated on the β subunit.

In light of the data obtained for the subunit structure of CF_1, the CF_1 inhibitor, the interaction of CF_1 with specific antibodies, and with NBD-Cl, a tentative model for the structure of CF_1 can be proposed. This model is schematically presented in Figure 1.

A different method for elucidating the mechanism of action of CF_1 is chemical modification in situ. Ryrie and Jagendorf (1971 a) observed that illumination of chloroplasts in the presence of sulfate, ADP and Mg^{2+} caused a maximum inhibition of about 50% for those reactions that involved CF_1. The studied reactions included photophosphorylation, ATPase activity of both bound and isolated CF_1, light-induced proton uptake and the $P/2e^-$ ratio. By removing the inhibited CF_1 from the membranes and subsequently recombining them with intact CF_1, proton uptake could be completely restored. A partial restoration of the photophosphorylation also occurred. The sulfate-induced inhibition could be prevented by the uncoupler NH_4Cl, by the energy transfer inhibitor phlorizin or by inclusion of either

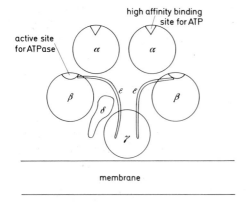

Fig. 1. Schematic model for the subunit structure of CF_1

phosphate or arsenate during the illumination period. Essentially similar results have been obtained in the same laboratory by DATTA et al. (1974) using permanganate instead of sulfate. Inhibition depends on the presence of the magnesium and ADP so that 5–10 μM ADP produces half-maximal inhibition. Permanganate treatment rendered the chloroplasts insensitive to ATP-induced stimulation of pH rise and concurrent ATP inhibition of electron transport. This treatment decreased the affinity of trypsin activated CF_1 toward ATP and the mode of ADP inhibition of its ATPase became of a simple competitive type without cooperative interactions in contrast to that found with untreated enzyme. MCCARTY et al. (1972) studied the effect of N-ethylmaleimide on isolated chloroplasts and found that upon illumination photophosphorylation and the Ca^{2+}-ATPase of activated CF_1 were about 50% inhibited. This treatment prevented the ATP-stimulation of H^+ uptake. Uncouplers abolished all the effects of N-ethylmaleimide treatment. Using labeled N-ethylmaleimide, MCCARTY and FAGAN (1973) clearly demonstrated that the modification of an SH group in the γ subunit of CF_1 caused all these effects.

A common feature of the inhibition of CF_1 in situ by several modifying agents such as sulfate, permanganate, N-ethylmaleimide and mercuric ions (IZAWA and GOOD, 1969) is that they all cause a maximal inhibition of about 50%. It is unlikely that 50% of the CF_1 molecules are differentially exposed to the reagents. Moreover, the fact that permanganate-modified CF_1 lost its allosteric properties suggests that all the CF_1 molecules were modified. When the subunit structure of CF_1 and the probable existence of a regulatory site on the α subunit (NELSON et al., 1973), are also considered, it seems likely that chemical modification of either the γ or α subunit imposes changes in the allosteric site of the α subunit. Indeed, one would expect from an enzyme with five different subunits wide variety of interactions, some of which might lead to similar effects.

H. Nucleotide Binding and the Mechanism of ATP Formation

As many as five nucleotide binding sites have been detected in the mitochondrial ATPase (F_1) (see BALTSCHEFFSKY and BALTSCHEFFSKY, 1974). Equilibrium dialysis experiments indicated the existence of two binding sites for ATP per CF_1 molecule (LIVNE and RACKER, 1969). Participation of two active sites in the ATPase activity of CF_1 molecule have also been deduced from kinetic studies (NELSON et al., 1972a). SHAHAK et al. (1973) proposed that there are two types of active sites upon CF_1, one of which is directly involved in the process of ATP formation. This proposal was based on the observation that a fluorescent ATP derivative was a poor substitute for ATP in various ATPase reactions, although a fluorescent ADP derivative quite effectively replaced ADP in photophosphorylation.

ROY and MOUDRIANAKIS (1971a) observed a very slow incorporation of $[^{14}C]ADP$ into purified CF_1. The bound ADP underwent a transphosphorylation reaction to give AMP and ATP. While AMP was easily separable, ATP remained tightly bound and only upon denaturation it could be liberated from the enzyme. Upon illumination of the chloroplasts in the presence of AMP and phosphate, ADP and a tight complex between ADP and CF_1 were formed. If this reaction can be shown to be fast enough, this might indicate that in photophosphorylation

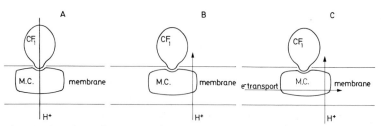

Fig. 2 A–C. Schematic proposal for the proton translocation in CF_1 and its membrane counterpart (MC)

AMP is an earlier acceptor for phosphate than ADP (see Roy and Moudrianakis, 1971 b). The formation of CF_1-ADP complex was sensitive to arsenate or sulfate, and exogenous ADP, GDP or ATP dissociate the complex even in darkness. Girault et al. (1973) extended these observations using circular dichroism studies. They similarly found fixation of two ADP molecules per CF_1. In addition the ADP binding was inhibited by phosphate, pyrophosphate and tripolyphosphate. However, they did not indicate if during complex formation a transphosphorylation reaction occurred. Hence it has been established that CF_1 can firmly bind nucleotides. It certainly seems desirable that further studies be conducted on nucleotide binding to CF_1, particularly those concerned with the effect of uncouplers. Such studies will hopefully provide insights into the relation between nucleotide binding and photophosphorylation.

On the basis of our current knowledge, we are still unable to propose a detailed mechanism for ATP formation. The Mitchell hypothesis (Mitchell, 1966) has seemingly been proved correct in the sense that electron transport in fact leads to the formation of a proton gradient which can in turn bring about the photophosphorylation of ADP to ATP (see Jagendorf and Uribe, 1966). The participation of reversible ATPase in the coupling of phosphorylation to electron transport has been widely demonstrated. However, it still remains to be shown if this ATPase is operating "chemiosmotically" (see Mitchell, 1974). Three schematic proposals for the transformation of electron transport energy to the chemical energy of ATP via reversible ATPase are illustrated in Figure 2. Only scheme A, in which the protons pass through CF_1 is in accord with the Mitchell hypothesis. In contrast the protons in scheme B pass through the membrane counterpart (MC) of CF_1 and energy might be transferred via conformational changes imposed on CF_1. In scheme C, either the proton gradient or electron transport itself lead to the conformational changes necessary for ATP formation. The observation that ATP increases proton uptake was interpreted by Telfer and Evans (1972) as proof that essentially scheme A is the operative one. However, ATP might affect the proton permeability properties of MC via CF_1 (see McCarty et al., 1971). The time seems ripe for experiments which will distinguish between these possibilities and elucidate the mechanism by which ATP is produced by photo-induced electron transport.

Acknowledgements. I am very grateful to Dr. J. Kuhn for correcting my English. The support during the preparation of this manuscript by the Advancement of Mankind Foundation is gratefully acknowledged.

References

Abrams, A., Smith, J.B.: In: The Enzymes. Boyer, P.D. (ed.). New York-London: Academic Press, 1974, Vol. X, pp. 395–429
Avron, M.: J. Biol. Chem. **237**, 2011–2017 (1962)
Avron, M.: Biochim. Biophys. Acta **77**, 699–702 (1963)
Avron, M., Neumann, J.: Ann. Rev. Plant Physiol. **19**, 137–166 (1968)
Baltscheffsky, H., Baltscheffsky, M.: Ann. Rev. Biochem. **43**, 871–897 (1974)
Carmeli, C., Racker, E.: J. Biol. Chem. **248**, 8281–8287 (1973)
Datta, D.B., Ryrie, I.J., Jagendorf, A.T.: J. Biol. Chem. **249**, 4404–4411 (1974)
Deters, D.W., Nelson, N.: Federation Proc. **32**, 516 (1973)
Deters, D.W., Racker, E., Nelson, N., Nelson, H.: J. Biol. Chem. **250**, 1041–1047 (1975)
Farron, F.: Biochemistry **9**, 3823–3828 (1970)
Farron, F., Racker, E.: Biochemistry **9**, 3829–3836 (1970)
Girault, G., Galmiche, J.M., Michel-Villaz, M., Thiery, J.: Europ. J. Biochem. **38**, 473–478 (1973)
Girault, G., Galmiche, J.M., Vermeglio, A.: In: Proc. 3rd Intern. Congr. Photosynthesis. Avron, M. (ed.). Amsterdam: Elsevier, 1975, pp. 839–847
Gregory, P., Racker, E.: In: 9th Intern. Congr. Biochemistry (Abstract). Stockholm: 1973, p. 238
Howell, S.H., Moudrianakis, E.N.: Proc. Nat. Acad. Sci. U.S. **58**, 1261–1268 (1967)
Izawa, S., Good, N.E.: In: Progress in Photosynthesis Research. Metzner, H. (ed.). Tübingen: Laupp, 1969, Vol. III, pp. 1288–1298
Jagendorf, A.T.: In: Bioenergetics of Photosynthesis. Govindjee (ed.). New York-San Francisco-London: Academic Press, 1975, pp. 414–492
Jagendorf, A.T., Uribe, E.: Brookhaven Symp. Biol. **19**, 215–245 (1966)
Kamienietzky, A., Nelson, N.: Plant Physiol. **55**, 282–287 (1975)
Karu, A.E., Moudrianakis, E.N.: Arch. Biochem. Biophys. **129**, 655–671 (1969)
Lien, S., Berzborn, R.J., Racker, E.: J. Biol. Chem. **247**, 3520–3524 (1972)
Lien, S., Racker, E.: Methods Enzymol. **23**, 547–555 (1971a)
Lien, S., Racker, E.: J. Biol. Chem. **246**, 4298–4307 (1971b)
Livne, A., Racker, E.: J. Biol. Chem. **244**, 1332–1338 (1969)
McCarty, R.E.: Methods Enzymol. **23**, 251–253 (1971)
McCarty, R.E., Fagan, J.: Biochemistry **12**, 1503–1507 (1973)
McCarty, R.E., Fuhrman, J.S., Tsuchiya, Y.: Proc. Nat. Acad. Sci. U.S. **68**, 2522–2526 (1971)
McCarty, R.E., Pittman, P.R., Tsuchiya, Y.: J. Biol. Chem. **247**, 3048–3051 (1972)
McCarty, R.E., Racker, E.: Brookhaven Symp. Biol. **19**, 202–214 (1966)
McCarty, R.E., Racker, E.: J. Biol. Chem. **242**, 3435–3439 (1967)
McCarty, R.E., Racker, E.: J. Biol. Chem. **243**, 129–137 (1968)
McEvoy, F.A., Lynn, W.S.: Arch. Biochem. Biophys. **156**, 335–341 (1973)
Mitchell, P.: Chemiosmotic Coupling in Oxidative and Photosynthetic Phosphorylation. Bodmin: Glynn, 1966
Mitchell, P.: FEBS Lett. **43**, 189–194 (1974)
Nelson, N., Deters, D.W., Nelson, H., Racker, E.: J. Biol. Chem. **248**, 2049–2055 (1973)
Nelson, N., Nelson, H., Racker, E.: J. Biol. Chem. **247**, 6506–6510 (1972a)
Nelson, N., Nelson, H., Racker, E.: J. Biol. Chem. **247**, 7657–7662 (1972b)
Penefsky, H.S.: In: The Enzymes. Boyer, P.D. (ed.). New York-London: Academic Press, 1974, Vol. X, pp. 375–394
Petrack, B., Lipmann, F.: Photophosphorylation and photohydrolysis in cell-free preparations of blue-green alga. In: Light and Life. McElroy, W.D., Glass, H.B. (eds.). Baltimore: John Hopkins, 1961, pp. 621–630
Pullman, M.E., Monroy, G.C.: J. Biol. Chem. **238**, 3762–3769 (1963)
Racker, E.: Federation Proc. **26**, 1335–1340 (1967)
Racker, E.: In: Membranes of Mitochondria and Chloroplasts. Racker, E. (ed.). New York-Cincinnati-Toronto-London-Melbourne: Van Nostrand Reinhold, 1970, pp. 127–171
Racker, E., Hauska, G., Lien, S., Berzborn, R.J., Nelson, N.: In: Proc. 2nd Intern. Congr. Photosynthesis. Forti, G., Avron, M., Melandri, A. (eds.). The Hague: Dr. Junk, 1972, Vol. II, pp. 1097–1113

Roy, H., Moudrianakis, E.N.: Proc. Nat. Acad. Sci. U.S. **68**, 464–468 (1971a)
Roy, H., Moudrianakis, E.N.: Proc. Nat. Acad. Sci. U.S. **68**, 2720–2724 (1971b)
Ryrie, I.J., Jagendorf, A.T.: J. Biol. Chem. **246**, 582–588 (1971a)
Ryrie, I.J., Jagendorf, A.T.: J. Biol. Chem. **246**, 3771–3774 (1971b)
Ryrie, I.J., Jagendorf, A.T.: J. Biol. Chem. **247**, 4453–4459 (1972)
Senior, A.E.: Biochim. Biophys. Acta **301**, 249–277 (1973)
Shahak, Y., Chipman, D.M., Shavit, N.: FEBS Lett. **33**, 293–296 (1973)
Shoshan, V., Shavit, N.: Europ. J. Biochem. **37**, 355–360 (1973)
Telfer, A., Evans, M.C.W.: Biochim. Biophys. Acta **256**, 625–637 (1972)
Tzagoloff, A., Meagher, P.: J. Biol. Chem. **246**, 7328–7336 (1971)
Vambutas, V.K., Racker, E.: J. Biol. Chem. **240**, 2660–2667 (1965)

8. Field Changes

B. RUMBERG

A. Introduction

I. Chemiosmotic Coupling Hypothesis

In his chemiosmotic coupling hypothesis MITCHELL (1961, 1966) stated that an electrochemical potential difference of protons across the thylakoid membrane should couple the processes of electron transport and ATP formation. This concept demands the existence of both a proton pump incorporated into the electron transport system and a translocator system for protons mediating their backflow which is able to funnel the free energy stored in the proton gradient into ATP.

The proton pump mechanism as proposed by MITCHELL is based on two postulates. First, the electron transport chain should have a sequence of alternating electron and hydrogen carriers. Second, the chain should have a zigzag arrangement within the membrane in order to achieve displacement of electrons one way (inside to outside) and hydrogen groups the other (outside to inside). Such mechanism enables a net translocation of protons (outside to inside) in addition to the transfer of electrons from water to $NADP^+$ (for details see JUNGE, Chapter II.1.a this volume). The resulting electrochemical potential difference for protons is made up of two components, the chemical one and the electrical one:

$$\Delta \bar{\mu} = RT \ln \frac{[H^+]_i}{[H^+]_o} + F \Delta \psi \tag{1}$$

where $[H^+]_i$, $[H^+]_o$ are the proton concentrations in the inner and outer phase respectively, and $\Delta \psi$ is the electrical potential difference between inner and outer phase (a $\Delta \psi$ of 58 mV corresponds to a $[H^+]_i/[H^+]_o$ ratio of 10).

II. Evidence for Field Changes

Since 1963 it has become clear that in fact a reversible uptake of protons during light-driven electron transport takes place, which is indicated by pH changes in a suspension of thylakoid vesicles (JAGENDORF and HIND, 1963; NEUMANN and JAGENDORF, 1964). These pH changes give information on the total amount of protons pumped in. They give, however, no direct information on the internal concentration of free protons $[H^+]_i$, because the bulk of protons pumped in binds to the inner surface of the thylakoid membrane (for details see ROTTENBERG, Chapter III.2 this volume). Neither is information included on $\Delta \psi$, because it is dependent on the imbalance of electrical charge due to transmembrane movements of both protons and secondary counterions:

$$\Delta\psi = \frac{1}{C}\sum \Delta Q_n \qquad\qquad (2)$$

where C is the electrical capacity of the thylakoid membrane, and ΔQ_n is the charge imbalance due to translocation of individual ion species.

The electrical field within the membrane is given as the electric potential gradient. For biological membranes the constant field approximation is usually applied (see DAINTY, 1962):

$$E = \frac{\Delta\psi}{a} \qquad\qquad (3)$$

where E is the electric field within the membrane, and a is the thickness of the insulating membrane layer.

Since 1967, evidence has accumulated for the existence of light-induced field and potential changes due to a charge displacement through the thylakoid membrane which loads the membrane positive on the inside. First, evidence came from the analysis of flash induced absorption changes at 515 nm which were interpreted as a response of chloroplast pigments to a light-induced electric field across the thylakoid membrane (JUNGE and WITT, 1968; first report on these results WITT, 1967). Analysis of those electrochromic absorption changes has currently been extended and refined since then and a lot of information on the extent and the kinetics of the electric field has been gathered (for reviews see WITT et al., 1968; WITT, 1971, 1975).

The results on the basis of the electrochromic absorption changes have been confirmed and completed by means of other independent techniques. Delayed light emission from chlorophyll has been proven as a further intrinsic probe of membrane potential changes (BARBER and KRAAN, 1970). Information by external probes has been obtained by three means: analysis of the light-induced redistribution of freely permeating ions (SCHRÖDER et al., 1972; ROTTENBERG et al., 1972), direct readings of a microelectrode inserted in a single giant chloroplast (BULYCHEV et al., 1972), and induction effects on macroscopic electrodes (FOWLER and KOK, 1974).

III. Evidence for Membrane Potential-Induced ATP Formation

From the very beginning of the understanding of the field-indicating character of the absorption shift at 515 nm it became clear that the electric field could serve as a driving force for ATP formation as proposed by MITCHELL. This was manifest from the observation that phosphorylation causes an accelerated decay of the electric field (RUMBERG and SIGGEL, 1968; JUNGE and WITT, 1968). ATP formation at the expense of the energy stored in the form of a membrane potential has further been proven by the artificial set up of a diffusion potential across the thylakoid membrane (URIBE, 1973).

B. Quantitative Results on Changes of Membrane Potential

I. Intrinsic Probes

A molecule located in the membrane or at its surface, and with spectroscopic properties which are influenced by changes of the electric field, could serve as an membrane potential indicator. WITT and his group have studied light-induced absorption shifts in chloroplasts and green algae and have concluded that a certain spectrum, normally followed at 515 nm, represents an electrochromic effect which is induced by the electric field on the carotenoids and the chlorophylls in the membrane. This photometric method has proven most appropriate for kinetic studies and will, therefore, be reviewed in some detail.

The interpretation is based on several arguments: (1) all of the pigments localized in the thylakoid membrane undergo absorption shifts attributable to field influence (EMRICH et al., 1969); (2) the spectral shifts are identical with those obtained when an electrical field is applied to artificial layers of chloroplast pigments (SCHMIDT et al., 1971, 1972); (3) the relaxation time of the absorption changes is accelerated by addition of ion permeability inducing agents (JUNGE and WITT, 1968); (4) relative amplitude and time course of the changes are identical with those of field-indicating signals obtained by an independent electrode technique (WITT and ZICKLER, 1973, 1974); similar absorption changes were observed when chromatophores of bacteria were subjected to an artificially induced diffusion potential (JACKSON and CROFTS, 1969).

These results factually prove field-indicating absorption changes which have the specification of a molecular voltmeter. The rise of the changes in the light indicates the formation of an electric field, and the decay in the dark indicates the breakdown of the field by field-driven ion fluxes. Quantitative studies in chloroplasts as well as in photosynthetic bacteria have revealed that these electrochromic absorption changes are a *linear* indicator of the voltage across the membrane. (SCHLIEPHAKE et al., 1968; REINWALD et al., 1968; JACKSON and CROFTS, 1969; WITT and ZICKLER, 1974). Calibration has been obtained using excitation with a short flash as a standard. The light-induced voltage per single short flash has been calculated on the basis of Eq. (2). Detailed studies on the electrochromic absorption changes at 515 nm showed:

1. The onset of the electric potential occurs within less than 20 nsec after flash excitation (WOLFF et al., 1969).

2. The electric field is a collective property of a unit containing at least 10^5 chlorophyll molecules which corresponds to the size of one thylakoid (JUNGE and WITT, 1968).

3. Both reaction centres contribute about equally to the electric potential on excitation with a short flash (SCHLIEPHAKE et al., 1968).

4. The flash-induced amplitude corresponds to a voltage of approx. 50 mV (SCHLIEPHAKE et al., 1968).

5. When permanent light is imposed a further increase until a maximum value of about 200 mV takes place within 10 msec, followed by a decrease in the order of seconds to the steady state level. For *Chlorella* a steady-state level of 100 mV has been determined (GRÄBER and WITT, 1974). Using isolated chloroplasts no unequivocal results on the steady state level could be obtained as yet, due to

large changes in scattering light superimposed to the electrochromic absorption changes under these conditions (see Thorne et al., 1975).

6. The decay in the dark is accelerated if the proton concentration in the inner phase is raised, which gives evidence that a field-driven efflux of *protons* takes place (Boeck and Witt, 1972).

Indication of electric potential changes has also been derived from the enhancement of delayed light emission from chlorophyll due to the recombination of electrons and electronholes across the membrane (Barber and Kraan, 1970). This method suffers from the influence of both, field changes and changes of internal pH, on the delayed light emission. Wraight and Crofts (1971) recognized a rapid potential rise when the light was turned on. By comparison with artificially induced diffusion potentials changing the external salt concentration it was deduced, that the light-induced steady-state potential in chloroplasts is between 75 and 105 mV with the inside positive (Barber, 1972).

II. External Probes

An ion which is in an electrochemical equilibrium across the membrane is distributed according to the Nernst equation. Therefore, the membrane potential can be read out from the ion distribution provided that the selected ion permeates freely in its charged form and that a true equilibrium distribution is established. The light-induced steady state potential from studies of the ion distribution using the Nernst equation and estimates of the internal volume came out to be 10 mV positive on the inside or even less (Schröder et al., 1972; Rottenberg et al., 1972). This result refers to the potential difference between inner and outer *bulk* phases of course.

Microelectrodes have also been used in an attempt to determine membrane potential changes in chloroplasts (Bulychev et al., 1972; Vredenberg and Tonk, 1975). Giant chloroplasts from *Peperomia metallica* have been selected for this purpose. Signals of the order of 10 mV positive inside have been obtained. When the light is turned on a rapid rise followed by a slow decrease to a steady state level takes place. The decline upon darkening is followed by a small undershoot. This technique, however, suffers from the uncertainty of the location of the electrode tip, and will therefore not be considered further.

Under special circumstances a direct electric response from two macroelectrodes inserted in a chloroplast suspension can be observed, which indicates directly a charge displacement across the thylakoid membrane. This response appears only if a nonsaturating flash of light is applied (Fowler and Kok, 1974). The effect results from a dipole field which is caused by differences of membrane polarization along the gradient of light intensity. Normally the response decays fast due to charge equilibration between paired membranes. The decay is retarded if the viscosity of the medium is increased, and finally becomes identical with that of the field-indicating electrochromic absorption changes (Witt and Zickler, 1973). Amplitude calibration in terms of the membrane potential has not be obtained as yet.

III. Conclusions

A unique result of all the different techniques employed is, that a membrane potential positive on the inside is created if chloroplasts are illuminated. However, quantitative results on the magnitude of the light-induced potential during steady state conditions obtained by different techniques are not in accord. Intrinsic probes reveal a value of approx. 100 mV, whereas studies of the ion distribution show on a low value of approx. 10 mV. In order to get insight into this discrepancy one has to consider that intrinsic probes respond to the potential difference between the inner and outer membrane surfaces, whereas salt distributions, when treated as bulk phase phenomena, deliver potential differences between inner and outer bulk phases. Disagreement between both methods signifies a potential difference between the regions of membrane surface and bulk phase. This leads to the influence of a membrane surface charge and an electric double layer. Details will be discussed in the next section.

C. Concept of Ion Transport Phenomena

I. General Model

The process of proton translocation is accompanied by the translocation of other ions. The general result is, that anions like chloride ions move inward and cations like potassium ions and magnesium ions move outward. A quasielectroneutrality relation holds, that is, the amount of translocated protons is nearly equal to the charge imbalance caused by all the other ions (SCHRÖDER et al., 1972; HIND et al., 1974):

$$\Delta H^+ \approx \sum z_n^- \, \Delta I_n^- - \sum z_n^+ \, \Delta I_n^+. \tag{4}$$

These results confirm the idea that the process of inward proton translocation is an active one, which is associated with the formation of a membrane potential. The events are depicted schematically in Figure 1.

At the onset of illumination very quickly a membrane potential is set up in connection with the active influx of protons into the thylakoid (stage a, for details see Sect. B.1).

The membrane potential gives rise to the secondary influx of permeant anions and efflux of permeant cations. These counterion fluxes occur until the ion gradients are equilibrated against the membrane potential (SCHRÖDER et al., 1972). As a consequence of the counterion fluxes the originally high membrane potential will be diminished and a large internal proton accumulation will take place (stage b). The counterion equilibrium in the dark as well as during stationary illumination is described by the Nernst equation:

$$\Delta \psi = \pm RT/F \ln \frac{[I^\mp]_{i,\,s}}{[I^\mp]_{o,\,s}} \tag{5}$$

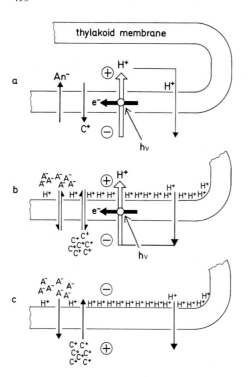

Fig. 1a–c. Scheme of ion translocating phenomena in chloroplasts. (a) Onset of illumination. (b) Steady state conditions. (c) Relaxation upon darkening

where $\Delta\psi$ is the electric potential across the membrane, and $[I]_{i,s}$, $[I]_{o,s}$ are the concentrations of monovalent ions at inner and outer membrane surface resp. The ion concentrations at membrane surface are related to those in the bulk solution through the membrane surface potential:

$$[I^{\pm}]_s = [I^{\pm}]_b \exp(F/RT\,\psi_s) \tag{6}$$

where $[I]_s$, $[I]_b$ are the ion concentrations at membrane surface and in bulk solution resp., and ψ_s is the electric surface potential.

In the dark the highly unbalanced proton gradient is removed through exchange with the counterions (stage c). During this process the fluxes of protons and counterions are coupled through a membrane diffusion potential which is described by the Goldman-Hodgkin-Katz equation (see DAINTY, 1962):

$$\Delta\psi_{\mathrm{diff}} = RT/F \ln \frac{p_{H^+}[H^+]_{o,s} + \sum p_{I^+}[I^+]_{o,s} + \sum p_{I^-}[I^-]_{i,s}}{p_{H^+}[H^+]_{i,s} + \sum p_{I^+}[I^+]_{i,s} + \sum p_{I^-}[I^-]_{o,s}} \tag{7}$$

where p_H and p_I are the permeability coefficients of protons and other ions resp.

The model bears specific consequences with respect to the kinetics of proton translocation, especially if the membrane permeability for protons and counterions resp. is changed. Experiments on this line have been performed which gave full confirmation (RUMBERG and MUHLE, 1976).

II. Role of Surface Charge

There is a great deal of evidence that biological membranes carry fixed charges on their surface regions which give rise to ionic double layers (see HAYDON and HLADKY, 1972). Consideration of the relationship between surface charge density and surface potential should be possible on the basis of the Gouy-Chapman theory (see DELAHAY, 1965):

$$\sinh(F/2RT\psi_s) = 136\,\sigma/\sqrt{c} \tag{8}$$

where σ is the surface charge density in units of electronic charges/Å^2, and c is the monovalent ion concentration in bulk solution in units of mol/l.

The charge density at the inner thylakoid surface under conditions of *darkness* has been estimated to be $5 \cdot 10^{-4}$ negative charges/Å^2 (RUMBERG and MUHLE, 1976). The corresponding surface potential depends on the ion concentration in bulk solution. It is -32 mV if an ion concentration of 10 mM is assumed. The thickness of the ionic double layer under these conditions is approx. 20 Å.

The protons which are pumped in when the light is turned on bind almost completely to the inner membrane surface thus giving rise to a surface charge increase. The amount pumped in under normal conditions at pH 8 is approx. 0.3 protons per chlorophyll molecule or 200 Å^2 resp. Therefore the surface charge density increases in the light to approx. $10 \cdot 10^{-4}$ positive charges/Å^2 which corresponds to a surface potential of $+58$ mV. The resulting light-induced *change* of the inner surface potential is (RUMBERG and MUHLE, 1976):

$$\Delta\psi_{s,\,light} = +90 \text{ mV.}$$

Investigation of the ion redistribution has revealed that the light-induced potential difference between the bulk phases during stationary conditions is negligible. Therefore, potential differences in the order of 100 mV as indicated by intrinsic probes should be due to surface potential differences only. This conclusion is in accord to the above made estimate on the surface potential change.

III. Change of Membrane Potential During Illumination Cycle

At the beginning of illumination at first a high membrane potential is created in connection with the active proton influx (see Fig. 2, stage a and Fig. 3, top).

Due to the onset of counterion fluxes protons accumulate in the inner phase of the thylakoid. The increased proton concentration in the inner phase causes an enhanced proton conductivity of the membrane which will give rise to a decrease of the membrane potential. On the other hand the binding of protons to the inner membrane causes an increase of the surface charge and as a consequence an increase of the inner surface potential. During steady-state conditions the potential difference across the membrane will nearly exclusively be due to a surface potential difference (see Fig. 2, stage 6 and Fig. 3, top). Corresponding to the condition of constant electrochemical potential throughout outer and inner phase resp. [Eq. (6)] changes of pH in the surface regions should take place corresponding to the changes of the electric potential (see Fig. 2 and Fig. 3, bottom).

If the light is turned off, the proton influx stops immediately. The uncompensated backflow of protons will cause a sudden drop of the membrane potential to the level of the diffusion potential. Analysis of the electrochromic absorption changes at 515 nm has shown, that the value of the remaining diffusion potential across the membrane is low (GRÄBER and WITT, 1974) (see Fig. 2, stage c and Fig. 3, top).

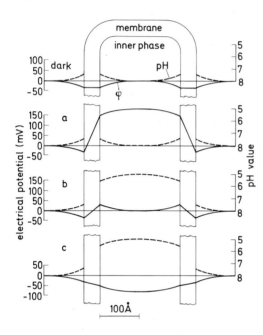

Fig. 2a–c. Profile of electrical potential and pH across the thylakoid. (a) Onset of illumination. (b) Steady state conditions. (c) Relaxation upon darkening. Calculation of inner double layer potential is described in the text. Outer double layer potential has been chosen arbitrary. Potential rise at the beginning of illumination is indicated by electrochromic absorption changes, inner bulk phase potential during steady state conditions is read out from counterion distribution. The membrane distance of 200 Å corresponds to an inner space volume of approx. 10 l/mol chlorophyll. See Figure 3 for complement

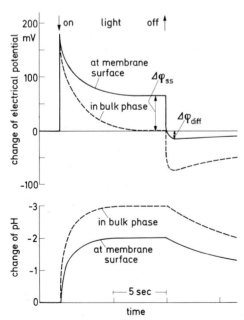

Fig. 3. *Top:* Proposed time course of light-induced potential changes between inner and outer membrane surfaces and between inner and outer bulk phases resp. *Bottom:* Corresponding changes of pH. A light-induced change of 3 units in the inner bulk phase has been observed by RUMBERG and SIGGEL (1969) and ROTTENBERG et al. (1972). See Figure 2 for complement

D. Relationship Between Membrane Potential and ATP Formation

I. Induction Phase

During the early stage of illumination the electrochemical proton potential difference will be poised towards the electrical component. Moreover, the membrane potential difference will not yet be balanced by counterion gradients. Two consequences arise supposed a chemiosmotic phosphorylation mechanism is at work. First, the field decay should be controlled by the ATPase system. Second, ATP production should be suppressed when the electric energy is dissipated by an artificially induced ion conductivity of the membrane. Both effects have been observed under illumination with short flashes or short flash groups. First, acceleration of the decay of the field-indicating absorption changes at 515 nm has been demonstrated which specifically depends on concomitant ATP production (RUMBERG and SIGGEL, 1968; JUNGE et al., 1970). Second, suppression of ATP production has been demonstrated in the presence of cation permeability inducing agents like valinomycin and gramicidin (JUNGE et al., 1970; BOECK and WITT, 1972). Quantitative titration with gramicidin has proven, that the functional unit of phosphorylation is of the minimum size of one thylakoid as it is for the electrical events (BOECK and WITT, 1972).

II. Steady-State Illumination

During steady-state conditions the protons are cyclically translocated between inner and outer phase. The magnitude of the electrochemical potential difference for protons results from the balance between proton influx and proton efflux down the gradient. The counterions are equilibrated across the membrane according to the light-induced electric potential difference. Therefore, during steady-state conditions all events should be completely independent from the membrane permeability of other ions than protons. With respect to phosphorylation three conclusions can be drawn, which all have been verified.

    1. $\Delta\psi$ as well as ΔpH should be diminished due to ATPase activity. Decrease of ΔpH is manifest from different methods which had been applied to estimate the internal pH (RUMBERG and SIGGEL, 1969; PICK et al., 1973). Decrease of $\Delta\psi$ on the other hand has also been demonstrated by analysis of the field-indicating absorption changes at 515 nm (BALTSCHEFFSKY and HALL, 1974).

    2. The rate of phosphorylation should depend on both ΔpH and $\Delta\psi$. This has in principle been demonstrated recently (GRÄBER and WITT, 1976). In these experiments ΔpH has been obtained from the amine distribution method. This method delivers information on the ΔpH between inner and outer bulk phases. However, the ATPase is subjected to the surface regions of the membrane and therefore, with respect to ΔpH and $\Delta\psi$, the surface values should be considered. Quantitative experiments under rigorously controlled conditions in this respect have not been carried out as yet.

    3. Suppression of ATP production by dissipation of the electrochemical potential difference should only be possible by addition of proton permeability inducing

agents (uncouplers). Enhancement of counterion permeability should be without effect. This prediction is also in accordance with the experimental results. It has been demonstrated that the potassium carrier valinomycin, which suppresses ATP production when applied during flash excitation, is without effect under continuous illumination (Junge et al., 1970).

III. Postillumination Phase

From Jagendorf's pioneer work it is known that chloroplasts preloaded with protons by light-induced proton uptake or by an acid-base transition are able to synthesize ATP subsequently (Jagendorf and Hind, 1963; Jagendorf and Uribe, 1966). Later on it was shown that the yield of ATP under conditions in which ΔpH is suboptimal is highly stimulated if a diffusion potential positive on the inside is artificially imposed by addition of KCl in the presence of valinomycin (Schuldiner et al., 1972, 1973; Uribe and Li, 1973).

In order to explain this stimulation effect one has to visualize that the yield of ATP is governed by the competition between the ATPase-coupled proton efflux and dissipative basal proton efflux. Gräber and Witt (1976) have shown that the proton efflux is poised towards the ATPase-coupled pathway if the electrochemical potential (ΔpH or $\Delta\psi$) is increased. Therefore, the relative yield of ATP should increase if the electrochemical potential difference for protons is enhanced by an additionally applied electric potential difference.

Experiments of this sort culminated in that of Uribe (1973), who showed that the synthesis of ATP by chloroplasts can be driven by the energy of an artificially produced membrane potential alone in the absence of a chemical component of the hydrogen ion gradient.

E. Summary

Evidence has accumulated that the pumps for protons which are incorporated in the photosynthetic electron transport system work in an electrogenic fashion generating an electrochemical proton gradient across the thylakoid membrane made up of both a chemical and an electrical component.

The electrical imbalance gives rise to passive counterion fluxes which, on their part, control the total amount of protons pumped in. The membrane potential exhibits a spike of 200 mV when the light is turned on and declines to a level of 100 mV which is exclusively caused by a change of the inner surface potential due to proton binding to the inner membrane.

There is ample evidence that ATP formation is driven at the expense of the electrical and the chemical component of the proton gradient as well.

References

Baltscheffsky, M., Hall, D.O.: FEBS Lett. **39**, 345–348 (1974)
Barber, J.: FEBS Lett. **20**, 251–254 (1972)
Barber, J., Kraan, G.P.B.: Biochim. Biophys. Acta **197**, 49–59 (1970)
Boeck, M., Witt, H.T.: In: Proc. 2nd Intern. Congr. Photosynthesis. Forti, G., Avron, M., Melandri, A. (eds.). The Hague: Dr. Junk, 1972, Vol. II, pp. 903–911
Bulychev, A.A., Andrianov, V.K., Kurella, G.A., Litvin, F.F.: Nature (London) **236**, 175–177 (1972)
Dainty, J.: Ann. Rev. Plant Physiol. **13**, 379–402 (1962)
Delahay, P.: Double Layer and Electrode Kinetics. New York: Interscience, 1965
Emrich, H.M., Junge, W., Witt, H.T.: Z. Naturforsch. **24 B**, 1144–1146 (1969)
Fowler, C.F., Kok, B.: Biochim. Biophys. Acta **357**, 308–318 (1974)
Gräber, P., Witt, H.T.: Biochim. Biophys. Acta **333**, 389–392 (1974)
Gräber, P., Witt, H.T.: Biochim. Biophys. Acta **423**, 141–163 (1976)
Haydon, D.A., Hladky, S.B.: Quart. Rev. Biophys. **5**, 187–282 (1972)
Hind, G., Nakatani, H.Y., Izawa, S.: Proc. Nat. Acad. Sci. U.S. **71**, 1484–1488 (1974)
Jackson, J.B., Crofts, A.R.: FEBS Lett. **4**, 185–189 (1969)
Jagendorf, A.T., Hind, G.: In: Photosynthetic Mechanism of Green Plants. Nat. Acad. Sci. Nat. Res. Council **1145**, 599–610 (1963)
Jagendorf, A.T., Uribe, E.: Proc. Nat. Acad. Sci. U.S. **55**, 170–177 (1966)
Junge, W., Rumberg, B., Schröder, H.: Europ. J. Biochem. **14**, 575–581 (1970)
Junge, W., Witt, H.T.: Z. Naturforsch. **23 B**, 244–254 (1968)
Mitchell, P.: Nature (London) **191**, 144–148 (1961)
Mitchell, P.: Biol. Rev. **41**, 445–502 (1966)
Neumann, J., Jagendorf, A.T.: Arch. Biochem. Biophys. **107**, 109–119 (1964)
Pick, U., Rottenberg, H., Avron, M.: FEBS Lett. **32**, 91–94 (1973)
Reinwald, E., Stiehl, H.H., Rumberg, B.: Z. Naturforsch. **23 B**, 1616–1617 (1968)
Rottenberg, H., Grunwald, T., Avron, M.: Europ. J. Biochem. **25**, 54–63 (1972)
Rumberg, B., Muhle, H.: J. Bioelectr. Bioenerget. (1976) (in press)
Rumberg, B., Siggel, U.: Z. Naturforsch. **23 B**, 239–244 (1968)
Rumberg, B., Siggel, U.: Naturwissenschaften **56**, 130–132 (1969)
Schliephake, W., Junge, W., Witt, H.T.: Z. Naturforsch. **23 B**, 1571–1578 (1968)
Schmidt, S., Reich, R., Witt, H.T.: Naturwissenschaften **58**, 414 (1971)
Schmidt, S., Reich, R., Witt, H.T.: In: Proc. 2nd Intern. Congr. Photosynthesis. Forti, G., Avron, M., Melandri, A. (eds.). The Hague: Dr. Junk, 1972, Vol. II, pp. 1087–1095
Schröder, H., Muhle, H., Rumberg, B.: In: Proc. 2nd Intern. Congr. Photosynthesis. Forti, G., Avron, M., Melandri, A. (eds). The Hague: Dr. Junk, 1972, Vol. II, pp. 919–930
Schuldiner, S., Rottenberg, H., Avron, M.: FEBS Lett. **28**, 173–176 (1972)
Schuldiner, S., Rottenberg, H., Avron, M.: Europ. J. Biochem. **39**, 455–462 (1973)
Thorne, S.W., Horvath, G., Kahn, A., Boardman, N.K.: Proc. Nat. Acad. Sci. U.S. **72**, 3858–3862 (1975)
Uribe, E.: FEBS Lett. **36**, 143–147 (1973)
Uribe, G., Li, B.: J. Bioenerget. **4**, 435–444 (1973)
Vredenberg, W.J., Tonk, W.J.M.: Biochim. Biophys. Acta **387**, 580–587 (1975)
Witt, H.T.: In: Fast Reactions and Primary Processes in Chemical Kinetics. Claesson, S. (ed.). New York-London-Sydney: Interscience, 1967, pp. 261–310
Witt, H.T.: Quart. Rev. Biophys. **4**, 365–477 (1971)
Witt, H.T.: In: Bioenergetics of Photosynthesis. Govindjee (ed.). New York-San Francisco-London: Academic Press, 1975, pp. 493–554
Witt, H.T., Rumberg, B., Junge, W.: In: Biochemie des Sauerstoffs. Hess, B., Staudinger, Hj. (eds.). Berlin-Heidelberg-New York: Springer, 1968, pp. 262–306
Witt, H.T., Zickler, A.: FEBS Lett. **37**, 307–310 (1973)
Witt, H.T., Zickler, A.: FEBS Lett. **39**, 205–208 (1974)
Wolff, C., Buchwald, H.E., Rüppel, H., Witt, K., Witt, H.T.: Z. Naturforsch. **24 B**, 1038–1041 (1969)
Wraight, C.A., Crofts, A.R.: Europ. J. Biochem. **19**, 386–397 (1971)

9. Acid Base ATP Synthesis in Chloroplasts

S. Schuldiner

A. Introduction

The ATPase system of chloroplasts catalyzes translocation of protons across the thylakoid membrane concomitantly with ATP hydrolysis (Carmeli, 1970). The finding that ATP formation is observed in chloroplasts following a transition from acid to base provided one of the first experimental demonstrations on the reversibility of the enzyme, i.e., ATP is synthesized from an artificially imposed pH differential of appropriate polarity across the membrane. This phenomenon strongly supports one of the basic tenets of the chemiosmotic theory for ATP formation in chloroplasts (Mitchell, 1966, 1968), i.e., that an electrochemical gradient of protons provides the energy for ATP synthesis.

Acid-base-induced ATP formation was first described by Hind and Jagendorf (1965). In this chapter, I will review the experimental evidence describing the properties of the system. The detailed molecular mechanism by which this ATP synthesis takes place is not yet clear and is beyond the scope of this chapter.

B. General Properties of the System

The standard procedure consists of two successive stages, both generally carried out at $0°$ C: (1) equilibration of chloroplasts at an acid pH (the acid stage), and (2) increase of the external pH with simultaneous addition of phosphorylating reagents (ADP, Mg^{2+} ions and phosphate). The quantity of ATP formed is markedly dependent on the size of the pH gradient and is strongly affected by the pK and permeability of the organic acid present in the acid stage (Jagendorf and Uribe, 1966; Uribe and Jagendorf, 1967b). Under optimal conditions, the total amount of ATP formed is quite large (up to 240 nmol can be formed per mg of chlorophyll) (Jagendorf and Uribe, 1966; Jagendorf, 1967).

ATP formation under these conditions is inhibited by known uncouplers of photophosphorylation. Thus ammonium salts, atebrin, carbonyl cyanide m-chlorophenylhydrazone (CCCP), and Triton X-100 are all effective at concentrations similar to those which inhibit photophosphorylation (Jagendorf and Uribe, 1966). Even more relevant to the question of whether the normal phosphorylation machinery is involved in this process is the finding that antibody specific to chloroplast ATPase inhibits both light dependent and acid-base dependent ATP formation (McCarty and Racker, 1966). Furthermore, the acid base transition and light aroused the same patterns of oxygen exchanges between ADP, P_i and H_2O (Shavit and Boyer, 1966). Moreover, it is now well established that acid base catalyzed

ATP formation does not involve electron transport. LYNN (1968) raised the possibility that the redox potentials of certain electron transport carriers might be altered due to alterations in pH. However, the high yields of ATP obtained rule out the possibility of a stoichiometric reaction; thus, for example, around 100 molecules of ATP are formed per molecule of cytochrome f. Furthermore, a broad range of inhibitors, specific for electron transport, have no effect on acid-base-induced phosphorylation. Thus, 3-(3,4-dichlorophenol)-1,1-dimethylurea (DCMU), 3-(p-chlorophenol)-1,1-dimethylurea (CMU), 2-n-heptyl-4-hydroxyquinoline-N-oxide (HOQNO), o-phenantroline, simazine and antimycin A, alone or in combination, inhibit acid-base phosphorylation less than 10% even at concentrations many times higher than those required to inhibit photophosphorylation (JAGENDORF and URIBE, 1966; MILES and JAGENDORF, 1970). Only n-butyl 3,5 diiodo 4-hydroxybenzoate inhibits both reactions at comparable concentrations (MILES and JAGENDORF, 1970); however, this compound was reported to act as an uncoupler as well as an electron transport inhibitor (GROMET-ELHANAN and AVRON, 1965). Other treatments that inhibit electron transport (i.e., chloride depletion) or agents which alter the redox potential of chloroplasts (i.e., exogenous electron acceptors) have no effect on acid base-catalyzed ATP formation (MILES and JAGENDORF, 1970). Moreover acid-base phosphorylation has been observed in chloroplasts completely depleted of plastocyanin (ANDERSON and McCARTY, 1969).

C. Dicarboxylic Acid Requirement

Acid-base-induced ATP synthesis is enhanced when certain dicarboxylic acids are present during the acid stage of the reaction (JAGENDORF und URIBE, 1966; URIBE and JAGENDORF, 1967a, b), and the specificity for dicarboxylic acids is very broad. The observations are consistent with the notion that the organic acid in its undissociated form penetrates the internal space of the chloroplast at the low pH, and then serves as a reservoir of protons when the chloroplasts are exposed to a high pH (see Fig. 1). This hypothesis stipulates only that the organic acid be freely permeable in its undissociated form and much less permeable in its unprotonated (i.e., charged) form. For maximal effectiveness, the acid should also have a pK intermediate between the pHs of the acid and basic stages because it should equilibrate rapidly across the membrane during the acid stage and effectively release protons during the alkaline stage of the reaction. Indeed, the most effective organic acids are C_4 and C_5 aliphatic dicarboxylic acids with one pK_a in the region of 4.2 to 4.4 and the second at 5.3 to 5.5 (URIBE and JAGENDORF, 1967b). Acids as diverse as succinic, fumaric, glycolic, lactic, adipic, mesaconic, itaconic, o-phtalic, p-phtalic and barbituric acid are all effective (URIBE and JAGENDORF, 1967b). Moreover, among a series of related dicarboxylic acids, effectiveness is related to the degree of protonation in the acid phase (i.e., pH 4.0). Modifications in structure which alter pK_a decrease the effectiveness of the acid. Succinic acid, which supports synthesis of high amounts of ATP, penetrates rapidly at pH 4.0; the rate of equilibration decreases markedly when the pH is increased, and at pH 6.5 where the acid is fully dissociated, there is little or no penetration (URIBE

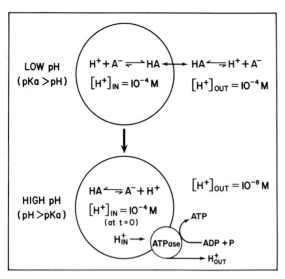

Fig. 1. Function of the internal buffers. At the low pH most of the acid is in its protonated form (HA) which can freely equilibrate across the membrane. When the pH on the outside is raised, protons will start moving outwards down their electrochemical gradient. The decrease in proton concentration on the inside will lead to dissociation of the acid ($HA \rightarrow H^+ + A^-$) therefore preventing a rapid increase in the pH for as long as enough protonated acid is present in the thylakoid

and Jagendorf, 1967a). Low permeability during the alkaline phase is apparently critical since acetic acid, which penetrates well both as the acid and the anion, is ineffective in supporting ATP synthesis (Uribe and Jagendorf, 1967a). Acids which penetrate very slowly in the acid stage, such as glutamic acid, also produce low yields of ATP (Uribe and Jagendorf, 1967a). It is noteworthy that certain amines which have very low pKs (i.e., aniline) can function as internal proton reservoirs and support acid base-induced ATP formation (Avron, 1972).

As postulated, the role of the internal buffer is to fulfill two functions: (1) the absolute quantity of internal protons available for ATP synthesis is increased significantly, and (2) the pH difference between the inside of the thylakoid and the medium is maintained for as long as possible. Regarding (1), the amount of internal succinate accumulated during acid stage correlates very well with the amount of ATP formed. This has been demonstrated either by changing the external concentration of succinate (Uribe and Jagendorf, 1967a) or by reducing the thylakoid volume in the presence of osmotically active agents such as sucrose (Uribe and Jagendorf, 1968). Sucrose is a potent inhibitor of acid-base ATP formation at concentrations which have no effect on photophosphorylation. The observations are consistent with the suggestion that sufficient internal volume is required for storage of as much organic acid as possible to act as a buffer. Regarding (2), acid base ATP formation, steady-state phosphorylation (Portis and McCarty, 1974; Pick et al., 1975) and postillumination ATP synthesis (Schuldiner et al., 1973) exhibit a threshold effect with respect to the pH gradient. These observations are discussed below.

D. The Electrochemical Gradient of Protons and ATP Synthesis

The yield of ATP as a function of pH in the acid stage exhibits an inverse relationship (JAGENDORF and URIBE, 1966), an observation which is consistent with the increasing amounts of free acid in the thylakoids at lower pH values. Moreover, at a given pH during the acid phase, more ATP is formed when the pH of the phosphorylation stage (i.e., the alkaline phase) is increased (JAGENDORF and URIBE, 1966). The pH curve for the phosphorylation step varies depending on the pH of the acid stage: as the pH of the acid stage is raised, both the pH optimum for phosphorylation and the minimum pH at which ATP synthesis is detected are also raised (Fig. 2). These effects probably reflect the need for a minimal pH differential between the two stages of the experiment, superimposed upon the pH requirements of the phosphorylation enzyme. The minimal pH differential for demonstrable ATP synthesis ranges from 2.2 to 2.9 pH units. This requirement is demonstrated more clearly in other systems. Thus, when postillumination (SCHULDINER et al., 1973) or steady-state (PORTIS and MCCARTY, 1974; PICK et al., 1975) phosphorylation is measured as a function of ΔpH (i.e., the difference in pH between the inside of the thylakoid and the medium), no ATP synthesis is detected when the ΔpH is below 2.5 to 3.0 pH units. Above this value there is a steep increase in the amount of ATP synthesized as a function of ΔpH. This "threshold" phenomenon is also apparent in acid/base experiments with submitochondrial particles. In this case a $\Delta\bar{\mu}_{H^+}$ of approximately 160 mV is necessary for significant rates of ATP synthesis (THAYER and HINKLE, 1975). The "thresholds" described probably represent the minimum proton gradients required to overcome the phosphate potential under the conditions described. The chemical gradient of protons can be replaced, partially at least, by an electrical

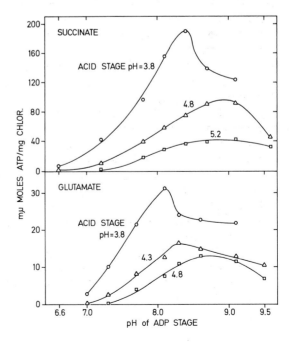

Fig. 2. The pH curve for phosphorylation following an acid-to-base transition. Note the higher pH optima for phosphorylation when the pH of the acid stage is higher. Note also the different optima for the two acids used. (From JAGENDORF and URIBE, 1966)

potential. Thus, if a diffusion potential is imposed across the membrane under conditions in which the pH difference is suboptimal, more ATP is synthesized than can be accounted for by the ΔpH (Schuldiner et al., 1972; Uribe, 1973). Small amounts of ATP are synthesized even in the absence of a pH differential (Uribe, 1973; Uribe and Li, 1973) when a diffusion potential is imposed across the chloroplast membrane. An internal buffer is required in these experiments also.

E. Kinetics

The amount of ATP synthesized in the basic stage is a function of incubation time in the acid stage and varies depending on the acid used, its concentration, and the pH of the acid stage (Jagendorf and Uribe, 1966). When glutamate is used a biphasic increase is detected: 85% of the maximum is reached in about 40 sec followed by a very slow increase which takes about 3 min. With 3 mM succinate at pH 3.8 the half-time is 8 sec, and with 10 mM succinate, 2 sec. At pH 4.5 the half-time is 4 sec at 10 mM succinate. As shown by Uribe and Jagendorf (1967a), these rates are in excellent agreement with the time course of acid penetration during the acid stage.

The time course of decay in ability to synthesize ATP at high pH was also estimated by omitting ADP and ^{32}P for varying lengths of time after the shift to alkaline. At pH 8 the rate of decay exhibits a half life which is less than 2 sec. After about 6 sec, ATP synthesis is no longer observed (Jagendorf and Uribe, 1966).

The kinetics of proton release during acid/base transition were studied by stopped flow (Nishizaki and Jagendorf, 1971; Nishizaki, 1972, 1973). Using bromocresol purple as a pH indicator, Nishizaki showed that the time course of proton release occurs in two phases, a fast phase ($t_{1/2} = 20-40$ msec) and a slow phase ($t_{1/2} = 1.5-3.0$ sec) (Nishizaki, 1973). A third minor component ($t_{1/2} = 5$ sec) was also detected, but was not apparently related to ATP formation.

ATP formation is largely finished after about 3 sec, i.e., when there is still a considerable amount of protons in the thylakoid; however, the ratio of ATP formed to protons released vary inversely with the rate of proton efflux. The faster the rate, the smaller the quantity of ATP formed (Nishizaki and Jagendorf, 1971; Nishizaki, 1972, 1973). ATP synthesis ceases before about 40% of the protons leave the thylakoid, probably because the gradient decays below the "threshold" value. These studies are consistent with the notion that the maintenance of an adequate H^+ gradient is a necessary prerequisite for ATP synthesis.

F. Activation of ATP Hydrolysis

Isolated chloroplasts have virtually no ATPase activity; however, ATP hydrolysis can be induced by light in the presence of dithiothreitol (see Chap. III.5 in this volume). Light is required for inducing a "high energy state" and its effect is prevented by agents which collapse this "high energy state". As with phosphorylation, the requirement for electron flow can be obviated by an acid-base transition (KAPLAN et al., 1967). Under all conditions, activity is dependent on the presence of Mg^{2+} ions and dithiothreitol, and in almost all other respects, induction of ATPase activity by a pH transition appears to be analogous to induction by light. In both cases, ATPase activity is inhibited by uncoupling reagents, and in both cases the ATPase activity continues following the termination of the activation; in some cases, continuous ATP hydrolysis was observed for up to 8 min although the acid-base induced ATP formation is over in 4 sec. This slow decay enabled KAPLAN and JAGENDORF (1968) to investigate two problems that can not be studied when ATP synthesis is investigated: (1) is activation due to the acidity of the acid stage or to the pH differential? and (2) is the driving force in the base stage reentry of cations other than protons? The results of the experiment designed to investigate point (1) are shown in Table 1. Chloroplasts were subjected to three different sets of conditions: (1) chloroplasts were brought from pH 5.9 to pH 8.4; (2) chloroplasts were brought from 5.9 to 10.1, a larger pH differential, but the base stage is above the pH optimum for ATPase, and (3) chloroplasts were taken from 5.9 to 10.1 for 5 sec only, then readjusted to pH 8.4. The amount of ATP hydrolyzed in condition (3) was three times higher than that observed under the other two sets of conditions. This result would not be expected if an acid pH is required for activation of the enzyme. The finding is consistent with the need for a large ΔpH in the activation mechanism. Regarding point (2), one alternative for the proposed mechanism of ATP formation is that the driving force in the alkaline phase may be reentry of cations other than protons which drive an ATP-dependent-cation pump in reverse. If this is the case, one would expect the mechanism to work only when an appropriate cation is supplied in the basic medium. The specificity towards a cation in the alkaline stage cannot be investigated when ATP synthesis is followed because ADP, phosphate and Mg^{2+} ions must be supplied in less than 2 sec to be able to trap the energy of the pH jump. However, the ATPase activation process itself does not require the simultaneous presence of ATP and Mg^{2+} ions. These may be added up to 15 sec later, with very little loss of activity in order to assess the extent of prior ATPase activation (KAPLAN et al., 1967). The data argue strongly that activation of ATPase during this intermediate base stage does not require the presence of

Table 1. ATPase activation by pH differential. Fresh chloroplasts were broken, washed, centrifuged, and resuspended in 40 mM dithiothreitol, pH 7.5. They were then incubated in the usual acid stage (0° C) at pH 5.9 for 20 sec and injected into one of three kinds of base stages. After 5 sec they were injected into base stage 2 and incubated for 115 sec before the reaction was stopped with 0.2 ml of 20% trichloracetic acid. (From KAPLAN and JAGENDORF, 1968)

Acid stage (20 sec) pH	Base stage 1 (5 sec) pH	Base stage 2 (115 sec) pH	ATP hydrolyzed mμmol mg^{-1} chlorophyll
5.9	8.4	8.4	77
5.9	10.1	10.1	77
5.9	10.1	8.4	222

a specific cation or series of cations. This suggests that the proton concentration gradient per se is the energetic component initiating activation.

More recent experiments point to the fact that the transmembrane pH differential leads to a change in conformation of ATPase, thereby permitting its hydrolytic subunits to become functional. Thus, changes in the reactivity of ATPase towards different agents (Datta et al., 1974; McCarty and Fagan, 1973; Oliver and Jagendorf, 1975) and variations in tritium exchange (Ryrie and Jagendorf, 1972) are induced by an acid-base transition. Hopefully, studies of these phenomena will lead to a more complete understanding of ATP synthesis on a molecular level.

Acknowledgements. I would like to thank Drs. P. Hinkle, A. Jagendorf and E. Uribe for providing their manuscripts prior to publication. In addition, I thank Dr. Jagendorf for his helpful suggestions and for his permission to reproduce Figure 2 and Table 1 in this chapter; and Drs. D. Gutnick, H.R. Kaback and G. Rudnick for editorial help in the preparation of this manuscript.

References

Anderson, M., McCarty, R.E.: Biochim. Biophys. Acta **189**, 193–206 (1969)
Avron, M.: In: Proc. 2nd Intern. Congr. Photosynthesis. Forti, G., Avron, M., Melandri, A. (eds.). The Hague: Dr. Junk, 1972, Vol. II, pp. 861–871
Carmeli, C.: FEBS Lett. **7**, 297–300 (1970)
Datta, D.B., Ryrie, I.J., Jagendorf, A.T.: J. Biochem. **249**, 4404–4411 (1974)
Gromet-Elhanan, Z., Avron, M.: Plant Physiol. **40**, 1053–1059 (1965)
Hind, G., Jagendorf, A.T.: J. Biol. Chem. **240**, 3195–3201 (1965)
Jagendorf, A.T.: Federation Proc. **26**, 1361–1369 (1967)
Jagendorf, A.T., Uribe, E.: Proc. Nat. Acad. Sci. U.S. **55**, 170–177 (1966)
Kaplan, J.H., Jagendorf, A.T.: J. Biol. Chem. **243**, 972–979 (1968)
Kaplan, J.H., Uribe, E., Jagendorf, A.: Arch. Biochem. Biophys. **120,**, 365–375 (1967)
Lynn, W.S.: Biochemistry **7**, 3811–3820 (1968)
McCarty, R.E., Fagan, J.: Biochemistry **12**, 1503–1507 (1973)
McCarty, R.E., Racker, E.: Brookhaven Symp. Biol. **19**, 202–214 (1966)
Miles, C.D., Jagendorf, A.T.: Biochemistry **9**, 429–434 (1970)
Mitchell, P.: Biol. Rev. **41**, 445–502 (1966)
Mitchell, P.: In: Chemiosmotic Coupling and Energy Transduction. Bodmin: Glynn, 1968
Nishizaki, Y.: Biochim. Biophys. Acta **275**, 177–181 (1972)
Nishizaki, Y.: Biochim. Biophys. Acta **314**, 312–319 (1973)
Nishizaki, Y., Jagendorf, A.T.: Arch. Biochem. Biophys. **133**, 255–262 (1969)
Nishizaki, Y., Jagendorf, A.T.: Biochim. Biophys. Acta **226**, 172–186 (1971)
Oliver, D.J., Jagendorf, A.T.: Federation Proc. **34**, 596 (1975)
Pick, U., Rottenberg, H., Avron, M.: FEBS Lett. **48**, 32–36 (1974)
Portis, A.R., McCarty, R.E.: J. Biol. Chem. **249**, 6250–6254 (1974)
Ryrie, I.J., Jagendorf, A.T.: J. Biol. Chem. **246**, 3771–3774 (1971)
Ryrie, I.J., Jagendorf, A.T.: J. Biol. Chem. **247**, 4453–4459 (1972)
Schuldiner, S., Rottenberg, H., Avron, M.: FEBS Lett. **28**, 173–176 (1972)
Schuldiner, S., Rottenberg, H., Avron, M.: Europ. J. Biochem. **39**, 455–462 (1973)
Shavit, N., Boyer, P.D.: J. Biol. Chem. **241**, 5738–5740 (1966)
Thayer, W.S., Hinkle, P.C.: J. Biol. Chem. **250**, 5330–5335 (1975)
Uribe, E.G.: FEBS Lett. **36**, 143–147 (1973)
Uribe, E.G., Jagendorf, A.T.: Plant Physiol. **42**, 697–705 (1967a)
Uribe, E.G., Jagendorf, A.T.: Plant Physiol. **42**, 706–711 (1967b)
Uribe, E.G., Jagendorf, A.T.: Arch. Biochem. Biophys. **128**, 351–359 (1968)
Uribe, E.G., Li, B.C.Y.: Bioenergetics **4**, 435–444 (1973)

10. Energy-Dependent Conformational Changes

R. KRAAYENHOF

A. Introduction

The three-dimensional structure of proteins has been shown to be of critical importance for their specific catalytic, regulating or translocating activities (KOSHLAND and NEET, 1968; KLOTZ et al., 1970; Cold Spring Harbor Symposium on Quantitative Biology, 1972). Over the past two decades protein biochemists have convincingly demonstrated that ligand-induced conformational rearrangements play a vital role in the dynamic control and coupling of metabolic reactions, in both soluble and membrane-bound proteins (see for instance WYMAN, 1968; GITLER, 1972; SINGER and NICOLSON, 1972). BOYER (1965) has plausibly proposed that the transduction of redox energy to the phosphorylation reaction might be mediated by high-energy protein conformations. This idea found little support among investigators of photosynthetic energy conservation who, with few exceptions, prefer the electrochemical coupling mechanism proposed by MITCHELL (1961) that may be more attractive from merely physical points of view.

In this chapter some attention will be focused on the conformational mechanism of energy transduction as it is visualized at present, followed by a brief description of light-dependent structural changes observed in chloroplasts. Finally, the observations of energy-dependent conformational changes of chloroplast membrane-bound ATPase will be summarized.

B. Conformational Mechanism of Energy Transduction

The original proposal of BOYER (1965), which has been further developed and experimentally supported by the groups of BOYER (1974) and SLATER (1971, 1974), can be formulated as follows. The energy set free by the redox reactions is primarily conserved in a conformationally strained high-energy form of the electron-transporting protein and is then "transmitted" to the ATP-synthesizing protein (ATPase) by means of one or more reversible transitions between strained and relaxed protein conformations. The essential difference from the chemiosmotic mechanism is that the conformational energy transduction is supposed to occur by a short-range interaction between the sequential components whereas in the chemiosmotic mechanism an electrochemical potential across the bulk membrane drives ATP synthesis by a long-range interaction (which is in fact one of the operational advantages of this latter mechanism since one needs neither direct contact nor stoichiometry between the component molecules).

Analogous to the conformational rearrangements in well-characterized proteins (see for instance PERUTZ, 1970) the interaction between the energy-transducing

proteins may be, at least partly, electrostatic in nature. Proximity changes of
basic and acidic groups of the peptides *must* lead to pK_a shifts by Coulombic
interaction and hence in local changes of proton binding and vice versa. On the
basis of such and other considerations Williams (1970, 1975) has argued that
a high local concentration of protons at low water activity in the membrane
proper can act as an intermediate step towards ATP synthesis. A similar (microche-
miosmotic) mechanism has been recently proposed by Izawa et al. (1974). Bennun
(1971) has designed a conformational type of hypothesis which contains elements
of the earlier proposals mentioned and emphasizes the possible electrostatic nature
of the conformational interaction between the ATPase and the redox enzymes.

The occurrence of different protein conformations with differences in both
"energy content" and proton affinity is nicely exemplified in oxy- and deoxyhemo-
globin (Perutz, 1970). This led Chance et al. (1970) to formulate the "membrane
Bohr effect" in order to explain a fast component of the light-dependent proton
uptake by *Chromatium* chromatophores.

A recent debate between the proponents of the chemiosmotic and conformational hypotheses (Mitchell, 1974, 1975; Boyer, 1975) on mechanistic details
of a supposedly proton-translocating ATPase has failed to resolve the essential
differences between the original proposals. Whether the *primary* energy-transducing
step involves a bulk delocalized transmembrane phenomenon or a localized phenomenon in or on the membrane proper (see also Williams, 1975) is still the
crucial problem that remains to be solved.

C. Energy-Dependent Structural Changes
in the Thylakoid Membrane

Different structural changes have been observed in response to energization of
chloroplasts using a variety of techniques, including absorption and light-scattering
measurements, electron microscopy, and application of fluorescent or paramagnetic
"probes", labeled inhibitors and chemical modifiers. No attempt will be made
here to cover all published material. Instead, a few observations that are or may
be relevant for the topic of this section will be discussed. Since the changes recorded
by these techniques in whole chloroplasts are not physically well-defined, the designation "conformational changes" in the strict sense may be premature in these
cases. However, since there is general agreement that such changes are involved,
either directly or indirectly, this term is also used here.

The gross conformational changes in chloroplasts, first observed by Packer
(1963), using light scattering, have been clearly correlated with fluxes of ions
and water (Dilley and Vernon, 1965) and may also be associated with the decrease
of membrane thickness revealed by densitometric scanning of electron micrographs
(Murakami and Packer, 1970). A light-induced modification of the hydrophobic
domains of thylakoid membranes is observed with a spin-labeled hydrocarbon
probe (Torres-Pereira et al., 1974). On the basis of available kinetic data these
"bulk" conformational changes cannot be considered to reflect "intermediary"
high-energy states preceding photophosphorylation. They are rather the result of
alternative energy-utilizing processes.

Earlier successful applications of fluorescent probes in proteins and model membranes have induced the search for high-energy conformations in membranes. At present (for review see KRAAYENHOF et al., 1975) it may be concluded that the process of energization is monitored only by fluorophores that bear an electrical charge and are amphiphilic in nature, like the naphthalene sulfonate and aminoacridine derivatives. The energy-linked responses are largely due to reversible binding of the probes to the organelles. In chloroplasts the energy-dependent binding of aminoacridines has been explained by a proton gradient-induced accumulation inside the thylakoid space (SCHULDINER et al., 1972) and by an electrostatic interaction with negative sites on the energized membrane surface (KRAAYENHOF and FIOLET, 1974). Experiments in favor of the latter possibility include the responses of aminoacridine derivatives that are irreversibly bound to the membrane, of derivatives that are both paramagnetic and fluorescent, and of atebrin conjugated to (nonpermeant) proteins or Sepharose (KRAAYENHOF and SLATER, 1974). If the binding of aminoacridines indeed reflects the creation of fixed negative membrane surface charges under energized conditions, we are dealing with a phenomenon that can be easily accommodated in a mechanism of energy transduction that involves localized charge separation at membrane interfaces and conformational transmission by electrostatic interaction. The observation of pK_a shifts of probes in the course of their interaction with differently charged model membranes and with mitochondrial membranes (MONTAL and GITLER, 1973) is in harmony with this view.

If the primary energy-conserving reaction is intimately coupled to the transfer of reducing equivalents one would expect the kinetics of these processes to be identical. Consequently, one may expect that probes that report concomitant conformational changes (e.g. by the associated charge displacements) will respond with kinetics similar to those of the redox changes. This is certainly not the case for the aminoacridines in chloroplasts (half-time response in the order of 2 sec), probably because the movement of the probe is rate-limiting. Recently, it was found that a covalently attached aminoacridine showed considerably faster responses to illumination (half-time between 40 and 100 msec), indicating that the energy-dependent exposure of negative sites is a rapid process (KRAAYENHOF, 1975), in contrast to the gross structural changes observed.

GIAQUINTA et al. (1973, 1974; see also Chap. II.18) have found that illumination causes a change in exposure of different groups on the chloroplast membrane surface (presumably amino- and carboxyl groups) using the water-soluble chemical modifiers p-(diazonium)-benzene sulfonic acid and N-cyclohexyl-N'-[2-(4-morpholinyl)-ethyl]-carbodiimide metho-p-toluene sulfonate. The binding of diazonium label is dependent on as yet unidentifiable redox changes around photosystem II and is uncoupler-insensitive.

A recent observation by HORTON and CRAMER (1974) that illumination of chloroplasts causes a change of accessibility of cytochromes c-554 ($=f$) and b-559 to ferricyanide suggests the occurrence of conformational rearrangements of the cytochromes relative to the membrane surface.

The above-mentioned structural rearrangements detected in the whole membrane assembly imply that conformational changes do occur in one or more membrane components but detailed mechanistic considerations will have to await the actual identification of the particular components that change their conformation.

D. Energy-Dependent Conformational Changes
in Chloroplast ATPase

The chloroplast membrane-bound ATPase is the only component of the energy-conserving apparatus to which one can specifically assign conformational rearrangements in response to energization. This is not very surprising because this enzyme is relatively loosely bound to the peripheral part of the thylakoid membrane and can be manipulated (see this vol. Chap. III.7) without drastic modification of the ATPase or the membrane.

RYRIE and JAGENDORF (1972) were the first to find a light-induced and uncoupler-sensitive conformational change in membrane-bound ATPase by measuring tritium incorporation after extraction of the ATPase. The inhibitory effect and binding to the ATPase of N-ethyl-maleimide is enhanced by previous energization of the chloroplasts (McCARTY and FAGAN, 1973; see also this vol. Chap. III.7). BAKKER-GRUNWALD and VAN DAM (1974) found that pretreatment of chloroplasts with dithioerythritol (which leads to release of one of the smaller subunits of the ATPase, called the ATPase inhibitor; see this vol. Chap. III.7) and subsequent washing yields a chloroplast preparation that needs only a short light trigger for activation of ATP hydrolysis. Apparently, this treatment removes a slow component in the activation process. Therefore, this preparation certainly offers promising possibilities for kinetic studies of the light-induced conformational changes in the ATPase under variable metabolic conditions. (For a detailed description of the relations between the ATPase conformations and its hydrolytic activity, see this vol. Chap. III.5). LUTZ et al. (1974) conclude from their experiments on the transfer of radioactive label via tightly-bound adenine nucleotides in *Rhodospirillum rubrum* chromatophores that conformational changes occur at the active site of the ATPase. HARRIS and SLATER (1975) recently showed that the tightly-bound nucleotides on chloroplast ATPase, which are not exchangeable in the dark, become rapidly exchangeable with added nucleotides in the light, indicating an energy-dependent exposure of the adenine nucleotide binding sites. The lack of inhibition of this phenomenon by the energy-transfer inhibitor Dio-9, however, does not allow explanations in terms of a single-step conformational change (HARRIS and SLATER, 1975).

The limiting success obtained with electrostatically or electrophoretically interacting probes is mainly due to confusing changes of binding parameters. Obviously, covalently attached reporter groups circumvent these problems and provide a better chance to detect primary energy-linked events. If applied to membrane components that can be easily isolated and reconstituted, like chloroplast ATPase, this approach seems a very promising one. Labeling of the ATPase and subsequent reconstitution will enable a direct and specific means to follow ATPase responses kinetically, provided the ATPase is still enzymatically active and indirect influences from other membrane changes can be ruled out. After finding optimal conditions for labeling and reconstitution of the ATPase and extensive screening of covalent fluorophores it was found that the amine-specific fluorogenic compound fluorescamine is highly suitable for detecting energy-dependent changes in the reconstituted ATPase (KRAAYENHOF and SLATER, 1974). Best results with regard to inhibitory effects of the label on either the recombination with the membrane or the restored

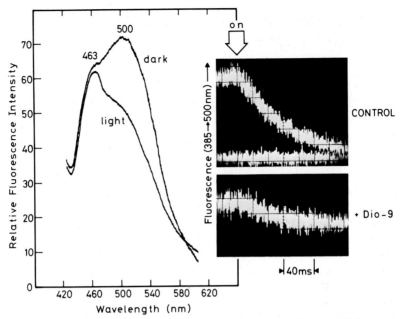

Fig. 1. Light-induced fluorescence emission change of chloroplasts reconstituted with fluoresc-amine-labeled ATPase. (For details see KRAAYENHOF and SLATER, 1974)

ATP synthesis activity were obtained if the labeling with fluorescamine was carried out in situ, i.e. on the membrane-bound ATPase. After the labeling the ATPase was extracted and rapidly reconstituted with freshly-depleted membranes. The fluorescamine-modified enzyme is able to restore ATP synthesis as well as the light-induced pH rise and atebrin-fluorescence quenching in ATPase-depleted membranes to a similar extent as unmodified enzyme. Figure 1 (taken from KRAAYENHOF and SLATER, 1974) shows the fluorescence emission spectra of this reconstituted system, taken in the dark and in the light. The spectra indicate that two components of the fluorophore with maxima at 463 and 500 nm contribute to the emission. These are assigned to hydrophobic and hydrophilic species, respectively (on the basis of model experiments). Illumination causes a partial removal of the hydrophilic species which indicates that peripheral amino groups on the ATPase are pulled inside the protein. The light-induced fluorescence decrease at 500 nm is rapid, with a half response time of 45 msec (the off-reaction has a half-time of 5–8 sec). An important observation is the inhibition of this phenomenon by the energy-transfer inhibitor Dio-9, since this rules out a number of trivial explanations (for instance changes in screening of the fluorophore during light-induced shrinkage). Uncouplers also inhibit this spectral change, but little effect was found from dithioerythritol, adenine nucleotides and inorganic phosphate. No decision is possible as to the intermediate steps between the redox changes and the rapid structural change in the ATPase but it is clear that these must show half-times between those of the redox change and about 40 msec.

A similar method employed for the more deeply buried membrane proteins will undoubtedly encounter severe hindrance by unavoidable modifications of the desired components as well as of the membrane residue. This is, nevertheless, a most challenging method of approach since it is almost the only way to resolve early energy-transducing events in the complicated membrane structure kinetically.

References

Bakker-Grunwald, T., Van Dam, K.: Biochim. Biophys. Acta **347**, 290–298 (1974)

Bennun, A.: Nature New Biology **233**, 5–8 (1971)

Boyer, P.D.: In: Oxidases and Related Redox Systems. King, T.E., Mason, H.S., Morrison, M. (eds.). New York: Wiley, 1965, Vol. II, pp. 994–1008

Boyer, P.D.: In: Dynamics of Energy-Transducing Membranes. Ernster, L., Estabrook, R.W., Slater, E.C. (eds.). Amsterdam: Elsevier, 1974, pp. 289–301

Boyer, P.D.: FEBS Lett. **50**, 91–94 (1975)

Chance, B., Crofts, A.R., Nishimura, M., Price, B.: Europ. J. Biochem. **13**, 364–374 (1970)

Cold Spring Harbor Symposium on Quantitative Biology: Vol. **36**: Cold Spring Harbor Lab. 1972

Dilley, R.A., Vernon, L.P.: Arch. Biochem. Biophys. **111**, 365–375 (1965)

Giaquinta, R., Dilley, R.A., Anderson, B.J.: Biochem. Biophys. Res. Commun. **52**, 1410–1417 (1973)

Giaquinta, R., Dilley, R.A., Anderson, B.J., Horton, P.: J. Bioenerget. **6**, 167–177 (1974)

Gitler, C.: Ann. Rev. Biophys. Bioeng. **1**, 51–92 (1972)

Harris, D.A., Slater, E.C.: Biochim. Biophys. Acta **387**, 335–348 (1975)

Horton, P., Cramer, W.A.: Biochim. Biophys. Acta **368**, 348–360 (1974)

Izawa, S., Ort, D.R., Gould, J.M., Good, N.E.: In: Proc. 3rd Intern. Congr. Photosynthesis. Avron, M. (ed.). Amsterdam: Elsevier, 1975, Vol. I, pp. 449–461

Klotz, I.M., Langerman, N.R., Darnall, D.W.: Ann. Rev. Biochem. **39**, 25–62 (1970)

Koshland, D.E., Neet, K.E.: Ann. Rev. Biochem. **37**, 359–410 (1968)

Kraayenhof, R.: Abstr. 5th Intern. Biophys. Congr. Copenhagen: 1975

Kraayenhof, R., Brocklehurst, J.R., Lee, C.P.: In: Concepts in Biochemical Fluorescence. Chen, R.F., Edelhoch, H. (eds.). Vol. 2. New York: Marcel Dekker, 1976, pp. 767–809

Kraayenhof, R., Fiolet, J.W.T.: In: Dynamics of Energy-Transducing Membranes. Ernster, L., Estabrook, R.W., Slater, E.C. (eds.) Amsterdam: Elsevier, 1974, pp. 355–364

Kraayenhof, R., Slater, E.C.: In: Proc. 3rd Intern. Congr. Photosynthesis. Avron, M. (ed.). Amsterdam: Elsevier, 1975, Vol. II, pp. 985–996

Lutz, H.U., Dahl, J.S., Bachofen, R.: Biochim. Biophys. Acta **347**, 359–370 (1974)

McCarty, R.E., Fagan, J.: Biochemistry **12**, 1503–1507 (1973)

Mitchell, P.: Nature (London) **191**, 144–148 (1961)

Mitchell, P.: FEBS Lett. **43**, 189–194 (1974)

Mitchell, P.: FEBS Lett. **50**, 95–97 (1975)

Montal, M., Gitler, C.: J. Bioenerget. **4**, 363–382 (1973)

Murakami, S., Packer, L.: J. Cell Biol. **47**, 332–351 (1970)

Packer, L.: Biochim. Biophys. Acta **75**, 12–22 (1963)

Perutz, M.F.: Nature (London) **228**, 726–739 (1970)

Ryrie, I.J., Jagendorf, A.T.: J. Biol. Chem. **247**, 4453–4459 (1972)

Schuldiner, S., Rottenberg, H., Avron, M.: Europ. J. Biochem. **25**, 64–70 (1972)

Singer, S.J., Nicolson, G.L.: Science **175**, 720–731 (1972)

Slater, E.C.: Quart. Rev. Biophys. **4**, 35–71 (1971)

Slater, E.C.: In: Dynamics of Energy-Transducing Membranes. Ernster, L., Estabrook, R.W., Slater, E.C. (eds.). Amsterdam: Elsevier, 1974, pp. 1–20

Torres-Pereira, J., Mehlhorn, R., Keith, A.D., Packer, L.: Arch. Biochem. Biophys. **160**, 90–99 (1974)

Williams, R.J.P.: In: Electron Transport and Energy Conservation. Tager, J.M., Papa, S., Quagliariello, E., Slater, E.C. (eds.). Bari: Adriatica Editrice, 1970, pp. 7–23

Williams, R.J.P.: FEBS Lett. **53**, 123–125 (1975)

Wyman, J.: Quart. Rev. Biophys. **1**, 35–80 (1968)

11. Uncoupling of Electron Transport from Phosphorylation in Chloroplasts

N.E. GOOD

In view of the limited space available, the author had to decide whether to present a comprehensive and documented list of uncouplers or to attempt an analysis of uncoupling. He chose the latter. Comprehensive lists of uncouplers are to be found elsewhere (GOOD et al., 1966a; IZAWA and GOOD, 1972; GOOD and IZAWA, 1973).

Most of the numerous undocumented statements contained in the text are based on unpublished observations by the author's associates. However, the author takes full responsibility for the interpretations presented, interpretations which are at best uncertain. The reader should keep in mind that the classification of uncouplers presented here is tentative since it rests on a number of unproven hypotheses.

A. The Concept of Uncoupling

Electron transport in the lamellae of chloroplasts (thylakoids) is usually associated with ATP formation if ADP and orthophosphate (P_i) are present. Since electron transport is required for, and to some extent depends on the phosphorylation of ADP, the two reactions are said to be coupled. "Uncoupling" implies a dissociation of this association. Any treatment which inhibits phosphorylation without causing a corresponding inhibition of electron transport, or increases electron transport without causing a corresponding increase in phosphorylation, must be thought of as "uncoupling" in the broadest sense of the term. However, the term used in this sense is not very useful since it encompasses very different phenomena with very different causes and very different consequences, phenomena which often have nothing to do with the mechanism of phosphorylation. At least two sequential reactions of chloroplasts result in phosphorylation: the photosystem I-dependent oxidation of plastohydroquinone (site I) and the photosystem II-dependent oxidation of water (site II). The rate of electron transport through site I often depends on phosphorylation whereas the rate of a electron transport through site II is usually independent of phosphorylation (GOULD and IZAWA, 1973). Therefore, in any system utilizing only site II, a phosphorylation inhibitor or even the omission of ADP or P_i would "uncouple" by the all-inclusive definition above. Moreover, electron transport in isolated chloroplasts need not involve either site I or site II; any substance which intercepts electrons before a phosphorylation site or donates electrons after a phosphorylation site, therefore, would seem to uncouple. Again the term as defined above is too inclusive.

To avoid these difficulties, the term "uncoupling" as used in the following discussion will be restricted to phenomena which represent malfunctions in normal energy conservation processes. It is assumed that electron transport through sites I and II somehow energizes the lamellar membrane so that it is able to carry out the highly endergonic phosphorylation of ADP. The author, therefore, proposes to define uncoupling as the destruction of the energized state of the lamellar membrane.

B. Criteria of Uncoupling

Typically uncoupling increases the rate of electron transport in the absence of ADP and P_i and decreases the rate of phosphorylation in the presence of ADP and P_i. However, for the reasons already given, neither effect is absolutely diagnostic of uncoupling; nor is an increase in electron transport a consequence of uncoupling when electron transport is itself primarily rate-determining (e.g. at low light intensities or in the presence of many inhibitors). A rigorous demonstration of uncou-

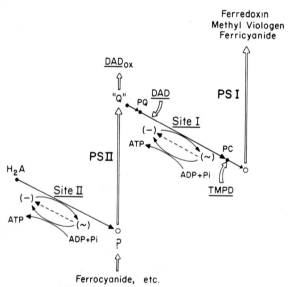

Fig. 1. Some phosphorylating and nonphosphorylating electron transport reactions which have been observed in isolated chloroplasts. *Straight arrows:* transfer of electrons. H_2A represents hydrogen donors such as water or catechol. *DAD:* lipophilic hydrogen donor, diaminodurene, and *DADox:* its oxidation product. Oxidations of hydrogen donors support phosphorylation, probably through the production of H^+ on the inner side of the lamellar membrane. Oxidations of electron donors such as N,N,N^1N^1-tetramethylphenylenediamine (*TMPD*) or ferrocyanide, which do not yield H^+, do not support phosphorylation (HAUSKA et al., 1973; IZAWA and ORT, 1974). *Curved arrows:* conservation of redox energy in the "energized" state of the membrane ($\sim$), concentration gradient energy which is used for the phosphorylation of ADP by the coupling factor. Uncoupling, as defined here: *dotted arrows*. By directly or indirectly dissipating ion activity gradients, it causes the energized state ($\sim$) to revert to the unenergized state ($-$). "*Q*": primary acceptor of electrons from photosystem II (PS II), *PQ:* plastoquinone, and *PC:* plastocyanine

pling involves three kinds of evidence. First, it must be shown that the increase in electron transport or decrease in phosphorylation result in a lowered efficiency of phosphorylation. Second, it must be shown that the lowered efficiency is not a result of an inhibition of phosphorylation per se (energy transfer inhibition). Third, it must be shown that the observed electron transport is not bypassing one or both sites of phosphorylation by following some of the truncated pathways illustrated in Figure 1.

If the putative uncoupler is a plausible acceptor or donor of electrons, the following additional tests can be made: (1) Circumvention of site I by a lipophilic acceptor such as DADox can be precluded either if the reaction being investigated is sensitive to inhibitors like KCN (OUITRAKUL and IZAWA, 1973) or poly-L-lysine (BRAND et al., 1972) which block photosystem I or if it is sensitive to dibromothymoquinone which prevents transport between the two photosystems (TREBST et al., 1970). (2) Electron transport can be separated from phosphorylation by preilluminating the chloroplasts in the presence of an electron acceptor, then adding ADP and P_i. The dark decay of the short-lived energized state, X_e, is speeded by uncouplers (HIND and JAGENDORF, 1963) but is unaffected by electron acceptors or electron donors. Thus true uncoupling can be demonstrated if the uncoupler is added after the light is turned off. (There is some danger in the use of this last criterion since we cannot be sure that the energized state responsible for delayed phosphorylation is the same as the energized state responsible for steady-state phosphorylation.)

C. Types of Uncoupling by Typical Uncouplers

In considering the possible mechanisms of uncoupling the author has assumed that electron transport produces hydrogen ion activity gradients and that these gradients are used by the well-characterized coupling factor to drive phosphorylation (MITCHELL, 1966). If we accept this general model of the coupling process, the plausibility of two quite different modes of uncoupling becomes obvious: either the coupling factor may malfunction by carrying ions without phosphorylating ADP, or other parts of the membrane may become "leaky" and unable to maintain the required gradients. There is a good deal of evidence implicating both modes of uncoupling, depending on the uncoupler used. However, many uncouplers cannot be thus classified as yet, either because they have not been sufficiently investigated or because their bevavior is imperfectly understood.

I. Malfunctions of the Coupling Factor

1. Basal (Nonphosphorylating) Electron Transport

Isolated chloroplasts are always somewhat loosely coupled in that they transport electrons at an appreciable rate in the absence of ADP and P_i and in the absence of uncouplers. Much of this basal electron transport seems to be related to the

Table 1. Some chloroplast uncouplers

Substance	Uncoupling concentration[a]	Special problems	References
Ammonium salts	10^{-3}–10^{-2} M	Light scattering can decrease. Inhibit water oxidation at high pH	KROGMANN et al. (1959) IZAWA and GOOD (1966) IZAWA et al. (1969)
Methylamine salts	10^{-3}–10^{-2} M	Light scattering can decrease	GOOD (1960)
Poly-L-lysine	10^{-5}–10^{-4} M[b]	Inhibits photosystem I by reacting with plasto-cyanin if chloroplast are swollen	DILLEY and PLATT (1968) BRAND et al. (1972)
Atebrin	10^{-6}–10^{-5} M	Light scattering can increase. Uncoupling not fully reversible	AVRON and SHAVIT (1963) DILLEY and VERNON (1965)
Carbonylcyanide 3-chlorophenyl-hydrazone (CCCP)	10^{-6}–10^{-5} M	Inhibits electron transport at low pH or high concentrations	HEYTLER (1963)
2,6-Dichloro-phenolindophenol	10^{-4} M	Ineffective at high pH. Absorbs actinic light	SAHA et al. (1971)
2,3′,6-Trichloro-phenolindophenol	$3 \cdot 10^{-5}$ M	Ineffective at high pH. Absorbs actinic light	KROGMANN and JAGENDORF (1959)
Desaspidin	10^{-7}–10^{-6} M	Uncoupler readily photo-xidized	GROMET-ELHANAN and AVRON (1966)
Nigericin	10^{-7}–10^{-6} M	Requires high K^+	KARLISH et al. (1969)
Gramicidin D	10^{-7}–10^{-6} M	None obvious (uses K^+ or Na^+)	KARLISH et al. (1969)
Arsenate	Higher than P_i	Requires ADP and active coupling factor	AVRON and SHAVIT (1965)

[a] These values are to be viewed with caution. For instance, amines are required at higher concentrations and indophenols at lower concentrations as the pH is lowered. Moreover the more active uncouplers tend to be strongly absorbed by the chloroplasts, in which case the amount of uncoupler relative to the chloroplast volume is more significant than the overall concentration in the reaction mixture
[b] In lysine monomer equivalents

activities of the coupling factor. The rate of the process has the same sharp pH dependence (optimum just above 8.0) as phosphorylation. In contrast, most kinds of deliberately uncoupled electron transport have lower and broader optima (GOOD et al., 1966a). Moreover the basal transport is markedly decreased by the presence of either ADP or ATP, both substrates of the coupling factor (AVRON et al., 1958; IZAWA and GOOD, 1969). Probably the basal transport results from some inadvertent uncoupling associated with the isolation procedure or with the presence of an abnormal medium surrounding the lamellae. Nevertheless, it is possible that the basal transport is a normal characteristic of electron transport in vivo, since the very large energies available from excited chlorophyll may make a portion of the transport essentially irreversible and, therefore, independent of the chemical potentials of the intermediate reactants.

2. Removal of the Coupling Factor

In the absence of cations, especially divalent cations, the coupling factor is released from the lamellae. Electron transport rates are then high whether or not ADP and P_i are present and no ATP is formed (AVRON, 1963). Clearly removal of the coupling factor leaves "holes" or other lesions in the membrane which allow the energized state to be dissipated, probably by ion leakage.

3. Phosphate Analogs Arsenate and Thiophosphate

There is no doubt whatsoever that these uncouplers cause the coupling factor to malfunction since uncoupling by them requires active coupling factor and the presence of ADP. Moreover, they are competitive with P_i (AVRON and JAGENDORF, 1959; AVRON and SHAVIT, 1965). It is reasonable to assume that arsenate and thiophosphate replace phosphate in the phosphorylation reaction, that unstable arsenylated or thiophosphorylated intermediates involving ADP are formed by the coupling factor, and that these spontaneously decompose.

II. Malfunctions of the Membrane

1. Neutralization of the Internal H^+ by Amines

Ammonia and other amines with lipid-soluble unprotonated and uncharged forms can freely permeate the lipoidal lamellar membranes. Once inside they take up accumulated hydrogen ions and are converted into the corresponding ammonium ions. The accumulation of ammonium ions inside requires an inward flux of anions in order to preserve electrical neutrality (KROGMANN et al., 1959; GOOD, 1960). Hence there can be a great accumulation of ammonium salts during electron transport if permeant anions are accessible in the medium (CROFTS, 1967; DEAMER and PACKER, 1969). This leads to an osmotic uptake of water and the vast swelling of thylakoids which is characteristic of amine uncoupling (IZAWA and GOOD, 1966). Small vesicles made by fragmenting chloroplasts or the photosynthetic membranes of bacteria are not much uncoupled by ammonia or amines. This observation has been rationalized by postulating that the vesicles are less permeable to anions so that a membrane potential (due to the uncompensated increase in cations) is preserved and with it a large part of the electrochemical driving force for phosphorylation (MCCARTY and COLEMAN, 1969). Amines are useful uncouplers of chloroplasts but at high pHs (above 8.0) they tend to inhibit electron transport by irreversibly inactivating the mechanism of water oxidation (IZAWA et al., 1969). Weak amines such as aniline and polar amines such as amino acids do not uncouple. The polar buffer tris (hydroxymethyl)-aminomethane uncouples only slightly (GOOD et al., 1966b).

2. Permeation of the Lamellar Membrane by H^+

Weak acids having anions which are relatively lipid soluble may uncouple. Presumably the protonatable anions dissolve in the material of the membrane where they serve as H^+ carriers. Nevertheless, it remains difficult to predict the uncoupling efficacy of weak acids on the basis of their physical properties. Carbonylcyanide phenylhydrazones, e.g. the 3-chlorophenylhydrazone (CCCP) and the 4-trifluoro-methoxyphenylhydrazone (FCCP), are particularly powerful uncouplers of chloroplasts and mitochondria (HEYTLER, 1963). In keeping with the proposed mechanism of uncoupling, it has been shown that they do indeed make phospholipid membranes permeable to H^+ (HOPFER et al., 1968). Unfortunately, they are very reactive substances and, in addition to uncoupling, they often inhibit electron transport.

Indophenols seem to uncouple by a similar mechanism. 2,6-dichlorophenolindophenol and 2,3',6-trichlorophenolindophenol are electron acceptors which have been used extensively in the study of electron transport because the disappearance of the blue color on reduction is so easily observed. However, these indophenols are fairly potent uncouplers (KROGMANN and JAGENDORF, 1959). As electron acceptors their apparent uncoupling action is suspect since they could accept electrons before site I, and to a limited extent they probably do. However, indophenols are true uncouplers since they inhibit the whole of phosphorylation, not just that part contributed by site I; moreover, they inhibit post-illumination ATP formation in the manner described above (HIND and JAGENDORF, 1963). Indophenols are not very satisfactory uncouplers since their uncoupling action depends on the existence of some of the dye in the protonated form. Therefore, the uncoupling depends on the pH and is weak above 8.0. Furthermore the intense colors of the dyes interfere with optical measurements and even with the absorption of light by chlorophyll.

3. Transmembrane Exchanges of H^+ for K^+

Neither 2,4-dinitrophenol, which makes membranes permeable to H^+ and uncouples mitochondria (HOPFER et al., 1968), nor valinomycin which makes membranes permeable to K^+, are by themselves good uncouplers of chloroplasts. However in combination they uncouple quite well. This observation suggests that the exchange of H^+ for K^+ across the membrane is particularly effective in uncoupling. The suggestion is further supported by the fact that nigericin and gramicidin D, ionophores which by themselves do catalyze such exchanges, are among the best uncouplers known (KARLISH et al., 1969). These latter uncouplers are useful not only because they uncouple so well but also because their action is so specific. In addition to abolishing phosphorylation when they are added at very low concentrations, they permit electron transport to proceed continuously at very high rates and they do not interfere with most assay systems.

III. Uncoupling by Unknown Mechanisms

Atebrin (also known as Quinacrin) represents a class of complex aliphatic amine uncouplers containing large heterocyclic rings. Included in this class are other physiologically active substances such as chlorpromazine (AVRON and SHAVIT,

1963), brucine (IZAWA and GOOD, 1968) and chloroquine (HIND et al., 1969). The mechanism by which they uncouple is obscure. In the presence of atebrin, electron transport causes a massive extrusion of salts and of water with a consequent collapse of the internal space of the thylakoid (DILLEY and VERNON, 1965). No convincing explanation of this phenomena has been proposed. Uncoupling by atebrin probably does not involve its movement across the membrane. Rather, it seems to become membrane-bound, since its fluorescence is quenched during electron transport (KRAAYENHOF et al., 1972), although perhaps not at uncoupling concentrations. Apparently atebrin acts simply by contact with the outer surface of the lamella; it seems to reversibly uncouple even if it is covalently bonded to large sepharose beads (KRAAYENHOF and SLATER, 1974). Most kinds of uncoupling speed the reequilibration of ions after the light is turned off and, therefore, hasten the decay of the pH gradient across the membrane but uncoupling by atebrin does not, at least not at pHs where phosphorylation is optimal. One is tempted to speculate that atebrin may interact with some portion of the coupling factor which is accessible from the medium, perhaps that portion which is accessible to ADP. On the other hand, atebrin causes chloroplasts to shrink during electron transport even after removal of the coupling factor (IZAWA and GOOD, 1966). This latter observation implies that the salt extrusion associated with atebrin uncoupling is independent of coupling factor, the only known site of involvement of ADP^+ in the lamellae.

D. A General Consideration of Mechanisms of Uncoupling

Early students of oxidative phosphorylation and photophosphorylation usually thought of uncoupling in terms of the catalysis of the breakdown of chemical intermediates. With the advent of the chemiosmotic theory of phosphorylation, it became clear that many kinds of uncoupling could be explained in terms of induced permeability of membranes to ions (MITCHELL, 1966). It should be pointed out, however, that the earlier concept of uncoupling by chemical catalysis is in no way incompatible with the chemiosmotic theory since the breakdown of a phosphorylation intermediate would, in terms of the chemiosmotic theory, simply constitute a malfunction of the coupling factor. The role of uncouplers as modifiers of ion transport phenomena cannot be taken as proof of a primary involvement of all uncouplers in changing permeabilities to ions if the energized state of the membrane, ion gradients and ATP formation are intimately and reversibly associated.

There are some striking differences in the behavior of different classes of uncouplers and, if all of these uncouplers act through ion diffusion, quite different kinds of "leaks" must be produced. For instance, there is a marked qualitative difference between the effects of low levels of carbonylcyanide phenylhydrazones and the effects of other uncouplers. CCCP totally inhibits phosphorylation at low light intensities, or during short illumination times, at concentrations which have little effect at higher intensities or longer times. Most other uncouplers inhibit phosphorylation equally at all light intensities (SAHA et al., 1970) and introduce a very small lag in the onset of phosphorylation, a few milliseconds at most.

There are also striking differences in the effects of uncouplers on the electron transport-dependent volume changes in chloroplasts (salt movements). Indeed there seems to be a hierarchy of effects: carbonylcyanide phenylhydrazones abolish all volume changes whether or not amines or atebrin are present. Amines cause massive swelling whether or not atebrin is present. Although removal of coupling factor does not of itself induce transport-related volume changes, it in no way interferes with amine-swelling or atebrin-shrinking (IZAWA and GOOD, 1966). As has already been pointed out, most uncouplers speed the decay of the electron transport-dependent pH gradient across the lamella membrane but, at phosphorylation pHs atebrin does not. None of the differences can be explained at present.

References

Avron, M.: Biochim. Biophys. Acta **77**, 699–702 (1963)
Avron, M., Jagendorf, A.T.: J. Biol. Chem. **234**, 967–972 (1959)
Avron, M., Krogmann, D.W., Jagendorf, A.T.: Biochim. Biophys. Acta **30**, 144–153 (1958)
Avron, M., Shavit, N.: Proc. Nat. Acad. Sci. U.S. **1145**, 611–618 (1963)
Avron, M., Shavit, N.: Biochim. Biophys. Acta **109**, 317–331 (1965)
Brand, J., Baszynski, T., Crane, F., Krogmann, D.: J. Biol. Chem. **247**, 2814–2819 (1972)
Crofts, A.R.: J. Biol. Chem. **242**, 3352–3359 (1967)
Deamer, D.W., Packer, L.: Biochim. Biophys. Acta **172**, 539–545 (1969)
Dilley, R.A., Platt, J.S.: Biochemistry **7**, 338–346 (1968)
Dilley, R.A., Vernon, L.P.: Arch. Biochem. Biophys. **111**, 365–375 (1965)
Good, N.E.: Biochim. Biophys. Acta **40**, 502–517 (1960)
Good, N.E., Izawa, S.: In: Metabolic Inhibitors. Hochster, R.M., Kates, M., Quastel, J.H. (eds.). New York-London: Academic Press, 1973, Vol. IV, pp. 179–214
Good, N.E., Izawa, S., Hind, G.: Current Topics Bioenerg. **1**, 75–112 (1966a)
Good, N.E., Winget, G.D., Winter, W., Connolly, T.N., Izawa, S., Singh, R.M.M.: Biochemistry **5**, 467–477 (1966b)
Gould, J.M., Izawa, S.: Biochim. Biophys. Acta **314**, 211–223 (1973)
Gromet-Elhanan, Z., Avron, M.: Plant Physiol. **41**, 1231–1236 (1966)
Hauska, G., Trebst, A., Draber, W.: Biochim. Biophys. Acta **305**, 632– (1973)
Heytler, P.G.: Biochemistry **2**, 357–361 (1963)
Hind, G., Jagendorf, A.T.: Z. Naturforsch. **18B**, 689–694 (1963)
Hind, G., Nakatani, H.Y., Izawa, S.: Biochim. Biophys. Acta **172**, 277–289 (1969)
Hopfer, U., Lehninger, A.L., Thompson, T.E.: Proc. Nat. Acad. Sci. U.S. **59**, 484–490 (1968)
Izawa, S., Good, N.E.: Plant Physiol. **41**, 533–543 (1966)
Izawa, S., Good, N.E.: Biochim. Biophys. Acta **162**, 380–391 (1968)
Izawa, S., Good, N.E.: In: Progress in Photosynthesis Research. Metzner, H. (ed.). Tübingen: Laupp, 1969, Vol. III, pp. 1288–1298
Izawa, S., Good, N.E.: Methods Enzymol. **24B**, 355–377 (1970)
Izawa, S., Heath, R.L., Hind, G.: Biochim. Biophys. Acta **180**, 388–398 (1969)
Izawa, S., Ort, D.R.: Biochim. Biophys. Acta **357**, 127–143 (1974)
Karlish, S.J.D., Shavit, N., Avron, M.: Europ. J. Biochem. **9**, 291–298 (1969)
Kraayenhof, R., Izawa, S., Chance, B.: Plant Physiol. **50**, 713–718 (1972)
Kraayenhof, R., Slater, E.C.: In: Proc. 3rd Intern. Congr. Photosynthesis. Avron, M. (ed.). Amsterdam: Elsevier, 1975, Vol. II, pp. 985–996
Krogmann, D.W., Jagendorf, A.T.: Plant Physiol. **34**, 277–282 (1959)
Krogmann, D.W., Jagendorf, A.T., Avron, M.: Plant Physiol. **34**, 272–277 (1959)
McCarty, R.E., Coleman, C.H.: J. Biol. Chem. **244**, 2492–2498 (1969)
Mitchell, P.: Biol. Rev. Cambridge Phil. Soc. **41**, 445–502 (1966)
Ouitrakul, R., Izawa, S.: Biochim. Biophys. Acta **305**, 105–118 (1973)
Saha, S., Izawa, S., Good, N.E.: Biochim. Biophys. Acta **233**, 158–164 (1970)
Saha, S., Ouitrakul, R., Izawa, S., Good, N.E.: J. Biol. Chem. **246**, 3204–3209 (1971)
Shavit, N., Dilley, R.A., San Pietro, A.: Biochemistry **7**, 2356–2363 (1968)
Trebst, A., Harth, E., Draber, W.: Z. Naturforsch. **25B**, 1157–1159 (1970)

12. Energy Transfer Inhibitors of Photophosphorylation in Chloroplasts

R.E. McCarty

A. Definition of Energy Transfer Inhibitors

The synthesis of ATP from ADP and P_i by chloroplasts (photophosphorylation) is coupled in a complex manner to light-dependent electron flow. Although nonidentified intermediates could exist, the evidence supports the concept (MITCHELL, 1961, 1966) that an electrochemical gradient in H^+ serves to couple phosphorylation to electron flow (for recent reviews, see HIND and McCARTY, 1973; JAGENDORF, 1975; McCARTY, 1976). Reagents which either prevent the formation of the H^+ gradient (inhibitors of electron flow) or promote its dissipation (uncouplers) inhibit photophosphorylation. In addition, a more terminal step in photophosphorylation may be inhibited without a *direct* effect on the rate of electron flow or on the magnitude of the H^+ gradient. Reagents which block phosphorylation in this manner are called energy transfer inhibitors.

A comparison of the effects of inhibitors of electron flow, uncouplers and energy transfer inhibitors on electron transport and photophosphorylation in isolated chloroplasts (thylakoids) is given in Figure 1. It may be seen that the three kinds of inhibitors can be readily distinguished even though all three inhibit photophosphorylation. The main distinguishing feature of an energy transfer inhibitor is that it decreases the rate of electron flow only under conditions where active phosphorylation can take place. Electron flow in the absence of P_i (basal electron flow) or in the presence of uncouplers is either not affected or only slightly inhibited.

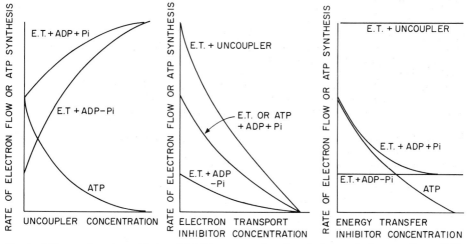

Fig. 1. Effects of various kinds of inhibitors on phosphorylation and electron flow. *E.T.:* electron transport rate, and *ATP:* rate of phosphorylation

In contrast, a direct inhibitor of electron flow would abolish electron flow under all conditions, whereas an uncoupler should enhance electron flow. An energy transfer inhibitor thus attenuates electron flow in an indirect manner. The rate of electron flow is low under nonphosphorylating conditions and the pH gradient across thylakoid membranes is high. Phosphorylation decreases the magnitude of the pH gradient (Pick et al., 1973; Portis and McCarty, 1974) and stimulates the rate of electron flow. By blocking a terminal step in ATP formation, energy transfer inhibitors also prevent the enhancement of the rate of electron flow due to phosphorylation. Since uncouplers reduce the magnitude of the pH gradient and overcome the rate limiting step of electron flow, energy transfer inhibitors would have no effect on uncoupled electron flow.

Good et al. (1966) and Izawa and Good (1972) reviewed energy transfer inhibition of photophosphorylation.

B. Energy Transfer Inhibitors Which Probably Exert Their Effects on Coupling Factor 1

Coupling factor 1 (CF_1) plays at least two roles in photophosphorylation: catalysis of the terminal phosphorylation reaction and the prevention of rapid leakage of H^+ from the thylakoids (McCarty and Racker, 1966, 1967). For a review on the function of CF_1 in photophosphorylation, see Jagendorf (1975). Reagents which interfere with the catalytic function of CF_1 would act as energy-transfer inhibitors.

I. Dio-9

Although oligomycin and aurovertin, inhibitors of energy transfer in oxidative phosphorylation by mitochondria, have little to no effect on photophosphorylation, Dio-9 inhibits photophosphorylation (McCarty et al., 1965) as well as oxidative phosphorylation (Guillory, 1964). The structure of Dio-9 is unknown and preparations contain several components (R.J. Guillory, pers. comm.). Although its usefulness suffers from lack of availability, Dio-9 is an almost ideal energy transfer inhibitor. It has little effect on basal electron flow, no effect at all on uncoupled electron flow but abolishes the stimulation of electron flow by phosphorylation. Furthermore, light-dependent H^+ uptake is either not inhibited (McCarty and Racker, 1966) or even stimulated by Dio-9 (Karlish and Avron, 1971; McCarty et al., 1971). Moreover, the ATPase activities of either chloroplasts or of trypsin-treated CF_1 (Vambutas and Racker, 1965) are sensitive to Dio-9.

II. Phlorizin and Related Compounds

Phlorizin (4,4',6'-trihydroxy-2'-glucosidodihydrochalcone) is present in the root back of certain Rosaceae, including that of apple, pear, plum and cherry trees.

Fig. 2. Structures of phlorizin and 4'-deoxyphlorizin

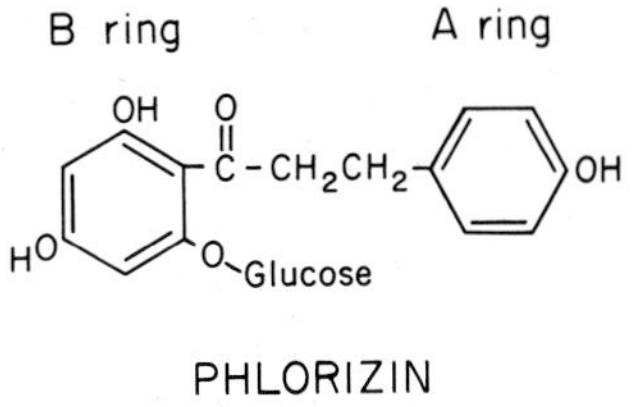

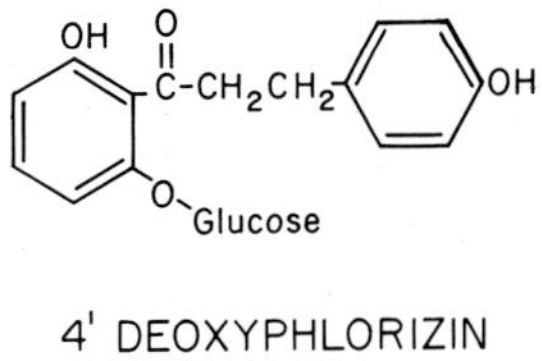

B ring   A ring

PHLORIZIN

4' DEOXYPHLORIZIN

The structure of phlorizin is shown in Figure 2. For over 30 years, phlorizin has been known to be an inhibitor of glycogen and starch phosphorylases as well as to elicit glucosuria in animals. Phlorizin is also an inhibitor of photophosphorylation (IZAWA et al., 1966). Phlorizin (0.05 to 2.0 mM) has no effect on nonphosphorylating electron flow (assayed in the presence of ADP, but absence of added phosphate) or uncoupled electron flow, but inhibits phosphorylating electron flow. It also inhibits (WINGET et al., 1969) the light- and SH-activated ATPase activity of chloroplasts (HOCH and MARTIN, 1963; MARCHANT and PACKER, 1963; PETRACK et al., 1965), a reaction which requires CF_1 (McCARTY and RACKER, 1966). Furthermore, the Ca^{2+}-dependent ATPase activity of purified CF_1 (VAMBUTAS and RACKER, 1965) is sensitive to phlorizin (50% inhibition at about 1 mM; McCARTY, unpublished).

If the rate of phosphorylation was reduced by decreasing either the light intensity or the ADP concentration, the sensitivity of phosphorylation to phlorizin was decreased, probably because the terminal reactions of phosphorylation are less rate-limiting under these conditions. In contrast, phosphorylation at low phosphate concentrations was more sensitive to phlorizin. The type of inhibition with respect to P_i concentration, as revealed by double reciprocal plots, was mixed. In view of these results and of the ability of phlorizin to inhibit other phosphate transfer reactions, phlorizin may inhibit photophosphorylation by interfering with P_i utilization in some manner.

Analogs of phlorizin were synthesized and their effects on phosphorylation and electron flow tested (WINGET et al., 1969). 4'-Deoxyphlorizin (see Fig. 2 for its structure) is also an excellent energy transfer inhibitor and is nearly ten-times more potent than phlorizin. Modifications of the A ring, generally resulted in compounds which were more effective energy transfer inhibitors than phlorizin. Although 4'-deoxyphloretin-2'-galactoside was more potent than 4'-deoxyphlorizin, phloretin itself inhibited all kinds of electron flow. However, at low concentrations (less than 100 μM) phloretin appears to have a much stronger effect on phosphorylating electron flow than on basal electron flow. At pH 7.0, electron flow was insensitive to phloretin (URIBE, 1970) and even at pH 8.0, basal electron flow was inhibited less than 50% at a concentration of phloretin which abolished phosphorylation.

Izawa and Good (1968) suggested that basal electron flow, which is phlorizin and Dio-9 insensitive, may represent a pathway which is not coupled to phosphorylation. If the basal electron flow rate were subtracted from the overall rate of electron flow under phosphorylating conditions, the calculated P/e_2 ratios approached 2. However, since pH gradients up to 3.5 units can be supported by basal electron flow, it hardly seems valid to subtract the basal flow. Some correction is perhaps justified. The rate of basal electron flow in chloroplasts supplemented with ATP is proportional to the internal H^+ concentration (Portis et al., 1975). Phosphorylation decreases ΔpH and, therefore, the internal H^+ concentration (Pick et al., 1973; Portis and McCarty, 1974). In consequence, the basal flux should be inhibited by phosphorylation. Using the internal H^+ concentration, we calculated the basal flux under phosphorylating conditions. By correcting the total electron flow rate for this basal flow, a P/e_2 value near 1.3 was obtained (A.R. Portis, Jr., unpublished).

III. Antisera to Coupling Factor I and its Subunits

A monospecific mouse antiserum to CF_1 inhibited the Ca^{2+}-dependent ATPase of CF_1 as well as all chloroplast reactions involving adenine nucleotides (McCarty and Racker, 1966). In contrast, H^+ uptake supported by pyocyanine-dependent cyclic electron flow was insensitive to the antiserum. Phosphorylating electron transport was also insensitive to the antiserum. It should be pointed out, however, that the chloroplasts used in these experiments had been swollen in 10 mM NaCl and, in these chloroplasts, phosphorylation did not enhance electron flow.

More recently, quite satisfactory antisera to CF_1 have been elicited in rabbits and mice are no longer used. Some rabbit sera inhibited H^+ uptake (McCarty et al., 1971) although phosphorylation was much more sensitive. However, the portion of H^+ uptake which is enhanced by ATP was quite sensitive to CF_1 antisera (McCarty et al., 1971; Nelson et al., 1973). Moreover, phosphorylating electron flow, but not basal electron flow, in nonswollen chloroplasts was sensitive to the antiserum (McCarty, unpublished).

CF_1 contains five components or subunits, denoted α, β, γ, δ and ε in order of increasing mobility (decreasing molecular weight) on acrylamide gel electrophoresis in the presence of sodium dodecyl sulfate (Racker et al., 1971). Noncrossreacting antisera to the purified subunits were elicited in rabbits. Photophosphorylation was sensitive only to the anti-α or anti-γ sera (Nelson et al., 1973). The anti-β serum, although it did not inhibit phosphorylation, did interact with CF_1 on the membrane since it caused agglutination of the chloroplasts. Interestingly, none of the antisera alone inhibited the Ca^{2+}-dependent ATPase of CF_1. However, a combination of anti-α and anti-γ was quite effective in inhibiting the ATPase. H^+ uptake was either unaffected or stimulated (anti-γ) by the antisera to the subunits. Anti-α, like anti-CF_1 abolished the stimulation of the extent of H^+ uptake by ATP.

IV. N-Ethylmaleimide

McCarty et al. (1971) suggested that the stimulation of the extent of H^+ uptake by ATP or ADP might be the consequence of the binding of these nucleotides to CF_1. The binding of ATP in the light was proposed to alter the conformation of CF_1 which in turn reduces the permeability of thylakoids to H^+. To test this hypothesis, a search was initiated for a convenient way in which to monitor interactions between chloroplast-bound CF_1 and adenine nucleotides.

RYRIE and JAGENDORF (1971 a, 1972) showed that CF_1 in chloroplasts undergoes energy-dependent conformational changes. Incorporation of ^3H into CF_1 was detected when chloroplasts were illuminated in the presence of ^3H_2O. ATP, however, had little effect on this ^3H-exchange. We reasoned that the conformational change could expose a group (or groups) to reaction with alkylating reagents such as N-ethylmaleimide (NEM). Furthermore, if we were very lucky, the reaction of this group with NEM would inhibit phosphorylation.

NEM (1 mM) partially inhibited phosphorylation only when chloroplasts were illuminated with this reagent. Incubation of chloroplasts with NEM in the dark had no effect (McCARTY et al., 1972). Uncoupled electron flow was not affected by the NEM and light treatment, but coupled electron flow was somewhat reduced. The extent of H^+ uptake and ΔpH (PORTIS et al., 1975) was also unaffected by NEM, except that the stimulation of uptake by ATP was decreased. The Ca^{2+}-dependent ATPase of CF_1 in NEM-inhibited chloroplasts was also inhibited by 20–40%.

The inhibition of phosphorylation by NEM was then correlated to its reaction with CF_1. Chloroplasts were first incubated with NEM in the dark and then exposed to ^3H-NEM in the light. A small amount of NEM (approximately 1 mol/mol CF_1) was incorporated into CF_1 in the light (McCARTY and FAGAN, 1973). Nearly all of the NEM was incorporated into the γ subunit. The incorporation of NEM into CF_1 and the inhibition of phosphorylation by this reagent responded to a variety of treatments or conditions in a very similar way. For example, both incorporation and inhibition had similar kinetics, similar dependence on NEM concentration and were abolished by uncoupling agents. It is thus very likely that the inhibition of phosphorylation by NEM is the consequence of its reaction of a group in the γ-subunit. Although this group is likely to be a SH group, NEM can react with the ε-NH_2 of lysine. However, dithio-bis-nitrobenzoic acid also inhibits phosphorylation in a light-dependent way, indicating that the NEM-reactive group could be an SH.

A number of other maleimide derivatives inhibit phosphorylation in a light-dependent manner. More hydrophobic maleimides, such as N-phenylmaleimide are slightly more effective than NEM. However, bifunctional maleimides, such as o-phenylenediaminedimaleimide, are about one-thousand times more potent than NEM and also react more quickly (McCARTY and WILTON, unpublished).

MAGNUSSON and McCARTY (1975) exploited the reaction of NEM with CF_1 to study the interactions of the membrane-bound enzyme with nucleotides. Low concentrations (5–30 µM) of ATP or ADP protect phosphorylation from inhibition by NEM and phosphate enhances their effectiveness. Thus, ATP and ADP can modify CF_1 in the light.

V. ATP

In their early studies on the coupling of phosphorylation to electron flow, AVRON et al. (1958) noted that low concentrations of ATP (1–20 µM) inhibited electron flow. Uncoupled electron flow was insensitive to ATP. Much later, it was reported (McCarty et al., 1971; Telfer and Evans, 1972) that similar low concentrations of ATP enhance the extent of H^+ uptake or the magnitude of ΔpH (Portis and McCarty, 1974). In view of the sensitivity of these effects of ATP to NEM or anti-CF_1 sera, CF_1 is involved, possibly by regulating the permeability of chloroplasts to H^+. It was proposed (Portis et al., 1975), that at ΔpH values of about 2.8 or higher, CF_1 undergoes conformational changes which allow leakage of H^+ and, therefore, faster rates of electron flow. ATP appeared to prevent these changes and consequently inhibited electron flow.

Shavit and Herscovici (1970) showed that ATP at much higher concentrations (1–100 mM) is an inhibitor of phosphorylation and phosphorylating electron flow. Half-maximal inhibition was achieved at about 3 mM ATP. Basal electron flow was much less sensitive to ATP and uncoupled electron flow was totally insensitive. Thus, ATP appears to act as an energy transfer inhibitor.

Since increasing the ADP (or P_i) concentration had no effect on the inhibition of electron flow by ATP, simple competition for nucleotide binding sites in CF_1 cannot explain the inhibition. However, high ATP concentrations in the medium could prevent the dissociation of the newly synthesized ATP from CF_1 thereby inhibiting phosphorylation. Boyer (1974) suggested that the dissociation of ATP from coupling factors may be the major energy-requiring step in phosphorylation. CF_1 could also contain a regulatory site which shuts down phosphorylation when the ATP concentration becomes too high.

C. Energy Transfer Inhibitors Whose Site of Action is Unknown

I. Dicyclohexylcarbodiimide

N,N'-Dicyclohexylcarbodiimide (DCCD), a useful reagent in organic synthesis, is also a useful inhibitor of oxidative phosphorylation (Beechey et al., 1966) and photophosphorylation (McCarty and Racker, 1967; Uribe, 1972). DCCD inhibited phosphorylation in chloroplasts 50% at about 20 µM. Coupled electron flow was also inhibited such that, at lower DCCD concentrations, the P/e_2 ratio was only slightly changed. Basal electron flow assayed in the absence of ADP and P_i can also be inhibited and this inhibition was not fully overcome by NH_4Cl (McCarty and Racker, 1967). Uribe (1972), however, reported that DCCD had no effect on basal electron flow (ferricyanide reduction) unless the chloroplasts were illuminated in the presence of DCCD prior to assay. However, it is not clear whether Uribe measured basal electron transport in the presence of ADP. It is possible that ADP could protect basal electron flow from inhibition by DCCD. Furthermore, it may be that more potent uncouplers than NH_4Cl could reverse the inhibition of basal electron flow by DCCD. In view of the fact (McCarty and Racker, 1967) that DCCD can stimulate H^+ uptake, one might expect that basal electron flow would be inhibited. DCCD is a potent inhibitor of the light- and SH-activated Mg^{2+}-ATPase of chloroplasts, but has no effect on the Ca^{2+}-ATPase of soluble CF_1. Moreover, hydrophylic carbodiimides, such as N-ethyl-N'-

(dimethylaminopropyl)-carbodiimide (EDAC), do not affect phosphorylation like DCCD (URIBE, 1972). EDAC is an inhibitor of electron transport (McCARTY, 1974) which is most effective when the chloroplasts are illuminated with the EDAC prior to assay. Perhaps the inhibition of basal electron flow by DCCD upon prolonged illumination is due to a slow reaction of DCCD with the EDAC-sensitive site. Although the site of action of DCCD is unknown, the lack of its effect on soluble ATPase and the fact that only lipophilic carbodiimides are effective suggest that some hydrophobic membrane component rather than CF_1 is involved.

The ability of chloroplasts to phosphorylate (AVRON, 1963) or to carry out light-dependent H^+ uptake (McCARTY and RACKER, 1966) was lost when part of the CF_1 was extracted by exposure of chloroplasts to dilute solutions of EDTA. Addition of CF_1 partially restored these activities. DCCD also restored H^+ uptake and enhances phosphorylation in EDTA-treated chloroplasts (McCARTY and RACKER, 1967; URIBE, 1972). The removal of CF_1 makes thylakoid membranes more permeable to H^+ and DCCD in some unknown manner blocks the flow of accumulated H^+ out of the chloroplasts. For DCCD to enhance phosphorylation in EDTA-treated chloroplasts, active CF_1 must be present on the thylakoids (McCARTY and RACKER, 1967).

II. Tri-n-butyltinchloride

Tri-n-butyltinchloride (TBT) inhibited phosphorylation in spinach and Euglena chloroplasts at very low concentrations. In spinach chloroplasts, one TBT per about 60 chlorophylls gave complete inhibition. TBT did not inhibit uncoupled electron flow, but did inhibit basal and coupled electron flow. TBT, like DCCD, restored H^+ uptake to EDTA-treated chloroplasts and gave some stimulation of phosphorylation (KAHN, 1968). TBT inhibited ATPase activity of chloroplasts (LYNN and STAUB, 1969) but not that of CF_1. Thus, TBT inhibits phosphorylation in a manner quite similar to that of DCCD.

III. Synthalin

The characteristics of the inhibition of photophosphorylation by synthalin (de-camethylenediguanidine) resemble those of the inhibition by Dio-9 or phlorizin (GROSS et al., 1968). Only electron flow coupled to phosphorylation was markedly inhibited by synthalin. H^+ uptake was not affected at concentrations of synthalin (1–2 mM) which inhibited phosphorylation 60 to 90%. Synthalin inhibition was unaffected by the phosphate concentration. The effects of synthalin on ATPase in chloroplasts or CF_1 were not tested.

IV. p-Chloromercuribenzoate

p-Chloromercuribenzoate (PCMB) is a phlorizin-like phosphorylation inhibitor (IZAWA and GOOD, 1969), although uncoupled electron flow appears to be slightly sensitive to PCMB. Fifty percent inhibition of phosphorylation was reached at approximately 50 μM PCMB. However, even at 700 μM PCMB, the phosphorylation rate was over 20% of the control. Since PCMB inhibited the Ca^{2+}-dependent ATPase of CF_1 (Vambutas and Racker, 1965), CF_1 could be the site of action of PCMB. Para-hydroxymercuribenzoate and phenylmercuric acetate had similar effects. Mercuric acetate or mercuric chloride (1–10 μM) partially inhibited ATP synthesis and coupled electron flow. Although electron flow in the presence of ADP, but absence of phosphate was unaffected by Hg^{2+}, electron flow assayed in the absence of ADP was stimulated. Bradeen et al. (1973) reported that Hg^{2+} did not inhibit post-illumination ATP synthesis, indicating that Hg^{2+} may not affect CF_1 in chloroplasts. Moreover, Hg^{2+} did not appear to inhibit phosphorylation linked to electron flow through photosystem II.

V. Kaempferol

In pea chloroplasts the flavonoid, kaempferol, inhibited phosphorylation nearly completely at 0.13 mM. It had phlorizin-like effects on electron flow and H^+ uptake. Strangely, 0.13 mM kaempferol had little affect on the light- and SH-activated ATPase in chloroplasts (Arntzen et al., 1974). It seems possible that differences in the conditions under which the phosphorylation and ATPase assay were run could account for the insensitivity of ATPase to kaempferol.

VI. Chloramphenicols

As a warning to those investigators who use chloramphenicol (0.1–0.3 mg/ml) as an inhibitor of protein synthesis, Wara-Aswapati and Bradbeer (1974) published that chloramphenicol acts as an energy transfer inhibitor. At higher concentrations, chloramphenicol also inhibited basal electron flow. D- and L-threochloramphenicols were equally effective although only the D-isomer inhibits protein synthesis.

VII. Tentoxin

The fungus, *Alternaria tenuis,* a plant pathogen which causes chlorosis and interferes with plastid development, excretes a cyclic tetrapeptide toxin into the culture medium. This toxin, called tentoxin, is a potent inhibitor of cyclic phosphorylation supported by N-methylphenazonium methosulfate-dependent cyclic electron flow (Arntzen, 1972). Full inhibition of cyclic phosphorylation was observed at 0.4 μM tentoxin, but tentoxin at this concentration stimulated rather than inhibited H^+ uptake. Only electron flow under phosphorylating conditions was sensitive to 0.4 μM tentoxin. Since tentoxin was said not to inhibit oxidative phosphorylation in corn shoot mitochondria, tentoxin may be an excellent inhibitor for the study of ATP synthesis in vivo.

VIII. Sulfate

The inhibition of photophosphorylation by sulfate is complex. When added to phosphorylation assay mixtures, sulfate inhibited phosphorylation in a manner competitive with phosphate (HALL and TELFER, 1969). Postillumination or acid-base induced ATP formation was also inhibited by sulfate with inhibition constants (K_i) of 2 to 3 mM. In addition, illumination of chloroplasts with sulfate in the presence or ADP and Mg^{2+} prior to the assay of phosphorylation resulted in an irreversible inhibition of phosphorylation (RYRIE and JAGENDORF, 1971 b). Chloroplasts treated with sulfate in this manner are actually uncoupled. Apparently the illumination of chloroplasts with ADP and sulfate causes some change in CF_1 (or a component to which CF_1 binds) so that the interactions between the thylakoids and CF_1 are destabilized. Thus, some CF_1 is lost from sulfate-inhibited chloroplasts which results in an uncoupling of phosphorylation (GREBANIER, 1975).

Sulfate has been reported to stimulate (HALL and TELFER, 1969), inhibit (PICK and AVRON, 1973) or have no effect (ASADA et al., 1968) on coupled electron flow. HALL and TELFER (1969), however, used prolonged periods of illumination (10 min) and it is quite possible that a coupling by ADP and sulfate could have been expressed. ASADA et al. (1968) used chloroplasts which had rather poor phosphorylation efficiencies. PICK and AVRON (1973) observed that sulfate inhibited coupled electron flow, but had little effect on basal or uncoupled electron flow. The inhibition was overcome by increasing the concentration of either phosphate or arsenate. Thus, it seems likely that sulfate can act as an energy transfer inhibitor by interfering with the utilization of phosphate.

D. Some Observations and Conclusions

Most of the energy transfer inhibitors have little effect on basal electron flow. However, AVRON (1971) reported that Dio-9, phlorizin and DCCD markedly inhibited basal electron flow in lettuce chloroplasts at pH values greater than about 7.5. The pH-dependence of basal electron flow in lettuce chloroplasts is markedly different from that in spinach chloroplasts. For example, the rate of basal electron flow in lettuce chloroplasts at pH 8.5 to 9 was as high or higher than phosphorylating electron flow (AVRON, 1971). In contrast, in spinach chloroplasts, phosphorylation stimulated electron flow by about two and one-half-fold in the same pH range (GOOD et al., 1966). Thus, lettuce chloroplasts are severely uncoupled at alkaline pH values. This uncoupling is probably related to changes in CF_1 which result in the leakage of H^+ from the thylakoid interior (PORTIS et al., 1975). Energy transfer inhibitors, which exert their effects on CF_1 or closely related components, may prevent these changes. This conclusion is supported by the observation that in lettuce chloroplasts energy transfer inhibitors markedly enhanced light-dependent H^+ uptake at alkaline pH values (KARLISH and AVRON, 1971).

The presence of ADP and P_i together may exert effects on chloroplast activities which are not due to ATP synthesis itself. For example, ADP and P_i inhibited light-dependent ^3H exchange in CF_1 in chloroplasts (RYRIE and JAGENDORF, 1972) and prevented the reaction of CF_1 with N-ethylmaleimide (MAGNUSSON and McCARTY, 1975). Yet, phlorizin did not reverse these effects of ADP and P_i, indicating that phosphorylation is not involved. In contrast, energy transfer inhibitors reversed the decrease in ΔpH in illuminated chloroplasts caused by ADP and P_i (Pick et al., 1973; Portis and McCarty, 1974). Thus, energy transfer inhibitors are useful reagents in the study of effects of nucleotides and of phosphorylation on chloroplast-catalyzed processes.

The mechanism(s) of the synthesis of ATP at the expense of ion gradients is unknown and is one of the more challenging problems in bioenergetics. By probing in depth into the mode of action of energy transfer inhibitors, it is hoped that more light can be shed on the mechanism of photophosphorylation.

References

Arntzen, C.J.: Biochim. Biophys. Acta **283**, 539–542 (1972)
Arntzen, C.J., Falkenthal, S.V., Bobick, S.: Plant Physiol. **53**, 304–306 (1974)
Asada, K., Deura, R., Kasai, Z.: Plant Cell Physiol. **9**, 143–146 (1968)
Avron, M.: Biochim. Biophys. Acta **77**, 699–702 (1963)
Avron, M.: In: Proc. 2nd Intern. Congr. Photosynthesis. Forti, G., Avron, M., Melandri, A. (eds.). The Hague: Dr. Junk, 1972, pp. 861–871
Avron, M., Krogmann, D.W., Jagendorf, A.T.: Biochim. Biophys. Acta **30**, 144–153 (1958)
Beechey, R.B., Holloway, C.T., Knight, I.G., Roberton, A.M.: Biochem. Biophys. Res. Commun. **23**, 75–80 (1966)
Boyer, P.D.: In: Dynamics of Energy-Transducing Membranes. Ernster, L., Estabrook, R., Slater, E.C. (eds.). Amsterdam: Elsevier, 1974, pp. 289–301
Bradeen, D.A., Winget, G.D., Gould, J.M., Ort, D.R.: Plant Physiol. **52**, 680–682 (1973)
Good, N.E., Izawa, S., Hind, G.: Current Topics Bioenerg. **1**, 75–112 (1966)
Grebanier, A.: Ph.D. Thesis, Cornell Univ., 1975
Gross, E., Shavit, N., San Pietro, A.: Arch. Biochem. Biophys. **127**, 224–228 (1968)
Guillory, R.J.: Biochim. Biophys. Acta **89**, 197–207 (1964)
Hall, D.O., Telfer, A.: In: Progress in Photosynthesis Research. Metzner, H. (ed.). Tübingen: Laupp, 1969, Vol. III, pp. 1281–1287
Hind, G., McCarty, R.E.: In: Photophysiology. Giese, A.C. (ed.). New York-London: Academic Press, 1973, Vol. VIII, pp. 113–156
Hoch, G., Martin, I.: Biochem. Biophys. Res. Commun. **12**, 223–228 (1963)
Izawa, S., Good, N.E.: Biochim. Biophys. Acta **162**, 380–391 (1968)
Izawa, S., Good, N.E.: In: Progress in Photosynthesis Research. Metzner, H. (ed.). Tübingen: Laupp, 1969, Vol. III, pp. 1288–1298
Izawa, S., Good, N.E.: Methods Enzymol. **23B**, 355–377 (1972)
Izawa, S., Winget, G.D., Good, N.E.: Biochem. Biophys. Res. Commun. **22**, 223–226 (1966)
Jagendorf, A.T.: In: Bioenergetics of Photosynthesis. Govindjee (ed.). New York: Academic Press, 1975, pp. 413–493
Kahn, J.S.: Biochim. Biophys. Acta **153**, 203–210 (1968)
Karlish, S.J.D., Avron, M.: Europ. J. Biochem. **20**, 51–57 (1971)
Lynn, W.S., Staub, K.D.: Proc. Nat. Acad, Sci. U.S. **63**, 540–547 (1969)
Magnusson, R.P., McCarty, R.E.: J. Biol. Chem. **250**, 2593–2598 (1975)
Marchant, R.H., Packer, L.: Biochim. Biophys. Acta **75**, 458–460 (1963)
McCarty, R.E.: Arch. Biochem. Biophys. **161**, 93–99 (1974)
McCarty, R.E.: In: Encyclopedia of Plant Physiology. Stocking, C.R., Heber, U. (eds.). Berlin-Heidelberg-New York: Springer, 1976
McCarty, R.E., Fagan, J.: Biochemistry **12**, 1503–1507 (1973)
McCarty, R.E., Fuhrman, J.S., Tsuchiya, Y.: Proc. Nat. Acad. Sci. U.S. **68**, 2522–2526 (1971)
McCarty, R.E., Guillory, R.J., Racker, E.: J. Biol. Chem. **240**, 4822–4823 (1965)
McCarty, R.E., Pittman, P.R., Tsuchiya, Y.: J. Biol. Chem. **247**, 3048–3051 (1972)
McCarty, R.E., Racker, E.: Brookhaven Symp. Biol. **19**, 202–214 (1966)
McCarty, R.E., Racker, E.: J. Biol. Chem. **242**, 3435–3439 (1967)
Mitchell, P.: Nature (London) **191**, 144–148 (1961)

Mitchell, P.: Biol. Rev. Cambridge **41**, 445–502 (1966)
Nelson, N., Deters, D.W., Nelson, H., Racker, E.: J. Biol. Chem. **248**, 2049–2055 (1973)
Petrack, B., Craston, A., Sheppy, F., Farron, F.: J. Biol. Chem. **240**, 906–914 (1965)
Pick, U., Avron, M.: Biochim. Biophys. Acta **325**, 297–303 (1973)
Pick, U., Rottenberg, H., Avron, M.: FEBS Lett. **32**, 91–94 (1973)
Portis, A.R., Jr., Magnusson, R.P., McCarty, R.E.: Biochem. Biophys. Res. Commun. **64**, 877–884 (1975)
Portis, A.R., Jr., McCarty, R.E.: J. Biol. Chem. **249**, 6250–6254 (1974)
Racker, E., Hauska, G.A., Lien, S., Berzborn, R.J., Nelson, N.: Resolution and reconstruction of the system of photophosphorylation. In: Proc. 2nd Intern. Congr. Photosynthesis. Forti, G., Avron, M., Melandri, A. (eds.). The Hague: Dr. Junk, 1972, pp. 1097–1113
Ryrie, I.J., Jagendorf, A.T.: J. Biol. Chem. **246**, 3771–3774 (1971a)
Ryrie, I.J., Jagendorf, A.T.: J. Biol. Chem. **246**, 582–588 (1971b)
Ryrie, I.J., Jagendorf, A.T.: J. Biol. Chem. **247**, 4453–4459 (1972)
Shavit, N., Herscovici, A.: FEBS Lett. **11**, 125–128 (1970)
Telfer, A., Evans, M.C.W.: Biochim. Biophys. Acta **256**, 625–637 (1972)
Uribe, E.G.: Biochemistry **9**, 2100–2106 (1970)
Uribe, E.G.: Biochemistry **11**, 4228–4235 (1972)
Vambutas, V.K., Racker, E.: J. Biol. Chem. **240**, 2660–2667 (1965)
Wara-Aswapati, O., Bradbeer, J.W.: Plant Physiol. **53**, 691–693 (1974)
Winget, G.D., Izawa, S., Good, N.E.: Biochemistry **8**, 2067–2074 (1969)

13. Photophosphorylation in vivo

H. GIMMLER

> "The concept of cyclic and noncyclic as two distinguishable types of photophosphorylation has survived for some 15 years. Neither stoichiometry nor rates of in vivo photophosphorylation have been made clear." [J. MYERS in: Conceptual developments in photosynthesis, 1924–1974 (1974)]

A. Introduction

As shown in the preceding chapters of this volume, the process of photophosphorylation (PP) is defined as light-induced synthesis of ATP from ADP and inorganic phosphate (P_i):

$$ADP + P_i \xrightarrow{hv} ATP. \tag{1}$$

PP is confined to the thylakoid systems of green plants and its driving force is a proton gradient, which is created by the photosynthetic electron transport (ET) across the thylakoid membranes. The operation of PP can be experimentally demonstrated in both intact cells (in vivo) or in thylakoid fragments isolated from green cells (in vitro). Most of our present knowledge about the mechanism of PP and its coupling to photosynthetic ET arises from in vitro studies with thylakoid membranes (cf. preceding chapters in this volume). Such preparations are for many reasons experimentally superior to the corresponding in vivo systems, if basic principles and properties of PP are investigated (Table 1).

However, as will be discussed in this chapter, several problems exist, which can be solved only by studies on intact cells. Such investigations require *physiological* methods rather than the *biochemical* methods characteristic of in vitro studies. Physiological methods, which are characterized by a "tendency to preserve the system" (MOHR, 1975), take into account the high cooperativity of intact cells. This provides the justification for allocating a separate chapter in this volume to in vivo PP. Rather than giving a complete survey of the subject, this chapter attempts to outline the present view of three main problems of in vivo PP. These are (1) the physiological significance of cyclic and pseudocyclic PP as sources of "extra-ATP" in intact cells (in addition to the well-established role of noncyclic PP as the main ATP-source in green cells); (2) the intracellular regulation of these three different types of PP, and (3) their relation to CO_2-assimilation.

For other aspects of in vivo PP we may refer to one of the following reviews. The discovery of PP in vivo and its early history was described by ARNON (1956). Later studies were reviewed by KANDLER (1960) and in more detail by SIMONIS (1960) in the first edition of this encyclopedia. Recent developments in this field were summarized by SIMONIS and URBACH (1973) in an article which also gives a good survey of the methods applied in PP in vivo and by HEBER (1976). Special problems of PP in algae are described in the reviews of WIESSNER (1970a, b) and RAVEN (1974), while some chapters of this volume also deal with some problems closely related to PP in photosynthetic microorganisms (this vol. Chaps. V.1, V.2, V.3). Reference to the proceedings of the international congresses on photosynthesis (METZNER, 1969; FORTI et al., 1972; AVRON, 1975) will provide specialized information on the progress in this scientific field.

The difference between PP in vivo and in vitro is mainly due to the removal of the chloroplast envelope in the preparations of chloroplasts from eukaryotes, while the plasmalemma is similarly ruptured during the isolation of thylakoid fragments from prokaryotes. In addition to causing a change in the physiological environment of the stroma matrix (and/or the cytosol) this destructive process also creates the following artifacts: (1) Some essential components of the natural ET chain, for example ferredoxin and $NADP^+$, are lost. Thus noncyclic ET from H_2O to the physiological electron acceptor $NADP^+$ is interrupted. Similarly the physiological cyclic ET, requiring ferredoxin, cannot proceed. Therefore, experiments with thylakoid fragments cannot yield information about reaction rates and relative efficiencies of these systems. (2) CO_2-assimilation and many other ATP-requiring, endogenous reactions of green plants are absent in broken chloroplasts. Thus, the high cooperativity typical of intact cells is missing. In vitro systems may only indicate the capacity for photophosphorylation and not the physiological ATP requirement in whole cells. (3) The phosphorylation potential, defined as [ATP]/[ADP] [P_i], is different in intact cells and in vitro preparations. The phosphorylation potential, however, is an important regulating factor of in vivo PP. Some further characteristics of PP in vivo and in vitro are summarized in Table 1.

We may define "photophosphorylation in vivo" as light-induced formation of ATP [cf. Eq. (1)] in intact plant cells or isolated intact chloroplasts, which are capable of reducing CO_2 at high rates. This definition includes light-induced phosphorylations of photosynthetic bacteria, algae, lichens, mosses, ferns, gymnosperms and angiosperms, but excludes PP of chlorophyll-free organisms. This restriction has more of a historical than a logical reason. The latter systems play only a minor role in plant metabolism in comparison to the fundamental role of PP in green plants. Because of the endogenous light-driven ATP synthesis in isolated, intact chloroplasts, which are also able to reduce CO_2 rapidly (type A chloroplasts, HALL, 1972) they must also be included within this definition (but compare SIMONIS and URBACH, 1973). However, the possibility of reduced cooperation and altered regulation in type A chloroplasts must be considered.

In in vivo work two different experimental approaches must be distinguished, results of which should be interpreted in different ways. If the investigated ATP synthesis is coupled to an ET, which proceeds exclusively via native electron carriers and does not need the addition of external electron acceptors or substrates for phosphorylation, the term PP in vivo is identical with the terms physiological PP (PPP) or endogenous PP (EPP). Unfortunately, the term "endogenous PP" is used also for chloroplast preparations, which require the addition of external ADP for ATP synthesis (NOBEL, 1967; MIGINIAC-MASLOW and MOISE, 1969; SCHÜRMANN et al., 1972). The term PP in vivo is by no means identical with physiological PP, if the addition of exogenous, mostly artificial electron acceptors or mediators (or substrates for phosphorylation) are required for ATP-synthesis, as these additives interfere with the endogenous metabolism. Investigations of either system give important insights into the properties of PP, but only results

Table 1. Differences between PP in intact cells or type A chloroplasts and thylakoid fragments

	Photophosphorylation	
	in vivo	in vitro
System	Intact cells or type A chloroplasts (Hall, 1972). Thylakoids are surrounded by an intact chloroplast envelope (eukaryotes) or by an intact plasmalemma (prokaryotes). Complexity of the system high, "closed system".	Isolated thylakoid fragments without a surrounding envelope or plasmalemma [type C, D, E, F chloroplasts, Hall (1972)]. Complexity limited, "open system".
Environment of the coupling factor	Physiological; pH and ionic content of the stroma and the thylakoid space only slightly influenced by the external medium.	Artificial; pH and ionic content of the medium according to the experimental conditions.
Requirements	CO_2 and/or O_2 and/or reducible anions, but no external ADP, P_i and Mg^{2+}. Adenylates do not penetrate the plasmalemma or the chloroplast envelope.	ADP, P_i, Mg^{2+}, external electron acceptors, mediators and donors according to the experimental conditions.
ADP and ATP	Limited amounts of ADP available (10^{-4}–10^{-3} M endogenous concentration). ADP/ATP ratio 0.1–5 (Krause, 1975). State III can be kept only for seconds, unless ATP consuming reactions occur.	ADP in excess ($2 \cdot 10^{-3}$ M in the medium). ADP/ATP ratio $\ll 0.1$. State III can be kept for minutes.
ATP-consumption	Yes, the rate of which is in the same order of magnitude as the ATP-formation. Rapid turnover of the ATP-pool.	No.
ET and PP	ET from H_2O via both photosystems to $NADP^+$ (noncyclic PP) or O_2 (pseudocyclic PP), cyclic ET via photosystem I (cyclic PP). With few exceptions PP of partial sequences of the ET chain cannot be investigated since most electron acceptors and donors used in studies of PP in vitro do not penetrate readily through the plasmalemma and/or the chloroplast envelope.	Noncyclic, pseudocyclic and cyclic ET according to the electron donors, mediators and donors applied. PP of partial sequences of the ET chain can be investigated.

Other phosphorylating systems	Yes, oxidative phosphorylation, glycolysis.	No.
Energy conserving sites	Native energy conserving sites.	PP by the operation of both native and artificial energy conservation sites (TREBST, 1974).
Regulating factors	Pool sizes of ADP, ATP, P_i, $NADP^+$, NADPH. CO_2 and O_2 pressures. Light intensity. Regulation between different systems of PP possible. ATP synthesis according to the physiological ATP-requirement.	Regulation limited.
Inhibitors, uncouplers	Accessibility limited, preferentially lipophilic compounds do penetrate the plasmalemma and/or the chloroplast envelope. Effects often less specific than in vitro due to the complexity of the system. Higher concentrations than in vitro are necessary for 50% inhibition.	All compounds reach at least the outer membrane face (stroma side), lipophilic compounds can penetrate into the intrathylakoid space. Effects more specific than in vivo, lower concentrations for 50% inhibition necessary.
Assays	(1) Direct approaches (measurements of changes of the pool sizes of ADP, P_i, ATP). (2) Indirect approaches (light-induced reactions depending upon a steady supply of ATP are taken as indicator reactions). Estimation of PP rates difficult, measured rates are lower than true rates. Direct assay of P/e_2 ratios not possible.	PP can be determined directly by the production of ATP or the disappearance of ADP, P_i or H^+. Rates of PP can be easily determined as well as P/e_2 ratios.

of the former systems (PPP) are relevant for the question of the physiological
ATP requirement of intact cells and the regulation of the different types of PP.
Experiments with the "artificial" systems in intact cells are more useful for prob-
lems such as the phosphorylation capacity or the localization of coupling sites.

B. Methods

I. Objects

Our present knowledge of in vivo PP arises from the investigation of only very
few plant genera. Blue green algae are represented by *Anacystis, Anabaena* and
Phormidium; the favored green algae are *Chlorella, Scenedesmus, Ankistrodesmus,
Euglena, Chlamydomonas, Dunaliella, Chlamydobothrys* and *Hydrodictyon; Nitella;
Chara* and *Tolypella* represent the coenocytic charophyta; the best investigated
genus of the rhodophyta is *Porphyridium,* while the photosynthetic bacteria *Chromatium, Rhodospirillum, Rhodopseudomonas* and *Chlorobium* have been subjects of
intensive studies (this vol. Chap. V.3). The standard objects for investigations
of PP in higher plants are the angiosperms *Spinacia, Pisum, Elodea* and *Phasaeolus.*
Little is known as yet about PP in pyrrophyta, chrysophyta, brown algae, lichens,
mosses, ferns and gymnosperms. Certainly it is not expected that the basic mechanism of PP will vary between different plant families, but rates, ratios of the
different types of PP and their regulatory mechanism may vary from family to
family and from species to species. Therefore, unwarranted generalizations, [e.g.
about the physiological significance of pseudocyclic or cyclic PP in vivo ("what's
good for spinach, is good for *Anacystis*")] must be avoided. In the future comparative aspects of in vivo PP deserve more consideration including their evolutionary
(Broda, 1970; Baltscheffsky, 1975), ontogenetic (Senger 1970a, b; Gimmler
et al., 1971; Oelze-Karow and Butler, 1971) and ecological aspects (Huber and
Edwards, 1975).

II. Test Reactions

For a variety of reasons, which are summarized in Table 1, in vivo PP cannot
be measured with the same methods used for studies of PP in vitro. In the former,
indicator reactions are measured, which reflect indirectly the rate of PP. These
indicator reactions were already extensively reviewed by Simonis (1960) and Simonis
and Urbach (1973) and are summarized in Table 2. These test reactions have
in common that they are (1) stimulated by light, (2) inhibited by inhibitors of
photosynthetic ET and (3) suppressed by uncouplers of PP. There are three further
requirements from an ideal test reaction for in vivo PP, but these are incompatible
with each other. They must (1) reflect the true rate of ATP-synthesis, (2) not
interfere with endogenous ATP-requiring reactions and (3) be independent of external additions. The selection of test reactions presented in Table 2 may be rather
arbitrary, but these methods have contributed most to the investigation of in
vivo PP.

The methods can be divided into two groups. *Direct approaches* and *indirect approaches*. The former monitor light-induced changes of the endogenous pools of the reactants in Eq. (1). These test reactions are independent of external additions and do reflect changes of the endogenous phosphorylation potential during dark-light transients. A disadvantage of these methods is that quantitative results can be obtained only during induction periods, whereas the comparison of steady-state pool sizes under different conditions give useful, but only qualitative, information (KYLIN and TILLBERG, 1967; BORNEFELD et al., 1972; BEDELL and GOVINDJEE, 1973; OKKEH and KYLIN, 1975; OKKEH et al., 1975). The interpretation of these changes is complicated by the recent observation that "steady-state levels" of ATP exhibit considerable oscillations in continuous light (LEWENSTEIN and BACHOFEN, 1972). Furthermore, the light-minus-dark differences in the ATP pool can be positive or negative, depending upon the experimental conditions, especially upon the occurrence of ATP-consuming reactions (URBACH and GIMMLER, 1970a; BEDELL and GOVINDJEE, 1973). The interpretation of these changes in vivo is facilitated if dark levels of ATP are kept low by anaerobisis or specific inhibitors of oxidative phosphorylation (URBACH and GIMMLER, 1970b; URBACH and KAISER, 1972; BEDELL and GOVINDJEE, 1973). Short time exposures to light (sec) and other physiological conditions, where the rate of photosynthetic CO_2-assimilation is limited, are useful in obtaining high rates of PP (BORNEFELD et al., 1972; BORNEFELD, 1976a; OWENS and KRAUSS, 1972).

III. Indirect Approaches

AVRON and NEUMANN (1968) distinguished three types of reactions, which depend upon a steady supply of ATP (Table 2).
  1. Intracellular reactions independent upon external additions.
  2. Uptake reactions of solutes, which are not metabolized.
  3. Combined systems, representing reactions where the ATP-dependent uptake of a compound is followed by the ATP-requiring metabolism of the compounds.

These indirect assays can be applied at steady-state conditions, but suffer from other disadvantages: namely, during uptake reactions limiting (dark) steps at the plasmalemma or the chloroplast envelope may occur. Also, the addition of a compound, the uptake of which and/or its later metablism requires ATP, introduces a new reaction competing for the ATP produced within the cell. This causes interference with other ATP-requiring endogenous reactions, if large amounts of ATP are needed. If only low amounts of ATP are required by the introduced system, this influence can be neglected, but then this reaction certainly cannot reflect the true rate of ATP synthesis in the cell. Further, the reason for the ATP-dependency of a test reaction (e.g. photokinesis) may be complex and its exact stoichiometry may not always be known. Finally, many of the indirect approaches can be applied only under special conditions (e.g. glucose uptake under anaerobisis) or to special objects (e.g. photokinesis in moving algae). No indirect assay, which can be applied universally to all organisms has been developed. The only indicator reaction for in vivo PP, which approximates such a univeral assay, is the light-induced incorporation of ^{32}P, because it can be applied at least to all aquatic plants and to type A chloroplasts regardless of their origin.

Table 2. Assays for photophosphorylation in vivo

Methods	References
A. Direct approaches	
Light-induced changes of the endogenous pools of the compounds taking part in the reaction $ADP + P_i \rightarrow ATP$ are taken as indicator reactions for PP.	1–3
1. Changes of the P_i-pool.	
a) Direct determination	4–6
b) After prelabeling with ^{32}P	9–10
2. Changes of the ATP-pool.	79
a) Conventional enzymic assays	11–13, 15
b) Firefly luciferase-luciferin test	16–17, 19–23, 77, 78
B. Indirect approaches	
Light-stimulated reactions, depending upon a steady supply of ATP, are taken as indicator reactions for PP.	3, 24
1. Endogenous reactions, independent of the uptake of solutes from the external medium. No limiting steps at the plasmalemma or the chloroplast envelope.	
a) Glucose conversion into starch	25, 39, 47
b) N_2-fixation in blue green algae	26–28
c) Photokinesis	29–32
d) Photoreduction of CO_2 in H_2-adapted algae	33–35
2. Reactions, in which the uptake of solutes from the external medium is involved. Limiting steps at the plasmalemma or the chloroplast envelope may occur.	36–38
A. Photoassimilation of organic compounds which can be metabolized	
a) Photoassimilation of glucose	36, 40–42
b) Photoassimilation of acetate	36, 43
B. Light-stimulated uptake of inorganic ions which can be metabolized	2
a) Uptake of phosphate	2, 44–46, 49–52
b) Uptake of phosphate and incorporation into organic compounds	2, 56–57
c) Uptake of phosphate and incorporation into polyphosphates	2, 8, 14, 58, 59
d) Photoassimilation of nitrate	60–62, 64
C. Light-stimulated uptake of ions which cannot be metabolized	37–38
a) Uptake of potassium	65–69, 71
b) Uptake of chloride	7, 69, 70, 72, 73
C. Miscellaneous	
1. Light-induced shrinkage (light scattering)	63, 75, 76
2. Photoinhibition of respiration	55, 74
3. Steady-state relaxation spectrophotometry (measurement of ET rather than of PP)	48, 53, 54

References see opposite page.

This test is applicable in the presence or absence of CO_2 or O_2. It is intimately linked to the phosphorylation process itself and therefore ascribed by some authors (AVRON and NEUMANN, 1968; RAVEN, 1974) even to the direct methods. However, because it involves an uptake reaction, we consider this test as an indirect assay, the results of which permit important conclusions, which however, are sometimes difficult to interpret quantitatively (JESCHKE et al., 1967; GIMMLER et al., 1968; GIMMLER et al., 1969; URBACH and GIMMLER, 1970b; KAISER and URBACH, 1974).

Nevertheless, the general validity of the methods listed in Table 2 is vindicated by the fact that in spite of their diversity, very similar results have been obtained. On the other hand, this methodological diversity has also created controversies in the field. In the following we attempt to outline the common denominator of all the results without regard to their methodical origin.

IV. Rates

Type A chloroplasts reduce CO_2 at rates between 50 and 200 µmol/mg chlorophyll^{-1} h^{-1} (HEBER, 1973a), the corresponding rates of intact leaves and green algae are in the same order of magnitude. Photosynthetic rates of blue green and red algae are even higher (up to 500 µmol CO_2 mg chlorophyll^{-1} h^{-1}) due to their low content of chlorophyll. Since it is generally assumed that 3 molecules ATP are required for the fixation of 1 molecule of CO_2 in the Calvin cycle, PP in intact leaves, type A chloroplasts and green algae should proceed at least with rates between 150 and 600, and in blue green and red algae with rates up to 1800 µmol ATP mg chlorophyll^{-1} h^{-1}.

These are minimal numbers, neglecting the requirement of ATP for many other reactions in the cell (cf. Table 2). Although the intrinsic difficulties in measur-

---

References: (1) KANDLER (1960), (2) SIMONIS (1960), (3) SIMONIS and URBACH (1973), (4) KANDLER (1950), (5) KANDLER (1957), (6) KLOB et al. (1972), (7) BARBER (1968a), (8) ULLRICH (1972a), (9) BRADLEY (1957), (10) SANTARIUS et al. (1964), (11) KANDLER et al. (1961), (12) FORTI and PARISI (1963), (13) SANTARIUS and HEBER (1965), (14) ULLRICH (1972b), (15) NOBEL et al. (1969), (16) STREHLER (1953), (17) STREHLER (1970), (18) ULLRICH (1970), (19) JOHN (1970), (20) BORNEFELD et al. (1972), (21) URBACH and KAISER (1972), (22) HEBER (1973a), (23) GIMMLER (1973), (24) AVRON and NEUMANN (1968), (25) MACLACHAN and PORTER (1959), (26) COX and FAY (1969), (27) FAY (1970), (28) BOTHE (1972), (29) NULTSCH (1967), (30) NULTSCH (1969), (31) NULTSCH (1970), (32) NULTSCH (1973), (33) BISHOP (1962), (34) BISHOP (1966), (35) BISHOP (1967), (36) WIESSNER (1970a), (37) WIESSNER (1970b), (38) MACROBBIE (1970), (39) GLAGOLEVA et al. (1972), (40) KANDLER (1954), (41) TANNER et al. (1969), (42) WIESSNER (1966), (43) WIESSNER (1965), (44) JESCHKE and SIMONIS (1965), (45) ULLRICH-EBERIUS and SIMONIS (1970), (46) ULLRICH-EBERIUS (1973a), (47) GLAGOLEVA and ZALENSKI (1970), (48) HOCH (1972), (49) KYLIN (1966), (50) SMITH (1966), (51) WEIGL (1967), (52) RAVEN (1974), (53) HOCH and RANDLES (1971), (54) RURAINSKI et al. (1970), (55) RIED (1968), (56) URBACH and GIMMLER (1970b), (57) WEICHART (1961), (58) WINTERMANS (1955), (59) ULLRICH and SIMONIS (1969), (60) KESSLER et al. (1970), (61) ULLRICH (1971), (62) ULLRICH (1974), (63) NOBEL (1968), (64) ULLRICH-EBERIUS (1973b), (65) MACROBBIE (1965), (66) RAVEN (1971), (67) BARBER (1968b), (68) JESCHKE (1970), (69) JESCHKE (1972a), (70) JESCHKE (1967), (71) JESCHKE (1970), (72) JESCHKE (1972b), (73) JESCHKE and SIMONIS (1969), (74) RIED (1969), (75) HEBER (1969), (76) GIMMLER (1973), (77) BORNEFELD (1976a), (78) BORNEFELD (1976b), (79) KRAUSE (1975).

ing the rates of in vivo PP are very well known the main criticism of the methods applied in this field are that they failed to demonstrate rates higher than 5 µmol ATP mg chlorophyll^{-1} h^{-1} (HEBER, 1969, 1973a, b; MYERS, 1974; KRAUSE, 1975). Therefore, as these rates are so low compared with the true phosphorylation rates and the cyclic or pseudocyclic PP themselves only represent a small proportion of that low rate, the physiological significance of the two processes have been queried. This criticism stimulated a hunt for higher rates of in vivo PP and resulted in a considerable increase in the reported rates. Up to 45 µmol ATP mg chlorophyll^{-1} h^{-1} were found for unicellular green algae (TANNER et al., 1969; URBACH and KAISER, 1972; GIMMLER, 1973), 20–60 µmol ATP mg chlorophyll^{-1} h^{-1} in type A chloroplasts (NOBEL, 1968; URBACH and KAISER, 1972; KAISER and URBACH, 1976) and 80–500 in blue green algae (OWENS and KRAUSS, 1972; BORNEFELD et al., 1972; BORNEFELD, 1976a, 1976b; BOTTOMLEY and STEWART, 1976): Nonetheless, these numbers account only for 3 to 10% of the "true rate" of in vivo PP, calculated on the basis of the rate of CO_2-fixation. These improved rates did not solve the principle problem, but they present a considerable improvement on older data.

C. Cyclic Photophosphorylation in vivo

There is now general agreement that cyclic PP not only exists in photosynthetic bacteria (FRENKEL, 1970; SYBESMA, 1970; BALTSCHEFFSKY et al., 1971; this vol. Chap. V.3) but also in algae and higher plants (SIMONIS and URBACH, 1973; RAVEN, 1974). However, in photosynthetic bacteria cyclic PP is the main source of ATP, whereas the contribution of cyclic PP to the total PP in the other organisms is subject to considerable controversy. Bacterial PP is described elsewhere in this volume (Chap. V.3). An evolutionary relationship between cyclic PP of photosynthetic bacteria and that of algae and higher plants is indicated by the following common properties. The cyclic PP in both bacteria and algae and higher plants requires only one light reaction (PS I) and is sensitive to O_2, antimycin and to DBMIB (BALTSCHEFFSKY, 1975). It contains possibly two energy conservation sites and is linked to an open chain ET system via a quinone.

The ability to perform a cyclic PP seems to be universal among algae and higher plants (but cf. SMITH and RAVEN, 1974). This can be readily demonstrated under experimental conditions, where other phosphorylations are suppressed. The quantitative role of cyclic PP in vivo is very difficult to assess. Since evidence for the operation of cyclic PP in vivo arises entirely from exclusion experiments, it can be always argued, that the exclusion of other phosphorylating systems induces the cyclic pattern rather than making the preexisting cyclic system visible. Such questions are characteristic for intact cells. Therefore, there is still disagreement, how far cells make use of their ability to perform cyclic PP, especially in the presence of O_2 and CO_2. From the evolutionary point of view it is interesting that cyclic PP in vivo may be more active in blue-green algae, which are closely related to the photosynthetic bacteria (BORNEFELD et al., 1972; BEDELL and GOVINDJEE, 1973; BORNEFELD, 1976a; BOTTOMLEY and STEWART, 1976), than in green algae and higher plants (HEBER, 1969, 1973a; GIMMLER, 1973). Thus it may be considered to be an evolutionary relic in higher plants, and to be of major physiological significance only in lower algae. Since cyclic PP precedes noncyclic PP during development (GIMMLER et al., 1971; OELZE-KAROW and BUTLER, 1971), it is tempting to speculate that the phylogenetic and ontogenetic behavior of cyclic and noncyclic PP in vivo confirm the biogenetic law, that the individual development of organisms recapitulates their phylogenetic development.

The characteristics of cyclic PP in algae and higher plants are summarized in Table 3. Cyclic PP requires the activity of only one light reaction (PS I), as revealed by its insensitivity to DCMU, its action spectra and by studies with mutants lacking eigher PS II or PS I. The sequence of electron-carriers may be arranged, according to the references given in Table 3, as follows: P-700, ferredoxin, cytochrome b-563, plastoquinone, cytochrome f, plastocyanine. The participation of ferredoxin in physiological cyclic PP is generally deduced from the severe inhibition of this reaction by antimycin and DSPD (cf. SIMONIS and URBACH, 1973; RAVEN, 1974; but compare BOTHE, 1969). Antimycin is a strong inhibitor of the ferredoxin-catalyzed cyclic PP in vitro (TAGAWA et al., 1963) and DSPD inhibits the photosynthetic reduction of ferredoxin (TREBST and BURBA, 1967). In addition the antimycin-sensitivity of cyclic PP in vivo confirms the involvement of cytochrome b-563 which was originally deduced from the oxidation and reduction of cytochrome b-563 by photosystem I (LEVINE, 1969; AMESZ et al., 1972; CRAMER, 1976). The role of plastoquinone is still unclear. The plastoquinone antagonist DBMIB, which blocks photosynthetic ET between plastoquinone and cytochrome f (TREBST et al., 1970; BÖHME et al., 1971; GIMMLER and AVRON, 1972) inhibits cyclic PP in algae (Table 3), suggesting the involvement of plastoquinone in cyclic PP. In type A chloroplasts the inhibitory effect of DBMIB on cyclic PP is still a matter of controversy (FORTI and ROSA, 1971; HUBER and EDWARDS, 1975; KAISER and URBACH, 1976). DBMIB does not influence the kinetics of the oxidation and reduction of cytochrome b-563 in *Porphyridium* (AMESZ et al., 1972; AMESZ, 1973), which implies that a significant part of the plastoquinone pool is not located within the cyclic pathway. However, the complete absence of this proton carrier would be very difficult to accept in view of its importance as a coupling site (TREBST, 1974). Evidence for the participation of cytochrome f, plastocyanine and P-700 in cyclic ET is discussed in the review of CRAMER (1976), KATOH (1976) and HOCH (1976) in this volume (II.4; II.12; II.14).

Cyclic PP in vivo requires, as indeed does PP in vitro (GROMET-ELHANAN, 1967) a precise poising by PS II, that is a slow but not completely inhibited electron transfer from PS II to PS I (MIGINIAC-MASLOW and MOYSE, 1969; VAN RENSEN, 1969; ARNON and CHAIN, 1975; KAISER and URBACH, 1976; RURAINSKI and HOCH, 1976). This is implied by the stimulation of cyclic ET or cyclic PP by DCMU under appropriate light intensities. Over-reduction, e.g. by high light intensities, prevents cyclic PP in vivo (HEBER, 1969).

The number of coupling sites in the physiological cyclic PP is still under debate. TANNER (1969), VAN RENSEN (1969) proposed two sites, whereas TREBST and HAUSKA (1974) favor a single energy conserving site. KYLIN et al. (1972) discuss two different cyclic PP systems containing one coupling site each. A quantum requirement of 2 Einstein per mol ATP was measured by WIESSNER (1965) and TANNER et al. (1968). One energy conservation site was localized between plastoquinone and cytochrome f (TREBST, 1974) and is identical with the corresponding site in noncyclic

Table 3. Characteristics of cyclic photophosphorylation in vivo

Photosystems	Only PS I needed. Evidence arises from: 1. Action spectra (1–7, 69), activity in far red light (cf. 8) 2. DCMU insensitivity (8, 61, 62, 68) 3. Experiments with mutants lacking either PS I or PS II (9–11, 22, 51) 4. Absence of enhancement (2, 12, 13)
Electron carriers involved	1. Cytochrome f and b-563 (14, 19, 21) 2. Plastoquinone (23–25, 52, 72, but compare 15, 29, 53) 3. Plastocyanine (16) 4. P-700 (17, 20–22) 5. Ferredoxin (18, 74)

Sequence of electron carriers

```
                antimycin                                              ~ (?)
         ┌──────────┊─────────────── Cyt b-563 ◄──────────────────────┐
         │          ┊                                                 │
   PQ ───────┊─────────────── ~ ──► Cyt f ──► PC ──► PS I ──► Fd
         │          ┊                                  ⌇hv            DSPD
               DBMIB
```

Inhibitors	1. Antimycin (33–43, 54, 56, 72, but compare 55) 2. DSPD (25, 30–32, 40) 3. DBMIB (23–28, 52, but compare 29, 53, 72)
Evidence for coupling to ATP synthesis	1. Stimulation of cyclic ET by uncouplers (20–22) 2. ATP synthesis under exclusion conditions and inhibition of test reactions by uncouplers (cf. 8)
Number of energy-conserving sites	Two (46, 62) or one (47) sites are discussed
Quantum requirement	2 Einstein/mol ATP (1, 48)
Light saturation	Low in comparison to noncyclic PP in vivo and cyclic PP in vitro (36, 40, 43, 57–61)
Rates	Up to 250 μmol ATP mg chlorophyll^{-1} h^{-1} in N_2 (49, 73) and 80 μmol ATP mg chlorophyl^{-1} h^{-1} (air, +DCMU) (50) were measured (conservative estimations). It is assumed that cyclic PP in vivo proceeds at least with the same rates as oxidative phosphorylation (44, 45, 66)
Regulation	Cyclic PP is favored by the absence of potential electron acceptors of noncyclic and pseudocyclic ET (CO_2, O_2, reducible anions, cf. 8) and needs a very precise redox balance (poising) by PS II (52, 54, 57, 70, 72, 74). Overreduction prevents cyclic PP (71)
Physiological function	Supply of "extra ATP" (e.g. for the photoassimilation of organic compounds, the uptake of ions, the synthesis of proteins (8, 45, 64, 65) especially under conditions where non-cyclic PP is impaired ($-CO_2$) and pseudocyclic PP is inhibited (N_2). Not obligatory for CO_2-assimilation (31, 32, 37, 40, 41, 63, 66, but compare 59, 67)

References see opposite page.

PP. If a second site does exist as it appears to do in photosynthetic bacteria, it should be localized between PS I and cytochrome b-563.

It is believed that the absence of potential acceptors of noncyclic or pseudocyclic ET (CO_2, O_2, reducible anions like SO_4^{2-}, NO_3^-) favor cyclic PP (ULLRICH, 1971; URBACH and SIMONIS, 1973; RAVEN, 1973; ULLRICH-EBERIUS, 1973a), although these effects may depend upon the plant subject and the experimental conditions. Thus, cyclic PP is suppressed by O_2 and CO_2 in intact spinach leaves (HEBER, 1969), in *Elodea* (Jeschke and SIMONIS, 1969) or in *Dunaliella* cells (GIMMLER, 1973), but not in *Anacystis* (BORNEFELD et al., 1972), *Ankistrodesmus* (URBACH and GIMMLER, 1970b), nor even in type A chloroplasts from spinach (KRAUSE and HEBER, 1971). A unique feature of cyclic PP in vivo is its low light saturation, which contrasts dramatically with the high light saturation of cyclic PP in vitro (Table 3).

Once the ability of intact cells to perform cyclic PP was established, many research laboratories tried with great enthusiasm to demonstrate the physiological significance of cyclic PP as an obligatory ATP source for CO_2-assimilation, but most of the evidence was negative. Treatments inhibiting cyclic PP in vivo very strongly (DSPD: GIMMLER et al., 1968; URBACH and GIMMLER, 1969; RAVEN, 1969b, 1970; antimycin: KANDLER and TANNER, 1966; TANNER et al., 1965; TANNER and KANDLER, 1969; salicylaldoxime: TANNER et al., 1969) exhibited only slight effects on the rate of photosynthetic CO_2-fixation, although the distribution pattern of ^{14}C was significantly influenced. The conclusion was drawn (cf. SIMONIS, 1967) that cyclic PP in vivo is not needed for CO_2-assimilation (but compare SCHÜRMANN et al., 1972). KLOB et al. (1972) discussed the possible function of cyclic PP in vivo as a priming reaction for CO_2-assimilation. They claimed that cyclic PP may be necessary for the induction period, but not for the steady state of photosynthesis.

References: (1) WIESSNER (1965), (2) GINGRAS (1966), (3) BISHOP (1967), (4) RAVEN (1969b), (5) FAY (1970), (6) NULTSCH (1970), (7) SIMONIS (1972), (8) SIMONIS and URBACH (1973), (9) BISHOP (1966), (10) TANNER et al. (1967), (11) LEVINE (1969), (12) BISHOP and GAFFRON (1963), (13) WIESSNER and GAFFRON (1964), (14) CRAMER (1976), (15) AMESZ (1976), (16) KATOH (1976), (17) HOCH (1976), (18) HALL (1976), (19) AMESZ et al. (1972), (20) TEICHLER-ZALLEN and HOCH (1967), (21) RURAINSKI et al. (1970), (22) TEICHLER-ZALLEN et al. (1972), (23) URBACH and KAISER (1972), (24) GRIMME (1972), (25) BOTHE (1972), (26) TREBST et al. (1970), (27) BÖHME et al. (1971), (28) GIMMLER and AVRON (1972), (29) AMESZ (1973), (30) TREBST and BURBA (1967), (31) GIMMLER et al. (1968), (32) URBACH and GIMMLER (1969), (33) TAGAWA et al. (1963), (34) URBACH and SIMONIS (1964), (35) SIMONIS (1964), (36) TANNER et al. (1965), (37) KANDLER and TANNER (1966), (38) RIED (1968), (39) RIED (1969), (40) RAVEN (1969a), (41) RAVEN (1970), (42) RAVEN (1971), (43) ULLRICH and SIMONIS (1969), (44) RAVEN (1972), (45) RAVEN (1973), (46) TANNER (1969), (47) TREBST and HAUSKA (1974), (48) TANNER et al. (1968), (49) OWENS and KRAUSS (1972), (50) BORNEFELD et al. (1972), (51) BISHOP (1962), (52) KAISER and URBACH (1976), (53) FORTI and ROSA (1971), (54) MIGINIAC-MASLOW and MOYSE (1969), (55) BOTHE (1969), (56) WIESSNER (1966), (57) VAN RENSEN (1969), (58) KANDLER (1957), (59) KLOB et al. (1972), (60) BEDELL and GOVINDJEE (1973), (61) URBACH and GIMMLER (1970b), (62) VAN RENSEN (1971), (63) TANNER and KANDLER (1969), (64) RAMIREZ et al. (1968), (65) BORNEFELD and SIMONIS (1975), (66) TANNER et al. (1969), (67) SCHÜRMANN et al. (1972), (68) KYLIN (1972), (69) HEBER (1969), (70) RURAINSKI and HOCH (1976), (71) HEBER (1969), (72) HUBER and EDWARDS (1975), (73) BOTTOMLEY and STEWART (1976), (74) ARNON and CHAIN (1975).

If cyclic PP is not essential for photosynthetic CO_2-assimilation, what then is its physiological significance? It should be kept in mind that the growth of cells need more ATP than used for CO_2-fixation (Raven, 1974). Ion uptake, photoassimilation of organic substrates, nitrogen fixation, photokinesis and photoreduction of CO_2 in H_2-adapted algae may be supplied with "extra-ATP" from cyclic PP (Table 2). Also, protein synthesis may be powered by cyclic PP (Ramirez et al., 1968; Bornefeld and Simonis, 1975). Cyclic PP in vivo may be of great significance when noncyclic PP is impeded by the lack of CO_2, e.g. in leaves with closed stomata (Arnon, 1967).

D. Pseudocyclic Photophosphorylation in vivo

Pseudocyclic ET is defined as noncyclic ET in which O_2 instead of $NADP^+$ is the terminal electron acceptor ("Mehler-reaction"). The characteristic of pseudocyclic PP in vivo are summarized in Table 4. The presence of the Mehler-reaction in vitro and its coupling can be easily demonstrated in broken chloroplasts by monitoring the light-induced O_2-uptake and its photosynthetic control by ADP (e.g. Whitehouse et al., 1971). Mass spectrometric gas exchange measurements with $^{18}O_2$ indicate that in vivo a light-induced O_2-uptake can also occur, but is masked normally by the photosynthetic O_2-evolution (Brown and Weis, 1959; Hoch et al., 1963; Jackson and Volk, 1969, 1970; Bunt and Heeb, 1971; Volk and Jackson, 1972). Only when photosynthetic evolution is impeded by exclusion of CO_2 can a light-stimulated net O_2-uptake be observed (e.g. Whitehouse et al., 1971). Unfortunately, no quantitative conclusion about the contribution of pseudocyclic ET in vivo could be drawn from these $^{18}O_2$-experiments, since in intact cells both dark respiration and photorespiration (glycolate oxidation) obscure the picture (Egneus et al., 1976). Measurements of different indicator reactions of in vivo PP with various plant subjects resulted in different interpretations of the physiological significance of pseudocyclic PP in vivo in the past. From studies of the light-induced incorporation of ^{32}P (Urbach and Gimmler, 1970; Ullrich, 1971) and the light-stimulated uptake of K^+ (Raven, 1969b) into unicellular green algae, the possible existence of pseudocyclic PP could not be excluded explicitly. However, a minor role was assigned to this reaction, especially with regard to the ATP requirements of CO_2-assimilation. Only under low light intensities in air-CO_2, a DCMU-sensitive PP could be observed (Urbach and Gimmler, 1970b). In contrast to these studies, the shrinkage experiments of Heber (1969; 1973b) and less convincing the ATP-measurements of Nobel (1968) implied a larger contribution of pseudocyclic PP to total PP in vivo under physiological conditions. Recent evidence confirm this view very strongly. Patterson and Myers (1973) demonstrated a considerable light-induced, DCMU-sensitive H_2O_2 formation in Anacystis, which could be stimulated by uncouplers. Similarly Glidewell and Raven (1976) and Raven and Glidewell (1976) showed in Hydrodictyon a light-induced uptake of $^{18}O_2$ parallel to photosynthetic O_2 evolution, which was accompanied by ATP synthesis. The light-stimulated O_2-uptake was sensitive to DCMU, but insensitive to concentrations of KCN which inhibited respiration as well as CO_2-assimilation. The coupling of this reaction to ATP-synthesis could

Table 4. Characteristics of pseudocyclic photophosphorylation in vivo

Photosystems	PS II and PS I needed. Evidence arises from: 1. Action spectra (1) and inactivity in far red light (1) 2. DCMU-sensitivity (1–5) 3. Experiments with mutants (1, 4)
Electron carriers involved	1. The electron carriers between PS II and PS I are identical with those in noncyclic ET from H_2O to NADP (cf. Table 5) 2. On the oxidizing site of PS I the participation of ferredoxin and an oxygen reducing factor (ORF) are discussed (15)

Sequence of electron carriers

$$H_2O \xrightarrow{} PS\ II \longrightarrow PQ \rightsquigarrow Cyt\ f \longrightarrow PC \longrightarrow PS\ I \longrightarrow Fd \longrightarrow ORF \dashrightarrow O_2$$

$$\uparrow_{hv} \quad DCMU \qquad DBMIB \qquad\qquad\qquad \uparrow_{h} \qquad DSPD$$

Inhibitors	1. DCMU (1–5) 2. DBMIB (16) 3. DSPD (??)
Requirements	O_2, high saturation in comparison to oxidative phosphorylation (3–5, 7)
Evidence for coupling to ATP-synthesis	1. ATP formation under exclusion conditions (7, 8) 2. Stimulation (or noninhibition) of pseudocyclic ET by uncouplers (2, 5) 3. Inhibition of indicator reactions for PP by uncouplers (1, 3)
Number of energy-conserving sites	Two, compare Table 5. P/e_2 ratio as in noncyclic PP
Quantum requirement	2 Einstein/mol equivalents (cf. Table 5)
Light saturation	Higher than that of cyclic PP (1, 4, 5), similar to that of CO_2-assimilation
Rates	Up to 24 μmol O_2 mg chlorophyll^{-1} h^{-1} (2, 4, 5, 6, 9) and 20 μmol ATP mg chlorophyll^{-1} h^{-1} (8) have been measured (conservative estimations)
Regulation	Low concentrations of O_2 inhibit (3–5, 7, 10), CO_2 stimulates (1, 6, 10). If the oxidation of the endogenous NADPH pool is limiting (e.g. because CO_2-fixation is limited by ATP), pseudocyclic ET is activated (6, 10, 11)
Physiological function	Supply of ATP for the uptake of ions (3, 12, 13, 17) or the formation of polyphosphates (14). Obligatory ATP-source for CO_2-assimilation (1, 6, 10, 11)?

References: (1) Heber (1969), (2) Glidewell and Raven (1976), (3) Raven and Glidewell (1976), (4) Heber and French (1968), (5) Patterson and Myers (1973), (6) Egneus et al. (1976), (7) Miginiac-Maslow and Moyse (1969), (8) Nobel (1968), (9) Whitehouse et al. (1971), (10) Heber (1973a), (11) Heber (1973b), (12) Jeschke (1970), (13) Raven (1970), (14) Ullrich and Simonis (1969), (15) Elstner and Heupel (1974), (16) Egneus (1975), (17) Ullrich (1972a).

be demonstrated for obvious reasons only under exclusion conditions. Oxidative phosphorylation was inhibited by the action of KCN and antimycin (Raven and Glidewell, 1976) cyclic PP by antimycin and noncyclic PP by KCN (via the inhibition of CO_2-assimilation). The remaining PP is sensitive to DCMU and requires O_2. It is strongly suppressed by uncouplers, whereas the corresponding pseudocyclic ET measured as light-stimulated O_2-uptake was not influenced or slightly stimulated. Even more striking is the demonstration that type A chloroplasts also exhibit a light-induced uptake of O_2 parallel to photosynthetic O_2-evolution (Egneus et al., 1976), because in this system dark respiration is absent and a correction can be made for glycolate oxidation. This O_2-reducing reaction was accompanied by H_2O_2-formation and occured only during CO_2-evolution with CO_2 as the electron acceptor, but not with PGA as acceptor. Since CO_2-assimilation (ATP/NADPH ratio 3/2) needs one more molecule than PGA-reduction (ATP/NADPH ratio 2/2), the pseudocyclic PP is considered to supply ATP under physiological conditions, where light and thereby ATP (Heber, 1973a) is rate-limiting (Egneus et al., 1976). The latter authors emphasize that the light-stimulated O_2-uptake was measured under low light intensities because the occurence of O_2-uptake under high light intensities could be interpreted as a valve-reaction for conditions of high electron pressures. The light saturation of pseudocyclic ET was found to be higher than that of cyclic PP (Heber and French, 1968; Heber, 1969; Patterson and Myers, 1973; but compare Urbach and Gimmler, 1970b). The absence of CO_2, restricting CO_2-assimilation, favors the pseudocyclic PP (Heber, 1969, 1973a,b; Raven and Glidewell, 1976). The oxygen saturation seems to be higher than that of oxidative phosphorylation.

The components of the electron transport chain of noncyclic and pseudocyclic PP between H_2O and PS I are essentially the same (cf. Tables 4 and 5), and it is also reasonable to believe that the same energy conservation sites are involved in both reactions. However, whereas in noncyclic PP the electrons are transferred from PS I via ferredoxin to $NADP^+$, the pathway of electrons from PS I to O_2 is still under investigation. Elstner and Heupel (1974) assume that the Mehler-reaction in vivo proceeds via ferredoxin and a special oxygen reducing factor ("ORF") which may be located in the vicinity of the $NADP^+$-ferredoxin oxidore-ductase. It should be mentioned that the interaction of O_2 with components of the ET chain after PS I are the most important, but not the only ones (Egneus, 1975).

The physiological function of pseudocyclic PP may be similar to that of cyclic PP, that is to produce "extra ATP" needed for the growth of the cell (Raven, 1974). However, Egneus et al. (1976) stress the function of pseudocyclic PP as an obligatory ATP-source for CO_2-assimilation. Since the availibility of ATP and not that of NADPH is probably rate-limiting in CO_2-fixation (Heber, 1973a), the endogenous pool of $NADP^+$ will soon be fully reduced during dark-light transients and $NADP^+$ reduction would be limiting, if electrons are not diverted to a second pathway. It may be suggested that pseudocyclic ET, coupled to ATP synthesis takes over and thereby removes the ATP limitation of CO_2 assimilation. Recently Jennings and Forti (1975) claimed that the pseudocyclic PP supplies CO_2 fixation with ATP during the induction period of photosynthesis, but not during the steady-state photosynthesis, a concept similar to that advanced for cyclic PP (Klob et al., 1972).

E. Noncyclic Photophosphorylation in vivo

Since the presence of noncyclic PP (ATP-formation coupled to photosynthetic ET from H_2O via PS II and I to $NADP^+$) in vivo is not questioned, it will only be reviewed briefly. As 2 NADPH and 3 ATP are required for the fixation of one molecule of CO_2 in the Calvin cycle (but compare also BEN-AMOTZ and AVRON, 1972) and as in vitro studies have shown that at least one molecule of ATP is formed in noncyclic ET per $NADP^+$ reduced, it may be assumed that at least the major part of ATP synthesis (2 ATP/CO_2), necessary for CO_2 assimilation, is coupled to noncyclic ET from H_2O to $NADP^+$. It should be kept in mind that from the experimental point of view it is difficult to discriminate between ATP formation by noncyclic and pseudocyclic ET. There are neither inhibitors, light qualities (as in cyclic PP) or other experimental conditions which allow pseudocyclic and noncyclic PP in vivo to be distinguished. Even in $N_2 + CO_2$ pseudocyclic PP may occur because O_2 is evolved in such a system. The properties of noncyclic PP in vivo are summarized in Table 5.

Most evidence from both in vivo and in vitro experiments support the requirement of two light reactions (PS II and PS I) in series for noncyclic PP (Z-scheme, but cf. ARNON et al., 1970, 1971; ARNON, 1971) and therefore, the discussion will be based on this particular scheme. Action spectra of noncyclic PP in vivo resemble those of photosynthetic CO_2-fixation (SIMONIS and MECHLER, 1963; RAVEN 1969a). Photosynthetic CO_2-assimilation exhibits the well known Emerson-enhancement effect (MYERS, 1971). The involvement of plastoquinone, cytochrome f, plastocyanine, P-700 and ferredoxin is demonstrated by both spectrophotometric measurements in vivo and in vitro as well as from extraction and replacement studies (cf. Table 5) with thylakoid fragments and is discussed in detail in the preceding chapters of this volume. Noncyclic PP in vivo is inhibited by DCMU, showing a requirement for photosystem II activity, while the inhibitory effect of DBMIB and DSPD (disalycylidenepropanediamine) provide specific in vivo evidence for the participation of plastoquinone and ferredoxin in this reaction, since DBMIB is a plastoquinone antagonist (TREBST et al., 1970) and DSPD an inhibitor of the photosynthetic reduction of ferredoxin (TREBST and BURBA, 1967). Contribution from in vivo experiments to the question of the number and the localization of energy conserving sites are rare. The main evidence in this respect is obtained by in vitro experiments. From experiments with broken chloroplasts evidence accumulates that two coupling sites are involved in noncyclic ET from H_2O to $NADP^+$ (TREBST and REIMER, 1973; IZAWA et al., 1973; REEVES and HALL, 1973; WEST and WISKICH, 1973; NYUNT and WISKICH, 1973; HEATHCOTE and HALL, 1974), one being located between the water-splitting apparatus and photosystem II, the other one between plastoquinone and cytochrome f. The latter coupling site was already postulated earlier from the stimulating effect of uncouplers on the rate of reduction of cytochrome f (AVRON and CHANCE, 1966; NISHIMURA, 1968). The P/e_2 ratios derived from in vitro studies varies between 1 and 2 (TREBST, 1974). Average numbers of 1.33 ATP/e_2 were found for the overall reaction of $NADP^+$-reduction and 0.66 ATP/e_2 if only PS II was functioning (TREBST and HAUSKA, 1974). P/e_2 ratios cannot be measured in intact cells and indirect estimations resulted in values "not much higher than 1" (HEBER, 1973a). The energy conservation site between the H_2O-splitting apparatus and PS II could be confirmed

Table 5. Characteristics of noncyclic photophosphorylation in vivo

Photosystems	PS II and PS I needed (but cf. 32–34). Evidence arises from: 1. Action spectra (1, 2) and nonactivity in far red light (3, 4) 2. DCMU sensitivity (3–8) 3. Experiments with mutants (9) 4. Enhancement of O_2 evolution or CO_2 fixation (10)
Electron carriers involved	Plastoquinone (11), cytochrome f (12), plastocyanine (13), P-700 (14), ferredoxin (15), ferredoxin-NADP$^+$-oxido-reductase (16), NADP$^+$
Sequence of electron carriers	

$$H_2O \xrightarrow{} PS\ II \xrightarrow{\qquad} PQ \xrightarrow{\qquad} Cyt\ f \to PC \to PS\ I \to Fd \to Fd\text{-}NADP^+\text{-}Red. \to NADP^+$$

hv DCMU   DBMIB hv DSPD

Inhibitors	1. DCMU (3–7) 2. DBMIB (17–19, 24, 26) 3. DSPD (20–23) 4. Indirect inhibition by conditions, which block photosynthetic CO_2-fixation ($-CO_2$, KCN, uncouplers, etc.)
Requirements	CO_2 (cf. 3, 4)
Evidence for coupling to ATP-synthesis	1. ATP synthesis (cf. 3, 4) 2. Inhibition of indicator reactions of PP by uncouplers (cf. 3)
Number of energy-conserving sites	Two sites (25–30), P/e_2 ratio 1–2 (25–30, 36), average P/e_2 ratio 1.33 (36)
Quantum requirement	2 Einstein/mol equivalents (31, 34, but cf. 35)
Light saturation	Similar as CO_2 fixation (7)
Rates	Up to 200 µmol ATP mg chlorophyll^{-1} h^{-1} have been measured (30, 38) (conservative estimations). True rates should be at least two times higher than the rate of CO_2 fixation
Physiological function	ATP for CO_2 fixation. It is improbable that noncyclic PP produces more ATP than needed for CO_2-fixation, probably less

References: (1) Simonis and Mechler (1963), (2) Raven (1969a), (3) Simonis and Urbach (1973), (4) Raven (1973), (5) Kylin et al. (1972), (6) van Rensen (1971), (7) Urbach and Gimmler (1970b), (8) Gingras and Lemasson (1965), (9) Levine (1969), (10) Myers (1971), (11) this vol. Chap. II.13, (12) this vol. Chap. II.12, (13) this vol. Chap. II.14, (14) this vol. Chap. II.4, (15) this vol. Chap. II.9, (16) this vol. Chap. II.11, (17) Gimmler and Avron (1972), (18) Bothe (1972), (19) Grimme (1972), (20) Trebst and Burba (1967), (21) Gimmler et al. (1968), (22) Urbach and Gimmler (1969), (23) Raven (1969b), (24) Trebst and Reimer (1973), (25) Izawa et al. (1973), (26) Gimmler (1973), (27) Reeves and Hall (1973), (28) Heathcote and Hall (1974), (29) West and Wiskich (1973), (30) Bornefeld et al. (1972), (31) Kok (1960), (32) Arnon et al. (1970), (33) Arnon (1971), (34) Arnon et al. (1971), (35) Heber (1973a), (36) Trebst and Hauska (1974), (37) Senger (1972), (38) Bottomley and Stewart (1976).

by in vivo measurements. As previously found in vitro in the presence of DBMIB a DCMU-sensitive ATP synthesis could be reactivated by the addition of benzoquinone in *Dunaliella* cells (Gimmler, 1973). This benzoquinone reduction was stimulated by uncouplers (Gimmler, 1972), whereas the coupled ATP synthesis was inhibited. Another coupling site could be demonstrated in the same system between plastoquinone and PS I, because DAD initiated a DCMU- and DBMIB-resistant ATP synthesis. However, one cannot be sure that this site is identical with the native site between plastoquinone and cytochrome f, since it might reflect an artificial coupling site (TREBST, 1974). From narrow interval titrations of PP in *Scenedesmus* with DCMU, KYLIN et al. (1972) proposed three different sites of energy conservation in noncyclic ET, but possibly these date are overinterpreted.

Beyond any doubt the physiological function on noncyclic PP is the supply of ATP for CO_2-fixation. Since the exact ATP/e_2 ratio in vivo is not known, it is uncertain whether noncyclic PP produces less ATP than required or more ATP than required for CO_2-fixation. As pointed out by HEBER and KIRK (1975) a ATP/e_2 ratio of 2 would be not much less of a problem than that of 1, if the ratios are constant. A flexible stoichiometry of the ATP/NADPH ratio is required to avoid accumulation of either ATP or NADPH (HEBER, 1976).

F. Regulation of Photophosphorylation in vivo

One of the outstanding features of in vivo systems is their ability for self-regulation, and this applies especially to the energy metabolism. Therefore, it is not surprising that in green cells the light-induced production of ATP is adapted in a very economic way to the actual requirements for ATP, which itself is governed to a large extent by the experimental conditions. Two mechanisms of regulation must be distinguished. The first operates via the phosphorylation potential. The consumption of ATP within the cell increases the rate of PP due to the photosynthetic control of ET by the adenylate system. The second mechanism involves the balance between ATP and reduction equivalents. This type of in vivo regulation is usually discussed on the basis of the Z-scheme which implies that noncyclic and pseudocyclic, as well as cyclic ET, are located in identical photosynthetic units in the thylakoid membranes (e.g. HEBER, 1973b, 1976). In a more comprehensive model, which takes into account some recent structural investigations, PARK and SANE (1971) proposed the existence of distinct photosystem I units in the stroma lamellae in addition to the PS II/PS I complex in the grana lamellae. These models are not mutually exclusive, but the regulatory mechanism suggested are rather different. In the latter system, the control of the ATP and NADPH balance required for CO_2-fixation would occur either via the phosphorylation potential (ATP is produced by the systems in the grana as well as in the stroma lamellae) or a soluble electron carrier, which mediates electron flow between plastocyanin of the noncyclic system in the grana lamellae and P-700 of the PS I units in the stroma lamellae (PARK and SANE, 1971). In the former, the redox levels of ferredoxin and plastoquinone are the critical factors, because ferredoxin is assumed to divert electrons from noncyclic electron transport to O_2 (pseudocyclic PP) or to cytochrome b (cyclic PP), whereas the electrons from the cyclic pathway reenter the noncyclic ET chain probably via plastoquinone. The redox level of ferredoxin is strongly dependent on the turnover of the $NADP^+/NADPH$ couple (ARNON and CHAIN, 1975). If CO_2 is absent, $NADP^+$ is completely reduced to NADPH, because there is no consumption of NADPH by the conversion of PGA to triose phosphate. The lack of reducible $NADP^+$ would lead to an overreduction of ferredoxin, if no other electron acceptors were available, and the ET from H_2O via both photosystems would stop. However, since in practice O_2 can be reduced by ferredoxin and the oxygen reducing factor (Elstner and Heupel, 1974) a pseudocyclic electron transport can take over and produce ATP. The ratio of pseudocyclic to noncyclic PP would be high in the absence of CO_2. However, even in the presence of CO_2 the pool size of reducible $NADP^+$ may limit noncyclic PP, if one assumes that the NADPH-oxidation during the conversion of PGA to triose phosphate is limited by the ATP required for this reaction (Heber, 1973a). Also in this case, some of the electrons must be diverted to O_2 in order to improve the ATP/NADPH balance. Nevertheless, the ratio between pseudocyclic and noncyclic PP would be much lower than in the absence of CO_2. If pseudocyclic PP is limited, e.g. under low O_2-pressures, it is still feasible that cyclic PP might replace or supplement pseudocyclic PP.

The operation of cyclic PP in intact cells requires that the pool of plastoquinone is not fully reduced, e.g. by high photosystem II activity and/or by low PS I activity (cf. Table 3). This follows from the stimulation of cyclic ET or PP by DCMU (Kaiser and Urbach, 1976; Rurainski and Hoch, 1976) in unicellular algae and type A chloroplasts. On the other hand the flow of electrons from PS II must not be completely inhibited. It seems that the absence of CO_2 and low pressures of O_2 produce conditions which favor cyclic PP in vivo.

G. Photophosphorylation in vivo and CO_2 Fixation

At present it is not known how much ATP can be formed during noncyclic PP in vivo. Only the maximal physiological ATP-requirement of the cells is known, which is governed by ATP-consumption during CO_2-fixation: three molecules of ATP are required for the fixation of one molecule of CO_2, corresponding to an ATP/e_2 ratio of 1.5. Since ATP/e_2 ratios cannot be measured directly in intact cells, such ratios estimated in thylakoid fragments are extrapolated to the in vivo situation (Heber, 1976). As already discussed these ratios vary between 1 and 2. A consequence of this unsatisfactory situation is that three different working hypotheses are still prevalent: noncyclic PP in vivo produces (1) less than 1.5 ATP/e_2, (2) exactly 1.5 ATP/e_2 and (3) more than 1.5 ATP/e_2. The first view is supported by indirect estimations of the ATP/e_2 ratio in vivo. The quantum requirement of PGA reduction in type A chloroplasts was found to be close to 4, whereas quantum requirement of CO_2 reduction was found to be 12 (Heber, 1973a, 1976). This was interpreted to indicate that ATP generated during the reduction of $NADP^+$ in vivo is insufficient to drive CO_2 reduction, corresponding to an ATP/e_2 ratio smaller than 1.5. This assumption is also supported by some theoretical considerations of Raven (1972). The latter view is favored by investigators who measure ATP/e_2 ratios of 2 in vitro (e.g. Heathcote and Hall, 1975) or a quantum requirement of 8 for photosynthetic CO_2 reduction in algae (Senger, 1972).

If the ATP/e_2 ratio in vivo is smaller than 1.5, either cyclic or pseudocyclic PP would be obligatory for CO_2 assimilation. Since cyclic PP can be inhibited completely without effecting CO_2 fixation as discussed above (cf. also Table 3), this suggests that cyclic PP is not an obligatory process for CO_2 fixation under normal conditions. Nonetheless, cyclic PP might supply ATP for CO_2 fixation, e.g. under conditions where the activity of other phosphorylating systems is limited. However, there is no experimental support for such a phenomenon as yet. Pseudocyclic PP, the other candidate for the role of the auxiliary ATP producer cannot be inhibited without affecting noncyclic PP in vivo. Therefore, it cannot be proven by inhibition experiments whether CO_2 assimilation needs supplementary ATP from pseudocyclic PP or not. However, in contrast to cyclic PP, the existence of which can be demonstrated only under exclusion conditions, pseudocyclic ET does occur parallel to photosynthetic CO_2 assimilation (PATTERSON and MYERS, 1973; GLIDEWELL and RAVEN, 1976; EGNEUS et al., 1976). Therefore, it is more likely that pseudocyclic PP provides the extra ATP for CO_2 fixation should this indeed be required.

If noncyclic PP produces exactly the amount of ATP required for CO_2 fixation ($ATP/e_2 = 1.5$) and no other ATP producing reactions exist in the light, no ATP would be available for other ATP-consuming reactions (Table 2). If noncyclic PP produces more ATP than needed for CO_2 assimilation ($ATP/e_2 = 2$), ATP would accumulate rapidly during CO_2 assimilation and NADPH would be limiting (HEBER and KIRK, 1975), unless ATP were consumed rapidly by other reactions. ATP consuming reactions having such high rates, other than CO_2 assimilation, have not been measured in intact cells. Therefore, it would be very desirable even under these favorable conditions to have an intracellular mechanism able to regulate the ATP/NADPH ratio (HEBER and KIRK, 1975). But this mechanism may not necessarily be that proposed by HEBER (1973a, b) (variation of the ATP/NADPH ratio by the operation of pseudocyclic PP). An alternative scheme has been offered by TREBST (1974). A flexible coupling could be caused also by changes of the proton conductivity of the thylakoids (chemiosmotic mechanism). Flexible coupling might be not only a passive consequence of ion conductivity of thylakoid membranes and the chloroplast envelope, but may be "regulated by special mechanisms" (HEBER, 1976).

H. Concluding Remarks

Since the appearance of the corresponding article about PP in vivo in the first edition of this encyclopedia 16 years ago (SIMONIS, 1960), a number of advances have been made, although in comparison to the progress observed in PP in vitro during this time, the progress in PP in vivo beyond any doubt lags behind. Nonetheless, it has been established that all green cells possess not only the ability to perform noncyclic PP, but also cyclic and pseudocyclic PP. The constituents of these different types of electron transport chains have been made clear, as well as their properties and requirements. However, it is, as yet, unclear how far cells make use of their ability to perform cyclic or pseudocyclic PP. So far the existence of cyclic PP can be demonstrated only under conditions which exclude other phosphorylating conditions and it is established that it is not obligatory for CO_2 assimilation. Since pseudocyclic ET does occur in parallel to CO_2 assimilation, it seems more probable that this is the reaction supplementing CO_2 fixation with "extra ATP." However, it should be remembered that the stoichiometry of noncyclic PP in vivo (ATP/NADPH ratio) is illdefined and thus it remains an open question whether CO_2 assimilation indeed requires ATP in addition to that produced by noncyclic PP. This is the main question which has to be solved in the future. We do not expect very much from a more extensive hunt for higher rates of in vivo PP, although measured rates still account only for maximally 10% of the true rates of endogenous PP. In the future *regulatory mechanisms* of in vivo PP should be studied more carefully. The interest of investigations should be focused also on *phylogenetic* and *ontogenetic* aspects of in vivo PP to avoid unwarranted generalizations.

References

Amesz, J.: Biochim. Biophys. Acta **301**, 35–51 (1973)

Amesz, J., Pulles, M.P.J., Visser, J.W.M., Sibbing, F.A.: Biochim. Biophys. Acta **275**, 442–452 (1972)

Arnon, D.I.: Ann. Rev. Plant Physiol. **7**, 325–354 (1956)

Arnon, D.I.: Physiol. Rev. **47**, 317–358 (1967)

Arnon, D.I.: Proc. Nat. Acad. Sci. U.S. **68**, 2883–2892 (1971)

Arnon, D.I., Chain, R.K.: Proc. Nat. Acad. Sci. U.S. **72**, 4961–4965 (1975)

Arnon, D.I., Chain, R.K., McSwain, B.D., Tsujimoto, H.Y., Knaff, D.B.: Proc. Nat. Acad. Sci. U.S. **67**, 1404–1409 (1970)

Arnon, D.I., Knaff, D.B., McSwain, B.D., Chain, R.K., Tsujimoto, H.Y.: Photochem. Photobiol. **14**, 397–425 (1971)

Avron, M. (ed.): Proc. 3rd Intern. Congr. Photosynthesis. Amsterdam: Elsevier, 1975, 3 Volumes

Avron, M., Chance, B.: Brookhaven Symp. Biol. **19**, 149–160 (1966)

Avron, M., Neumann, J.: Ann. Rev. Plant Physiol. **19**, 137–166 (1968)

Baltscheffsky, H.: Proc. 3rd Intern. Congr. Photosynthesis. Avron, M. (ed.). Amsterdam: Elsevier, 1975, Vol. III, pp. 2061–2066

Baltscheffsky, H., Baltscheffsky, M., Thore, A.: Curr. Top. Bioenerg. **4**, 273–325 (1971)

Baltscheffsky, M.: Proc. 3rd Intern. Congr. Photosynthesis. Avron, M. (ed.). Amsterdam: Elsevier, 1975, Vol. I, pp. 799–806

Barber, J.: Nature (London) **217**, 876–877 (1968a)

Barber, J.: Biochim. Biophys. Acta **163**, 141–149 (1968b)

Bedell, G.W., Govindjee: Plant Cell Physiol. **14**, 1081–1097 (1973)

Ben-Amotz, A., Avron, M.: Plant Physiol. **49**, 244–248 (1972)

Bishop, N.I.: Nature (London) **195**, 55–57 (1962)

Bishop, N.I.: Ann. Rev. Plant. Physiol. **17**, 185–208 (1966)

Bishop, N.I.: Photochem. Photobiol. **6**, 621–628 (1967)

Bishop, N.I.: In: Photophysiology. Giese, A.C. (ed.). New York: Academic Press, 1973, Vol. VIII, pp. 65–96

Bishop, N.I., Gaffron, H.: In: Photosynthetic Mechanism of Green Plants. Kok, B., Jagendorf, A.T. (eds.). Washington: Nat. Acad. Sci., 1963, pp. 441–451

Böhme, H., Reimer, S., Trebst, A.: Naturforsch. **268b**, 341–352 (1971)

Bornefeld, T.: Biochem. Physiol. Pflanzen **170**, 333–344 (1976a)

Bornefeld, T.: Biochem. Physiol. Pflanzen **170**, 345–353 (1976b)

Bornefeld, T., Domanski, J., Simonis, W.: In: Proc. 2nd Intern. Congr. Photosynthesis. Forti, G., Avron, M., Melandri, A. (eds.). The Hague: Dr. Junk, 1972, Vol. II, pp. 1379–1386

Bornefeld, T., Simonis, W.: In: Proc. 3rd Intern. Congr. Photosynthesis. Avron, M. (ed.).
 Amsterdam: Elsevier, 1975, Vol. II, pp. 1557–1565
Bothe, H.: Z. Naturforsch. **24 B**, 1574–1582 (1969)
Bothe, H.: In: Proc. 2nd Intern. Congr. Photosynthesis. Forti, G., Avron, M., Melandri,
 A. (eds.). The Hague: Dr. Junk, 1972, Vol. III, pp. 2169–2177
Bottomley, P.J., Stewart, W.D.P.: Arch. Microbiol. **108**, 249–258 (1976)
Bradley, D.F.: Arch. Biochem. Biophys. **68**, 172–185 (1957)
Broda, E.: Progr. Biophys. Mol. Biol. **21**, 143–208 (1970)
Brown, A.H., Weis, D.: Plant Physiol. **34**, 224–234 (1959)
Bunt, J.S., Heeb, M.A.: Biochim. Biophys. Acta **226**, 354–359 (1971)
Cox, R.M., Fay, P.: Proc. Roy. Soc., Ser. B **172**, 357–366 (1969)
Egneus, H.: Physiol. Plantarum **33**, 203–213 (1975)
Egneus, H., Heber, U., Matthiesen, U.: Biochim. Biophys. Acta **408**, 252–268 (1975)
Elstner, E.F., Heupel, A.: Z. Naturforsch. **29 C**, 564–571 (1974)
Fay, P.: Biochim. Biophys. Acta **216**, 353–356 (1970)
Forti, G., Avron, M., Melandri, B.A. (eds.): Proc. 2nd Intern. Congr. Photosynthesis. The
 Hague: Dr. Junk, 1972, 3 Vol
Forti, G., Parisi, B.: Biochim. Biophys. Acta **71**, 1–6 (1963)
Forti, G., Rosa, L.: FEBS Lett. **18**, 55–58 (1971)
Frenkel, A.W.: Biol. Rev. **45**, 569–616 (1970)
Gimmler, H.: In: 6th Intern. Congr. Photobiology, Book of Abstracts. Schenck, G.O. (ed.).
 Bochum: 1972, p. 283
Gimmler, H.: Z. Pflanzenphysiol. **68**, 289–307 (1973)
Gimmler, H., Avron, M.: In: Proc. 2nd Intern. Congr. Photosynthesis. Forti, G., Avron,
 M., Melandri, A. (eds.). The Hague: Dr. Junk, 1972, Vol. I, pp. 789–800
Gimmler, H., Neimanis, S., Eilmann, I., Urbach, W.: Z. Pflanzenphysiol. **64**, 358–366 (1971)
Gimmler, H., Simonis, W., Urbach, W.: Naturwissenschaften **56**, 371–372 (1969)
Gimmler, H., Urbach, W., Jeschke, W.D., Simonis, W.: Z. Pflanzenphysiol. **58**, 353–364
 (1968)
Gingras, G.: Physiol. Vegetal. **4**, 1–65 (1966)
Gingras, G., Lemasson, C.: Biochim. Biophys. Acta **109**, 67–78 (1965)
Glagoleva, T., Chulanovskaya, M.V., Zalenskii, O.V.: Photosynthetica **6**, 354–363 (1972)
Glagoleva, T., Zalenskii, O.V.: Photosynthetica **4**, 15–20 (1970)
Glidewell, S.M., Raven, J.A.: J. Exp. Botany **26**, 479–488 (1975)
Grimme, L.H.: In: Proc. 2nd Intern. Congr. Photosynthesis. Forti, G., Avron, M., Melandri,
 A. (eds.). The Hague: Dr. Junk, 1972, Vol. III, pp. 2009–2019
Gromet-Elhanan, Z.: Biochim. Biophys. Acta **131**, 526–537 (1967)
Hall, D.O.: Nature New Biol. **235**, 125–126 (1972)
Heathcote, P., Hall, D.O.: Biochem. Biophys. Res. Commun. **56**, 767–774 (1974)
Heathcote, P., Hall, D.O.: In: Proc. 3rd Intern. Congr. Photosynthesis. Avron, M. (ed.).
 Amsterdam: Elsevier, 1975, Vol. I, pp. 463–471
Heber, U.: Biochim. Biophys. Acta **180**, 302–319 (1969)
Heber, U.: Biochim. Biophys. Acta **305**, 140–152 (1973a)
Heber, U.: Ber. Deut. Botan. Ges. **86**, 187–195 (1973b)
Heber, U.: J. Bioenerg. Biomembranes **9**, 157–171 (1976)
Heber, U., French, C.S.: Planta (Berl.) **79**, 99–112 (1968)
Heber, U., Kirk, M.: Biochim. Biophys. Acta **376**, 136–150 (1975)
Hoch, G.E.: Methods Enzymol. **24 B**, 297–303 (1972)
Hoch, G.E., Owens, O.V.H., Kok, B.: Arch. Biochem. Biophys. **101**, 171–180 (1963)
Hoch, G.E., Randles, J.: Photochem. Photobiol. **14**, 435–449 (1971)
Huber, S.C., Edwards, G.E.: FEBS Lett. **58**, 211–214 (1975)
Izawa, S., Gould, J.M., Ort, D.R., Felker, P., Good, N.E.: Biochim. Biophys. Acta **305**,
 119–128 (1973)
Jackson, W.A., Volk, R.J.: Nature (London) **222**, 269–271 (1969)
Jackson, W.A., Volk, R.J.: Ann. Rev. Plant Physiol. **21**, 385–432 (1970)
Jennings, R.C., Forti, G.: In: Proc. 3rd Intern. Congr. Photosynthesis. Avron, M. (ed.).
 Amsterdam: Elsevier, 1975, Vol. I, pp. 735–742
Jeschke, W.D.: Planta (Berl.) **73**, 161–174 (1967)
Jeschke, W.D.: Planta (Berl.) **91**, 111–128 (1970)

Jeschke, W.D.: Planta (Berl.) **103**, 164–180 (1972a)
Jeschke, W.D.: Z. Pflanzenphysiol. **66**, 409–419 (1972b)
Jeschke, W.D., Gimmler, H., Simonis, W.: Plant Physiol. **42**, 380–386 (1967)
Jeschke, W.D., Simonis, W.: Planta (Berl.) **67**, 6–32 (1965)
Jeschke, W.D., Simonis, W.: Planta (Berl.) **88**, 157–171 (1969)
John, J.B.S.: Anal. Biochem. **37**, 409–416 (1970)
Kaiser, W., Urbach, W.: Ber. Deut. Botan. Ges. **86**, 213–226 (1973)
Kaiser, W., Urbach, W.: Ber. Deut. Botan. Ges. **87**, 145–153 (1974)
Kaiser, W., Urbach, W.: Biochim. Biophys. Acta **423**, 91–102 (1976)
Kandler, O.: Z. Naturforsch. **5B**, 423–437 (1950)
Kandler, O.: Z. Naturforsch. **9B**, 624–644 (1954)
Kandler, O.: Z. Naturforsch. **12B**, 271–280 (1957)
Kandler, O.: Ann. Rev. Plant Physiol. **11**, 37–54 (1960)
Kandler, O., Liesenkoetter, I., Oaks, B.B.: Z. Naturforsch. **16B**, 50–61 (1961)
Kandler, O., Tanner, W.: Ber. Deut. Botan. Ges. **79**, 48–57 (1966)
Kessler, E., Hofmann, A., Zunft, W.: Arch. Mikrobiol. **72**, 23–26 (1970)
Klob, W., Tanner, W., Kandler, O.: In: Proc. 2nd Intern. Congr. Photosynthesis. Forti,
    G., Avron, M., Melandri, A. (eds.). The Hague: Dr. Junk, 1972, Vol. III, pp. 1998–2010
Kok, B.: In: Handbuch der Pflanzenphysiologie. Ruhland, H. (ed.). Berlin: Springer, 1960,
    Vol. V, pp. 566–633
Krause, C.: Photosynthetica **9**, 412–453 (1975)
Krause, G.H., Heber, U.: In: Proc. 1st Europ. Biophys. Congr. Broda, E., Locher, A., Springer-
    Lederer, H. (eds.). Vienna: Wiener Med. Akad., 1971, pp. 79–84
Kylin, A.: Physiol. Plantarum **19**, 644–649 (1966)
Kylin, A., Sundberg, J., Tillberg, J.E.: Physiol. Plantarum **27**, 376–383 (1972)
Kylin, A., Tillberg, J.E.: Z. Pflanzenphysiol. **58**, 165–174 (1967)
Levine, R.P.: Ann. Rev. Plant. Physiol. **20**, 523–540 (1969)
Lewenstein, A., Bachofen, R.: Biochim. Biophys. Acta **267**, 80–85 (1972)
MacLachan, G.A., Porter, H.K.: Proc. Roy. Soc., Ser. B **150**, 460–473 (1959)
MacRobbie, E.A.C.: Biochim. Biophys. Acta **94**, 64–73 (1965)
MacRobbie, E.A.C.: Quart. Rev. Biophys. **3**, 251–294 (1970)
Metzner, H. (ed.): Progress in Photosynthesis Research. Tübingen: Intern. Union Biol. Sci.,
    1969, 3 Vols.
Miginiac-Maslow, M., Moyse, A.: In: Progress in Photosynthesis Research. Metzner, H. (ed.).
    Tübingen: Laupp, 1969, Vol. III, pp. 1203–1212
Mohr, H.: Naturwiss. Rundschau **28**, 154–160 (1975)
Myers, J.: Ann. Rev. Plant Physiol. **22**, 289–312 (1971)
Myers, J.: Plant Physiol. **54**, 420–426 (1974)
Nishimura, M.: Biochim. Biophys. Acta **153**, 838–847 (1968)
Nobel, P.S.: Plant Physiol. **42**, 1389–1394 (1967)
Nobel, P.S.: Biochim. Biophys. Acta **153**, 170–182 (1968)
Nobel, P.S., Chang, D., Wang, C.T., Smith, S.S., Barcus, D.E.: Plant Physiol. **44**, 655–661
    (1969)
Nultsch, W.: Z. Pflanzenphysiol. **56**, 1–11 (1967)
Nultsch, W.: Photochem. Photobiology **10**, 119–123 (1969)
Nultsch, W.: In: Photobiology of Microorganisms. Halldal, P. (ed.). London: Wiley, 1970,
    pp. 213–251
Nultsch, W.: In: Primary Events in Photobiology. Checcuci, A., Weale, R.A. (eds.). Amster-
    dam: Elsevier, 1973, pp. 245–273
Nyunt, U.T., Wiskich, J.T.: Plant Cell Physiol. **14**, 1099–1106 (1973)
Oelze-Karow, H., Butler, W.L.: Plant Physiol. **48**, 621–625 (1971)
Okkeh, A., Kylin, A.: Physiol. Plantarum **33**, 118–123 (1975)
Okkeh, A., Tillberg, J.E., Kylin, A.: Physiol. Plantarum **33**, 124–127 (1975)
Owens, O., Krauss, R.W.: Plant Physiol. Suppl. **49**, 52 (1972)
Park, R.B., Sane, P.V.: Ann. Rev. Plant Physiol. **22**, 395–430 (1971)
Patterson, P.C.O., Myers, J.: Plant Physiol. **51**, 104–109 (1973)
Ramirez, J.U., Del Campo, F.F., Arnon, D.I.: Proc. Nat. Acad. Sci. U.S. **59**, 606–612 (1968)
Raven, J.A.: New. Phytologist **68**, 45–62 (1969a)
Raven, J.A.: New Phytologist **68**, 1089–1113 (1969b)

Raven, J.A.: J. Exp. Botany **21**, 1–16 (1970)
Raven, J.A.: J. Exp. Botany **22**, 420–433 (1971)
Raven, J.A.: New Phytologist **71**, 227–247 (1972)
Raven, J.A.: In: Algal Physiology and Biochemistry. Stewart, W.D.P. (ed.). Oxford: Blackwell Scientific, 1974, pp. 391–423
Raven, J.A.: J. Exp. Botany **25**, 221–229 (1974)
Raven, J.A., Glidewell, S.M.: New Phytologist **75**, 197–204 (1975)
Reeves, S.G., Hall, D.O.: Biochim. Biophys. Acta **314**, 66–78 (1973)
Rensen, J.J.S. van: In: Progress in Photosynthesis Research. Metzner, H. (ed.). Tübingen: Laupp, 1969, Vol. III, pp. 1769–1776
Rensen, J.J.S. van: Thesis. Landbouwhogeschool Wageningen, Holland. Wageningen: Veenman & Zonen, 1971, pp. 1–80
Ried, A.: Biochim. Biophys. Acta **153**, 653–663 (1968)
Ried, A.: In: Progress in Photosynthesis Research. Metzner, H. (ed.). Tübingen: Laupp, 1969, Vol. I, pp. 521–530
Rurainski, H.J., Hoch, G.E.: Z. Naturforsch. **30C**, 761–770 (1975)
Rurainski, H.J., Randles, J., Hoch, G.E.: Biochim. Biophys. Acta **205**, 254–262 (1970)
Santarius, K.A., Heber, U.: Biochim. Biophys. Acta **102**, 39–54 (1965)
Santarius, K.A., Heber, U., Ullrich, W., Urbach, W.: Biochem. Biophys. Res. Commun. **15**, 139–146 (1964)
Schürmann, P., Buchanan, B.B., Arnon, D.I.: Biochim. Biophys. Acta **267**, 111–124 (1972)
Senger, H.: Planta (Berl.) **90**, 243–266 (1970a)
Senger, H.: Planta (Berl.) **92**, 327–346 (1970b)
Senger, H.: In: Proc. 2nd Intern. Congr. Photosynthesis. Forti, G., Avron, M., Melandri, A. (eds.). The Hague: Dr. Junk, 1972, Vol. I, pp. 723–730
Simonis, W.: In: Handbuch der Pflanzenphysiologie. Ruhland, H. (ed.). Berlin: Springer, 1960, Vol. V, pp. 966–1007
Simonis, W.: Ber. Deut. Botan. Ges. **77**, 5–13 (1964)
Simonis, W.: Ber. Deut. Botan. Ges. **80**, 395–402 (1967)
Simonis, W.: In: Theoretical Foundations of the Photosynthetic Productivity. Nichiporovich, A.A. (ed.). Moscow: Nauska, 1972, pp. 75–83
Simonis, W., Mechler, E.: Biochem. Biophys. Res. Commun. **13**, 241–245 (1963)
Simonis, W., Urbach, W.: Ann. Rev. Plant Physiol. **24**, 89–114 (1973)
Smith, F.A.: Biochim. Biophys. Acta **126**, 94–99 (1966)
Smith, F.A., Raven, J.A.: New Phytologist **73**, 1–12 (1974)
Strehler, B.L.: Arch. Biochem. Biophys. **43**, 67–79 (1953)
Strehler, B.L.: In: Methoden der enzymatischen Analyse. Bergmeyer, H.U. (ed.) Weinheim: Chemie, 1970, Vol. II, pp. 2036–2050
Sybesma, C.: In: Photobiology of Microorganisms. Halldal, P. (ed.). London: Wiley-Interscience, 1970, pp. 57–93
Tagawa, K., Tsujimoto, H.Y., Arnon, D.I.: Proc. Nat. Acad. Sci. U.S. **49**, 567–572 (1963)
Tanner, W.: Biochem. Biophys. Res. Commun. **36**, 278–283 (1969)
Tanner, W., Dächsel, L., Kandler, O.: Plant Physiol. **40**, 1151–1156 (1965)
Tanner, W., Kandler, O.: In: Progress in Photosynthesis Research. Metzner, H. (ed.). Tübingen: Laupp, 1969, Vol. III, pp. 1217–1223
Tanner, W., Löffler, M., Kandler, O.: Plant Physiol. **44**, 422–428 (1969)
Tanner, W., Loos, E., Klob, W., Kandler, O.: Z. Pflanzenphysiol. **59**, 301–303 (1968)
Tanner, W., Zinecker, U., Kandler, O.: Z. Naturforsch. **22B**, 358–359 (1967)
Teichler-Zallen, D., Hoch, G.E.: Arch. Biochem. Biophys. **120**, 227–230 (1967)
Teichler-Zallen, D., Hoch, G.E., Bannister, T.T.: In: Proc. 2nd Intern. Congr. Photosynthesis. Forti, G., Avron, M., Melandri, A. (eds.). The Hague: Dr. Junk, 1972, Vol. I, pp. 643–647
Trebst, A.: Ann. Rev. Plant Physiol. **25**, 423–458 (1974)
Trebst, A., Burba, M.: Z. Pflanzenphysiol. **57**, 419–433 (1967)
Trebst, A., Harth, E., Draber, W.: Z. Naturforsch. **25B**, 1157–1159 (1970)
Trebst, A., Hauska, G.: Naturwissenschaften **61**, 308–316 (1974)
Trebst, A., Reimer, S.: Biochim. Biophys. Acta **305**, 129–139 (1973)
Ullrich-Eberius, C.I.: Planta (Berl.) **115**, 25–36 (1973a)
Ullrich-Eberius, C.I.: Planta (Berl.) **109**, 161–176 (1973b)
Ullrich-Eberius, C.I., Simonis, W.: Planta (Berl.) **93**, 214–226 (1970)

Ullrich, W.R.: Planta (Berl.) **90**, 272–285 (1970)
Ullrich, W.R.: Planta (Berl.) **100**, 18–30 (1971)
Ullrich, W.R.: Planta (Berl.) **102**, 37–54 (1972a)
Ullrich, W.R.: Arch. Mikrobiol. **87**, 323–339 (1972b)
Ullrich, W.R.: Planta (Berl.) **116**, 143–152 (1974)
Ullrich, W.R., Simonis, W.: Planta (Berl.) **84**, 358–367 (1969)
Urbach, W., Gimmler, H.: In: Progress in Photosynthesis Research. Metzner, H. (ed.). Tübingen: Laupp, 1969, Vol. III, pp. 1274–1280
Urbach, W., Gimmler, H.: Ber. Deut. Botan. Ges. **83**, 439–442 (1970a)
Urbach, W., Gimmler, H.: Z. Pflanzenphysiol. **62**, 276–286 (1970b)
Urbach, W., Kaiser, W.: In: Proc. 2nd Intern. Congr. Photosynthesis. Forti, G., Avron, M., Melandri, A. (eds.). The Hague: Dr. Junk, 1972, Vol. II, pp. 1401–1411
Urbach, W., Simonis, W.: Biochem. Biophys. Res. Commun. **17**, 39–45 (1964)
Volk, R.J., Jackson, W.A.: Plant Physiol. **49**, 218–223 (1972)
Weichart, G.: Planta (Berl.) **56**, 262–289 (1961)
Weigl, J.: Planta (Berl.) **75**, 327–342 (1967)
West, K.R., Wiskich, J.T.: Biochim. Biophys. Acta **292**, 197–205 (1973)
Whitehouse, D.G., Ludwig, L.J., Walker, D.A.: J. Exp. Botany **22**, 772–791 (1971)
Wiessner, W.: Nature (London) **205**, 56–57 (1965)
Wiessner, W.: Nature (London) **212**, 403–404 (1966)
Wiessner, W.: In: Photobiology of Microorganisms. Halldal, P. (ed.). London: Interscience, 1970a, pp. 95–133
Wiessner, W.: In: Photobiology of Microorganisms. Halldal, P. (ed.). London: Interscience, 1970b, pp. 135–163
Wiessner, W., Gaffron, H.: Federation Proc. **23**, 226 (1964)
Wintermans, J.F.G.M.: Medelingen. Landbouwhogeschool Wageningen (Netherlands) **55**, 69–126 (1955)

14. Delayed Luminescence

S. Malkin

A. General

I. Introduction

Photosynthetic organisms are able to emit light for some time after an exposure to illumination. This delayed luminescence (d.l.) is to be distinguished from prompt fluorescence which is emitted during illumination and decays almost instantaneously, when the exciting light is extinguished. Delayed luminescence is a general phenomenon manifested by all intact photosynthetic systems investigated so far: microorganisms, leaves of higher plants, photosynthetic bacteria and active preparates such as chloroplasts and chromatophores. According to its spectrum d.l. originates from chlorophyll a (bacteriochlorophyll a in purple bacteria).

II. Relevance to Photosynthesis

Delayed luminescence from photosynthetic organisms was discovered quite accidentally by Strehler and Arnold, during an attempt to measure photophosphorylation by the luciferin-luciferase luminescence method (Strehler and Arnold, 1951). Following the discovery, the close connection to the photosynthetic process was established by studying the influence of factors which also affect the photosynthetic process (Arthur and Strehler, 1957; Strehler and Lynch, 1957). It was suggested that d.l. is a result of a reverse process of the primary photochemical event of photosynthesis. Further studies confirmed that indeed d.l. is intimately connected with the photosynthetic process. It involves the pigment system, the charge separation process (cf. Sect. D.I) and is considerably influenced (a surprising result of the last decade of research) by characteristic membrane parameters, such as pH gradients and electric potential difference (cf. Sect. C.III.5).

It appears that d.l. may be used to probe a variety of parameters which change during adjustment to steady-state photosynthesis. In itself, however, it reflects an insignificant loss of energy, compared to the other processes leading to photosynthesis (Table 1a).

III. Origin of Delayed Luminescence in General Systems
(cf. Parker, 1964)

Emission originates from molecules in an electronically excited state: such molecules return to the ground state by either radiative transition (photon emission) or by non-radiative transition (losing the extra energy by heating the surrounding medium). *Prompt fluorescence* emission is due to excited singlet states which were formed directly by the light-absorption process. *Phosphorescence* is a relatively long-lived emission originating from excited triplet states, which are produced from the excited singlet state (such emission was not detected from photosynthetic organisms). *Delayed luminescence* is observed when there is at first a production of relatively stabilized products from the excited state, followed by a back-reaction which again forms the excited state. This can be summarized by the following equations:

$$M + h\nu \rightarrow {}^1M^* \qquad \text{Absorption} \tag{1}$$

$${}^1M^* \rightarrow M + h\nu' \quad \text{Emission} \tag{2}$$

$${}^1M^* \rightleftarrows M' \tag{3}$$

$${}^1M^* + A \rightleftarrows M + A' \tag{3a}$$

Mono or bimolecular photochemical transformations

d.l. is the result of the reversal of Eq. (3) or Eq. (3a) followed by Eq. (2).

The details of reactions Eqs. (3) or (3a) give rise to the classification of d.l. in several ways. The first possibility is concerned with excited-state physics only. Thus, M' may be long lived excited triplet state ($^3M^*$). This is a very common type for organic molecules in solution, and is further classified according to the way $^1M^*$ is produced again from $^3M^*$ (either directly: $^3M^* \rightarrow {}^1M^*$, or by interaction with another triplet: $2\,^3M^* \rightarrow {}^1M^* + M$) (PARKER, 1964).

The second possibility is a photochemical production of molecular fragments (radicals) which after recombination lead back to an excited state. Aromatic amines, acriflavin and related compounds, for example, at low temperatures and rigid media give rise to a photochemical reaction in which there is an electron ejection to the medium (solvated electron) and formation of a positive ion ("hole") ($R^* \rightarrow R^+ + e^-$) leading to d.l. by the reverse reaction (LINSCHITZ et al., 1954; LIM and SWENSON, 1962; LIM and WENG, 1963). In these examples d.l. is produced when the system is warmed.

Similar mechanisms as outlined above operate also in solid systems, such as organic semiconductors. In such systems, cooperative interactions introduce new aspects, such as the mobility of the excitation, and also of the electrons and holes formed in a photoionization process (GUTMAN and LYONS, 1967).

B. Methods

I. General

Essentially, d.l. measurements require a fast shutter to stop the pre-illumination light and a sensitive sensor (a photomultiplier operated at its maximum sensitivity). In addition, the weakness of d.l. requires protection of the photomultiplier from the several fold stronger, scattered light and fluorescence during the pre-illumination time. These factors call for some sophistication in the experimental system. Several methods are outlined below:

(1) Flow system: the sample flows in a closed circuit. Through the flow line there are segments for irradiation and for observation of delayed light (STREHLER and ARNOLD, 1951). (2) Becquerel type phosphoroscope: the sample is irradiated by light which is chopped by a rotating wheel containing closed and opened segments. During irradiation, another rotating wheel placed between the sample and the sensor, screens the sensor, but opens the way to d.l. when the sample is not illuminated (ARTHUR and STREHLER, 1957). This technique was also utilized for single flash experiments (ZANKEL, 1971). (3) Double shutter techniques: the sample is irradiated through one shutter and the sensor receives d.l. through another shutter, which is opened only when the pre-illumination light stops (BERTSCH and AZZI, 1965; MALKIN and HARDT, 1971). (4) Fast electronic gating built in the sensor which is synchronized to shut off the sensor response during pre-illumination, thus preventing the need for screening (RUBY, 1968; STACY et al., 1971).

There is always a finite "dead" time which elapses, in all methods, between the end of irradiation to commencement of observation. In an orthodox use of the above methods, the minimal time may be in the order of 0.1–10 msec. It is possible, however, to achieve transit time of about 250 nsec by the use of a focused continuous laser beam and fast chopping by a rotating sector. This is about the minimal time necessary for d.l. measurements. A faster chopping (25 nsec) was achieved by electro-optical modulation of the laser beam (HAUG et al., 1969).

II. Induction and Transient Effects

Delayed luminescence is influenced by several photosynthetic parameters (e.g. concentrations of reduced electron acceptor, oxidized electron donor, S-state of the oxygen system, membrane parameters) which change continuously during illumination and tend towards a steady-state value. In principle, when one stops the pre-illumination at certain times and observe d.l., the resulting behavior of d.l. (i.e. its extrapolated initial value, as well as its entire decay kinetics) is dependent on the value of all these photosynthetic parameters. This change of d.l. vs. pre-illumination time (t_l) is called *induction* and is reflecting the combined effects of the photosynthetic parameters on d.l. as they change during t_l.

A common way to study such induction is by the Becquerel phosphoscope method, fixing a certain observation time (t_L)—quite often in the msec time range. d.l. changes gradually as the excitation with the chopped light continues. The exciting light serves the double purpose of (slow) actinic action (changes of properties of the system), and measuring action, imposing relatively fast processes of excitation, which are monitored by d.l.

The relaxation of the system back to the dark-steady-state is measured indirectly by the same method. After switching off the exciting light for a given period (t_d), the light is switched on again and d.l. measurement is repeated. The initial value of d.l. at the moment of switching on defines d.l. as a function of t_d. By extrapolation to long values of t_d one can estimate the steady-state "dark" value of L.

For full utilization of the method it is incorrect to assume that only the value of d.l. is changed but not its entire kinetics. In fact, this is not true (ITOH and

Murata, 1974). The correct representation of induction requires several measurements with different times t_L. This fact has been frequently ignored.

This method is particularly suited to study transient responses and changes in steady-state values of d.l. caused by the responses' addition of various reagents (some representative examples are: Mayne, 1967; Clayton, 1969; Bertsch et al., 1969; Barber, 1972).

C. Phenomenology

I. Emission and Excitation Spectrum

Almost all d.l. phenomena described below have emission spectrum of chlorophyll a in vivo (or bacteriochlorophyll a in photosynthetic bacteria), and cannot be distinguished, on this basis, from prompt fluorescence (Strehler and Arnold, 1951; Arnold and Davidson, 1954; Hardt and Malkin, 1974). The excitation spectrum corresponds in general to the action spectrum of photosynthesis, particularly to that of photosystem II (Arnold and Thompson, 1956; Goedheer, 1962; Mayne, 1968). For exceptional cases see Section D.I.4.

II. Decay of Delayed Luminescence

From the end of irradiation, d.l. is observed as a decaying function in time. The decay is complex: it can be divided into segments or phases ranging over many orders of magnitude. The decay is sometimes represented mathematically at least in a formal way, by a sum of decaying exponentials (Ruby, 1968; Lavorel, 1973).

A clear distinction between the decay of prompt fluorescence and the first phase of d.l. has been achieved (Haug et al., 1972). This allows for a comparison between d.l. intensity at any time and the level of steady-state prompt fluorescence (Table 1b). The contribution of d.l. in relation to total fluorescence is small, in confirmation of previous conclusions by Müller and Lumry (1965). This resolves in a negative fashion, the suggestion by Arnold and Davidson (1963) that a major part of the observed fluorescence is in fact delayed light.

Most of d.l. decays in a very short time (< 100 µsec). The millisecond range component, which was extensively studied by the phosphoroscope method, represents, in itself, a small fraction of d.l. intensity.

Integration of the d.l. in time to give the total light emitted from time zero to any desired time ($_0\int {}^tL\, dt'$), shows, however, that each phase contributes comparable amounts of emitted light (Table 1c). This emphasizes the significance of each phase. If an exact stoichiometry existed between each emitted photon and the amount of entities which enter into the reaction generating d.l., one could conclude that the amount of precursors for each phase are comparable, and only the rate-limiting steps for different precursors are widely different.

Very often d.l. measurement is performed at a particular point in time during the decay (e.g. msec d.l.).

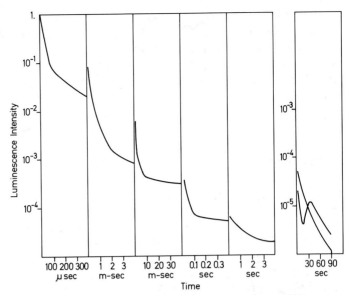

Fig. 1. Representative results of d.l. decay through several phases. (After LAVOREL, 1975a — time range <40 msec; TOLLIN et al., 1958 — time range <0.4 sec; STREHLER and ARNOLD, 1951 — time range <4 sec; RUBIN et al., 1966 — time range <90 sec). Vertical positions for each subfigure were adjusted to each other in an arbitrary but suggestive way. In the <90 sec sub-figure the curves correspond to white light pre-illumination: relatively weak light (regular decay) and strong light (curve with the maximum). Illumination with far-red light (>700 nm) produces similar curve with the maximum displaced to longer time, or even the appearance of two maxima

Table 1. Some parameters of delayed luminescence

a) Absolute intensity

Quanta emitted/quanta absorbed	$4 \cdot 10^{-7} - 4 \cdot 10^{-6}$ (x:2–3)	TOLLIN et al. (1958)
Quanta emitted/reaction centers		
35 µsec component	10^{-4}	ZANKEL (1971)
acid-base triggered d.l.	10^{-4}	HARDT and MALKIN (unpublished)

b) Ratio of d.l. at various decay times to prompt fluorescence (HAUG et al., 1972)

Time	0 (extrapolation)	1 µsec	10 µsec	1 msec
Ratio to fluorescence	1/165	1/168	1/256	1/2,900

c) Total light output of d.l. at various times from 1 msec (light integral L). Total light integral emitted from 1 msec to 10 sec is taken as 100%[a] (TOLLIN et al., 1958)

Material	1–10 msec	10–100 msec	0.1–1 sec	1–10 sec
Chlorella	3.7	7.1	27.2	61.5
Spinach chloroplasts	10.8	16.3	28.3	44.6

[a] The contribution of faster d.l. (<1 msec) is roughly equal to the contribution from 1–10 msec (estimation from Fig. 1)

III. Activation of Delayed Luminescence

As described before, d.l. appears spontaneously after pre-illumination. There are other types of d.l. which are activated by proper chemical or physical treatment, and require pre-illumination before the treatment:

1. Thermoluminescence

a) Glow Curves

By slow heating of leaves or chloroplasts after per-illumination at some chosen low temperature in the range between 77 K to room temperature, one or several peaks of emission appear as the temperature increases (Arnold and Sherwood, 1957, 1959; Arnold, 1966; Arnold and Azzi, 1971; Rubin and Venediktov, 1969; Shuvalov and Litvin, 1969; Desai et al., 1975). The graph of luminescence vs time (or temperature) is called a *glow curve*. This type of d.l. presumably reflects the need for activation energy in the luminescence process of products which were formed in the light at a low temperature. There are more peaks when pre-illumination is started at room temperature and the sample allowed to cool down during illumination to the starting temperature, compared to the case where dark adapted material was pre-illuminated at the starting temperature (Rubin and Venediktov, 1969). This is understood when considering that, at low temperature, only the primary photochemical steps are possible, while more processes are available at higher temperature. Glow curves were observed also from photosynthetic bacteria (Fleischman, 1971a).

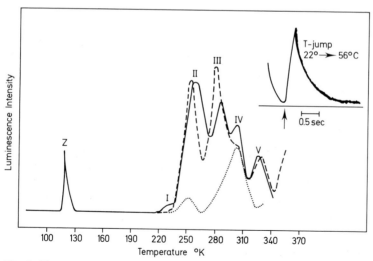

Fig. 2. Thermoluminescence. *Main figure:* glow curves. *Solid line:* after Desai et al. (1975) (their peak designation is followed) for pre-illumination at 77 K. *Broken lines:* after Rubin and Venediktov (1969); ---- pre-illumination during cooling, ... pre-illumination only at the low temperature (223 K). *Insert:* T-jump pulsed thermoluminescence, as obtained in the author's lab. (See Hardt and Malkin, 1973)

b) Temperature-Jump

Delayed luminescence is also obtained by sudden heating when pre-illumination was done at room temperature (MAR and GOVINDJEE, 1971). It appears as a pulse of emission immediately following the T-jump (MALKIN and HARDT, 1973).

These methods yield, at least in principle, the activation energy required for d.l. Most estimates give values roughly around 0.6 eV, which agrees roughly with the difference of energy between the chlorophyll a excited singlet state and the products of the primary photoact (ARNOLD and AZZI, 1971, for glow curves; MALKIN and HARDT, 1973, for T-jump; JURSINIC and GOVINDJEE, 1972, and HOP-KINS and BARBER, 1975, for the change of d.l. as a function of the temperature).

SHUVALOV and LITVIN (1969) presented a more complex analysis in which activation energies for forward and back reactions were considered. They estimated a range of activation energies from 0.15 eV to 0.9 eV for the various phases of d.l. and suggested correspondence between the glow curve peaks and the various decaying components of regular d.l. Their analysis seems too speculative compared to the available data.

2. Electric Field

An external a.c. field of several hundred V/cm causes a considerable enhancement of regular d.l. (ARNOLD and AZZI, 1971). Such enhancement was observed at 10–20 kHz, and was voltage dependent in an apparently increasing exponential function. The significance of this type of d.l. has not yet been evaluated critically. LAVOREL (1975b) suggested that the field acts in the same way as the induced electric membrane potential (cf. Sect. C.III.5).

3. Solvent-Induced Delayed Luminescence

Sudden addition of organic solvents (e.g. methanol, acetone, dimethylsulfoxide, to mention only a few) to chloroplast suspension causes a burst of d.l. (Fig. 3). This phenomenon was attributed to the increase of the membrane dielectric constant by the solvent (HARDT and MALKIN, 1972a).

4. Chemiluminescence, by Oxidation or Reduction

Luminescence is generated by adding various oxidizing or reducing agents e.g. hydrosulfite to chloroplasts (MAYNE, 1966, 1969); oxygen, hydrosulfite, ferricyanide to chromatophores (FLEISCHMAN, 1969). This type of d.l. apparently involves chemiluminescent reactions of the added compounds which probably interact with the chlorophyll of the reaction center.

5. Activation of Membrane Energization

The connection between d.l. and membrane energization was discovered by the observation (MAYNE, 1967) of the inhibitory effect of phosphorylation-uncouplers on d.l. This observation was extended to photosynthetic bacteria by FLEISCHMAN and CLAYTON (1968). At about the same time MAYNE and CLAYTON (1966) dis-

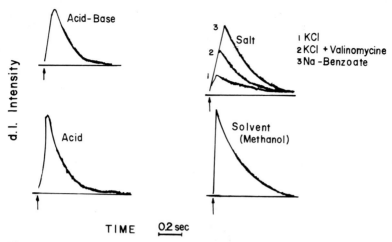

Fig. 3. Triggered d.l., as obtained by sudden (~ 50 msec) addition of base (after prior acidification), or of acid, salt or solvent to chloroplasts. Typical details are: Pre-illumination light 510–600 nm, 0.1 sec, 35 nano Einstein/cm^2 sec. Waiting period 2 sec. Injections: 0.25 ml succinic acid 20 mM and then 0.25 ml Tris base 0.1 M (acid-base). 0.25 ml HCl (0.1 M) (acid). 0.25 ml KCl (2 M)$\pm$ Valinomycin (10^{-7} M) or sodium benzoate (1.75 M) (salt). 0.25 ml methanol. Sample (1 ml) contains low buffered chloroplasts (50 µg chlorophyll)

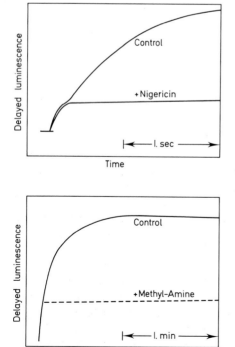

Fig. 4. msec delayed luminescence, as function of time in continuous chopped illumination, during the induction time. (After Wraight and Crofts, 1971, *upper figure*. After Itoh et al., 1971b, *bottom*.) Effect of preaddition of uncouplers is shown

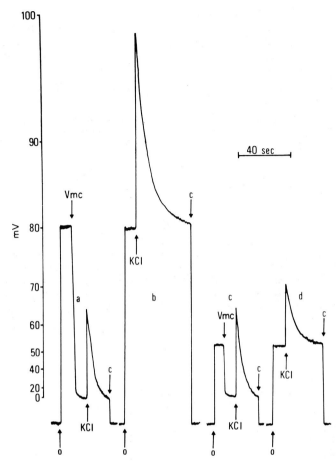

Fig. 5. Calibration of protonmotive force by delayed light during steady-state, by salt pulses. (From BARBER, 1972.) ↑ start of illumination and measurements. ↓Vmc addition of Valinomycin ↑KCl addition of KCl ↑0

covered that a pH transition (acid to base) in chloroplast suspension caused the emission of a burst of light. Since such a transition could also make ATP it was thought initially that perhaps energy from a "high energy" state (with an energy content of about 10 kcal/mol) which is a precursor of ATP, could somehow become "concentrated" in a chlorophyll electronic excited state ($\sim$40 kcal/mol). This view was questioned on the basis of energic reason and soon rejected by the observation that pre-illumination is absolutely required (MAYNE, 1968, 1969). The pre-ilumination presumably supplies most of the energy for d.l.; the acid to base transition serves only as a further "push" or triggering. Further support to this latter concept came from the discovery that similar triggering is observed by sudden salt addition (MILES and JAGENDORF, 1969; BARBER and KRAAN, 1970) as well as by a reverse pH shift (base to acid) (MILES and JAGENDORF, 1969); treatments which by themselves do not produce ATP (MILES and JAGENDORF, 1969).

The significance of membrane parameters was evaluated by the work of KRAAN (1971), KRAAN et al. (1970), BARBER and KRAAN (1970), BARBER and VARLEY (1972) and BARBER (1972). When a salt is injected in to a chloroplast suspension an electric potential gradient is established between the two sides of the (closed) thylakoid membrane, as the less permeable anion lags behind the cation in the diffusion of the salt from outside to inside. Qualitatively, the intensity of the triggered d.l. was related to the membrane potential, being larger as the potential was increased. This was especially seen when the permeability of the cation was made even larger by the addition of ionophorous compounds (cation carriers). Addition of valinomycin, for example, which is a specific K^+ carrier, increased the d.l. caused by KCl injection several fold. Addition of Na Benzoate, the anion of which is presumably much less permeable than Cl^- causes triggered light several fold stronger compared to the effect of KCl (Fig. 3). Quantitative studies suggested that d.l. is proportional to the exponential of the induced membrane potential.

The effect of the pH shifts may be explained by a superposition of the following factors. (1) Creation of electric diffusion potential formed by the addition of H^+. (2) Participation of the d.l. precursors in protonation equilibria which modulate the active forms of the precursors (KRAAN et al., 1970; CROFTS et al., 1971). (3) Reverse electron flow as demonstrated by SHAHAK et al. (1975) which may generate the d.l. precursors (MAYNE, 1968).

The effect of the membrane parameters on d.l. is also demonstrated during steady-illumination, although the separation of all the contributing factors is not "clean":

Illumination of coupled chloroplasts causes a general increase of msec d.l., from zero to a steady level. The details of this rise depend presumably on the exciting light intensity. In general, there is a fast phase followed by a slow increase (ITOH et al., 1971a; WRAIGHT and CROFTS, 1971). The extent of the slow phase is increased considerably with the addition of electron acceptors, and is abolished by uncouplers and by DCMU (WRAIGHT and CROFTS, 1971; ITOH et al., 1971a, b; BERTSCH et al., 1971). A positive correlation of the d.l., during the slow phase to the extent of the pH transient was found by EVANS and CROFTS (1973). The fast phase was correlated with a rapid establishment of a membrane potential (WRAIGHT and CROFTS, 1971). The effect of DCMU and uncouplers is probably caused by the elimination of the pH gradient.

CROFTS, WRAIGHT and FLEISCHMAN (1971) considered the energetics of d.l. production from the point of view of the recombination hypothesis (cf. Sect. D.II.1). The difference between the energy of the chlorophyll excited-state and the primary oxidized and reduced products (around 0.6 eV) seems to be too large to account for the magnitude of d.l. They reasoned, that the energized membrane may contribute to decrease this gap. An essential point in their picture is that the recombining species (P^+A^-) are situated on opposite sides of the membrane. In this case the membrane electric potential difference ΔV introduces a term in the energy of P^+A^-, proportional to ΔV. The equation for the rate of recombination in this case is (FLEISCHMAN, 1971b):

$$J = N \exp \left(-\frac{\Delta E - F \Delta V}{RT} \right). \tag{4}$$

N is the amount of the recombining molecules. ΔE is the activation energy without the membrane factor, F is the Faraday constant, R is the gas constant and ΔV is the membrane potential difference.

BARBER (1972) provided experimental results which indeed suggested a positive exponential dependence of d.l. on ΔV, as expressed by Eq. (4).

It was also possible to include the pH gradient in the same theoretical framework, by assuming that the recombining species participate in protonation equilibria which modulate their active form in d.l. production. The resulting equation is a modified form of Eq. (4)

$$J = N \exp - \frac{\Delta E - (F\Delta V - (\log_{10} e\, RT(\Delta pH)))}{RT}.$$
(5)

The term in brackets is proportional to the protonmotive force, assumed to be the driving energy term in photophosphorylation according to the chemiosmotic model (MITCHELL, 1966). Indeed, msec d.l. response to the phosphorylation reagents, uncouplers, energy transfer inhibitors, agrees qualitatively well with this model (NEUMANN et al., 1973). It was also possible to calibrate d.l. in terms of protonmotive force (this calibration is subject to uncertainty of constant factor and gives minimal values), by creating artificial membrane potential as a result of salt injection (BARBER, 1972). This showed protonmotive force in the range of about 100 mV during steady illumination.

IV. Delayed Luminescence and the S-States
(cf. Chap. II.8)

Delayed luminescence depends considerably on the S-state of the oxygen evolution precursor, as defined by O_2 evolution in saturating flashes (Fig. 6). This was shown for the various phases of the decaying d.l. [For the µsec to msec region by ZANKEL (1971); for d.l. from Chlorella at 240 msec by BARBIERI et al. (1970)], and for the triggered d.l. (HARDT and MALKIN, 1973). The pattern of the oscillation at shorter times (less than about 6 msec in *Chlorella*) of the d.l. is very similar to the O_2 yield pattern. For longer times of d.l. there is a positive phase shift by 1 flash. This suggests that the main influencing factor is S_3 for the longer times, and S_4 for the shorter times of d.l., before it reacts with water to give oxygen (see LAVOREL, 1975b Sect. 4.2). The participation of S_4 is also indicated by isolated chloroplasts at low temperatures, in the range of $-40°$ C to 0 (VELTHUIS and AMESZ, 1975), where the exciting flashes, except the last one, were given at room temperature.

The stimulating influence of higher S-states is further confirmed by the specific effect of reagents which accelerate their decay [Reagents of Substances which accelerate the deactivation reaction of the water-splitting system Y (ADRY); RENGER, 1972)]. It was shown that ADRY reagents completely inhibited both d.l. (at least in the > 50 msec region) and triggered d.l. (SIDERER et al., 1975).

The oscillatory pattern in d.l. was used as a tool to identify the site of inhibition of a photosynthetic mutant (*Chlamydomonas reinhardi* lfd-13 mutant), which is probably in the formation of S_4 from S_3 (HARDT et al., 1975). In a very similar

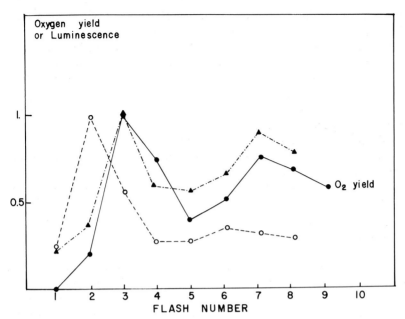

Fig. 6. Oscillations of d.l. in response to S-states. ○ delayed light at 30 μsec after excitation (Zankel, 1971). △ delayed light at 240 msec after excitation (Barbieri et al., 1970). ○ oxygen evolution yield (Barbieri et al., 1970)

way the inhibitory effect of NH_4Cl, low temperature and high pH on oxygen evolution was explained by a slow-down of S_4 reaction with water (Velthuys, 1974, Velthuys and Amesz, 1975).

D. Origin of Delayed Luminescence in Photosynthetic Systems

I. Delayed Luminescence from Plants

1. Relation to Photosystem II

Delayed luminescence at times shorter than a few seconds and triggered d.l. origi-nate in photosystem II. The following is a list of evidence supporting this statement: d.l. missing from photosystem I particles (Lurie et al., 1972; Vernon et al., 1972); mutants which lack photosystem II activity (Bertsch et al., 1967; Lavorel and Levine, 1968; Haug et al., 1972). The relatively slower phases (20 msec) of d.l. and triggered d.l. are inhibited by treatments which inhibit photosystem II (Lurie and Malkin, 1974). msec d.l. is less affected but changes its decay pattern in a way suggesting, according to Bertsch and Lurie (1971) and Cohn et al. (1975), that parasitic cyclic electron flow around the reaction center of photosystem II is induced by these treatments. The emission spectrum of both regular d.l. and triggered d.l. looks similar to the spectrum of the in vivo chlorophyll a fluorescence (cf. Sect. C.I), originating in the main part from photosystem II. The action spectrum for d.l. is similar to the action spectrum for photosystem II reaction (cf. Sect. C.I). Addition of far-red light, absorbed mainly in PS I, causes a decrease in delayed light which is excited by PS II light, showing that a product of PS II reaction which can be used by PS I is responsible for delayed light (BERTSCH, 1962; GOEDHEER, 1963). Finally, d.l. oscillates in response to the S-states of the oxygen evolution system (cf. Sect. C.IV).

These observations show that d.l. is tied with *active photosystem II units* and requires essentially the primary photoact of photosystem II. This answers in a positive way the fundamental question of whether d.l. is a result of processes involving the photosynthetic machinery or whether it merely represents insignificant portions of nonfunctional chlorophyll. This last conclusion is extended to photosynthetic bacteria, where the requirement for intact photosynthetic units was shown (CLAYTON and BERTSCH, 1965).

2. Relation to Photosynthetic Units

MAYNE showed a saturation of the acid-base triggered d.l. with respect to the requirement of pre-illumination light (MAYNE, 1968). He determined that the number of absorbed quanta required for saturation is approximately equivalent to 1:200 chlorophyll molecules, roughly corresponding to the size of the photosynthetic unit. Similar saturation properties were observed for d.l. in the msec region (ARTHUR and STREHLER, 1957) and also with other types of triggered d.l. (HARDT and MALKIN, 1971). This shows that formation of precursors for d.l. is limited to photosynthetic units.

3. Delayed Luminescence and Prompt Fluorescence

A correlation between the fluorescence yield changes and d.l. changes was observed in a number of cases: (1) (WRAIGHT, 1972) during the induction period (cf. Sect. B.II), where both fluorescence yield and d.l. increase. In order to eliminate the effect of membrane parameters this rise was studied in uncoupled chloroplasts (it corresponds to the fast rise phase of coupled chloroplasts—cf. Sect. C.III.5). When the exciting light intensity is not too strong, a clear correlation exists between d.l. and the fluorescence rise which indicates the reduction of the primary acceptor Q of photosystem II (Fig. 7). (2) CLAYTON (1969) observed parallel effects upon

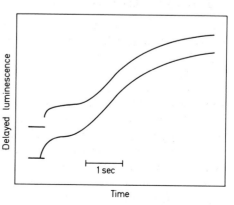

Fig. 7. Fluorescence and msec d.l. rise, as a function of time, in continuous chopped illumination, during the induction time, in *uncoupled* chloroplasts. (After WRAIGHT, 1972.) *Horizontal lines:* zero level for each experiment

addition of $K_3Fe(CN)_6$ and DCMU on both prompt fluorescence and d.l. (3) The decay of d.l. in the second range after pre-illumination, and the decrease of the fluorescence yield of chloroplasts inhibited by DCMU, are parallel (BEN-NOUN, 1970; LAVOREL, 1975b).

The conclusion from such experiments may fall into two alternatives. (1) That the emission yield controls also the intensity of d.l. (LAVOREL, 1968, 1969, 1975b). In this case d.l. would depend only indirectly on the state of the reaction center, while there would not necessarily be any mechanistic connection between the two (CLAYTON, 1969). (2) That the emission yield for d.l. is in fact constant, and that d.l. depends directly on reduced Q as a substrate for recombination (WRAIGHT, 1972).

LAVOREL (1968, 1975b) expressed the view that d.l. depends on the product of both factors; the macroscopic fluorescence yield and the concentration of reduced Q: (the first is the yield factor while the second is the rate factor). This view could be acceptable if the excitation after recombination migrated freely between many photosynthetic units (LAVOREL and JOLIOT, 1972). Otherwise the emission yield should be that of an "open" photosynthetic unit, which corresponds to the fluorescence yield of a "dark adapted" state. The analysis of WRAIGHT (1972) favors possibility (2) above. Close inspection of the dependence of luminescence on the fluorescence changes (CLAYTON, 1969; BENNOUN, 1970) gives more support to this alternative. This would indicate, in turn, that the photosynthetic units are largely independent, in terms of excitation transfer between them, in agreement with flashing light experiments (MALKIN, 1974).

4. Exceptional Delayed Luminescence; Delayed Luminescence from Photosystem I

Thermoluminescence glow curves from leaves and chloroplasts reveal certain d.l. peaks which are exceptional in their general behavior (SHUVALOV and LITVIN, 1969; ARNOLD and AZZI, 1971; DESAI et al., 1975). The peak at around 120 K was claimed to be excited by blue light only (between approximately 350 and 550 nm). However, DESAI et al. claim that it can be excited by red light, which is absorbed by chlorophyll a, as well. It has an emission spectrum with maximum beyond 730 nm, and is somewhat resistant to heating which destroys photosystem II activities. Similar d.l. thermoluminescence was detected also from chlorophyll in vitro (SANE et al., 1974) and claimed to be chlorophyll a phosphorescence, arising from the triplet state; thus it is not relevant to the photosynthetic photoact. The luminescence peaks at 283, 198 and 321 K were assigned as originating from photosystem I (DESAI et al., 1975). The one at 321 K was directly connected to the photoact of photosystem I, the mechanism of others included interaction with photosystem II. The only peak which is assigned to photosystem II proper was the peak at 260 K (DESAI et al., 1975). The evidence for these assignments does not seem conclusive yet.

Long-lived d.l. in the region of decay time from a few seconds to minutes apparently seems to be preferentially excited by photosystem I (BERTSCH and AZZI, 1965; RUBIN et al., 1966; BJÖRN, 1971) and requires oxygen (BJÖRN, 1971). This type of d.l. grows to a maximum and then decays (cf. Fig. 1). It is perhaps possible to consider it as an oxygen-dependent chemiluminescence, which is not actually a true reversal of the photosynthetic events.

BJÖRN and SIGFRIDSSON (1971) described unusual, very long-lived d.l. from leaves and isolated chloroplasts, which is initiated by adding small amounts of alcohols and aldehydes. The relation to the fast pulse also obtained with organic solvents (cf. Sect. C.III.3) is not clear yet.

II. Models for the Mechanism of Delayed Luminescence in Photosynthetic Systems

1. Direct Recombination Model

Based on the original suggestion of ARTHUR and STREHLER (1957), with the knowledge that d.l. can be produced by recombination of charged fragments, LAVOREL suggested (1968, 1969, 1971) that d.l. is a direct expression of a radiative back reaction of the primary photoact. Denoting the electron donor by P (P-680 or chlorophyll a_{II} in photosystem II; P-870 or P-890 in bacterial systems) and the electron acceptor by A (Q in photosystem II and X in bacterial systems) the reactions leading to d.l. can be formulated as follows:

$$PA \xrightarrow{h\nu} P^*A \longrightarrow P^+A^- \qquad \text{Excitation; primary charge separation,} \qquad (6)$$

$$P^+A^- \longrightarrow P^*A \longrightarrow PA + h\nu \quad \text{Recombination; delayed light.} \qquad (7)$$

Presumably, the excitation on P*, can migrate back through the antennae pigments, and then lead to emission.

According to this model d.l. must be related directly to the momentary concentrations of the recombining species. At very short times from excitation d.l. should follow the decay of P^+ and A^-, at longer times one may expect the intervention of secondary electron carriers which feed electrons, or positive charges, back into the primary carriers. This would be responsible for the slower components of the decaying d.l.

VAN GORKOM and DONZE (1973) interpreted the 35 μsec decay phase by the electron donation to P^+ from a secondary electron donor, which takes place during this time. The 200 μsec phase is due to fewer units where the secondary donor is already oxidized, and the rate limiting step is displaced.

The following main points contribute support for the above model: (1) Correlation to the fluorescence changes (see Sect. D.I.3). (2) d.l. correlation to the S-states (see Sect. C.V.5) points towards direct participation of higher S-states. A possible explanation for this participation is a local electrostatic field exerted on the reaction center by the charge storing species which influences in a similiar way as the membrane electric potential (VAN GORKOM and DONZE, 1973). In the case of the longer phases of d.l. and the triggered d.l., one may also consider a back pressure of positive charges, placing a certain fraction of positive charges on P, e.g.

$$Z^{+n}P \rightleftharpoons Z^{+(n-1)}P^+$$

which promotes recombination (Z is the electron carrier capable of storing the charges of the S-state). Delayed luminescence at longer times on the decay scale is inhibited by hydroxyl-amine, which keeps P in a reduced form by electron donation (BENNOUN, 1970; MAR and GOVINDJEE, 1971).

In bacterial systems d.l. was correlated directly with the concentration of oxidized P-870 (FLEISCHMAN, 1971b, 1974).

The feasibility of d.l. production by the above mechanisms is supported by observation of chemiluminescence from chemically oxidized chlorophyll a in solution as the result of its reduction by suitable reductants (GOEDHEER and VEGT, 1962).

In the investigation of triggered luminescence, it was found that the T-jump and the membrane excited-triggered d.l. behave in a different way, especially with regard to their decay as a function of the dark-time between the end of pre-illumination and the triggering act (HARDT and MALKIN, 1971). It was suggested that the precursors for the two types are different. This view was further supported by the fact that the two types of d.l. could be triggered in succession, with the same pre-illumination.

Examination of the data suggests that very likely P^+ participates in most d.l. phenomena. The participation of A^- is not proved conclusively for all cases, and alternative electron sources may be considered, for each case.

2. Semi-Conductor Model

The production of d.l. at a very low temperature (77 K) led some workers (TOLLIN and CALVIN, 1957; TOLLIN et al., 1958) to deny the participation of enzymic processes and to suggest a model which places much significance on semi-conductor phenomena. This model was suggested earlier for photosynthesis by KATZ (1949).

Motivated by such phenomena as thermoluminescence and observation of luminescence and photoconductance of dried photosynthetic materials (all of which are very typical of semiconducting systems) this view was particularly adopted by ARNOLD et al. (ARNOLD and SHERWOOD, 1957, 1959; ARNOLD and CLAYTON, 1960). Later a very special version of the semiconductor model was suggested (ARNOLD and AZZI, 1971; BERTSCH, 1969) to explain certain specific phenomena, such as the enhancement of d.l. upon addition of electron acceptors to isolated chloroplasts (BERTSCH et al., 1969), the dependence of d.l. on the square of the intensity of a non-saturating pre-illumination flash (JONES, 1967), and the fact that at low temperatures (77 K) d.l. is observed only after the first flash (ARNOLD and AZZI, 1971). However, these phenomena are explained in a much better way (LAVOREL, 1975b) in terms of the direct recombination hypothesis with consideration of the effects of the S-states and membrane parameters. It is also known that limited electron transfer reactions can take place at very low temperatures (BUTLER et al., 1973). From these considerations the semiconductor model seems unnecessary and is indeed not generally accepted.

Some phenomena indicate mobility of the recombining charges (T-jump d.l. — MALKIN and HARDT, 1973; recombination of P^+ and Q^- which is second-order in Q^- — BENNOUN, 1970). Solvent-induced d.l. was interpreted as a detrapping of electrons caused by the increase of dielectric constant (HARDT and MALKIN, 1972). It is therefore possible that semiconductor phenomena play some role, and are involved perhaps in the details of the recombination process (see also: WRAIGHT and CROFTS, 1971).

3. Photophysical Phenomena

The importance of triplet excited-state mechanisms in producing d.l. in photosynthesis was never critically investigated, although some speculations were expressed in the literature (PARKER and JOYCE, 1966; SHUVALOV and LITVIN, 1969; STACY et al., 1971; MALKIN and HARDT, 1971). The life-time of the triplet state, if it is produced at all, is probably very short (MORAW and WITT, 1961). It may thus play a role only in producing the very fast phase of d.l., and also perhaps as a transitory intermediate. According to SANE et al. (1974) one of the thermoluminescence peaks is due directly to triplet state phosphorescence, but is not tied particularly with photosynthetic reactions.

References

Arnold, W.: Science **154**, 1046–1049 (1966)
Arnold, W., Azzi, J.: Photochem. Photobiol. **14**, 233–240 (1971)
Arnold, W., Clayton, R.K.: Proc. Nat. Acad. Sci. U.S. **46**, 769–776 (1960)
Arnold, W., Davidson, J.B.: J. Gen. Physiol. **37**, 677–684 (1954)
Arnold, W., Davidson, J.B.: In: Photosynthetic mechanisms of green plants, pp. 698–700. Publ. 1145, Nat. Acad. Sci. Nat. Res. Counc., Washington, D.C. (1963)
Arnold, W., Sherwood, H.K.: Proc. Nat. Acad. Sci. U.S. **43**, 105–114 (1957)
Arnold, W., Sherwood, H.K.: J. Phys. Chem. **63**, 1–4 (1959)
Arnold, W., Thompson, J.: J. Gen. Physiol. **39**, 311–318 (1956)
Arthur, W.E., Strehler, B.L.: Arch. Biochem. Biophys. **70**, 507–526 (1957)
Barber, J.: Biochim. Biophys. Acta **275**, 105–116 (1972)
Barber, J., Kraan, G.P.B.: Biochim. Biophys. Acta **197**, 49–95 (1970)
Barber, J., Varley, W.J.: J. Exp. Botany **23**, 216–228 (1972)
Barbieri, G., Delosme, R., Joliot, P.: Photochem. Photobiol. **12**, 197–206 (1970)
Bennoun, P.: Biochim. Biophys. Acta **216**, 357–363 (1970)
Bertsch, W.F.: Proc. Nat. Acad. Sci. U.S. **48**, 2000–2004 (1962)
Bertsch, W.F.: In: Progress in Photosynthesis Res. Metzner, H. (ed.). Tübingen: Laupp, 1969, Vol. II, pp. 996–1005
Bertsch, W.F., Azzi, J.R.: Biochim. Biophys. Acta **94**, 15–26 (1965)
Bertsch, W.F., Azzi, J.R., Davidson, J.B.: Biochim. Biophys. Acta **143**, 129–143 (1967)
Bertsch, W.F., Lurie, S.: Photochem. Photobiol. **14**, 251–260 (1971)
Bertsch, W.F., West, J., Hill, R.: Biochim. Biophys. Acta **172**, 525–538 (1969)
Bertsch, W.F., West, J., Hill, R.: Photochem. Photobiol. **14**, 241–250 (1971)
Björn, L.O.: Photochem. Photobiol. **13**, 5–20 (1971)
Björn, L.O., Sigfridsson, B.: Physiol. Plantarum **25**, 308–315 (1971)
Butler, W.L., Visser, J.W.M., Simons, H.L.: Biochim. Biophys. Acta **292**, 140–151 (1973)
Clayton, R.K.: Biophys. J. **9**, 61–76 (1969)
Clayton, R.K., Bertsch, W.F.: Biochem. Biophys. Res. Commun. **18**, 415–419 (1965)
Cohn, D.E., Cohen, W.S., Lurie, S., Bertsch, W.F.: In: Proc. 3rd Int. Congr. Photosynthesis. Avron, M. (ed.). Amsterdam: Elsevier, 1975, Vol. I, pp. 65–74
Crofts, A.R., Wraight, C.A., Fleischman, D.E.: FEBS Lett. **15**, 89–100 (1971)
Desai, T.S., Sane, P.V., Tatake, V.G.: Photochem. Photobiol. **21**, 345–350 (1975)
Evans, E.H., Crofts, A.R.: Biochim. Biophys. Acta **292**, 130–139 (1973)
Fleischman, D.E.: In: Progress in Photosynthesis Res. Metzner, H. (ed.). Tübingen: Laupp, 1969, Vol. II, pp. 952–955
Fleischman, D.E.: Photochem. Photobiol. **14**, 65–70 (1971a)
Fleischman, D.E.: Photochem. Photobiol. **14**, 277–286 (1971b)
Fleischman, D.E.: Photochem. Photobiol. **19**, 59–68 (1974)
Fleischman, D.E., Clayton, R.K.: Photochem. Photobiol. **8**, 287–298 (1968)

490   S. Malkin:

Goedheer, J.C.: Biochim. Biophys. Acta **64**, 294–308 (1962)
Goedheer, J.C.: Biochim. Biophys. Acta **66**, 61–71 (1963)
Goedheer, J.C., Vegt, G.R.: Nature (London) **193**, 875–876 (1962)
Gorkom, H.J. van, Donze, M.: Photochem. Photobiol. **17**, 333–342 (1973)
Gutman, F., Lyons, L.E.: Organic semiconductors (1967) Wiley, New York
Hardt, H., Malkin, S.: Photochem. Photobiol. **14**, 483–492 (1971)
Hardt, H., Malkin, S.: Biochim. Biophys. Acta **267**, 588–594 (1972)
Hardt, H., Malkin, S.: Photochem. Photobiol. **17**, 433–440 (1973)
Hardt, H., Malkin, S.: FEBS Lett. **42**, 293–295 (1974)
Hardt, H., Malkin, S., Epel, B.: In: Proc. 3rd Intern. Congr. Photosynthesis. Avron, M.
    (ed.). Amsterdam: Elsevier, 1975, Vol. I, pp. 75–82
Haug, A., Jorquet, D.D., Beall, H.: Biochim. Biophys. Acta **283**, 92–99 (1972)
Haug, A., Kohler, B.E., Priestley, E.B., Robinson, G.W.: Rev. Sci. Instr. **40**, 1439–1444
    (1969)
Hopkins, M.E., Barber, J.: In: Proc. 3rd Intern. Congr. Photosynthesis. Avron, M. (ed.).
    Amsterdam: Elsevier, 1975, Vol. I, pp. 101–114
Itoh, S., Katoh, S., Takamiya, A.: Biochim. Biophys. Acta **245**, 121–128 (1971a)
Itoh, S., Murata, N.: In: Proc. 3rd Intern. Congr. Photosynthesis. Avron, M. (ed.). Amsterdam:
    Elsevier, 1975, Vol. I, pp. 115–126
Itoh, S., Murata, N., Takamiya, A.: Biochim. Biophys. Acta **245**, 109–120 (1971b)
Jones, L.W.: Proc. Nat. Acad. Sci. U.S. **58**, 75–80 (1967)
Jursinic, P., Govindjee: Photochem. Photobiol. **15**, 331–348 (1972)
Katz, E.: In: Photosynthesis in Plants. Franck, J., Loomis, W.E. (eds.). Ames, Ia: Iowa
    State College Press, 1949, pp. 267–292
Kraan, G.P.B.: Thesis, State Univ. Leiden, The Netherlands, 1971
Kraan, G.P.B., Amesz, J., Velthuys, B.R., Steemers, R.G.: Biochim. Biophys. Acta **223**,
    129–145 (1970)
Lavorel, J.: Biochim. Biophys. Acta **153**, 727–730 (1968)
Lavorel, J.: In: Progress Photosyntheis Res. Metzner, H. (ed.). Tübingen: Laupp, 1969, Vol. III,
    pp. 883–898
Lavorel, J.: Photochem. Photobiol. **14**, 261–275 (1971)
Lavorel, J.: Biochim. Biophys. Acta **325**, 213–229 (1973)
Lavorel, J.: Photochem. Photobiol. **21**, 331–343 (1975a)
Lavorel, J.: Luminescence. In: Bioenergetics of Photosynthesis. Govindjee (ed.). New York-
    London: Academic Press, 1975b, pp. 223–317
Lavorel, J., Joliot, P.: Biophys. J. **12**, 815–831 (1972)
Lavorel, J., Levine, R.P.: Plant Physiol. **43**, 1049–1055 (1969)
Lim, E.C., Swenson, G.W.: J. Chem. Phys. **36**, 118–122 (1962)
Lim, E.C., Weng, Y.: J. Chem. Phys. **39**, 847–848 (1963)
Linschitz, H., Berry, M.G., Schweitzer, D.: J. Am. Chem. Soc. **76**, 5833–5839 (1954)
Lurie, S., Cohen, W., Bertsch, W.F.: In: Proc. 2nd Intern. Congr. Photosynthesis Res. Forti, G.,
    Avron, M., Melandri, A. (eds.). The Hague: Dr. Junk, 1972, Vol. I, pp. 197–206
Lurie, S., Malkin, S.: In: Proc. 3rd Intern. Congr. Photosynthesis. Avron, M. (ed.). Amsterdam:
    Elsevier, 1975, Vol. I, pp. 83–91
Malkin, S.: Biophys. Chem. **2**, 327–337 (1974)
Malkin, S., Hardt, H.: In: Proc. 2nd Intern. Congr. Photosynthesis. Res. Forti, G., Avron, M.,
    Melandri, A. (eds.). The Hague: Dr. Junk, 1972, Vol. I, pp. 253–269
Malkin, S., Hardt, H.: Biochim. Biophys. Acta **305**, 292–301 (1973)
Mar, T., Govindjee: Biochim. Biophys. Acta **226**, 200–203 (1971)
Mayne, B.C.: Brookhaven Symp. Biol. **19**, 460–466 (1966)
Mayne, B.C.: Photochem. Photobiol. **6**, 189–197 (1967)
Mayne, B.C.: Photochem. Photobiol. **8**, 107–113 (1968)
Mayne, B.C.: In: Progress of Photosynthesis Res. Metzner, H. (ed.). Tübingen: Laupp, 1969,
    Vol. II, pp. 947–955
Mayne, B.C., Clayton, R.K.: Proc. Nat. Acad. Sci. U.S. **55**, 494–497 (1966)
Miles, C.D., Jagendorf, A.T.: Arch. Biochem. Biophys. **129**, 711–719 (1969)
Mitchell, P.: Biol. Rev. **41**, 445–502 (1966)
Moraw, R., Witt, H.T.: Z. Physikal. Chem. **29**, 25–42 (1961)

Müller, A., Lumry, R.: Proc. Nat. Acad. Sci. U.S. **54**, 1479–1485 (1965)
Neumann, J., Barber, J., Gregory, P.: Plant Physiol. **51**, 1069–1073 (1973)
Parker, C.A.: In: Advances in Photochemistry. Noyes, W.A., Hammond, G.S., Pitts, J.N. (eds.). New York: Interscience, 1964, Vol. II, pp. 305–383
Parker, C.A., Joyce, T.A.: Nature (London) **210**, 701–703 (1966)
Renger, G.: Biochim. Biophys. Acta **256**, 428–439 (1972)
Rubin, A.B., Fokht, A.S., Venediktov, P.S.: Biofizika (English Transl.) **11**, 341–348 (1966)
Rubin, A.B., Venediktov, P.S.: Biofizika (English Transl.) **14**, 105–109 (1969)
Ruby, R.H.: Photochem. Photobiol. **8**, 299–308 (1968)
Sane, P.V., Tatake, V.G., Desai, T.S.: FEBS Lett. **45**, 290–298 (1974)
Shahak, Y., Hardt, H., Avron, M.: FEBS Lett. **54**, 151–154 (1975)
Shuvalov, V.A., Litvin, F.F.: Mol. Biol. **3**, 45–56 (1969)
Siderer, Y., Hardt, H., Malkin, S.: (1975). Abst. FEBS 10th Meeting p. 1203
Stacy, W.T., Mar, T., Swenberg, C.E., Govindjee: Photochem. Photobiol. **14**, 197–219 (1971)
Strehler, B.L., Arnold, W.: J. Gen. Physiol. **34**, 809–829 (1951)
Strehler, B.L., Lynch, V.: Arch. Biochem. Biophys. **70**, 527–545 (1957)
Tollin, G., Calvin, M.: Proc. Nat. Acad. Sci. U.S. **43**, 895–908 (1957)
Tollin, G., Gujimori, E., Calvin, M.: Proc. Nat. Acad. Sci. U.S. **44**, 1035–1046 (1958)
Velthuys, B.R.: In: Proc. 3rd Intern. Congr. Photosynthesis. Avron, M. (ed.). Amsterdam: Elsevier, 1975, Vol. I, pp. 93–100
Velthuys, B.R., Amesz, J.: Biochim. Biophys. Acta **376**, 162–168 (1975)
Vernon, L.P., Klein, S., White, F.G., Shaw, E.R., Mayne, B.C.: In: Proc. 2nd Intern. Congr. Photosynthesis Res. Forti, G., Avron, M., Melandri, A. (eds.). The Hague: Dr. Junk, 1972, pp. 801–812
Wraight, C.A.: Biochim. Biophys. Acta **283**, 247–258 (1972)
Wraight, C.A., Crofts, A.R.: Europ. J. Biochem. **19**, 386–397 (1971)
Zankel, K.L.: Biochim. Biophys. Acta **245**, 373–385 (1971)

15. Exchange Reactions

C. CARMELI

A. Introduction

Exchange reactions have been successfully used in the study of the mechanisms of biochemical catalysis (DIXON and WEBB, 1964). This method, which is based on the use of isotopic tracers makes it possible to distinguish between two types of mechanisms of enzyme catalyzed reactions. One involves a covalently bound enzyme intermediate, while in the other the reaction occurs between substrates at the active site of the enzyme but does not proceed through the formation of such an intermediate. The study of the rate of exchange reactions and the requirement of such reactions for various substrates might also lead to the formulation of the rate limiting step in a multistep reaction. Reactions devoid of a covalently bound enzyme intermediate are expected to proceed at an optimal rate near equilibrium conditions. However, such reactions when coupled to an energy-conserving system might proceed at considerable rate even when they are far from equilibrium. The recognition of this fact led to the use of exchange reactions as a powerful tool in the investigation of the mechanism of energy conservation.

BOYER (1967) thoroughly reviewed the use of exchange reactions in the study of oxidative and photophosphorylation. It is the purpose of this chapter to review briefly the development of this field and to concentrate on the recent contributions of the study of exchange reactions to the understanding of the mechanism of energy coupling in photophosphorylation.

B. The Development of the Study of Exchange Reactions in Photophosphorylation

Exchange reactions can take place only in a process which is kinetically reversible. In their early work with chloroplasts AVRON and JAGENDORF (1959) did not detect the thermodynamically favorable hydrolysis of ATP which was expected to reflect the reversal of phosphorylation. First indication for the reversibility of the system came from the work of AVRON et al. (1965) who found a light-dependent ATP-HOH exchange. They found that the oxygen bridge between the β-γ phosphoryl groups in ATP was contributed by ADP during photophosphorylation, and observed with the use of ^{18}O an exchange of oxygen between water and ATP which occurred during net synthesis of ATP; however, the exchange also occurred in the absence of net synthesis. The occurrence of ATP-HOH exchange was confirmed by SHAVIT et al. (1967) who also detected a slow light dependent P_i-HOH exchange.

A major development in the field were the findings by PETRACK et al. (1965) that chemical modification of chloroplasts induced reversibility of the terminal steps of the process of photophosphorylation. They observed that illumination of chloroplasts in the presence of −SH reagents, such as cysteine, induced a rapid hydrolysis of ATP. Exchange reactions which required kinetic reversibility were expected to be found. It was soon found that light-triggered chloroplasts concomitantly catalyzed ATPase and P_i-ATP exchange activities (CARMELI and AVRON, 1966; RIENITS, 1967; McCARTY and RACKER, 1968). The rate of other exchange reactions was also found to be accelerated under the conditions which induced ATP hydrolysis. SKYE et al. (1967) demonstrated high rates of both ATP-HOH and P_i-HOH exchange under these conditions.

The existence of ADP-ATP exchange which is related to the catalysis of ATP formation in chloroplasts is yet to be established. AVRON (1960) could not find any such exchange in chloroplasts isolated from Swiss chard leaves. However, KAHN and JAGENDORF (1961) demonstrated an ADP-ATP exchange reaction in spinach chloroplasts. They also purified an enzyme which catalyzed the activity. It was, however, demonstrated that this enzyme is not related to the mechanism of photophosphorylation (BEN-YEHOSHUA and AVRON, 1964). Isolated coupling factor from chloroplasts (CF_1) does not catalyze this exchange. No significant ADP-ATP exchange could be demonstrated in washed lettuce chloroplasts under conditions in which they catalyzed both ATPase and P_i-ATP exchange activities (CARMELI and AVRON, 1967). However, some light-induced ADP-ATP exchange could be demonstrated in spinach chloroplasts (STEWART and RIENITS, 1968), but it was not sensitive to the inhibitors of photophosphorylation.

Various experimental results can be considered to support a relation between exchange reactions and the coupling mechanism of photophosphorylation. Convincing evidence is the observed inhibition of P_i-ATP exchange by uncouplers and by energy transfer inhibitors of photophosphorylation (CARMELI and AVRON, 1967; RIENITS, 1967). Moreover, the exchange is catalyzed by the same enzyme which is involved in ATP formation since an antibody against CF_1 inhibits both the exchange and the photophosphorylation activities (McCARTY and RACKER, 1968). The recognition of the requirement for energy in exchange reactions led to the use of these reactions as indicators for energy conversion. In resolution and reconstitution studies it was observed that the purified coupling factor could not catalyze exchange reactions unless attached to the chloroplast membrane (McCARTY and RACKER, 1968). In other resolution experiments vesicles were obtained which were reconstituted from a fraction of the chloroplast membrane deficient in chlorophyll and in electron transport carriers (CARMELI and RACKER, 1973). The formation of membrane vesicles to which CF_1 was attached was essential for restoration of P_i-ATP exchange activity, thus, supporting the suggestion (MITCHELL, 1968) that the coupling apparatus is composed of two separate energy generating systems coupled by a closed membrane. One of these systems is probably the reversible ATPase which when attached to the membrane of closed vesicles can conserve energy as indicated by the presence of P_i-ATP exchange activity in the reconstituted vesicles.

C. Mechanisms of Exchange Reactions

One of the major problems concerning the mechanism of phosphorylation involves the question of whether the energy which is generated in the process of oxidation is transduced through a formation of covalently bound intermediates. According to the chemical hypothesis of oxidative and photophosphorylation (SLATER, 1953), energy is conserved by formation of a covalent bond between an electron transport carrier and an intermediate. According to the chemiosmotic hypothesis (MITCHELL, 1968) the primary energy conserving event involves formation of an electrochemical proton gradient which is used in the synthesis of ATP. However, both hypotheses do not preclude the formation of a phosphorylated intermediate at one of the steps leading to the synthesis of ATP. The possible existence of a phosphorylated intermediate was pursued intensively in numerous works. One of the ways to identify an enzyme-bound intermediate is to use exchange reactions. This method makes it possible to distinguish between two mechanisms of enzyme catalyzed reactions. One mechanism involves the catalysis of interaction between molecules attached to the surface of the enzyme. The second mechanism involves an actual transfer of a group from one of the reactants to form a covalent bond with the enzyme as an intermediate step. The overall reaction of ATP synthesis can be written as:

$$\text{ADP-OH} + \text{P-OH} \underset{}{\overset{E}{\rightleftharpoons}} \text{ADP-O-P} + H_2O. \tag{1}$$

In order to follow the atoms and bonds involved in the reaction the β-γ oxygen bridge in ATP is shown as P-O-P and one of the hydroxyls in ADP and in orthophosphate is shown as ADP-OH and P-OH respectively. The forward reaction requires energy which is provided by oxidation. A reaction which involves a phosphorylated intermediate will proceed through two steps:

$$\text{P-OH} + \text{E-H} \rightleftharpoons \text{E-P} + H_2O, \tag{2}$$

$$\text{ADP-OH} + \text{E-P} \rightleftharpoons \text{E-H} + \text{ADP-O-P}. \tag{3}$$

First, an energy requiring phosphorylation of an enzyme in which phosphoryl is covalently bound to the enzyme (E-P) and a molecule of water is formed, and then a transfer of the phosphoryl to ADP which results in the formation of ATP.

If kinetically reversible, at equilibrium both mechanisms are expected to catalyze exchange reactions. Thus, if Eq. (1) represents the true mechanism, exchange reaction will take place only if all reactants are present. Equations (2) and (3), on the other hand, require only one substrate for exchange. P_i-HOH exchange, e.g. can take place according to Eq. (2) in the presence of orthophosphate alone if adequate energy is provided for the formation of the enzyme intermediate. There is no need to add ADP as the final acceptor for the exchange reaction to occur. A small amount of E-P will be reversibly formed [Eq. (2)] at an amount stoichiometrically equivalent to the enzyme. Addition of isotopically labeled $H^{18}OH$ will result in the incorporation of ^{18}O into P-OH when Eq. (2) is followed by the reversed Eq. (4).

$$E\text{-}P + H^{18}OH \rightarrow E\text{-}H + P\text{-}^{18}OH. \tag{4}$$

The process will proceed till isotopic equilibrium will be reached.

Another exchange that can be catalyzed following the addition of only one substrate to the enzyme is an ADP-ATP exchange. As is apparent from Eq. (3) the addition of ATP will result in a reversible formation of a small amount of ADP. In the presence of a radioactively labeled ADP the exchange will result in the formation of labeled ATP. The reaction would not require the addition of orthophosphate although a phosphoryl enzyme intermediate is involved in the exchange.

It follows that the two mechanisms of enzyme catalyzed reactions can be distinguished according to their requirements for substrates. A reaction which proceeds through the formation of a covalently bound intermediate will not require all substrates and product for exchange, while a reaction which does not involve a formation of a bound intermediate will require all substrates in order to catalyze an exchange. The latter will be a result of a dynamic reversal of catalysis and is expected to proceed at optimal rates near the thermodynamic equilibrium. In both cases, the occurrence and the relative rates of various exchanges is expected to be dependent also on the rate constants of the association and dissociation of the substrate to the enzyme. Thus in phosphorylation a low dissociation constant for binding of ADP will result in a slow ADP-ATP exchange or in its absence but would not necessarily effect the rate of other exchanges such as P_i-ATP or ATP-HOH. It should be pointed out that the formation of a phosphorylated intermediate is only one example for a possible covalently bound intermediate which can be formed during ATP synthesis. A covalently bound ADP could also be a proper intermediate.

D. Requirement for Substrates

A requirement for a substrate in an exchange reaction can be experimentally demonstrated by the effects of either its addition or by its depletion from the reaction system. An increase in the rate of ATP-HOH exchange by addition of P_i was demonstrated (SKYE et al., 1967) to be catalyzed in the dark by light triggered chloroplasts. ATP-HOH exchange catalyzed in the light was however inhibited by either ADP or P_i (AVRON et al., 1965). BOYER (1967) suggested that the inhibition of the exchange in the light reflects the requirement for these substrates in the reaction. Since phosphorylation proceeds at high rates in the light, addition of ADP will result in phosphorylation of P_i to ATP and P_i will be depleted from the system. The addition of P_i will deplete the system from ADP. BOYER, therefore, proposed that both dark and light experiments demonstrated the requirement for all substrates for ATP-HOH exchange. These facts support his suggestion that the exchange is a result of a dynamic reversal of ATP formation and not of a covalently bound intermediate.

The rate of P_i-HOH exchange catalyzed in the dark by light-triggered chloroplasts was greatly stimulated by addition of ATP and P_i (SKYE et al., 1967). Although these results could indicate a requirement for ATP and P_i for P_i-HOH exchange they should be regarded with caution. It was demonstrated that the light-triggered state, which renders kinetic reversibility to CF_1, decays unless energy is provided for its maintenance either from light or from ATP (PETRACK et al., 1965; CARMELI, 1969). Thus, the light-triggered state of CF_1 which is essential for P_i-HOH exchange probably decayed in the dark in the absence of ATP. It is therefore difficult to decide whether ATP stimulated the exchange by preservation of the triggered state or because it was required for the catalysis. P_i was also found (CARMELI and AVRON, 1967) to stabilize the triggered state in the absence of any source of energy and therefore did not necessarily stimulate the catalysis of P_i-HOH exchange.

Since ATP is present during the assay of P_i-ATP exchange it seems likely that the stimulation of the exchange by P_i (CARMELI and AVRON, 1967) reflects its requirement for catalysis. The P_i-ATP exchange was also found to be dependent on ADP (CARMELI and LIFSHITZ, 1969). Addition of ADP stimulated the exchange while its removal by phosphorylation of ADP with pyruvate kinase and phospho-enolpyruvate almost completely inhibited the exchange. An analog of ADP $1,N^6$-ethenoadenosine diphosphate (εADP) also stimulated P_i-ATP exchange but εATP competitively inhibited the exchange (SHAHAK et al., 1973). These results tend to support the suggestion that the exchanges are a result of a concerted reaction involving all substrates. It is, therefore, possible that ATP formation does not proceed through a formation of a covalently bound enzyme intermediate.

E. The Relations Between Exchange Reactions and the Mechanism of Photophosphorylation

I. Inhibition by Antibody Against CF_1

The inhibition of P_i-ATP exchange by antibody against CF_1 presents the best evidence for the relation of this reaction to photophosphorylation (McCARTY and RACKER, 1968). Moreover, it suggests that CF_1 which probably catalyzes the terminal step in ATP synthesis also catalyzes the exchange. Recently however, SHOSHAN et al. (1974) have shown that a monovalent fragment of the antibody against CF_1 inhibited, under certain conditions, ATP formation but did not inhibit ATPase and P_i-ATP exchange. These results are similar to the results obtained by NELSON et al. (1973) who studied the effect of antibodies against the individual subunits of CF_1 on various reactions of photophosphorylation. They found that although all antibodies interacted with CF_1 their inhibitory effects also depended on the conformation state of CF_1. It is therefore possible that the monovalent fragment of the antibody against CF_1 inhibited the conformation changes which are required to confer kinetic reversibility to the CF_1. Thus the antibody prevented CF_1 from catalyzing ATPase and P_i-ATP exchange but did not prevent ATP synthesis which proceeded without these changes.

The role of CF_1 in the catalysis of this exchange was also demonstrated by CARMELI and AVRON (1967) and McCARTY and RACKER (1968) who showed that partial removal of CF_1 from chloroplasts by treatment with EDTA caused inhibition of P_i-ATP exchange. This treatment was shown to uncouple photophosphorylation (AVRON, 1963). Energy transfer inhibitors such as Dio-9 and phlorizin which inhibit phosphorylation also inhibited the P_i-ATP exchange (CARMELI and AVRON, 1967; RIENITS, 1967; McCARTY and RACKER, 1968).

II. Effects of Light

The observed light dependence of ATP-HOH, P_i-HOH and P_i-ATP exchange reactions (AVRON et al., 1965; SHAVIT et al., 1967) serves as a strong indication for the relation between these reactions and the light dependent process of photophosphorylation. Moreover, the requirement for light in the presence of an $-SH$ reagent for stimulation of the ATPase and the exchange reactions gives further support to this suggestion. Since light drives electron transport, these results might be interpreted to indicate that the oxidation is directly linked to the catalysis of exchange reactions. However, the light triggered ATPase and exchange reactions can proceed in the dark (PETRACK et al., 1965; CARMELI and AVRON, 1967; SKYE et al., 1967). It is more likely, therefore, that the energy released during light induced electron transport, possibly in the form of electrochemical potential, induces a conformation change in CF_1 which renders kinetic reversibility to the enzyme. Indeed, RYRIE and JAGENDORF (1971) have demonstrated changes in conformation of CF_1 during illumination. The combination of an electrochemical potential and $-SH$ reagents enhanced the changes in conformation and therefore stimulated the reversible activity.

The role of the electrochemical gradient in conferring kinetic reversibility to CF_1 was demonstrated under conditions where the potential was formed in the dark in the absence of oxidation reactions. JAGENDORF and URIBE (1966) formed a proton gradient in the dark by transferring chloroplasts from acid to base. The artificially formed proton gradient drove ATP synthesis in the dark. This electrochemical gradient was shown to induce P_i-HOH exchange (SHAVIT and BOYER, 1966). Addition of an $-SH$ reagent to chloroplasts during acid-base transition induced P_i-ATP exchange reaction (BACHOFEN and SPECHT-JURGENSEN, 1967; KAPLAN and JAGENDORF, 1968). Thus, it was demonstrated that artificially-formed electrochemical gradient in the dark induced effects similar to those carried by light induced electron transport. It follows that the process of oxidation-reduction is not compulsory for induction of reversibility which is probably brought about by a change in conformation of CF_1. Although the conformation change requires energy it is probably not an intermediate which leads to ATP synthesis, since both the formation and the decay of the light induced electrochemical gradient (HIND and JAGENDORF, 1963) are faster than those of the changes which confer reversibility (PETRACK et al., 1965). Moreover, phosphate stabilizes the conformational changes while ADP accelerates their decay (CARMELI and LIFSHITZ, 1972) but these reagents do not alter the electrochemical potential.

Similar to photophosphorylation all the exchange reactions were dependent on Mg^{2+} ions. However some incorporation of P_i into ATP was found to be catalyzed by chloroplasts in the presence of Ca^{2+} ion during continuous illumination (BAKKER-GRUNWALD, 1974). This incorporation was interpreted as indicative for the catalysis of P_i-ATP exchange.

F. Energy Requirements

The energy requirements for induction of kinetic reversibility to CF_1 should be distinguished from the thermodynamic considerations. As indicated in Section D the exchanges probably do not proceed through the formation of a covalently bound intermediate. The rates of the exchanges should be, therefore, optimal near equilibrium. However, most of the exchanges were measured, at considerable rates, at substrate concentrations which were far from the thermodynamic equilibrium. These results can be explained in view of the fact that the reactions catalyzed by CF_1 are coupled to the energy transduction apparatus through the chloroplast membrane. The inhibition of P_i-ATP exchange by uncouplers such as carbonyl cyanide p-trifluoromethoxyphenylhydrazone (FCCP) and NH_4Cl, which dissipate the electrochemical gradient, supports this suggestion (CARMELI and AVRON, 1967; RIENITS, 1967; MCCARTY and RACKER, 1968). Although the effect of uncouplers was not tested on reactions which involve oxygen exchange there are some indications that these reactions also require energy. Thus, the requirement for ATP in the P_i-HOH exchange (SKYE et al., 1967) could partially be due to the requirement for energy which is provided during the concomitant hydrolysis of ATP. In the case of ATP-HOH exchange the energy requirements have to be determined with the aid of uncouplers since ATP is present and can provide the energy to support the reaction. The study of energy requirement is of a special interest in view of the finding of an uncoupler insensitive P_i-HOH exchange in mitochondria (BOYER et al., 1973). These authors suggested that the dissimilar sensitivity of various exchanges to uncouplers can be explained if it is assumed that water formation during ATP synthesis at the catalytic site does not require energy. Energy is only required for the release of ATP which is tightly bound to the enzyme.

G. Reconstitution of Vesicles Catalyzing P_i-ATP Exchange

The accumulated data indicate that energy is required for P_i-ATP exchange. Therefore, this reaction could be an indicator for energy conservation. In attempts to reassemble the energy conversion system from its elementary components the reconstitution of the complete phosphorylation apparatus either was not desired or was not attained. In this type of work P_i-ATP exchange can serve as a marker for partial reconstitution. MITCHELL (1968) suggested that the oxido-reduction and the hydro-dehydro systems are separately located in the membrane. The proton motive force couples the two systems. The coupling factor is probably the reversible ATPase of the suggested hydro-dehydro system since CF_1 which is required for ATP synthesis also catalyzes ATP hydrolysis. The hydrolysis is coupled to energy requiring processes such as proton transduction (CARMELI, 1969) and P_i-ATP exchange. Recently partial separation between electron transport and the ATP synthesis apparatus was achieved. The energy transducing system was reconstituted from the resolved fraction (CARMELI and RACKER, 1973).

In this work the chloroplast membrane was solubilized with the aid of cholate. A fraction containing lipids, CF_1 and some other proteins but deficient in chloro-

Table 1. ATPase and P_i-ATP activities in chloroplasts and in vesicles reconstituted from a resolved fraction of the chloroplast membrane. Chloroplasts were pretreated with 50 mM dithiothreitol in order to induce ATPase and P_i-ATP activities. Then treated with 2% cholate and centrifuged at 160,000·g for 1 h. The supernatant was either used as such or supplemented with purified chloroplast lipids. In order to form vesicles the supernatant was passed on a Biogel P-4 column which removed the detergent

Preparation	$nmol·mg\ chlorophyll^{-1}/10\ min$	
	P_i-ATP exchange	ATPase activity
Chloroplasts	1,490	15,957
Vesicles	329[a]	3,021[a]
Vesicles + 5 mg digalactosyl diglycerides	531[a]	3,052[a]
Vesicles + 5 µM FCCP	39[a]	5,141[a]

[a] Activities are given in nmol P_i per vesicles obtained from chloroplasts containing mg chlorophyll per 10 min

phyll and cytochromes was separated. On the removal of cholate either by dialysis or by gel filtration (CARMELI and LIFSHITZ, 1974) vesicles were formed. These vesicles catalyzed ATP hydrolysis accompanied by an uncoupler sensitive P_i-ATP exchange.

The resolution is indicated from the fact that the reconstituted system was deficient in components of the oxidation reduction apparatus. The vesicles contained no chlorophyll and had only 12% of the chloroplast content in cytochrome b_6 and f. However, about 25% of ATPase and 40% of the P_i-ATP exchange activities were recovered in the vesicles. The importance of the membrane in energy conversion is indicated by the fact that the vesicles were seen in electron micrographs as closed membranous formations. The fact that addition of lipids during the reconstitution increased the rate of the exchange also supports the suggested role of the membrane in energy conversion. It is interesting to note that the vesicles catalyzed exchange even more efficiently than the chloroplasts. This is apparent when the rate of the exchange is compared to the rate of the concomitant ATP hydrolysis in each of the systems (Table 1). From these results, it is apparent that in chloroplasts as in mitochondria (KAGAWA and RACKER, 1971) the coupling device which is responsible for the utilization of the oxidation reduction energy for the formation of ATP can be partially separated from the electron transport chain.

References

Avron, M.: Biochim. Biophys. Acta **40**, 257–272 (1960)
Avron, M.: Biochim. Biophys. Acta **77**, 699–702 (1963)
Avron, M., Grisaro, V., Sharon, N.: J. Biol. Chem. **240**, 1381–1386 (1965)
Avron, M., Jagendorf, A.T.: J. Biol. Chem. **234**, 967–972 (1959)
Bachofen, R., Specht-Jurgensen, I.: Z. Naturforsch. **22B**, 1051–1054 (1967)
Bakker-Grunwald, T.: Biochim. Biophys. Acta **347**, 141–143 (1974)

Ben-Yehoshua, S., Avron, M.: Biochim. Biophys. Acta **82**, 67–73 (1964)
Boyer, P.D.: Current Topics Bioenerg. **2**, 99–149 (1967)
Boyer, P.D., Cross, R.L., Momsen, W.: Proc. Nat. Acad. Sci. U.S. **70**, 2837–2839 (1973)
Carmeli, C.: Biochim. Biophys. Acta **189**, 256–266 (1969)
Carmeli, C., Avron, M.: Biochem. Biophys. Res. Commun. **24**, 923–928 (1966)
Carmeli, C., Avron, M.: Europ. J. Biochem. **2**, 318–326 (1967)
Carmeli, C., Lifshitz, Y.: FEBS Lett. **5**, 227–230 (1969)
Carmeli, C., Lifshitz, Y.: Biochim. Biophys. Acta **267**, 86–95 (1972)
Carmeli, C., Lifshitz, Y.: In: Proc. 3rd Intern. Congr. Photosynthesis. Avron, M. (ed.). Amster-
    dam: Elsevier, 1975, pp. 849–857
Carmeli, C., Racker, E.: J. Biol. Chem. **248**, 8281–8287 (1973)
Dixon, M., Webb, E.C.: Enzymes. 5ondon: Longman, 1964, pp. 282–286
Hind, G., Jagendorf, A.T.: Proc. Nat. Acad. Sci. U.S. **49**, 715–722 (1963)
Jagendorf, A.T., Uribe, E.: Proc. Nat. Acad. Sci. U.S. **55**, 170–177 (1966)
Kagawa, Y., Racker, E.: J. Biol. Chem. **246**, 5477–5487 (1971)
Kahn, J.S., Jagendorf, A.T.: J. Biol. Chem. **236**, 940–943 (1961)
Kaplan, J.H., Jagendorf, A.T.: J. Biol. Chem. **243**, 972–979 (1968)
McCarty, R.E., Racker, E.: J. Biol. Chem. **243**, 129–137 (1968)
Mitchell, P.: Chemiosmotic Coupling and Energy Transduction. Bodmin: Glynn, 1968
Nelson, N., Deters, D.W., Nelson, H., Racker, E.: J. Biol. Chem. **248**, 2049–2055 (1973)
Petrack, B., Carston, A., Sheppy, F., Farron, F.: J. Biol. Chem. **240**, 906–914 (1965)
Rienits, K.G.: Biochim. Biophys. Acta **143**, 595–605 (1967)
Ryrie, I.J., Jagendorf, A.T.: J. Biol. Chem. **246**, 3771–3774 (1971)
Shahak, Y., Chipman, D.M., Shavit, N.: FEBS Lett. **33**, 293–296 (1973)
Shavit, N., Boyer, P.D.: J. Biol. Chem. **241**, 5738–5740 (1966)
Shavit, N., Skye, G.E., Boyer, P.D.: J. Biol. Chem. **242**, 5125–5130 (1967)
Shoshan, V., Tel-Or, E., Shavit, N.: In: Proc. 3rd Intern. Congr. Photosynthesis. Avron,
    M. (ed.). Amsterdam: Elsevier, 1975, pp. 831–838
Skye, G.E., Shavit, N., Boyer, P.D.: Biochem. Biophys. Res. Commun. **28**, 724–729 (1967)
Slater, E.C.: Nature (London) **172**, 975–978 (1953)
Stewart, B.W., Rienits, K.G.: Biochim. Biophys. Acta **153**, 907–909 (1968)

IV. Structure and Function

1. Introduction to Structure and Function of the Photosynthesis Apparatus

K. MÜHLETHALER

In the light microscope a chloroplast can be differentiated into a clear structureless stroma and a number of small green discs. The latter have been termed "grana" and were thought to contain exclusively the pigments with photosynthetic activity. Investigations with the electron microscope have shown that the grana are composed of a stack of lamellae which have been termed thylakoids by MENKE (1961). These structures are the basic elements of the photosynthetic apparatus, from bacteria up to the chloroplasts of higher plants.

A. The Membrane Components

It is now well established that the thylakoids contain all the chlorophyll and that they represent the structural framework for the photosynthetic primary reaction. The chemical composition of the thylakoid membranes is complex as they contain lipids, pigments, and proteins. Carbohydrates are only present in small quantities (VAN WYK, 1966). About half of the membrane mass consists of lipids and the outer half of protein. The lipids represent a complex mixture. They appear to differ from those of other plant structures in having a high proportion of glycolipids and certain phospholipids, and a low triglyceride content (VAN DEENEN, 1965). Quantitatively, the most important lipids are the glycolipids, e.g. mono- and digalactosyldiglycerides. They constitute about 40% by weight of the total lipid. The pigments account for about 20%, the remainder being phospholipids, quinones, and other components (WINTERMANS, 1967).

As shown by MENKE and RUPPEL (1971) and by NOLAN and PARK (1975), the protein fraction can be separated into at least 15 different proteins. These are either associated with the membrane or are an intrinsic part of it. The loosely bound fraction may be readily released by mild treatments on changing the ionic strength or pH of the medium, or by chelating the divalent cations that bind some of the proteins to the membrane. The isolation of these *peripheral (extrinsic) membrane proteins* is greatly facilitated by their solubility in water. The gentle treatment preserves in most cases their enzymatic activities. It has been shown that most of them bear a net negative charge at neutral pH (STECK and FOX, 1972). Their attachement to the membrane surface takes place either through divalent cation bridges or through specific membrane proteins.

The proteins forming the basic framework are termed *integral (intrinsic) membrane proteins*. In the early days of biochemical analysis, these proteins were solubilized by organic solvents such as butanol, pentanol, or 2-chloroethanol. However, the bulk of the protein was then denatured. This led many authors to the conclusion that the membranes are composed of two different classes of proteins: the stable "structural proteins" and the enzymatic active proteins (CRIDDLE, 1966). If these two classes really exist, we should expect to find some proteins which have a helical conformation, as described for many enzymes, and some which have a β-conformation, as in fibrous proteins. Measurements of fractions of the membrane proteins by means of infrared absorption, circular dichroism, and optical rotatory dispersion have indicated that β-structure is rarely found (WALLACH and ZAHLER, 1966). MENKE (1962), on the other hand, obtained a yield of lamellar structural protein up to 54% of the total weight of the chloroplast preparation. Depending on the method of fractionation, the amount of "structural" and "enzymatic" protein may vary considerably. Due to the complex character of the binding forces between the different membrane substances, a uniform solubilization of the components without denaturation is very difficult. The membrane components may interact with polar groups (via hydrogen bonds) and/or with nonpolar groups (via hydrophobic bonds). Partial and selective protein solubilization can be obtained by a number of methods, involving chelating agents, detergents, and the manipulation of ionic strength and pH. The interactions of detergents with membrane proteins will differ from those seen with water-soluble proteins, such as serum albumin, due to differences in conformational structure of the two protein types. However, the amino acid composition of water-soluble proteins is very similar to the water-insoluble lamellar structure proteins (WEBER, 1962, 1963).

It is unreasonable to characterize membrane proteins as a class ("structural protein" versus "soluble proteins" etc.). Their composition and function is as diverse as any other proteins present in different parts of the cell. It is likely that membrane proteins will have distinct characteristics only in those cases where the protein is involved in strong interactions with the membrane lipids.

Accordingly, it could be envisaged that some membrane proteins would be so designed as to interact with the apolar portions of the membrane lipids. Thus, we may distinguish between polar side chain residues exposed to the outer surface of the membrane, and extensive apolar regions oriented towards the membrane interior. Such proteins could then intercalate an appreciable portion of their structure into the hydrophobic layer. The interactions between the membrane lipids and proteins would be stabilized by hydrophobic forces, as well as through surface ion-dipole, dipole-dipole, and ion-pair formation (GITLER, 1972). The changes occurring in the membrane components during solubilization are complex. As shown by MACHOLD (1975), the protein moieties of the pigment-protein complexes from thylakoids exhibit altered electrophoretic mobilities after the removal of the pigments by high concentrations of sodium dodecylsulfate or by acetone, as compared with the undenatured complexes in which most of the pigment-protein bonds are preserved. This indicates that the native complexes can disintegrate into subunits. The formation of protein aggregates into multi-enzyme complexes is an important feature in the structural and functional hierarchy of enzyme systems. ATP-synthetase, for instance, was found to consist of five subunits (NELSON et al., 1973). The water-soluble, bacterial chlorophyll protein was found to have a molecular weight of 152,000 and to consist of four identical subunits, each containing five bacteriochlorophylls (OLSON et al., 1969). The pigment-protein complexes of higher plants, which are believed to be from photosystems I and II may also be constructed from subunits (ARNTZEN et al., 1969). The molecular organization of these photosystems is not yet determined. The ordered arrangement of different subunits into a larger complex may allow directional or vectorial enzymatic reactions through or across the membrane. In each complex, several catalytic sites could be brought into a position such that the entering substrate molecule would be processed into metabolic substances. Since the size of these multi-enzyme-complexes may be up to 200 Å, their position in the membrane can be determined in the electron microscope.

B. Ultrastructure of Thylakoid Membranes

I. General Aspects

The stroma regions generally contain single thylakoids, whereas the grana region contain closely packed stacks of thylakoids (Fig. 1). A single membrane measures 75 Å in thickness, but variations between 50 to 100 Å are possible. In the grana region, two adjacent membranes are fused to form a membrane pair called a "partition". The distance between membrane pairs in the unswollen condition amounts to about 230 Å (WEHRLI, 1975). The thickness of a membrane pair varies between 100–200 Å and in very thin sections the profile appears undulated.

For about a decade between 1958–1968 it was believed that the so-called "unit membrane" structure, postulated by ROBERTSON (1958), would also be applicable to chloroplast membranes. According to this concept, which originated from the earlier hypothesis proposed by DANIELLI and DAVSON (1935), it was generally believed that the thylakoid membrane was composed of three strata, containing on each side a monolayer of "structure protein" molecules separated by a intermediate lipid bilayer. In addition to the electron microscopic results, the trilamellar concept was supported by X-ray diffraction work (FINEAN, 1962). The view that the dark strata represent protein layers, as advocated by STOECKENIUS and MAHR (1965), was not in agreement with the earlier work by CRIEGEE (1936), who showed that osmium tetroxide forms cyclic monoesters with unsaturated compounds, such as unsaturated fatty acids. In a subsequent reaction, dioles may be formed by hydrolysis and they in turn may form stable diesters with a neighboring cyclic monoester. As also suggested by WIGGLESWORTH (1957), osmium tetroxide may stabilize lipids by bridging neighboring molecules. According to KORN (1966), 40% of the oleic acids in the plasmalemma of *Amoeba* react with osmium. Similarly a quantitative investigation by HAYES et al. (1963) showed that a direct relationship exists between the uptake of osmium tetroxide and the lipid content of lipoproteins. With an elegant new experimental approach, KOPP (1972) showed that, in the membrane, osmium reacts primarily with the lipids. To trace the reaction of osmium tetroxide, he compared membranes before and after the removal of double bonds. In addition, the staining characteristics of membranes before and after lipid extraction were studied. If the double bonds were blocked before osmium-fixation, the contrast in membranes was lost. These reactions were tested with spinach chloroplasts. Control experiments with membranes before and after lipid extraction in thin section, and after freeze-etching, led to the conclusion that the dark strata

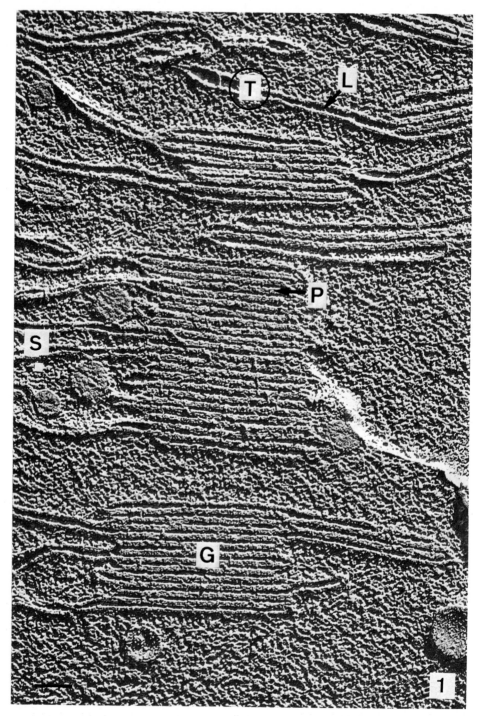

Fig. 1. Cross-sectional view of a spinach chloroplast showing the arrangement of grana and stroma thylakoids. Thylakoid (*T*), Lumen (*L*), Partition (*P*), Grana (*G*), Stroma (*S*). × 100,000 (Wehrli, 1975)

seen in osmium fixed membranes corresponds with layers having a high content of unsatturated lipids. Therefore, contrary to the earlier theories, the membrane surface must be covered with lipids, occasionally interrupted with peripheral protein complexes.

However, it should be noted that FLEISCHER et al. (1967) described a distinct unit-membrane structure in mitochondria after lipid extraction with acetone, followed by osmium fixation and uranyl acetate and lead citrate staining. This result does not comply with either the trilamellar DANIELLI and DAVSON model or the result of KOPP. In accordance with KOPP's (1972) result, ONGUNG et al. (1968) had earlier noted a loss of structure in chloroplasts after chloroform-methanol extraction. However, the original membrane system remained when fixation with osmium tetroxide or potassium permanganate was accomplished before lipid extraction. According to ONGUNG et al. (1968), the glycolipids and phospholipids in chloroplasts cannot be extracted with acetone or chloroform-methanol if fixation with osmium tetroxide precedes. All these investigations show that the results reported by FLEISCHER et al. (1967) are not in agreement with similar studies of other authors, and are so far inexplicable. It is not clear why a total removal of the membrane lipids, which constitute approximately half of the membrane structure, does not cause a change of the trilamellar pattern. More recent experiments have shown that osmium tetroxide itself may cause drastic changes in the membrane, due to denaturation and solubilization of membrane components. As shown by MCMILLAN and LUSTIG (1973), up to 85% of the proteins in human erythrocyte ghosts can be solubilized from osmium tetroxide treated samples. In comparison only 10% are removed from glutaraldehyde treated ghosts. This again confirms the generally accepted view that glutaraldehyde stabilizes the protein components, and osmium tetroxide the lipid fraction of the membrane. In spite of the improvements achieved in recent fixation, embedding, and sectioning techniques, contributions to a better understanding of membrane structure are very limited.

More structural details could be seen in isolated thylakoids which were simply dried on a specimen support and shadowcast. With this technique FREY-WYSSLING and STEINMANN (1953) found that isolated granum discs consisted of flat vesicles with a globular substructure. Based on these results they postulated a new model composed of globular units. In repeating these experiments PARK and PON (1963) proposed the quantasome model. The elementary particles termed "quantasomes", with a thickness of 100 Å and a diameter of 200 Å, were thought to be the morphological expression of the photosynthetic unit as formulated by EMERSON and ARNOLD in 1933. With the introduction of the freeze-fracture technique (MOOR et al., 1961), a new method was created for the investigation of chemically unaltered cell structures. The first studies on chloroplast membranes confirmed the presence of particles and allowed a detailed study of their size, structure, and arrangement over large areas of the thylakoid (MÜHLETHALER et al., 1965). With model lipid membrane systems, BRANTON (1966), concluded that the fracture process of frozen specimens splits the membrane along the hydrophobic zone and thus exposes the internal part of the thylakoid structure. This finding could be confirmed with the double fracture technique, which enables replication of both sides of the cleaved plane (WEHRLI et al., 1969). This technique can be combined with the "etching" procedure, which reveals the surfaces of the membrane. The combined use of these methods allows studies of the inner and outer surfaces, as well as both sides of the cleaved hydrophobic middle zone. As a result, it became apparent that all the four membrane faces thus revealed contain particles.

The nature of the globular particles was at first a matter of controversy. During the electron microscopy of negatively stained lipid preparations, LUCY and GLAUERT (1964) observed globular subunits which they interpreted as specific complex assemblies of small globular micelles of lipids. They speculated that similar macromolecular assemblies of lipids could be present in biological membranes. X-ray diffraction studies of model systems involving lipids, proteins and water revealed a remarkable polymorphism which was highly sensitive to temperature and water concentration (LUZZATI, 1968). For this reason, it was thought that a possible rearrangement of the membrane lipids could occur during the freeze-fracture process. At higher temperatures the lipid chains behave like liquids and are less ordered than at low temperatures, where their conformation is stiff but with frequent disorder of the c-chain conformation. An X-ray diffraction analysis by DEAMER et al. (1970), where the preservation of various lipids which were also freeze-etched was studied, indicated that the X-ray diffraction patterns of dry, or nearly dry, phospholipid samples before and after freezing were quite similar. The 4.5 Å diffuse band typical of liquid paraffins was found in both samples. This demonstrates that the liquid structure was preserved after freezing and that crystallization of the paraffin chains was avoided. From these experiments, we can conclude that freezing of the lipids does not detectably alter the lamellar phase normally present in membranes and seen as a small layer at the base of the fracture face after freeze-etching. The possibility that the particles seen in freeze-fractured membranes could represent globular lipid micelles can be excluded with certainty. That they must therefore be related to protein structures has also been shown by biochemical analysis combined with freeze-fracture studies (ARNTZEN et al., 1969); by antibody labeling (PINTO DA SILVA et al., 1971; BERZBORN et al., 1974) and by enzymatic degradation (BAMBERGER and PARK, 1966).

II. The Outer (Matrix Side) Surface (OS)

In general, only the outer thylakoid membrane of a granum stack is accessible for replication, because in the rest of the granum the thylakoid membranes are fused. The inner side of the thylakoid membrane can best be replicated if the vesicles are swollen by osmotic treatment.

The outer surface (OS), i.e. the surface next to the matrix of the chloroplast, reveals two populations of particles with an average size of 100 Å and 140 Å respectively (Fig. 2). The larger ones can be removed with EDTA and, as shown by KANNANGARA et al. (1970), BERZBORN et al. (1974), GARBER and STEPONKUS (1974), MILLER and STAEHELIN (1974), and OLESKO and MOUDRIANAKIS (1974), they represent the coupling factor particles CF_1. In freeze-fractured preparations it became evident, that these particles were missing along the contact zone of neighboring grana thylakoids (partition). In the larger stroma thylakoids which traverse a grana stack, particle-free areas are also seen (WEHRLI, 1975). From this observation we conclude that ATP-synthesis is restricted to stroma membranes.

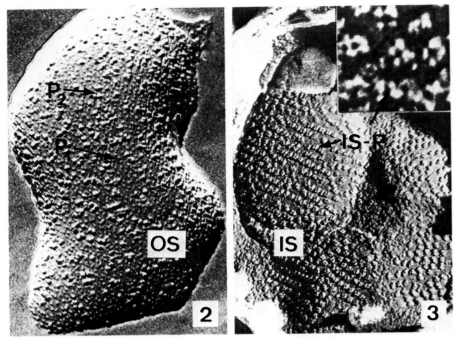

Fig. 2. Outer surface structure of a thylakoid membrane of a spinach (*S. oleracea*) chloroplast. Deep-etching. Small particles (P_1), large particles (P_2). $\times 100,000$ (WEHRLI, 1975)

Fig. 3. Inner surface structure of a thylakoid membrane (*S. oleracea*). Deep-etching. Inner surface particles (*IS-P*). $\times 100,000$ (WEHRLI, 1975). Inset $\times 500,000$

III. The Inner (Lumen Side) Surface (IS)

The lumen side of the grana thylakoids are also covered with particles (Fig. 3). Here they exhibit a characteristic rectangular shape, with an average edge length of 100 by 150 Å (WEHRLI, 1975). There are 8 particles μm^{-2}. With increased resolution it can be seen that the complexes are built up of four subunits with an average diameter of 50 Å. Very often it is not possible to distinguish each of them, because two particles may merge together forming a short rod. Their presence was seen in grana thylakoids. Here, a smaller population of 40 to 60 Å particles is also present. They are difficult to see because they do not rise far enough above the background to provide sufficient contrast in the replica (WEHRLI, 1975).

The particles attached at the inner and outer surfaces of the thylakoids are also easily detectable with negative staining. Here they appear as bright components which coincide in size, form, and distribution with those seen in frozen specimens. Since after negative staining, images are projected from both thylakoid surfaces simultaneously, it becomes difficult to determine on which side the particles are located. The staining medium is not able to penetrate into the inner hydrophobic zone and for this reason the integral particles cannot be detected.

IV. The Inner Zone of the Thylakoid Membrane

As mentioned earlier, the hydrophobic middle layer can be analyzed after fracturing the membrane in the frozen state. With the double fracturing technique, both sides of the cleaved thylakoid membrane are accessible for replication. In Figures 4 and 5 a complementary pair of micrographs from a fractured region containing a granum stack is shown. In one picture the fracture face of the outer half (OFF) viewed from the luman side is shown (Fig. 4), while the other picture (Fig. 5) represents the same fracture plane viewed from the matrix side. The latter is the inner membrane half and is termed the inner fracture face (IFF) (Mühlethaler, 1971).

1. The Inner Fracture Faces (IFF)

The arrangement and number of particles, or holes if these complexes are torn out during the fracturing process, is different in the grana region from that in the stroma region. In the grana areas, the number of particles is considerably higher than in the stroma, whereas the number of holes is lower (Fig. 5). Wehrli (1975) counted 1,250 particles μm^{-2} and 1,700 holes μm^{-2} within a grana region, compared with 450 particles μm^{-2} and 3,700 holes μm^{-2} in the stroma area. The size of the particles present in the grana region is larger than those in the stroma areas. According to Wehrli (1975) the grana particles have a rectangular form with a length of 150 Å and a width of 110 Å. In addition, a small number of 100 Å particles can be seen. In the stroma region the particles are rectangular with an average diameter of 100 Å. The particles are composed of two subunits. Similar measurements were recorded by Goodenough and Staehelin (1971).

The large grana particles (150×110 Å) have been found so far only in photosynthetic membranes, which indicates that their activity must be related with the photosynthetic primary process. Due to their size we can relate them to the "quantasomes" (Park and Biggins, 1964). From specimens where the angle of evaporation is known, it could be determined that the particles, *present in fused pairs of membranes,* approximately correspond to the thickness of the outer half of the thylakoid membrane which is 50–55 Å (Fig. 9). This observation does not confirm the view proposed by Goodenough and Staehelin (1971), that these complexes are penetrating through adjacent grana membranes.

2. The Outer Fracture Faces (OFF)

In the stroma area, Wehrli (1975) counted 3,700 particles μm^{-2} (IFF:450 p μm^{-2}) in comparison with 4,500 μm^{-2} for the grana region (IFF:1,250 p μm^{-2}). As described for the ones attached to the inner fracture face, these particles are also composed of two subunits with a dimer size of 100 Å. One third of the total particle number is only 50 Å in diameter and forms a population of its own.

The picture (Fig. 4) of the grana region shows a very dense particle population, in which free spaces between the particles are generally not present. Particle size and form corresponds to those observed in the stroma region. There are some indications that the particles are more deeply embedded in the membrane matrix

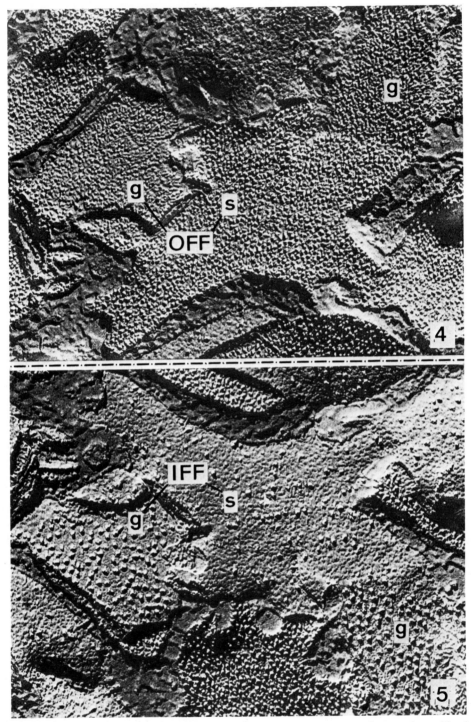

Figs. 4 and 5. Two complementary replicas showing corresponding freeze-fracture faces. Fig. 4 shows the outer fracture face (*OFF*); Fig. 5 the inner fracture face (*IFF*). Grana (*g*) and stroma (*s*) region. × 100,000 (WEHRLI, 1975)

of the grana thylakoids than those present in the stroma thylakoids. For this reason the outer fracture faces appear less granular than in the stroma membranes. From observations of complementary replicas it can be concluded that the 150 Å complexes present at the inner fracture face of the grana thylakoid are torn out of the outer half of the membrane. In many cases, the particles are split and remnants remain in the outer membrane layer. The molecules aggregated into these clusters must have an ordered arrangement similar to that described for multi-enzyme complexes (Lehninger, 1966).

C. The Relations Between Peripheral and Integral Particles

Based on the result obtained with the freeze-fracture and freeze-etch technique, the approximate structural arrangement of the discernible membrane components becomes evident. The two parts of the split membrane are different in thickness. Wehrli (1975) found that the outer layer of the membrane is considerable thicker than the inner one. He estimated a ratio of 2.5 to 1. Given a total membrane thickness of 75 Å, fracturing would give rise to an outer layer of 50–55 Å and an inner one, measuring 20–25 Å in thickness. In view of the results obtained by Kopp (1972) using osmium tetroxide, we must conclude that the inner lamella represents a lipid monolayer. The particles standing out of the inner fracture layer of the stroma region are higher than the thickness of the outer part of the thylakoid (Fig. 9). This means that the peaks of the 150 Å particles must penetrate the surface. The small particle population seen at the thylakoid surface between the CF_1 complexes must be the peaks of particles located within the membrane. A structural continuity between the peripheral four-subunit complexes at the lumen side and the adjacent integral particles is very likely.

Some additional observations can be obtained if chemically unchanged thylakoids are compared with lipid depleted membranes. According to Ji et al. (1968), the bulk of the galactolipids, about one third of the phosphotidylglycerol, all xanthophylls, but no chlorophyll, can be extracted with 45% acetone. After this treatment, the membrane particles aggregate and the whole membrane structure collapses. For morphological studies with lipid depleted thylakoids, the membranes must be fixed previously with glutaraldehyde. After the extraction, the matrix substance of the outer membrane zone diminishes and the particle structure of the internal proteins becomes visible (Wehrli, 1975). In contrast to the effects on the membrane matrix of the outer zone, acetone extraction does not attack the inner matrix layer. Since chlorophyll is not extracted by this treatment, it might be concluded that this pigment is mainly located at the lumen side of the thylakoid. As discussed later, this view is not in agreement with results derived from antibody labeling tests (Radunz et al., 1971). The result of these extraction studies gives clear evidence that an asymmetric distribution of lipids exist. Wehrli (1975) postulates that the outer matrix layer contains mainly monogalactolipids, sulfolipids, and phosphatdidylglycerol, and the inner layer contains digalactolipids and the bulk of the chlorophylls.

D. Mobility of Membrane Particles

Experiments have shown that the thylakoid membrane is not a stable structure. The lipid molecules are in a fluid state (LUZZATI, 1968) and therefore in continuous motion. It was also observed that the thickness of a pair of fused grana membranes decreases by 25% after illumination. This process is reversible (MURAKAMI and PACKER, 1970). This structural change was related to the capacity of illuminated chloroplasts to carry out a light-dependent protonation across the membrane and the formation of a pH gradient. WANG and PACKER (1973) used the freeze-fracture technique to investigate the influence of light and ionic environement on membrane organization. They found that particles seen in the hydrophobic core of grana membranes can change their orientation within the membrane, upon illumination or treatment by divalent cations. DEAMER et al. (1967) have shown that a swelling effect results from the extremely low internal pH, caused by light dependent H^+ uptake into chloroplasts. This decrease in pH results in membrane disorganization. The configuration of the normal grana structure is pulled apart, the spacing between the thylakoids becomes larger, and the chloroplasts as a whole undergo an increase in volume due to the uptake of water and solutes. This swelling is accompanied by a decrease in the number of particles per μm^2 by 32% and the average particle diameter is decreased by 40 Å (WANG and PACKER, 1973). As an explanation of these changes the authors suggest that protein components are penetrating further into the interior of the membrane in response to illumination, or that they are being withdrawn from the hydrophobic domains of the membrane during darkness.

Serological reactions also indicate that the membrane particles are mobile. As shown by BERZBORN et al. (1974) and by MILLER and STAEHELIN (1974), the coupling factor particles at the thylakoid surface form clusters if the thylakoids are incubated with antisera containing only antibodies agains CF_1 (Fig. 6). These results indicate that the aggregates are formed by lateral movement of the particles at the membrane surface. Since the mean distance between the particles in the untreated control was greater than the distance which can be bridged by an antibody, BERZBORN et al. (1974) proposed that the antibody first reacts through only one of its binding sites with a particle, that this complex moves laterally within the surface of the thylakoid until it meets another particle, and that the antibody then reacts through its second binding site. There is every reason to believe that the reaction with the antibodies does not by itself induce the capacity of lateral movement at the thylakoid surface, but that this capacity is a property of the 140 Å particles in vivo. The phenomenon of macroscopic agglutination shows that the antigen is rather tightly bound to the membrane, because the thylakoids cannot be easily shaken apart. As shown by WANG and PACKER (1973) and OJAKIAN and SATIR (1974) integral particles are also mobile.

E. The Identification of Membrane Constituents

It is a main concern of electron microscopists to correlate the structural details seen in the specimens with biochemical data, and to locate and identify certain components such as enzymes, lipids, or carbohydrates etc. in sections or replicas. A number of histochemical reactions, using heavy metal compounds which introduce a higher contrast, have been proposed. In most cases these substances must be attached at a specific site with a chemical reaction. This involves changes in temperature, pH, and salt concentration which are likely to destroy the labile structural framework of the native specimen. So far, the best results have been obtained, as shown in the previous paragraph (Fig. 6), with serological methods. An antibody molecule is, of course, not visible with the ordinary methods. Enough contrast, however, can be introduced if ferritin is first bound to the antibody. This marker, is however, very large and cannot be introduced into the living cell or into a membrane. Methods are needed to overcome these difficulties.

Valuable information concerning the biochemical composition of the thylakoid surface can be obtained with macroscopic agglutination tests. This phenomenon takes place if the bifunctional antibody molecules react with two stroma-freed chloroplasts. In the case of the antigenic determinant being located in a depression of the thylakoid surface, an antibody molecule cannot form a bridge between two plastids and is simply absorbed at the specific site. For details see this vol. Chapter II.17.

With this serological technique a number of proteins involved in the photosynthetic pathway have been located. It was found that the following proteins are accessible to antibodies at the thylakoid surface:

1. $NADP^+$-ferredoxin reductase (BERZBORN et al., 1966; BERZBORN, 1969).
2. The coupling factor CF_1 (KANNANGARA et al., 1970; BERZBORN et al., 1974).
3. Ferredoxin (HIEDEMANN-VAN WYK and KANNANGARA, 1971).
4. Ferredoxin-reducing substance (BERZBORN et al., 1966).
5. Carboxydismutase (KANNANGARA et al., 1970).
6. Plastocyanin (SCHMID et al., 1975).

Antibodies against a protein exhibiting photosystem I activity were found to react with a site located on the lumen side and another on the matrix side of the thylakoid membrane. An inhibition site for photosystem II was also found to be located at the surface of the thylakoids (KOENIG et al., 1972; MENKE, 1973). BRIANTAIS and PICAUD (1972) prepared antibodies against enriched photosystems I and II proteins, and found a precipitation of isolated thylakoids with antibodies against photosystem I, and to a lesser extent also against photosystem II. They concluded that photosystem I is in the membrane and accessible to antibodies. The authors take their results as proof that photosystem I particles are located in the outer part of the thylakoid membrane on top of the photosystem II particles in the inner membrane layer. This arrangement had been previously proposed by ARNTZEN et al. (1969), based on freeze-fracture studies.

As described by RADUNZ and BERZBORN (1970) galacto- and sulfolipids are also orientated towards the thylakoid surface. Tests for pigments indicated that lutein and chlorophyll are associated with these lipids (RADUNZ and SCHMID, 1973).

At the lumen side of the thylakoids the following proteins have been detected by serological tests:

1. Cytochrome f (RACKER et al., 1971).
2. Plastocyanin (HAUSKA et al., 1971; RACKER et al., 1971). [In a recent paper, SCHMID et al. (1975) have shown that plastocyanin could be detected at the matrix side of the thylakoid.]

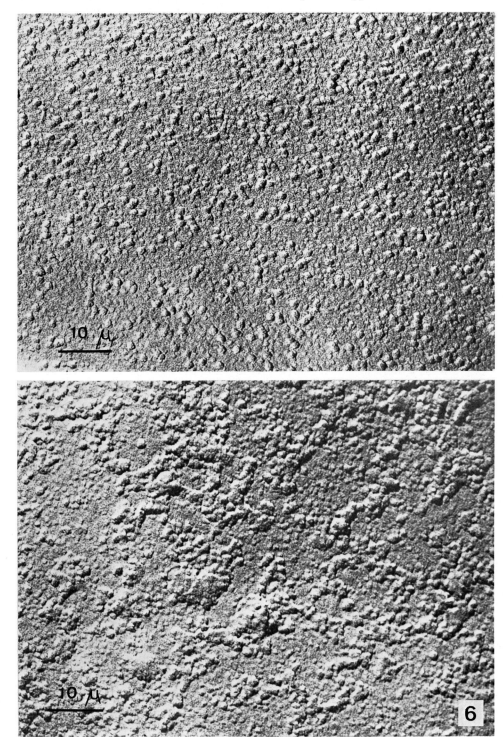

Fig. 6. Thylakoid outer surface of spinach chloroplast, before and after treatment with antiserum against CF_1. The particles are aggregated to clusters. $\times 140,000$ (BERZBORN et al., 1974)

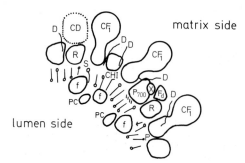

Fig. 7. Proposed model of the thylakoid membrane, based on serological tests with antibodies against: carboxydismutase (*CD*), ferredoxin (*Fd*), coupling factor (*CF₁*), glycolipopeptide (*D*), primary acceptor (*X*), P-700 protein (*P-700*), ferredoxin-NADP⁺-reductase (*R*), chlorophyll (*chl a*), sulfolipid (*S*), phosphatidylglycerol (*P*), plastocyanin (*PC*) and cytochrome f (*f7*), according to BERZBORN (1972)

Based on such serological studies, BERZBORN (1972) modified the earlier membrane model suggested by KREUTZ (1970) (Fig. 7). In this bilamellar model the proteins would be located in the outer (matrix) half of the membrane and the lipids along the lumen side. As previously described, the staining reaction with osmium tetroxide (KOPP, 1972) and the results obtained by freeze-fracturing and etching (BRANTON, 1969; MÜHLETHALER, 1972; WEHRLI, 1975) do not support this bilamellar asymmetric model. As outlined earlier, the serological tests also indicate the presence of lipids on both sides of the thylakoid membrane.

A schematic model for vectorial electron flow has been proposed by HAUSKA (1972) (Fig. 8). This scheme is based on serological and functional considerations and does not claim to be in accordance with the ultrastructural organization.

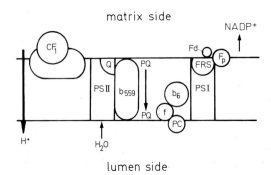

Fig. 8. Attempt to correlate the photosynthetic electron transport process with the results of antigen labeling. Modell suggested by HAUSKA (1972). Abbreviations: photosystem I (*PS I*) and II (*PS II*), quencher (*Q*), cytochrome b-559 and b-6, plastoquinone (*PQ*), ferredoxin-NADP⁺-reductase (*Fp*), reducing site of PS I (*FRS*)

F. Correlation Between Ultrastructural and Serological Studies

Based on his freeze-etch and fracture studies, WEHRLI (1975) proposed a thylakoid membrane model with an asymmetric structure (Fig. 9). The lipid matrix would be present in two layers, whereby the outer leaflet would contain monogalactolipids (RADUNZ, 1972), sulfolipids (RADUNZ and BERZBORN, 1970), phosphatidylglycerol (RADUNZ, 1971), lutein and plastoquinone (RADUNZ and SCHMID, 1973), together

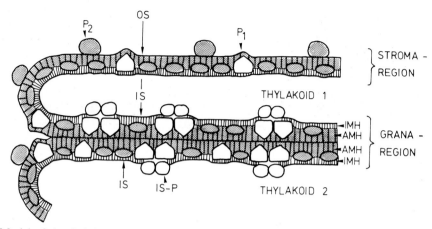

Fig. 9. Model of the thylakoid membrane based on freeze-fracture studies. Inner membrane half (*IMH*), outer membrane half (*AMH*), inner surface (*IS*), outer surface (*OS*), particle (*P*), inner surface particles (*IS-P*) small particles (P_1), large particles (P_2) (WEHRLI, 1975)

with some chlorophyll (RADUNZ et al., 1971). The lumen side leaflet would consist of digalactolipids and chlorophyll. In the lipid matrix, up to 6,000 protein particles per μm^2 with a variable size between 40 to 200 Å can be counted. Assuming an average particle cross section of 100 $Å^2$, about 50% of the stroma and 70% of the grana area are covered with particles (WEHRLI, 1975). The distribution of integral particles is different in each half of the cleaved membrane. The ratio of the number of particles per μm^2 between the outer- and inner layer amounts to 7/1 for the stroma region and 3/1 for the grana area. This does not imply that the proteins are predominantly located along the outer part of the membrane. From our studies on cleaved membranes we have to conclude that the particles are located in the middle of the hydrophobic zone. During the fracturing process the particles are torn out to some extent into the outer or inner membrane layer, depending on their binding strength to associated components, such as lipids, pigments etc. The freeze-fracture work indicate a sidewise packing instead of a binary top-on-top orientation as proposed by ARNTZEN et al. (1969). According to these authors, material isolated by digitonin treatment consisted of membrane fragments with 175 Å particles with enriched photosystem II activity, and smaller 110 Å particles with photosystem I activity. According to GOODENOUGH and STAE-HELIN (1971), however, the large complexes should contain both systems in the same complex. As discussed, the distribution of the 100–110 Å and 150–175 Å particles is different in the grana and stroma regions. The smaller (PS I) particles are found in both regions, while the large ones (PS II) are only seen in grana thylakoids. As shown by STAEHELIN and MILLER (1974) and WEHRLI (1975) the four subunit particles at the lumen side are present in the grana region. This indicates a functional relationship between the large 150 Å particles in the membrane and the peripheral four subunit complexes, which may perhaps represent cytochrom f according to serological tests. As mentioned previously, the coupling factor particles are only present at the matrix surface of stroma thylakoids and at the end-grana lamellae, but not in the partitions in the grana stack.

G. The Relationship Between Structure and Function

Trebst (1974) discussed the present evidence for vectorial electron flow and correlated the familiar zig-zag scheme of photosynthetic electron flow with the membrane
structure. The biochemical evidence (Trebst and Hauska, 1974) is in agreement
with the earlier biophysical studies by Witt (1971) and the hypothesis by Mitchell
(1961) that electron transfer across the membrane is vectorial in such a way that
the acceptor sites of photosystem I and II are at the outside of the thylakoid
and the donor sites are inside. As suggested by Junge and Witt (1968), transfer
is perpendicular to the membrane. The two photosystems are connected with a
plastoquinone loop which pumps protons from the outside to the thylakoid lumen.
Phosphorylation is coupled to electron flow via the discharge of the electrically
energized membrane and the field driven proton gradient (Witt 1971).

As discussed by Park and Sane (1971) there is now evidence which supports
the occurrence of two PS I reactions and one PS II reaction in chloroplasts. The
PS I reaction located in the stroma lamellae seems to be physically separated
from PS II, and may be primarily involved in cyclic photophosphorylation. The
PS I in the grana may be largely involved in noncyclic electron transport to
$NADP^+$. Freeze-fracture studies favor this view because only grana thylakoids
contain complexes which can be related to PS I and II, whereas stroma membranes
contain only PS I particles and CF_1 complexes.

The complicated electron and proton transport processes taking place in the
thylakoid membrane are possible because the numerous enzymes involved in this
pathway are organized into an assembly wherein each subunit has its proper place
in the whole functional chain.

H. Conclusions

1. The membrane system of chloroplasts is differentiated into grana and stroma
lamellae, which are termed thylakoids. In the grana stack the membranes of adjacent
thylakoids are fused to form partitions.

2. The membranes contain particles, which can be seen after freeze-fracture
(integral or intrinsic particles), or freeze-etching (peripheral or extrinsic particles).

3. These particles are fully or partly embedded in the lipid matrix, containing
mainly mono- and digalactolipids, sulfolipids, and pigments.

4. The fluid-like consistency of the lipids, which are found on both surfaces,
allows a free movement of the particles in the membrane.

5. With the method of antibody labeling some of the enzymes involved in
the electron transport chain could be located at the membrane surfaces.

6. Based on freeze-etch results, it must be concluded that stroma and grana
thylakoids are different in their structure and function. Photophosphorylation seems
to be restricted to the stroma thylakoids, because the CF_1-particles are absent
in grana thylakoids. In contrast, the large complexes expected to contain PS II
are only seen in grana thylakoids.

7. The membrane organization indicates that the metabolic active proteins are assembled into large complexes which catalyze multi-stage vectorial enzymatic reactions across the membrane. These complexes are inserted into, or attached to, membrane structure at specific, recurring intervals in the form of globular clusters. All these numerous multi-enzyme complexes produce the photosynthesis products simultaneously over the membrane area.

I. The Freeze-Fracture Nomenclature Used for Studies of the Thylakoid Membrane

MÜHLETHALER (1971) WEHRLI (1975)		BRANTON (1969)	GOODENOUGH and STAEHELIN (1971)
Outer surface (matrix side)	OS	"A" face	A'
Outer fracture face	OFF	"C" face	Cs, Cu
Inner fracture face	IFF	"B" face	Bs, Bu
Inner surface (lumen side)	IS	"D" face	D
The index s stands for stroma, g for grana (OFFs; OFFg etc.)			s stands for stacked; u for unstacked

A new nomenclature to describe and label the fracture faces and surfaces was recently proposed by BRANTON et al. (1975). It is suggested that for any membrane that can be split, the half closest to the cytoplasm, nucleoplasm, chloroplast stroma, or mitochondrial matrix be designated the "protoplasmic" half, abbreviated P; the half closest to the extracellular space, exoplasmic space, or endoplasmic space be designated the extracellular, exoplasmic or endoplasmic half, abbreviated E. The exoplasmic space would also include the thylakoid lumen. Once the half-membrane is labeled P or E, the particular aspect seen in the electron microscope can then be identified as a true surface of the membrane or as a fracture face. The surface, to be designated either PS or ES, the fracture face, PF or EF. If the new nomenclature were used to label the thylakoid freeze-etched membranes, the outer surface (OS) has to be termed PS, the inner surface (IS) would be ES. The outer fracture face (OFF) is identical with PF and the inner fracture face (IFF) with EF.

Acknowledgements. The outhor wishes to thank Dr. GREGORY R. PARISH for his correction of the English text and to Miss ERIKA ABÄCHERLI and Mrs. RUTH RÄBER for their help with the manuscript.

References

Arntzen, C.J., Dilley, R.A., Crane, F.L.: J. Cell Biol. **43**, 16–31 (1969)
Bamberger, E.S., Park, R.B.: Plant Physiol. **41**, 1591–1600 (1966)
Berzborn, R.: Z. Naturforsch. **24B**, 436–446 (1969)

Berzborn, R.: Habilitationsschrift, Univ. Bochum, 1972, pp. 1–149
Berzborn, R., Kopp, F., Mühlethaler, K.: Z. Naturforsch. **29C**, 694–699 (1974)
Berzborn, R., Menke, W., Trebst, A., Pistorius, E.: Z. Naturforsch. **21B**, 1057–1059 (1966)
Branton, D.: Proc. Nat. Acad. Sci. U.S. **55**, 1048–1056 (1966)
Branton, D.: Ann. Rev. Plant Physiol. **20**, 209–238 (1969)
Branton, D. et al.: Science **190**, 54–56 (1975)
Briantais, J.-M., Picaud, M.: FEBS Lett. **20**, 100–104 (1972)
Criddle, R.S.: In: Biochemistry of Chloroplasts. Goodwin, T.W. (ed.). New York: Academic Press, 1966, Vol. I, pp. 203–231
Criegee, R.: Ann. Chem. **522**, 75–96 (1936)
Danielli, J.F., Davson, H.: J. Cell Physiol. **5**, 495–508 (1935)
Deamer, D.W., Crofts, A.R., Packer, L.: Biochim. Biophys, Acta **131**, 81–96 (1967)
Deamer, D.W., Leonard, R., Tardieu, A., Branton, D.: Biochim. Biophys. Acta **219**, 47–60 (1970)
Deenen, L.L.M. van: In: Progress in the Chem. of Fats and other Lipids. Hofman, R.T. (ed.). Oxford: Pergamon Press, 1965, Vol. VIII, pp. 1–127
Emerson, R., Arnold, W.: J. Gen. Physiol. **16**, 191–205 (1933)
Finean, J.B.: Circulations **26**, 1151–1162 (1962)
Fleischer, S., Fleischer, B., Stoeckenius, W.: J. Cell Biol. **32**, 193–208 (1967)
Frey-Wyssling, A., Steinmann, E.: Vierteljahrschr. Naturforsch. Ges. Zürich **98**, 20–29 (1953)
Garber, M., Steponkus, P.L.: J. Cell Biol. **63**, 24–34 (1974)
Gitler, C.: Ann. Rev. Biophys. Bioeng. **1**, 51–92 (1972)
Goodenough, U.W., Staehelin, L.A.: J. Cell Biol. **48**, 594–619 (1971)
Hauska, G.A.: Angew. Chem. **84**, 123–124 (1972)
Hauska, G.A., McCarty, R.E., Berzborn, R.J., Racker, E.: J. Biol. Chem. **246**, 3524–3531 (1971)
Hayes, T.L., Lindgren, F.T., Gofman, J.W.: J. Cell Biol. **19**, 251–255 (1963)
Hiedemann-Van Wyk, D., Kannangara, C.G.: Z. Naturforsch. **26B**, 46–50 (1971)
Ji, T.H., Hess, J.L., Benson, A.A.: Biochem. Biophys. Acta **150**, 676–685 (1968)
Junge, W., Witt, H.T.: Z. Naturforsch. **23B**, 244–254 (1968)
Kannangara, C.G., Van Wyk, D., Menke, W.: Z. Naturforsch. **25B**, 613–618 (1970)
Koenig, F., Menke, W., Craubner, H., Schmid, G.H., Radunz, A.: Z. Naturforsch. **27B**, 1225–1238 (1972)
Kopp, F.: Cytobiologie **6**, 287–317 (1972)
Korn, E.D.: Biochim. Biophys. Acta **116**, 325–335 (1966)
Kreutz, W.: Advan. Botan. Res. **3**, 53–169 (1970)
Lehninger, A.L.: Naturwissenschaften **53**, 57–63 (1966)
Lucy, J.A., Glauert, A.M.: J. Mol. Biol. **8**, 727–748 (1964)
Luzzati, V.: In: Biological Membranes. Chapman, D. (ed.). London-New York: Academic Press, 1968, pp. 71–124
Machold, O.: Biochim. Biophys. Acta **382**, 494–505 (1975)
McMillan, P.N., Lustig, R.B.: Proc. Nat. Acad. Sci. U.S. **70**, 3060–3064 (1973)
Menke, W.: Z. Naturforsch. **16B**, 334–336 (1961)
Menke, W.: Ann. Rev. Plant Physiol. **13**, 27–44 (1962)
Menke, W.: Physiol. Veg. **11**, 231–238 (1973)
Menke, W., Ruppel, H.G.: Z. Naturforsch. **26B**, 825–831 (1971)
Miller, K.R., Staehelin, A.: 8th Intern. Congr. Elec. Microsc., Canberra, (1974) Vol. **2**, 204–205
Mitchell, P.: Nature (London) **191**, 144–148 (1961)
Moor, H., Mühlethaler, K., Waldner, H., Frey-Wyssling, A.: J. Biophys. Biochem. Cytol. **10**, 1–13 (1961)
Mühlethaler, K.: Studies on freeze-etching of cell membranes. In: Intern. Rev. Cytology. Bourne, G.H., Danielli, J.F. (eds.). New York-London: Academic Press, 1971, Vol. XXXI, pp. 1–19
Mühlethaler, K.: Freeze-etching studies on chloroplast thylakoids. In: Proc. 2nd Intern. Congr. Photosynthesis. Forti, G., Avron, M., Melandri, A. (eds.). The Hague: Dr. Junk, 1972, Vol. III, pp. 1423–1429
Mühlethaler, K., Moor, H., Szarkowski, J.M.: Planta (Berl.) **67**, 305–323 (1965)
Murakami, S., Packer, L.: J. Cell Biol. **47**, 332–351 (1970)

Nelson, N., Deters, D.W., Nelson, H., Racker, E.: J. Biol. Chem. **248**, 2049–2055 (1973)
Nolan, W.G., Park, R.B.: Biochim. Biophys. Acta **375**, 406–421 (1975)
Ojakian, G.K., Satir, P.: Proc. Nat. Acad. Sci. U.S. **71**, 2052–2056 (1974)
Oleszko, S., Moudrianakis, E.N.: J. Cell Biol. **63**, 936–948 (1974)
Olson, J.M., Thornber, J.P., Koenig, D.F., Ledbetter, M.C., Olson, R.A., Jennings, W.H.: In: Progress in Photosynthesis Research. Metzner, M. (ed.). Tübingen: Laupp, 1969, Vol. I, pp. 217–225
Ongung, A., Thomson, W.W., Mudd, J.B.: J. Lipid Res. **9**, 416–424 (1968)
Park, R.B., Biggins, J.: Science **144**, 1009–1010 (1964)
Park, R.B., Pon, N.G.: J. Mol. Biol. **6**, 105–114 (1963)
Park, R.B., Sane, P.V.: Ann. Rev. Plant Physiol. **22**, 395–430 (1971)
Pinto da Silva, P., Douglass, S.D., Branton, D.: Nature (London) **232**, 194–195 (1971)
Racker, E., Hauska, G.A., Lien, S., Berzborn, R.J., Nelson, N.: In: Proc. 2nd Intern. Congr. Photosynthesis. Forti, G., Avron, M., Melandri, A. (eds.). The Hague: Dr. Junk, 1972, Vol. II, pp. 1097–1113
Radunz, A.: Z. Naturforsch. **26 B**, 916–919 (1971)
Radunz, A.: Z. Naturforsch. **27 B**, 822–826 (1972)
Radunz, A., Berzborn, R.J.: Z. Naturforsch. **25 B**, 412–419 (1970)
Radunz, A., Schmid, G.H.: Z. Naturforsch. **28 C**, 36–44 (1973)
Radunz, A., Schmid, G.H., Menke, W.: Z. Naturforsch. **26 B**, 435–446 (1971)
Robertson, J.D.: In: Proc. 4th Intern. Congr. Elec. Microsc. Bargmann, W., Peters, D., Wolpers, C. (eds.). Berlin-Göttingen-Heidelberg: Springer, 1960, pp. 159–171
Schmid, G.H., Radunz, A., Menke, W.: Z. Naturforsch. **30 C**, 201–212 (1975)
Staehelin, L.A., Miller, K.R.: In: 8th Intern. Congr. Elec. Microsc. Canberra, (1974), Vol. II, pp. 202–203
Steck, Th.L., Fox, C.F.: Membrane Proteins. In: Membrane Molecular Biology. Fox, C.F., Keith, A.D. (eds.). Stamford: Sinauer, 1972, pp. 27–75
Stoeckenius, W., Mahr, S.C.: Lab. Invest. **14**, 458–469 (1965)
Trebst, A.: Ann. Rev. Plant Physiol. **25**, 423–458 (1974)
Trebst, A., Hauska, G.: Naturwissenschaften **61**, 308–316 (1974)
Wallach, D.F.H., Zahler, P.H.: Proc. Nat. Acad. Sci. U.S. **56**, 1552–1559 (1966)
Wang, A.Y., Packer, L.: Biochim. Biophys. Acta **305**, 488–492 (1973)
Weber, P.: Z. Naturforsch. **17 B**, 683–688 (1962)
Weber, P.: Z. Naturforsch. **18 B**, 1105–1110 (1963)
Wehrli, E.: Strukturen der Thylakoidmembranen. Diss. ETH Nr. 5571, 1975
Wehrli, E., Mühlethaler, K., Moor, H.: Exp. Cell Res. **59**, 336–339 (1969)
Wigglesworth, V.B.: Proc. Roy. Soc. (London) Ser. B, **147**, 185–199 (1957)
Wintermans, J.F.G.M.: In: Le Chloroplaste. Sironval, C. (ed.). Paris: Masson, 1967, pp. 86–90
Witt, H.T.: Quart. Rev. Biophys. **4**, 365–477 (1971)
Wyk, D. van: Z. Naturforsch. **21 B**, 700–703 (1966)

2. The Topography of the Thylakoid Membrane of the Chloroplast

P.V. SANE

A. Introduction

In an earlier chapter (this vol. IV.1) MÜHLETHALER has discussed in detail the ultrastructural characteristics of the higher plant chloroplast membranes, as studied by several techniques. In this chapter the relation between structure and function has been considered. In the first section we shall consider the overall distribution of the photosystems in the lamellar structure, and examine in brief the support and apparent contradictions for the PARK and SANE (1971) model. In the next section, the reactivity in the partition and nonpartition regions of the membrane is considered. There we shall consider the pertinent data for localizing the two most important enzymes, associated with the photosynthetic electron transport chain viz. the $NADP^+$ reductase and ATPase. The role of grana in a higher plant chloroplast will then be discussed in view of the distribution of photosystem I and photosystem II, as well as $NADP^+$ reductase and ATPase in the lamellar structure of chloroplasts.

The evidence for the sidedness of the membrane is considered in the last section. In view of the vectorial electron transport taking place in the membrane the topography across and along the membrane will be examined. Finally a possible arrangement of the different components of the two photosystems in the membrane will be discussed.

In view of a large number of papers published covering the area to be discussed in the article, it was decided to refer to a limited number of the most pertinent papers. In this article an attempt has been made to reinterpret some of the data in order to reconcile the apparent contradictions observed in the published material.

B. The Distribution of Photosystems in the Chloroplast Lamellar Structure

I. The Model

The studies on the fractionation of chloroplast membranes using detergents or mechanical methods as discussed by JACOBI (this vol. Chap. IV.3) have shown that a heavy fraction sedimenting between 1,000 g and 10,000 g is enriched in PS II but also contains PS I, and a lighter fraction sedimenting between 50,000 g and 160,000 g contains predominantly, or only, PS I. These fractions have been fully characterized biochemically as well as biophysically, and a detailed discussion of this has been provided by JACOBI (this vol. Chap. IV.3).

The ultrastructural characterization of the fractionated material has been reported by only a few workers. Since digitonin and triton released the PS I fraction from the chloroplast membranes easily, ANDERSON and BOARDMAN (1966) and BRIANTAIS (1969) proposed that the PS I may be located on the outside of the membrane. ARNTZEN et al. (1969) for the first time carried out in detail the ultra-structural studies of PS I and PS II enriched fractions obtained by digitonin incubation. They proposed a binary model for the chloroplast membrane, in which the outer half of the membrane consists of PS I and the inner half of PS II. This model is inconsistent with several observations and has been critisized by PARK and SANE (1971).

The ultrastructural characterization of the detergent-treated material is difficult and less convincing, since detergents are known to release and replace lipids of the membranes with the unavoidable result of changed characteristics in staining and fracturing. In view of this, ultrastructural studies of fractions obtained by mechanical methods are considered more meaningful. JACOBI and LEHMANN (1968, 1969) developed a mild sonication technique to obtain PS I and PS II fractions from chloroplast membranes. The thin sectioning studies of different fractions led them to suggest that the PS I fraction may have arisen from the intergrana area or stroma lamellae. These studies, however, did not show whether PS II was absent from intergrana areas.

The freeze fracturing, deep etching, thin sectioning and biochemical studies of SANE et al. (1970) on the PS II enriched fraction (10 K) and PS I fraction (160 K), obtained by passing the chloroplasts through a french pressure cell at appropriate pressures, provided convincing evidence to suggest that the PS I (160 K) fraction originated from the stroma lamellae, and possibly end grana membranes, whereas the PS II (10 K) fraction arose from grana regions. The crucial evidence was based on the presence of only small particles in the PS I fraction and of both small and large particles in the PS II fraction observed on freeze fracturing. In a class I freeze-fractured chloroplast stroma lamellae and end grana membranes contained only small particles, whereas the partition regions of the grana contained both small and large particles. On the basis of these studies, and of those of GOODCHILD and PARK (1971), PARK and SANE (1971) proposed a model for the distribution of photosystems in the lamellar structure of a higher plant chloroplast. A schematic representation of this model is shown in Figure 1.

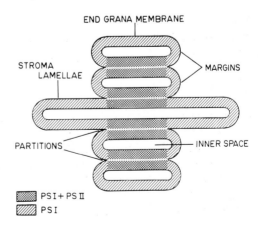

Fig. 1. A schematic representation of the distribution of PS I and PS II in the lamellar structure of chloroplasts

PS I + PS II
PS I

In this model PS II is restricted to the partition regions of the grana, whereas PS I is present in both stroma and grana lamellae. The stroma lamellae is devoid of PS II. Thus, there are two kinds of PS I in the higher plant chloroplast, one present in the grana and physically close to PS II, and another present in the stroma and physically separated from the PS II. It is possible that the major function of the stroma PS I is cyclic photophosphorylation. However, the presence of as much as 50% of $NADP^+$ reductase needed for maximum rates of $NADP^+$ reduction in stroma lamellae suggest that stroma lamellae may also be involved in $NADP^+$ reduction. It was suggested that the PS II in the grana may be connected with PS I of the stroma lamellae by a small molecular weight electron carrier (SANE et al., 1970). SANE and PARK (1971a) subsequently wondered if this may be responsible for the enhancement observed in $NADP^+$ reduction by isolated chloroplasts.

On the basis of similarities observed in the biochemical and biophysical characteristics of PS I and PS II fractions (SANE et al., 1970; PARK et al., 1971) obtained by different procedures, PARK and SANE (1971) argued that all fractionation procedures possibly result in the separation of stroma lamellae from grana membranes to yield PS I and PS II enriched fractions. In the case of digitonin this is borne out by the EM studies of WEHRMEYER (1962) and WESSELS and VOORN (1972). Further, all the EM studies conducted so far show that the fractions are not "particles" as designated previously, but are vesicles or grana fragments. The fragmentation of grana by detergents into PS I and PS II has now been achieved and is dealt with by JACOBI (this vol. Chap. IV.3).

An interesting characteristic of all the PS I fractions is that they show a decline in quantum yield for PS I reaction at shorter wavelengths (see PARK and SANE, 1971). This was interpreted by SANE and PARK (1971b, 1972) and PARK and SANE (1971) in terms of the presence of some inactive chlorophylls in stroma lamellae that may be precursors of PS II. It was suggested that this PS II may be completed on folding to form grana. In examining this proposal HENRIQUES and PARK (1974), using radio tracer techniques, demonstrated that the stroma lamellae do not give rise to grana membranes. The evidence of these workers is valid for mature chloroplasts only. The present author feels that stroma lamellae of only the young chloroplasts give rise to grana membranes during development, and that this may not be true for the stroma lamellae of the matured chloroplast.

PARK and SANE (1971) explained a great deal of data on the basis of the model shown in Figure 1. We shall, therefore, not go into details of this but would consider in brief the data supporting this model as well as those contradicting it.

II. The Supporting Evidence

The distribution of photosystems in the lamellar structure as shown in Figure 1 is consistent with most observations in the development of structure and function in greening chloroplasts (see this vol. Chap. IV.6). The studies on greening chloroplasts have shown that PS I activity precedes the formation of PS II activity. The structural studies conducted simultaneously demonstrate the presence of only stroma lamellae when PS I activity is present. The formation of grana or appressed regions during greening is associated with the appearance of PS II. The freeze fracturing and biochemical studies of PHUNG-NHU-HUNG and her co-workers (1970a, b) have shown that primary membranes possess only small particles and are the sites of PS I activity. The formation of grana is associated with the appearance of both the small and large particles as expected from the model.

The chloroplasts of plants undergoing a physiological stress, such as mineral deficiency, have been studied by CRANE and his co-workers for their photochemical activities and ultrastructural appearance (BARR et al., 1972; BASZYNSKY et al., 1972; HALL et al., 1972). These studies have demonstrated that increased grana stacking in mineral deficient plants is associated with increased PS II activity.

The model is also consistent with the observations made on mutants of higher plant chloroplasts (see PARK and SANE, 1971). The studies on mutant chloroplasts show that the mutant chloroplasts with a high chlorophyll a/b ratio possess higher PS I activity. Such chloroplasts also contain predominantly stroma lamellae. The formation of grana in these mutants is correlated with PS II activity (HORAK and ZALIK, 1973; BAZZAZ et al., 1974).

In C-4 plants there are two types of chloroplasts. In malate formers of Graminae the bundle sheath chloroplasts are deficient in grana and possess only stroma lamellae (WOO et al., 1970; DOWNTON and co-workers, 1970, 1971). These chloroplasts, as would be predicted from the model, possess only PS I activity.

A perusal of the above data indicates that during greening, in mutants and in plants growing under mineral deficiency, as well as in normal higher plant chloroplasts, there is an association between stroma lamellae and PS I (but not PS II), whereas the presence or development of grana is associated with the non-cyclic electron transport i.e. PS I + PS II. Thus PS II is exclusively associated with grana membranes.

III. The Contradictions

SMILLIE et al. (1972), examining the relationship between the presence of grana and PS II activity in agranal chloroplasts of bundle sheath cells of C-4 plants, concluded that the formation of grana is not correlated with the appearance of functional PS II. These authors did not provide the EM pictures of these chloroplasts. Since the physiological conditions under which the plants are grown can change the extent of grana formation, it is essential to provide electron micrographs of the chloroplasts claimed to be agranal. Further, as appropriately pointed out by LAETSCH (1974), the chloroplasts of all bundle sheath cells are not agranal, as some of them do possess rudimentary grana. Hence, the arguments of SMILLIE et al. are not totally acceptable.

HALL et al. (1971) used cytochemical techniques to localize the site of copper ferricyanide photoreduction in the chloroplast membranes. Since in a short incubation period ferrocyanide appeared mainly on the stroma lamellae, they proposed that stroma lamellae must be containing PS II. This proposal is questionable, since in class II chloroplasts ferricyanide is known to be reduced primarily by PS I (see TREBST, 1974). Thus, their data suggest that stroma lamellae possess PS I rather than PS II. Now if stroma lamellae do not contain PS II, where do the reducing equivalents come from? SANE et al. (1970) have postulated that the stroma lamellae PS I may be in contact with PS II located in the grana through a small molecular weight electron carrier. If this is true, it is not difficult to explain the results of HALL et al. (1971) on the basis of the model proposed for the distribution of PS I and PS II in the chloroplast membranes.

Although the PS II is associated with the presence and formation of grana, the PS II activity is not totally lost on losing the partition regions. Thus, EDTA-treated chloroplasts that have lost partition regions still show good PS II activity (HOWELL and MOUDRIANAKIS, 1967; PARK and PFEIFHOFER, 1968). Similarly, the big particles observed only in the grana are not lost from the higher plant chloroplasts on swelling. This suggests that the formation of grana may be necessary only for the development of PS II and its coupling to PS I during development.

C. Reactivity in the Partition and Nonpartition Regions

In the previous section we have considered the overall distribution of the two photosystems in the grana containing chloroplasts of higher plants. In this section we shall look into the distribution and localization of the two most important enzymes related to the electron transport viz., the ATPase and the $NADP^+$ reductase. Finally, we shall attempt to assess the role of grana in view of the distribution of the two photosystems and enzymes in the lamellar structure of chloroplasts.

As discussed earlier the grana contains the components of the noncyclic electron transport chain, whereas the stroma lamellae contain only the functional components of photosystem I. In grana itself we have two morphologically distinct membrane regions, namely the partitions and the limiting membranes. The term limiting membrane in this Article will include the end grana membranes and the margins. It was shown earlier that the end grana membranes on freeze fracturing reveal small particles on the B face, whereas this face in the partition regions contains large particles (SANE et al., 1970). On this basis we had proposed that, in addition to stroma lamellae, the end grana membranes also contain only PS I. Although it is not known whether the margins of the grana contain only PS I, or both PS I as well as PS II, it appears at present that this nonappressed membrane region of the grana contains only PS I. That the partition regions of the grana contain both PS I and PS II is evident from the cytochemical studies of WEIER et al. (1966) and HALL et al. (1971). It is tempting to suggest that in grana, partition regions may contain only PS II, whereas the limiting membranes of the grana may house only PS I. This, however, does not seem to be true as several studies have shown that spillover of energy can take place from one photosystem to another (AVRON and BEN-HAYYIM, 1969; MURATA, 1969; SUN and SAUER, 1971; VANDERMEULEN and GOVINDJEE, 1974). This could happen only if the antenna chlorophylls of both the systems are not separated from each other by a distance of as much as 1–2 microns. Thus, one must conclude that both PS I and PS II are located in the partition regions of the grana. This brings up the question of (1) whether the terminal electron acceptor, $NADP^+$, of the noncyclic electron transport is reduced in the hydrophobic environment present in the partitions and, (2) whether the ATP synthesis also occurs in the partition regions. One way to look at the situation would be to find out exactly where the two enzymes $NADP^+$ reductase and ATPase are located in the membranes.

I. The Localization of NADP⁺ Reductase

The location of the reductase has been studied by immunological technique. The work of BERZBORN (1968) showed that the antisera against the reductase did not agglutinate the chloroplast lamellae directly. Subsequently BERZBORN (1969), using indirect agglutination techniques, showed that $NADP^+$ reductase is located on the outer surface of the chloroplast membrane but in a depression formed by the knobs of ATPase. Although these studies showed that reductase is present on the outer surface, they did not show that it was not present in the partition regions. One should also consider the possibility that if, during isolation of chloroplasts or lamellae used for agglutination, the grana are swollen due to low ionic strength, it would expose even the components that are located in the partition regions. From the conditions used by BERZBORN it is difficult to judge whether the stacks had remained intact. For a relevant discussion see SCHMID and RADUNZ (1974).

In their experiments KEISTER et al. (1962) and SAN PIETRO (1963) observed more or less complete inhibition of the $NADP^+$ reduction by their antibody. The chloroplast preparation of these workers must have preserved the grana stacks, since sufficient ionic strength was present in the isolation medium. The reaction mixture also contained sufficient salts to maintain thylakoids in the stacked condition. This inference has been drawn on the basis of studies conducted by IZAWA and GOOD (1966a, b). Since they have observed a more or less complete inhibition of $NADP^+$ reductase by the antibody under conditions where the grana stacks were preserved, one can conclude that all the sites of $NADP^+$ reduction are accessible to the antibodies. Because the hydrophilic antibody cannot penetrate the hydrophobic partition regions of grana, all the sites of $NADP^+$ reduction must be present on the outer surface of thylakoid exposed to the matrix. This study shows that $NADP^+$ reductase could not be present in the partition region. Of course the reduced NADP is utilized in the stroma of the chloroplast for CO_2 reduction and, hence, it seems logical that this cofactor is reduced only on the outer surface of the membrane. The proposal that $NADP^+$ cannot be reduced in the partitions brings up the problem of how the reducing equivalents are transferred from partition regions to the surface of the membrane. Are there small molecular weight electron carriers? Theoretical calculations suggest that the diffusion of a small molecular weight compound over 2–3 micron distance will not impose any limitation for the kinetics of CO_2 fixation (SANE et al., 1970). Other consequences of this proposal are discussed later.

II. The Localization of ATPase

Although the enzyme ATPase has been the subject of investigation of several workers, its precise location is yet to be confirmed. We will analyse the data obtained so far using different approaches and try to arrive at a most likely location for this important enzyme in the thylakoid system. Since the noncyclic electron transport is present in the partition regions, it is logical to expect the ATPase to be located in the partition regions. Besides, the immunological studies to be discussed later have demonstrated the presence of this enzyme on the outer surface of the membranes. The view that ATPase may also be present in the partition regions is strengthened from two recent ultrastructural studies. Garber and Steponkus (1974) used negative staining and freeze fracturing techniques to identify the coupling factor. It may be recalled that several workers, by using only negative staining technique, considered ATPase knobs observed on the outer surface of the chloroplast membrane to be identical with the quantasome, a designation proposed by Park for the morphological expression of a photosynthetic unit capable of carrying out H_2O-$NADP^+$ noncyclic electron transport (see Park and Biggins, 1964). The position regarding this has been virtually cleared up and has been dealt with in some detail by Park and Sane (1971). It is sufficient to say that quantasome has now been identified with the big B face particles and its back side appearing on the D surface by freeze etch techniques (Park and Pfeifhofer, 1969a). The studies of Garber and Steponkus (1974) showed that washing of thylakoids with 0.8% STA (silicotungstate) removed the 150 Å particles from the A' surface (OS in their terminology) by freeze etching. These particles were identifiable as the coupling factor. In negatively stained preparations these particles on the outer surface were 90 Å and lost on STA treatment. This result is in agreement with the work of Lien and Racker (1971) who have done detailed work on the STA treated chloroplast membranes. These results are also in agreement with the earlier work of Howell and Moudrianakis (1967) and Park and Pfeifhofer (1968), who have used EDTA to remove the 90 Å particles from the outer surface of lamellae. An interesting observation made by Garber and Steponkus (1974) is that the inner fracture face (corresponding to B face according to Park and Pfeifhofer, 1969a) in untreated thylakoids contains two types of particles—100 Å and 150 Å. The frequency of these particles was found to be 1,650 particles μm^{-2}. On STA treatment this frequency decreased to 710 particles μm^{-2} and moreover, this fracture face was devoid of any 150 Å particles.

These workers also carried out reconstitution experiments in which isolated coupling factor was added back to the STA-treated membranes. On negative staining the 90 Å particles reappeared on the surface of the membranes. In freeze-fractured preparations the 150 Å particles on the inner fracture face also reappeared, although the concentration of these particles was approximately 30% less than in untreated thylakoids. These results point to the possibility that the big 150 Å particles, which are lost on STA treatment and reappear on reconstitution, may be related to the partially buried coupling factor. Since these particles are present only in the partition regions these studies indicate the presence of coupling factor in the partition region of chloroplast membranes.

Garber and Steponkus (1974) did consider the other possibility that the loss and reappearance of 150 Å particles from the inner fracture face may be due to a change in the membrane brought about by the STA treatment. In this connection it should be remembered that Lien and Racker (1971) have already shown that the treatment of chloroplast membranes with as low as 0.3% STA completely inhibited PS II and PS I activity. Apparently STA has many other effects on the membrane. Further, Park and Pfeifhofer (1969b) consider 100–120 Å particles observed on the outer surface of chloroplast membrane on deep etching as identical to coupling factor. Also in their studies, Park and Pfeifhofer (1968), using EDTA to remove coupling factor particles, did not see disappearance of B face particles.

Another point that should be considered is that the big B face particles are supported by a very thin, probably 1.5 nm, layer (see PARK and PFEIFHOFER, 1969a for a detailed discussion). It is because of this thin support that B face big particles are not persistent during deep etching. In the case of STA treated membranes it appears that this thin support is further affected, resulting in the loss of B face particles on freeze fracturing.

Another piece of evidence pointing to the location of ATPase in the partitions comes from the work of OLESZKO and MOUDRIANAKIS (1974). Using negative staining they have shown that the coupling factor can be seen on the stromal lamellae extending into the matrix. They also provide an EM picture indicating occasional appearance of coupling factor on grana lamellae, wherever fused membranes appear slightly separated. These authors also show a vesicle with 90 Å knobs on all its surface. They argue that the disc preparation they obtained should have come from the grana, since in spinach chloroplasts the membranes in situ are predominantly arranged in grana. This argument is difficult to accept because, although in general it is true that chloroplast membranes are arranged predominantly in grana, a picture of a chloroplast shown by the authors in Figure 7 of their article clearly indicates a preponderance of stromal lamellae. Such a chloroplast on swelling should yield predominantly vesicles arising from stroma lamellae. In such studies it is difficult to discern whether a given disc or vesicle has come from the grana stacks or from stromal lamellae. The evidence would have been convincing if they had started with a purified preparation of grana stacks, for which methods are now available, and then demonstrated the appearance of particles related to coupling factor on all the surface of the vesicles obtained from them. In freeze fractured chloroplasts the 120 Å particles located on the C face (outer fracture face) do not seem to be identical with coupling factor, as these are not lost by treatments that remove the coupling factor (PARK and PFEIFHOFER, 1969b; GARBER and STEPONKUS, 1974).

In view of the above, the author feels that the EM data so far available do not lend unequivocal support to the idea that the ATPase is present in the partitions in addition to the outer surface of the whole thylakoid membranes. We will now consider immunological and biochemical data.

The earlier studies of MCCARTY and RACKER (1966) have shown that an antiserum against ATPase could inhibit photophosphorylation, as well as acid-base induced phosphorylation (JAGENDORF and URIBE, 1966), indicating that ATPase on the membrane is accessible to the hydrophilic antibody. On the basis of their experimental conditions it is difficult to decide whether ATPase could be present in the partitions in addition to the outer surface of the thylakoids. However, subsequent studies of MCCARTY and co-workers (1971) are more revealing in this regard. In these studies an antiserum from rabbit showed as much as 80% inhibition of photophosphorylation. In these experiments a considerable amount of ionic strength (50 mM NaCl in addition to 5 mM $MgCl_2$) was maintained. On the basis of the work of IZAWA and GOOD (1966a, b) it can be concluded that under conditions used by MCCARTY and co-workers the grana must have remained intact. Thus, it appears that under conditions when partitions are retained intact the hydrophilic antibody can bind on to most of the ATPase. Since the antibody is unlikely to penetrate the hydrophobic partition regions it is logical to conclude that no ATPase is present in the partition regions.

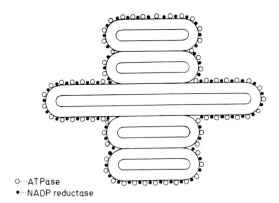

Fig. 2. A schematic diagram showing distribution of NADP⁺ reductase and ATPase in the lamellar structure of chloroplasts. Note the absence of these enzymes from the partition regions

o···ATPase
•···NADP reductase

The studies of KANNANGARA et al. (1970), in which they have used tobacco aurea mutant chloroplasts isolated under conditions where the partition regions are retained intact i.e. high ionic strength, showed almost 100% inhibition of ATP synthesis by an antibody. These results also suggest that practically all of the ATPase is present only on the outer surface of the thylakoid. Recent studies by NELSON et al. (1973) further support the conclusion that coupling factor must be present only on the outer surface of the thylakoid and not in the partition regions, since under conditions where partitions are retained intact antibodies against different subunits of coupling factor inhibited both cyclic and noncyclic photophosphorylation.

The data from the studies with lactoperoxidase have also suggested the location of the coupling factor on the outside (ARNTZEN et al., 1974a). It is difficult to infer from their results whether ATPase is present at all in partitions. The DABS studies of GIAQUINTA et al. (1974b) have shown that this chemical reagent severely affected the photophosphorylation, but not the electron transport from H_2O to methylviologen. The use of ^{35}S DABS (diazonium benzene sulfonic acid) enabled them to observe binding of DABS to the coupling factor, indicating that coupling factor was accessible to this nonpenetrating water soluble chemical reagent. The conditions of the experiment suggest that partitions must have been preserved. Thus, one could conclude from these studies that in a thylakoid system where partitions are retained, practically all of the coupling factor is accessible. Therefore, the coupling factor of ATPase may not be present in the partition regions.

Considering the above discussion one comes up with a model for distribution of NADP⁺ reductase and ATPase as shown in Figure 2. What are the consequences of localization of NADP⁺ reductase and ATPase on the outer surfaces of limiting membranes of grana and stroma lamellae?

III. The Consequences of the Model

Since the ATPase and NADP⁺ reductase are located only on the outer surface of the limiting membranes of the grana and stroma lamellae, it would be expected that the stroma lamellae fraction would be relatively enriched in proteins, whereas the grana would be relatively enriched in chlorophylls.

If $NADP^+$ reduction can only take place on the surface and not in the partitions, the reducing equivalents generated by PS I must somehow be transferred to a place where the reductase is located. This could happen if a small molecular weight component reduced by the primary acceptor of PS I shuttled back and forth between the partition region and the outer surface of the membrane. An alternate possibility could be that the components reduced by PS II in the partitions could diffuse to the PS I located at the margins and oxidized there. In this case the PS I present in the partition regions could function as a cyclic system.

If ATPase is not located in the partition regions the ratio of coupling factor to chlorophyll will change depending upon the extent of grana present in the chloroplasts. This has been observed by BERZBORN and MÜLLER (personal communication). They have noted that the chloroplasts containing higher amounts of stroma lamellae contained higher amount of coupling factor. This also explains the discrepancy in the coupling factor particles per μ^2 or number of chlorophylls observed by different workers (MURAKAMI, 1968; STROTMANN et al., 1973; OLESZKO and MOUDRIANAKIS, 1974). STROTMANN et al. (1973), in fact, have already speculated that grana membranes may lack coupling factor particles where they are in contact with each other (partition regions), and occupied by coupling factor particles only where they are exposed to the matrix.

Since the stroma lamellae possess a relatively larger surface area exposed to the matrix as compared to grana, one would expect a higher ATPase activity per mg chlorophyll in the stroma lamellae (PS I) fraction than in the grana fraction. This has been already observed by ARNTZEN et al. (1969) and HAUSKA and SANE (1972), using digitonin and french press procedures to isolate stroma lamellae and grana lamellae fractions.

LOCKSHIN et al. (1971) have reported that maize etioplasts from which the chloroplasts develop contained most, or probably all, of the coupling factor activity subsequently present in the chloroplasts. It was suggested that the photosynthetic membranes were formed in stages. If this is so, then, during greening when stroma lamellae fold to form grana, the coupling factor will have to be excluded and kept outside of the partition. This seems possible, as BERZBORN et al. (1974) have shown that coupling factor together with its base is mobile. This mobility of the coupling factor could also explain the observation of OLESZKO and MOUDRIANAKIS (1974) regarding the occasional presence of coupling factor on grana lamellae whenever the fused membranes appear slightly separated.

If ATPase is absent in the partition regions, how is energy conservation during electron transport in terms of ATP brought about? One possibility is that the high energy intermediate ($\sim$), be it a chemical or a conformational entity, will have to traverse a long distance. Alternately, if one accepts that the vectorial electron transport with the plastoquinone loop (see TREBST, 1974) brings about the establishment of the proton gradient or pH difference between the inner space and outer space, it is simple to visualize that ATPase situated anywhere on the vesicle will be able to utilize this to synthesize ATP. In this case, it is presumed that each vesicle of the grana thylakoid acts as a single unit. This is a valid presumption in view of the results obtained by WITT's group. WITT and his co-workers (see WITT, 1971), using gramicidin, have suggested that the functional unit of phosphorylation is one thylakoid.

IV. The Role of Grana

The models (Figs. 1 and 2) discussed above indicate that, in the chloroplasts, the stroma lamellae are membranes that are rich in enzymes such as $NADP^+$ reductase, ATPase and possibly RUDP (ribulose-1,5-bisphosphate) carboxylase, whereas the partitions are pigment-rich enzyme-deficient membrane regions. Although partition formation seems essential during greening for the development of PS II and the coupling of PS II and I, the extent of grana formed relative to total membrane area may vary in different plants growing under different conditions. One can consider a situation in which plants have to grow in very low light intensities. In these plants (the shade plants) the limiting factor for photosynthesis is light and the plant, therefore, must utilize whatever light is available most efficiently. This is possible, if the concentration of light-harvesting pigments is increased. The formation of extensive grana stacks in these plants provides such a possibility. Under these conditions, however, the amount of $NADP^+$ reductase and ATPase need not be high since the rates of electron transport per unit time are relatively low. Thus, the need of the plants growing under low light intensities is high pigment concentration per unit area and low enzyme content. The extensive grana formation in the shade plants, as observed by ANDERSON et al. (1973), can be readily explained on the basis of this argument. These workers proposed that grana formation may be a means of achieving a higher density of light harvesting assemblies for more efficient collection of light quanta. On the basis of the models proposed (Figs. 1 and 2) it could be additionally said that by increasing the grana, the enzymes not needed in higher amounts are simultaneously kept at a minimum. If ATPase and $NADP^+$ reductase were to be a part of the partition regions it would have been wasteful for the plant to increase this enzyme concentrations along with increase in pigments.

Another situation that supports the above statement is the case of C-4 plants. In these plants the CO_2 fixation pathway requires additional ATP than does the C-3 pathway. This, it appears, is met by the bundle sheath chloroplasts that have few grana. The extensive network of stroma lamellae in these chloroplasts provides for higher rates of ATP synthesis.

Thus, the role of grana may be to provide the flexibility for the plant to adjust to the light conditions under which it has to grow. This suggestion predicts that a given plant, if grown under low light intensity, will possess relatively higher grana, less $NADP^+$ reductase and ATPase per mg chlorophyll, as compared to the plant grown under high light intensity. It may be pointed out that an increase in grana content should decrease the chlorophyll a/b ratio, resulting in increasing the range of quality and quantity of light absorbed by the pigments. This is advantageous, as it allows more efficient collection of the light.

Are there other roles for grana? So far we have no data to suggest other roles. What is the reason for increased stacking in plants subjected to grow in a media deficient in certain minerals? Why are most chlorotic mutants deficient in grana? Such mutants do not saturate at low light intensities. Is this characteristic related to the very few appressed regions observed in these mutants? Answers to these questions could only be given after a very careful and thorough study of plants grown under different physiological conditions (high and low light intensities, different temperatures, mineral deficiency, and possibly different CO_2 concentrations), and after studying relations between the grana content and several photosynthetic activities.

D. The Asymmetry of the Membrane

In the earlier sections we considered the distribution of the photosystems and the two enzymes associated with the electron transport chain in the chloroplast lamellar structure. In this section we shall consider the possible arrangement of the different electron transport components, keeping in view the vectorial electron transport taking place in the membrane. The localization of the donor and acceptor sites of the two photosystems has been attempted using immunological, biochemical, and biophysical techniques. These studies have demonstrated that the chloroplast membrane is functionally and structurally asymmetric. An excellent review has appeared recently in which TREBST (1974) has considered all the relevant data on this aspect. Although the studies in the past five years have brought out the functional asymmetry of the membrane, the ultrastructural studies using freeze fracturing and deep etching has earlier provided the morphological evidence for the asymmetry of the membrane. We shall, therefore, briefly consider the morphological evidence first.

I. The Morphological Evidence for the Asymmetry of the Membrane

If one considers the individual thylakoid of the granum as a vesicle on freeze fracturing and deep etching, it shows two fractured faces and two surfaces. On the basis of studies conducted by PARK and BRANTON (1966), BRANTON and PARK (1967), PARK and PFEIFHOFER (1969a) a model for the distribution of particles on different faces and surfaces observed on freeze fracturing and deep-etching was provided by PARK and PFEIFHOFER (1969a). This model is shown in Figure 3. Although previously MÜHLETHALER and his colleagues favored the view that fractures occurred at the interface between the surface of the membrane and the surrounding medium, these workers now agree with the interpretation of the Berkeley group (BRANTON, 1966; BRANTON and PARK, 1967; BRANTON and SOUTHWORTH, 1967; PARK and PFEIFHOFER, 1969a), that the fractures occurred along the hydrophobic regions, thus exposing two matching faces through a single break (WEHRLI et al., 1970). MÜHLETHALER (this vol. Chap. IV.1) has considered the details of this aspect.

The two fracture faces resulting from the splitting open of the membrane are termed B and C faces. The B face is covered with large particles approximately 175 Å in diameter and 90 Å in height. It may be mentioned that the fracture occurs along the hydrophobic region such that the B face has a very thin support, whereas the C face is underlaid and strengthened by almost the entire membrane thickness (PARK and PFEIFHOFER, 1969a). The D surface is the interior or inner surface of the vesicle and forms the back portion of the B face. The C face is covered with particles about 120 Å in diameter and about 90 Å in height.

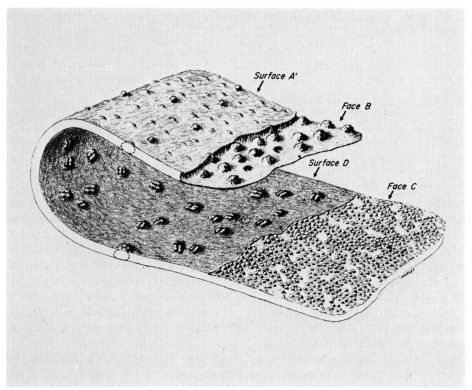

Fig. 3. A diagrammatic representation of the faces and surfaces observed in a thylakoid from partition region. (Adopted from Park and Pfeifhofer, 1969 a)

The surface A′ is the outer surface of the vesicle. There is some confusion regarding the nature of this surface. In the stack an additional face is exposed from the fracture occurring along the partition. This face has been termed A face and should be considered as the outer surface of that part of the granum vesicle taking part in partition formation. The remainder of the outer surface of the granum vesicle viz, at the margin, should be identical to A′ surface covered with a number of proteins such as ATPase, reductase, and the carboxylase. The outer surface of the end grana membrane will be identical to A′ surface. Thus, the outer surface of the vesicle from the partition regions seems to be differentiated along its length—the hydrophobic part involved in partition formation identical to A face; the hydrophilic part that is exposed to the matrix being identical to A′ surface. The big B face particles are presumably protruding through the A face and are exposed on unstacking.

Thus, the two parts (not halves) of the thylakoid membrane are not identical. The two surfaces are also very different morphologically, showing the asymmetry of the thylakoid membrane.

What is the situation in the stroma lamellae? The studies of Sane et al. (1970) have shown that the stroma lamellae contain only small particles. In a stroma lamellae vesicle both the fracture faces namely, B and C, are covered with only small 120 Å diameter particles. The outer surface in this case is identical to A′ surface; there is no A face. What does the inner surface (related to D surface of the granum vesicle) look like? It is known that it is not covered with enzymes such as ATPase, NADP$^+$ reductase, or carboxylase. To this extent the two surfaces of the stroma lamellae are not identical.

II. The Topography Across the Membrane

1. The Vectorial Electron Transport

The flashing light studies of WITT and his co-workers (WITT, 1971), JUNGE and AUSLÄNDER (1973), AUSLÄNDER and JUNGE (1974), and FOWLER and KOK (1974) have suggested that during photoact II the oxidation of water takes place at the inner surface of the membrane, releasing protons in the inner space. The electron in this photoact traverses across the membrane as evidenced by the reduction of Q, which seems to be located towards the outer side. Q then reduces the plastoquinone which picks up two protons from the outer aqueous phase, as plastoquinone is a H-carrier. The reduced plastoquinone traverses from the outer side of the membrane towards the inner side, where it is oxidized by PS I via cytochrome f or plastocyanin releasing the protons in the inner space from the reduced plastoquinone. In the photoact I the electron once again traverses from inside to outside, reducing X located towards the outer surface. Subsequently, NADP$^+$ is reduced, which picks up protons from the outer aqueous phase. Thus, during the two photoacts, electrons travel across the membrane from inside to outside. The two photoacts are coupled by a H-transport in the opposite direction i.e. from outside to inside. This vectorial electron flow provides for an alternate electron and H-transport, resulting in the deposition of the protons in the inner space and withdrawl of protons from the outer aqueous phase. This establishes a pH difference between the inside and outside of the membrane vesicle, which, according to MITCHELL's hypothesis (1966) could drive ATP synthesis. This type of vectorial electron transport could be predicted on the basis of MITCHELL's hypothesis. Although the biophysical studies of the Berlin group supported this type of electron transport across the membrane, the biochemical support for it came with the introduction of DBMIB as a plastoquinone analog (see TREBST, 1972). The experiments employing this analog first established the functioning of the plastoquinone loop during the electron transport between the two photosystems. In the presence of this compound, it was possible to study the electron transport and photophosphorylation associated with PS II. Since in the presence of DBMIB the coupling site between plastoquinone and plastocyanin is blocked, the observation (see TREBST 1974, 1975) that ATP is produced during electron transport from water to ferricyanide in the presence of DBMIB could only be explained by an additional energy conserving site at PS II (TREBST and REIMER, 1973; IZAWA et al., 1973). The acceptance of chemiosmotic hypothesis was essential for justifying this additional energy conserving site. TREBST suggested that the oxidation of water liberates two protons inside the inner space and that these protons contribute to the pH difference necessary for ATP synthesis. The existence of the energy conserving site at PS II also explains the variable P/e$_2$ ratios reported in the literature. The details of the vectorial electron transport have been considered in Chapter II.1.a of this volume.

The essential feature of the vectorial electron flow is that the donor and the acceptor sites of the two photosystems must be on the opposite sites of the membrane. This would mean that the components of the donor site of PS II and PS I should be located towards the inner surface, while their acceptors Q and X must be located towards the outer side of the membrane vesicle. We shall now examine if the data so far obtained are consistent with this type of arrangement.

2. The Distribution of the Components of the Two Photosystems in the Membrane

a) The Donor Site of PS II

Although the flash photolysis data have demonstrated convincingly the presence of a PS II donor site at or near the inner surface of the membrane, the immunological and biochemical evidence suggests that the PS II, and particularly the water oxidation site, may be exposed to outside. This contradiction proves to be only apparent if one critically examines the immunological and biochemical results.

The immunological studies of RADUNZ et al. (1971), BRAUN and GOVINDJEE (1972), BRIANTAIS and PICAUD (1972) and RADUNZ and SCHMID (1973) have shown inhibition of PS II by an antibody prepared against the PS II complex, or one of its components. This was taken to suggest an external location of PS II. Further support for the external location of PS II, or part of it, came from the DABS studies of DILLEY and his co-workers (DILLEY et al., 1972; GIAQUINTA et al., 1973; GIAQUINTA et al., 1974a) and the lactoperoxidase investigations of ARNTZEN et al. (1974a, b). In all these studies it was assumed that the thylakoid system was not altered during isolation or experimental procedures, and that addition of antibodies or chemicals had no indirect effects on the membrane. These assumptions are not valid. Firstly, in most experiments no attempts were made to ensure that unstacking did not take place. In fact, the studies of SELMAN et al. (1973), using trypsin, demonstrated that unstacking renders the membranes more vulnerable to trypsin. ARNTZEN et al. (1974b) also argue that in stacked membrane iodination of PS II is blocked, suggesting unstacking results in exposure of PS II. Secondly, in some experiments it appears that partial inhibition observed could be due to partial damage to PS II, resulting from experimental procedure — e.g. the chloroplast used by BRAUN and GOVINDJEE were frozen and thawed at least once prior to use. Such a procedure is known, from the observations of SANE et al. (1975), to alter the characteristics of the chloroplast membranes. Thirdly, ZILINKAS (1975), on the basis of her studies with trypsin, chaotropic agents, and membrane fixatives, has suggested that addition of antisera or trypsin brings about macroconformational changes, resulting in the repositioning of the specific site of the PS II complex. This, she explains, permits an access to impermeable inhibitors. Conformational changes around PS II have also been observed by GIAQUINTA et al. (1974a).

In view of this, the author is of the opinion that partial inhibition observed in some of these studies may be due to exposure of part of PS II, as a result of unstacking or damage to PS II during experimental procedure. Thus, under tightly stacked conditions, PS II is not accessible to hydrophilic antibodies or chemicals. This is consistent with its location in the hydrophobic partition regions.

It has been shown that the inhibition of PS II observed (probably on unstacking) is between H_2O and DPC donation (BRAUN and GOVINDJEE, 1972; GIAQUINTA et al., 1973). Thus, it seems that the reaction centre of PS II is not exposed on unstacking, but rather a component associated with H_2O oxidation and involved in a conformational change may be located towards the outside. Could this exposed part of PS II complex be identical with the protruding portion of the big B face particles through the outer surface of the membrane? It is tempting to suggest this since the B face particles are considered as PS II markers. The exposure of part of PS II complex on unstacking is clearly against the binary model of the membrane.

It thus appears that the donor site of PS II is located towards the inner surface of the membrane, as suggested from flash photolysis work.

b) The Acceptor Site of PS II

The flash photolysis studies supporting the vectorial electron transport indicate that the acceptor of PS II, the Q, is towards the outer side of the membrane. This is in agreement with studies on the photoreductions by PS II, which show the acceptor of PS II to be situated towards the outside (see TREBST, 1974). ZILINKAS (1975) also favours the localization of the primary acceptor of PS II close to the outer surface, on the basis of her observation that silicomolybdate and silico-tungstate can accept electrons directly from Q. Further confirmation for the location of Q towards the outer side comes from the observation that chloroplasts on fragmentation (which possibly exposes the acceptor site of PS II) actively reduce ferricyanide or DCIP.

It may be mentioned here that although Q is situated towards the outer surface of the membrane, it is still embedded in the hydrophobic region of partitions. This is evident from the studies of SAHA et al. (1971), who have shown that lipophilic acceptors are reduced by PS II.

c) The Donor Site of PS I

Although it is now agreed that there are two distinct PS I in the chloroplasts, no attempts have been made to treat these two PS I differently by the workers.

The earlier studies of RACKER et al. (1972) showed that an antibody against P-700 complex agglutinated chloroplast particles and inhibited PS I reaction. It was suggested that P-700 may be located on the outer surface. However, the plastocyanin, a primary donor of electrons to P-700 seems to be located inside or towards the inside of the membrane. This conclusion is based on the studies using antibody against plastocyanin conducted by HAUSKA et al. (1970), HAUSKA et al. (1971), RACKER et al. (1972), and SANE and HAUSKA (1972). A detailed discussion of this is provided by BERZBORN and LOCKAU (this vol. Chap. II.17).

The location of plastocyanin at or near the inner surface of the membrane as proposed by HAUSKA and his co-workers has been questioned by SELMAN et al. (1975), on the basis of their work with DABS and CDIS + GEE (glycine ethylester). They observed that the plastocyanin associated with disrupted membrane accumulated three to four times more label from the hydrophilic modifiers than the plastocyanin associated with intact membranes. They, therefore, postulate that plastocyanin is located in the hydrophilic cleft in the membrane. This would be consistent with the polylysine inhibition of PS I between cytochrome f and P-700 (BRAND et al., 1972a, b). SCHMID et al. (1975) also report on an accessibility of plastocyanin to its antibody.

That the donor site of PS I is located towards the inside, as indicated by flash photolysis studies, is further supported by the investigations of HAUSKA and co-workers (HAUSKA, 1972; HAUSKA et al., 1973) with artificial electron donors (see this vol. Chap. II.15). These studies demonstrate that PS I can only oxidize lipophilic uncharged compounds in the reduced form. It was explained that the sulfonated PMS or pyocyanine that have similar redox properties as the parent compounds cannot catalyze cyclic photophosphorylation, because they are unable to penetrate the lipophilic regions of the chloroplast membranes. These studies therefore suggest that the oxidizing site of PS I is located inside the lipid barrier of the chloroplast membranes. Further support for the location of donor site towards the inside comes from the cytochrome f antibody studies of RACKER et al. (1972).

The present author favours the view that the donor site of PS I and the P-700 are on the inner surface or towards the inner side. The accessibility of plastocyanin to such compounds as polylysine, DABS is difficult to explain unless one considers that these compounds have some additional effects on the membrane — a possibility that is difficult to rule out at present.

d) The Acceptor Site of PS I

As discussed in the previous section the reduction of $NADP^+$, which is the terminal acceptor of PS I, takes place on the outside. The enzyme $NADP^+$ reductase is located on the outerside but not in the partition regions. The immunological studies of BRIANTAIS and PICAUD (1972) and KOENIG et al. (1972) suggest external location of PS I. Specifically, ferredoxin, which is the physiological acceptor of PS I, is also located on the outside since the antibody prepared against ferredoxin agglutinates chloroplasts (HIEDEMANN-VAN WYK and KANNANGARA, 1971). Further one of the components involved in $NADP^+$ reduction also seems to be located on the outside, as the work of REGITZ et al. (1970) shows. There is no confusion regarding the external location of the terminal acceptor of PS I.

It thus appears that the donors of PS II and I are located towards the inside, whereas their acceptors are located towards the outside.

III. The Topography Along the Membrane

In view of tbe fact that PS II, and therefore the noncyclic electron transport, is located in the partition regions, it appears that the molecular architecture of the partition region or grana will be different from that of the stroma lamellae which contains only PS I. Thus, in the grana the PS II donor site with its reaction centre P-682 seems to be situated towards the inner surface. The oxidation of water probably takes place at the inner surface, releasing protons in the inner space. It appears, however, that part of the PS II complex is exposed to the outside in unstacked membranes. The Q is situated towards the outside. Thus,

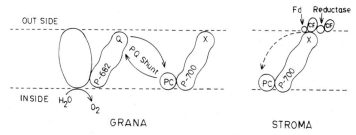

Fig. 4. A schematic representation of the arrangement of different components of the electron transport chain in the grana membranes and stroma lamellae. *CF*: coupling factor; *Fd*: ferredoxin; *PC*: plastocyanin; *PQ*: plastoquinone; *Q*: primary acceptor of PS II; *X*: primary acceptor of PS I

the PS II is extending across the membrane. In the partitions this PS II is alternated by a PS I complex with its reaction centre P-700 and plastocyanin at, or close to, the inner surface. The acceptor X is located towards the outer side. The two photosystem complexes are joined by the plastoquinone loop. This type of arrangement satisfies the requirements of the vectorial electron flow with alternating electron and H-transport in the grana. The reduction of hydrophilic acceptors by PS I of grana may be mediated by a diffusable electron carrier. As one proceeds from the partition regions to the stroma lamellae, or limiting membranes, the topography of the membrane changes. In these nonappressed regions probably only the components of PS I are present. The PS I complex extends across the thickness of the membrane with the donor site situated towards the inner side. The acceptor X is towards the outer side. The whole surface of these membranes is covered with ferredoxin, NADP$^+$-reductase, and the ATPase. Thus, the acceptor of PS I of the grana is embedded in the lipophilic region in the partitions, whereas the PS I acceptor of stroma lamellae is probably accessible to hydrophilic acceptors. This difference between the two would be shown in studies involving kinetics of reduction of lipophilic and hydrophilic acceptors of PS I.

A schematic representation of the possible arrangement of the different electron components in the grana (partitions) and stroma membranes is shown in Figure 4. Basically it is similar to the one proposed by TREBST (1974, 1975), but further brings out the differences that may exist as one proceeds along the membrane.

E. Concluding Remarks

In this chapter we have considered the overall distribution of the two photosystems and the two enzyme (NADP$^+$-reductase and ATPase) associated with the electron transport chain in the chloroplast membranes. It is argued that those regions of the inner membrane of the chloroplast that are exposed to the matrix (containing enzymes of CO_2 fixation) possess only PS I and are covered on the outer surface with NADP$^+$-reductase and ATPase. The partition regions that contain both PS I and PS II are devoid of these enzymes.

As a further step in understanding the topography of the membrane, data obtained by using several techniques have been considered in order to arrive at the most likely location of the different components of the two photosystems in the membrane. It has been shown that the chloroplast membrane is asymmetric in nature and it is felt that in conducting future studies on the topography of the membrane one must take into account the state of the membrane viz, whether it is stacked or unstacked. It is hoped that further researches in the area of identifying, characterizing, and localizing the protein and lipid components of the membrane would enhance our understanding of the molecular architecture of the chloroplast membrane. Such investigations are already being pursued in some laboratories (ANDERSON and LEVINE, 1974a, b; BERG et al., 1974).

Acknowledgements. The author wishes to thank Prof. A. TREBST of the Ruhr University, Bochum (West Germany) for suggestions in preparing the manuscript. Thanks are also due to Dr. R. BERZBORN of the Ruhr University, Bochum for discussions in preparing a part of this manuscript.

References

Anderson, J.M., Boardman, N.K.: Biochim. Biophys. Acta **112**, 403–421 (1966)
Anderson, J.M., Goodchild, D.J., Boardman, N.K.: Biochim. Biophys. Acta **325**, 573–585 (1973)
Anderson, J.M., Levine, R.P.: Biochim. Biophys. Acta **333**, 378–387 (1974a)
Anderson, J.M., Levine, R.P.: Biochim. Biophys. Acta **357**, 118–126 (1974b)
Arntzen, C.J., Armond, P.A., Zettinger, C.S., Vernotte, C., Briantais, J.M.: Biochim. Biophys. Acta **347**, 329–339 (1974a)
Arntzen, C.J., Dilley, R.A., Crane, F.L.: J. Cell Biol. **43**, 16–31 (1969)
Arntzen, C.J., Vernotte, C., Briantais, J.M., Armond, P.: Biochim. Biophys. Acta **368**, 39–53 (1974b)
Ausländer, W., Junge, W.: Biochim. Biophys. Acta **357**, 285–298 (1974)
Avron, M., Ben-Hayyim, G.: In: Progress in Photosynthesis Research. Metzner, H. (ed.). Tübingen: Laupp, 1969, Vol. I, pp.1185–1196
Barr, R., Hall, J.D., Baszynski, T., Brand, J., Crane, F.L.: Proc. Indiana Acad. Sci. **81**, 114–120 (1972)
Baszynski, T., Brand, J., Barr, R., Krogmann, D.W., Crane, F.L.: Plant Physiol. **50**, 410–411 (1972)
Bazzaz, M.B., Govindjee, Paolillo, D.J.: Z. Pflanzenphysiol. **72**, 181–192 (1974)
Berg, S., Dodge, S., Krogmann, D.W., Dilley, R.A.: Plant Physiol. **53**, 619–627 (1974)
Berzborn, R.J.: Z. Naturforsch. **23B**, 1096–1104 (1968)
Berzborn, R.J.: Z. Naturforsch. **24B**, 436–446 (1969)
Berzborn, R.J., Kopp, F., Mühlethaler, K.: Z. Naturforsch. **29C**, 694–699 (1974)
Brand, J., Baszynski, T., Crane, F.L., Krogmann, D.W.: J. Biol. Chem. **247**, 2814–2819 (1972a)
Brand, J., San Pietro, A., Mayne, B.C.: Arch. Biochem. Biophys. **152**, 426–428 (1972b)
Branton, D.: Proc. Nat. Acad. Sci. **55**, 1048–1056 (1966)
Branton, D., Park, R.B.: J. Ultrastruct. Res. **19**, 283–303 (1967)
Branton, D., Southworth, D.: Exp. Cell Res. **47**, 648–653 (1967)
Braun, B.Z., Govindjee: FEBS Lett. **25**, 143–146 (1972)
Briantais, J.M.: Physiol. Veg. **7**, 135–180 (1969)
Briantais, J.M., Picaud, M.: FEBS Lett. **20**, 100–104 (1972)
Dilley, R.A., Peters, G.A., Shaw, E.R.: J. Membrane Biol. **8**, 163–180 (1972)

Downton, W.J., Berry, J.A., Tregunna, E.B.: Z. Pflanzenphysiol. **63**, 194–198 (1970)
Downton, W.J., Pyliotis, N.A.: Can. J. Botany **49**, 179–180 (1971)
Fowler, C.F., Kok, B.: Biochim. Biophys. Acta **357**, 308–318 (1974)
Garber, M., Steponkus, P.L.: J. Cell Biol. **63**, 24–34 (1974)
Giaquinta, R.T., Dilley, R.A., Anderson, B.J.: Biochem. Biophys. Res. Commun. **52**, 1410–1417 (1973)
Giaquinta, R.T., Dilley, R.A., Selman, B.R., Anderson, B.J.: Arch. Biochem. Biophys. **162**, 200–209 (1974a)
Giaquinta, R.T., Selman, B.R., Bering, C.L., Dilley, R.A.: J. Biol. Chem. **249**, 2873–2878 (1974b)
Goodchild, D.J., Park, R.B.: Biochim. Biophys. Acta **226**, 393–399 (1971)
Hall, D.O., Edge, H., Kalina, M.: J. Cell Sci. **9**, 289–303 (1971)
Hall, J.D., Barr, R., Al-Abbas, A.H., Crane, F.L.: Plant Physiol. **50**, 404–409 (1972)
Hauska, G.A.: FEBS Lett. **28**, 217–220 (1972)
Hauska, G.A., McCarty, R.E., Berzborn, R.J., Racker, E.: J. Biol. Chem. **246**, 3524–3531 (1971)
Hauska, G.A., McCarty, R.E., Racker, E.: Biochim. Biophys. Acta **197**, 206–218 (1970)
Hauska, G.A., Sane, P.V.: Z. Naturforsch. **27 B**, 938–942 (1972)
Hauska, G.A., Trebst, A., Draber, W.: Biochim. Biophys. Acta **305**, 632–641 (1973)
Henriques, F., Park, R.: Plant Physiol. **54**, 386–391 (1974)
Hiedemann-Van Wyk, D., Kannangara, C.G.: Z. Naturforsch. **26 B**, 46–50 (1971)
Horak, A., Zalik, S.: Plant Physiol. **51** (suppl.), 66 (1973)
Howell, S.H., Moudrianakis, E.N.: J. Mol. Biol. **27**, 323–333 (1967)
Izawa, S., Good, N.E.: Plant Physiol. **41**, 533–543 (1966a)
Izawa, S., Good, N.E.: Plant Physiol. **41**, 544–552 (1966b)
Izawa, S., Gould, J.M., Ort, D.R., Felker, P., Good, N.E.: Biochim. Biophys. Acta **305**, 119–128 (1973)
Jacobi, G., Lehmann, H.: Z. Pflanzenphysiol. **59**, 457–476 (1968)
Jacobi, G., Lehmann, H.: In: Progress in Photosynthesis Research. Metzner, H. (ed.). Munich: Goldman, 1969, pp. 159–173
Jagendorf, A.T., Uribe, E.: Proc. Nat. Acad. Sci. U.S. **55**, 170–177 (1966)
Junge, W., Ausländer, W.: Biochim. Biophys. Acta **333**, 59–70 (1973)
Kannangara, C.G., Van Wyk, D., Menke, W.: Z. Naturforsch. **25 B**, 613–618 (1970)
Keister, D.L., San Pietro, A., Stolzenbach, F.E.: Arch. Biochem. Biophys. **98**, 235–244 (1962)
Koenig, F., Menke, W., Craubner, H., Schmid, G.H., Radunz, A.: Z. Naturforsch. **27 B**, 1225–1238 (1972)
Laetsch, W.M.: Ann. Rev. Plant Physiol. **25**, 27–52 (1974)
Lien, S., Racker, E.: J. Biol. Chem. **246**, 4298–4307 (1971)
Lockshin, A., Falk, R.H., Bogorad, L., Woodcock, C.L.F.: Biochim. Biophys. Acta **226**, 366–382 (1971)
McCarty, R.E., Fuhrmann, J.S., Tsuchiya, Y.: Proc. Nat. Acad. Sci. U.S. **68**, 2522–2526 (1971)
McCarty, R.E., Racker, E.: Brookhaven Symp. Biol. **18**, 202–214 (1966)
Mitchell, P.: Biol. Rev. **41**, 445–502 (1966)
Murakami, S.: In: Comparative Biochemistry and Biophysics of Photosynthesis. Shibata, K., Takamiya, A., Jagendorf, A.T., Fuller, R.C. (eds.). Tokyo: Tokyo Univ., 1968, pp. 82–98
Murata, N.: Biochim. Biophys. Acta **172**, 242–251 (1969)
Nelson, N., Deters, D.W., Nelson, H., Racker, E.: J. Biol. Chem. **248**, 2049–2055 (1973)
Oleszko, S., Moudrianakis, E.N.: J. Cell Biol. **63**, 936–948 (1974)
Park, R.B., Biggins, J.: Science **144**, 1009–1011 (1964)
Park, R.B., Branton, D.: Brookhaven Symp. Biol. **19**, 341–352 (1966)
Park, R.B., Pfeifhofer, A.O.A.: Proc. Nat. Acad. Sci. U.S. **60**, 337–343 (1968)
Park, R.B., Pfeifhofer, A.O.A.: J. Cell Sci. **5**, 299–311 (1969a)
Park, R.B., Pfeifhofer, A.O.A.: J. Cell Sci. **5**, 313–319 (1969b)
Park, R.B., Sane, P.V.: Ann. Rev. Plant Physiol. **22**, 395–430 (1971)
Park, R.B., Steinback, K., Sane, P.V.: Biochim. Biophys. Acta **253**, 204–207 (1971)
Phung-Nhu-Hung, S., Hoarau, A., Moyse, A.: Z. Pflanzenphysiol. **62**, 245–258 (1970a)
Phung-Nhu-Hung, S., Lacourly, A., Sarda, C.: Z. Pflanzenphysiol. **62**, 1–16 (1970b)

Racker, E., Hauska, G.A., Lien, S., Berzborn, R.J., Nelson, N.: In: Proc. 2nd Intern. Congr. Photosynthesis. Forti, G., Avron, M., Melandri, A. (eds.). The Hague: Dr. Junk, 1972, pp. 1097–1113

Radunz, A., Schmid, G.H.: Z. Naturforsch. **28 C**, 36–44 (1973)

Radunz, A., Schmid, G.H., Menke, W.: Z. Naturforsch. **26 B**, 435–446 (1971)

Regitz, G., Berzborn, R., Trebst, A.: Planta (Berl.) **91**, 8–17 (1970)

Saha, S., Ouitrakul, R., Izawa, S., Good, N.E.: J. Biol. Chem. **246**, 3204–3209 (1971)

San Pietro, A.: Ann. N.Y. Acad. Sci. **103**, 1093–1105 (1963)

Sane, P.V., Desai, T.S., Tatake, V.G.: Indian J. Biochem. Biophys. **12**, 38–42 (1975)

Sane, P.V., Goodchild, D.J., Park, R.B.: Biochim. Biophys. Acta **218**, 162–178 (1970)

Sane, P.V., Hauska, G.A.: Z. Naturforsch. **27 B**, 932–938 (1972)

Sane, P.V., Park, R.B.: Biochem. Biophys. Res. Commun **44**. 491–496 (1971a)

Sane, P.V., Park, R.B.: Biochim. Biophys. Acta **253**, 208–212 (1971b)

Sane, P.V., Park, R.B.: In: Proc. 2nd Intern. Congr. Photosynthesis. Forti, G., Avron, M., Melandri, A. (eds.) The Hague: Dr. Junk, 1972, pp. 826–832

Schmid, G.H., Radunz, A.: Z. Naturforsch. **29 C**, 384–391 (1974)

Schmid, G.H., Radunz, A., Menke, W.: Z. Naturforsch. **30 C**, 201–212 (1975)

Selman, B.R., Bannister, T.T., Dilley, R.A.: Biochim. Biophys. Acta **292**, 566–581 (1973)

Selman, B.R., Johnson, G.L., Dilley, R.A., Voegeli, K.K.: In: Proc. 3rd Intern. Congr. Photosynthesis. Avron, M. (ed.). Amsterdam: Elsevier, 1975

Smillie, R.M., Bishop, D.G., Anderson, K.S.: In: Proc. 2nd Intern. Congr. Photosynthesis. Forti, G., Avron, M., Melandri, A. (eds.). The Hague: Dr. Junk, 1972, pp. 770–788

Strotmann, H., Hesse, H., Edelman, K.: Biochim. Biophys. Acta **314**, 202–210 (1973)

Sun, A.S.K., Sauer, K.: Biochim. Biophys. Acta **234**, 399–414 (1971)

Trebst, A.: In: Proc. 2nd Intern. Congr. Photosynthesis. Forti, G., Avron, M., Melandri, A. (eds.). The Hague: Dr. Junk, 1972, pp. 399–417

Trebst, A.: Ann. Rev. Plant Physiol. **25**, 423–458 (1974)

Trebst, A.: In: Proc. 3rd Intern. Congr. Photosynthesis. Avron, M. (ed.). Amsterdam: Elsevier, 1975, pp. 439–448

Trebst, A., Reimer, S.: Biochim. Biophys. Acta **305**, 129–139 (1973)

Trebst, A., Reimer, S.: Biochem. Physiol. Planzen **168**, 225–232 (1975)

Vandermeulen, D.L., Govindjee.: Biochim. Biophys. Acta **368**, 61–70 (1974)

Wehrli, E., Mühlethaler, K., Moor, H.: Exp. Cell Res. **59**, 336–339 (1970)

Wehrmeyer, W.: Z. Naturforsch. **17 B**, 54–57 (1962)

Weier, T.E., Stocking, C.R., Shumway, L.K.: Brookhaven Symp. Biol. **19**, 353–374 (1966)

Wessels, J.S.C., Voorn, G.: In: Proc. 2nd Intern. Congr. Photosynthesis. Forti, G., Avron, M., Melandri, A. (eds.). The Hague: Dr. Junk, 1972, pp. 835–845

Witt, H.T.: Quart. Rev. Biophys. **4**, 365–477 (1971)

Woo, K.C., Anderson, J.M., Boardman, N.K., Downton, W.J., Osmond, C.B.: Proc. Nat. Acad. Sci. U.S. **67**, 18–25 (1970)

Zilinkas, B.A.: Ph.D. thesis. Univ. Illinois, Urbana, 1975

3. Subchloroplast Preparations

G. JACOBI †

A. Introduction

The investigations on chloroplast fragmentation reported up to the beginning of the last decade were essentially concerned with the isolation of the "photosynthetic unit" as originally defined by EMERSON and ARNOLD (1932). The major interest was the relationship of particle size to photochemical activity, which during this time was solely characterized by the Hill activity. Some of these classical experiments have been discussed by THOMAS (1960) and by CLENDENNING (1960) in Volume V.1 of the Encyclopedia of Plant Physiology. As photosynthesis became understood in terms of two cooperating light reactions, PS I and PS II, studies in many laboratories were concerned with the preparation of subchloroplast particles enriched in these activities. The increasing interest in this topic is documented by the publication of some excellent review articles (BOARDMAN, 1970; PARK and SANE, 1971; BROWN, 1973; ARNTZEN and BRIANTAIS, 1975) and by a workshop on special problems of chloroplast fragmentation (JACOBI, 1972). Some laboratories succeeded recently in the isolation of highly purified reaction centers, which are discussed in detail by WESSELS (this vol. Chap. IV.4).

The fact that definite subunits are enclosed in the chloroplast lamellae raised a number of essential problems which were both stimulating and provocative for biochemists and morphologists. The existence of globular, photochemically active entities ran particularly counter to the long accepted membrane model, which postulated distinct inner layers of lipids and pigments covered by regularly arranged proteins at the surface (KREUTZ, 1964; MENKE, 1966). Although the reaction centers are at present generally assumed to be integral membrane entities, they constitute only a minor fraction of the thylakoid system. It is one intention of this article to compare rather large fragments in which the reaction centers are still associated with building blocks, and to evaluate the effect of various preparation procedures on the removal of certain electron carriers. Although these investigations provided strong evidence that the chloroplast lamellae include definite subunits with different photochemical activities, they threw little light upon the quantity and the distribution of the photosystems in the entire thylakoid system. However, a discussion of these problems cannot be meaningful without a simultaneous consideration of structure and activity. Therefore, this article will further emphasize a combination of electron microscopy and biochemistry in the characterization of various fractions obtained by different methods. Moreover, since the optimum conditions for certain photochemical reactions are drastically altered with the degree of disruption, it is necessary to pay attention to these conditions as indicators for membrane integrity.

B. The Fractionation Pattern

From the comparison of various subchloroplast fragments described in the literature, it became evident that the isolation of a certain type of particle is strongly
dependent on the conditions of chloroplast pretreatment, and on the method used
to break down the thylakoid system. Therefore, if one attempts to survey the
numerous fractions, it is necessary to correlate the manner of fragmentation with
the function of the resulting subunits.

The architecture of the thylakoid system has been well established from electron
microscopic investigations. With some exceptions, most of the chloroplasts in higher
plants show the common feature of stacking in certain regions. These "grana
stacks" are connected by a number of single membranes which are designated

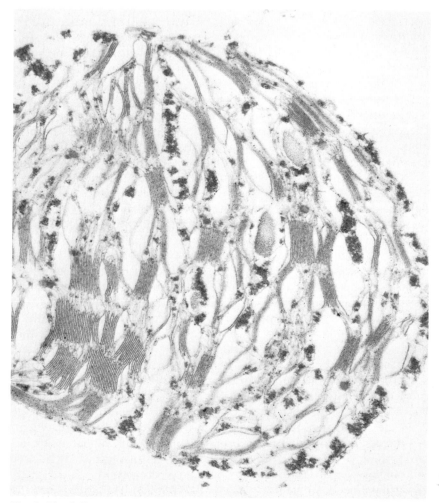

Fig. 1. The influence of ionic strength on the configuration of the thylakoid system. Once
washed chloroplasts were suspended in a buffered solution of 2% NaCl. Note the stabilization
of grana contact. (From Jacobi and Lehmann, 1968, 1969)

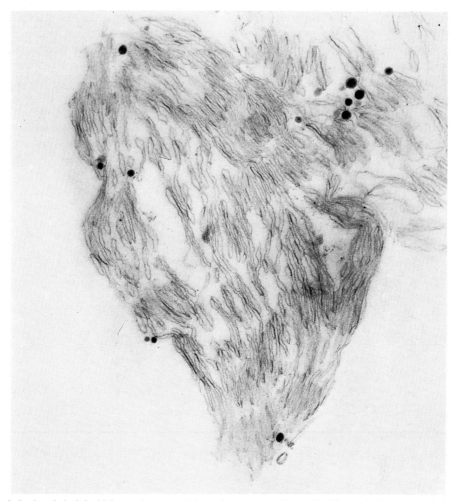

Fig. 2. Isolated thylakoid from the same chloroplast preparation as in Figure 1, but suspended in 0.2% NaCl. Note the dissociation of the grana membranes. (From JACOBI and LEHMANN, 1968, 1969)

as stroma lamellae. However, the grana cannot be viewed as a single pile of close discs. As demonstrated in electron microscopic investigations, the thylakoid system has to be considered as a membrane continuum (HESLOP-HARRISON, 1963; WEHRMEYER, 1963; WEIER et al., 1963, 1965; MÜHLETHALER, 1971). Although the internal space seems not to be separated by a border, the question arises as to whether the photosynthetic apparatus is identical in the two parts and whether the degree of stacking is correlated with a definite photosynthetic event.

Particular interest in fragmentation studies has been focused upon the configuration of the thylakoid system as influenced by the preparation conditions. As demonstrated by IZAWA and GOOD (1966), the contact of membranes in the grana region is lost when the chloroplasts are isolated in media of low ionic strength. Though the effect of salts upon stacking is not yet fully understood, the stabilization of the grana area by high-salt media can be used to separate the stacks from the interconnecting stroma lamellae by fragmentation. The influence of increasing salt concentration on the configuration of the thylakoid system is shown in Figures 1 and 2. Evidence for the separation of grana stacks from smaller vesicles came from several laboratories using different methods such as sonication (Gross and Packer, 1967; Jacobi and Lehmann, 1968, 1969), the French-press (Sane et al., 1970; Goodchild and Park, 1971; Arntzen et al., 1972), or detergents (Wehr-meyer, 1962; Anderson and Vernon, 1967; Arntzen et al., 1971; Wessels and Voorn, 1972). The preservation of grana stacks after a short sonication shock is illustrated in Figure 3. The results obtained by other methods showed a similar tendency.

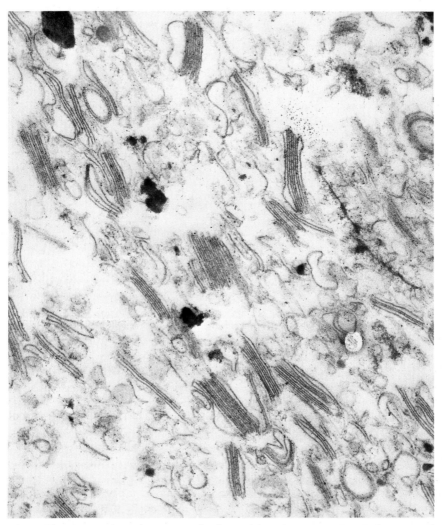

Fig. 3. The preservation of grana stacks from chloroplasts. The thylakoid preparation was suspended in 2% NaCl and 15 sec treated with sonic oscillation. (From Jacobi and Lehmann, 1968, 1969)

The various subchloroplast fragments were separated by differential centrifugation or by density gradient centrifugation. The heavy material which is present in the pellet after centrifugation at 10,000·g has been shown to consist mainly of grana stacks. A subsequent centrifugation at 50,000 or 80,000·g led to the sedimentation of vesicles which are probably derived from disrupted stacks. The remainder of the supernatant is further centrifuged for several hours at 144,000 or 170,000·g. Only the pellet of this fraction seems to represent the intergrana material. The supernatant contains released proteins and pigments which are especially solubilized by detergents.

Although the pattern of thylakoid breakage caused by physical methods and by detergents seems to be similar, the amounts of fragments in various fractions after differential centrifugation are different. Moreover, as shown with experiments using density gradient centrifugation, the size of particles released from the intact structure varies largely when different methods are used. MICHEL and MICHEL-WOLWERTZ (1969a, b) reported on a three-band pattern after disruption with a French press. In contrast, five distinct bands were described by KOK and RURAINSKI (1966) after treatment with digitonin.

In order to explain these variabilities, it is necessary to compare carefully the experimental conditions which differed largely from laboratory to laboratory. Regardless of the method used, the composition of the isolation and the fragmentation media, the period of exposure and the concentration of detergents were not the same. Therefore, some conflicting results noticed later are easily explained by different conditions which are often of greater importance than the fragmentation method itself.

C. The Distribution of Photosystems in the Grana and in the Intergrana Region of Chloroplasts from Higher Plants

BOARDMAN and ANDERSON (1964) were the first to report that the light material separated from chloroplasts treated with 0.5% digitonin after differential centrifugation was inactive in the Hill reaction, but nevertheless catalyzed a photoreduction of $NADP^+$ if provided with an electron donor couple. The small particles contained a significantly higher amount of chlorophyll a, whereas the heavier material was enriched in chlorophyll b. This basic observation was generally confirmed in several laboratories by the use of different fragmentation techniques providing an approximate ratio of 2/1 for chlorophyll a to b in the heavy fraction. The corresponding values for the light material were somewhat variable and ranged from 4 to 7.

Based on these observations, several investigations were undertaken to correlate the release of fragments with the distribution of photosystems in the intact thylakoid entity. Since the heavy fraction was shown to be composed mainly of grana stacks, it became apparent that chloroplyll b and photosystem II are concentrated in this region. In contrast, the light material which was assumed to derive from the intergrana area, catalyzed exclusively reactions related to photosystem I and was depleted in chlorophyll b.

Since the heavy fraction produced by various fragmentation procedures always retained a certain portion of photosystem I activity, it became apparent that photosystem I is also an actual constituent of the grana. The insufficient separation of photosystem I and photosystem II activities led JACOBI and LEHMANN (1968, 1969), and thereafter SANE et al. (1970) and ARNTZEN et al. (1971, 1972), to conclude that the two photosystems are closely associated in the grana region. The photosystem I present in the intergrana area was assumed to catalyze in vivo cyclic phosphorylation which would account for the extra ATP required for the CO_2-assimilation and other synthetic events. This proposal was made more evident by the findings of NELSON et al. (1970), ARNTZEN et al. (1971), HAUSKA and SANE (1972), who demonstrated that the light material produced by digitonin, by French press treatment, or even by sonication retained their property to catalyze cyclic phosphorylation.

An opposite conclusion was reached by HALL et al. (1971, 1972), who presented electron micrographs showing deposits of copper ferrocyanide distributed in both grana and intergrana lamellae. Since these complexes were formed under conditions which were thought to be characteristic for PS II activity, a uniform distribution of PS II was proposed. However, since ferricyanide is known to be reduced by almost all electron carriers, their conclusion is not convincing. Moreover, by the use of tetrazolium salts, WEIER et al. (1967) found the hydrophobic reduced formazan preferentially in the partition.

Although the concept of a functional differentiation within the thylakoid system has been widely accepted, it has raised a number of problems concerning the quantitative distribution of pigments and electron carriers. In spite of the similar breakage pattern caused by different methods, the relative portion of particles present in the centrifugal fractions varies largely. Differences are especially due to an additional fragmentation of grana and to the solubilization of pigments and membrane lipids by detergents.

The recovery of chlorophylls from the centrifugal fractions after separation has been well-established and was compared for different fragmentation procedures by BOARDMAN (1972a). The relative amount of chlorophyll present in the last supernatant was shown to be higher after treatment with digitonin than with the French press. As demonstrated by GOODCHILD and PARK (1971), digitonin continues to break the grana during the time of exposure, which could be stopped when glutaraldehyde was added.

Other detergents such as Triton X-100 and sonic oscillation seem less useful to break completely the thylakoid system without impairing the grana. Though a separation of subunits with significantly different pigment composition was achieved after short exposure to sonication (JACOBI and LEHMANN, 1968, 1969), only 1–2% of the total chlorophyll was collected in the light fraction with an a/b ratio of about 5. However, this amount is obviously too low to account for the total portion of intergranal material. After prolonged sonication, smaller vesicles are formed from the grana. Under these conditions, the pigment composition of the light material was almost the same as in intact chloroplasts. Moreover, the Hill-activity was found to be equally distributed, including in the fraction of smallest particles, which were originally thought to represent quantasomes (PARK and PON, 1962).

The most efficient method to break the thylakoid system, but leaving the grana almost intact, seems to be the French press. This method was first employed by MICHEL and MICHEL-WOLWERTZ (1969a, b), who found about 20% of the total chlorophyll in the light fraction after density gradient centrifugation. However, this value seems too high to account for the amount of chlorophyll in the intergrana region. In expanded experiments by SANE et al. (1970) and by ARNTZEN et al. (1971), a value of 5–7% was determined for the light fraction. An optimal release of the intergranal material was proposed by ARNTZEN et al. (1972) after two or three passes through the French pressure cell. In spite of the differences described, and with regard to the insufficient separation by differential centrifugation, approximately 10–15% of the total chlorophyll can be attributed to the intergranal area. The observed differences are most likely due to the use of presses with different valves and to the age of leaf material (see discussions in JACOBI, 1972, pp. 7 and 51).

In attempting to correlate the pigment distribution with individual chlorophyll forms, several groups have investigated the light and the heavy fractions by means of fluorescence emission and absorption spectra at low temperature. Additional information on the distribution of the photosystems has been derived more specifically from measurements of P-700 as a marker for PS I. Moreover, the determination of certain electron carriers, which became known to be related to redox events of either photosystem, completed the concept of a different location of photosynthetic assemblies along the thylakoid membrane.

The most sensitive method to evaluate the separation of the photosystems is the determination of fluorescence emission. At liquid nitrogen temperature, the large maximum at 730 nm can be ascribed to PS I, whereas the emission at 683–685 nm with a shoulder at 695 nm is a criterion for PS II. Using these characteristics, the intergranal particles produced by various methods show typical PS I spectra. In contrast, the ratio of the emission maxima at 683 and 730 nm increased significantly in the grana fraction (CEDERSTRAND and GOVINDJEE, 1966; KOK and RURAINSKI, 1966; ANDERSON, 1967; BROWN and PRAGER, 1969; MICHEL and MICHEL-WOLWERTZ, 1969a, b; BOARDMAN and THORNE, 1969; BOARDMAN, 1970; GOVINDJEE, 1972; JACOBI, 1972).

Additional information for the absence of PS II in the stroma lamellae came from measurements of variable fluorescence at room temperature (BOARDMAN, 1970; PARK et al., 1971; SANE and PARK, 1972). This property is thought to be related to the amount of oxidized reaction center of PS II. When assayed by the addition of dithionite or more directly by a light-induced increase, this portion was found to be negligible in the light fraction but concentrated in the grana. Finally, the energy transfer from PS II to the long wavelength absorbing species was only observed in the heavy fraction.

The absorption spectra of fractions as reported from several laboratories show good coincidence, demonstrating a decrease of chlorophyll b absorption at 650 nm, associated with a shift of the maximum towards the species absorbing at longer wavelength in the light material. Although low temperature absorption spectra were presented from various laboratories, the most comprehensive information came from the Carnegie group in Stanford, who investigated the light and the heavy fractions in a large number of lower and higher plants by curve analysis (FRENCH, 1970; FRENCH et al., 1971). On the whole, especially the long wavelength-absorbing chlorophyll a forms chl 691 and 705 were shown to be enriched in the light material, which was released after the first breakage. Of interest is the recent finding by BROWN et al. (1972) that chl 691 is completely absent in fraction II from *Dunaliella*.

Various procedures have been employed to measure the P-700 content as an indicator of PS I. Though the values measured after chemical oxidation and by light-induced absorbance changes are not directly comparable (BOARDMAN, 1972a, b), the data obtained by one method led to an estimation of the relative portion in the two fractions. There is fairly good agreement concerning the enrichment in the light fraction, showing a chlorophyll/P-700 ratio of 100/200 as compared with about 450 for intact chloroplasts (ANDERSON et al., 1966; OHKI and TAKAMIYA, 1970; SANE et al., 1970; ARNTZEN et al., 1971; BOARDMAN, 1972a; SANE and PARK, 1972; WESSELS and VOORN, 1972). However, considerable differences were reported for the heavy fraction, which are with certainty due to the additional solubilization of chlorophyll by detergents, and to a continuous disintegration of grana stacks. The values ranged from 650 up to 3,000, dependent upon the method used and on the enrichment of PS II. In summary, if one assumes approximately 15% of the toal chlorophyll to be present in the intergrana region, about 30% of P-700 is calculated to account for this region.

Similar calculations can be made for the distribution of various electron carriers. As reported by BOARDMAN and ANDERSON (1967), the molar ratio of cytochrome f/cytochrome b-6/cytochrome b-559 is approximately 1/2/2 per 430 chlorophylls in intact chloroplasts. Although the cytochromes are partly solubilized by detergents, it became apparent that cytochrome b-559 is almost absent in the light fraction (BOARDMAN et al., 1971). Together with C-550, cytochrome b-559 seems to be exclusively associated with photosystem II, which is consistent with their proposed relation to photochemical events at the level of photosystem II (KE et al., 1972; KITAJIMA and BUTLER, 1973).

The distribution of various membrane constituents in the grana and in the intergranal area is summarized in Table 1. However, these values are only approximate estimates and may vary with regard to the different extent of grana, as related to the age of the leaf material and influenced by growth conditions.

Table 1. Distribution of chlorophylls and electron carriers in grana and integranal membranes of isolated spinach chloroplasts

	chl a/b	%	chl / P-700	chl / cyt f	chl / cyt b-6 + b-559	cyt b-6 + b-559 / cyt f
Chloroplasts	2.8–3.1	100	400– 500	400–500	100–150	4
Grana	1.7–2.3	80–85	650–1000	600–750	100–120	6
Intergrana	5.0–7.0	15–20	100– 200	350–400	250–300	2

D. The Fractionation of Grana Stacks

The proposed concept that photosystem I occurs in the grana and in the intergrana region raised the fundamental question whether the reaction centers localized in different parts of the thylakoid system are equivalent. In an effort to elucidate this problem, some fractionation studies with isolated grana stacks were undertaken (JACOBI and LEHMANN, 1968; JACOBI, 1969, 1970; WESSELS and VOORN, 1972; ARNT-ZEN et al., 1972). However, the preparation of a pure photochemical subunit from grana is not easily obtained because of their more complex composition. A number of experimental difficulties have to be overcome to separate completely the two closely connected photosystems from each other (VERNON et al., 1971). Therefore, the individual mode of fragmentation methods used leads to the release of a puzzling number of sub-granal particles which are different in composition and activity. The majority of preparations described so far are rather complex and rarely represent purified photosystems.

In order to estimate the tendency of separation, several criteria have been used to characterize the subunit release. Though the best evidence seems to derive from measurements of photochemical activities, a quantitative calculation is often impaired by inactivation and by the alteration of the reaction properties as sum-marized in the next chapter. Other instructive information comes from the determi-nation of certain components involved in photochemical events and from reconstitu-tion experiments.

From the fragmentation of chloroplasts into larger units, it became clear that the enrichment of PS II is associated with an increase in chlorophyll b. However, as demonstrated by VERNON et al. (1971, 1972) and by WESSELS and VOORN (1972), the highly purified reaction centers of both PS I and PS II possess a similar high chlorophyll a/b ratio of 7 (summarized by WESSELS, this vol. Chap. IV.4). Although the simple measurement of the a/b ratio is therefore not conclusive for the isolation of reaction centers themselves, it is a fairly good indicator for the degree of separa-tion into subunits of larger size. Since chlorophyll b and xanthophylls are specific constituents of the accessory complex of PS II (VERNON et al., 1971, 1972), the lowering of the a/b ratio is indicative of the isolation of more complex PS II units (HUZISIGE et al., 1969; BOARDMAN, 1972 a, b).

Following VERNON's concept that the reaction center of PS II and the accessory complex are linked by lipids, a separation should only succeed by detergents, but not after mechanical breakage. In a series of experiments, JACOBI (1970) and JACOBI et al. (1972) used the French press and sonication for grana fragmentation. Even the smallest particles retained their capacity to catalyze activities related to both PS I and PS II. A crucial point in the preparation of smaller particles was their disintegration in salt-free media. At higher ionic strength, the vesicles tended to aggregate and became inactive. Similar results were reported by ARNTZEN et al. (1972).

In spite of the more effective solubilization caused by detergents, most of the PS I particles released from the grana retained some PS II activity (VERNON et al., 1969). Only a very small portion seemed to be uncontaminated and was recently shown to be completely identical with those PS I units which were released from the intergrana lamellae (WESSELS, this vol. Chap. IV.4). Though a number of fractions were reported to be free of PS II, they had an a/b ratio of 3.5/4.6, thus indicating that they were still contaminated with non-specific pig-ments (JACOBI, 1969; WESSELS and VOORN, 1972; ARNTZEN et al., 1972; BROWN et al., 1972). A further insufficient separation derived from curve analysis of in vivo chlorophyll forms, as has been described by BROWN et al. (1973). With the exception of a somewhat reduced chlorophyll b-650 concentration, the chlorophyll a forms resembled largely those of the grana fraction, rather than an intergrana PS I.

One fact of considerable interest in the investigations by BROWN et al. (1972) is the absence of chlorophyll a-705 in the total grana fraction, which surprisingly appeared again in PS I prepared from this material. The observed lack of P-700 raised the question as to whether in certain preparations this form is obscured by the relatively high portion of collector pigments. This problem is of fundamental interest since ARNON et al. (1971) and MALKIN (1971) reported on the isolation of a fraction which contained no P-700, but retained the capacity to photoreduce $NADP^+$. MALKIN determined P-700 chemically and by light-induced absorbance changes and found this pigment absent in the heavy fraction, but concentrated in the intergranal material. However, these results were questioned by ESSER (1974), who measured a light-induced P-700 signal in MALKIN's preparations after addition of plastocyanin.

Partially enriched grana PS I prepared by digitonin have generally a chlorophyll/ P-700 ratio of 170 (ARNTZEN et al., 1971) or 250 (WESSELS and VOORN, 1972), compared to a value of 120 in the intergrana material. Though this ratio can be reduced to 30 (VERNON et al., 1971) by further fractionation with Triton X-100, or even to 7 after treatment with ethanol (IKEGAMI and KATOH, 1975), any purification is accompanied by an inactivation. These investigations raised in particular the principle question concerning a specific criterion, which should account for the characterization of PS I after grana fragmentation. In general, if the enrichment of PS I from grana is followed under the aspect of photochemical activity, the critical size of particles to be expected is relatively large and more complex in composition than a highly purified reaction center.

An entirely different approach to distinguish PS I from grana and stroma lamellae came from reconstitution experiments by combining PS II with PS I, which were derived from the two regions (ARNTZEN et al., 1972). Under certain conditions, only the grana PS I was effective in reconstituting the electron transport from DPC to $NADP^+$. The fact that stroma PS I did not reconstitute, led them to conclude that vesicles released from the intergrana region are deficient in a factor necessary to link them PS II fragments. However, a similarly plausible explanation is that the fragmentation procedure causes a different accessibility of plastocyanin to PS I.

E. The Alteration of Reaction Properties and the Diversity of Chloroplast Fragments

Among the several lines of evidence purported to characterize the quality of a chloroplast fragment, the photochemical activity is undoubtedly of greatest importance. In spite of the progress that has been made in recent years to measure partial reactions of the electron transport chain (TREBST, 1972), the rates which were determined in the various fragments did not necessarily testify to their actual activity. In principle, the activities measured in the fractions after separation are simple presumed to attain those values which were determined in the original chloroplast suspension. However, such a calculation is only valid if the reaction properties of the disrupted material correspond to those of the original, which turns out to be true only for a small number of fragments prepared under certain conditions.

In order to distinguish uncoupling from inactiviations caused by various treatments, GOOD and IZAWA (1965) compared the Hill activity in several centrifugal fractions, with and without methylamine, and concluded an increase of inactivation with decreasing particle size. However, their estimates did probably not account for the actual capacity of the subunits, because of the pronounced shift of the pH-optimum of Hill activity after fragmentation, which has been overlooked. As demonstrated by KATOH and SAN PIETRO (1966), and more extensively by JACOBI (1966) and JACOBI and LEHMANN (1968, 1969), the reduction of ferricyanide and of $NADP^+$ have their own pH-optimum at pH 6 in particles of every size produced by sonication. This result explained also some conflicting observations reported from different laboratories, which noticed either a decrease (IZAWA and GOOD, 1965; GRESSEL and AVRON, 1965) or even a stimulation (BECKER et al., 1965), when the Hill activity before and after sonication was measured at pH 7.8 or 6.8 respectively.

However, to compare the various fragments described, one should be cautious in correlating the change of pH-optimum with the effect of a certain method. The breakage is similarly influenced by the homogenization media as demonstrated by McCARTY (1968) and by HAUSKA and SANE (1972), who reported that fragments produced by sonication also retain some phosphorylating capacity. These particles were derived from thylakoids suspended in sucrose, whereas those prepared by JACOBI and LEHMANN (1968) came from chloroplasts which were sonicated in salt media. Therefore, the subunits released by sonication are, under certain conditions, also similar to those produced by French press treatment, and resemble in some properties an intact membrane with the same pH-optimum of electron transport reactions (MURATA and BROWN, 1970).

A better estimate for the quality of particles produced by any method is the requirement of electron carriers to be added for obtaining maximum rates of activities related to PS I and PS II. However, these criteria are not necessarily correlated to the release of enzymes involved in certain photochemical events. For instance, the coupling factor has been shown to be attached to those subunits which, to a great extent, lose their phosphorylating capacity (HAUSKA and SANE, 1972; SCHOPF et al., 1975).

The following synopsis is an attempt to classify the large and confusing number of particles described in the literature and extends the classification of chloroplast preparations of HALL (1972) with special regard to the properties of fragments. Some of these resemble largely an intact membrane of Hall's type D, whereas others come more close to the purified reaction centers.

Class P-I. This class includes phosphorylating particles with an intact electron transport system. The larger fragments represent grana stacks. The two photosystems are well-coordinated and catalyze an electron transport from water to $NADP^+$. To obtain maximum rates, it is necessary to add ferredoxin and sometimes ferredoxin-$NADP^+$-oxidoreductase. However, plastocyanin has no effect. The smaller intergrana particles show only cyclic phosphorylation and $NADP^+$-reduction mediated by an electron donor couple. The particles of this class are generally prepared by lower concentrations of digitonin as first described by Wessels (1966a, b). Wessels et al. (1967); Hauska et al. (1970) and Arntzen et al. (1972) presented electron micrographs showing small knobs attached at the surface of the phosphorylating digitonin particles which were absent in fragments prepared at higher concentrations of the detergent. The ATP-formation in the grana stacks was shown to be accompanied by the formation of a proton gradient (Arntzen et al., 1972; Hauska and Sane, 1972). In contrast, the light fraction did not show a proton gradient and could not be uncoupled by methylamine (Arntzen et al., 1972; McCarty, 1968).

Class P-II. A considerable portion of larger fragments are uncoupled, but sustain most of the electron transport activities. The essential property of this type is an intact oxygen-evolving system tightly associated with PS II. Therefore, a normal Hill activity with various acceptors is observed. However, the $NADP^+$ reduction often becomes dependent upon the addition of plastocyanin as an indicator for a break at the level of cytochrome f. If this is the case, the pH-optimum of the Hill reaction drops to about pH 6. This type is produced by detergents, by certain French presses, or by sonication and is enriched in the heavy fractions obtained after centrifugation.

Class P-III. An indicator for this type is the damage to the water-splitting machinery which occurs by more vigorous disruption. The reduction of electron acceptors catalyzed by PS II requires the addition of certain donors, such as DPC (Vernon et al., 1969). These particles are generally smaller than those of class P-II and are especially derived from fragmented grana stacks.

Class P-IV. The majority of fractions which contain partially purified photosystems belong to this category. They differ from pure reaction centers in that they are still more complex in composition. The reaction centers are associated with other membrane constituents and particularly with accessory pigments as indicated by a chlorophyll a/b ratio much smaller than 7. As summarized before, a great number of PS II subunits, particularly those released by mechanical breakage of grana stacks, contain certain amounts of PS I and obviously represent the smallest complex photosynthetic entities.

Class P-V. Purified reaction centers (see Wessels, this vol. Chap. IV.4). Although the overwhelming number of fragments produced by various methods can be distinguished by these criteria, the classification is still incomplete with regard to the altered kinetic properties of certain electron carriers, as described in the next section. However, regardless of these details, it becomes clear that the recovery is hardly reflected by the activities measurd in the centrifugal fractions which are often not comparable in their degree of disruption. Therefore, the measured rate of a photochemical reaction displays merely the quality of a subunit, rather than its actual capacity. The problem always involved in the interpretation of altered activity is that two opposite events occur during the fragmentation: a stimulation of activity caused by the favored accessibility of electron donors or acceptors and the decrease due to inactivations. A remarkable increase of $NADP^+$-reduction after fragmentation of grana and of the photooxidation of cytochrome c by the treatment with Triton has been demonstrated by Jacobi et al. (1972)

and by BROWN (1971). The problem of estimating the extent of inactivation is particularly impaired by the stimulation of activity and by the above-noticed alteration of reaction optima. In addition to the irreversible damage, the reactivity of a fragment is also affected by an inhibition of certain membrane constituents, which are released during disruption. Among the substances responsible for the inhibition, free fatty acids must be especially considered. The onset of lipolysis accompanied by the degradation of unsaturated fatty acids during fragmentation has recently been shown by HEISE and JACOBI (1973) and by SCHOPF et al. (1975).

F. The Disorientation of Electron Carriers and the Effect of Plastocyanin

The proposed classification in Section E is considered a useful approach to recognizing a certain type of fragment by means of biochemical methods. However, since the measurements are generally carried out under steady-state conditions, they do not evaluate the disorientation of electron carriers which become established by kinetic studies. The following considerations deal particularly with the dissociation of electron carriers at the donor site of PS I. In spite of the fact that the breakage pattern at the level of PS I can be characterized by kinetic measurements of light-induced absorbance changes of cytochrome f and P-700, the interpretations are still confusing with regard to plastocyanin. Although a critical review on the function of the copper enzyme is submitted by KATOH (this vol. Chap. II.14) a survey on the different properties of subunits would be incomplete without a discussion on the discrepancies concerning the content and the action of plastocyanin in certain fragments.

Plastocyanin is a membrane-bound enzyme which is not released when the thylakoids are washed in hypotonic media, or treated with EDTA generally used to remove proteins from the outer surface. The ratio chlorophyll/plastocyanin determined in several laboratories agrees fairly well and ranges for intact chloroplasts between 220 and 600 (KATOH et al., 1961; PLESNICAR and BENDALL, 1970; BASZYNSKI et al., 1971; ARNTZEN et al., 1972; SANE and HAUSKA, 1972; MURATA and FORK, 1973). Although the liberation of the enzyme by fragmentation has been reported by these groups, the literature displays severe contradictions concerning its retention by the particles. Among the fragmentation methods compared, Triton X-100 and sonic oscillation were shown to be most effective in liberating the protein (SANE and HAUSKA, 1972; MURATA and FORK, 1973). However, HAUSKA et al. (1970, 1971) found the amount within the fragments unchanged when the chloroplasts were sonicated in the presence of soluble plastocyanin. Furthermore, since plastocyanin in situ was not accessible to an antibody, they speculated on a localization in the inner space of the thylakoids. Discrepancies arose especially in the findings of KNAFF and ARNON (1970) and of MURATA and FORK (1971 b, 1973), who described particles produced by mild treatment with the French press which were almost depleted in plastocyanin, but retained their electron transport system in its native state. The photoactive French press fragments prepared by the Stanford group had a chlorophyll/plastocyanin ratio of 3,000 and 7,000 in

PS I and PS II particles, respectively. This result was questioned by Baszynski et al. (1971) and by Sane and Hauska (1972), who noticed a retention when the chloroplasts were fragmented in 0.4 M sucrose. However, since the efficiency of various types of presses to release plastocyanin is different, such a comparison seems inconclusive.

Regardless of whether the French press particles retained a small amount of plastocyanin or not, direct evidence for its participation in photochemical events in depleted particles is still lacking. The measurement of light-induced absorbance changes is impeded by the relatively weak absorbancy of plastocyanin, and was therefore only reported in few cases in intact chloroplasts by the use of highly sensitive methods (R. Malkin and Bearden, 1973; Haehnel, 1975). As a consequence of these limitations, the role of the enzyme in electron transport is derived mostly from its requirement for some reactions after breakage. Other conclusions were reached from its influence on the photooxidation and the photoreduction of cytochrome f and P-700 in fragments.

As shown by Katoh and Takamiya (1963), Davenport (1965), Katoh and San Pietro (1966), Jacobi (1966, 1967), Arnon et al. (1968), Elstner et al. (1968), and thereafter in several other laboratories, the reduction of NADP$^+$ is strongly dependent upon the addition of plastocyanin in particles produced by sonication. Similar results were obtained with fragments prepared after treatment with Triton X-100 (Vernon and Shaw, 1965; Vernon et al., 1966), Tween 20 (Kok et al., 1964), and digitonin (Trebst and Elstner, 1965; Wessels, 1966b, 1969). However, the subunits produced by mild treatment with the French press as described by Knaff and Arnon (1970), Fork and Murata (1971b, 1972), and more recently by Knaff (1973), represent a completely different type. These fragments were shown to be depleted of plastocyanin, but to catalyze the reduction of NADP$^+$ and of methylviologen with high rates independent of the presence of the copper enzyme. Therefore, these particles resemble largely an intact thylakoid membrane, showing no response to added plastocyanin.

Although the effect of the copper enzyme serves as a sensitive indicator of the degree of fragmentation, it is difficult to come to confident conclusions concerning the actual role of plastocyanin and its exact position in the sequence of electron carriers. From the accelerated photooxidation of cytochrome f in chloroplasts treated with either Triton X-100 or sonication by added plastocyanin, Hind (1968) and Avron and Shneyour (1971) discussed a sequence in which the enzyme is located between the cytochrome and P-700. However, since the oxidation rate of cytochrome f was about half of the reduction rate of P-700, and the reduction of cytochrome was slow, Hind (1968) concluded that "this pigment cannot participate significantly in continuous electron flow between ascorbate and viologen". Therefore, he postulated an additional pathway in which plastocyanin is a redox carrier functioning parallel to cytochrome. This idea had previously been suggested by Kok and Rurainski (1965).

Since the double beam spectrophotometer used by Hind and by Avron and Shneyour allows only the measurement of slow changes, the observed effects of plastocyanin are possibly not related to primary photochemical events which occur in the range of μ- or milliseconds. Therefore, the rapid changes analyzed by Fork and Jacobi (1970) in grana, and by Fork and Murata (1971a, b, 1972) in smaller particles, are not comparable with the data reported by Hind (1968).

This more rapid method allowed the comparison of fast changes which are characteristic for an intact membrane with those alterations occurring after structural damage. The absorbance changes in the French press particles were shown to be entirely identical with those of the original chloroplasts — i.e. the photooxidation of cytochrome f and the photoreduction of P-700 in the presence of DCMU were significantly accelerated by artificial donors such as DPIP, DAD, and PMS. It is important to note that the addition of plastocyanin or of cytochromes prepared from *Porphyra* and *Euglena,* as well as mammalian cytochrome c, had no effect. However, the situation changed drastically when the particles were treated with Triton X-100. In this state, the fast oxidation of cytochrome f was completely abolished, but the natural proteins became highly efficient donors for P-700. In contrast, artificial donors, with the exception of PMS, were inefficient and could not overcome the effect of proteins. These results led FORK and JACOBI (1970), FORK and MURATA (1971, 1971 b), and JACOBI (1970) to conclude that after disruption with detergents or with sonication, cytochrome f is "uncoupled" from P-700 and plastocyanin is allowed direct access to the reaction center.

A modification of PS I would best explain the various effects described for the different fragments. It furthermore corresponds with the concept of an in vivo charge transfer complex between cytochrome f and P-700, as proposed by KOK et al. (1964) and KOK and RURAINSKI (1965). If this complex is broken, the maximum rate of $NADP^+$ reduction is obtained solely by the addition of plastocyanin plus ascorbate, but is not enhanced by artificial donors as shown for the intergranal particles released by sonication (JACOBI, 1969).

In view of the close association of the reaction center with its natural donor and their dissociation under certain conditions, the reconstitution experiments carried out by HUZISIGE et al. (1969), ARNTZEN et al. (1972), and others must be discussed in a different manner. ARNTZEN et al. (1972) compared PS I particles from the intergranal region with those released from the grana in their capacity to photoreduce $NADP^+$ from DPC, by a reconstitution with grana PS II units in the presence of plastocyanin. Since the intergrana particles, and not the grana PS I, were completely inefficient in $NADP^+$ reduction, it was assumed that the reaction centers localized in the two structural parts are different. However, the particles from the intergrana membranes which were released by the action of the French press represent more intact entities, showing no response to plastocyanin. In contrast, the PS I units from the grana were liberated by the action of digitonin, thus resulting in a partially broken complex.

In the same manner, some effects described by ARNON et al. (1971), MALKIN (1971), and KNAFF (1973) are reasonably explained by the different accessibility of plastocyanin to PS I. From a series of biochemical and biophysical investigations, the Berkeley group assumed that three light reactions are involved in the entire electron transport system (ARNON, 1971; ARNON et al., 1968; KNAFF and ARNON, 1971; KNAFF and MCSWAIN, 1971). The essential prediction was that P-700 and cytochrome f do not participate in the noncyclic electron flow from water to $NADP^+$ which was suggested to be driven by two wavelength-absorbing centers, designated as chl-IIb and chl-IIa. Since the action spectra showed a similar red drop for the photooxidation of cyt b-599 and C-550, and since the photooxidation of these components was similarly effected by plastocyanin, the two reaction centers were thought to be connected by these three carriers. The third light reaction was suggested to operate in parallel and was said to represent an individual system which is responsible for cyclic phosphorylation. This photosystem is assumed to catalyze the long-wavelength-absorbing reactions including the photooxidations of cytochrome f and b-6. P-700 is placed into the reaction center of this photosystem, which was named PS I.

Although the effect of plastocyanin and the absorption changes of cytochromes are simply explained by the breakage of a charge-transfer complex, some observations are still inconsistent with the sequence of electron carriers in the generally accepted Z-scheme. For instance, MALKIN (1971) reported that high rates of NADP$^+$ reduction in the donor couple including plastocyanin were measured with actinic light at 664 nm, whereas 715 nm was practically ineffective. Furthermore, since P-700 was not detected in these preparations, the results are hardly explained on the basis of the conventional concept. At this point, it is worthwhile to mention that RURAINSKI et al. (1971); RURAINSKI and HOCH (1972) proposed an alternative mechanism. By the use of steady-state relaxation spectroscopy, the authors found in chloroplasts that P-700 under certain conditions does not participate in the reduction of NADP$^+$. Although this observation is not yet fully understood, the direct measurement of the rates of electron flow through the intermediates, as demonstrated in these investigations, provides a future problem of characterizing in a more precise manner the actual state of the carriers after fragmentation. This kind of experiment may also help to clarify the role of plastocyanin in situ, which is better characterized in terms of electron flux rather than by its endogenous concentration. Though the apparent concentration of plastocyanin required to saturate the NADP$^+$ reduction is correlated to the degree of disruption (JACOBI, 1969), its actual effect is governed by the ratio of $[PCy]_{red}/[PCy]_{total}$ (KOK and RURAINSKI, 1965).

G. Prospect

The fragmentation studies described above provide compelling evidence for a functional differentiation within the thylakoid membrane system. It must be recognized, however, that this conclusion is derived almost exclusively from results obtained with chloroplasts of certain "standard" plants, particularly spinach. It is beyond the scope of this article to review those experiments which have dealt with the fragmentation of chloroplasts having an unusual structure. Those thylakoids with an extremely high extent of stacking (BJÖRKMAN et al., 1972; ANDERSON et al., 1973) and others having no grana, such as certain mutants and bundle sheath chloroplasts, bring out the significance of stacking for photosynthesis (WOO et al., 1970; ANDERSON et al., 1971a, b; BISHOP et al., 1971, 1972; GOODENOUGH and STAEHELIN, 1971). This problem has recently been reviewed in a more comprehensive manner by ARNTZEN and BRIANTAIS (1975). It must be pointed out that some results correspond exactly to those obtained with spinach, whereas others seem to be inconsistent with the concept of a disparate distribution of PS II. However, the contradiction is largely solved if one considers membrane contact as not being a prerequisite for the Hill-activity itself. Therefore, the lack of stacking is not necessarily an indicator for the absence of PS II. However, it is reasonable to assume that PS II is not randomly distributed, but localized in certain regions of the thylakoid system. If the model of grana formation as a spirocyclic overlap is correct, PS II should be localized only in the peripheral regions (WEHRMEYER, 1963). In fact, a separation of photosystems by fragmentation has been demonstrated as yet only with chloroplasts having grana. On the other hand, the unsuccessful separation of photosystems described so far for agranal thylakoids, which have the capacity to carry out the Hill reaction, may be simply explained by the inability to separate and recognize individual fragments of such thylakoid systems.

At present, the significance of stacking is best discussed with regard to the concentration of collecting pigments (ANDERSON et al., 1973). From this viewpoint, the alteration of structure and function during development (ARNTZEN et al., 1971; HILLER and BOARDMAN, 1971), caused by ecological factors (BJÖRKMAN, 1973), or induced by physiological variations, opens a wide field for future experiments. Chemical analysis during the various stages may further help to recognize those components which are responsible for membrane contact. Based on this premise it appears reasonable to expect a better understanding of the apparent light-induced conformation changes (PACKER et al., 1970) as the next level of investigations.

Acknowledgements. I am grateful to Dr. H.J. RURAINSKI for critical reading of the manuscript and valuable comments. My special thanks go to Dr. D. ROBINSON for the revision of the English text.

References

Anderson, J.M.: Carnegie Inst. Year Book **65**, 479–481 (1967)

Anderson, J.M., Boardman, N.K.: Biochim. Biophys. Acta **112**, 403–421 (1966)

Anderson, J.M., Boardman, N.K., Spencer, D.: Biochim. Biophys. Acta **245**, 253–258 (1971a)

Anderson, J.M., Fork, D.C., Amesz, J.: Biochem. Biophys. Res. Commun. **23**, 874–879 (1966)

Anderson, J.M., Goodchild, N.H., Boardman, N.K.: Biochim. Biophys. Acta **325**, 573–585 (1973)

Anderson, J.M., Vernon, L.P.: Biochim. Biophys. Acta **143**, 363–376 (1967)

Anderson, J.M., Woo, K.C., Boardman, N.K.: Biochim. Biophys. Acta **245**, 398–408 (1971b)

Arnon, D.I.: Proc. Nat. Acad. Sci. U.S. **68**, 2883–2892 (1967)

Arnon, D.I., Knaff, D.B., McSwain, B.D., Chain, R.K., Tsujimoto, H.Y.: Photochem. Photobiol. **14**, 397–425 (1971)

Arnon, D.I., Tsujimoto, H.Y., McSwain, B.D., Chain, R.K.: In: Comparative Biochemistry and Biophysics of Photosynthesis. Shibata, K., Takamiya, A., Jagendorf, A.T., Fuller, R.C. (eds.). State College, Pa.: Univ. Park Press, 1968, pp. 113–132

Arntzen, C.J., Briantais, J.-M.: In: Bioenergetics of Photosynthesis. Govindjee (ed.). New York: Academic Press, 1975, pp. 52–113

Arntzen, C.J., Dilley, R.A., Neumann, J.: Biochim. Biophys. Acta **245**, 409–424 (1971)

Arntzen, C.J., Dilley, R.A., Peters, G.A., Shaw, E.R.: Biochim. Biophys. Acta **256**, 85–107 (1972)

Avron, M., Shneyour, A.: Biochim. Biophys. Acta **226**, 498–500 (1971)

Baszynski, T., Brand, J., Krogmann, D.W., Crane, F.L.: Biochim. Biophys. Acta **234**, 537–540 (1971)

Becker, M.J., Shefner, A.M., Gross, J.A.: Plant Physiol. **40**, 243–250 (1965)

Bishop, D.G., Andersen, K.S., Smillie, R.M.: Biochem. Biophys. Res. Commun. **42**, 74–81 (1971)

Bishop, D.G., Andersen, K.S., Smillie, R.M.: Plant Physiol. **49**, 467–470 (1972)
Björkman, O.: In: Photophysiology. Giese, A.C. (ed.). New York-London: Academic Press, 1973, Vol. VIII, pp. 1–63
Björkman, O., Boardman, N.K., Anderson, J.M., Thorne, S.W., Goodchild, D.J., Pyliotis, N.A.: Carnegie Inst. Year Book **71**, 115–135 (1972)
Boardman, N.K.: Ann. Rev. Plant Physiol. **21**, 115–140 (1970)
Boardman, N.K.: Fractionation and properties of the photosystems. In: Chloroplast Fragments. Jacobi, G. (ed.). Göttingen: Workshop, 1972a
Boardman, N.K.: Biochim. Biophys. Acta **283**, 469–482 (1972b)
Boardman, N.K., Anderson, J.M.: Nature (London) **203**, 166–167 (1964)
Boardman, N.K., Anderson, J.M.: Biochim. Biophys. Acta **143**, 187–203 (1967)
Boardman, N.K., Anderson, J.M., Hiller, R.G.: Biochim. Biophys. Acta **234**, 126–136 (1971)
Boardman, N.K., Thorne, S.W.: Biochim. Biophys. Acta **189**, 294–297 (1969)
Brown, J.S.: Carnegie Inst. Year Book **70**, 499–504 (1971)
Brown, J.S.: In: Photophysiology. Giese, A. (ed.). New York-London: Academic Press, 1973, Vol. VIII, pp. 97–112
Brown, J.S., Gasanov, R.A., French, C.S.: Carnegie Inst. Year Book **72**, 351–359 (1972)
Brown, J.S., Prager, L.: Carnegie Inst. Year Book **67**, 516–520 (1969)
Cederstrand, C.N., Govindjee: Biochim. Biophys. Acta **120**, 177–180 (1966)
Clendenning, K.A.: In: Encyclopedia of Plant Physiology. Ruhland, W. (ed.). Berlin-Heidelberg-New York: Springer, 1960, Vol. V, pp. 736–768
Davenport, H.E.: The role of soluble protein factors in chloroplast electron transport. In: Non-Heme Iron Proteins. San Pietro, A. (ed.). Yellow Springs, Ohio: Antioch Press, 1965, pp. 115–135
Elstner, E., Pistorius, E., Böger, P., Trebst, A.: Planta (Berl.) **79**, 146–161 (1968)
Emerson, R., Arnold, W.: J. Gen. Physiol. **16**, 191 (1932)
Esser, A.F.: Biochim. Biophys. Acta **275**, 199–207 (1974)
Fork, D.C., Jacobi, G.: Carnegie Inst. Year Book **69**, 690–695 (1970)
Fork, D.C., Murata, N.: Carnegie Inst. Year Book **69**, 682–690 (1971a)
Fork, D.C., Murata, N.: Photochem. Photobiol. **13**, 33–44 (1971b)
Fork, D.C., Murata, N.: In: Proc. 2nd Intern. Congr. Photosynthesis. Forti, G., Avron, M., Melandri, A. (eds.). The Hague: Dr. Junk, 1972, pp. 847–858
French, C.S.: Carnegie Inst. Year Book **68**, 578–587 (1970)
French, C.S., Brown, J.S., Lawrence, M.C.: Carnegie Inst. Year Book **70**, 487–495 (1971)
Goodchild, D.J., Park, R.B.: Biochim. Biophys. Acta **226**, 393–399 (1971)
Goodenough, U.W., Staehelin, L.A.: J. Cell Biol. **48**, 594–619 (1971)
Govindjee: Discussions on chlorophyll fluorescence in vivo: techniques and applications. In: Chloroplast Fragments. Jacobi, G. (ed.). Göttingen: Workshop, 1972, pp. 17–45
Gressel, J., Avron, M.: Biochim. Biophys. Acta **94**, 31 (1965)
Gross, E.L., Packer, L.: Arch. Biochem. Biophys. **121**, 779–789 (1967)
Haehnel, W.: In: Proc. 3rd Intern. Congr. Photosynthesis. Avron, M. (ed.). Amsterdam, Elsevier, 1975, pp. 557–568
Hall, D.O.: New Biol. **235**, 125–126 (1972)
Hall, D.O., Edge, H., Kalina, M.: J. Cell Sci. **9**, 289–303 (1971)
Hall, D.O., Edge, H., Reeves, S.G., Stocking, C.R., Kalina, M.: In: Proc. 2nd Intern. Congr. Photosynthesis. Forti, G., Avron, M., Melandri, A. (eds.). The Hague: Dr. Junk, 1972, pp. 701–721
Hauska, G.A., McCarty, R.E., Berzborn, R.J., Racker, E.: J. Biol. Chem. **246**, 3524–3531 (1971)
Hauska, G.A., McCarty, R.E., Racker, E.: Biochim. Biophys. Acta **197**, 206–218 (1970)
Hauska, G.A., Sane, P.V.: Z. Naturforsch. **27B**, 938–942 (1972)
Heise, K.P., Jacobi, G.: Z. Naturforsch. **28C**, 120–127 (1973)
Heslop-Harrison, J.: Planta (Berl.) **60**, 243–260 (1963)
Hiller, R.G., Boardman, N.K.: Biochim. Biophys. Acta **253**, 449–458 (1971)
Hind, G.: Biochim. Biophys. Acta **153**, 235–240 (1968)
Huzisige, H., Usiyama, H., Kikuti, T., Azi, T.: Plant Cell Physiol. **10**, 441–445 (1969)
Ikegami, I., Katoh, S.: Biochim. Biophys. Acta **376**, 588–592 (1975)
Izawa, S., Good, N.E.: Biochim. Biophys. Acta **109**, 372–381 (1965)

Izawa, S., Good, N.E.: Plant Physiol. **41**, 544–552 (1966)
Jacobi, G.: Ber. Deut. Botan. Ges. **79**, 72–81 (1966)
Jacobi, G.: Z. Pflanzenphysiol. **57**, 255–268 (1967)
Jacobi, G.: Z. Pflanzenphysiol. **61**, 203–217 (1969)
Jacobi, G.: Ber. Deut. Botan. Ges. **83**, 451–463 (1970)
Jacobi, G.: Chloroplast Fragments. Göttingen: Workshop, 1972
Jacobi, G., Lehmann, H.: Z. Pflanzenphysiol. **59**, 457–476 (1968)
Jacobi, G., Lehmann, H.: In: Progress in Photosynthetic Research. Metzner, H. (ed.).
    Tübingen: Laupp, 1969, Vol. I, pp. 159–173
Jacobi, G., Murakami, S., Heise, K.P.: In: Proc. 2nd Intern. Congr. Photosynthesis. Forti,
    G., Avron, M., Melandri, A. (eds.). The Hague: Dr. Junk, 1972, pp. 813–824
Katoh, S., San Pietro, A.: In: Biochemistry of Copper. Preisach, J., Aisen, P., Blumberg,
    W.E. (eds.). New York: Academic Press, 1966, pp. 407–422
Katoh, S., Suga, I., Shiratori, I., Takamiya, A.: Arch. Biochem. Biophys. **94**, 136–141 (1961)
Katoh, S., Takamiya, A.: Plant Cell Physiol. **4**, 335–347 (1963)
Ke, B., Vernon, L.P., Chaney, T.H.: Biochim. Biophys. Acta **256**, 345–357 (1972)
Kitajima, M., Butler, W.L.: Biochim. Biophys. Acta **325**, 558–564 (1973)
Knaff, D.B.: Biochim. Biophys. Acta **292**, 186–192 (1973)
Knaff, D.B., Arnon, D.I.: Biochim. Biophys. Acta **223**, 201–204 (1970)
Knaff, D.B., Arnon, D.I.: Biochim. Biophys. Acta **226**, 400–408 (1971)
Knaff, D.B., McSwain, B.D.: Biochim. Biophys. Acta **245**, 105–108 (1971)
Kok, B., Rurainski, H.J.: Biochim. Biophys. Acta **94**, 588–590 (1965)
Kok, B., Rurainski, H.J.: Biochim. Biophys. Acta **126**, 584–587 (1966)
Kok, B., Rurainski, H.J., Harmon, E.A.: Plant Physiol. **39**, 513–520 (1964)
Kreutz, W.: Z. Naturforsch. **19B**, 441–446 (1964)
Malkin, R.: Biochim. Biophys. Acta **253**, 421–427 (1971)
Malkin, R., Bearden, A.J.: Biochim. Biophys. Acta **292**, 169–185 (1973)
McCarty, R.E.: Biochem. Biophys. Res. Commun. **32**, 37–43 (1968)
Menke, W.: In: Biochemistry of Chloroplasts. Goodwin, T.W. (ed.). London-New York:
    Academic Press, 1966, Vol. I, pp. 3–18
Michel, J.M., Michel-Wolwertz, M.R.: Carnegie Inst. Year Book **67**, 508–514 (1969a)
Michel, J.M., Michel-Wolwertz, M.R.: In: Progress in Photosynthetic Research. Metzner, H.
    (ed.). Tübingen: Laupp, 1969b, Vol. I, pp. 115–121
Mühlethaler, K.: In: Structure and Function of Chloroplasts. Gibbs, M. (ed.). Berlin-Heidel-
    berg-New York; Springer, 1971
Murata, N., Brown, J.S.: Plant Physiol. **45**, 360–361 (1970)
Murata, N., Fork, D.C.: Biochim. Biophys. Acta **245**, 356–364 (1971a)
Murata, N., Fork, D.C.: Carnegie Inst. Year Book **70**, 468–472 (1971b)
Murata, N., Fork, D.C.: Carnegie Inst. Year Book **72**, 376–384 (1973)
Nelson, N., Drechsler, Z., Neumann, J.: J. Biol. Chem. **245**, 143–151 (1970)
Ohki, R., Takamiya, A.: Biochim. Biophys. Acta **197**, 240–249 (1970)
Packer, L., Murakami, S., Mehard, C.W.: Ann. Rev. Plant Physiol. **21**, 271–304 (1970)
Park, R.B., Sane, P.V.: Ann. Rev. Plant Physiol. **22**, 395–430 (1971)
Park, R.B., Steinback, K.E., Sane, P.V.: Biochim. Biophys. Acta **253**, 204–207 (1971)
Plesničar, M., Bendall, D.S.: Biochim. Biophys. Acta **216**, 192–199 (1970)
Rurainski, H.J., Hoch, G.E.: In: Proc. 2nd Intern. Congr. Photosynthesis. Forti, G., Avron,
    M., Melandri, A. (eds.). The Hague: Dr. Junk, 1972, pp. 133–141
Rurainski, H.J., Randles, J., Hoch, G.E.: FEBS Lett. **13**, 98–100 (1971)
Sane, P.V., Goodchild, D.J., Park, R.B.: Biochim. Biophys. Acta **216**, 162–178 (1970)
Sane, P.V., Hauska, G.A.: Z. Naturforsch. **27B**, 932–938 (1972)
Sane, P.V., Park, R.B.: In: Proc. 2nd Intern. Congr. Photosynthesis. Forti, G., Avron, M.,
    Melandri, A. (eds.). The Hague: Dr. Junk, 1972, pp. 825–832
Schopf, R., Heise, K.P., Schmidt, B., Jacobi, G.: In: Proc. 3rd Intern. Congr. Photosynthesis.
    Avron, M. (ed.). Amsterdam: Elsevier, 1975, pp. 911–919
Thomas, J.B.: In: Encyclopedia of Plant Physiology. Ruhland, W. (ed.). Berlin-Heidelberg-New
    York: Springer, 1960, Vol. V, pp. 511–566
Trebst, A.: Methods Enzymol. **24**, 146–165 (1972)
Trebst, A., Elstner, E.: Z. Naturforsch. **20B**, 925–926 (1965)

Vernon, L.P., Ke, B., Mollenhauer, H.H., Shaw, E.R.: In: Progress in Photosynthesis Research. Tübingen: Laupp, 1969, Vol. I, pp. 137–148

Vernon, L.P., Klein, S., White, F.G., Shaw, E.R., Mayne, B.C.: In: Proc. 2nd Intern. Congr. Photosynthesis. Forti, G., Avron, M., Melandri, A. (eds.). The Hague: Dr. Junk, 1972

Vernon, L.P., Shaw, E.R., Ogawa, T., Raveed, D.: Photochem. Photobiol. **14**, 343–357 (1971)

Wehrmeyer, W.: Z. Naturforsch. **17 B**, 54–57 (1962)

Wehrmeyer, W.: Planta (Berl.) **59**, 280–295 (1963)

Weier, T.E., Stocking, R.C., Bracker, C.E., Risley, E.B.: Am. J. Botany **52**, 339–352 (1965)

Weier, T.E., Stocking, C.R., Shumway, L.K.: Brookhaven Symp. Biol. **19**, 353–374 (1967)

Weier, T.E., Thomson, W.W., Drewer, H.: J. Ultrastruct. Res. **8**, 122–143 (1963)

Wessels, J.S.C.: In: Currents in Photosynthesis. Thomas, J.B., Goedheer, J.C. (eds.). Rotterdam: A.D. Donker, 1966a, pp. 129–139

Wessels, J.S.C.: Biochim. Biophys. Acta **126**, 581–583 (1966b)

Wessels, J.S.C.: Biochim. Biophys. Acta **153**, 497–500 (1968)

Wessels, J.S.C.: In: Progress in Photosynthesis. Research. Metzner, H. (ed.). Tübingen: Laupp, 1969, Vol. I, pp. 128–136

Wessels, J.S.C., Dorsman, A., Luitingh, A.J.: In: Le Chloroplaste. Sironval, C. (ed.). Paris: Masson and Cie, 1967

Wessels, J.S.C., Voorn, G.: In: Proc. 2nd Intern. Congr. Photosynthesis. Forti, G., Avron, M., Melandri, A. (eds.). The Hague: Dr. Junk, 1972, pp. 833–846

Woo, K.C., Anderson, J.M., Boardman, N.K., Downtown, W.J.S., Osmond, C.B., Thorne, S.W.: Proc. Nat. Acad. Sci. U.S. **67**, 18–25 (1970)

4. Fragmentation

J.S.C. WESSELS

A. Introduction

There is now ample evidence that photosynthetic electron transport from water to NADP$^+$ is driven by two light reactions operating in series. The light reactions are catalyzed by two different pigment assemblies, termed photosystem I and photosystem II, each containing a photochemical reaction center and light-harvesting pigment molecules.

The photosystems are located in the internal membrane system of the chloroplast, which is surrounded by an embedding matrix referred to as the stroma portion of the chloroplast. The internal membrane system consists of numerous flattened sacs, which have been named lamellae or thylakoids. In the mesophyll chloroplasts of higher plants the lamellae are arranged in densely packed regions, the grana, and in a more loosely organized system of membranes which interconnect the grana stacks. These two types of lamellae are known as the grana lamellae and the stroma lamellae, respectively. Little if any stacking of lamellae is found in the chloroplasts of algae and of bundle-sheath cells of several plants having the C$_4$-dicarboxylic acid pathway of CO$_2$ fixation (LAETSCH, 1974).

Disruption of the chloroplast lamellar system is a prerequisite for the separate isolation of the photosystems I and II. The methods which have been applied to membrane fragmentation comprise those based on the generation of shearing forces, such as sonication and the French pressure cell, and those making use of detergents.

At extremely low concentrations, all detergents exist in water in monomeric form, but at a critical micelle concentration, different for each detergent, specific aggregates (micelles) are formed. This critical micelle concentration is lowest for the non-ionic detergents (e.g. Triton X-100, digitonin), and highest for the more strongly ionized, protein-denaturing detergents (e.g. sodium dodecylsulphate). The larger critical micelle concentration of ionic detergents is caused by electrical repulsion between the ionized hydrophilic groups.

The solubilizing action of detergents is based on their ability to disrupt the hydrophobic interactions existing between protein and lipid in biological membranes. The hydrophobic part of the detergent can compete with lipids for the hydrophobic area of the protein, and the hydrophilic head group of the detergent may confer solubility in water on the resulting protein–detergent complex. At the same time detergents may be expected to form mixed micelles with lipids, thereby creating a situation favorable for separation of the protein from the lipids with which it was originally associated.

The features governing the binding of detergents to proteins have recently become more clear (MAKINO et al., 1973; COLEMAN, 1974; HELENIUS and SIMONS, 1975). At concentrations below the critical micelle concentration, the detergent molecules primarily bind hydrophobically to high-affinity binding sites, which are specific and located at or near the surface of the protein molecule. At higher detergent concentrations, non-specific low-affinity binding comes into play, which ordinarily results in a disruption of the native conformation of the protein and loss of biological activity. In the case of non-ionic detergents this concentration is never attained, since the critical micelle concentration is reached and micelle formation intervenes. Thus many of the proteins solubilized by non-ionic detergents retain a conformation which is able, or potentially able, to ensure biological activity. In the case of denaturing ionic detergents, however, the low-affinity binding sites start to be filled before the critical micelle concentration is reached. A characteristic feature of the low-affinity binding is that it is generally cooperative; the high charge results in unfolding of the protein molecule, allowing the access of further detergent molecules to previously buried hydrophobic side chains. Thus sodium dodecylsulphate, though occasionally used in small amounts for preparing solubilized proteins and lipoprotein complexes, is normally used in larger amounts for size analysis of polypeptides.

Membrane proteins solubilized by means of a detergent may contain large quantities of the detergent. If they are exposed to detergent-free solutions, the detergent separates and many of the proteins polymerize or form insoluble aggregates as a result of interaction between the hydrophobic regions of the protein molecules.

B. Differentiation of the Photosystems

Action spectra have shown that several different pigments and forms of chlorophyll are preferentially active in photosystem I or II (Fork and Amesz, 1969). Since chlorophyll b has been shown to be more active in photosystem II than in photosystem I, the value of the chlorophyll a/b ratio has often been considered an indication of the relative amount of each photosystem in chloroplast fragments. However, the isolation from spinach chloroplasts of a small photosystem II subchloroplast particle which has a high chlorophyll a/b ratio (Vernon et al., 1971, 1972; Wessels et al., 1973) should lead to some concern about the significance of this value.

A more reliable differentiation method is the measurement of light-induced electron transport. A survey of the different types of photoreduction in chloroplasts has been given by Trebst (1972). The DCMU-sensitive photoreduction of DCIP (2,6-dichlorophenolindophenol) is most frequently used to assess the activity of photosystem II. Since subchloroplast particles have normally lost the ability to evolve oxygen, artificial electron donors for photosystem II have to be applied. Yamashita and Butler (1969) have introduced a number of compounds that can be oxidized by photosystem II. One of the most convenient electron donors in chloroplast fragments is 1,5-diphenylcarbazide, as it is effective and specific for photosystem II, and its dark oxidation by DCIP is rather slow (Vernon and Shaw, 1969a, b). The photoreduction of DCIP can be measured spectrophotometrically at 590 nm.

A recently introduced method for the detection of photosystem II in chloroplast fragments is the measurement of the photooxidation of P-680, the reaction center chlorophyll of photosystem II (VAN GORKOM et al., 1974; VAN GORKOM et al., 1975). This reaction is most easily followed by monitoring the reversible, light-induced absorbance change at 435 nm in the presence of ferricyanide. P-680 bleaching is favored by a low pH (5–5.5) and strong actinic light.

Photosystem I activity in chloroplast fragments can be assayed with $NADP^+$ as electron acceptor and reduced DCIP or phenylenediamines as electron donor. The donor is kept reduced by excess ascorbate. $NADP^+$ photoreduction is measured by the increase in absorbance at 340 nm. In photosystem I particles, the photoreduction of $NADP^+$ is usually dependent on the addition of ferredoxin, ferredoxin-$NADP^+$ reductase and plastocyanin. Alternative reactions used for the assay of photosystem I activity are the light-induced oxygen uptake in the presence of reduced DCIP, plastocyanin, and methyl viologen as an autooxidizable electron acceptor, and the disproportionation of 1,5-diphenylcarbazone. The latter reaction was also found to be specifically dependent on photosystem I, and can be measured by the photo-induced absorbance change at 485 nm in the presence of plastocyanin (SHNEYOUR and AVRON, 1971; VERNON and SHAW, 1972).

The reaction requiring the smallest number of intact components around photosystem I is the light-induced absorbance change at 700 nm due to electron transfer from the reaction center, P-700, to the primary electron acceptor (MARSHO and KOK, 1971). Usually some ascorbate is required to bring P-700 to the reduced state at the beginning of the illumination period. Addition of methyl viologen minimizes the back flow of electrons from the reduced primary electron acceptor to P-700. The P-700 content of chloroplast fragments may also be determined from the ferricyanide oxidized-minus-ascorbate reduced difference spectrum.

C. Fragmentation of Chloroplasts

Partial separation of the photosystems has been achieved with the aid of detergents and mechanical devices such as the French pressure cell and sonicators. Many of the results can be explained by a model postulated for the first time by JACOBI and LEHMANN (1968, 1969), in which stroma lamellae contain only photosystem I and grana lamellae contain both photosystem I and photosystem II. Since stroma lamellae are more sensitive to the action of detergents and to French press and sonication treatments than the grana area, differential centrifugation of the disrupted chloroplasts may allow the isolation of a light fraction having only photosystem I activity and a heavy fraction containing both photosystem I and II.

The disruption has to be relatively gentle in order not to break the grana into fragments similar in size to the fragments of stroma lamellae. The fractionation methods are most successful when the grana are stabilized by a high ionic strength in the incubation medium. According to the model, there could be two kinds of photosystem I in chloroplasts. One of these is located in the stroma lamellae, and the other type of photosystem I, which may be closely associated with photosystem II, is located in the grana region.

Much relevant material about the methods used to disrupt the chloroplast membrane system and the structure and properties of the resulting fragments has been reviewed in detail by Boardman (1970), Park and Sane (1971), and Jacobi (this vol. Chap. IV.3). Detailed information about isolation procedures may also be found in Methods in Enzymology, Vol. XXIII (Boardman, 1971; Vernon and Shaw, 1971; Jacobi, 1971; Shibata, 1971). The present article will be concerned mainly with the isolation and properties of photosystem I and photosystem II subchloroplast *particles*.

D. Digitonin Subchloroplast Particles

Treatment of spinach chloroplasts with low concentrations of the non-ionic detergent digitonin disrupts the stroma lamellae to yield vesicles, which can be separated from the grana lamellae by fractional centrifugation between 10,000 and 100,000·g (Wessels and van Leeuwen, 1971; Wessels and Voorn, 1972). The stroma vesicles have only photosystem I activity and are still capable of performing cyclic photophosphorylation. It should be noted that the term "low concentrations of digitonin" also includes a low digitonin/chlorophyll ratio in the incubation medium, as solubilization by detergents is normally even more dependent on the detergent/protein ratio than on the final detergent concentration.

At high concentrations of digitonin not only stroma lamellae but also grana lamellae are disrupted, producing vesicles which partly further disintegrate to yield, among other things, chlorophyll-containing particles. These particles are the main constituent of the 100,000·g supernatant. The vesicles derived from the grana lamellae show both photosystem I and photosystem II activity, and are even able to reduce DCIP with water as electron donor. Though the 100,000·g supernatant is enriched in photosystem I, it has also photosystem II activity, but in this fraction the latter activity is dependent on diphenylcarbazide as electron donor.

Wessels and co-workers (Wessels and van Leeuwen, 1971; Wessels and Voorn, 1972; Wessels et al., 1973) have shown that by density-gradient centrifugation of the 100,000·g supernatant three kinds of chlorophyll-containing particles can be obtained, as illustrated in Figure 1. Below a colorless zone and a wide,

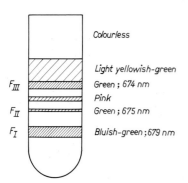

Fig. 1. Location of green bands in gradient tube after density-gradient centrifugation of 100,000 · g supernatant of digitonin-treated chloroplasts (Wessels et al., 1973)

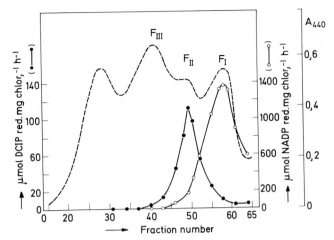

Fig. 2. Photosystem I and II activities of different fractions obtained from zonal rotor after sucrose density-gradient centrifugation of subchloroplast particles (WESSELS et al., 1973). *Broken line:* absorbance at 440 nm

light yellowish-green colored zone of solubilized pigments, the gradient tube exhibits a green band with an absorption maximum at 674 nm (F_{III}), a minor green band with an absorption maximum at 675 nm (F_{II}), and a main, bluish-green colored band with an absorption maximum at 679 nm (F_I). Between the two green bands a light pink band is observed, which contains the cytochromes f and b-6 in a molar ratio of 1/2.

Solubilization of grana lamellae subsequent to the removal of most of the stroma material yields relatively more of the middle green band (F_{II}), and a further increase in yield and purity of this fraction can be accomplished by using a zonal rotor. Figure 2 represents the distribution of the photosystem I and II activities over the different gradient fractions obtained from a zonal rotor after density-gradient centrifugation of subchloroplast particles mainly derived from the grana lamellae. It is seen that the photosystem II activity is concentrated in F_{II} and the photosystem I activity in F_I. Table 1 summarizes some properties of the particles after further purification by means of chromatography on a DEAE-cellulose column. The upper green band of the density gradient (F_{III}), which has no photochemical activity and is characterized by a low chlorophyll a/b ratio, has been considered to be an "accessory" or "light-harvesting" complex.

Table 1. Some properties of purified digitonin subchloroplast particles. The activities are expressed in μmol mg chlorophyll^{-1} h^{-1}. Asc, ascorbate; DPC, diphenylcarbazide; Chl, chlorophyll; PS, photosystem (WESSELS and BORCHERT, 1974)

Fraction	Abs. max. (nm)	Chl a/b	P-700/Chl	PS I activity (Asc-DCIP → NADP$^+$)	PS II activity (DPC → DCIP)
F_I	680	7.5–10	1/110–1/130	1,700–2,000	0
F_{II}	674–675	7–9	0	0	200–300
F_{III}	674–675	1.3–1.5	0	0	0

When stroma vesicles are disrupted by a high concentration of digitonin, density-gradient centrifugation of the $100,000 \cdot g$ supernatant yields only the blue-green band having photosystem I activity (F_I) and the pink band containing the cytochromes f and b-6 (Wessels and Voorn, 1972). The photosystem I particles prepared in this way have been designated stroma photosystem I particles to distinguish them from the photosystem I particles derived from the grana lamellae. Previously some differences were found between grana and stroma photosystem I particles, particularly in chlorophyll a/b ratio and P-700 content (Wessels and Voorn, 1972; Arntzen et al., 1972; Gasanov and French, 1973). Since during the isolation of grana photosystem I, contrary to that of stroma photosystem I, much of the accessory complex is released too, contamination of the former by F_{III} may occur. This results in an apparently lower chlorophyll a/b ratio and P-700 content in grana photosystem I particles. However, further purification of the particles shows unambiguously that there is essentially no significant difference between photosystem I particles derived from stroma and grana lamellae (Wessels and Borchert, 1974; Elgersma and Voorn, 1974).

The observation that chlorophyll b is more active in photosystem II than in photosystem I, and stroma lamellae do not contain F_{III}, may suggest that the accessory complex mainly transfers the absorbed light energy to photosystem II.

In addition to the properties summarized in Table 1, the photosystem I particles are characterized by a high content of β-carotene, the absence of significant amounts of cytochromes, which are separated by the density-gradient centrifugation, and a weak fluorescence (Wessels, 1968, 1969). At liquid-nitrogen temperature, an emission band is observed at about 730 nm. The $NADP^+$ photoreduction is strictly dependent on ferredoxin, ferredoxin-$NADP^+$ reductase and plastocyanin, which serves as the electron donor for photosystem I, and partially dependent on DCIP. After removal of the digitonin by precipitation of the complex with 40% ethanol, the particles are found to contain about 75% protein and 25% lipid, the latter made up of 15% chlorophyll, 2% carotenoids and 8% non-pigment lipids. Electron micrographs show the photosystem I particles to be rod-shaped, with a diameter of 6–6.5 nm and a length of 15–16 nm (Wessels and Voorn, 1972).

Wessels, van Alphen-van Waveren and Voorn (1973) have shown that the photosystem II activity of F_{II} is inhibited by DCMU, but the DCMU-sensitivity of the particles is somewhat less than is usually observed with chloroplasts. The pH optimum of the DCIP photoreduction is 6.0. The fluorescence emission maxima at room temperature and at liquid-nitrogen temperature are located at 681 nm and at 685 nm, respectively. Photosystem II particles contain about 10 mg of protein per mg chlorophyll, and are enriched in P-680, the components Q and C-550, which have been related to the primary electron acceptor of photosystem II, cytochrome b-559 and β-carotene. The molar ratio chlorophyll/cytochrome b-559 is 40–50. The chlorophyll/Q ratio, as determined from the light-induced absorbance change at 320 nm (presumably due to the formation of plastosemiquinone anion), ranges from 20 to 70 in different batches (van Gorkom et al., 1975). At liquid-nitrogen temperature the particles show a photoreduction of C-550, which is accompanied by a photooxidation of cytochrome b-559 if ascorbate is added prior to freezing (Wessels et al., 1973). That this photoreaction is dependent on β-carotene, as shown by Okayama and Butler (1972) and by Sofrova and Bendall (1972), fits in with the observation that the photosystem II particles are enriched in this carotenoid.

The cytochrome b-559 in the particles is mainly present in the oxidized state. This may be due to the fact that digitonin, like a variety of other disruptive treatments (OKAYAMA and BUTLER, 1972; WADA and ARNON, 1971; ERIXON et al., 1972; COX and BENDALL, 1972), converts the high-potential form of cytochrome b-559 to a low-potential, autooxidizable form. Actually most of the cytochrome b-559 is present in the dithionite-reducible form; only a part of it is reducible by ascorbate and none by hydroquinone. Presumably, it is the ascorbate-reducible part of cytochrome b-559 that can be oxidized by the primary electron donor of photosystem II at 77 K. The photosystem II particles show no photoreduction of cytochrome b-559 under a variety of conditions.

The accessory complex F_{III} contains a relatively high concentration of chlorophyll b and xanthophylls, but only a little β-carotene (WESSELS, 1968). The F_{III} particles are highly fluorescent; at liquid-nitrogen temperature the fluorescence emission spectrum exhibits a maximum at 682 nm and a shoulder at 695 nm (WESSELS and BORCHERT, 1974). Electron micrographs show the diameter of the particles to be about 6 nm (WESSELS and VAN LEEUWEN, 1971; WESSELS and VOORN, 1972).

E. Triton Subchloroplast Particles

Particles with either photosystem I or photosystem II activity have also been prepared by disruption of spinach chloroplasts with high concentrations of Triton X-100. In the extensive studies of VERNON and co-workers (VERNON et al., 1971, 1972; VERNON and SHAW, 1971) the fragments are separated by an initial centrifugation at 10,000·g, followed by successive centrifugations at 144,000·g for 1, 5 and, after dilution with an equal volume of water, 10 h. The sediment obtained after 1 h centrifugation at 144,000·g is enriched in photosystem II and is designated TSF-2. The light fragments which sediment after 10 h centrifugation at 144,000·g, and have been named TSF-1, closely resemble the digitonin photosystem I particles in most of their properties.

TSF-1 particles are very active in photosystem I reactions; the photoreduction of $NADP^+$ with the donor system ascorbate-DCIP requires ferredoxin, ferredoxin-$NADP^+$ reductase and plastocyanin. There is an enrichment of TSF-1 in chlorophyll a (chlorophyll a/b = 5.7), β-carotene, plastoquinone and P-700 (the ratio of chlorophyll/P-700 is approximately 100). The particles contain cytochrome f and cytochrome b-6, but these cytochromes may co-sediment with TSF-1 and not be an integral part of the photoactive particle (VERNON et al., 1969). The absorption maximum of TSF-1 is at 675 nm (VERNON et al., 1966), which is lower than that of digitonin photosystem I particles.

If the treatment with Triton X-100 is carried out on carotenoid-extracted, freeze-dried chloroplasts, small particles are obtained which are enriched in P-700 relative to total chlorophyll, showing a ratio of chlorophyll/P-700 of 30 (VERNON and SHAW, 1971). These particles are designated HP-700 particles. As compared to TSF-1, HP-700 particles show a lower chlorophyll a/b ratio (4.0) and a marked decrease in the ratio of the fluorescence intensities at 735 nm to 685 nm at liquid-nitrogen temperature. The absorption maximum is 676 nm. The particles contain both cytochrome f and cytochrome b-6. Their activity for $NADP^+$ photoreduction is very low, however, and the protein content (3.2 mg per 100 nmol of total chlorophyll) is more than three times that of TSF-1. The origin of the HP-700 particle is not clear, but the high protein content suggests that a lot of chlorophyll has been removed during the isolation procedure, and that the high P-700/chlorophyll ratio of the particle is due to P-700 being relatively more resistant to solubilization than the bulk pigment.

The fractions prepared by the action of Triton X-100 on chloroplast membranes have characteristic shapes when examined in the electron microscope. TSF-2 appears as a relatively smooth membrane, while TSF-1 preparations consist of small rod-shaped particles which show a marked tendency to aggregate. The individual rods are 7–8 nm in diameter. The HP-700 preparations also contain small particles which are rod-to-ellipsoidal in shape (about 15·6 nm).

When the TSF-2 fraction is extracted with a solution 0.25 M in sucrose and 0.02 M in Tris buffer, pH 8.0, some chlorophyll-containing material is released. By subjecting the 105,000·g supernatant of the suspension to sucrose density-gradient centrifugation, or passing it through a Bioglass 2,500 column, a particle, designated TSF-2a, has been prepared which is very active in DCIP photoreduction with diphenylcarbazide and shows no photosystem I activity (Vernon et al., 1971, 1972). With preparations made from summer-grown spinach, and using the adsorption step with Bioglass, activities of 1,000 µmol DCIP reduced mg chlorophyll^{-1} h^{-1} can be obtained.

In many respects the TSF-2a particles resemble the digitonin photosystem II particles (F$_\infty$). The absorption and fluorescence spectra and the content of cytochrome b-559, which is tightly bound to the particles, are similar. Both types of particles have a high chlorophyll a/b ratio, are devoid of P-700, and contain relatively more β-carotene than xanthophylls. In TSF-2a, the DCMU-inhibition of the DCIP photoreduction is also somewhat less than in whole chloroplasts, whereas the photoreduction of ferricyanide by diphenylcarbazide is even resistant to this inhibitor. It is conceivable that in TSF-2a and F$_\infty$ ferricyanide is able to interact with the particle before the DCMU site of action. Electron microscopic studies of the TSF-2a fraction show the presence of discrete, small particles, approximately 10 nm in diameter, which have a great tendency to aggregation. Ke et al. (1974) have detected C-550 and P-680 in TSF-2a by light-induced difference spectra at 77 K. The presence of P-680 has further been corroborated by a free-radical signal revealed by low-temperature EPR spectroscopy. The enrichment in P-680, C-550 and cytochrome b-559 is five- to ten-fold relative to unfractionated chloroplasts.

Digitonin and Triton-fractionated photosystem II particles differ, however, in the photochemical activity of cytochrome b-559, even though this cytochrome is present largely in the oxidized form in both types of particles. In TSF-2a no photooxidation of cytochrome b-559 is observed at liquid-nitrogen temperature even when it is pre-reduced chemically. At room temperature the cytochrome undergoes a DCMU-resistant photoreduction, both in the presence and in the absence of an electron donor such as diphenylcarbazide (Ke et al., 1972). The different behavior of cytochrome b-559 in the two particles may be caused by a difference in location of this compound with respect to the reaction center.

By recombination of TSF-1 and TSF-2a particles in the presence of lecithin, Ke and Shaw (1972) have been able to reconstitute light-induced electron transport from diphenylcarbazide to $NADP^+$. The reconstituted photochemical activity requires plastocyanin and is sensitive to DCMU. $NADP^+$-photoreduction rates of up to 67 µmol mg $chlorophyll^{-1}$ h^{-1} have been obtained. Recombination of highly purified digitonin subchloroplast particles has not been successful as yet, though some reconstitution can be obtained with impure preparations (WESSELS and BORCHERT, 1974). It is possible that purification of the digitonin particles involves the loss of a factor that is necessary for reconstitution of this photochemical activity.

F. Protein Composition of Subchloroplast Particles

Currently work is being carried out on the protein composition of fractions representing the photosystems (REMY, 1971; LEVINE et al., 1972; KLEIN and VERNON, 1974a, 1974b; ANDERSON and LEVINE, 1974a; WESSELS and BORCHERT, 1974; NOLAN and PARK, 1975). Obvious differences can be seen in the electrophoresis patterns of these fractions and further characterization of the different polypeptides is in progress.

The main components observed after sodium dodecylsulphate-polyacrylamide gel electrophoresis of washed, acetone-extracted chloroplast lamellae are found in the 20–30 and 60 kilodalton ranges. It has been shown that the chloroplast membrane polypeptides with molecular weights in the 60 kilodalton range are associated with a fraction enriched in photosystem I. Polypeptides in the 20–30 kilodalton range, labeled IIa, b and c by LEVINE et al. (1972), predominate in fractions enriched in photosystem II. ANDERSON and LEVINE (1974a) have shown that the polypeptides IIb and c are lacking in the unstacked lamellae of chloroplasts from chlorophyll-deficient mutant strains of barley and pea, and from the bundle-sheath cells of maize. Since both the barley and pea mutant chloroplasts and the maize bundle-sheath chloroplasts possess photosystem II activities, it has been proposed that these polypeptides are required for membrane stacking in higher plant chloroplasts. Recently it has been demonstrated that in digitonin photosystem II particles (F_{II}) a prominent band is observed in the region characteristic of polypeptide IIa. The polypeptides IIb and c, on the other hand, which predominate in the accessory complex F_{III}, are missing in F_{II} (WESSELS and BORCHERT, 1974). These results suggest that polypeptide IIa is associated with photosystem II activity, and that the accessory complex may be required for stacking.

When chlorophyll-containing chloroplast lamellae are treated with an anionic detergent such as sodium dodecylsulphate or sodium dodecyl-benzenesulphonate, three green bands are obtained upon electrophoresis in sodium dodecylsulphate polyacrylamide gels (OGAWA et al., 1966; THORNBER et al., 1967; THORNBER and OLSON, 1971). The green zone of lowest electrophoretic mobility and the zone of intermediate mobility are termed complex I and complex II, respectively. The zone of fastest mobility, termed complex III, is a mixture of free pigments solubilized by the detergent. Though the complexes I and II are photochemically inactive, it has been suggested that they are related to photosystem I and II, respectively. If subchloroplast fragments prepared by the action of digitonin are treated with sodium dodecyl-benzenesulphonate, the heavy digitonin fraction is found to be enriched in complex II and the light digitonin fraction in complex I. Complex I has a high, and complex II a low chlorophyll a/b ratio. Recently, a P-700–chlorophyll a–protein complex has been purified from several higher plants by hydroxylapatite chromatography of Triton X-100-dissociated chloroplast membranes (SHIOZAWA et al., 1974). This complex exhibits a red wavelength maximum at 677 nm and a chlorophyll/P-700 ratio of 40–50, and yields complex I upon sodium dodecylsulphate-polyacrylamide gel electrophoresis. Complex I is also obtained by gel electrophoresis of digitonin photosystem I particles (F_I) (WESSELS, unpublished). These experiments indicate that complex I is actually derived from photosystem I.

THORNBER and HIGHKIN (1974) have shown that complex II, which has been previously termed photosystem II chlorophyll-protein, is not present in a mutant of barley lacking chlorophyll b. As this mutant can live photoautotrophically, a more appropriate name, light-harvesting chlorophyll a/b protein, is proposed for this pigment-protein complex. Since the formation of appressed lamellae is decreased in the barley mutant chloroplasts, it is suggested that this complex is required for membrane stacking. Gel electrophoresis of the digitonin subchloroplast particle F_{III} yields two chlorophyll-containing zones, a fast migrating zone containing free pigment and a zone of intermediate mobility characteristic of complex II (WESSELS and BORCHERT, 1974). This result indicates that the accessory complex F_{III} may be identical with the light-harvesting chlorophyll a/b protein described by THORNBER and HIGHKIN, and is consistent with the view that this complex is required for membrane stacking in chloroplasts. ANDERSON and LEVINE (1974b) have recently compared the polypeptide profiles obtained on electrophoresis of green, lipid-containing chloroplast-membrane preparations with those of lipid-extracted membrane preparations. Their data also show that the polypeptides IIb and c are part of the light-harvesting chlorophyll-protein complex.

If digitonin photosystem II particles (F_{II}) are subjected to sodium dodecylsulphate-polyacrylamide gel electrophoresis, most of the pigment is solubilized by the detergent and only a weakly green-colored zone is observed with a mobility intermediate to those of the complexes I and II (WESSELS, unpublished).

References

Anderson, J.M., Levine, R.P.: Biochim. Biophys. Acta **333**, 378–387 (1974a)
Anderson, J.M., Levine, R.P.: Biochim. Biophys. Acta **357**, 118–126 (1974b)
Arntzen, C.J., Dilley, R.A., Peters, G.A., Shaw, E.R.: Biochim. Biophys. Acta **256**, 85–107 (1972)
Boardman, N.K.: Ann. Rev. Plant Physiol. **21**, 115–140 (1970)
Boardman, N.K.: Methods Enzymol. **23A**, 268–276 (1971)
Coleman, R.: Biochem. Soc. Transact. **2**, 813–816 (1974)
Cox, R.P., Bendall, D.S.: Biochim. Biophys. Acta **283**, 124–135 (1972)
Elgersma, O., Voorn, G.: In: Proc. 3rd Intern. Congr. Photosynthesis. Avron, M. (ed.). Amsterdam: Elsevier, 1975, Vol. III, pp. 1943–1949
Erixon, K., Lozier, R., Butler, W.L.: Biochim. Biophys. Acta **267**, 375–382 (1972)
Fork, D.C., Amesz, J.: Ann. Rev. Plant Physiol. **20**, 305–328 (1969)
Gasanov, R.A., French, C.S.: Proc. Nat. Acad. Sci. U.S. **70**, 2082–2085 (1973)

Gorkom van, H.J., Pulles, M.P.J., Wessels, J.S.C.: Biochim. Biophys. Acta **408**, 331–339 (1975)

Gorkom van, H.J., Tamminga, J.J., Haveman, J., van der Linden, I.K.: Biochim. Biophys. Acta **347**, 417–438 (1974)

Helenius, A., Simons, K.: Biochim. Biophys. Acta **415**, 29–79 (1975)

Jacobi, G.: Methods Enzymol. **23A**, 289–296 (1971)

Jacobi, G., Lehmann, H.: Z. Pflanzenphysiol. **59**, 457–476 (1968)

Jacobi, G., Lehmann, H.: In: Progress in Photosynthesis Research. Metzner, H. (ed.). Tübingen: Laupp, 1969, Vol. I, pp. 159–173

Ke, B., Sahu, S., Shaw, E., Beinert, H.: Biochim. Biophys. Acta **347**, 36–48 (1974)

Ke, B., Shaw, E.R.: Biochim. Biophys. Acta **275**, 192–198 (1972)

Ke, B., Vernon, L.P., Chaney, T.H.: Biochim. Biophys. Acta **256**, 345–357 (1972)

Klein, S.M., Vernon, L.P.: Photochem. Photobiol. **19**, 43–49 (1974a)

Klein, S.M., Vernon, L.P.: Plant Physiol. **53**, 777–778 (1974b)

Laetsch, W.M.: Ann. Rev. Plant Physiol. **25**, 27–52 (1974)

Levine, R.P., Burton, W.G., Duram, H.A.: Nature New Biol. **237**, 176–177 (1972)

Makino, S., Reynolds, J.A., Tanford, C.: J. Biol. Chem. **248**, 4926–4932 (1973)

Marsho, T.V., Kok, B.: Methods Enzymol. **23A**, 515–522 (1971)

Nolan, W.G., Park, R.B.: Biochim. Biophys. Acta **375**, 406–421 (1975)

Ogawa, T., Obata, F., Shibata, K.: Biochim. Biophys. Acta **112**, 223–234 (1966)

Okayama, S., Butler, W.L.: Plant Physiol. **49**, 769–774 (1972)

Park, R.B., Sane, P.V.: Ann. Rev. Plant Physiol. **22**, 395–430 (1971)

Remy, R.: FEBS Lett. **13**, 313–317 (1971)

Shibata, K.: Methods Enzymol. **23A**, 296–302 (1971)

Shiozawa, J.A., Alberte, R.S., Thornber, J.P.: Arch. Biochem. Biophys. **165**, 388–397 (1974)

Shneyour, A., Avron, M.: Biochim. Biophys. Acta **253**, 412–420 (1971)

Sofrova, D., Bendall, D.S.: In: Proc. 2nd Intern. Congr. Photosynthesis. Forti, G., Avron, M., Melandri, A. (eds.). The Hague: Dr. Junk, 1972, Vol. I, pp. 561–567

Thornber, J.P., Gregory, R.P.F., Smith, C.A., Bailey, J.L.: Biochemistry **6**, 391–396 (1967)

Thornber, J.P., Highkin, H.R.: Europ. J. Biochem. **41**, 109–116 (1974)

Thornber, J.P., Olson, J.M.: Photochem. Photobiol. **14**, 329–341 (1971)

Trebst, A.: Methods Enzymol. **24**, 146–165 (1972)

Vernon, L.P., Ke, B., Mollenhauer, H.H., Shaw, E.R.: In: Progress in Photosynthesis Research. Metzner, H. (ed.). Tübingen: Laupp, 1969, Vol. I, pp. 137–148

Vernon, L.P., Klein, S., White, F.G., Shaw, E.R., Mayne, B.C.: In: Proc. 2nd Intern. Congr. Photosynthesis. Forti, G., Avron, M., Melandri, A. (eds.). The Hague: Dr. Junk, 1972, Vol. I, pp. 801–812

Vernon, L.P., Shaw, E.R.: Biochem. Biophys. Res. Commun. **36**, 878–884 (1969a)

Vernon, L.P., Shaw, E.R.: Plant Physiol. **44**, 1645–1649 (1969b)

Vernon, L.P., Shaw, E.R.: Methods Enzymol. **23A**, 277–289 (1971)

Vernon, L.P., Shaw, E.R.: Plant Physiol. **49**, 862–863 (1972)

Vernon, L.P., Shaw, E.R., Ke, B.: J. Biol. Chem. **241**, 4101–4109 (1966)

Vernon, L.P., Shaw, E.R., Ogawa, T., Raveed, D.: Photochem. Photobiol. **14**, 343–357 (1971)

Wada, K., Arnon, D.I.: Proc. Nat. Acad. Sci. U.S. **68**, 3064–3068 (1971)

Wessels, J.S.C.: Biochim. Biophys. Acta **153**, 497–500 (1968)

Wessels, J.S.C.: In: Progress in Photosynthesis Research. Metzner, H. (ed.). Tübingen: Laupp, 1969, Vol. I, pp. 128–136

Wessels, J.S.C., van Alphen-van Waveren, O., Voorn, G.: Biochim. Biophys. Acta **292**, 741–752 (1973)

Wessels, J.S.C., Borchert, M.T.: In: Proc. 3rd Intern. Congr. Photosynthesis. Avron, M. (ed.). Amsterdam: Elsevier, 1975, Vol. I, pp. 473–484

Wessels, J.S.C., van Leeuwen, M.J.F.: In: Energy Transduction in Respiration and Photosynthesis. Quagliariello, E., Papa, S., Rossi, C.S. (eds.). Bari: Adriatica Editrice, 1971, pp. 537–550

Wessels, J.S.C., Voorn, G.: In: Proc. 2nd Intern. Congr. Photosynthesis. Forti, G., Avron, M., Melandri, A. (eds.). The Hague: Dr. Junk, 1972, Vol. I, pp. 833–845

Yamashita, T., Butler, W.L.: Plant Physiol. **44**, 435–438 (1969)

5. The Organization of Chlorophyll in vivo[1]

J.P. Thornber and R.S. Alberte

A. Introduction

It has become increasingly apparent during the last decade that chlorophyll is organized into several specific pigment-protein complexes in a photosynthetic organism. Some of these complexes function as light-harvesting components feeding their absorbed light energy to other chlorophyll-protein(s), the photochemical reaction center complex(es) (cf. Thornber, 1975). This knowledge has largely arisen from the application of known membrane fractionation techniques to chlorophyll-containing membranes. Detergents, particularly SDS (Sodium dodecyl sulfate), have been added to photosynthetic membranes, and chlorophyll-protein complexes of a size expected for a fully molecularly dispersed protein have been solubilized. The necessity to titrate accurately the plant membranes with anionic detergents to achieve complete solubilization of the membranes while maintaining specific associations of chlorophyll and protein has been found to be an essential consideration. From such solubilized extracts, homogenous chlorophyll-proteins have been isolated, thereby permitting the specific complexes to be characterized biochemically and biophysically. The aim of this article is to summarize briefly our present knowledge of the major chlorophyll-proteins in plants. A more extensive treatise on the early development of the subject can be found in Kupke and French (1960), and a much more detailed account of developments to the present time in Thornber (1975).

B. Existence of Multiple Chlorophyll-Proteins in Higher Plants

Ogawa et al. (1966) and Thornber et al. (1966, 1967) used anionic detergents to obtain chloroplast membrane extracts which upon electrophoresis in SDS-containing polyacrylamide gels were resolved into three major chlorophyll-containing zones. The two zones of lower electrophoretic mobilities were shown to be complexes of chlorophyll and protein, whereas the fastest moving zone was found to be detergent-complexed, protein-free chlorophyll and carotenoid (free pigment). It was also observed that the two chlorophyll-proteins not only accounted for a large proportion of the total photosynthetic pigment in the organism, but that the majority of the protein in chloroplast lamellae was involved in the organization

---

[1] This review was completed in May 1975. Further and more recent information is available in J.P. Thornber, R.S. Alberte, F.A. Hunter, J.A. Shiozawa and K. Kan, Brookhaven Symp. Biol. **28**, in press (1977).

of this chlorophyll. Differences between the pigment and amino acid compositions (cf. THORNBER, 1975) of the two pigment-proteins led to the conclusion that they were two distinct membrane-bound entities. Following these few basic observations on the organization of chlorophyll in vivo, much further data on these two complexes have accumulated during recent years and will be described in the next section.

C. P-700-Chlorophyll a-Protein

One of the two complexes is now termed the P-700-chlorophyll a-protein (DIETRICH and THORNBER, 1971). Recent work has shown this to be a more appropriate name than the others by which it has been variously known — (pigment-protein) complex I, component I, photosystem I chlorophyll-protein or CPI.

I. Isolation

The complex can be isolated as a stable, homogenous and functional entity by hydroxylapatite chromatography of (a) SDS extracts of blue-green algae (THORNBER, 1969), or (b) Triton X-100 extracts of eucaryotic photosynthetic organisms (SHIOZAWA et al., 1974). Addition of SDS to the Triton-prepared complex changes its spectrum and P-700 content; thus, the chlorophyll-protein observed upon polyacrylamide gel electrophoresis of SDS extracts of eucaryotic plant chloroplasts is a slightly altered form of this component (SHIOZAWA et al., 1974).

II. Characteristics

1. Chemical Composition

Chlorophyll a and β-carotene (molar ratio 20–30/1) account for essentially all of the pigment in the complex. Although galacto- and phospholipids are present in trace amounts, these probably do not represent specific lipid associations with the complex; however, quinones appear to be an integral part of the complex. The amino acid composition shows that the protein moiety contains a high proportion of non-polar amino acid residues. There are 14 chlorophyll a molecules per 90,000 g protein in the isolated material. Further details of the chemical composition can be found in THORNBER (1975).

2. Physical Properties

The characteristic absorption spectrum of the complex (Fig. 1) with its red wavelength maximum at 677 nm is a useful criterion for the spectral purity of a preparation of this chlorophyll-protein. The following spectral forms of chlorophyll a have been observed in the complex: Chl a 662, 669, 677 and 686 (BROWN et al.,

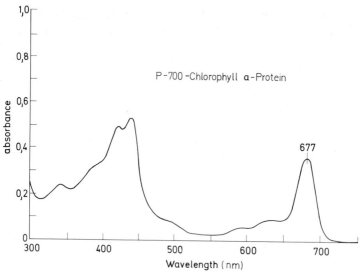

Fig. 1. Room temperature absorption spectrum of P-700-chlorophyll a-protein of tobacco chloroplasts

1975). This chlorophyll-protein is particularly enriched in Chl a 686 compared to intact chloroplast lamellae. Even longer wavelength forms are sometimes encountered in preparations of this complex (cf. Thornber, 1975). Fluorescence emission spectra (77° K) show a peak at 676 nm with relatively little emission beyond 700 nm. The sedimentation coefficient of the complex is 9 S.

3. Size

The molecular size of the native complex is 110,000 daltons (Thornber, 1975). There are 14 chlorophyll a molecules for every 110,000 dalton unit. However, if the ratio of chlorophyll a/P-700 in the complex is 40/1, then not every 110,000 dalton unit can contain a P-700 entity. Rationalization of these data has led to a proposed model in which three essentially identical 110,000 dalton units contain a total of about 40 chlorophyll a molecules. One of these three units contains, or has attached to it, a P-700 entity. If the P-700 is located within one of every three units, this may prevent P-700 from being isolated in a more enriched form than occurs in the P-700-chlorophyll a-protein. But, alternatively, if P-700 is located in a different yet distinct protein that is tightly attached to one of every three units, then hope remains that P-700 can be isolated in the same highly enriched state as has been accomplished for the bacterial reaction center.

III. Function

The isolated complex shows light-induced oxidation and dark reduction of P-700 (Dietrich and Thornber, 1971; Shiozawa et al., 1974). The close stoichiometric equivalence of P-700 and β-carotene in the chlorophyll-protein indicates that the β-carotene in this component may function to protect P-700 from photochemical damage. The P-700-chlorophyll a-protein preparation provides P-700 in the most enriched form currently obtainable. In higher plants and green algae the complex represents about a 10-fold enrichment of the reaction center. In most organisms studied the molar ratio of chlorophyll/P-700 in the complex is 40/1. This component thus represents the heart of photosystem I, and is found, therefore, to occur in relatively high concentrations in isolated photosystem I fractions. Such fractions, however, also contain many other components (redox and pigmented proteins) in addition to the P-700-chlorophyll a-protein.

IV. Occurrence

The complex has been observed in mono- and dicotyledonous angiosperms, gymnosperms, all the algal phyla, and in blue-green algae (cf. THORNBER, 1975). It is apparently ubiquitous in photosynthetically competent plants (cf. BROWN et al., 1975). But interestingly, several mutant plants which lack or may lack P-700 also lack this chlorophyll-protein (cf. THORNBER, 1975). The complex represents 4–30% of the total chlorophyll in all organisms examined with 10–18% being typical of its content in higher plants (BROWN et al., 1975).

D. The Light-Harvesting Chlorophyll a/b-Protein

This component is the other chlorophyll-protein complex observed in SDS extracts of green plants. THORNBER and HIGHKIN (1974) proposed the above name for this major chlorophyll-protein complex because it provided a more accurate description of its function than all of its former names—(pigment-protein or chlorophyll-protein) complex II, component II, the photosystem II chlorophyll-protein or CP II.

I. Isolation

A homogenous, stable preparation of the complex is obtained by hydroxylapatite chromatography of SDS extracts of chloroplast lamellae of chlorophyll b-containing organisms (KUNG and THORNBER, 1971; KAN and THORNBER, 1976).

II. Characteristics

1. Chemical Composition

The complex contains equimolar quantities of chlorophylls a and b. Every carotenoid (chlorophyll/carotenoid=3–5/1; mol/mol) in whole chloroplasts occurs in the isolated chlorophyll-protein but not in the same stoichiometric ratio; lutein and

β-carotene generally represent the major proportion of the carotenoid. Galacto- and phospholipids occur in trace amounts; again, probably indicating a non-specific lipid association. Its amino acid composition shows a high proportion of non-polar amino acid residues as would be expected for a membrane-bound protein. A more detailed description of the chemical composition of this complex can be found in the review by THORNBER (1975).

2. Physical Properties

The room temperature absorption spectrum of the complex shows a double peak in the red wavelength region with maxima at 672 and 653 nm (Fig. 2), as would be expected for a component containing equimolar quantities of chlorophylls a and b. Several spectral forms (Chl b 650, Chl a 662, 670, 677 and 684) have been detected in the complex; Chl a 684 is very much reduced and Chl b 650 is greatly enriched in comparison to the proportion of these spectral forms present in whole chloroplasts (BROWN et al., 1975). The complex exhibits a single fluores- cence emission peak at 77 K around 680 nm. The component sediments with a boundary of 2.3–3.1 S (THORNBER, 1975).

3. Size

The molecular size of the native pigment-protein complex has been reported by several groups to be 27,000–35,000 daltons (cf. THORNBER, 1975); its apoprotein has a size 3,000 daltons lower than the complex (GENGE et al., 1974). The most recent data indicate that the complex is very likely composed of a single polypeptide chain (cf. THORNBER, 1975) of 29,500 daltons and six chlorophyll molecules (3 chlo- rophyll a + 3 chlorophyll b) (KAN and THORNBER, 1976).

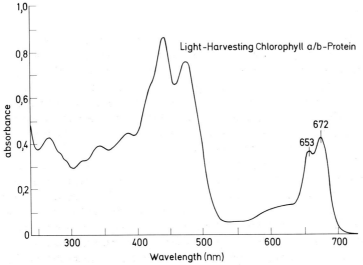

Fig. 2. Room temperature absorption spectrum of light-harvesting chlorophyll a/b-protein of green alga, *Chlamydomonas reinhardii*

III. Function

The observation (ANDERSON and LEVINE, 1974; GENGE et al., 1974; THORNBER and HIGHKIN, 1974) that this complex, which contains the majority of the pigment in a green plant, was absent from a photosynthetically competent barley mutant indicated that this chlorophyll-protein can function only as the *major* light-harvesting component of the photosynthetic apparatus, and not as a component of the photosynthetic electron transport chain. It is thought (ALBERTE and THORNBER, 1974; BROWN et al., 1975; cf. also GENGE et al., 1974; THORNBER, 1975) that *all* of the chlorophyll b in an organism together with an equivalent amount of chlorophyll a occurs in this complex. Thus the percentage of total chlorophyll associated with this complex in a plant or subchloroplastic fraction can be readily calculated from the chlorophyll a/b ratio of the material. It has been postulated that this chlorophyll-protein also functions to maintain appressed lamellae in contact with each other, while others have proposed that the complex is responsible for causing grana formation; however, it now appears very unlikely in the light of recent observations (summarized in THORNBER, 1975) that this latter proposal is correct.

IV. Occurrence

The complex is almost certainly present in all chlorophyll b-containing plants. It has been observed in many groups of angiosperms, gymnosperms and green algae in which it has generally accounted for 40–60% of the total chlorophyll.

E. Other Chlorophyll-Proteins in the Plant Kingdom

I. Detergent-Soluble Complexes

At the most 75% of the chlorophyll in green plants can be accounted for by pigment in the two major SDS-soluble chlorophyll-proteins, while in some algal phyla only 4% can be accounted for in these components. The organization of the uncategorized chlorophyll is for the most part unknown: in most plants such chlorophyll is found in the free pigment zone after electrophoresis of SDS extracts (BROWN et al., 1975). The presence of other chlorophyll-proteins has been reported. In some cases the additional electrophoretic zones were almost certainly polymers of the two known SDS-soluble complexes (cf. THORNBER, 1975). GENGE et al. (1974) have presented some of the only evidence for the existence of a third, distinct chlorophyll-protein in higher plants and green algae — its size is intermediate between those of the two known pigment-proteins. This third component is less stable to anionic detergents than the other two complexes, and consequently during electrophoresis pigments are removed from this component, and subsequently move into the free pigment zone. It therefore appears very probable that most of the free pigment observed in SDS extracts arises from complete dissociation of this third complex rather than from partial dissociation of the other two components as originally thought (cf. THORNBER, 1975).

II. Water-Soluble Complexes

Certain groups of higher plants contain low concentrations of a water-soluble chlorophyll a- and b-containing complex. The presence of such a component has been known for some time (cf. TAKAMIYA, 1971), but its function remains an enigma. Some dinoflagellates contain a water-soluble peridinin-chlorophyll a-protein. This component occurs in a greater concentration than the other water-soluble entity, and it apparently functions in the antenna of photosystem II (cf. THORNBER, 1975).

F. Content of Chlorophyll-Proteins in Higher Plants

In an SDS extract of higher plant chloroplast lamellae 12–18% of the total chlorophyll is accounted for in the P-700-chlorophyll a-protein complex while 48–55% is typically observed in the light-harvesting chlorophyll a/b-protein; the remaining 33% is found to be free pigment.

Naturally occurring examples of alterations in the proportions of the chlorophyll-proteins in photosynthetic membranes have been observed: chloroplasts in the bundle sheaths of C-4 plants have a much reduced content of the light-harvesting chlorophyll a/b-protein compared to mesophyll chloroplasts of C-3 and C-4 plants (GENGE et al., 1974). Also, grana lamellae contain a preponderance of this chlorophyll-protein while the majority of the chlorophyll in stroma lamellae is associated with the P-700-chlorophyll a-protein (cf. BROWN et al., 1975).

Environmental factors and mutations also influence the proportion of at least one of the chlorophyll-proteins in the chloroplast: light intensity, salinity conditions, tissue water status, mineral deficiency and some mutations alter the lamellar content of the major chlorophyll a/b-protein. It thus appears that the plasticity of this component is a highly adaptive feature of the photosynthetic apparatus (cf. THORNBER, 1975).

G. Summary and Concluding Remarks

It has long been recognized (cf. KUPKE and FRENCH, 1960) that the conversion of light energy into chemical energy in photosynthetic organisms will be fully understood only when the molecular architecture describing the relationship of chlorophyll molecules with each other and with other substances in the membrane is known. During the last decade studies directed at an elucidation of these relationships have substantiated the early notion that chlorophyll is conjugated with protein(s) in photosynthetic membranes. The task that has been undertaken is that of obtaining the smallest possible, yet homogenous chlorophyll-containing entities that retain properties (spectrum and function) which reflect some of those characteristics observed in intact membrane systems. Much of the success in this area has come from the use of ionic detergents to solubilize the membranes, and from the use of hydroxylapatite chromatography of detergent extracts to isolate such homogenous chlorophyll-protein complexes. Other methods for dissociating and fractionating photosynthetic membranes (e.g. non-ionic detergents, physical methods) yield, for the most part, heterogenous preparations of chlorophyll-proteins in which any one such complex is enriched, but in which other pigment-protein complexes or colorless proteins in addition to lipids are also present.

The earlier skepticism that chlorophyll-protein complexes might be formed by a non-specific detergent-induced co-solubilization of chlorophyll and thylakoid membrane protein(s) has now been essentially eliminated. The fact that the two membrane-bound complexes which have been extensively described in this article exhibit spectral characteristics (absorption, fluorescence, circular dichroism), and spectral forms that precisely reflect these same characteristics in intact chloroplast lamellar membranes argues very strongly that little or no changes have occurred within these pigment-protein complexes during their isolation. But perhaps the most persuasive evidence is that not only have mutants been found which lack one but not the other chlorophyll-protein, but in each instance the effect of the mutation on photosynthetic activity of the organism is exactly what would be predicted if either the P-700-chlorophyll a-protein or the light-harvesting chlorophyll a/b-protein were absent.

No knowledge is available on the three-dimensional organization of the chlorophylls within their protein framework, or on the interaction between similar and dissimilar chlorophyll-protein moieties, or between pigment-proteins and other membrane components. But, obviously, the various chlorophyll-proteins must be arranged in a specific and ordered manner. The authors favor an arrangement similar to the detailed model of SEELY (1973) or the tripartite model of BUTLER and KITAJIMA (1975). In other words, we support the idea of one continuous array of pigments in a photosynthetic unit. The chlorophylls in the major light-harvesting chlorophyll a/b-protein feed energy preferentially to the reaction center of photosystem II and therefore are most probably located adjacent to this reaction center component. Energy absorbed in this component can "spill over" to the reaction center of photosystem I to maintain an even distribution of energy to the reaction centers of both photosystems. Such an organization, rather than a separate package model, would explain how the amount of the chlorophyll a/b-protein can be varied substantially without detriment to the functioning of the photosynthetic apparatus, and would explain why isolated photosystem II fractions are particularly enriched in this component. We envisage that the third chlorophyll-protein proposed feeds energy preferentially to the reaction center of photosystem I. Little can be added at this time to the description in KIRK (1971) on the site of location of the chlorophyll-proteins in the thylakoid membrane; the only new data are on the differing concentrations of the two major chlorophyll-proteins in grana and stroma lamellae.

SDS-polyacrylamide gel electrophoresis has revealed much about the organization of chlorophylls in plants, and such a technique is a rapid and accurate method for obtaining data on the proportions of the known chlorophyll-proteins in plants or fractions thereof. But, in order to account in full for the organization of *all* the chlorophylls in photosynthetic membranes, it is essential that new techniques be applied or developed for dissociating the chlorophyll-proteins from the membranes, and for maintaining them in a stable form during their isolation as homogeneous entities. It is obvious that apart from the two major chlorophyll-proteins, other chlorophyll-protein complexes (e.g. the third, proposed chlorophyll-protein and the reaction center of photosystem II) are much less stable to ionic detergents; consequently, we know virtually nothing of the organization of 33% of the chlorophyll in higher plants and green algae, and of 90% of the chlorophyll in other algal phyla. Although it is possible that some of the photosynthetic pigment in an organism may not be conjugated with protein, an alternative arrangement would make it very difficult to envisage how biosynthesis and deposition of such chlorophylls into the membrane would be controlled.

Acknowledgements. The senior author's research since 1971 referred to in this article, and the preparation of this article, were largely supported by funds from the National Science Foundation, Grant number GB 31207.

References

Alberte, R.S., Thornber, J.P.: Plant Physiol. **53**, 47 (1974)

Anderson, J.M., Levine, R.P.: Biochim. Biophys. Acta **333**, 378–387 (1974)

Brown, J.S., Alberte, R.S., Thornber, J.P.: In: Proc. 3rd Intern. Congr. Photosynthesis. Avron, M. (ed.). Amsterdam: Elsevier, pp. 1951–1962, 1975

Butler, W.L., Kitajima, M.: In: Proc. 3rd Intern. Congr. Photosynthesis. Avron, M. (ed.). Amsterdam: Elsevier, pp. 13–24, 1975

Dietrich, W.E., Jr., Thornber, J.P.: Biochim. Biophys. Acta **245**, 482–493 (1971)

Genge, S., Pilger, D., Hiller, R.G.: Biochim. Biophys. Acta **347**, 22–30 (1974)

Kan, K., Thornber, J.P.: Plant Physiol. **57**, 47–52 (1976)

Kirk, J.T.O.: Ann. Rev. Biochem. **40**, 161–196 (1971)

Kung, S.D., Thornber, J.P.: Biochim. Biophys. Acta **253**, 285–289 (1971)

Kupke, D.W., French, C.S.: In: Encyclopedia of Plant Physiology. Berlin: Springer, 1960, Vol. V, pp. 298–322

Ogawa, T., Obata, F., Shibata, K.: Biochim. Biophys. Acta **112**, 223–234 (1966)

Seely, G.R.: J. Theoret. Biol. **40**, 189–199 (1973)

Shiozawa, J.A., Alberte, R.S., Thornber, J.P.: Arch. Biochem. Biophys. **165**, 388–397 (1974)

Takamiya, A.: Methods Enzymol. **23**, 603–613 (1971)

Thornber, J.P.: Biochim. Biophys. Acta **172**, 230–241 (1969)

Thornber, J.P.: Ann. Rev. Plant Physiol. **26**, 127–158 (1975)

Thornber, J.P., Gregory, R.P.F., Smith, C.A., Bailey, J.L.: Biochemistry **6**, 391–396 (1967)

Thornber, J.P., Highkin, H.R.: Europ. J. Biochem. **41**, 109–116 (1974)

Thornber, J.P., Smith, C.A., Bailey, J.L.: Biochem. J. **100**, 14 p. (1966)

6. Development of Chloroplast Structure and Function

N.K. Boardman

An important aspect of the study of photosynthesis is the correlation of photochemical and biochemical activities of the chloroplast thylakoid membrane with its structure. Earlier chapters in this volume have considered the structure and topography of the thylakoid membrane of the mature, fully-greened chloroplast in relation to function, and the properties of subchloroplast fragments obtained by disruption of the mature chloroplast with detergents and mechanical methods. Another approach to the study of chloroplast structure and function is to examine the sequence of biochemical and structural changes during the maturation of the chloroplast. Ideally such a study should be carried out on material which is in synchrony. In the case of higher plants, such as *Phaseolus,* synchrony is approached by growing etiolated seedlings in the dark for a number of days and then transferring them to light. The use of appropriate etiolated plants also has the advantage that sufficient material is available for biochemical analyses.

This section is devoted to the development of chloroplast structure and function in the leaves of higher plants, but it should be mentioned that *Euglena gracilis* (SCHIFF, 1975) and mutants of *Chlamydomonas reinhardi* (OHAD et al., 1972) and *Scenedesmus obliquus* (SENGER et al., 1975) also have been employed extensively in studies of chloroplast development.

A. Ultrastructural Changes During Greening

When angiosperm seedlings are germinated and grown in the dark, then, during the differentiation of the meristematic tissue, the relatively undifferentiated proplastids (approximately 1 μm in diameter) develop into the etioplasts (3–5 μm in diameter) of the etiolated leaf cells (VON WETTSTEIN, 1958; GUNNING and JAGOE, 1967; WEIER and BROWN, 1970). The characteristic structural feature of the etioplast is the presence of one or more quasi-crystalline three-dimensional tubular structures, termed prolamellar bodies (Figs. 1 and 2). A few lamellae usually extend outwards from a prolamellar body, and some of these may connect adjacent prolamellar bodies. The etioplasts are devoid of chlorophyll, but they contain a small amount of protochlorophyllide. Fluorescence microscopy, combined with electron microscopy, indicates that the protochlorophyllide is localized in centres which correspond in size with the prolamellar bodies (BOARDMAN and WILDMAN, 1962; SCHNEPF, 1964; KAHN, 1968a). Protochlorophyllide is bound in some way to a protein, termed holochrome, which is essential for the photoconvertibility of protochlorophyllide to chlorophyllide a (BOARDMAN, 1966).

The ultrastructural changes which occur in the etioplast on illumination of dark-grown seedlings have been studied extensively. In the early studies (von Wett-stein, 1958; Klein et al., 1964) leaf material was fixed with permanganate, but subsequent work suggests that the permanganate procedure may give rise to arte-facts during the early stages of greening (Kahn, 1968b). When glutaraldehyde and osmium tetroxide are used as fixatives, the following sequence of structural change is observed on illumination of dark-grown *Avena* plants with white light of reasonably high intensity (e.g. 600–1,000 foot candles) (Gunning and Jagoe, 1967) (Figs. 3–8). Similar observations have been made with bean and barley seedlings.

The first structural change which is observed is a loss of regularity of the crystalline prolamellar body, although the continuity of the membrane surface is retained. Then follows a complete dispersal of the prolamellar body into sheets of perforated membranes, which give rise to the thylakoids. After a few hours of illumination, depending on the species and age of the seedlings, there is a fusion and elaboration of the thylakoids in certain regions to form grana, which are the characteristic feature of mature mesophyll chloroplasts of higher plants. There appears to be continuity of membrane structures during the early stages of plastid development, which suggests that the membranes of the etioplast are the building blocks for the formation of photosynthetically-active membranes in the early stages of chloroplast development.

The proplastids of dark-grown cells of *Euglena* are about 1–2 μm in size. They contain prolamellar bodies, but these are smaller and less highly organized than

---

Figs. 1–8. The greening process in *Avena sativa*

Fig. 1. Etioplast with prolamellar body and a few lamellae. × 22,000

Fig. 2. Detail of another prolamellar body showing how tetrahedrally branched tubules inter-connect to create the quasi-crystalline lattice. At the periphery, the tubules connect with lamellae passing out into the stroma. The stroma contains ribosomes, also to be seen within the prolamellar body lattice. × 58,000

Figs. 3 and 4. After 2 h illumination there are numerous lamellae (primary thylakoids) spaced throughout the stroma. At higher magnification, and especially in face views (Fig. 4), the primary thylakoids are seen to be perforated, the perforations (e.g. *arrows*) perhaps being relics of the spaces in the prolamellar body lattice such as those occupied by ribosomes in Figure 2. Polyribosome configurations are now visible (*bracket*). Fig. 3 × 24,000; Fig. 4 × 52,000

Figs. 5 and 6. After 4 h illumination overgrowths on the primary thylakoids visible (*vertical arrows*) in profile (Fig. 5) and face view (Fig. 6) show that granum formation has commenced. Polyribosome configurations are now common both in the stroma and (*horizontal arrows*) on the developing thylakoid system. Fig. 5 × 64,000; Fig. 6 × 36,000

Figs. 7 and 8. Granum formation is considerably more advanced after 10-h greening, with 2 partitions per granum on average (Fig. 7 and *insert*). Polyribosomes are now very prominent on the thylakoids [*arrows* in insert and the face view (Fig. 8) of granum discs that have developed upon the primary thylakoids]. The collection of plastoglobuli in Figure 8 probably represents a remnant of the prolamellar body. Fig. 7 × 24,000; insert × 36,000; Fig. 8 × 36,000

Figures kindly provided by B.E.S. Gunning; Figs. 2–8 reproduced by permission from Gun-ning and Steer, Ultrastructure and the Biology of Plant Cells, Arnold, London, 1975

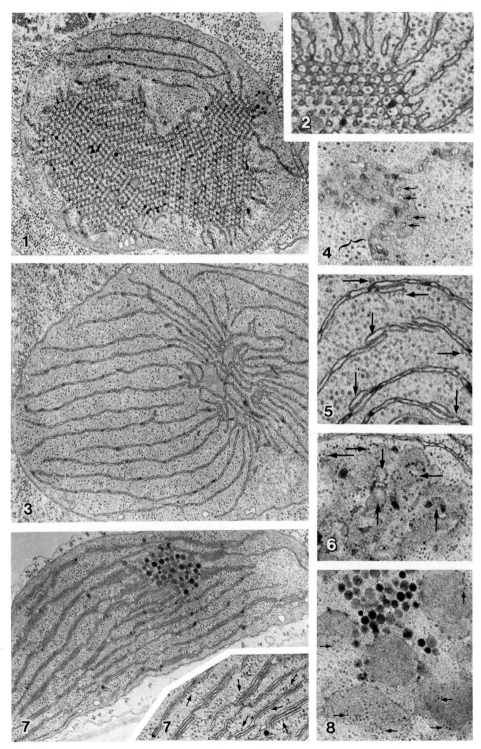

Figs. 1–8. Legends see opposite page

those in the etioplasts of higher plants (Klein et al., 1972). Klein and Schiff (1972) have pointed out that the proplastids of very young 3–4-day-old dark-grown bean seedlings do not contain prolamellar bodies and they conclude that prolamellar body formation is a result of prolonged etiolation and, therefore, not an obligate step in chloroplast development. Prolamellar bodies, however, are evident in some monocots (e.g. maize and sugar cane) even when the seedlings are grown in the light from germination (Laetsch and Price, 1969).

Certain plants with the C_4-dicarboxylic acid pathway of CO_2 fixation, such as maize and sugar cane, contain chloroplasts of two morphological types (Hodge et al., 1955; Laetsch et al., 1966). The chloroplasts of the vascular bundle sheath cells of sugar cane and maize are agranal i.e. they either lack grana or show extremely poor granal development; while the mesophyll cell chloroplasts contain well-developed grana. Developmental studies of sugar cane (Laetsch and Price, 1969) and maize (Horvath et al., 1975) indicate a temporary granal stage during the maturation of the agranal bundle sheath chloroplast, particularly if the plants are grown in the light from the beginning of germination. In dark-grown seedlings, prolamellar bodies are found in the etioplasts of both mesophyll and bundle sheath cells (Laetsch and Price, 1969; Montes and Bradbeer, 1975).

Additional information about the relationship between chloroplast structure and function has been obtained by greening dark-grown seedlings under unusual illumination conditions. Grana formation is inhibited if etiolated seedlings are greened in continuous far-red light (De Greef et al., 1971), by exposure to 1 msec flashes every 15 min (Sironval et al., 1968) or alternating periods of 2 min light and 98 min dark (Akoyunoglou and Michelinaki-Maneta, 1975). Under the first two conditions, the chloroplasts contain long parallel stacks of nonappressed primary thylakoids. Young five-day-old etiolated bean seedlings greened under the alternating light-dark regime also have primary thylakoids, but with older seedlings (12-day-old) some appression of thylakoids is observed.

B. Spectroscopic Changes During Greening

Spectroscopic studies of etiolated leaves or isolated prolamellar body membranes indicate three spectroscopic forms of protochlorophyll(ide) with absorption maxima at 628 nm, 637 nm and 650 nm (Shibata, 1957; Kahn et al., 1970). Two fluorescence forms of protochlorophyll(ide) are observed at 77 K with maximum at 630–633 nm and 655 nm (Litvin and Krasnovsky, 1957; Kahn et al., 1970). Measurements of excitation spectra of fluorescence indicate that the 655 nm emission is activated by light absorbed by PChl-650 and PChl-637. Light absorbed by PChl-637 is transferred to PChl-650 with high efficiency, which accounts for the absence of fluorescence emission from PChl-637. The fluorescence emission at 630 nm originates from PChl-628. The relative amounts of the three forms of protochlorophyllide may vary among plant species, and depend on the age of the seedlings. For example, PChl-650 and PChl-637 are present in approximately equal amounts in 10–14-day-old etiolated bean leaves and together they account for 85–90% of the total protochlorophyllide of the leaf. Younger seedlings contain a higher proportion of PChl-628 and PChl-637 (BOARDMAN et al., 1971; THORNE, 1971; KLEIN and SCHIFF, 1972). The molecular basis for the difference in spectroscopic properties of the protochlorophyll(ides) is not established. Different states of aggregation (KRASNOVSKY and KOSOBUTSKAYA, 1953; SELISKAR and KE, 1968) or different modes of binding of the pigment to protein (BOARDMAN, 1966) have been suggested. In the etioplast the protochlorophyllide molecules are organized into energy-transferring units containing at least 4 protochlorophyllide molecules (KAHN et al., 1970; VAUGHAN and SAUER, 1974) and there is some evidence that the units may contain as many as 20 chromophores (THORNE, 1971).

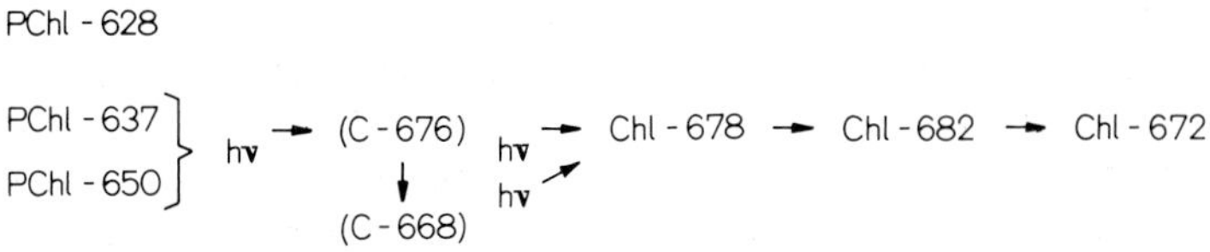

Fig. 9. Spectroscopic intermediates in the conversion of protochlorophyllide to chlorophyll

A number of spectroscopic intermediates are observed when dark-grown leaves are illuminated and these are summarized in Figure 9. The review by KIRK (1970) should be consulted for a fuller discussion and reference to the original literature. PChl-637 and PChl-650 are photoconverted to a form of chlorophyll absorbing at 678 nm (Chl-678), which is converted within 30 sec at room temperature to Chl-682. Then follows a slower spectral shift to 672 nm (SHIBATA, 1957), which takes 10–30 min to complete, depending on the species of plant and the age of the seedlings. Finally, with the accumulation of additional chlorophyll in the leaf the absorption peak gradually shifts to 678 nm, which corresponds with the chlorophyll absorption maximum in the fully-greened leaf.

Intermediate pigment forms absorbing at 676 nm (C-676) and 668 nm (C-668) are observed after very short flashes of high intensity light or illuminations with low intensity light (LITVIN and BELYAEVA, 1968, 1971; THORNE, 1971). C-668 appears to be derived from C-676 and is stable in the dark. It may represent either a small fraction of protochlorophyll(ide) which is converted more rapidly than the majority of the protochlorophyllide, or a one-photon intermediate in the conversion of protochlorophyllide to chlorophyllide a (THORNE, 1971).

On the basis of circular dichroism and fluorescence spectral changes, SCHULTZ and SAUER (1972) and MATHIS and SAUER (1972, 1973) suggested that the photoactive protochlorophyllide is present as dimers in the etiolated leaf, and photoconversion involves a two step light reaction to produce a dimeric form of chlorophyllide a, which then dissociates into monomeric chlorophyll(ide) a. They proposed that C-676 is a dimer of chlorophyllide a and protochlorophyllide, which dissociates in the dark to give C-668.

In young bean leaves (3–4-day-old), which do not have prolamellar bodies, most of the transformable pigment, consisting both of protochlorophyllide and protochlorophyll, absorbs at 635 nm and on illumination it is converted directly to chlorophyll(ide)-672 (KLEIN and SCHIFF, 1972; SCHIFF, 1975). *Euglena* resembles the young bean leaves in its spectroscopic changes on illumination (SCHIFF, 1975).

If only part of the protochlorophyllide of an etiolated leaf is converted to chlorophyllide by a short irradiation, then energy transfer is observed between the remaining protochlorophyllide and the various spectroscopic forms of chlorophyll(ide) (Boardman et al., 1971; Thorne, 1971; Vaughan and Sauer, 1974). This indicates that in the initial stages of greening some of the newly formed chlorophyll(ide) remains in close proximity to the protochlorophyllide. The kinetics of photoconversion of protochlorophyllide do not obey first-order kinetics, as might be expected, since protochlorophyllide is the photoreceptor for its own conversion. The kinetics which approximate to second-order are explainable in terms of a competition between the photoconversion process and transfer of excitation energy from protochlorophyllide to chlorophyllide a (Thorne and Boardman, 1972; Nielsen and Kahn, 1973). The energy transfer rate increases as chlorophyllide a is formed and this decreases the rate of photoconversion.

C. Chlorophyll Formation in Relation to Ultrastructural and Spectroscopic Changes

Several studies have indicated that the photoconvertible forms, PChl-637 and PChl-650, of the higher plant etioplast are protochlorophyllide, whereas the nonconvertible form, PChl-628, is esterified protochlorophyllide (Wolff and Price, 1957; Virgin, 1960; Ogawa et al., 1975). Some recent evidence suggests that the esterifying alcohol of protochlorophyllide is geranylgeraniol and not phytol (Liljenberg, 1974). In *Euglena* and very young bean leaves, however, it would appear that both protochlorophyllide and protochlorophyll are photoconvertible to chlorophyllide a and chlorophyll a respectively (Schiff, 1975).

It has been proposed that the loss in regularity of the prolamellar body correlates with the photoconversion of protochlorophyllide to chlorophyllide a (Kahn, 1968b; Henningsen and Boynton, 1969), but this conclusion was questioned by Treffry (1970) on the basis that some esterification of chlorophyllide a cannot be precluded during the time needed to fix the tissue with glutaraldehyde. Treffry (1970) observed that there was no loss of regularity of the prolamellar body if the illumination is carried out at 0°, where esterification of chlorophyllide a is inhibited, but photoconversion occurs. A correlation of the loss of regularity with the spectral shift from 678 nm to 682 nm also seems unlikely, since this spectral change is completed in about 5 min at 0° (Gassman et al., 1968). Treffry considers that some phytylation of chlorophyllide a accompanies the loss of regularity of the prolamellar body.

Phytylation of chlorophyllide a has been implicated in the dispersal of the transformed prolamellar body into the sheets of perforated membranes and the Shibata spectral shift from 682 nm to 672 nm. In the studies of Boardman (1967) and Akoyunoglou and Michalopoulos (1971), the time courses of the spectral shift and phytylation did not coincide, but in the more recent studies of Henningsen and Thorne (1974) there was a good correlation between phytylation and the spectral shift in both bean and barley seedlings at a number of temperatures. Mutants of barley which did not show esterification of chlorophyllide lacked the spectral shift (HENNINGSEN and THORNE, 1974), which is in agreement with the early studies of SMITH et al. (1959) with corn mutants. The dispersal of the prolamellar body membranes is inhibited at 0° and it shows the same temperature dependence as the Shibata spectral shift and phytylation. There appears, therefore, to be a causal relationship between the three phenomena, although the exact sequence of events is still not resolved.

Gel filtration studies of the isolated protochlorophyll(ide) and chlorophyll(ide) holochromes from barley show an approximate halving of the apparent molecular weight of the complex, which correlates with the Shibata spectral shift to 672 nm (HENNINGSEN et al., 1974). This may represent the dissociation of the chlorophyll-protein complex into a colorless photoenzyme (containing the site of photoconversion of protochlorophyllide) and a chlorophyll a carrier protein, which transports the chlorophyll a to its sites on the developing thylakoid membrane (cf. BOARDMAN, 1967; BOGORAD et al., 1968). The photoenzyme appears to be used repeatedly as the site for protochlorophyllide accumulation and photoconversion (NADLER and GRANICK, 1970; SÜZER and SAUER, 1971; THORNE, 1971).

On illumination of dark-grown seedlings, there is a lag period before chlorophyll(ide) a is synthesized, other than that formed by photoconversion of preexisting protochlorophyllide. The length of the lag varies from about 5 min to several hours, depending on the species of plant and the age of the seedlings. Figure 10

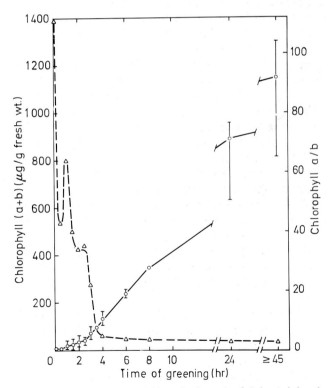

Fig. 10. Accumulation of chl a + chl b (———), and chl a/chl b ratio (-----) of 6-day etiolated barley seedlings on illumination with white light. (From HENNINGSEN and BOARDMAN, 1973)

shows the accumulation of chlorophyll on illumination of six-day-old dark-grown barley seedlings with continuous white light of 450 foot candles. Following the lag phase of 30 min, there is a slow accumulation of chlorophyll for the first 2.5 h and then a more rapid synthesis. The ratio of chlorophyll a to chlorophyll b is also shown. Chlorophyll b cannot be detected immediately after the photoconversion of protochlorophyllide, but it is formed during the lag phase and the ratio of chl a/chl b falls rapidly to about 40/1. After about 2 h the chl a/chl b again decreases rapidly to reach a value of 3/1 at about 6 h of illumination. During this period grana are formed. Similar time courses of chlorophyll b synthesis were reported for pea (THORNE and BOARDMAN, 1971) and maize (SHLYK et al., 1972). The enzymic steps involved in the formation of chlorophyll b are unknown, but radiotracer studies suggest that chlorophyll b is formed from chlorophyll a (SHLYK et al., 1972).

D. Composition of Developing Thylakoids

Studies with pea, bean, and barley seedlings indicate that there is little change in the amounts of the complex lipids and fatty acids during the first 6 h of greening (ROUGHAN and BOARDMAN, 1972; TEVINI, 1972), which supports the view that the membranes of the etioplast are the direct precursors of the primary thylakoids. With the rapid formation of grana, however, additional membrane synthesis occurs, and this correlates with increases in the level of the galactolipids (monogalactosyl diglyceride and digalactosyl diglyceride) and phosphatidyl glycerol (cf. TREMOLIÈRES and LEPAGE, 1971; LEECH et al., 1972; HEISE and JACOBI, 1973).

Etioplast membranes contain 7–15 polypeptides, extractable with sodium dodecylsulphate and separable by SDS-polyacrylamide gel electrophoresis (REMY, 1973b; COBB and WELLBURN, 1973; GUIGNERY et al., 1974; LÜTZ, 1975). There are no significant changes in the gel pattern of lamellar membranes for the first few hours of greening in continuous light. During the period where grana are formed, there is rapid synthesis of a polypeptide of molecular weight of about 24,000, so that in the green leaf this peptide is the major constituent of thylakoids (REMY, 1973b; LÜTZ, 1975). Plants greened in intermittent light give a pattern similar to that of the etioplast (REMY, 1973b).

ALBERTE et al. (1972) examined the appearance of SDS-extractable chlorophyll-protein complexes during the rapid greening of etiolated jack bean leaves. Complex II (chl a/chl b ratio of 1.2) was first detected after 2 h of illumination and it increased markedly during the period of chlorophyll b synthesis and grana formation. Complex I (chl a/chl b > 8) appeared after 6 h of greening and correlated with the observation of light-induced oxidation of P-700. Chloroplasts from bean seedlings exposed to alternating periods of light and dark have a very high chl a/chl b ratio and appear to lack complex II (ARGYROUDI-AKOYUNOGLOU et al., 1971; HILLER et al., 1973).

Cytochromes f, b-559$_{LP}$ (low potential form) and b-563, which are associated structurally with photosystem I (BOARDMAN and ANDERSON, 1967; ANDERSON and BOARDMAN, 1973) are found in the etioplast of dark-grown seedlings (BENDALL

Table 1. Cytochrome content of greening barley leaves. (From Henningsen and Boardman, 1973)

Time of greening (h)	Cyt f (nmol/g fr.wt.)	Cyt b-559$_{LP}$+ Cyt b-563 (nmol/g fr.wt.)	Cyt b-559$_{HP}$ (nmol/g fr.wt.)
0	0.73	2.0	0
2	0.71	2.4	0.05
4	0.73	2.0	0.33
6	1.15	2.9	0.38
8	0.81	2.0	0.76
24	1.16	2.7	1.8

et al., 1971). Cytochrome b-559$_{HP}$ (high potential form) which is associated with photosystem II is absent from etioplasts and it is formed during greening (Board-man, 1968; Plesničar and Bendall, 1972; Whatley et al., 1972). During the early stages of greening of barley seedlings, when photochemical activity develops, there appears to be little net synthesis of cytochrome f, b-559$_{LP}$ and b-563 (Henning-sen and Boardman, 1973) (Table 1). The mature chloroplast, however, contains substantially more of the cytochromes than does the etioplast, showing that all cytochromes are synthesized during the period of massive membrane synthesis and new thylakoid formation (Whatley et al., 1972). Plastocyanin, ferredoxin, and ferredoxin-NADP$^+$ reductase are also present in the dark-grown leaf and like-wise show substantial increases in amounts on illumination and maturation of the chloroplast (Plesničar and Bendall, 1972; Whatley et al., 1972; Haslett and Cammack, 1974).

Etiolated leaves contain the coupling factor of photophosphorylation, and elec-tron microscopy shows that it is attached to the membranes of the etioplast (Horak and Hill, 1971; Lockshin et al., 1971). It is similar to the coupling factor of chloroplasts, except that its ATPase activity is not light-triggered (Gregory and Bradbeer, 1975). Light-triggering of bean ATPase, and a simultaneous fall in dark activity was observed if the dark-grown seedlings were illuminated for 30 min before isolation of the plastids. On a per plastid basis, the light-triggered ATPase activity of bean plastids remained remarkably constant between 30 min and 48 h of greening. Gregory and Bradbeer (1975) estimated the ATPase activity per m^2 of thylakoid membrane; there was an 83% decline in the ATPase activity over 48 h of greening. They concluded that the dark-grown bean leaf synthesizes its full complement of chloroplast ATPase, which contrasts with the net synthesis of cytochromes and other components during greening.

E. Development of Photochemical Activity

The time-course of development of photochemical activity on illumination of dark-grown seedlings varies markedly with the age of the seedlings, the species of

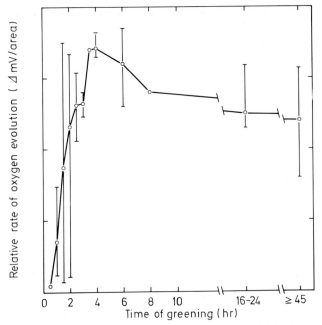

Fig. 11. Photosynthetic O_2 evolution of 6-day etiolated barley leaves at various times of greening in white light. (From Henningsen and Boardman, 1973)

plant, and conditions of growth. It is difficult, therefore, to compare the relative times of appearance of different photosynthetic activities unless measurements are made on plants of comparable physiological age. The early studies of Anderson and Boardman (1964) and Gyldenholm and Whatley (1968) were carried out with bean seedlings, which varied in age from 10–14 days and 14–16 days respectively. The development of photochemical activity was slow compared with that reported in more recent studies, using young bean or barley seedlings. The older plants had one advantage in that the slower time scale made it easier to differentiate the sequential appearance of different photochemical activities. These early studies suggested that the appearance of an active photosystem I (PS I) precedes that of an active photosystem II (PS II) [cf. review by Park and Sane (1971) for earlier work].

The sequential appearance of PS I and PS II is illustrated here by reference to more recent work with greening barley seedlings. Photosynthetic O_2 evolution from a lightly abraded six-day-old leaf is detected after 30 min of illumination (Fig. 11) (cf. Smith, 1954, and Egnéus et al., 1972). At 2 to 2.5 h, the rate of O_2 evolution per g fresh weight of leaf is as high as that from a green leaf at 45 h. If the photosynthetic rate of O_2 evolution is related to the chlorophyll content of the leaf, it is seen that the rate is 80-fold greater after 90 min than after 45 h (Fig. 12). It would seem, therefore, that PS I and PS II are active at an early stage of greening, but the photosynthetic units are small compared with those of the mature chloroplast. This conclusion is supported by the high light

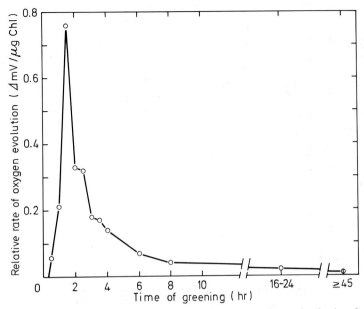

Fig. 12. Photosynthetic O_2 evolution by greening barley leaves expressed on the basis of their chlorophyll content. (From HENNINGSEN and BOARDMAN, 1973)

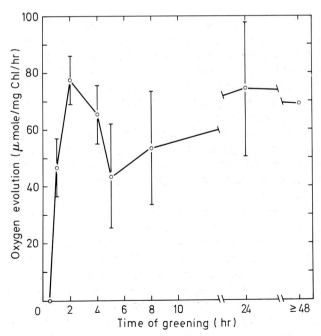

Fig. 13. Photochemical O_2 evolution by barley plastids at various stages of greening. (From HENNINGSEN and BOARDMAN, 1973)

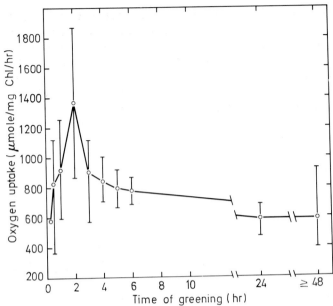

Fig. 14. Photosystem I activity of barley plastids at various stages of greening. (From HENNING-SEN and BOARDMAN, 1973)

intensities required to saturate photosynthetic O_2 evolution in the early stages of greening (KIRK and GOODCHILD, 1972). Isolated barley plastids show substantial photochemical oxygen evolution with ferricyanide as oxidant after 1 h of seedling greening and a peak of activity is observed at 2 h of greening (Fig. 13). A comparison of the time-courses of O_2 evolution for leaves and plastids suggests some inactivation of the developing plastids during isolation. Substantial PS I activity of plastids, measured by the rate of O_2 uptake with reduced N,N,N',N'-tetramethylphenylenediamine as electron donor and autoxidizable methyl viologen as acceptor, is detected as early as 15 min (Fig. 14). PLESNICAR and BENDALL (1972) and EGNÉUS et al. (1972) observed a similar early appearance of PS I activity in plastids from greening barley seedlings. Cyclic phosphorylation with phenazine methosulphate as cofactor was observed very soon after the exposure of the plants to light, but noncyclic phosphorylation with ferricyanide as oxidant could not be detected before 2 h of illumination (PLESNIČAR and BENDALL, 1972). EGNÉUS et al. (1972) examined oxygen exchange from developing barley seedlings, illuminated alternatively with red and far-red light, and concluded that there is cooperation between the two photosystems at 3 h of greening. O_2 uptake in 710 nm light was taken as a measure of PS I activity. PHUNG NHU HUNG et al. (1970a) observed a small photoreduction of $NADP^+$ in plastids isolated from etiolated barley seedlings illuminated for 2 h, although in the earlier studies with beans $NADP^+$ photoreduction lagged behind ferricyanide photoreduction (ANDERSON and BOARDMAN, 1964; GYLDENHOLM and WHATLEY, 1968).

RHODES and YEMM (1966) examined CO_2 fixation by greening barley seedlings. A light-dependent CO_2 exchange was observed at 3 h of greening, but the compensation point was not reached until 30–40 h. WOLF (1971) reported that the CO_2 compensation point was reached in $3^1/_2$ h with wheat seedlings grown under high light intensity. The onset of CO_2 fixation in relation to O_2 evolution in greening barley seedlings has recently been reinvestigated by ROBERTSON and LAETSCH (1974) and LÜTTGE et al. (1974). CO_2 fixation developed more slowly than O_2 evolution, the CO_2/O_2 ratio being <0.2 for several hours after the onset of O_2 evolution. One obvious explanation for such an imbalance of CO_2 and O_2 is that the Calvin carbon reduction cycle is not active in the early stages of greening. Since it is very likely that carbon compounds are imported into the plastid from the rest of the cell at this time, it is not difficult to envisage a substrate for the electron from O_2 evolution, other than CO_2. In some recent work with isolated plastids from greening pea seedlings O_2 evolution was observed at 3 h of greening either with 3-phosphoglycerate or oxaloacetate as substrate, but not with CO_2 (HEBER et al., 1976).

Plastids isolated from seedlings greened by exposure to flashes of light (1 msec flash every 15 min) do not evolve O_2 even after 300 flashes, but they are active in cyclic phosphorylation (PHUNG NHU HUNG et al., 1970a). This suggests that PS I, but not PS II, is active in the primary thylakoids of flashed beans (cf. DUJAR-DIN et al., 1970). However, REMY (1973a) observed PS II activity if plastids from flashed leaves were supplied with an artificial donor to PS II. He concluded that PS II reaction centers were present, but there was a defect in the water-splitting sequence of reactions. Oxygen evolution is induced in the flashed seedlings by exposure to continuous light for just a few minutes (STRASSER and SIRONVAL, 1972), or by flashes given at short intervals apart (INOUE et al., 1975). Photoactiva-tion of O_2 evolution may involve an activation of Mn, which is known to be involved in water-splitting (CHENIAE and MARTIN, 1971).

Plants greened in far-red light (DE GREEF et al., 1971), or in alternating periods of 2 min white light and 98 min dark (HILLER et al., 1973; AKOYUNOGLOU and MICHELINAKI-MANETA, 1975), slowly develop an active PS II, with the ability to evolve O_2. On the other hand, OGAWA and SHIBATA (1973) reported that etiolated wheat seedlings greened under a low intensity white light develop an active PS I, but have a very low rate of O_2 evolution.

F. Cytochrome and P-700 Redox Changes in Developing Plastids

Evidence for the sequential appearance of PS I and PS II has also come from measurements of light-induced redox changes of cytochrome f. In the mature chlo-roplast, cytochrome f is oxidized by PS I and reduced by PS II, and cytochrome b-563 is reduced by PS I. In the greening bean leaf a photooxidation of cytochrome f by PS I was observed after 90 min of greening, and photoreduction (by PS II) could be demonstrated about 30 min later (BONNER and HILL, 1963; HILLER and BOARDMAN, 1971). If dark-grown bean leaves are illuminated for 3 min, returned to darkness for $2^1/_2$ h and illuminated again, cytochrome f oxidation is observed immediately at the second illumination, and cytochrome f reduction (by PS II) is observed within a period of a few minutes (HILLER and BOARDMAN, 1971). The ability of the leaves to photooxidize cytochrome f develops in the dark period following the initial illumination, and requires no chlorophyll except that formed from protochlorophyllide at the first photoconversion. The ability to photoreduce cytochrome f, which develops during the second illumination, is probably due to the activation of electron flow from water. These redox changes of cytochrome f support the view that the reaction centers of PS I and PS II are formed at an early stage of greening. The size of the light-harvesting assemblies, both of PS I and PS II, is small compared with those of the green leaf, and chlorophyll synthesis during the first few hours following the lag period serves mainly to increase the size of the light-harvesting assemblies. With greening pea seedlings, a photooxidation of cytochrome f and a photoreduction of cytochrome b-563 is observed as early as 20 min. PS II becomes active after illumination for 1–2 h (BOARDMAN et al., 1972).

In contrast with the early appearance of cytochrome f oxidation, a photooxidation of P-700 is not observed in the early stages of greening. BOARDMAN et al. (1972) detected some oxidation of P-700 at 3 h of greening of pea seedlings, while ALBERTE et al. (1972) could not observe P-700 photooxidation for at least 6 h with greening bean seedlings. In barley seedlings, P-700 photooxidation was observed at 1 h (BOARDMAN, unpublished). Because of the slow appearance of P-700 photooxidation, ALBERTE et al. (1972) concluded that PS I activity lags behind PS II. BOARDMAN et al. (1972) suggested that the reaction center chlorophyll of PS I in the early stages of greening may not have the absorption properties of P-700. The appearance of the band at P-700 may correspond to the accumulation of additional chlorophyll molecules around the reaction center chlorophyll. An alternate explanation can be advanced to explain the failure to detect a light-induced bleaching of P-700 in the early stages of greening. The efficiency of photooxidation of P-700 may be poor at this stage of the development of the PS I units, so that the re-reduction is fast compared with its photooxidation.

The redox state of the primary electron acceptor (Q) of PS II is reflected in the fluorescence yield of chlorophyll a of leaves and plastids (DUYSENS and SWEERS, 1963). In mature chloroplasts, red light absorbed by PS II causes a reduction of Q and increases the fluorescence yield. A sequential illumination with far-red light, absorbed by PS I oxidizes Q^- and decreases the fluorescence yield. BUTLER (1965) examined fluorescence yield changes in red and far-red light in leaves of greening bean seedlings. Light-induced changes in fluorescence yield were detected after 2 h of illumination, indicating the appearance of active PS I and PS II, in agreement with the redox changes of cytochrome f.

G. Correlation of Ultrastructural Changes with Function

The earlier studies on the greening of dark-grown seedlings under continuous illumination suggested a close correlation between the formation of grana and the appearance in isolated plastids of electron flow from water and noncyclic phosphorylation (BOARDMAN and ANDERSON, 1964; ANDERSON and BOARDMAN, 1964; RHODES and YEMM, 1966; GYLDENHOLM and WHATLEY, 1968). The appearance during greening of the large particles seen in freeze fracture electron micrographs of chloroplasts (cf. this vol. Chap. IV/1) also correlated with the formation of grana (PHUNG NHU HUNG et al., 1970b). Later studies, particularly those with intact leaves of greening seedlings, indicated that photosynthetic O_2 evolution and redox changes associated with photochemical activity of PS II are observed ahead of the time when grana formation occurs. Since PS I is active ahead of PS II, it must be concluded that neither photosystem is dependent on the formation of grana for activity. This conclusion is substantiated by the greening studies under flashing light, alternating regimes of light and dark, and continuous illumination in far-red light. These latter studies further show that formation of chlorophyll b and chlorophyll-protein complex II are not essential for active PS I and PS II in developing plastids. Experiments with mutants of higher plants also indicate that there is no correlation between the formation of grana and PS I and PS II activity (SMILLIE et al., 1975), or between the presence of chlorophyll b and photochemical activity (HIGHKIN and FRENKEL, 1962). It is impossible, however, to dismiss a possible functional role for small regions of paired membranes, as distinct from well-recognized grana, during the early stages of chloroplast development. Even etioplasts of some plants contained some paired regions which remain during early greening (HENNINGSEN and BOYNTON, 1969).

References

Akoyunoglou, G., Michalopoulos, G.: Physiol. Plantarum **25**, 324–329 (1971)

Akoyunoglou, G., Michelinaki-Maneta, M.: Development of photosynthetic activity in flashed bean leaves. In: Proc. 3rd Intern. Congress Photosynthesis. Avron, M. (ed.). Amsterdam: Elsevier, 1975, Vol. III, pp. 1885–1886

Alberte, R.S., Thornber, J.P., Naylor, A.W.: J. Exp. Botany **23**, 1060–1069 (1972)

Anderson, J.M., Boardman, N.K.: Australian J. Biol. Sci. **17**, 93–101 (1964)

Anderson, J.M., Boardman, N.K.: FEBS Lett. **32**, 157–160 (1973)

Argyroudi-Akoyunoglou, J.H., Feleki, Z., Akoyunoglou, G.: Biochem. Biophys. Res. Commun. **45**, 606–614 (1971)

Bendall, D.S., Davenport, H.E., Hill, R.: Methods Enzymol. **23A**, 327–344 (1971)

Boardman, N.K.: Protochlorophyll. In: The Chlorophylls. Vernon, L.P., Seely, G.R. (eds.). New York: Academic Press, 1966, pp. 437–479

Boardman, N.K.: Chloroplast structure and development. In: Harvesting the Sun. San Pietro, A., Greer, F.A., Army, T.J. (eds.). New York: Academic Press, 1967, pp. 211–230

Boardman, N.K.: Cytochromes of developing chloroplasts. In: Comparative Biochemistry and Biophysics of Photosynthesis. Shibata, K., Takamiya, A., Jagendorf, A.T., Fuller, R.C. (eds.). State College, Pa.: Univ. Park, 1968, pp. 206–213

Boardman, N.K., Anderson, J.M.: Australian J. Biol. Sci. **17**, 86–92 (1964)

Boardman, N.K., Anderson, J.M.: Biochim. Biophys. Acta **143**, 187–203 (1967)

Boardman, N.K., Anderson, J.M., Hiller, R.G., Kahn, A., Roughan, P.G., Treffry, T.E., Thorne, S.W.: Biosynthesis of the photosynthetic apparatus during chloroplast development in higher plants. In: Proc. 2nd Intern. Congr. Photosynthesis. Forti, G., Avron, M., Melandri, A. (eds.). The Hague: Dr. Junk, 1972, Vol. III, pp. 2265–2287

Boardman, N.K., Anderson, J.M., Kahn, A., Thorne, S.W., Treffry, T.E.: Formation of photosynthetic membranes during chloroplast development. In: Autonomy and Biogenesis of Mitochondria and Chloroplasts. Boardman, N.K., Linnane, A.W., Smillie, R.M. (eds.). Amsterdam: North Holland, 1971, pp. 70–84

Boardman, N.K., Wildman, S.G.: Biochim. Biophys. Acta **59**, 222–224 (1962)

Bogorad, L., Laber, L., Gassman, M.: Aspects of chloroplast development: transitory pigment-protein complex and protochlorophyllide regeneration. In: Comparative Biochemistry and Biophysics of Photosynthesis. Shibata, K., Takamiya, A., Jagendorf, A.T., Fuller, R.C. (eds.). State College, Pa.: Univ. Park, 1968, pp. 299–312

Bonner, W., Hill, R.: Light induced optical changes in green leaves. In: Photosynthetic Mechanisms of Green Plants. Nat. Acad. Sci. Nat. Res. Council. **1145**, 82–90 (1963)

Butler, W.L.: Biochim. Biophys. Acta **102**, 1–8 (1965)

Cheniae, G.M., Martin, I.F.: Biochim. Biophys. Acta **253**, 167–181 (1971)

Cobb, A.H., Wellburn, A.T.: Planta (Berl.) **114**, 131–142 (1973)

Dujardin, E., de Kouchkovsky, Y., Sironval, C.: Photosynthetica **4**, 223–227 (1970)

Duysens, L.N.M., Sweers, H.E.: Mechanism of two photochemical reactions in algae as studied by means of fluorescence. In: Microalgae and Photosynthetic Bacteria. Tokyo: Univ. Tokyo, 1963, pp. 353–372

Egnéus, H., Reftel, S., Sellden, G.: Physiol. Plantarum **27**, 48–55 (1972)

Gassman, M., Granick, S., Mauzerall, D.: Biochem. Biophys. Res. Commun. **32**, 295–300 (1968)

Greef, J. de, Butler, W.L., Roth, T.F.: Plant Physiol. **47**, 457–464 (1971)

Gregory, P., Bradbeer, J.W.: Biochem. J. **148**, 433–438 (1975)

Guignery, G., Luzzati, A., Duranton, J.: Planta (Berl.) **115**, 227–243 (1974)

Gunning, B.E.S., Jagoe, M.P.: The prolamellar body. In: Biochemistry of Chloroplasts. Goodwin, T.W. (ed.). London-New York: Academic Press, 1967, Vol. II, pp. 655–676

Gyldenholm, A.O., Whatley, F.R.: New Phytologist **67**, 461–468 (1968)

Haslett, B.G., Cammack, R.: Biochem. J. **144**, 567–572 (1974)

Heber, U., Boardman, N.K., Anderson, J.M.: Biochim. Biophys. Acta **423**, 275–292 (1976)

Heise, K.P., Jacobi, G.: Planta (Berl.) **111**, 137–148 (1973)

Henningsen, K.W., Boardman, N.K.: Plant Physiol. **51**, 1117–1126 (1973)

Henningsen, K.W., Boynton, J.E.: J. Cell Sci. **5**, 757–793 (1969)

Henningsen, K.W., Thorne, S.W.: Physiol. Plantarum **30**, 82–89 (1974)

Henningsen, K.W., Thorne, S.W., Boardman, N.K.: Plant Physiol. **53**, 419–425 (1974)

Highkin, H.R., Frenkel, A.W.: Plant Physiol. **37**, 814–820 (1962)

Hiller, R.G., Boardman, N.K.: Biochim. Biophys. Acta **253**, 449–458 (1971)

Hiller, R.G., Pilger, D., Genge, S.: Plant Sci. Lett. **1**, 81–88 (1973)

Hodge, A.J., McLean, J.D., Mercer, F.V.: J. Biophys. Biochem. Cytol. **1**, 605–613 (1955)

Horak, A., Hill, R.D.: Can. J. Biochem. **49**, 207–209 (1971)

Horvath, G., Garab, G.I., Halász, N., Faludi-Daniel, A.: Maturation of thylakoids in mesophyll and bundle sheath chloroplasts of maize. In: Proc. 3rd Intern. Congr. Photosynthesis. Avron, M. (ed.). Amsterdam: Elsevier, 1975, Vol. III, pp. 1925–1932

Inoue, Y., Ichikawa, T., Kobayashi, Y., Shibata, K.: Multiple flash activation of the water-splitting system in wheat leaves grown under intermittent illumination. In: Proc. 3rd Intern. Congr. Photosynthesis. Avron, M. (ed.). Amsterdam: Elsevier, 1975, Vol. III, pp. 1833–1840

Kahn, A.: Plant Physiol. **43**, 1769–1780 (1968a)

Kahn, A.: Plant Physiol. **43**, 1781–1785 (1968b)

Kahn, A., Boardman, N.K., Thorne, S.W.: J. Mol. Biol. **48**, 85–101 (1970)

Kirk, J.T.O.: Ann. Rev. Plant Physiol. **21**, 11–42 (1970)

Kirk, J.T.O., Goodchild, D.J.: Australian J. Biol. Sci. **25**, 215–241 (1972)

Klein, S., Bryan, G., Bogorad, L.: J. Cell Biol. **22**, 433–442 (1964)

Klein, S., Schiff, J.A.: Plant Physiol. **49**, 619–626 (1972)

Klein, S., Schiff, J.A., Holowinsky, A.: Develop. Biol. **28**, 253–273 (1972)

Krasnovsky, A.A., Kosobutskaya, L.M.: Dokl. Akad. Nauk. SSSR **91**, 343–346 (1953)

Laetsch, W.M., Price, I.: Am. J. Botany **56**, 77–87 (1969)

Laetsch, W.M., Stetler, D.A., Vlitos, A.J.: Z. Pflanzenphysiol. **54**, 472–476 (1966)

Leech, R.M., Rumsby, M.G., Thomson, W.W., Crosby, W., Wood, P.: Lipid changes during plastid differentiation in developing maize leaves. In: Proc. 2nd Intern. Congr. Photosynthesis. Forti, G., Avron, M., Melandri, A. (eds.). The Hague: Dr. Junk, 1972, Vol. III, pp. 2479–2488

Liljenberg, C.: Physiol. Plantarum **32**, 208–213 (1974)
Litvin, F.F., Belyaeva, O.B.: Biokhimiya **33**, 928–936 (1968)
Litvin, F.F., Belyaeva, O.B.: Photosynthetica **5**, 200–209 (1971)
Litvin, F.F., Krasnovsky, A.A.: Dokl. Akad. Nauk. SSSR **117**, 106 (1957)
Lockshin, A., Falk, R.H., Bogorad, L., Woodcock, C.L.F.: Biochim. Biophys. Acta **226**, 366–382 (1971)
Lüttge, V., Kramer, D., Ball, E.: Z. Pflanzenphysiol. **71**, 6–21 (1974)
Lütz, C.: Z. Pflanzenphysiol. **75**, 346–359 (1975)
Mathis, P., Sauer, K.: Biochim. Biophys. Acta **267**, 498–511 (1972)
Mathis, P., Sauer, K.: Plant Physiol. **51**, 115–119 (1973)
Montes, G., Bradbeer, J.W.: The biogenesis of the photosynthetic membranes and the development of photosynthesis in greening maize leaves. In: Proc. 3rd Intern. Congr. Photosynthesis. Avron, M. (ed.). Amsterdam: Elsevier, 1975, Vol. III, pp. 1867–1876
Nadler, K., Granick, S.: Plant Physiol. **46**, 240–246 (1970)
Nielsen, O.F., Kahn, A.: Biochim. Biophys. Acta **292**, 117–129 (1973)
Ogawa, T., Bovey, F., Inoue, Y., Shibata, K.: Early stages in greening in etiolated bean leaves. In: Proc. 3rd Intern. Congr. Photosynthesis. Avron, M. (ed.). Amsterdam: Elsevier, 1975, Vol. III, pp. 1829–1832
Ogawa, T., Shibata, K.: Physiol. Plantarum **29**, 112–117 (1973)
Ohad, I., Jennings, R.C., Goldberg, I., Bar-Nun, S., Wallach, D.: Biogenesis of chloroplast membranes in *Chlamydomonas reinhardi*. In: Proc. 2nd Intern. Congr. Photosynthesis. Forti, G., Avron, M., Melandri, A. (eds.). The Hague: Dr. Junk, 1972, Vol. III, pp. 2563–2584
Park, R.B., Sane, P.V.: Ann. Rev. Plant Physiol. **22**, 395–430 (1971)
Phung Nhu Hung, S., Hoarau, A., Moyse, A.: Z. Pflanzenphysiol. **62**, 245–258 (1970a)
Phung Nhu Hung, S., Lacourly, A., Sarda, C.: Z. Pflanzenphysiol. **62**, 1–16 (1970b)
Plesničar, M., Bendall, D.S.: The development of photochemical activities during greening of etiolated barley. In: Proc. 2nd Intern. Congr. Photosynthesis. Forti, G., Avron, M., Melandri, A. (eds.). The Hague: Dr. Junk, 1972, Vol. III, pp. 2367–2374
Remy, R.: Photochem. Photobiol. **18**, 409–416 (1973a)
Remy, R.: FEBS Lett. **31**, 308–312 (1973b)
Rhodes, M.J.C., Yemm, E.W.: New Phytologist **65**, 331–341 (1966)
Robertson, D., Laetsch, W.M.: Plant Physiol. **54**, 148–159 (1974)
Roughan, P.G., Boardman, N.K.: Plant Physiol. **50**, 31–34 (1972)
Schiff, J.A.: The control of chloroplast differentiation in *Euglena*. In: Proc. 3rd Intern. Congr. Photosynthesis. Avron, M. (ed.). Amsterdam: Elsevier, 1975, Vol. III, pp. 1691–1717
Schnepf, E.: Planta (Berl.) **61**, 371–373 (1964)
Schultz, A., Sauer, K.: Biochim. Biophys. Acta **267**, 320–340 (1972)
Seliskar, C.J., Ke, B.: Biochim. Biophys. Acta **153**, 685–691 (1968)
Senger, H., Bishop, N.I., Wehrmeyer, W., Kulandaivelu, G.: Development of structure and function of the photosynthetic apparatus during light-dependent greening of a mutant of *Scenedesmus obliquus*. In: Proc. 3rd Intern. Congr. Photosynthesis. Avron, M. (ed.). Amsterdam: Elsevier, 1975, Vol. III, pp. 1913–1923
Shibata, K.: J. Biochem. **44**, 147–173 (1957)
Shlyk, A.A., Rudoi, A.B., Vezitsky, A.Y.: Metabolism of chlorophyll pigments at centers of biosynthesis during the initial stages of greening of etiolated seedlings. In: Proc. 2nd Intern. Congr. Photosynthesis. Forti, G., Avron, M., Melandri, A. (eds.). The Hague: Dr. Junk, 1972, Vol. III, pp. 2289–2308
Sironval, C., Bronchart, R., Michel, J.M., Brouers, M., Kuyper, Y.: Bull. Soc. Franc. Physiol. Végétale **14**, 195–225 (1968)
Smillie, R.M., Nielsen, N.C., Henningsen, K.W., von Wettstein, D.: Ontogeny and environmental regulation of photochemical activity in chloroplast membranes. In: Proc. 3rd Intern. Congr. Photosynthesis. Avron, M. (ed.). Amsterdam: Elsevier, 1975, Vol. III, pp. 1841–1860
Smith, J.H.C.: Plant Physiol. **29**, 143–148 (1954)
Smith, J.H.C., Durham, L.J., Wurster, C.F.: Plant Physiol. **34**, 340–345 (1959)
Strasser, R.J., Sironval, C.: FEBS Lett. **28**, 56–60 (1972)
Süzer, S., Sauer, K.: Plant Physiol. **48**, 60–63 (1971)
Tevini, M.: The formation of lipids following illumination of etiolated seedlings. In: Proc.

2nd Intern. Congr. Photosynthesis. Forti, G., Avron, M., Melandri, A. (eds.). The Hague: Dr. Junk, 1972, Vol. III, pp. 2471–2488

Thorne, S.W.: Biochim. Biophys. Acta **226**, 113–127 (1971)

Thorne, S.W., Boardman, N.K.: Plant Physiol. **47**, 252–261 (1971)

Thorne, S.W., Boardman, N.K.: Biochim. Biophys. Acta **267**, 104–110 (1972)

Treffry, T.: Planta (Berl.) **91**, 279–284 (1970)

Tremoliéres, A., Lepage, M.: Plant Physiol. **47**, 329–334 (1971)

Vaughan, G.D., Sauer, K.: Biochim. Biophys. Acta **347**, 383–394 (1974)

Virgin, H.I.: Physiol. Plantarum **13**, 155–164 (1960)

Weier, T.E., Brown, D.L.: Am. J. Bot. **57**, 267–275 (1970)

Wettstein, D. von: Brookhaven Symp. Biol. **11**, 138–159 (1958)

Whatley, F.R., Gregory, P., Haslett, B.G., Bradbeer, J.W.: Development of electron transport intermediates in greening chloroplasts. In: Proc. 2nd Intern. Congr. Photosynthesis. Forti, G., Avron, M., Melandri, A. (eds.). The Hague: Dr. Junk, 1972, Vol. III, pp. 2375–2381

Wolf, F.T.: Z. Pflanzenphysiol. **64**, 124–129 (1971)

Wolff, J.B., Price, L.: Arch. Biochem. Biophys. **72**, 293–301 (1957)

V. Algal and Bacterial Photosynthesis

1. Eukaryotic Algae

W. URBACH

A. Introduction

Plant physiologists and biochemists have made manifold use of algae for the study of photosynthesis. Since the introduction of *Chlorella* as a standard object of photosynthetic research in 1919 by WARBURG, experiments, especially with eukaryotic algae, have provided important contributions to our present knowledge on photosynthesis. Recently an excellent survey has been presented by MYERS (1974) on the progress in photosynthesis research in the last half century.

The following article on algal photosynthesis is added to the preceding survey on the photosynthetic electron transport and photophosphorylation, because those findings are based mainly on in vitro studies with isolated chloroplasts and thylakoid fragments of higher plants. It considers the properties of some photosynthetic processes in algae and underlines to what extent studies with algae have contributed to present knowledge on the operation of electron flow pathways and photophosphorylating systems. Since, however, eukaryotic algae, unlike blue-green algae and photosynthetic bacteria, possess electron transport systems rather similar to those of the chloroplasts of higher plants, references will frequently be made to the corresponding articles in this volume to avoid repetition.

This article cannot give complete coverage to the extensive field of photosynthesis in eukaryotic algae and will attempt primarily to outline the function of the photosynthetic apparatus and the photosystems, the electron transport process, and photophosphorylation reactions of these algae. Even these subjects cannot be reviewed thoroughly and reference will be made to other specialized articles. A detailed survey on photophosphorylation in vivo, based mainly on algal studies is given by GIMMLER (1977) in this volume, Chapter III.13. Aspects of the relationship of algal photosynthesis to carbon metabolism will be discussed extensively in several reviews in the second volume on photosynthesis of this encyclopedia.

Since the publication of the first edition of this encyclopedia in 1960, which includes many detailed reviews on algal photosynthesis, progress has been made in several subjects especially by using new approaches for measuring separate photosynthetic reactions in algae. Various aspects of this development in algal photosynthesis have been discussed in the following reviews: SMITH and FRENCH, 1963 (pigments); BLINKS, 1964 (pigments); DUYSENS, 1964 (general); HIND and OLSON, 1968 (electron transport); FORK and AMESZ, 1969, 1970 (energy transfer and electron transport); LEVINE, 1969a; BISHOP, 1973 (mutants); CHENIAE, 1970 (O_2 evolution); HALLDAL, 1970 (photobiology of microorganisms); BISHOP, 1971 (electron transport); MYERS, 1971 (enhancement); GOVINDJEE and PAPAGEORGIOU, 1971; GOEDHEER, 1972 (fluorescence); SIMONIS and URBACH, 1973 (photophosphorylation); RAVEN, 1974 (electron transport and photophosphorylation); GOVINDJEE and BRAUN, 1974 (absorption spectra and pigment systems); GIMMLER, 1977 (photophosphorylation). The proceedings of the international congresses on photosynthesis research (Metzner, 1969; Forti et al., 1972; Avron, 1975) contain many articles on photosynthetic processes in eukaryotic algae.

B. Objects

Eukaryotic algae offer many advantages in the investigation of photosynthesis. In contrast to higher plants they are simple systems, especially in the case of unicellular algae. Some of them possess a similar and comparable composition of pigments (e.g. Chlorophyceae, Euglenophyceae) which allows to extend conclusions of experimental results to higher plants. On the other hand, the variation of the pigment composition either by the lack of some components (e.g. chlorophyll b in Cryptophyceae, Dinophyceae, Chrysophyceae, Bacillariophyceae, Xanthophyceae, Phaeophyceae, Rhodophyceae) or by the presence of additional pigments like phycobilins (e.g. Rhodophyceae) or fucoxanthin (e.g. Phaeophyceae) offers the possibility of studying primary photosynthetic reactions in relation to the pigment systems. The considerable ability to change pigment composition in response to various environmental conditions favors the use of algae for experiments on special problems of photosynthesis (cf. Halldal, 1970).

Another form of adaptation, typical for several eukaryotic algae, is the ability to metabolize molecular hydrogen. Since these algae possess a hydrogenase system, they can use H_2 as an electron acceptor. Studies with these algae have provided many contributions to the understanding of the function of photosynthetic reactions (see review by Kessler, 1974). Much progress has been made in the analysis of photosynthesis by studies with mutants, for which some eukaryotic algae are very suitable (Levine, 1969a, 1974; Bishop, 1973).

Unicellular algae are very useful experimental organisms for investigations in the field of photosynthesis. They can be cultured easily under controlled conditions in the laboratory and provide a simple experimental material. Diverse methods of culturing unicellular algae have been described by Myers (1960) and Starr (1971). The culture conditions (e.g. for Chlorella, Scenedesmus, Ankistrodesmus, Dunaliella, Euglena, Porphyridium and some diatoms) have been improved by using methods for synchronizing the cultures of these algae (see reviews by Pirson and Lorenzen, 1966; Tamiya, 1966; Lorenzen, 1970; Lorenzen and Hesse, 1974; cf. San Pietro, 1971). Synchronized cultures of unicellular algae provide homogenous cell material in a definite developmental stage with high growth rates, which yield optimal conditions for physiological experiments on photosynthesis. Investigations of the photosynthetic capacity, quantum yield, fluorescence, enhancement effect, the activities of the two photosystems, and photophosphorylation carried out with synchronized algae show that there occur distinct changes in these photosynthetic activities during the algal life cycle (Senger and Bishop, 1967, 1969; Schor et al., 1970; Senger, 1970; Bishop and Senger, 1971; Gimmler et al., 1971; Senger, 1975; Senger and Frickel-Faulstich, 1975).

An increasing number of investigations of photosynthetic processes in vitro have been done in the last years using preparations of chloroplasts from eukaryotic algae (BÖGER, 1971b; GRAHAM and SMILLIE, 1971) or chemically treated algae (PARK, 1971). Active particles or chloroplasts have been extracted with success from *Chlamydomonas* (e.g. GIVAN and LEVINE, 1967; SATO et al., 1971; BRAND et al., 1975; CURTIS et al., 1975; GARNIER and MAROC, 1975), *Euglena* (e.g. CHANG and KAHN, 1970; SHNEYOUR and AVRON, 1970, 1975; WILDNER and HAUSKA, 1975), *Scenedesmus* (e.g. PRATT and BISHOP, 1968; POWLS et al., 1969; BERZBORN and BISHOP, 1973), *Dunaliella* (BEN-AMOTZ and AVRON, 1972a), *Acetabularia* (STROTMANN and BERGER, 1969), *Hymenomonas* (JEFFREY et al., 1966), *Bumilleriopsis* (BÖGER, 1969b) and have been used for measuring partial reactions of electron transport or photophosphorylation in vitro. Treatment of algae with formaldehyde, glutaraldehyde, or benzoquinone yields a better penetration of redox reagents and allows measurements of various redox reactions in this algal material (cf. LUDLOW and PARK, 1969; PARK, 1971; GIMMLER and AVRON, 1971). In comparison with the results obtained from investigations of higher plant chloroplasts it seems desirable to measure photosynthetic activities also in algal chloroplasts. Some of the results from in vitro studies with algae are discussed by RAVEN (1974).

For studies of photosynthetic processes in eukaryotic algae mainly members of the following major algal groups have been used: Chlorophyceae (e.g. *Chlorella, Ankistrodesmus, Scenedesmus, Hydrodictyon, Chlamydomonas, Chlamydobotrys, Dunaliella, Ulva, Acetabularia, Caulerpa, Chara, Nitella, Tolypella*), Euglenophyceae (*Euglena gracilis*), Rhodophyceae (e.g. *Porphyridium cruentum, Porphyridium aeruginium, Porphyra, Iridaea, Cryptopleura, Schizymenia, Drouetia, Cryptonemia*), Phaeophyceae (e.g. *Pheostrophion, Endarachne*). Moreover some interesting and experimentally useful members of following groups of eukaryotic algae have been employed for photosynthesis research: Xanthophyceae (e.g. *Botrydiopsis, Bumilleriopsis*), Haptophyceae (e.g. *Hymenomonas*), Chrysophyceae (e.g. *Ochromonas*), and Bacillariophyceae (e.g. *Phaeodactylum*).

It is a continous task for plant physiologists to look for other organisms within the fascinating variety of eukaryotic algae, which may contribute to the elucidation of the complex process of photosynthesis in the future.

C. Pigments and Pigment Systems

The eukaryotic algae contain various pigments for absorption of light from the visible spectrum. These pigments may be classified into three main groups: chlorophylls, carotenoids, and phycobilins. *Chlorophylls* (Chl) are the basic photosynthetic pigments in algae as well as in higher plants and photosynthetic bacteria. Chlorophyll a is the essential pigment for the conversion of light energy to chemical energy. It is present in all algae, whereas the other algal chlorophylls (Chl b, c, and d) are not universally distributed (see reviews by EGLE, 1960; FRENCH, 1960; ALLEN, 1966; MEEKS, 1974; GOVINDJEE and BRAUN, 1974). Chlorophyll b is absent in many groups of eukaryotic algae but is found particularly in Euglenophyceae and Chlorophyceae, where it functions as a light-harvesting pigment transferring its absorbed light energy to Chl a in a manner comparable to higher plants. Chlorophyll c acts as an accessory pigment in Dinophyceae, Chrysophyceae,

Bacillariophyceae, Xanthophyceae, and Phaeophyceae as concluded from experiments with diatoms (Mann and Myers, 1968) and brown algae (Fork, 1963; Goedheer, 1970). While the Chl a/Chl b ratio generally ranges from 2/1 to 3/1 (Strain et al., 1971) the Chl a/Chl c ratio varies between 1.2/1 and 5.5/1 (Jeffrey, 1972). Chlorophyll d has been found in extracts of many Rhodophyceae (O'Heocha, 1971) but its photosynthetic function is still unknown. Eukaryotic algae, as well as higher plants, contain different forms of chlorophyll a (Chl a-660, Chl a-680, Chl a-685, Chl a-690, Chl a-705) (Brown, 1972; French, 1971; Govindjee and Braun, 1974). Also, two forms of Chl b absorbing at 640 and 650 nm may exist in green algae (Thomas, 1971). Although the chemical nature of these different forms of chlorophyll in vivo is still unknown, it is postulated that many forms may have evolved to increase the efficiency of transfer of excitation energy (see Seely, 1972). The energy transfer from Chl b to Chl a in algae and also from the various forms of Chl a to the reaction center is very efficient (Cho and Govindjee, 1970).

A greater variety of *carotenoids* is found in algae than in higher plants (Hager and Stransky, 1970a, b; Goodwin, 1974). The distribution of carotenoids in algae and their function have been reviewed recently (Goedheer, 1970; Goodwin, 1971a, 1974; Bisalputra, 1974; Govindjee and Braun, 1971). Only β-carotene enjoys an ubiquitous distribution (Goedheer, 1970). From the numerous types of xanthophylls in algae, beside the high content of lutein especially in Chlorophyceae, the occurrence of fucoxanthin primarily in Phaeophyceae, Chrysophyceae, and Bacillariophyceae is of importance. The energy transfer from carotenoids to Chl a in green algae is about 50% (Emerson and Lewis, 1943; Cho and Govindjee, 1970) and from fucoxanthin to Chl a about 70% (Tanada, 1951).

The *phycobilins,* phycoerythrin and phycocyanin, generally occur only in Cyanophyceae and Rhodophyceae. Since these algae contain only Chl a, the phycobilins are important for light absorption and energy transfer. The transfer of excitation energy from phycoerythrin to Chl a is very efficient (French and Young, 1952; Brody and Brody, 1959). The distribution and function of the phycobilins are discussed in many excellent reviews (see Egle, 1960; O'Heocha, 1965, 1971; Goodwin, 1974; Bogorad, 1975).

The great variety of pigment composition in algae and its flexibility by adaptation to various environmental factors (see Halldal, 1970; Govindjee and Braun, 1974) favor the use of these organisms for photosynthesis research. Besides blue-green algae, red algae in particular show alterations in pigment ratios and photosynthetic efficiencies of Chl a and phycobilins (Rabinowitch, 1951; Yocum and Blinks, 1958; Brody and Emerson, 1959 b) if grown at different light intensities and light qualities. In diatoms and brown algae, as well as in green algae, variations in pigment compositions can be induced by light of varying intensities (Brown and Richardson, 1968; Halldal, 1970). An increase of the Chl b/Chl a ratio in green algae grown at low light intensities (as known from shade plants) and high amounts of photosynthetically inactive carotenoids in algae grown at high light intensities have been observed (Brown and Richardson, 1968; Oquist, 1969).

The concept of two light reactions operating in photosynthesis led to the assumption that each of the two photoreactions is functioning in a separated pigment system. Investigations of the composition of the two pigment systems have shown that most of the pigments are present in both systems but their composition is different in regard to concentrations. In green algae, as well as in higher plants, pigment system I has a higher Chl a/Chl b ratio than pigment system II. Red algae contain a larger amount of phycobilins in pigment system II.

From the different forms of Chl a, probably present in all algae (with the possible exception of red and blue-green algae) and in higher plants, Chl a-670 and Chl a-678 are found in both pigment systems with a somewhat greater proportion of Chl a-670 in pigment system II (GOVINDJEE et al., 1967; GOVINDJEE and BRAUN, 1974). The distribution of Chl a-660 between both pigment systems is not yet clear. The long wavelength Chl a forms (Chl a-685, Chl a-690, Chl a-705) are predominantly present in pigment system I. It is commonly accepted that Chl b is present in both pigment systems with a larger portion in pigment system II, but there are only a few reports about the distribution of the two Chl b forms. In eukaryotic algae lacking Chl b, the distribution described for this pigment may be true for that of Chl c. Investigations on the carotenoid distribution in the two pigment systems generally show that the carotenes (especially β-carotene) are present mainly in pigment system I, while the xanthophylls are found predominantly in pigment system II. In the phycobilin-containing algae a large proportion of chlorophyll a forms is found in pigment system I and a greater amount of phycobilins is present in pigment system II. The pigment systems of these algae are not yet well understood but it is known that the phycobilins are located in special bodies — the phycobilisomes.

Much of our knowledge on the distribution of pigments between both pigment systems in algae has been obtained from investigations which tried to fractionate algae physically by several procedures in order to separate the two systems. In a similar way to that used with higher plants (see BOARDMAN, 1970), generally two fractions e.g. from green algae (*Chlorella, Scenedesmus, Chlamydomonas, Euglena*) have been obtained indicating a partial separation of the pigment systems (ALLEN et al., 1963; BROWN et al., 1965; BROWN, 1969; FRENCH, 1971). Fractionation experiments with red algae (*Porphyridium cruentum, Porphyra*) and the diatom *Phaeodactylum tricornutum* are more difficult (OGAWA et al., 1968; BROWN, 1969). The two fractions found in these algae are interesting, since one of them contains more long wavelength forms of Chl a together with carotenes (β-carotene), while in the other fraction a larger proportion of Chl a-670 and more xanthophylls are present. In *Phaeodactylum* the fraction comparable to photosystem II-enriched particles of higher plants and green algae contains more Chl c than Chl b (OGAWA et al., 1968). In general, the composition of the two pigment systems of eukaryotic algae shows many similarities with that of higher plants.

Estimations of the number of Chl molecules necessary to evolve one molecule of O_2 using cells of Chlorella with varying Chl concentrations and bright short flashes for excitation of the photosystems led to the first formulation of the concept of the photosynthetic unit (PSU) by EMERSON and ARNOLD (1932). They found that in Chlorella about 2,400 molecules of Chl are capable of evolving one molecule of O_2 or reducing one molecule of CO_2. Since about eight quanta of light must be absorbed for the evolution of one molecule O_2, the size of the photosynthetic unit for each of the primary photoacts was calculated to be 250–300 molecules of Chl. These findings were supported by many investigations (cf. SCHMID and GAFFRON, 1968). Whereas in higher plants the PSU varies between 300 and 500 Chl/CO_2 fixed, it is rather more constant in algae (MYERS and GRAHAM, 1971;

SCHMID and GAFFRON, 1971) even during their life cycle (MYERS and GRAHAM, 1975). In chlorophyll deficient mutants of Chlorella, however, smaller sizes of the PSU have been estimated (400–900 Chl/O_2 evolved) (WILD and EGLE, 1968; WILD et al., 1971; DUBERTRET and JOLIOT, 1974), which are developed during the greening process in a characteristic manner (HERRON and MAUZERALL, 1972). Studies of the greening process of Chlorella mutants have shown that during the development of the PSU the enzymes of the reaction centers and the electron transfer agents are assembled first and then more antenna chlorophyll and the carotenoids are added (HERRON and MAUZERALL, 1972). From investigations of the chloroplast structure of Chlorella mutants with largely undeveloped chloro-plasts, as well as from observations of red and blue-green algae it is concluded that highly organized chloroplasts are not required for an effective photosynthetic apparatus. Investigating the formation of a photosystem II unit during the greening of a dark-grown Chlorella mutant, the presence of an active photosystem II at the initial stages could be shown. It consists of a relative constant amount of Chl a, and during the greening it is complemented by the addition of Chl b, which causes a two-fold increase in the size of the unit (DUBERTRET and JOLIOT, 1974).

The biosynthesis of chlorophylls and other pigments has very often been studied with algae. The great advantage of the availability of special algal mutants (GRA-NICK, 1971), and of algae growing heterotrophically in the dark, provide additional information on the biosynthetic pathway of chlorophylls (GOODWIN 1971 b; MEEKS, 1974).

The close relation between the biosynthesis of the chlorophylls and its accumula-tion in the chloroplast membranes and the development of thylakoids and their photosynthetic activities have been studied extensively in eukaryotic algae, espe-cially in *Euglena* (SCHIFF, 1971, 1975), different strains of *Chlamydomonas reinhardi* (OHAD et al., 1972), and Clorella (cf. HASE, 1971).

For better understanding of the function of the photosynthetic apparatus in algae a detailed study on the structure and ultrastructure of algal plastids and chloroplast membranes is neces-sary. Chloroplasts of eukaryotic algae possess thylakoids arranged in various ways. In red algae the thylakoids may remain separated from each other along their length with phycobili-somes located on their surfaces, and in brown algae they are arranged in 3-thylakoid bands (GIBBS, 1970). Stacked thylakoids similar to those in chloroplasts of higher plants are character-istic for the chloroplast of Euglenophyceae and Chlorophyceae (WEIER et al., 1966; GIBBS, 1970; GOODENOUGH and STAEHELIN, 1971).

For information on the comparative ultrastructure of algal chloroplasts the reader is referred to several excellent reviews (GRANICK, 1961; MANTON, 1966; ECHLIN, 1970; BISALPU-TRA, 1974; DODGE, 1974).

D. Electron Transport and Photophosphorylation

I. General Aspects

Photosynthetic energy conversion is initiated by light absorption of the various pigments present in photosynthesizing organisms. During the process of light ab-sorption by the light-harvesting pigments the light energy is transfered to the reaction centers of the pigment systems. Here the energy can be utilized for photochemical reactions which can drive electron transport systems and form "high energy phosphate". If absorbed energy is not used for photosynthetic reactions it can be dissipated by emission of light quanta in the form of fluorescence and delayed light emission.

Measurements of absorption spectra in connection with action spectra and diverse fluorescence spectra of various algae have provided a great deal of information on the photosynthetic apparatus of these organisms and its function (EMERSON and LEWIS, 1943; HAXO, 1960; FORK, 1963; BLINKS, 1964; FORK and AMESZ, 1969, RIED, 1972).

Absorption spectra are also needed for calculations of quantum yields of photosynthesis and its various processes. Measurements of quantum yields of photosynthesis in Chlorella have been a subject of a long controversy between WARBURG and EMERSON (cf. KOK, 1960). The value of 8 or more quanta per O_2 molecule found by EMERSON (1958) is now generally accepted and is in correspondence with the present scheme of photosynthesis. It was in early experiments with Chlorella that the "red drop" phenomenon i.e. the decrease in quantum yield in the long wavelength region of the spectrum was discovered (EMERSON and LEWIS, 1943; EMERSON et al., 1957; DAS and GOVINDJEE, 1967; DAS et al., 1968). This effect can today be explained by the function of two pigment systems in photosynthesis, since in the far red region one pigment system predominantely absorbs, resulting in inefficient photosynthesis.

In red as well as in blue-green algae, the red drop begins before the decline of the Chl a absorption band in the red and is accompanied by a drop in the quantum yield in the blue region (BRODY and EMERSON, 1959a; HAXO, 1960; HOCH and KOK, 1961; FORK, 1963). These declines of the quantum yield can be avoided if supplementary light, which can be absorbed by the accessory pigments (e.g. Chl b in green algae, the phycobilins in red and blue-green algae, and fucoxanthin in brown algae), is used in addition to the light absorbed predominantely by Chl a (EMERSON et al., 1957; EMERSON, 1958; EMERSON and RABINOWITCH, 1960; FORK, 1963). This effect known as the "Emerson enhancement effect" has been demonstrated in many different algal strains and in chloroplasts of higher plants and is now accepted as a general characteristic of photosynthetic organisms which are able to evolve oxygen (MYERS, 1971). It suggests the necessity of a simultaneous excitation of Chl a and the accessory pigments for high quantum yields (EMERSON, 1958). The action spectra of the enhancement effect in various algae containing different accessory pigments show peaks at the absorption maxima of these pigments and also around 670 nm (GOVINDJEE and RABINOWITCH, 1960), which indicates the effectivity of the accessory pigments and of the shorter wavelength forms of Chl a (e.g. Chl a-670). The enhancement effect has also been observed when the far red and the supplementary light were given successively (MYERS and FRENCH, 1960; MYERS, 1971).

Related to the enhancement effect was the observation of BLINKS (1957) of chromatic transients during his measurements of photosynthesis of the red alga Porphyra (BLINKS, 1960; MYERS, 1971). Conclusions drawn from these and other results, mostly obtained by experiments with various algae, again provided much of the evidence for the postulation of the concept of two pigment systems.

The discovery of P-700 as the reaction center of pigment system I by measuring the decrease of absorption at 700 nm upon illumination with far red light and its restoration by red light, led to the first proposal for the separated light reactions (KOK, 1957, 1961). Several other experimental observations in the early 1960s have been presented, which indicated strongly the necessity of two light reactions operating in series (HILL and BENDALL, 1960; KAUTSKY et al., 1960; DUYSENS et al., 1961; WITT et al., 1961).

In one of the first direct demonstrations of the existence of two light reactions it was shown that in red algae illumination with red light, which is absorbed by Chl a, causes an oxidation of cytochrome f, while green light, which is absorbed by the phycobilins, causes a reduction of this cytochrome (DUYSENS et al., 1961). Similar experiments were carried out with blue-green algae concerning P-700 (KOK and HOCH, 1961) and with Chlorella and chloroplasts measuring redox reactions of cytochrome and plastoquinone (WITT et al., 1961; RUMBERG et al., 1964). Studies of the action spectra of cytochrome oxidation and reduction provided informations on the contribution of active pigments and its composition in the two photosystems (DUYSENS and AMESZ, 1962). Thus, at first a great deal of evidence for the characterization of the two photosystems and the electron transport pathways were derived from experiments with algae, using absorption difference spectrophotometry (see reviews by WITT, 1960; DUYSENS, 1964; FORK and AMESZ, 1970). A first summary of studies on light-induced absorption changes of intact algae (Chlorella, Scenedesmus, Nitzschia, Nostoc, Porphyridium, Porphyra) has already been presented by KOK (1957). In the meantime, a large body of experimental data has been accumulated concerning the localization and function of electron carriers in the electron transport chain, and the nature of endogenous electron acceptors of the photosystems, mainly obtained from experiments with isolated chloroplasts and thylakoid membranes of higher plants (see review by TREBST, 1974 and preceding articles in this volume).

In general the composition of the two photosystems and the arrangement of the electron carriers in the electron transport systems of eukaryotic algae are similar to that of chloroplasts of higher plants. But due to the greater variety of algae e.g. in relation to pigment composition, and to the high variability obtained by changing growth and environmental conditions, there are some differences and peculiarities in the nature of electron acceptors and carriers, their sequences and requirements in the electron transport systems in comparison to that of higher plants. The differences between algal photosynthesis and photosynthesis in higher plants, however, are much more relevant with regard to secondary processes of photosynthesis e.g. carbon metabolism, and will not be discussed in this article.

At the present time it seems well-established that several endogenous photosynthetic electron transport systems exist in algae and higher plants (see reviews by SIMONIS and URBACH, 1973; RAVEN, 1974; TREBST, 1974; GIMMLER, this vol. Chap. III.13). There is a large body of evidence to show that at least two different native electron transport systems must be distinguished: a noncyclic and a cyclic electron flow. Additionally a pseudocyclic electron transport system in vivo as is known in isolated chloroplasts (Mehler-Reaction), has been increasingly discussed in the last years (see GIMMLER, this vol. Chap. III.13).

The arrangement and functioning of components in the electron transport systems in algae were studied using diverse experimental approaches. In addition to absorption difference spectrophotometry, including inhibitors or mutants lacking in special components of the photosynthetic systems (cf. LEVINE, 1969a; BISHOP, 1973), measurements of fluorescence phenomena have been extensively used. The successful application of fluorescence studies to elucidate the composition and operation of the photosystems in algae has been discussed thoroughly by GOVINDJEE and PAPAGEORGIOU (1971) and GOEDHEER (1972).

II. Photosystems

The chemical nature of the two photoreactions in algae has been much less investigated than that in higher plants. The first investigated component of a primary reaction in photosynthesis of algae is P-700 which is involved in photosystem I (KOK and HOCH, 1961). After its discovery in algae it has been investigated primarily in higher plants (see this vol. Chap. II.4). Much less is known about the nature of the primary electron acceptor of photosystem I. There is some evidence obtained by flash spectrophotometry that P-430 is the long-sought electron acceptor of photosystem I observed in spinach, as well as in blue-green algal particles (HIYAMA and KE, 1971a, b; KE, 1975; cf. this vol. Chap. II.7). Even the chemical nature of photosystem II is still unknown. The component P-690, detected by light-induced absorbance changes only in photosystem II-enriched particles of higher plant chloroplasts (DÖRING et al., 1967), might be a component of the primary reaction in photosystem II, paralleling P-700 in photosystem I (cf. KOK et al., 1975) but it has not yet been observed in algae. The primary electron acceptor of photosystem II called Q (as a quencher of fluorescence) has been characterized by kinetic studies of O_2 evolution in algae also (see review by CHENIAE, 1970). Associated with the unknown electron acceptor Q might be the component C-550, discovered by KNAFF and ARNON (1969), which has also been observed in green algae (BENDALL and SOFROVA, 1971). Recently the existence of a special electron acceptor of photosystem II is postulated in photoheterotrophically cultivated Chlamydomonas stellata (MENDE and WIESSNER, 1975).

Experiments on the separation of the photosystems in order to enable one to investigate more directly the composition of these systems, which have been done with higher plants (see reviews by BOARDMAN, 1970; PARK and SANE, 1971 and preceding articles of this volume), are much more difficult with algae than with chloroplasts of higher plants, as already mentioned (see BROWN, 1973).

The significance of chlorophyll proteins as light-harvesting and reaction center components studied in chloroplasts of algae and higher plants has been extensively reviewed by THORNBER (1975).

III. Noncyclic Electron Transport

In comparison to the components in the photosystems, the chemical nature of possible electron carriers and acceptors of the electron transport systems and their location and function in the chain are better known (see preceding articles of this volume). Most of these intermediates, like cytochrome f, cytochrome b-559, cytochrome b-563, plastocyanin, plastoquinone, and ferredoxin have also been detected in algae (Amesz and Vredenberg, 1966; Bendall and Hill, 1968; Ikegami et al., 1968; Levine, 1969a; Barr and Crane, 1971; Buchanan and Arnon, 1971; Yakushiji, 1971; Levine and Armstrong, 1972).

On the basis of the generally accepted Z-scheme of electron transport in photosynthesis, the noncyclic electron transport system from H_2O to $NADP^+$ includes in the region between the two light reactions at least two cytochromes e.g. cytochrome f and cytochrome b-559 (probably in its low potential form), plastocyanin, and plastoquinone. The properties of cytochrome f of various eukaryotic algae (Yakushiji, 1971) e.g. *Euglena* (Perini et al., 1964; Böger and San Pietro, 1967; Mitsui, 1971) *Chlamydomonas* (Gorman and Levine, 1966b), *Chlorella* (Grimme and Boardman, 1975), *Bumilleriopsis* (Lach et al., 1973; Lach and Böger, 1974), *Porphyra* (Katoh, 1960a) and its arrangement and function in noncyclic electron transport in relation to the photosystems and the other intermediates of this electron transport chain has been investigated in red algae (Duysens and Amesz, 1962; Duysens, 1964; Nishimura and Takamiya, 1966; Amesz and Fork, 1967; Nishimura, 1968; Amesz et al., 1972a, b; Gimmler and Avron, 1972; Biggins, 1973) and in green algae (Levine and Armstrong, 1972 (*Clamydomonas*), Katoh and San Pietro, 1967 (*Euglena*), Fork and Urbach, 1965 (*Ulva*), Fork and Urbach, 1966 (*Chlorella*), Powls et al., 1969 (*Scenedesmus*), Wildner and Hauska, 1974, 1975 (*Euglena*), Senger and Frickel-Faulstich, 1975 (*Chlamydomonas, Scenedesmus*).

These experiments generally show that in eukaryotic algae cytochrome f is functioning at the oxidizing site of photosystem I. Studies with mutants of *Chlamydomonas* (Gorman and Levine, 1965, 1966c; Levine and Gorman, 1966) support these findings, although the exact position of cytochrome f in the electron transport chain of algae, especially in relation to plastocyanin, seems to be still unclear (Fork and Amesz, 1970; Kunert and Böger, 1975). There are some indications that cytochrome f is involved also in the cyclic electron transport system (see below).

The role and localization of b-type cytochromes in the noncyclic electron transport system of algae is difficult to define. Cytochrome b-559, which exists in a high and in a low potential form, is most likely functioning between the two photosystems in certain algae (Levine and Gorman, 1966; Ikegami et al., 1970; cf. Hind and Olson, 1968; Bishop, 1971). It has been argued that probably the low potential form of cytochrome b-559 is located in the noncyclic electron transport chain, whereas the high potential form is functioning under special conditions in a cyclic electron flow around photosystem II (cf. Bendall and Sofrova, 1971; Epel and Butler, 1971; Epel et al., 1972; Levine, 1974; Garnier and Maroc, 1975; Maroc and Garnier, 1975). Cytochrome b-563 is most likely an intermediate of the cyclic electron flow with photosystem I (see below).

Plastocyanin, as an apparently ubiquitous photosynthetic electron carrier in algae and higher plants (see Katoh, this vol. Chap. II.14), has been isolated also from several eukaryotic algae e.g. from *Chlorella* (Katoh, 1960b; Kunert and Böger, 1975), *Scenedesmus* (Powls et al., 1969; Kunert and Böger, 1975) and *Chlamydomonas* (Gorman and Levine, 1966c).

Its redox potential, and also a number of properties of this component, are similar to those of cytochrome f in algae but not in higher plants, which apparently make these two electron carriers interchangeable in certain photosynthetic reactions (cf. ELSTNER et al., 1968; BISHOP, 1971). The eukaryotic alga *Bumilleriopsis* shows a lack of plastocyanin which can be replaced by cytochrome f, functioning as a direct electron donor for photosystem I (KUNERT and BÖGER, 1975). Probably due to this fact, and to some difficulties in measuring the redox reactions of plastocyanin by light-induced absorbance changes (KOUCHKOVSKY and FORK, 1964), the localization of this component in the electron transport chain is still an open question (see reviews by HIND and OLSON, 1968; FORK and AMESZ, 1970; BISHOP, 1971). Whereas experiments with *Ulva* (FORK and URBACH, 1965) and *Chlorella* (FORK and URBACH, 1966) indicate that plastocyanin is functioning before cytochrome f, studies on wild-type and mutant strains of *Chlamydomonas* (GORMAN and LEVINE, 1965, 1966c; LEVINE and GORMAN, 1966) suggest the opposite sequence in the noncyclic electron transport chain. Additionally, a parallel arrangement of both electron carriers has been discussed (KOK et al., 1964). Since the observed redox reactions of plastocyanin in vivo are relatively slow, it has been argued that electron flow cannot be completely carried by plastocyanin, and may also be reduced by a cyclic electron flow, as discussed for cytochrome f (see below).

The existence and function of a large pool of plastoquinone near photosystem II (see AMESZ, this vol. Chap. II.13) has been investigated in great detail in eukaryotic algae also (SUN et al., 1968; CHENIAE, 1970; FORK and AMESZ, 1970; BISHOP, 1971). Comparative studies of the reactions of plastoquinone, mainly by measuring UV absorption changes in relation to the other intermediates of the electron transport chain and to fluorescence and oxygen evolution of algae (RUMBERG et al., 1964; STIEHL and WITT, 1968; AMESZ et al., 1972a, b, d; AMESZ, 1973; SENGER and FRICKEL-FAULSTICH, 1975), support the findings in chloroplasts of higher plants regarding the role of this component in the noncyclic transport system (see AMESZ, this vol. Chap. II.13). Additionally it could be shown that plastoquinone may also be involved in the cyclic electron transport of algae (see below).

Absorption difference spectroscopy has revealed a pronounced absorbance change at 515 to 520 nm, similar to that observed in higher plants, which has been found in several green algae e.g. *Chlorella* (DUYSENS, 1954; RUBINSTEIN and RABINOWITCH, 1964; GOVINDJEE and GOVINDJEE, 1965; FORK and URBACH, 1966; KOUCHKOVSKY, 1969), *Ulva* (KOUCHKOVSKY and FORK, 1964; FORK and KOUCHKOVSKY, 1966), *Scenedesmus* (PRATT and BISHOP, 1968), *Chlamydomonas* (CHUA and LEVINE, 1969). This complex absorbance change, which has been attributed to a number of components (e.g. chlorophyll b, carotenoids) functioning in the electron transport chain near photosystem II, may partly indicate a membrane potential (see reviews by FORK and AMESZ, 1970; WITT, 1971). In this connection it should be mentioned that a number of red and brown algae (FORK and AMESZ, 1967) and a yellow-green alga (FORK, 1969) show light-induced shifts in the absorption spectrum of carotenoids comparable to the absorption change at 515 nm, which indicate the participation of carotenoids also in photosynthesis of algae (cf. FORK and AMESZ, 1970).

Ferredoxin as a substantial component of the photosynthetic electron transport system (see HALL and RAO, this vol. Chap. II.9) has also been isolated from several eukaryotic algae (BÖGER and SAN PIETRO, 1967; MATSUBARA, 1968; BÖGER, 1970; BUCHANAN and ARNON, 1971). Its function in the electron transport chain in relation to ferredoxin-NADP$^+$-oxidoreductase and in certain ferredoxin catalyzed reactions has been intensively investigated (Powls et al., 1969; Böger, 1969a, 1971a; cf. Bishop, 1971; Böger et al., 1973; Trebst, 1974). In some algae the flavoprotein phytoflavin partly substitutes for ferredoxin in iron-deficient cultures (cf. Smillie and Entsch, 1971; Zumft and Spiller, 1971). In addition to its function in the reduction of $NADP^+$, the generally accepted terminal electron acceptor in noncyclic electron transport path, ferredoxin may also play an essential role in cyclic electron flow in vivo (see below).

In studies with the halophilic green alga *Dunaliella parva*, the significance of $NADP^+$ as an obligatory intermediate in photosynthesis has been questioned (Ben Amotz and Avron, 1972b). Within the O_2-evolving steps or at another site before photosystem II, manganese seems to play an important role in the electron transport chain in algae also (see reviews by Kok and Cheniae, 1966; Cheniae, 1970).

IV. Cyclic Electron Transport

The existence of a cyclic electron transport system of photosynthesis in eukaryotic organisms in vivo is mainly inferred from experiments with various algae, which are described in detail in several reviews over the last few years (Simonis and Urbach, 1973; Raven, 1974; Gimmler, this vol. Chap. III.13). Most of these data are obtained from measurements of photophosphorylation by more or less indirect approaches (Urbach and Simonis, 1964; Kandler and Tanner, 1966; Simonis, 1967; Tanner et al., 1969; Wiessner, 1970a; Ried, 1970; Raven, 1971), but direct measurements of cyclic electron flow in algae have also been carried out by absorption difference spectroscopy (Levine, 1969b; Amesz et al., 1972a, b; Biggins, 1973, 1974) and steady-state relaxation spectrophotometry (Teichler-Zallen and Hoch, 1967; Rurainski et al., 1970; Hoch and Randles, 1971; Rurainski and Hoch, 1975). Also studies with mutants of *Chlamydomonas* and *Scenedesmus* (Teichler-Zallen et al., 1972, see reviews by Levine, 1969a, 1974; Bishop, 1973, and inhibitor studies (cf. Gimmler, this vol. Chap. III.13) were helpful in elucidating the function of cyclic electron transport in vivo and the intermediates involved in this process.

Cyclic electron flow depends preferentially on the activity of photosystem I, but also in intact algae it needs apparently a precise redox balance (poising) by photosystem II (Van Rensen, 1969; Rurainski and Hoch, 1975) as has been observed in intact chloroplasts (Kaiser and Urbach, 1976). Investigations of the cyclic electron transport system in eukaryotic algae support results obtained with other photosynthetic organisms (Hind and Olson, 1968; Böhme and Cramer, 1972) that cytochrome b-563 is most likely involved in the cyclic electron flow pathway of these algae (Levine, 1969b; Ikegami et al., 1970; Amesz et al., 1972b, c). This conclusion has already been drawn from inhibitor studies with antimycin (Urbach and Simonis, 1964; Tanner et al., 1965), which blocks cyclic photophosphorylation in algae (see reviews by Simonis and Urbach, 1973; Gimmler, this vol. Chap. III.13). The inhibition of reactions in algae depending on cyclic photophosphorylation by antimycin and DSPD (Gimmler et al., 1968; Urbach and Gimmler, 1969) also suggests the participation of ferredoxin in cyclic electron flow, since the ferredoxin-catalyzed cyclic photophosphorylation in vitro is blocked by antimy-cin (TAGAWA et al., 1963) and DSPD inhibits only ferredoxin-dependent reactions in isolated chloroplasts (TREBST and BURBA, 1967). But, in comparison with the results obtained in vitro, more information is necessary in order to elucidate the role of ferredoxin in algal cyclic electron transport systems.

Evidence for the involvement of plastoquinone in cyclic electron flow of euka-ryotic algae (URBACH and KAISER, 1972; BIGGINS, 1974) and of intact chloroplasts of higher plants (KAISER and URBACH, 1976) has been obtained from experiments with the plastoquinone antagonist DBMIB (TREBST et al., 1970), but there have also been reported opposite observations with *Porphyridium* (AMESZ et al., 1972c; AMESZ, 1973) and chloroplasts (FORTI and ROSA, 1971). The functioning of a proton carrier-like plastoquinone in an endogenous cyclic transport system, how-ever, would be in accordance with the concept of TREBST (1974, 1975), in view of the necessity of a native energy conserving site in such a cyclic electron flow. Several data, mainly obtained from studies with mutants of *Chlamydomonas* (GOR-MAN and LEVINE, 1966c; LEVINE, 1969a; TEICHLER-ZALLEN et al., 1972) and *Scene-desmus* (POWLS et al., 1969; BISHOP, 1972, 1973) and spectrophotometric measure-ments in red and green algae (TEICHLER-ZALLEN and HOCH, 1967; RURAINSKI et al., 1970; BIGGINS, 1973) strongly indicate the participation of cytochrome f, and also of plastocyanin in the cyclic electron transport system of these algae.

V. Pseudocyclic Electron Transport

Since the discovery of a pseudocyclic electron flow with O_2 as the terminal electron acceptor (Mehler reaction) in isolated chloroplasts of higher plants attempts have been made to prove the existence of this reaction also in algae (HOCH et al., 1963; RAVEN, 1970; URBACH and GIMMLER, 1970b; BUNT and HEEB, 1971; ULL-RICH, 1971). Recently, in the green alga *Hydrodictyon* a light-induced $^{18}O_2$ uptake and a pseudocyclic photophosphorylation could be demonstrated (GLIDEWELL and RAVEN, 1975; RAVEN and GLIDEWELL, 1975), similar to that shown earlier in experiments with leaves of higher plants (HEBER, 1969), *Anacystis* (PATTERSON and MYERS, 1973), and intact chloroplasts of higher plants (HEBER, 1973; EGNEUS et al., 1975). The role and magnitude of the pseudocyclic electron transport system as an essential part of photosynthesis in vivo is still under investigation. Preliminary measurements of the rates of cyclic and pseudocyclic photophosphorylation in intact spinach chloroplasts indicate that the rates of the pseudocyclic photophos-phorylation reach maximally one half of the rates of cyclic photophosphorylation (KAISER and URBACH, unpublished).

The intermediates of the pseudocyclic electron transport system should be most likely the same as those of the noncyclic electron transport chain with the exception that, instead of $NADP^+$ oxygen acts as the terminal electron acceptor, and the electron pathway after photosystem I is probably somewhat different. Evidence for the existence and significance of pseudocyclic electron flow and photophosphor-ylation in vivo are discussed in detail in the review of GIMMLER in this vol. Chap. III.13.

VI. Regulation of Electron Transport Systems

If we accept that in algae and probably in all intact cells three different photosynthetic electron transport systems exist, the question arises as to how their function is regulated. There are many indications in the studies on photosynthetic reactions in algae under various conditions (e.g. light intensities, developmental stages) that a fairly strong interaction between the different electron transport systems in vivo occurs (Urbach and Simonis, 1964; Van Rensen, 1969; Urbach and Gimmler, 1970a; Senger, 1970; Gimmler et al., 1971; Raven, 1971; Urbach and Kaiser, 1972; Biggins, 1973). In particular, the changes in different photosynthetic activities during the life cycle of synchronized algae (Senger, 1970; Gimmler et al., 1971; Senger, 1975) may be caused by some factors regulating the electron transport systems (Schor et al., 1970; Senger and Frickel-Faulstich, 1975).

The participation of some cofactors, especially of plastoquinone and ferredoxin, in several electron transport systems implies that probably the redox levels of these components may be decisive in whether electrons flow to $NADP^+$ (noncyclic), to oxygen (pseudo-cyclic) or reenter the electron transport chain via cytochrom b-6 and plastoquinone in a cyclic electron pathway. Therefore it seems also reasonable to assume that e.g. the partial or complete absence or presence of CO_2 and/or O_2 favors one or other electron flow path in vivo (Heber, 1969; Jeschke and Simonis, 1969; Heber, 1973).

As already found in vitro (Tagawa et al., 1963; Gromet-Elhanan, 1967), the function of the cyclic electron transport system in vivo requires a slow electron flow from photosystem II, maintaining a precise redox balance of the cyclic system (poising). This has been concluded from its stimulation by certain concentrations of DCMU (Rurainski and Hoch, 1975; Kaiser and Urbach, 1976). More details, including some models of the regulation of the electron transport and photophosphorylation systems in vivo are discussed in the review by Raven (1974) and by Gimmler (this vol. Chap. III.13). The interactions between photosynthesis and respiration in algae is another important aspect of regulation in vivo (cf. Ried, 1968).

VII. Special Electron Acceptors

It seems worthwhile to mention that several eukaryotic algae are able to use various electron acceptors (e.g. H_2, NO_3^-, NO_2^-) when the electron flow is functioning under special conditions (e.g. H_2-atmosphere, anaerobiosis). Studies on hydrogen metabolism have provided many contributions to our knowledge on photosynthetic reactions and nitrate reduction. The ability to metabolize molecular hydrogen depends on the presence of a hydrogenase which has been found in certain bacteria and in about 50% of the algae from most major groups examined so far (see review by Kessler, 1974). Since the discovery by Gaffron (1940), the *photoreduction* of H_2 in hydrogen-adapted algae has been investigated in great detail, although the mechanism is still not clear. It is widely accepted that only photosystem I and cyclic photophosphorylation participate, and that the final reduction of CO_2 occurs with the hydrogen activated by hydrogenase via ferredoxin (Arnon et al., 1961; but see Kessler, 1974).

The mechanism of the *photoproduction* of H_2, which again occurs in algae with a hydrogenase system under anaerobic conditions, is also still a subject of some controversy. The results of many investigations, especially from GAFFRON and his co-workers (HEALY, 1970; STUART, 1971; STUART and GAFFRON, 1972a, b), seem to indicate that this process may be due to a noncyclic electron flow from photosystem II, or from organic substances through photosystem I to hydrogenase depending on the organism and experimental conditions used. For detailed information on hydrogen metabolism in algae, the reader is referred to the reviews by BISHOP (1966) and KESSLER (1974).

Reducible anions (e.g. NO_3^-, NO_2^-) can also serve as electron acceptors in algae. The requirement for ATP, however, e.g. for the light-dependent nitrate assimilation, is still an open question (cf. KESSLER, 1964; MORRIS and AHMED, 1969; ULLRICH, 1971, 1974; THACKER and SYRETT, 1972; MORRIS, 1974).

VIII. Photophosphorylation

Photosynthetic electron transport systems are normally coupled to synthesis of ATP, which is used in CO_2-fixation and in other energy-requiring processes in the cell. Therefore, investigations of the electron flow pathways in eukaryotic algae discussed in the preceding chapters of this review allow some conclusions on the properties and functions of photophosphorylation reactions in these organisms. On the other hand, studies of energy-requiring reactions in vivo, which are influenced by light, provide information on the complex process of photophosphorylation in intact cells. Generally, such indirect measurements of light-dependent reactions (e.g. photoassimilation of organic compounds (glucose, acetate), uptake of ions (phosphate, potassium, chloride), incorporation of ^{32}P into organic compounds and polyphosphates, photokinesis, photoreduction of CO_2 in H_2-adapted algae, photoinhibition of respiration, light-scattering changes) have been used to investigate the different reactions of photophosphorylation in vivo (cf. SIMONIS and URBACH, 1973). In addition to noncyclic and pseudocyclic photophosphorylation, the existence of a cyclic photophosphorylation could be demonstrated with these indirect approaches. Since these investigations, including several other aspects on photophosphorylation in vivo, have recently been reviewed in detail (SIMONIS and URBACH, 1973; RAVEN, 1974 and in the article by GIMMLER, this vol. Chap. III.13), the reader is referred to these articles for information on photophosphorylation in eukaryotic algae. Earlier studies have been discussed in reviews by KANDLER (1960), SIMONIS (1960), and WIESSNER (1970a, b). Various aspects of photophosphorylation in vitro with chloroplasts and particles isolated from several eukaryotic algae are described by RAVEN (1974).

References

Allen, M.B.: In: The Chlorophylls. Vernon, L.P., Seely, G.R. (eds.). New York-London: Academic Press, 1966, pp. 511–519

Allen, M.B., Murchio, J.C., Jeffrey, S.W., Bendix, S.A.: In: Studies on Microalgae and Photosynthetic Bacteria. Tokyo: Univ. Tokyo, 1963, pp. 407–412

Amesz, J.: Biochim. Biophys. Acta **301**, 35–51 (1973)

Amesz, J., van den Ergh, G.J., Visser, J.W.M.: In: Proc. 2nd Intern. Congr. Photosynthesis. Forti, G., Avron, M., Melandri, A. (eds.). The Hague: Dr. Junk, 1972d, Vol. I, pp. 419–430

Amesz, J., Fork, D.C.: Photochem. Photobiol. **6**, 903–912 (1967)

Amesz, J., Pulles, M.P.J., Visser, J.W.H., Subbing, F.A.: Biochim. Biophys. Acta **275**, 442–452 (1972c)

Amesz, J., Visser, J.W.H., van den Ergh, G.J., Dirks, M.P.: Biochim. Biophys. Acta **256**, 370–380 (1972a)

Amesz, J., Visser, J.W.H., van den Ergh, G.J., Pulles, M.P.J.: Physiol. Veg. **10**, 319–328 (1972b)

Amesz, J., Vredenberg, W.J.: In: Biochemistry of Chloroplasts. Goodwin, T.W. (ed.). New York-London: Academic Press, 1966, Vol. II, pp. 593–600

Arnon, D.I., Losada, M., Nozaki, M., Tagawa, K.: Nature (London) **190**, 601–606 (1961)

Avron, M. (ed.): Proc. 3rd Intern. Congr. Photosynthesis, 3 vols. Amsterdam: Elsevier, 1975, pp. 1–2194

Barr, R., Crane, F.L.: In: Methods in Enzymology. San Pietro, A. (ed.). New York-London: Academic Press 1971, **23A**, 372–408

Ben-Amotz, A., Avron, M.: Plant Physiol. **49**, 240–243 (1972a)

Ben-Amotz, A., Avron, M.: Plant Physiol. **49**, 244–248 (1972b)

Bendall, D.S., Hill, R.: Ann. Rev. Plant Physiol. **19**, 167–186 (1968)

Bendall, D.S., Sofrova, D.: Biochim. Biophys. Acta **234**, 371–380 (1971)

Berzborn, R.J., Bishop, N.I.: Biochim. Biophys. Acta **292**, 700–714 (1973)

Biggins, J.: Biochemistry **12**, 1165–1169 (1973)

Biggins, J.: FEBS Lett. **38**, 311–314 (1974)

Bisalputra, T.: Plastids. In: Algal Physiology and Biochemistry. Stewart, W.D.P. (ed.). Oxford: Blackwell Scientific, 1974, Vol. X, pp. 124–160

Bishop, N.I.: Ann. Rev. Plant Physiol. **17**, 185–208 (1966)

Bishop, N.I.: Ann. Rev. Biochem. **40**, 197–226 (1971)

Bishop, N.I.: In: Proc. 2nd Intern. Congr. Photosynthesis. Forti, G., Avron, M., Melandri, A. (eds.). The Hague: Dr. Junk, 1972, Vol. I, pp. 459–468

Bishop, N.I.: In: Photophysiology. Giese, A.C. (ed.). New York-London: Academic Press, 1973, Vol. VIII, pp. 65–96

Bishop, N.I., Senger, H.: In: Methods in Enzymology. San Pietro, A. (ed.). New York-London: Academic Press, 1971, **23**, 53–66

Blinks, L.R.: In: Research in Photosynthesis. Gaffron, H. (ed.). New York: Interscience, 1957, pp. 444–449

Blinks, L.R.: Proc. Nat. Acad. Sci. U.S. **46**, 327–333 (1960)

Blinks, L.R.: In: Photophysiology. Giese, A.C. (ed.). New York-London: Academic Press, 1964, Vol. I, pp. 199–221

Boardman, N.K.: Ann. Rev. Plant Physiol. **21**, 115–140 (1970)

Böger, P.: Z. Pflanzenphysiol. **61**, 447–461 (1969a)

Böger, P.: Z. Pflanzenphysiol. **61**, 85–97 (1969b)

Böger, P.: Planta (Berl.) **92**, 105–128 (1970)

Böger, P.: Planta (Berl.) **99**, 319–338 (1971a)

Böger, P.: In: Methods in Enzymology. San Pietro, A. (ed.). New York-London: Academic Press, 1971b, **23A**, 242–248

Böger, P., Lien, S.S., San Pietro, A.: Z. Naturforsch. **28c**, 505–510 (1973)

Böger, P., San Pietro, A.: Z. Pflanzenphysiol. **58**, 70–75 (1967)

Böhme, H., Cramer, W.A.: Biochim. Biophys. Acta **283**, 302–315 (1972)

Bogorad, L.: Ann. Rev. Plant Physiol. **26**, 369–401 (1975)

Brand, J.J., Curtis, V.A., Togasaki, R.K., San Pietro, A.: Plant Physiol. **55**, 187–191 (1975)

Brody, M., Emerson, R.: J. Gen. Physiol. **43**, 251–264 (1959a)

Brody, M., Emerson, R.: Am. J. Botany **46**, 433–440 (1959b)

Brody, S.S., Brody, M.: Arch. Biochem. Biophys. **82**, 161–178 (1959)

Brown, J.S.: Biophys. J. **9**, 1542–1552 (1969)

Brown, J.S.: Ann. Rev. Plant Physiol. **23**, 73–86 (1972)

Brown, J.S.: In: Photophysiology. Giese, A.C. (ed.). New York-London: Academic Press, 1973, Vol. VIII, pp. 97–112

Brown, J.S., Bril, C., Urbach, W.: Plant Physiol. **40**, 1086–1090 (1965)
Brown, T.E., Richardson, F.T.: J. Phycol. **4**, 38–54 (1968)
Buchanan, B.B., Arnon, D.I.: In: Advances in Enzymology. San Pietro, A. (ed.). New York-London: Academic Press, 1971, **23A**, 413–440
Bunt, J.S., Heeb, M.A.: Biochim. Biophys. Acta **226**, 354–359 (1971)
Chang, I.C., Kahn, J.S.: J. Protozool. **17**, 556–564 (1970)
Cheniae, G.M.: Ann. Rev. Plant Physiol. **21**, 467–498 (1970)
Cho, F., Govindjee: Biochim. Biophys. Acta **216**, 139–150 (1970)
Chua, N.-H., Levine, R.P.: Plant Physiol. **44**, 1–6 (1969)
Curtis, V.A., Brand, J.J., Togasaki, R.K.: Plant Physiol. **55**, 183–186 (1975)
Das, M., Govindjee: Biochim. Biophys. Acta **143**, 570–576 (1967)
Das, M., Rabinowitch, E., Szalay, L.: Biophys. J. **8**, 1131–1137 (1968)
Dodge, J.D.: Sci. Progr. **61**, 257–274 (1974)
Döring, G., Stiehl, H.H., Witt, H.T.: Z. Naturforsch. **22b**, 639–644 (1967)
Dubertret, G., Joliot, P.: Biochim. Biophys. Acta **357**, 399–411 (1974)
Duysens, L.N.M.: Science **120**, 353–354 (1954)
Duysens, L.N.M.: Progr. Biophysics **14**, 1–100 (1964)
Duysens, L.N.M., Amesz, J.: Biochim. Biophys. Acta **64**, 243–260 (1962)
Duysens, L.N.M., Amesz, J., Kamp, B.M.: Nature (London) **190**, 510–511 (1961)
Echlin, P.: In: Organization and Control in Prokaryotic and Eukaryotic Cells. Charles, H.P., Knight, B.C.J.G. (eds.). Cambridge: Cambridge Univ., 1970, pp. 221–248
Egle, K.: In: Encyclopedia of Plant Physiology. Ruhland, W. (ed.). Berlin-Heidelberg-New York: Springer, 1960, Vol. V, pp. 444–496
Egneus, H., Heber, U., Matthiesen, U., Kirk, M.: Biochim. Biophys. Acta **408**, 252–268 (1975)
Elstner, E., Pistorius, E., Böger, P., Trebst, A.: Planta (Berl.) **79**, 146–161 (1968)
Emerson, R.: Ann. Rev. Plant Physiol. **9**, 1–24 (1958)
Emerson, R., Arnold, W.: J. Gen. Physiol. **16**, 191–205 (1932)
Emerson, R., Chalmers, R., Cederstrand, C.: Proc. Nat. Acad. Sci. U.S. **43**, 133–143 (1957)
Emerson, R., Lewis, C.M.: Am. J. Botany **30**, 165–178 (1943)
Emerson, R., Rabinowitch, E.: Plant Physiol. **35**, 477–485 (1960)
Epel, B.L., Butler, W.L.: Biophys. J. **12**, 922–929 (1971)
Epel, B.L., Butler, W.L., Levine, R.P.: Biochim. Biophys. Acta **275**, 395–400 (1972)
Fork, D.C.: In: Photosynthetic Mechanism of Green Plants. Kok, B., Jagendorf, A.T. (eds.). Washington: Nat. Acad. Sci., 1963, pp. 352–361
Fork, D.C.: In: Progress in Photosynthesis Res., Metzner, H. (ed.). Tübingen: Laupp, 1969, Vol. II, pp. 800–810
Fork, D.C., Amesz, J.: Photochem. Photobiology **6**, 913–918 (1967)
Fork, D.C., Amesz, J.: Ann. Rev. Plant Physiol. **20**, 305–328 (1969)
Fork, D.C., Amesz, J.: In: Photophysiology. Giese, A.C. (ed.). New York-London: Academic Press, 1970, Vol. V, pp. 97–126
Fork, D.C., Kouchkovsky, Y.: Photochem. Photobiol. **5**, 609–619 (1966)
Fork, D.C., Urbach, W.: Proc. Nat. Acad. Sci. U.S. **53**, 1307–1315 (1965)
Fork, D.C., Urbach, W.: In: Currents in Photosynthesis. In: Proc. 2nd Western-European Conf. Photosynthesis. Rotterdam: A.D. Donker, 1966, pp. 293–303
Forti, G., Avron, M., Melandri, B.A. (eds.): Proc. 2nd Intern. Congr. Photosynthesis 3 vols. The Hague: Dr. Junk, 1972, pp. 1–2752
Forti, G., Rosa, L.: FEBS Lett. **18**, 55–58 (1971)
French, C.S.: In: Encyclopedia of Plant Physiology. Ruhland, W. (ed.). Berlin-Heidelberg-New York: Springer, 1960, Vol. V, pp. 252–297
French, C.S.: Proc. Nat. Acad. Sci. U.S. **68**, 2893–2907 (1971)
French, C.S., Young, V.M.K.: J. Gen. Physiol. **35**, 873–890 (1952)
Gaffron, H.: Am. J. Botany **27**, 273–283 (1940)
Garnier, J., Maroc, J.: In: Proc. 3rd Intern. Congr. Photosynthesis. Avron, M. (ed.). Amsterdam: Elsevier, 1975, Vol. I, pp. 547–556
Gibbs, S.P.: Ann. N.Y. Acad. Sci. **175**, 454–473 (1970)
Gimmler, H., Avron, M.: Z. Naturforsch. **26b**, 585–588 (1971)
Gimmler, H., Avron, M.: In: Proc. 2nd Intern. Congr. Photosynthesis. Forti, G., Avron, M., Melandri, A. (eds.). The Hague: Dr. Junk, 1972, Vol. I, pp. 789–800

Gimmler, H., Neimanis, S., Eilmann, I., Urbach, W.: Z. Pflanzenphysiol. **64**, 358–366 (1971)
Gimmler, H., Urbach, W., Jeschke, W.D., Simonis, W.: Z. Pflanzenphysiol. **58**, 353–364 (1968)
Givan, A.C., Levine, R.P.: Plant Physiol. **42**, 1264–1268 (1967)
Glidewell, S.M., Raven, J.A.: J. Ex. Botany **26**, 479–488 (1975)
Goedheer, J.C.: Photosynthetica **4**, 97–106 (1970)
Goedheer, J.C.: Ann. Rev. Plant Physiol. **23**, 87–112 (1972)
Goodenough, U.W., Staehelin, L.A.: J. Cell Biol. **48**, 594–619 (1971)
Goodwin, T.W.: In: Aspects of Terpenoid Chemistry and Biochemistry. Goodwin, T.W. (ed.). New York-London: Academic Press, 1971a, pp. 315–356
Goodwin, T.W.: In: Structure and Function of Chloroplasts. Gibbs, M. (ed.). Berlin-Heidelberg-New York: Springer, 1971b, pp. 215–276
Goodwin, T.W.: In: Algal Physiology and Biochemistry. Stewart, W.D.P. (ed.). Oxford-London-Edinburgh-Melbourne: Blackwell Scientific, 1974, Vol. X, pp. 176–205
Gorman, D.S., Levine, R.P.: Proc. Nat. Acad. Sci. U.S. **54**, 1665–1669 (1965)
Gorman, D.S., Levine, R.P.: Plant Physiol. **41**, 1637–1642 (1966a)
Gorman, D.S., Levine, R.P.: Plant Physiol. **41**, 1643–1647 (1966b)
Gorman, D.S., Levine, R.P.: Plant Physiol. **41**, 1648–1656 (1966c)
Govindjee, Braun, B.Z.: In: Algal Physiology and Biochemistry. Stewart, W.D.P. (ed.). Oxford: Blackwell Scientific, 1974, pp. 346–390
Govindjee, Govindjee, R.: Photochem. Photobiol. **4**, 793–801 (1965)
Govindjee, Munday, J.C., Papageorgiou, G.: In: Energy Conversion by the Photosynthetic Apparatus. Brookhaven Symp. Biol. **19**, 434–445 (1967)
Govindjee, Papageorgiou, G.: In: Photophysiology. Giese, A.C. (ed.). New York-London: Academic Press, 1971, Vol. VI, pp. 1–46
Govindjee, Rabinowitch, E.: Biophysics **1**, 73–89 (1960)
Graham, D., Smillie, R.M.: In: Methods of Enzymology. San Pietro, A. (ed.). New York-London: Academic Press, 1971, **23A**, 228–242
Granick, S.: In: The Cell, Biochemistry, Physiology, Morphology. Brachet, J., Mirsky, A.E. (eds.). New York-London: Academic Press, 1961, pp. 489–602
Granick, S.: In: Methods in Enzymology. San Pietro, A. (ed.). New York-London: Academic Press, 1971, **23A**, 162–168
Grimme, L.H., Boardman, N.K.: In: Proc. 3rd Intern. Congr. Photosynthesis. Avron, M. (ed.). Amsterdam: Elsevier, 1975, Vol. III, pp. 2115–2124
Gromet-Elhanan, Z.: Biochim. Biophys. Acta **131**, 526–537 (1967)
Hager, A., Stransky, H.: Arch. Mikrobiol. **72**, 68–83 (1970a)
Hager, A., Stransky, H.: Arch. Mikrobiol. **73**, 77–89 (1970b)
Halldal, P.: In: Photobiology of Microorganisms. Halldal, P. (ed.). London-New York-Sydney-Toronto: John Wiley-Interscience, 1970, pp. 17–55
Hase, E.: In: Autonomy and Biogenesis of Mitochondria and Chloroplasts. Boardman, N.K., Linnane, A.W., Smillie, R.M. (eds.). Amsterdam: North Holland, 1971, pp. 434–446
Haxo, F.T.: In: Comparative Biochemistry of Photoreactive Systems. Allen, M.B. (ed.). New York-London: Academic Press, 1960, pp. 339–360
Healey, F.P.: Planta (Berl.) **91**, 220–226 (1970)
Heber, U.: Biochim. Biophys. Acta **180**, 302–319 (1969)
Heber, U.: Ber. Deut. Botan. Ges. **86**, 187–195 (1973)
Herron, H.A., Mauzerall, D.: Plant Physiol. **50**, 141–148 (1972)
Hill, R., Bendall, F.: Nature (London) **186**, 136–137 (1960)
Hind, G., Olson, J.M.: Ann. Rev. Plant Physiol. **19**, 249–282 (1968)
Hoch, G.E., Kok, B.: Ann. Rev. Plant Physiol. **12**, 155–194 (1961)
Hoch, G.E., Owens, O.H., Kok, B.: Arch. Biochem. Biophys. **101**, 171–180 (1963)
Hoch, G.E., Randles, J.: Photochem. Photobiol. **14**, 435–449 (1971)
Hiyami, T., Ke, B.: Proc. Nat. Acad. Sci. U.S. **68**, 1010–1013 (1971a)
Hiyami, T., Ke, B.: Arch. Biochem. Biophys. **147**, 99–108 (1971b)
Ikegami, I., Katoh, S., Takamiya, A.: Biochim. Biophys. Acta **162**, 604–607 (1968)
Ikegami, I., Katoh, S., Takamiya, Y.: Plant and Cell Physiol. **11**, 777–791 (1970)
Jeffrey, S.W.: Biochim. Biophys. Acta **279**, 15–33 (1972)
Jeffrey, S.W., Ulrich, J., Allen, M.B.: Biochim. Biophys. Acta **112**, 35–44 (1966)

Jeschke, W.D., Simonis, W.: Planta (Berl.) **88**, 157–171 (1969)
Kaiser, W., Urbach, W.: Biochim. Biophys. Acta **423**, 91–102 (1976)
Kandler, O.: Ann. Rev. Plant Physiol. **11**, 37–54 (1960)
Kandler, O., Tanner, W.: Ber. Deut. Botan. Ges. **79**, 48–57 (1966)
Katoh, S.: Nature (London) **186**, 138–139 (1960a)
Katoh, S.: Nature (London) **186**, 533–534 (1960b)
Katoh, S., San Pietro, A.: Arch. Biochem. Biophys. **119**, 488–496 (1967)
Kautsky, H., Appel, W., Amman, H.: Biochem. Z. **332**, 277–292 (1960)
Ke, B.: In: Proc. 3rd Intern. Congr. Photosynthesis. Avron, A. (ed.). Amsterdam: Elsevier, 1975, Vol. I, pp. 373–382
Kessler, E.: Ann. Rev. Plant Physiol. **15**, 57–71 (1964)
Kessler, E.: In: Algal Physiology and Biochemistry. Stewart, W.D.P. (ed.). Oxford: Blackwell Scientific, 1974, Vol. X, pp. 456–473
Knaff, D.B., Arnon, D.I.: Proc. Nat. Acad. Sci. U.S. **64**, 715–722 (1969)
Kok, B.: Acta Botan. Neerl. **6**, 316–336 (1957)
Kok, B.: In: Encyclopedia of Plant Physiology. Ruhland, W. (ed.). Berlin-Heidelberg-New York: Springer, 1960, Vol. V, pp. 566–633
Kok, B.: Biochim. Biophys. Acta **48**, 527–533 (1961)
Kok, B., Cheniae, G.M.: In: Current Topics Bioenergy. Sanadi, D.R. (ed.). New York: Academic Press, 1966, **1**, 1–47
Kok, B., Hoch, G.: In: Light and Life. McElroy, W.D., Glass, B. (eds.). Baltimore: Johns Hopkins, 1961, pp. 397–423
Kok, B., Radmer, R., Fowler, C.F.: In: Proc. 3rd Intern. Congr. Photosynthesis. Avron, M. (ed.). Amsterdam: Elsevier, 1975, Vol. I, pp. 485–496
Kok, B., Rurainski, H.J.: Biochim. Biophys. Acta **94**, 588–590 (1965)
Kok, B., Rurainski, H.J., Harmon, E.A.: Plant Physiol. **39**, 513–520 (1964)
Kouchkovsky, Y.: In: Progress in Photosynthesis Res. Metzner, H. (ed.). Tübingen: Laupp, 1969, Vol. II, pp. 959–970
Kouchkovsky, Y., Fork, D.C.: Proc. Nat. Acad. Sci. U.S. **52**, 232–239 (1964)
Lach, H.J., Böger, P.: Z. Pflanzenphysiol. **72**, 427–442 (1974)
Lach, H.J., Ruppel, H.G., Böger, P.: Z. Pflanzenphysiol. **70**, 432–451 (1973)
Levine, R.P.: Ann. Rev. Plant Physiol. **20**, 523–540 (1969a)
Levine, R.P.: In: Progress in Photosynthesis Res. Metzner, H. (ed.). Tübingen: Laupp, 1969b, Vol. II, pp. 971–977
Levine, R.P.: In: Algal Physiology and Biochemistry. Stewart, W.D.P. (ed.). Oxford: Blackwell Scientific, 1974, Vol. X, pp. 424–433
Levine, R.P., Armstrong, J.: Plant Physiol. **49**, 661–662 (1972)
Levine, R.P., Gorman, D.S.: Plant Physiol. **41**, 1293–1300 (1966)
Lorenzen, H.: In: Photobiology of Microorganisms. Halldal, P. (ed.). London-New York: Wiley-Interscience, 1970, pp. 187–212
Lorenzen, H., Hesse, M.: In: Algal Physiology and Biochemistry. Stewart, W.D.P. (ed.). Oxford: Blackwell Interscience, 1974, Vol. X, pp. 894–908
Ludlow, C.J., Park, R.B.: Plant Physiol. **44**, 540–543 (1969)
Mann, J.E., Myers, J.: Plant Physiol. **43**, 1991–1993 (1968)
Manton, J.: In: Biochemistry of Chloroplasts. Goodwin, T.W. (ed.). London-New York: Academic Press, 1966, Vol. I, pp. 23–47
Maroc, J., Garnier, J.: Biochim Biophys. Acta **387**, 52–68 (1975)
Matsubara, H.: J. Biol. Chem. **243**, 370–375 (1968)
Meeks, J.C.: In: Algal Physiology and Biochemistry. Stewart, W.D.P. (ed.). Oxford: Blackwell Scientific, 1974, Vol. X, pp. 161–175
Mende, D., Wiessner, W.: In: Proc. 3rd Intern. Congr. Photosynthesis. Avron, M. (ed.). Amsterdam: Elsevier, 1975, Vol. I, pp. 505–513
Metzner, H. (ed.): Progress in Photosynthesis Res. 3 vols. Tübingen: Laupp, 1969, pp. 1–1807
Mitsui, A.: In: Methods in Enzymology. San Pietro, A. (ed.). New York-London: Academic Press 1971, **23A**, 368–371
Morris, J.: In: Algal Physiology and Biochemistry. Stewart, W.D.P. (ed.). Oxford: Blackwell Scientific, 1974, Vol. X, pp. 583–609
Morris, J. Ahmed, J.: Physiol. Plantarum **22**, 1166–1174 (1969)

Myers, J.: In: Encyclopedia of Plant Physiology. Ruhland, W. (ed.). Berlin-Heidelberg-New York: Springer, 1960, Vol. V, pp. 211–233
Myers, J.: Ann. Rev. Plant Physiol. **22**, 289–312 (1971)
Myers, J.: Plant Physiol. **54**, 420–426 (1974)
Myers, J., French, C.S.: Plant Physiol. **35**, 963–969 (1960)
Myers, J., Graham, J.R.: Plant Physiol. **48**, 282–286 (1971)
Myers, J., Graham, J.R.: Plant Physiol. **55**, 686–688 (1975)
Nishimura, M.: Biochim. Biophys. Acta **153**, 838–847 (1968)
Nishimura, M., Takamiya, A.: Biochim. Biophys. Acta **120**, 45–56 (1966)
Ogawa, T., Kanai, R., Shibata, K.: In: Comparative Biochemistry and Biophysics of Photosynthesis. Shibata, K., Takamiya, A., Jagendorf, A.T., Fuller, R.C. (eds.). Tokyo: Univ. Tokyo, 1968, pp. 22–35
Ohad, I., Eytan, E., Jennings, R.C., Goldberg, J., Bar-Nun, S., Wallach, D.: In: Proc. 2nd Intern. Congr. Photosynthesis. Forti, G., Avron, M., Melandri, A. (eds.). The Hague: Dr. Junk, 1972, pp. 2563–2584
O'Heocha, C.: In: Chemistry and Biochemistry of Plant Pigments. Goodwin, T.W. (ed.). New York-London: Academic Press, 1965, pp. 175–196
O'Heocha, C.: Oceanogr. Mar. Biol. Ann. Rev. **9**, 61–82 (1971)
Oquist, G.: Physiol. Plantarum **22**, 516–528 (1969)
Park, R.B.: In: Methods in Enzymology. San Pietro, A. (ed.). New York-London: Academic Press, 1971, **23A**, 248–250
Park, R.B., Sane, P.V.: Ann. Rev. Plant Physiol. **22**, 395–430 (1971)
Patterson, P.C.O., Myers, J.: Plant Physiol. **51**, 104–109 (1973)
Perini, F., Kamen, M.D., Schiff, J.A.: Biochim. Biophys. Acta **88**, 74–90 (1964)
Pirson, A., Lorenzen, H.: Ann. Rev. Plant Physiol. **17**, 439–458 (1966)
Powls, R., Wong, J., Bishop, N.J.: Biochim. Biophys. Acta **180**, 490–499 (1969)
Pratt, L., Bishop, N.I.: Biochim. Biophys. Acta **162**, 369–379 (1968)
Rabinowitch, E.: Photosynthesis and Related Processes. New York: Interscience, 1951, Vol. II, pp. 603–828
Raven, J.A.: J. Exp. Botany **21**, 1–16 (1970)
Raven, J.A.: J. Exp. Botany **22**, 420–433 (1971)
Raven, J.A.: In: Algal Physiology and Biochemistry. Stewart, W.D.P. (ed.). Oxford: Blackwell Scientific, 1974, pp. 391–423
Raven, J.A., Glidewell, S.M.: New Phytologist **75**, 197–204 (1975)
Rensen, J.J.S. van: In: Progress in Photosynthesis Res. Metzner, H. (ed.). Tübingen: Laupp, 1969, Vol. III, pp. 1769–1776
Ried, A.: Biochim. Biophys. Acta **153**, 653–663 (1968)
Ried, A.: In: Proc. IBP/PP Technical Meeting, Trebon. Wageningen: Centre Agric., 1970, pp. 231–246
Ried, A.: In: Proc. 2nd Intern. Congr. Photosynthesis. Forti, G., Avron, M., Melandri, A. (eds.). The Hague: Dr. Junk, 1972, Vol. I, pp. 763–772
Rubinstein, D.: Biochim. Biophys. Acta **109**, 41–47 (1965)
Rubinstein, D., Rabinowitch, E.: Biophys. J. **4**, 107–113 (1964)
Rumberg, B., Schmidt-Mende, P., Witt, H.T.: Nature (London) **201**, 466–468 (1964)
Rurainski, H.J., Hoch, G.E.: Z. Naturf. **30c**, 761–770 (1975)
Rurainski, H.J., Randles, J., Hoch, G.E.: Biochim. Biophys. Acta **205**, 254–262 (1970)
Sato, V.L., Levine, R.P., Neumann, J.: Biochim. Biophys. Acta **253**, 437–448 (1971)
Schiff, J.A.: In: Autonomy and Biogenesis of Mitochondria and Chloroplasts. Boardman, N.K., Linnane, A.W., Smillie, R.M. (eds.). Amsterdam: North Holland, 1971, pp. 98–118
Schiff, J.A.: In: Proc. 3rd Intern. Congr. Photosynthesis. Avron, M. (ed.). Amsterdam: Elsevier, 1975, Vol. III, pp. 1691–1717
Schmid, G.H., Gaffron, H.: J. Gen. Physiol. **52**, 211–239 (1968)
Schmid, G.H., Gaffron, H.: Photochem. Photobiol. **14**, 451–464 (1971)
Schor, S., Siekevitz, Ph., Palade, G.E.: Proc. Nat. Acad. Sci. U.S. **66**, 174–180 (1970)
Seely, G.R.: In: Proc. 2nd Intern. Congr. Photosynthesis. Forti, G., Avron, M., Melandri, A. (eds.). The Hague: Dr. Junk, 1972, Vol. I, pp. 341–348
Senger, H.: Planta (Berl.) **92**, 327–346 (1970)

Senger, H.: In: Les Cycles Cellulaires et leur Blocage chez Plusieurs Protistes. Lefort-Tran, M. (ed.) 1975
Senger, H., Bishop, N.I.: Nature (London) **214**, 140–142 (1967)
Senger, H., Bishop, N.I.: Nature (London) **221**, 975 (1969)
Senger, H., Frickel-Faulstich, B.: In: Proc. 3rd Intern. Congr. Photosynthesis. Avron, M. (ed.). Amsterdam: Elsevier, 1975, Vol. I, pp. 715–721
Shneyour, A., Avron, M.: FEBS Lett. **8**, 164–166 (1970)
Shneyour, A., Avron, M.: Plant Physiol. **55**, 137–141 (1975)
Simonis, W.: In: Encyclopedia of Plant Physiology. Ruhland, W. (ed.). Berlin-Heidelberg-New York: Springer, 1960, Vol. V, pp. 266–1007
Simonis, W.: Ber. Deut. Botan. Ges. **80**, 395–402 (1967)
Simonis, W., Urbach, W.: In: Studies on Microalgae and Photosynthetic Bacteria. Tokyo: Univ. Tokyo, 1963, pp. 597–611
Simonis, W., Urbach, W.: Ann. Rev. Plant Physiol. **24**, 89–114 (1973)
Smillie, R.M., Entsch, B.: In: Methods in Enzymology. San Pietro, A. (ed.). New York-London: Academic Press, 1971, **23 A**, 504–514
Smith, J.H.C., French, C.S.: Ann. Rev. Plant Physiol. **14**, 181–224 (1963)
Starr, R.C.: In: Methods in Enzymology. San Pietro, A. (ed.). New York-London: Academic Press, 1971, **23 A**, 29–53
Stiehl, H.H., Witt, H.T.: Z. Naturforsch. **23b**, 220–224 (1968)
Strain, H.H., Cope, B.T., Svec, W.A.: In: Methods in Enzymology. San Pietro, A. (ed.). New York-London: Academic Press, 1971, **23 A**, 452–476
Strotmann, H., Berger, S.: Biochem. Biophys. Res. Commun. **35**, 20–26 (1969)
Stuart, T.S.: Planta (Berl.) **96**, 81–92 (1971)
Stuart, T.S., Gaffron, H.: Planta (Berl.) **106**, 91–100 (1972a)
Stuart, T.S., Gaffron, H.: Planta (Berl.) **106**, 101–112 (1972b)
Sun, E., Barr, R., Crane, F.L.: Plant Physiol. **43**, 1935–1940 (1968)
Tagawa, K., Tsujimoto, H.Y., Arnon, D.I.: Proc. Nat. Acad. Sci. U.S. **49**, 567–572 (1963)
Tamiya, H.: Ann. Rev. Plant Physiol. **17**, 1–26 (1966)
Tanada, T.: Am. J. Botany **38**, 276–283 (1951)
Tanner, W., Dächsel, L., Kandler, O.: Plant Physiol. **40**, 1151–1156 (1965)
Tanner, W., Löffler, M., Kandler, O.: Plant Physiol. **44**, 422–428 (1969)
Teichler-Zallen, D., Hoch, G.E.: Arch. Biochem. Biophys. **120**, 227–230 (1967)
Teichler-Zallen, D., Hoch, G.E., Bannister, T.T.: In: Proc. 2nd Intern. Congr. Photosynthesis. Forti, G., Avron, M., Melandri, A. (eds.). The Hague: Dr. Junk, 1972, Vol. I, pp. 643–647
Thacker, A., Syrett, P.J.: New Phytologist **71**, 423–433 (1972)
Thomas, J.B.: FEBS Lett. **14**, 61–64 (1971)
Thornber, J.P.: Ann. Rev. Plant Physiol. **26**, 127–158 (1975)
Trebst, A.: Ann. Rev. Plant Physiol. **25**, 423–458 (1974)
Trebst, A.: In: Proc. 3rd Intern. Congr. Photosynthesis. Avron, M. (ed.). Amsterdam: Elsevier, 1975, Vol. I, pp. 439–447
Trebst, A., Burba, M.: Z. Pflanzenphysiol. **57**, 419–433 (1967)
Trebst, A., Harth, E., Draber, W.: Z. Naturforsch. **25b**, 1157–1159 (1970)
Ullrich, W.R.: Planta (Berl.) **100**, 18–30 (1971)
Ullrich, W.R.: Planta (Berl.) **116**, 143–152 (1974)
Urbach, W., Gimmler, H.: In: Progress in Photosynthesis Res. Metzner, H. (ed.). Tübingen: Laupp, 1969, Vol. III, pp. 1274–1280
Urbach, W., Gimmler, H.: Ber. Deut. Botan. Ges. **83**, 439–442 (1970a)
Urbach, W., Gimmler, H.: Z. Pflanzenphysiol. **62**, 276–286 (1970b)
Urbach, W., Kaiser, W.: In: Proc. 2nd Intern. Congr. Photosynthesis. Forti, G., Avron, M., Melandri, A. (eds.). The Hague: Dr. Junk, 1972, Vol. II, pp. 1401–1411
Urbach, W., Simonis, W.: Biochem. Biophys. Res. Commun. **17**, 39–45 (1964)
Warburg, O.: Biochem. Z. **103**, 188–217 (1919)
Weier, T.E., Bisalputra, T., Harrison, A.: J. Ultrastruct. Res. **15**, 38–56 (1966)
Wiessner, W.: In: Photobiology of Microorganisms. Halldal, P. (ed.). London: Wiley-Interscience, 1970a, pp. 95–133
Wiessner, W.: In: Photobiology of Microorganisms. Halldal, P. (ed.). London: Wiley-Interscience, 1970b, pp. 135–163

Wild, A., Egle, K.: Beitr. Biol. Pflanzen **45**, 213–241 (1968)
Wild., A., Zickler, H.O., Grahl, H.: Planta (Berl.) **97**, 208–223 (1971)
Wildner, G.F., Hauska, G.: Arch. Biochem. Biophys. **164**, 127–135 (1974)
Wildner, G.F., Hauska, G.: In: Proc. 3rd Intern. Congr. Photosynthesis. Avron, M. (ed.).
    Amsterdam: Elsevier, 1975, Vol. I, pp. 525–534
Witt, H.T.: In: Encyclopedia of Plant Physiology. Ruhland, W. (ed.). Berlin-Heidelberg-New
    York: Springer, 1960, Vol. V, pp. 634–674
Witt, H.T.: In: Nobel Symposium V, Almquist and Wiksell Stockholm. New York-London-
    Sidney: Interscience Publishers, 1967, pp. 261–316
Witt, H.T.: Quart. Rev. Biophys. **4**, 365–477 (1971)
Witt, H.T., Müller, A., Rumberg, B.: Nature (London) **191**, 194–196 (1961)
Yakushiji, E.: In: Methods in Enzymology. San Pietro, A. (ed.). New York-London: Academic
    Press, 1971, **23A**, 364–368
Yocum, C.S., Blinks, L.R.: J. Gen. Physiol. **41**, 1113–1117 (1958)
Zumft, W.G., Spiller, M.: Biochem. Biophys. Res. Commun. **45**, 112–118 (1971)

2. Blue-Green Algae

D.W. KROGMANN

A. Introduction

The blue-green algae represent a unique stage in the evolution of plant life. The cyanophyta are clearly procaryotic organisms and so they might well be called blue-green bacteria. However they are the only procaryotes able to produce oxygen from water so their photosynthetic activity is like that of the higher plants (KROGMANN, 1973). Before making any generalizations about blue-green algal photosynthesis it is important to note that there are between one and two thousand species of blue-green algae and very few have been examined in the laboratory. The taxonomy of these algae is not at all clear since their morphology is simple and their physiology is difficult. Thus there is as yet no clear choice for the best blue-green algae to use for studying photosynthesis nor any assurance that the best organism would be representative of all blue-greens. At present all results should be attributed only to the species from which they have been obtained and even then further qualified in that these results may be limited to a particular strain, culture or set of growth conditions. One might have been tempted to conclude that some blue-green algae contained no ferredoxin prior to the observation that this protein disappears promptly from iron deficient cells. The ease of extraction of a given enzyme might vary greatly depending on some yet to be recognized properties of the membrane. Presently it appears that plastocyanin can be isolated from only a few blue-greens but this may be due either to the absence of the proper gene or to some difficulties that confound the isolation procedure.

There are promises of distinct experimental advantage in studying photosynthetic electron transport and photophosphorylation in blue-green algae. As with other procaryotic organisms, growth in a precisely controlled environment allows easy manipulation of the physiology of these cells. There is an opportunity to use the powerful genetic techniques which have been applied to *Escherichia coli*. Some species of blue-green algae will grow in an unicellular mode in axenic culture. Slowly the requirements for dark, heterotrophic growth of these organisms are being recognized so that one may hope to grow cells with specific genetic deletions in the photosynthetic apparatus. Induced mutations in the blue-greens are being made and the discovery of viruses — the cyanophages — which use blue-green algae as their host makes transduction and genetic mapping of the photosynthetic characters possible. Finally these most primitive of the oxygen evolving plants seem to have a more loosely organized photosynthetic membrane so that some of the photosynthetic machinery is easily released from these structures. The major accessory light absorbing pigment can be removed either by washing the isolated membranes or by subjecting the growing cell to nitrogen deficiency. Cytochromes are released from these membranes with much greater ease than is the case with chloroplasts from most eucaryotes.

There are at least two important aspects in which the photosynthetic membranes of blue-green algae differ from that of most higher plants. In the blue-green algae, the photosynthetic membranes are not organized within an organelle and phycobiliproteins are present instead of chlorophyll b.

B. Membrane Structure

All oxygen-producing plants except blue-green algae have a chloroplast structure which contains the photosynthetic enzymes. In the blue-green algae, the chlorophyll and other photosynthetic constituents are organized on thylakoids. The thylakoids may be arranged in concentric shells near the periphery of the cell or they may pervade all areas of the cell in a rather disorganized network (LANG and WHITTON, 1973; FOGG et al., 1973). These thylakoids may show a direct connection to the plasma membrane of the cell as they do in purple bacteria (BISALPUTRA, 1974). The thylakoids remain separated from one another along their entire length in some species while in other species they can be tightly appressed or narrowly separated from adjacent thylakoids. There is no tendency for the thylakoids of blue-green algae to run in pairs, threes or fives as they do in the chloroplasts of eucaryotes. The thylakoid shows a tripartite membrane structure and the membranes are usually closely appressed to give a flattened vesicle with no obvious internal lumen. We have not been able to observe osmotic induced volume changes in isolated membranes by either electron microscopy or packed volume measurements.

The blue-green algae membranes must also differ in their internal construction. JOST (1965) used the freeze etch technique to examine the thylakoid membranes of *Oscillatoria rubescens* and distinguished a 50 to 70 Å diameter particle on one fracture face and a 100 to 200 Å diameter particle on the other. This is quite different from the freeze etch data from chloroplasts. On the external surface or stroma side of the thylakoid membrane in many blue-greens, one can see the very large phycobilisome particle — 280 to 550 Å in diameter at the base (WILDMAN and BOWEN, 1974).

OGAWA et al. (1969) attempted a detergent fractionation of photosynthetic membranes from *Anabaena variabilis*. Membrane fragments were prepared by sonic disruption of the cells and were washed free of phycobiliproteins. The membrane fragments were next exposed to the detergent Triton X-100 which is known to produce two types of particles from spinach chloroplasts. With spinach chloroplasts a small particle is released from the membrane by the detergent and this small particle is enriched for photosystem I activity while the residual membrane contains the components of photosystem II. The opposite result was obtained with *Anabaena* membrane fragments. Photosystem I remained bound to large membranous sheets while photosystem II components were associated with small irregular strands of material released from the membranes by detergent. The large photosystem I-enriched membrane fragments showed many 100 Å particles on their surface but these particles did not behave like the chloroplast coupling factor which they resemble. While digitonin is routinely used to split higher plant chloroplast membranes into small photosystem I and large photosystem II subparticles, this mild detergent apparently has no disruptive effect on membrane fragments of *Nostoc* (ARNON et al., 1974). These observations suggest some important chemical difference between chloroplast and blue-green algal membranes. Gross compositional analyses have revealed no striking differences in the make up of chloroplast and blue-green membranes. KLEIN and VERNON (1974) have reported on the general qualitative similarity seen after disc gel electrophoresis of membrane proteins from *A. variabilis* and *Chlamydomonas reinhardi*. The *Anabaena* protein pattern is distinct in that it lacks polypeptides in the 20 to 30,000 molecular weight range which are associated with chlorophyll b in higher plants. It could be that a few proteins in photosystem II including the chlorophyll b protein might cause radical differences in membrane structure but so little is known about this area that no conclusions are possible.

C. Major Accessory Pigments

The blue-green algae contain no chlorophyll b. Instead there are biliproteins in which a linear tetrapyrole is covalently held to a protein. These accessory pigments can be considered from two viewpoints — the molecular properties of the pigments and the organization of these pigment molecules on the photosynthetic membrane.

I. Molecular Properties of Phycobiliproteins

The red and blue bile pigment-protein complexes are found in the blue-green, red and some of the cryptomonad algae (CHAPMAN, 1973). Phycocyanin and allophycocyanin contain the same linear tetrapyrrole chromophore and are believed to be universally present in the biliprotein containing algae. Phycoerythrin bears a chromophore of a different structure which causes a pronounced shift in its absorption maximum to shorter wavelengths. Phycoerythrin is found in some blue-green and most red algae. Table 1 gives a few representative data to illustrate some of the physical characteristics of the pigment-protein complexes of blue-green algae. Measurement of molecular weights of these proteins usually reveals a family of sizes — monomer, trimer, hexamer, dodecomer, even aggregates of 18 subunits. At high protein concentration and with proper handling the hexamer of phycocyanin seems to predominate. However the aggregation state of these molecules in solution is only an indirect indication of the in vivo condition since these molecules exist on the membrane as tightly packed bundles — the phycobilisomes. EISERLING and GLAZER (1974) used electron microscopy to recognize the predominant hexamer form of phycocyanin of *Synechococcus* sp. and they found higher assembly forms in the shape of short rods and ordered bundles which may relate to the structure in the intact phycobilisome.

Phycocyanin, allophycocyanin and phycoerythrin, even from a single species source, show no common antigenic determinants indicating that each has a distinct genetic origin. However there is an excellent immunological cross reactivity for

Table 1. Composition of phycobiliproteins

	Molecular weight	Chromophore	λ_{max}
Phycoerythrin *Aphanocapsa* sp.[a]	α 20,000; β 22,000	$1/\alpha$; $2/\beta$	565 nm
Phycocyanin *Synechococcus* sp.[a]	α 15,900; β 19,100	$1/\alpha$; $2/\beta$	622 nm
Allophycocyanin *Synechococcus* sp.[a]	α 15,200; β 17,250	—	652 nm
Allophycocyanin B *Synechococcus* sp.[b]	16,000; 17,000	—	671 nm
Chlorophyll-protein complex *Phormidium luridum*[c]	350,000 containing 6 subunits of 35,000	5/subunit	677 nm
Chromoprotein *Anabaena cylindrica*[d]	50,000	—	695 nm

[a] Glazer et al. (1971); [b] Glazer and Bryant (1975); [c] Thornber (1969); [d] Pjon and Fujita (1974)

each type of biliprotein regardless of the algal source. Thus antibody prepared against phycocyanin from a blue-green alga will cross react with phycocyanin from a red or a cryptomonad alga (Bennett and Bogorad, 1971; Glazer et al., 1971). This conservation of structure is documented even more convincingly in the partial amino acid sequences that have been done for these proteins. Troxler et al. (1971) examined the sequence of the N-terminal regions of alpha and beta subunits of phycocyanin from the red alga *Cyanidium caldarium*. The two different subunits showed a 60% homology within the first 27 residues and residues 10 through 27 were almost identical. Williams et al. (1974) found a similar high degree of homology in the N-terminal regions of the alpha and beta subunits of phycocyanin from *Synechococcus* sp. When the alpha and beta chain sequences of *Synchecoccus* phycocyanin were compared with those of *C. caldarium*, there was also a high degree of homology. In comparing the red with the blue-green algal proteins, 22 of 32 positions were identical and another five positions showed substitutions representing single base changes in the codons. Thus, the alpha and beta subunits of phycocyanin are coded on highly conserved products of an ancestral gene doubling event. Further these genes are conserved in the evolution of red algae from blue-greens.

In contrast to chlorophyll-protein complexes, the phycobiliproteins have a covalent attachment of the chromophore to the polypeptide. The pigment is attached either through an ester or a thioether bond to the polypeptide. The biliproteins show a rather high content of acidic and non-polar amino acid residues. Hydrophobic interactions seem to play an important role in the monomer-hexamer equilibrium (MacColl et al., 1971). Aggregation behavior depends on ionic strength, pH, denaturing agents and, perhaps most important, the presence of other proteins. Glazer et al. (1973) made an excellent spectroscopic characterization of purified phycocyanin. Aggregation of phycocyanin at high protein concentration results in a red shift of the absorption maximum and a large increase in the extinction coefficient. Circular dichroism measurements indicate increased intermolecular interactions between the chromophores as the biliprotein aggregates. There is evidence that interactions of phycoerythrin with phycocyanin which allow efficient energy transfer do not cause major changes in the conformation of these proteins (FRAN-KOWIAK and GRABOWSKI, 1973). These interactions of the phycobiliproteins at high protein concentration are all germaine to the in vivo condition since these proteins are packed within the phycobilisomes on the membrane surface.

The overlapping fluorescence and absorption bands at the red end of the spectrum make the phycobiliproteins an excellent array of light harvesting and energy transmitting pigments. The absorption maxima listed in Table 1 indicate the range of light harvesting capacity of these pigments. For efficient energy transfer the fluorescence spectrum of the energy donor pigment overlaps with the absorption spectrum of energy acceptor and the pigments are held in close proximity in the phycobilisome to maximize energy transfer. Thus it is possible to demonstrate

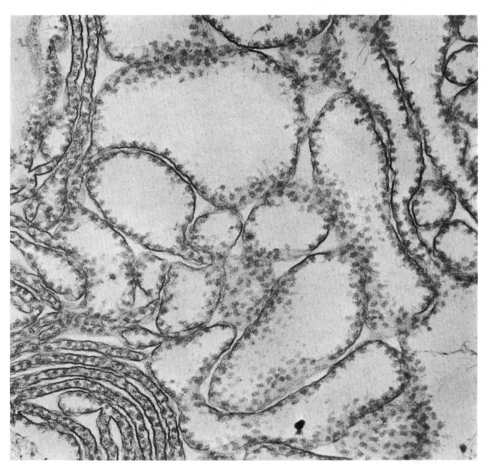

Fig. 1. Chloroplast section from the red alga *Porphyridium cruentum* showing phycobilisomes attached to the chloroplast lamellae. Numerous blue-green algae have been found to have similar phycobilisome structures. Phycobilisomes are composed of phycoerythrin, phycocyanin and allophycocyanin and serve as major light harvesting pigments. Fixed in 4% gluteraldehyde and post fixed in 1% osmium tetroxide. Magnification ×67,000. (Contributed by E. GANTT)

energy transfer through the sequence phycoerythrin → phycocyanin → allophyco-cyanin → chlorophyll. An intermediate in the sequence might also serve as an efficient light harvesting pigment (Lemasson et al., 1973). Glazer and Bryant (1975) have recently reported on a new allophycocyanin B whose absorption maximum at 671 nm makes it a very attractive candidate as the immediate energy donor to chlorophyll a.

II. Phycobilisomes

The phycobiliproteins are probably all contained within the large phycobilisome particle visible on the surface of photosynthetic membranes of both blue-green and red algae (Gantt and Lipschultz, 1974; Wildman and Bowen, 1974). These are broad based structures attached to the outer or stroma surface of the membrane. The phycobilisomes vary in different species of blue-green algae from triangular to semicircular shapes. They range in diameter at the base from 280 to 550 Å but there is no clear correlation between the size of the phycobilisome and the age or species of the cell in which they are found. Phycobilisomes can be isolated with little fragmentation and there is good evidence for energy transfer through the phycobiliproteins within the structure (Gantt and Lipschultz, 1973; Gray et al., 1973; Gray and Gantt, 1975). Gray and Gantt (1975) have suggested a model where allophycocyanin forms the base of the phycobilisome which is attached to the photosynthetic membrane; above this there may be a layer of phycocyanin which in turn has a phycoerythrin layer above it forming the outermost layer of the phycobilisome. The precise organization of the proteins in the phycobilisome promises to be a fascinating problem in supramolecular assembly.

D. Photosystem II Reaction Centers

While the phycobiliproteins collect energy which is transferred into a chlorophyll a-like molecule of photosystem II, very little is known of the biochemical nature of the reaction center where the absorbed light energy is made to do chemical work. Fujita and his colleagues have isolated a chromoprotein complex from *Anabaena cylindrica* which they suggest may be the reaction center pigment for photosystem II. As the photosynthetic membranes of *A. cylindrica* are incubated in dilute buffer, photosystem II activity is lost and the chromoprotein complex is released (Fujita et al., 1974). The purified chromoprotein has a molecular weight of 50,000 and shows absorption maxima at 275, 450, 475, 630 and 690 nm (Pjon and Fujita, 1974). The chromophore of this protein is not easily released by organic solvent extraction so it is very different from the usual chlorophyll-protein associations. The chromoprotein complexes readily in vitro with phycocyanin, but not phycoerythrin or several other proteins tested. This may reflect a natural association of a reaction center pigment with light harvesting pigment but one would rather see an association which allowed energy transfer from phycocyanin to allophycocyanin to chlorophyll a and then finally to a reaction center chromoprotein since this would afford the optimal spectral overlaps for efficient energy migration. Phycocyanin adsorbed to the chromoprotein can transfer energy to it and the light energized chromoprotein can catalyze the reduction of oxygen or methyl red by the reduced forms of dichloroindophenol, tetramethylphenylene diamine or cytochrome c (Tsuji and Fujita, 1971).

E. Photosystem I

Chlorophyll a is the principal light absorber for photosystem I and this pigment is firmly held in the membrane. Thornber (1969) treated the photosynthetic membranes of *Phormidium luridum* with the detergent sodium dodecyl sulfate and released a chlorophyll-protein complex in amounts that accounted for 75% of the chlorophyll in the membranes. The very hydrophobic pigment protein complex has a molecular weight of approximately 150,000 and contains four identical subunits of molecular weight 35,000. Each polypeptide subunit contains five molecules of chlorophyll. One fifth of the molecules in this preparation were found to contain one P-700 plus approximately 20 light-harvesting chlorophyll a molecules (Dietrich and Thornber, 1971). The other 80% of the molecules in the preparation contain only chlorophyll a. The P-700-containing molecules could not be separated from those containing only chlorophyll a. The midpoint potential for P-700 in this complex lies between 0.38 and 0.42 V (Thornber and Olson, 1971). The preparation catalyzes electron transfer from reduced cytochrome c to oxygen with good quantum efficiency.

Ogawa and Vernon (1970) have made another approach to the photosystem I reaction center complex. *A. variabilis* was grown in diphenylamine containing media to diminish its carotenoid content. The photosynthetic membranes were isolated from these cells and treated with the detergent Triton X-100. Sucrose density gradient centrifugation allowed for the separation of a membrane fragment enriched in P-700 (one P-700 per 30 chlorophyll a molecules in the fragment) and purified with respect to total protein content. When supplemented with the appropriate enzymes, this fragment catalyzed a rapid photoreduction of $NADP^+$ by reduced indophenol. Sodium dodecyl sulfate – disc gel electrophoresis analysis of this fragment might lead to the recognition of proteins which are neighbors of P-700 in the membrane.

F. Electron Transport from Photosystem I to $NADP^+$

Light energy collected by the antennae chlorophyll of photosystem I is funneled to a reaction center where it is used to push electrons from P-700 ($E'_0 \cong +0.43$) to an electron acceptor of strongly negative redox potential. P-700 was recognized in blue-green algae coincident with its discovery in higher plants. Hiyama and Ke (1972) have obtained difference spectra and extinction coefficients for P-700 in *A. variabilis* which lend support to the notion that P-700 is a special form of chlorophyll a. P-700 is currently thought to photoreduce a non-heme iron protein — a special type of ferredoxin tightly bound to the photosynthetic membrane. HIYAMA and KE (1971) observed light induced spectral changes suggestive of ferredoxin in preparations of *Plectonema boryanum* and *A. variabilis* and these changes showed kinetic properties expected of a primary photoreductant. EVANS et al. (1973) found the light induced electron spin resonance signal which they attributed to iron reduction in this photoreductant in *A. cylindrica*. VISSER et al. (1974) have shown an excellent correlation between the oxidation of P-700 and the appearance of the spin reasonance signal in *Anacystis nidulans*. Their results strongly support the notion that a bound ferredoxin is the primary electron acceptor of photosystem I. One hopes for the purification and chemical characterization of both P-700 and the new ferredoxin.

Electrons from the primary reductant produced by photosystem I have several alternative routes in blue-green algae. The most likely course is for the electron to pass to the conventional ferredoxin which is easily released from the membrane and has been characterized from many algae and higher plants. Several recent reports describe large-scale preparations of ferredoxin from blue-green algae (RAO et al., 1972; HALL et al., 1972; MITSUI and SAN PIETRO, 1973). These proteins appear to be more stable than ferredoxins from higher plants and the larger crystals reported by MITSUI and SAN PIETRO (1973) suggest that these molecules may be especially well suited for X-ray structural analysis. In terms of size, iron content and spectroscopic properties, the ferredoxins of blue-green algae are similar to ferredoxins from higher plants and not at all like the ferredoxins found in the green and purple photosynthetic bacteria. Consistent with the above are the immunological data of TEL-OR and AVRON (1974). Antibodies prepared against Swiss chard ferredoxin would cross react with ferredoxin from *Phormidium persicinum* but not the one from *Clostridium*.

There have been no detailed studies of the flavoprotein ferredoxin: $NADP^+$ reductase from blue-green algae since its first purification by SUSOR and KROGMANN (1966).

An alternative to the sequence through ferredoxin, then ferredoxin: $NADP^+$ reductase to $NADP^+$ appears in algae that have become deficient in iron. Flavodoxin appears in iron deficient cells where it replaces both ferredoxin and ferredoxin: $NADP^+$ reductase. Flavodoxin has a very negative redox potential and forms a stable semiquinone radical (ENTSCH and SMILLIE, 1972). The characterization of the free radical form of this enzyme allowed a convincing demonstration that flavodoxin does function in vivo (NORRIS et al., 1972).

Yet another route of reducing power generated by photosystem I leads to the reduction of oxygen. PATTERSON and MYERS (1973) found a light dependent production of hydrogen peroxide by several, but not all of the blue-green algae they examined. This reaction occurs in intact cells and seems to reflect the production of a low potential autooxidizable reductant which cannot be drained off fast enough by CO_2 reduction. Earlier, FUJITA and MYERS (1966, 1967) had demonstrated the photosystem I dependent accumulation of a reductant in the lamellae of blue-green algae which they called the cytochrome reducing substance. This reductant is either on the path from the primary photoreductant to ferredoxin or in equilibrium with it. The reductant is autooxidizable and probably is the same as the oxygen reducing substance studied by HONEYCUTT and KROGMANN (1970, 1972). While the attempts to isolate the cytochrome-reducing substance and the oxygen-reducing substance may have been vitiated by the creation of an iron-EDTA artifact (OETTMEIER and LOCKAU, 1973), there is strong evidence that large quantities of low potential autooxidizable substance can accumulate in the illuminated membranes. The blue-green membrane seems to react with O_2 more readily than do higher plant systems. The autooxidation of reduced ferredoxin is known to generate the superoxide anion radical. This very destructive radical might also be produced by autooxidation of other reducing substances. Blue-green algae are known to contain superoxide dismutase (LUMSDEN and HALL, 1974). Using *A. nidulans* ABELIOVICH et al. (1974) demonstrated that the synthesis of superoxide dismutase is induced by high O_2 concentration and that the enzyme functions to protect the organism from photooxidative death under conditions of CO_2 deprivation.

G. Reactants Linking the Photosystems

The sequence photosystem II → C-550 → plastoquinone → cytochrome b-559 → cytochrome f → plastocyanin → P-700 seems to be the best contemporary assignment of electron carriers between the two photosystems but this is very tentative and worthy of much debate. C-550 is known only from spectroscopic observation as a substance which undergoes reduction in photosystem II light even at liquid nitrogen temperature. BUTLER (1973) has found C-550 in *Anacystis marina* and KNAFF (1973) has described C-550 photoreduction in *Nostoc* at 77 K. Electrons from photoreduced C-550 may then pass through plastoquinone to cytochrome b-558. The chemical identity of plastoquinone and its participation in sequence is known with certainty but its exact site — before or after cytochrome b-558 — is less secure. Using Nostoc membrane fragments, KNAFF found that 2,5-dibromo-3-methyl-6-isopropyl-p-benzoquinone (DBMIB), which is thought to be a plastoquinone antagonist, inhibits the reduction of cytochrome b-558 by photosystem II light but not the oxidation of this cytochrome by photosystem I light. BÖHME and CRAMER (1972) observed the opposite result — DBMIB inhibition of cytochrome b-559 photooxidation by photosystem I but not its reduction by photosystem II — in spinach chloroplasts. Either Nostoc has a different order of electron carriers than spinach or the assumptions on which these interpretations rest need revision. KNAFF estimated the redox potential of cytochrome b-558 (synonymous with b-559) in Nostoc to be the same as that of cytochrome f ($E_m = +0.35$). FUJITA (1974) obtained a very similar redox potential value for the cytochrome b-559 in *A. variabilis*. In these experiments cytochrome reduction by photosystem II was inhibited by EDTA which may later prove to be a useful diagnostic tool. Cytochrome b-559 oxidation by photosystem I was inhibited by $HgCl_2$ as would be expected of plastocyanin dependent reaction. APARICIO et al. (1974) have found the high potential form of cytochrome b-559 in *Nostoc muscorum* and as in higher plants, this form can be photooxidized at 77 K by photosystem II. The meaning of this phenomenon is unclear. Cytochrome f is believed to mediate electron transfer between cytochrome b-558 and plastocyanin on the basis of spectroscopic measurements in situ and reconstitution experiments. The determination of the amino acid sequence of cytochrome f from *Spirulina maxima* by AMBLER and BARTSCH (1975) provides highly detailed evidence of the similarity of this protein to cytochrome f from eucaryotic algae. The algal cytochrome f is a small monomeric acidic protein which may differ significantly from the higher plant protein which appears to be oligomeric and rather hydrophobic.

While plastocyanin can not be readily isolated from every species of blue-green alga, it has been identified in several species. The tenacity of plastocyanin binding to the membrane may be the variable among species which accounts for its absence from the soluble proteins of some cells. VISSER et al. (1974) described the electron paramagnetic resonance signal of plastocyanin in intact *A. nidulans*. This work demonstrated the presence of plastocyanin in an alga which has not yet yielded plastocyanin by extraction and fractionation methods and further demonstrated plastocyanin function in vivo. Plastocyanin from *A. variabilis* shows the opposite net charge from higher plant plastocyanins (LIGHTBODY and KROGMANN, 1967) and TSUJI and FUJITA (1972) have shown that the plastocyanin binding site in algal membrane requires basic proteins while the analogous site in spinach chloroplast membranes will accept electrons only from acidic proteins.

H. Water Splitting, Integrated Function and Phosphorylation

Very little is known about the reactions which transfer electrons from water to photosystem II in blue-green algae. GERHARDT and WIESSNER (1967) had shown that manganese is required in this region by *A. nidulans* just as it is in eucaryotic systems. Recently TEL-OR and AVRON (1975) reported on the purifications of a small, heat-stable manganese-containing factor which restores Hill reaction activity to osmotically shocked membranes from *Phormidium luridum*. The site of action of this factor is between water and the site of electron donation to photosystem II. SOFROVA et al. (1974) have shown that a number of electron donors will restore photosystem II activity to *P. boryanum* preparations inactivated by high concentrations of Tris buffer and thus established another point of similarity between blue-green algae and higher plants.

There is ample evidence that the integrated function of all the components of the electron transport sequence in blue-green algae is very sensitive to the osmotic environment in which the membranes are suspended (SUZUKI and FUJITA, 1972; WARD and MYERS, 1972). This indicates a more fragile construction than the higher plant chloroplast membrane and should continue to provide an advantage in dissecting the photosynthetic apparatus.

There have been studies (BEDELL and GOVINDJEE, 1973; BORNEFELD and SIMONIS, 1974) which, through measurement of cellular ATP levels, indicate the functioning of both non-cyclic and cylic photophosphorylation in blue-green algae. Little has been added to the details of photophosphorylation in cell-free preparations from blue-green algae (see KROGMANN, 1973) however the author has found that *A. variabilis* membranes show proton gradient formation with much the same characteristics as phosphorylating chloroplasts.

References

Abeliovich, A., Kellenberg, D., Shilo, M.: Photochem. Photobiol. **19**, 379–382 (1974)

Ambler, R.P., Bartsch, R.G.: Nature (London) **253**, 285–288 (1975)

Aparicio, P.J., Ando, K., Arnon, D.I.: Biochim. Biophys. Acta **357**, 246–251 (1974)

Arnon, D.I., McSwain, B.D., Tsujimoto, H.Y., Wada, K.: Biochim. Biophys. Acta **357**, 231–245 (1974)

Bedell, G.W., Govindjee: Plant Cell Physiol. **14**, 1081–1087 (1973)

Bennett, A., Bogorad, L.: Biochemistry **10**, 3625–3634 (1971)

Bisalputra, T.: In: Algal Physiology and Biochemistry. Stewart, W.D.P. (ed.). Berkeley-Los Angeles: Univ. Calif., 1974, pp. 124–160

Böhme, H., Cramer, W.A.: Biochemistry **11**, 1155–1160 (1972)

Bornefeld, T., Simonis, W.: Planta (Berl.) **115**, 309–318 (1974)

Butler, W.L.: Accounts Chem. Res. **6**, 177–184 (1973)

Chapman, D.: In: The Biology of Blue-Green Algae. Carr, N.G., Whitton, B.A. (eds.). Oxford: Blackwell, 1973, pp. 162–185

Dietrich, W.E., Thornber, J.P.: Biochim. Biophys. Acta **245**, 482–493 (1971)

Eiserling, F.A., Glazer, A.N.: J. Ultrastruct. Res. **47**, 16–25 (1974)

Entsch, B., Smillie, R.M.: Arch. Biochem. Biophys. **151**, 378–386 (1972)

Evans, M.C.W., Reeves, S.G., Telfer, A.: Biochem. Biophys. Res. Commun. **51**, 593–596 (1973)

Fogg, G.E., Stewart, W.D.P., Fay, P., Walsby, A.E.: The Blue-Green Algae. London-New York: Academic Press, 1973

Frackowiak, D., Grabowski, J.: Photosynthetica **7**, 305–310 (1973)

Fujita, Y.: Plant Cell Physiol. **15**, 861–874 (1974)

Fujita, Y., Myers, J.: Arch. Biochem. Biophys. **113**, 730–737 (1966)

Fujita, Y., Myers, J.: Arch. Biochem. Biophys. **119**, 8–15 (1967)

Fujita, Y., Pjon, C., Suzuki, R.: Plant Cell Physiol. **15**, 779–787 (1974)

Gantt, E., Lipschultz, C.A.: Biochim. Biophys. Acta **292**, 858–861 (1973)

Gantt, E., Lipschultz, C.A.: Biochemistry **13**, 2960–2966 (1974)

Gerhardt, B., Wiessner, W.: Biochem. Biophys. Res. Commun. **28**, 958–964 (1967)

Glazer, A.N., Bryant, D.A.: Arch. Microbiol. **104**, 15–22 (1975)

Glazer, A.N., Cohen-Bazire, G., Stanier, R.Y.: Proc. Nat. Acad. Sci. U.S. **68**, 3005–3008 (1971)

Glazer, A.N., Fang, S., Brown, D.M.: J. Biol. Chem. **248**, 5679–5685 (1973)

Gray, B.H., Gantt, E.: Photochem. Photobiol. **21**, 121–128 (1975)

Gray, B.H., Lipschultz, C.A., Gantt, E.: J. Bact. **116**, 471–478 (1973)

Hall, D.O., Rao, K.K., Cammack, R.: Biochem. Biophys. Res. Commun. **47**, 798–802 (1972)

Hiyama, T., Ke, B.: Proc. Nat. Acad. Sci. U.S. **68**, 1010–1013 (1971)

Hiyama, T., Ke, B.: Biochim. Biophys. Acta **267**, 160–171 (1972)

Honeycutt, R.C., Krogmann, D.W.: Biochim. Biophys. Acta **197**, 267–275 (1970)

Honeycutt, R.C., Krogmann, D.W.: Biochim. Biophys. Acta **256**, 467–476 (1972)

Jost, M.: Arch. Mikrobiol. **50**, 211–245 (1965)

Klein, S.M., Vernon, L.P.: Plant Physiol. **53**, 777–778 (1974)

Knaff, D.B.: Biochim. Biophys. Acta **325**, 284–296 (1973)

Krogmann, D.W.: In: The Biology of Blue-Green Algae. Carr, N.G., Whitton, B.A. (eds.). Oxford: Blackwell, 1973, pp. 80–98

Lang, N.J., Whitton, B.A.: In: The Biology of the Blue-Green Algae. Carr, N.G., Whitton, B.A. (eds.). Oxford: Blackwell, 1973, pp. 66–79

Lemasson, C., Tandeau de Marsac, N., Cohen-Bazire, G.: Proc. Nat. Acad. Sci. U.S. **70**, 3130–3133 (1973)

Lightbody, J.J., Krogmann, D.W.: Biochim. Biophys. Acta **131**, 508–515 (1967)

Lumsden, J., Hall, D.O.: Biochem. Biophys. Res. Commun. **58**, 35–41 (1974)

MacColl, R., Berns, D.S., Koven, N.L.: Arch. Biochem. Biophys. **146**, 477–482 (1971)

Mitsui, A., San Pietro, A.: Plant Sci. Lett. **1**, 157–163 (1973)

Norris, J.R., Crespi, H.L., Katz, J.J.: Biochem. Biophys. Res. Commun. **49**, 139–146 (1972)

Oettmeier, W., Lockau, W.: Z. Naturforsch. **28C**, 717–721 (1973)

Ogawa, T., Vernon, L.P.: Biochim. Biophys. Acta **197**, 292–301 (1970)

Ogawa, T., Vernon, L.P., Mollenhauer, H.H.: Biochim. Biophys. Acta **172**, 216–229 (1969)

Patterson, P.C.O., Myers, J.: Plant Physiol. **51**, 104–109 (1973)

Pjon, C., Fujita, Y.: Plant Cell Physiol. **15**, 789–797 (1974)

Rao, K.K., Smith, R.V., Cammack, R., Evans, M.C.W., Hall, D.O.: Biochem. J. **129**, 1159–1162 (1972)

Sofrova, D., Slechta, V., Leblova, S.: Photosynthetica **8**, 34–39 (1974)

Susor, W.A., Krogmann, D.W.: Biochim. Biophys. Acta **120**, 65–72 (1966)

Suzuki, R., Fujita, Y.: Plant Cell Physiol. **13**, 427–436 (1972)

Tel-Or, E., Avron, M.: Europ. J. Biochem. **47**, 417–421 (1974)

Tel-Or, E., Avron, M.: In: Proc. 3rd Intern. Congr. Photosynthesis. Avron, M. (ed.). Amsterdam: Elsevier, 1975

Thornber, J.P.: Biochim. Biophys. Acta **172**, 230–241 (1969)

Thornber, J.P., Olson, J.M.: Photochem. Photobiol. **14**, 329–341 (1971)

Troxler, R.F., Brown, A., Foster, J.A., Franzblau, C.: Federation Proc. **33**, 1258 Abs. (1974)

Tsuji, T., Fujita, Y.: Plant Cell Physiol. **12**, 807–811 (1971)

Tsuji, T., Fujita, Y.: Plant Cell Physiol. **13**, 93–99 (1972)

Visser, J.W.M., Amesz, J., van Gelder, B.F.: Biochem. Biophys. Acta **333**, 279–287 (1974a)

Visser, J.W.M., Rijgersberg, K.P., Amesz, J.: Biochim. Biophys. Acta **368**, 235–246 (1974b)

Ward, B., Myers, J.: Plant Physiol. **50**, 547–550 (1972)

Wildman, R.B., Bowen, C.C.: J. Bact. **117**, 866–881 (1974)

Williams, V.P., Freidenreich, P., Glazer, A.: Biochem. Biophys. Res. Commun. **59**, 462–466 (1974)

3. Electron Transport and Photophosphorylation in Photosynthetic Bacteria

Z. Gromet-Elhanan

A. Introduction

Photosynthetic bacteria are unique among the eubacteria in their ability to use light as the ultimate energy source (Engelmann, 1888; van Niel, 1932). They are also unique among other photosynthetic organisms in their inability to use water as an ultimate reductant, and consequently they do not evolve oxygen and require the addition of other reductants. Their ability to grow photosynthetically under strictly anaerobic conditions led bacterial systematists to set them off in a special major subgroup.

Aside from their common property of photosynthetic growth the photosynthetic bacteria display a remarkable diversity and versatility, not only in their morphology and growth conditions but also in their pigment systems and photosynthetic metabolism. This diversity led to their division into two major groups on the basis of their pigmentation: the green and purple bacteria. The purple bacteria are further subdivided into sulfur and nonsulfur bacteria on the basis of their source of reducing power: the purple sulfur and all the green bacteria can use reduced inorganic sulfur compounds, whereas the purple nonsulfur bacteria require reduced organic compounds. A summary of the main properties of the three groups is given in Table 1.

Table 1. General description of the three main groups of photosynthetic bacteria

Microorganisms	Green sulfur bacteria	Purple bacteria	
		Sulfur	Nonsulfur
Old classification[a]	Chlorobacteriaceae	Thiorodaceae	Athiorodaceae
New classification[b]	Chlorobiaceae	Chromatiaceae	Rhodospirillaceae
Representative genera	*Chlorobium* *Chloropseudomonas*	*Chromatium* *Thiospirillum*	*Rhodospirillum* *Rhodopseudomonas*
Pigments	Bacteriochlorophyll c or d[c] Monocyclic carotenoides	Bacteriochlorophyll a or b Acyclic carotenoids	
Source of reducing power	H_2S; S; $S_2O_3^=$ or H_2	H_2S; S; $S_2O_3^=$; H_2 and organic compounds	Organic compounds (H_2 in some species)
Relation to oxygen	Obligate anaerobes	Obligate anaerobes	Many species facultative aerobes
Growth in the dark	None	None	Yes, aerobic

[a] van Niel (1957); [b] Pfennig and Trüper (1974); [c] Jensen et al. (1964)

The striking deviations of the photosynthetic bacteria from the general pattern of photosynthesis as well as the marked differences among them aroused a vast interest in the study of their photosynthetic metabolism, but also complicated the issue. Because there are differences not only between the three groups but also within each group, there is no representative organism for study and, in spite of the temptation to generalize, results obtained with one species cannot be attributed without further tests to any other species. FRENKEL (1970) has emphasized this situation by drawing attention to the multiplicity of electron transport reactions in bacterial photosynthesis. Although the differences might be variations on a general theme, we are at present still searching for it.

Cell-free membrane preparations termed chromatophores (SCHACHMAN et al., 1952), which contain the photosynthetic pigments and carry out light-induced electron transport and photophosphorylation, have been isolated from various photosynthetic bacteria. The variability observed with whole cells extends also to the chromatophores, where differences in structure and function of particles isolated from different bacteria have been reported. The development, composition and structural organisation of the chromatophores in relation to the cell membrane have been reviewed recently by LASCELLES (1968) and OELZE and DREWS (1972) and will not be dealt with here. This chapter will cover mainly the function of chromatophores in photosynthetic electron transport and energy conversion reactions. Within the limited scope of this chapter it is impossible to cover in detail every aspect of these topics. Therefore, only an overall picture will be presented, including the major deviations from it as well as the unsolved problems, together with references to more specialized recent reviews.

B. Photosynthetic Electron Transport

I. General

The study of light-induced electron transport in photosynthetic bacteria began with the observation of DUYSENS (1952) that illumination of a suspension of *Chromatium vinosum* cells caused a small reversible decrease in an absorption band at 890 nm, which was accompanied by a blue shift in another band near 800 nm. This decrease in absorption was later ascribed to photooxidation (DUYSENS et al., 1956) of a special minor bacteriochlorophyll component, the reaction center bacteriochlorophyll (RC Bchl). Further studies indicated that light absorbed by the bulk light-harvesting, or antenna, bacteriochlorophyll is very efficiently transferred to this RC Bchl which functions as the initiator of the electron transport reactions peculiar to photosynthesis.

Although the initiation of photosynthetic electron transport follows a similar pattern in chloroplasts (cf. this vol. Chap. II.1) and chromatophores (see below) the further bacterial type of electron transport is quite different. Since the first observation of FRENKEL (1954) that illumination of chromatophores prepared from *Rhodospirillum rubrum* led to rapid synthesis of ATP, it has been established that photophosphorylation occurs in chromatophores in the complete absence of added electron donors or acceptors. It therefore follows that the photosynthetic electron transport to which ATP formation is coupled, is cyclic in nature in the chromatophores and involves no net oxidation or reduction. The components participating in this cyclic electron transport and their proposed sequence, the number and possible location of the sites of the coupled phosphorylation will be outlined below.

II. Reaction Centers and Primary Events

The nature of the reaction center and the primary photochemical process in bacterial photosynthesis were studied extensively in the last few years and have recently been reviewed in detail (CLAYTON, 1973; PARSON and COGDELL, 1975). These studies have gained momentum especially since the first successful isolation of a reaction center from *Rhodopseudomonas spheroides* by REED and CLAYTON (1968).

Reaction center preparations that contain the RC Bchl but are free of most or all of the bulk antenna Bchl have already been isolated from chromatophores of several species of purple non sulfur bacteria, although special preparative procedures were required for each species. The purple sulfur and green bacteria have so far yielded preparations which are enriched in RC Bchl but still contain a substantial number of antenna Bchl molecules per active center. The reaction center preparations generally contain, in addition to the protein-pigment complex of the RC Bchl, also bacteriopheophytin, ubiquinone and nonheme iron. Some larger preparations contain also copper and b and c-type cytochromes. Summaries of the various preparations and their composition have been published (OELZE and DREWS, 1972; PARSON, 1974; SAUER, 1975).

The spectral properties characteristic to RC Bchl in intact cells and chromatophores were observed also in reaction center preparations. Thus, illumination of a reaction center causes bleaching of the major band of the RC Bchl, which varies between 860 nm to 890 nm among the different species containing Bchl a, and even more so when species with Bchl b are included. For simplicity they are referred to by one common designation: P-870. A blue shift near 800 nm occurs also in reaction centers in conjunction with P-870. The bleaching of P-870 is very rapid, proceeds with a quantum yield near unity and is virtually temperature-independent. These properties have been observed both in *Chromatium* and *R. rubrum* chromatophores (PARSON, 1968) and in a reaction center preparation from *Rps. spheroides* (WRAIGHT and CLAYTON, 1974). Therefore, the bleaching of P-870 seems to be the primary reaction in photosynthetic bacteria in general. Since the bleaching was shown to be due to photooxidation it appears that the primary photochemical event is the transfer of an electron from the RC Bchl (the primary donor) to a primary acceptor: $P\text{-}870\ X \rightarrow P\text{-}870^+\ X^-$.

The identity of the primary electron acceptor is still eluding us, although it is one of the most active areas of current research. A variety of components have been proposed for the role, including quinones, nonheme iron proteins and other components identified only in terms of absorbance changes or EPR signals. The finding that photochemically active reaction center preparations contain nonheme iron and ubiquinone has focused the attention of them. But the investigations designed to discover whether either of them or a complex of both function as the primary acceptor have yielded conflicting results.

The observations could be explained by suggesting that only one of these components (although we do not know which one) is the physiological acceptor, but that P-870 might be flexible enough so that if the native primary acceptor is removed or inactivated a substitute can take its place. There is, however, another possible explanation, namely that there may be several different acceptors for a given donor, each of them functioning at a different set of conditions. This suggested multiplicity of primary acceptors ties in with the reported existence, in a number of photosynthetic bacteria, of more than one functional c-type cytochrome, which can donate electrons to P-870 but under different conditions (see Sect. B.III). These observations might reflect a situation of parallel pathways operating on both the acceptor and donor sides of P-870 (see FRENKEL, 1970). In this connection it might be worth while to mention earlier suggestions on the possible existance of more than one photochemical reaction center in some photosynthetic bacteria especially *Chromatium* (CUSANOVICH et al., 1968; SCHMIDT and KAMEN, 1971) and *R. rubrum* (SYBESMA, 1969; FOWLER and SYBESMA, 1970; OKAYAMA et al., 1970). These suggestions were based on the observation that illumination of intact cells, chromatophores and reaction centers resulted in a number of infrared absorbance changes besides P-870, which could be due to multiple types of RC Bchl. Each of these types might in turn react with a different acceptor and donor resulting in completely independent but parallel electron transport chains. Recently, the view that there is only one type of reaction center present in each species has become more popular (PARSON, 1974). But this view is based mainly on data with *Chromatium* (THORNBER, 1970; PARSON and CASE, 1970; DUTTON et al., 1971; CASE and PARSON, 1973) and should therefore await critical experimentation in a number of other species before it can be accepted as a general rule.

The chemical potential produced by the absorbed photon is defined by the difference between the electrochemical potential of the oxidized primary donor and the reduced primary acceptor. KUNTZ et al. (1964) measured the reversible midpoint potential for P-870/P-870$^+$ in *R. rubrum* chromatophores and found it to be +0.44 V. In *Chromatium* a value of +0.49 V was reported by CUSANOVICH et al. (1968) and numbers varying within these narrow limits were later reported in various other species, except for the green sulfur bacteria in which lower values ranging between +0.24 V and +0.33 V were reported (FOWLER et al., 1971; KNAFF et al., 1973).

Attempts to assay the midpoint potential of the primary acceptor resulted, however, in large differences not only among various species but even between different measurements in the same species (PARSON and COGDELL, 1975). These variations are at least partially due to the limitations of the indirect methods used. Thus, the redox titrations of X/X$^-$ were based on measurements of P-870 assuming that a chemical reduction of the primary acceptor will result in loss of ability of P-870 to transfer electrons upon illumination. Such measurements gave values ranging between −0.03 V to −0.16 V. If, however, the electron transport between the primary and secondary acceptor is very rapid, then this method might reflect the midpoint potential of the secondary rather than the primary acceptor or an average of both. Indeed, by using flash spectroscopy following brief laser pulses (rather than continuous illumination) DUTTON (1971) and SEIBERT and DE VAULT (1971) found midpoint potentials near −0.14 V for *Rps. gelatinosa* and *Chromatium,* but they also obtained evidence that a small portion of the photoreaction was operative down to −0.35 V. A similar low value was recently reported by LOACH et al. (1975) in *R. rubrum.* Indications for even lower values were obtained by GOVINDJEE et al. (1974) and by SILBERSTEIN and GROMET-ELHANAN (1974) who observed a stimulation of the dark decay of photoinduced absorbance changes in *R. rubrum* chromatophores by benzyl and methyl viologen.

The accurate determination of the midpoint potential of the primary acceptor is of utmost importance in the study of photosynthetic electron transport in bacteria, because it will determine whether the reducing power generated in the light is enough to mediate the direct reduction of NAD^+ (see Sect. B.V).

III. Components of the Electron Transport Chain

Besides the primary electron donor and acceptor the cytochromes are the most prominent components of the bacterial photosynthetic electron transport. Since the first report on the presence of cytochrome c in *R. rubrum* (VERNON, 1953) and the demonstration of its reversible photooxidation (DUYSENS, 1954), every examined species of photosynthetic bacteria was found to contain at least one c-type cytochrome, which appears to function as the direct electron donor to oxidized RC Bchl. Until very recently the purple and green sulfur bacteria were considered to contain only cytochrome c, whereas in the purple nonsulfur species both c- and b-type cytochromes were demonstrated. But in two independent recent reports (FOWLER, 1974; KNAFF and BUCHANAN, 1975) b-type cytochromes were observed in *Chromatium* as well as in two species of green bacteria: *Chlorobium limicola* and *Chl. thiosulfatophilum.*

Besides these common features the photosynthetic bacteria display, however, a remarkable diversity in the number and properties of their cytochromes. At our present state of knowledge it is impossible to decide whether this diversity is due to basic differences in the photosynthetic electron transport chains or is rather a result of the use of different experimental approaches in the examination of various species. This section will, therefore, describe the main findings reported for each of the most extensively studied species. KAMEN and HORIO (1970) and HORIO and KAMEN (1970) have reviewed in detail the structure and function of bacterial cytochromes in general. Light-induced cytochrome reactions in photosynthetic bacteria were reviewed by FRENKEL (1970) and some more recent data were summarized by PARSON (1974).

Chromatium. The large magnitude of the absorbance changes due to cytochromes and the minimal interference by carotenoids in *Chromatium* chromatophores made them particularly useful in this study. Two c-type cytochromes were shown to be photooxidized by these chromatophores (CUSANOVICH et al., 1968). C-555 (Soret band at 422 nm) has an Em of +0.33 V and is photooxidized with a half time of 2 μsec (PARSON and CASE, 1970) but only at high redox potentials and high light intensity. Its re-reduction in the dark requires 50 msec and it was found to be a component of the cyclic electron transport from X^- to P^+ which is coupled to photophosphorylation (see Sect. B.IV). The other cytochrome, C-552 (sometimes called C-553, with a Soret at 423 nm) has an Em of +0.1 V and its photooxidation (half time 1 μsec) is nearly temperature-independent to at least

77 K, where C-555 oxidation stops completely (CHANCE and NISHIMURA, 1960; DUTTON, 1971). Under reducing conditions, when both cytochromes should be able to donate electrons, only C-552 is photooxidized. Its re-reduction is, however, very slow (> 10 sec) and it seems therefore to have no connection to energy storage and has been implicated in the noncyclic electron transport from reduced sulfur compounds to NAD^+, although there is as yet no experimental evidence for it.

Since conditions for photooxidation of both cytochromes are mutually exclusive each of them must be able to donate electrons directly to RC Bchl, and it seems to be the same RC Bchl at least in *Chromatium* chromatophores (see Sect. B.II).

KNAFF and BUCHANAN (1975) have recently demonstrated the presence of a b-type cytochrome (with an α-band at 560 nm and Em of -0.05 V) in *Chromatium* chromatophores. By poising the chromatophores at $+0.2$ V, C-555 was photooxidized with no observable change in b-560. But upon addition of antimycin A (which is a known inhibitor of electron transport between cytochrome b and c, see Sect. B.IV) b-560 was photoreduced. They, therefore, proposed that b-560 is the electron donor to C-555 in the cyclic electron transport chain.

Chlorobium. Very little work has been done with the green sulfur bacteria and recent reports on cytochrome photoreactions in *Chlorobium* chromatophores are so contradictory that further clarification will be required before any overall picture can be drawn. In both *Chl. limicola* (FOWLER et al., 1971)[1] and *Chl. thiosul-fatophilum* (KNAFF et al., 1973) photooxidation of high potential c-type cytochrome, C-551 Em between $+0.17$ V or $+0.22$ V was observed. KNAFF et al., have also observed that C-551 photooxidation was attenuated to zero in two one electron steps with Em of $+0.03$ V and -0.13 V and suggested that the first might be an additional, low potential c-type cytochrome and the second a primary (or secondary) acceptor. Very recently b-types cytochromes have been found in chromatophores from both species (FOWLER, 1974; KNAFF and BUCHANAN, 1975). But, while FOWLER observed photooxidation of b-562, KNAFF and BUCHANAN reported on its photoreduction. This photoreduction was stimulated by antimycin A, which was shown to inhibit the dark oxidation of b-564 but had no effect on C-551 photooxidation.

Rhodopseudomonas. Here chromatophores or reaction center preparations from many species have been studied including *Rps. viridis, gelatinosa, capsulata* and *spheroides.* The first two species are similar to *Chromatium* and contain two membrane-bound c-type cytochromes, a high- and a low-potential. Both appear to feed electrons directly to $P-870^+$ by parallel pathways (THORNBER et al., 1969; CASE et al., 1970) and the photooxidation of both is very fast at 77 K. (KIHARA and CHANCE, 1969; DUTTON, 1971).

Rps. spheroides and *Rps. capsulata* on the other hand, are more similar to *R. rubrum.* They have no membrane bound low-potential c-type cytochrome which can donate electrons to $P-870^+$ (DUTTON and JACKSON, 1972; EVANS and CROFTS, 1974b) and their reactive c-type cytochrome does not undergo low temperature reactions (KIHARA and CHANCE, 1969). Also in both species, as in *R. rubrum,* there are membrane-bound b-type cytochromes which undergo photoreduction, but are reoxidized by way of the high-potential c-type cytochrome and not directly by $P-870^+$.

---

[1] This report deals with *Chloropseudomonas ethylica,* which has later been found to be a mixture of *Chl. limicola* and another non-photosynthetic bacterium (GRAY et al., 1973).

Working with carotenoid deficient mutants, so that characterization of α-bands of both b- and c-type cytochromes was possible, DUTTON and JACKSON (1972) in *Rps. spheroides* and EVANS and CROFTS (1974b) in *Rps. capsulata* report different observations which cannot as yet be reconciled with one scheme for both species. Thus, in *Rps. spheroides* only one c-type cytochrome was potentiometrically measurable, the high potential C-551 (Em +0.29 V) which is the electron donor to oxidized RC Bchl, but three b-type cytochromes were measured with Em +0.15 V, +0.05 V and −0.09 V (the first two appear to reduce C-551). Similar redox titrations in *Rps. capsulata* revealed on the other hand, three c-type components with Em of +0.34 V, +0.12 V and 0.0 V, only the high potential one being photooxidized through RC Bchl, and only one b-type cytochrome b-561 (Em +0.06 V). Moreover, b-561 did not seem to donate electrons directly to the high-potential photooxidizable cytochrome c (Em +0.34 V) and required an intermediary unidentified carrier Z.

Rhodospirillum. As early as 1963 NISHIMURA observed in whole *R. rubrum* cells light-induced absorbance changes which appeared to indicate photooxidation of cytochrome c and photoreduction of cytochrome b, but only when he added the inhibitors HOQNO (2-*n*-heptyl-4-hydroxyquinoline-N-oxide) or antimycin A. In chromatophores of *R. rubrum,* however, light-induced cytochrome reactions are difficult to follow, because of their small magnitude and the large interference in the α-band by carotenoids and in the Soret band by the remarkably large light-induced absorption peak at 430 nm (SMITH and BALTSCHEFFSKY, 1959). There is only one report by PARSON (1967) on light-induced cytochrome changes in chromatophores. By extrapolation of flash-induced absorbance changes, again in the presence of the inhibitor antimycin A, he found an indication for the photooxidation of C-551 (Soret 418) and photoreduction of b-563 (Soret 431).

Dark potentiometric titrations of α-bands of cytochromes were recently carried out in chromatophores of the carotenoidless mutant G9 (KAKUNO et al., 1971; DUTTON and JACKSON, 1972). Both groups found a C-552 with Em of +0.29 V to +0.31 V which is similar to the well known purified C$_2$ (BARTSCH et al., 1971). As in *Rps. spheroides* more than one b-type component was resolved by the potentiometric titration: KAKUNO et al. found two b-types b-562 Em +0.02 V and −0.16 V and DUTTON and JACKSON observed even three components with Em +0.16 V; −0.05 V and −0.10 V. At present, however, no investigation on the correlation of these components with the light-induced cytochrome reactions has been done.

Ubiquinones have been found in large amounts in photosynthetic bacteria and from their presence in reaction center preparations (see Sect. B.II) they seem to participate in the photosynthetic electron transport. The extraction of ubiquinone-10 from *R. rubrum* chromatophores greatly diminished ATP formation which was reactivated upon addition of ubiquinone-10 (OKAYAMA et al., 1968). In *Chromatium* chromatophores ubiquinone-7 has been demonstrated (CUSANOVICH and KAMEN, 1968a).

The exact location of the ubiquinone in the cyclic electron transport is, however, not clear and it was included in a different position in almost every published scheme: in place of the secondary acceptor, Y (EVANS and CROFTS, 1974b) after Y (FRENKEL, 1970) or in between two b-type cytochromes (DUTTON and JACKSON, 1972). In some cases it was even proposed to be the primary acceptor (see Sect. B.II).

Ferredoxins have also been isolated from a number of photosynthetic bacteria

(YOCH and VALENTINE, 1972) but have not been shown to participate directly in the cyclic photosynthetic electron transport. There is only one report on a ferredoxin-dependent photoreduction of NAD^+ by *Chlorobium* chromatophores (BUCHANAN and EVANS, 1969; see also Sect. B.V). It is not clear whether there is any connection between the isolated ferredoxins and the nonheme iron found in reaction center preparations. In a recent report the soluble ferredoxins isolated from *R. rubrum* were found to be distinct from bound nonheme iron (YOCH et al., 1975).

IV. Sites of Coupled Phosphorylation

Photophosphorylation coupled to the light-induced electron transport was observed in chromatophores (FRENKEL, 1954) at about the same time as in chloroplasts (ARNON et al., 1954). The number of phosphorylating sites and their location has since been constantly under investigation, but their elucidation is still far from complete. The research was hindered by the uncertainty regarding the number of components participating in the electron transport chain and their exact sequence (see Sect. B.III), by the absence of specific inhibitors for every electron transport step and by the cyclic nature of the system, in which no direct measurement of the number of coupling sites per electron transferred is possible.

The information obtained up to now is summarized in the scheme of Figure 1. It is based on the use of those inhibitors which have been shown, by a number of approaches, to inhibit the electron transport at specific sites as well as on the use of electron transport carriers, which when added to the inhibited chromatophores, caused resumption of their phosphorylation (see this vol. Chap. II.15).

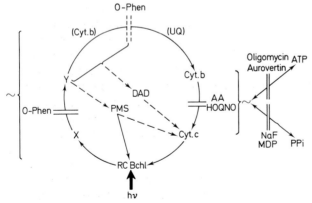

Fig. 1. Schematic presentation of the light-induced cyclic electron transport and the sites of coupled phosphorylation in *R. rubrum* chromatophores. —— Electron transport reaction which was followed spectrophotometrically. ---- suggested electron transport reaction. ( ) indicates a component observed as yet only in dark redox titrations. =: site of action of inhibitor. ===: suggested site of inhibition. ∼: high energy state. *AA*: antimycin A; *MDP*: methylene disphosphate; *O-Phen*: orthophenanthroline; *UQ*: ubiquinone. For further explanations see text

Practically all the research in this field has been carried out with chromatophores from these species: *R. rubrum, Rps. capsulata* and *Chromatium*.

Photophosphorylation in *R. rubrum* chromatophores was inhibited by low concentrations of antimycin A or HOQNO (SMITH and BALTSCHEFFSKY, 1959; GELLER and LIPMANN, 1960). Since they are well known inhibitors of electron transport between cytochromes b and c in mitochondria, it was indicative of a similar site of inhibition in chromatophores. This assumption was corroborated in further investigations where, by applying the inhibitors to whole cells (NISHIMURA, 1963) or to chromatophores (PARSON, 1967), it became possible to demonstrate the photooxidation of cytochrome c and photoreduction of cytochrome b. Evidence for the location of a site of phosphorylation between cytochromes b and c was obtained by BALTSCHEFFSKY (1967) in experiments based on the crossover theorem of CHANCE et al. (1955). BALTSCHEFFSKY found that addition of either ATP or PP$_i$ (inorganic pyrophosphate) in the dark caused the oxidation of cytochrome c simultaneously with the reduction of cytochrome b (see Fig. 1). This energy-induced reverse eletron transport is indicative of the existence of a coupling site between these cytochromes. Unfortunately the effect of antimycin A or HOQNO on this reverse electron transport has not been reported.

The inhibition of phosphorylation by antimycin A or HOQNO could be overcome by the addition of PMS to *R. rubrum* chromatophores (BALTSCHEFFSKY and BALTSCHEFFSKY, 1958; GELLER and LIPMANN, 1960). This indicated that PMS can "by-pass" the electron transport step between cytochrome b and c and suggested the existence of at least one other site of phosphorylation, which is operative in the PMS by-pass of the inhibited site. These data raised the question of the number of phosphorylation sites on this cyclic electron transport. BALTSCHEFFSKY and ARWIDSSON (1962) have reported that the antibiotic valinomycin inhibited endogenous ATP formation to the extent of only 50% and had no effect on PMS phosphorylation. They, therefore, suggested that there are only two sites of phosphorylation in *R. rubrum* chromatophores, one sensitive to valinomycin and the other (the site operating with PMS) resistant to the antibiotic. In further experiments this site-specificity of valinomycin was not confirmed, since the endogenous phosphorylation was found to be inhibited by over 80% (THORE et al., 1968; GROMET-ELHANAN, 1970)[2]. Further experiments will therefore be required before a clear-cut decision regarding the total number of coupling sites can be reached. The site of energy conservation drawn in Figure 1 between RC Bchl and cytochrome b might represent a number of phosphorylating sites whose locations is still undecided. CROFTS et al. (1974) have proposed that in *Rps. spheroides* one coupling site is associated with the photochemical reaction. This possibility has not been tested in *R. rubrum*.

The exact extent of the PMS by-pass is also not known. COST et al. (1966) and PARSON (1967) presented evidence that the reduced form of PMS can donate electrons directly to oxidized RC Bchl in *R. rubrum* chromatophores. But the exclusion of cytochrome c from the phosphorylating PMS by-pass is still uncertain (see Fig. 1). Regarding the direct electron donor to PMS OKAYAMA et al. (1968) have reported that extraction of ubiquinone from *R. rubrum* chromatophores

---

[2] This effect of valinomycin was independent of the presence of K$^+$ and is different from its effect on K$^+$ permeability (see Sect. C.III).

greatly diminished ATP formation even in the presence of PMS, thus suggesting that ubiquinone should be included in the PMS by-pass. Horio and Yamashita (1964) tested the effect of o-phenanthroline on various phosphorylating systems, and found most of them to be 50% inhibited at concentrations below 0.1 mM, but the PMS by-pass was much more resistant, reaching 50% inhibition only above 1 mM. These results could indicate that at low concentrations o-phenanthroline inhibits one site, whereas at 1 mM inhibition of a second, a less sensitive site sets in. Clayton (1973) has, indeed, suggested that o-phenanthroline has two sites of action in *Rps. spheroides* reaction centers, one between the primary and secondary acceptors (X and Y in Fig. 1) and the other inhibiting directly the primary photoact. If this would be the case also in *R. rubrum* it would mean that ubiquinone must be the primary acceptor and the PMS by-pass might just be a stimulation of the back reaction from X to RC Bchl. In *Chromatium,* however, Parson and Case (1970) could not observe inhibition of the primary photoact even by very high o-phenanthroline concentrations. It, therefore, seems more plausible that the less sensitive site is between X and Y, while the more sensitive one can be between Y and cytochrome b. In that case the scheme for the PMS by-pass would be as presented in Figure 1, assuming that ubiquinone can be the secondary acceptor Y. It should, however, be emphasized that this minimal scheme does not include the information on dark redox titrations (see Sect. B.III) which indicated that there might be additional steps between Y and cytochrome b, involving an additional b-type cytochrome and may be more than one ubiquinone component.

Besides PMS other compounds were reported to by-pass partially or completely the segment inhibited by antimycin A or HOQNO. These include ascorbate-DCIP and ascorbate-DAD, both being effective only in the presence of air or other electron acceptors such as NAD^+ or methyl viologen (see Sect. B.V). High concentrations of DAD by itself formed a by-pass which, as with PMS, markedly stimulated the rate of phosphorylation (Gromet-Elhanan, 1970), indicating that the electron transport between cytochrome b and c is rate limiting in the endogenous system. Unlike with PMS, there is as yet no information about the extent of the DAD by-pass (see Fig. 1).

Since photophosphorylation is in chromatophores cyclic in nature the relationship between electron transport and the coupled phosphorylation, which is usually measured by the respiratory control or the stimulation of electron transport by uncouplers (see this vol., Chap. III.11), cannot be investigated here. Both uncouplers and energy transfer inhibitors (see this vol. Chap. III.12) block cyclic phosphorylation in all systems. They can, therefore, be differentiated from the electron transport inhibitors discussed above, but not from each other. Feldman and Gromet-Elhanan (1971) described, however, a light-induced noncyclic electron transport from ascorbate-DCIP to oxygen in *R. rubrum* chromatophores. This system was stimulated by phosphorylating conditions and afforded a clear differentiation between uncouplers, such as atebrin and CCP, which stimulated oxygen uptake and the energy transfer inhibitor oligomycin which inhibited it. Differentiation between the two types of inhibitors was also obtained by following their effects on such reactions as PP_i formation (see below), NAD^+ reduction (see Sect. B.V) and proton uptake (see Sect. C.II). But these systems, unlike the coupled noncyclic oxygen uptake, were inhibited by uncouplers and stimulated by oligomycin (for details see the above mentioned sections).

In the absence of ADP a light-induced phosphorylation of P_i to PP_i has been observed in *R. rubrum* chromatophores in the endogenous system as well as in the PMS + HOQNO by-pass (BALTSCHEFFSKY and VON STEDINGK, 1966). PP_i formation resembled ATP formation in sensitivity to electron transport inhibitors and to uncouplers, but not to oligomycin. At oligomycin concentrations which block ATP formation PP_i formation was slightly stimulated. This was one of the first indications for the function of oligomycin as an energy transfer inhibitor blocking a specific terminal step in the synthesis of ATP rather than interfering with the high energy state (see Sect. C.IV). Further experiments have suggested that two different and independent coupling factor proteins are involved in the synthesis of ATP and PP_i (see Sect. C.V).

Chromatium. Chromatophores capable of photophosphorylation were first isolated from *Chromatium* by NEWTON and KAMEN (1957). Extensive studies of their phosphorylation process were, however, initiated much later by CUSANOVICH and KAMEN (1968b), HOCHMAN and CARMELI (1973) and KNAFF and BUCHANAN (1975). Most of the reactions which have been studied up to now were found to be very similar to those reported in *R. rubrum*. In the absence of added carriers the photophosphorylation was inhibited by the uncouplers CCP, gramicidin and atebrin as well as by the electron transport inhibitors antimycin A and HOQNO. In the presence of PMS, the slightly stimulated rate of photophosphorylation was resistant to both inhibitors, indicating the operation of the PMS by-pass in these chromatophores. As mentioned earlier (see Sect. B.III) these inhibitors were shown to inhibit electron transport between cytochrome b and c also in *Chromatium*.

A marked difference was, however, observed in the sensitivity of *Chromatium* and *R. rubrum* chromatophores to energy transfer inhibitors. HOCHMAN and CARMELI (1973) found that in *Chromatium*, as in chloroplasts (see this vol. Chap. III.12), Dio-9 and phlorizin acted as energy-transfer inhibitors, whereas oligomycin was inactive. In *R. rubrum*, on the other hand, oligomycin is an energy transfer inhibitor whereas Dio-9 and phlorizin were reported to act as uncouplers (GROMET-ELHANAN, 1969).

The only other species in which photophosphorylation was studied in detail is *Rps. capsulata*. Although it has more similarity to *R. rubrum* than *Chromatium* even in its cytochrome content (see Sect. B.III), its phosphorylating activity is, in some aspects at least, surprisingly different. Thus, PMS does not by-pass the site of antimycin A inhibition (KLEMME and SCHLEGEL, 1968). Moreover, HOQNO, unlike antimycin A, does not inhibit any of the studied phosphorylating systems nor does it affect the photoreduction of cytochrome b, which is stimulated by antimycin A (EVANS and CROFTS, 1974b). KLEMME (1969) has also reported that PP_i could not replace ATP as an energy donor in *Rps. capsulata* (see Sect. B.V). The ineffectiveness of PMS does not mean that there is in *Rps. capsulata* only one site of coupled phophorylation, since DAD was recently found to increase markedly the endogenous rate of phosphorylation and to partially by-pass the antimycin A site of inhibition (GROMET-ELHANAN and GEST, unpublished observations). These results suggest that the inability of PMS to overcome the inhibition by antimycin A might be due to some specific difference in an immediate electron transport donor or acceptor in *Rps. capsulata* which makes it incapable of reacting with PMS. But further studies will be required to clarify this difference.

V. Photoreduction of NAD$^+$

The mechanism for generation of reducing power in general and NAD$^+$ reduction in particular has been a subject of active research and discussion for almost two decades. An excellent, extensive review of the field has recently appeared (Gest, 1972). In this section only the main results together with the different theoretical interpretations will be summarized.

Frenkel (1958) was the first to observe photoreduction of pyridine nucleotides in *R. rubrum* chromatophores. Unlike chloroplasts (see this vol. Chap. I.2) this photoreduction was found to be specific for NAD$^+$, dependent on the addition of an electron donor and inhibited under phosphorylating conditions, e.g. when ADP, P$_i$ and Mg^{2+} were added (Frenkel, 1959). These early results have already suggested that the photoreduction of NAD$^+$ cannot be coupled with phosphorylation, but rather competes with the phosphorylation coupled to the cyclic electron transport. This competition is not in favor of NAD$^+$ reduction, since much more ATP was formed than NAD$^+$ reduced. The competition could, in theory, exist at various levels:

1. At the level of the photoreactions. If two electron transport systems, the phosphorylating cyclic and a nonphosphorylating noncyclic one leading to NAD$^+$ reduction, function in chromatophores and are driven via different RC Bchl participating in independent parallel pathways. The competition can then be a reflection of a regulatory mechanism which allows full activity of NAD$^+$ reduction only when the cyclic pathway is partially or completely inactive. This suggestion will require the demonstration of such parallel pathways (see Sect. B.II).

2. At the level of electron-transport. The cyclic and noncyclic systems might compete for electrons coming from a common carrier, either RC Bchl or the primary or secondary acceptors. As summarized in Section B.III there are indications for parallel independent pathways feeding electrons into a common Rc Bchl and these could be separated again after the primary or secondary acceptor. Unfortunately, however, most of this evidence comes from spectrophotometric observations of cytochromes in *Chromatium, Rps. viridis* and *Rps. gelatinosa* where NAD$^+$ reduction has not been studied in detail.

3. At the level of energy transfer. If, as has been suggested by Gest (1966), NAD$^+$ is reduced via the reverse electron flow demonstrated in mitochondria by Chance and Hollunger (1961), it will be dependent on an adequate supply of energy and will have to compete with ATP synthesis for the high energy state formed during the cyclic electron flow.

The last possibility has become the most favored one in recent years (Gest, 1972), because it could easily explain almost all of the rather confusing results, and also because some of its predictions have been verified. However, in the present author's opinion the reverse electron flow hypothesis has not been established beyond any doubt, and some of the accumulated information cannot be explained by it without invoking further assumptions. The results will therefore be evaluated here in light of all possible interpretations.

The most important observations are the following:

1. Inhibition of NAD$^+$ photoreduction by phosphorylation could be overcome by the addition of oligomycin (Keister and Yike, 1967a) or of aurovertin (Gromet-Delhanan, 1974a). Both compounds even slightly stimulated NAD$^+$ reduction

when added in the absence of ADP and P_i. This effect is completely different from that of uncouplers such as CCP, gramicicin or atebrin, that inhibited NAD^+ reduction in the absence or presence of ADP and P_i (KEISTER and YIKE, 1967a). It was therefore concluded that oligomycin (and aurovertin), both well known mitochondrial energy transfer inhibitors, function in a similar way also in chromatophores (see Fig. 1).

The different effects of uncouplers and energy-transfer inhibitors were taken as an indication for reverse electron flow (KEISTER and YIKE, 1967a). Uncouplers inhibit ATP synthesis by dissipating the high energy state (see this vol. Chap. III.11) and should therefore inhibit also the energy-dependent reverse electron flow, whereas energy transfer inhibitors block ATP synthesis without affecting or even stimulating the high energy state (see this vol. Chap. III.12). These effects can, however, be explained also by competition for electrons, because uncouplers stimulate the coupled cyclic electron transport and so increase its effectiveness in shunting electrons away from the noncyclic NAD^+ reducing pathway, whereas energy transfer inhibitors would inhibit the cyclic pathway and consequently favor NAD^+ reduction.

2. NAD^+ was shown to be reduced in complete darkness at the expense of ATP, and oligomycin inhibited the reduction in this case (KEISTER and YIKE, 1967a; KLEMME, 1969). In *R. rubrum* PP_i could replace ATP as energy donor for NAD^+ reduction (KEISTER and YIKE, 1967a) with a similar efficiency but in *Rps. capsulata* PP_i was inactive (KLEMME, 1969).

Although this demonstration fulfilled one of the basic predictions of the reverse electron flow hypothesis, it could not be accepted as unequivocal proof that all the light-induced NAD^+ reduction occurs via this mechanism, because of two problems: (1) In both bacterial species the rate of NAD^+ reduction with ATP was at best only about 20% of the light-induced rate, although in other ATP-driven reactions, such as energy-linked transhydrogenase, the rate with ATP was similar to the light-induced one (KEISTER and YIKE, 1967b). (2) The 20% activity with ATP was observed in both species with succinate as the electron donor, whereas with other electron donors the ATP-driven rate was only 5% or less of the light-induced one. This was observed with H_2 which was as efficient a donor as succinate in *Rps. capsulata* (KLEMME, 1969) and with electron donors such as ascorbate-DAD, which was twice as effective as succinate in the light-induced NAD^+ reduction in *R. rubrum* (KEISTER and MINTON, 1969).

3. The observations mentioned in Section B.II, dealing with the primary acceptor and its redox potential were also regarded as supportive of the reverse electron flow, because the redox potential of the postulated primary acceptor, as measured by the available indirect methods, was found to be insufficient for NAD^+ reduction. More recent experiments have, however, indicated that the redox potential of the primary acceptor in *R. rubrum* may be negative enough to allow it to reduce NAD^+ in the dark, since it was found to react with methyl viologen (see Sect. B.II). In this case the noncyclic electron flow would become much more probable than the rather wasteful reverse electron flow. The observed ATP-driven reverse electron flow from succinate to NAD^+ can then be envisaged as a respiratory rather than photosynthetic activity, especially since both species in which this activity has been demonstrated can grow aerobically in the dark (see Table 1).

In the green sulfur obligate anaerobe, *Chlorobium*, BUCHANAN and EVANS (1969) have reported a ferredoxin dependent light-induced reduction of NAD^+ with either sodium sulfide or mercaptoethanol serving as the electron donors. This electron transport was nonphosphorylative and was assumed to operate via a noncyclic electron transport acting in parallel with the cyclic phosphorylating system. With sulfide as donor NAD^+ reduction was sensitive to antimycin A, whereas with mercaptoethanol it was not (KNAFF and BUCHANAN, 1975). In *R. rubrum* NAD^+ reduction with succinate as donor is sensitive to the inhibitor whereas with ascorbate-DCIP it is not (NOZAKI et al., 1961; BOSE and GEST, 1963). These results suggest that succinate (and sulfide) donate electrons to the level of cytochrome b but ascorbate-DCIP (and mercaptoethanol) can donate to cytochrome c.

From the data summarized above it is clear that further experiments will be required to clarify the mechanism of NAD^+ reduction in chromatophores. A fruitful new approach to this rather old problem might be provided by experiments with mutants defective in specific photosynthetic reactions which are now beginning to be isolated (see Sect. D).

C. Energy Conservation

I. General

The current hypotheses dealing with the mechanism of photophosphorylation coupled to electron transport have been evaluated in this volume, Chapter III.1. They will be discussed here only in connection with the special characteristics of energy conservation as well as energy conversion reactions observed in chromatophores, which are at variance with the situation encountered in chloroplasts. These include:

1. The relatively large membrane potential formed in the light across the chromatophore membrane, its association with photophosphorylation and its effect on postillumination and acid-base phosphorylation.

2. The easily demonstrable dark activity of ATPase and pyrophosphatase, which do not require any light-triggering and enable ATP (or PP_i) to function as energy donors in a large number of energy-dependent reactions, such as energy-linked transhydrogenase and reverse electron transport.

3. The properties of the coupling factor(s) isolated from a number of bacterial species which have been shown to participate in the terminal steps of ATP synthesis or breakdown, but are not involved in the formation (or utilization) of the light-induced high energy state.

Energy transformation reactions in photosynthetic bacteria have been reviewed in detail by BALTSCHEFFSKY et al. (1971). Some of the more recent results reported in chromatophores have been evaluated by JAGENDORF (1975) in his excellent essay on the mechanism of photophosphorylation.

II. Proton Uptake, pH Gradient and Membrane Potential

A reversible light-induced proton uptake was first observed in chromatophores by VON STEDINGK and BALTSCHEFFSKY (1966). They also reported that in chromatophores, as in chloroplasts (see this vol. Chap. III.1 and III.2), proton uptake was inhibited by uncouplers but not by the energy transfer inhibitor oligomycin which rather stimulated it. With another energy transfer inhibitor, aurovertin, no inhibition but also no stimulation of the proton uptake was found (GROMET-ELHANAN, 1974a). Both coupling sites (see Sect. B.IV) were shown to be involved in proton uptake, since addition of antimycin A or HOQNO inhibited it and the inhibition was overcome by addition of PMS (VON STEDINGK, 1967; GROMET-ELHANAN and BRILLER, 1969).

The net light-induced proton uptake could be drastically inhibited in chromatophores without affecting ATP formation. This differential effect was obtained with nigericin + KCl (SHAVIT et al., 1968) or with NH$_4$Cl (BRILLER and GROMET-EL-HANAN, 1970). In chloroplasts both compounds were effective uncouplers (see this vol. Chaps. III.1 and III.11). The difference between chloroplasts and chromatophores was ascribed to a lower permeability of the chromatophore membrane to Cl$^-$ (JACKSON et al., 1968). Therefore, the positive electric potential formed during the light-induced proton uptake cannot be neutralized in chromatophores by the movement of Cl$^-$ and the electrochemical proton gradient will be composed of a membrane potential in addition to the pH gradient. In the presence of NH$_4$Cl, or nigericin + KCl, the pH gradient will collapse (for a detailed explanation see this vol. Chap. III.2). But the H$^+$-induced membrane potential will be exchanged with an NH$_4^+$- or K$^+$-induced one, which might even be larger, since the total amount of NH$_4^+$ or K$^+$ taken up will not be limited by the availability of internal buffering groups. According to the chemiosmotic hypothesis this increased membrane potential can be the driving force for the normal rate of phosphorylation observed in the presence of these compounds in chromatophores.

Experimental support for the suggested development of a membrane potential was obtained by testing the effect of valinomycin type ionophores on chromatophores. These compounds are known to facilitate an electrogenic movement of K$^+$ or NH$_4^+$ across biological membranes, resulting in a discharge of the membrane potential (see this vol. Chap. III.2). If the membrane potential was limiting the rate of additional proton uptake, its dissipation might result in an increased proton uptake. Indeed, these ionophores were found to stimulate the initial rate and extent of light-induced proton uptake in the presence of K$^+$ in chromatophores, again without affecting ATP formation (VON STEDINGK and BALTSCHEFFSKY, 1966; THORE et al., 1968).

Another kind of ion flux, which should dissipate the membrane potential and consequently stimulate proton uptake in chromatophores, is that resulting from the diffusion of permeable anions. Investigating a group of synthetic anions, ISAEV et al. (1970) found that phenyl dicarbaundecaborane accumulated inside R. rubrum chromatophores in the light together with the accumulation of protons. A stimulation of both the rate and extent of proton uptake by the addition of SCN$^-$ and ClO$_4^-$ (but not of SO$_2^-$) was reported by GROMET-ELHANAN and LEISER (1973). Moreover, the sodium salts of these anions did not affect ATP formation, but their ammonium salts behaved as uncouplers, inhibiting ATP formation as well as NAD$^+$ reduction (GROMET-ELHANAN, 1972a; GROMET-ELHANAN and LEISER, 1973). Such uncoupling of photophosphorylation by addition of a compound which dissipates the membrane potential in combination with one which can collapse the pH gradient, has also been observed when valinomycin was added together with nigericin + KCl (THORE et al., 1968; JACKSON et al., 1968) or with NH$_4$Cl (BRILLER and GROMET-ELHANAN, 1970).

These effects drew attention to the presumably large membrane potential developed in the light in chromatophores. Further investigations were therefore designed to find means for demonstrating this membrane potential and also for measuring it quantitatively in comparison with the pH gradient. A number of internal as well as external probes were found to reflect the light-induced membrane potential in chromatophores. These include delayed light emission (FLEISCHMAN and CLAYTON, 1968); band shifts of carotenoids (JACKSON and CROFTS, 1969); enhancement of ANS (8-anilinonaphtalene 1-sulfonate) fluorescence (GROMET-ELHANAN, 1972b; LEISER and GROMET-ELHANAN, 1976) and distribution of SCN$^-$ (at 10 µM) across the chromatophore membrane (SCHULDINER et al., 1974). In all cases valinomycin + KCl or SCN$^-$ (at 10 mM) inhibited the light-induced change, whereas nigericine + KCl or NH$_4$Cl did not, or even stimulated it. Delayed light emission and the enhancement of ANS fluorescence were also reported to decrease on addition of ADP, P$_i$ and Mg^{2+} and oligomycin abolished this decrease, thus indicating the close association between the membrane potential and photophosphorylation in chromatophores. A similar decrease of the pH gradient could be obtained only in the presence of SCN$^-$ or valinomycin + KCl, when the membrane potential was eliminated (GROMET-ELHANAN and LEISER, 1973).

III. Quantitative Estimation of the Light-Induced Electrochemical Proton Gradient in Relation to the Phosphate Potential

An obligate requirement of the chemiosmotic hypothesis is that the overall electrochemical proton gradient developed across the membrane during electron transport should be quantitatively sufficient to account for the observable equilibrium phosphate potential sustainable by photophosphorylating chromatophores (MITCHELL, 1968). For this comparison an accurate quantitative estimation of the overall electrochemical proton gradient is required. The relatively easy demonstration of both components of this gradient in chromatophores made them suitable systems for such determinations.

It should, however, be emphasized that, although we have at our disposal a number of methods for measuring the pH gradient (see this vol. Chap. III.2) and the membrane potential (see this vol. Chap. III.8) none of them is direct, accurate and completely free of complications. Therefore any reported quantitative determinations should be evaluated with the proper reservations, expecially those relying on only one method for measurement of each parameter.

A typical example of the problems encountered in the quantitative determination can be illustrated by the first method developed for measuring the membrane potential—that of the carotenoid band shifts. JACKSON and CROFTS (1969) have observed that the addition of valinomycin + KCl in the dark caused identical shifts to those observed in the light, and the extent of the change was proportional to the logarithm of the external KCl concentration. They, therefore, concluded that the height of the light-induced change can be calibrated by the dark K$^+$ diffusion potential. This calibration is, however, based on the assumption that the light-induced cartenoid band shift is due solely to the light-induced membrane potential and does not register any other effect. The assumption was questioned recently by CROFTS et al. (1974). They have conducted a detailed examination of the magnitude of light-induced carotenoid effect in a mutant of *Rps. spheroids* which has a single major carotenoid rather than two or more. With this simplified test system they found that only an undecided part of the light-induced carotenoid effect can be accounted for by an electrochromic effect related to the membrane potential. Therefore, any calibration of the light-induced effect by the dark valino-mycin-induced K^+ diffusion potential must lead to an overestimation of the light-induced membrane potential.

From all the determinations of membrane potential reported to date in chroma-tophores, those based on the cartenoid band shifts in *Rps. spheroids* and *Rps. capsulata* gave the highest estimates ranging between 280 mV to 420 mV (JACKSON and CROFTS, 1969; EVANS and CROFTS, 1974a; CASADIO et al., 1974). A number of other methods were employed for determination of the membrane potential in *R. rubrum*. These include the distribution of radioactive thiocyanate following cen-trifugation in the light (SCHULDINER et al., 1974); the enhancement of ANS fluores-cence (LEISER and GROMET-ELHANAN, 1976) and the enhancement of 3,3-dipentylox-acarbocyanine fluorescence (PICK and AVRON, 1976), both calibrated by the dark K^+ diffusion potential. With these methods the estimated light-induced membrane potential ranged between 89 mV to 110 mV.

Measurements of the pH gradient in *Rps. capsulata* by the quenching of 9-amino acridine fluorescence (CASADIO et al., 1974) and in *R. rubrum* by the quenching of either atebrin fluorescence (LEISER and GROMET-ELHANAN, 1975) or 9-amino acridine fluorescence (LEISER and GROMET-ELHANAN, 1976) gave rather similar estimates, ranging between 2.3 to 3.0 pH units. In *R. rubrum* the pH gradient was measured also by the distribution of radioactive methylamine following centrifuga-tion in the light (SCHULDINER et al., 1974). The value obtained was about 20 to 30% lower than those cited above, but this could be due to an underestimation (see JAGENDORF, 1975).

Recent estimates of the phosphate potential maintained under steady state conditions in *Rps. capsulata* (CASADIO et al., 1974) and *R. rubrum* (LEISER and GRO-MET-ELHANAN, 1976) are also quite similar when measured under similar conditions. By comparing this similar phosphate potential with the very different overall electro-chemical proton gradient measured in these two species (due to the large difference in the estimation of their membrane potential), one can obviously reach dissimilar conclusions regarding the applicability of the chemiosmotic hypothesis. It is impor-tant, therefore, to stress the point that the experimental evidence is as yet too scarce and uncertain to justify any kind of conclusion.

IV. The High-Energy State and its Utilization (Postillumination and Acid-Base Phosphorylation)

All the available theories on the mechanism of coupled phosphorylation implicate some kind of high energy intermediate or state between the electron transport and the ATP synthesis apparatus (see this vol. Chaps. III.1 and III.4). This high energy state can be used to drive energy requiring reactions.

An energy-linked transhydrogenase catalysing a redution of NADP$^+$ by NADH has been demonstrated in chromatophores of *R. rubrum* (KEISTER and YIKE, 1966) and of *Rps. spheroides* (ORLANDO et al., 1966). This reaction could be induced either by light or by ATP in the dark (see Sect. C.V). The energy linked nature of the light-induced reaction was demonstrated by its sensitivity to uncouplers and resistance to oligomycin (KEISTER and YIKE, 1966, 1967b). FISHER and GUILLORY (1969b) were able to remove from *R. rubrum* chromatophores a soluble protein factor which was necessary for the transhydrogenase activity and have also purified it extensively (FISHER and GUILLORY, 1971).

In chloroplasts the high energy state was demonstrated by postillumination ATP synthesis (see this vol. Chap. III.6). In this system a complete separation between the light-induced electron transport and the dark steps leading to ATP synthesis was achieved. A high yield of postillumination ATP synthesis was recently reported also in *R. rubrum* chromatophores (LEISER and GROMET-ELHANAN, 1975), and was found to be dependent on the presence of SCN$^-$ or valinomycin+KCl in the light stage. This dependence suggested that the light-induced membrane potential could not be used for postillumination ATP synthesis, and substantial amounts of ATP can be formed only when this membrane potential is exchanged into an increased pH gradient (see Sect. C.II). It should, however be emphasized that this membrane potential was rather closely correlated with ATP synthesis under continuous illumination as in the presence of NH$_4$Cl or nigericin+KCl (see Sect. C.II).

Further evidence for the close correlation of postillumination ATP synthesis with the pH gradient was obtained by testing the pH optimum of the light stage (LEISER and GROMET-ELHANAN, 1975). This was found to be around pH 8.0 and not at pH 6.0 as in chloroplasts (see this vol. Chap. III.6). This pH optimum was rather surprising, especially since the dark decay rate of the high energy state was much faster at this pH. The reason for this pH optimum was found to be due to the dependence of pH gradient on the external pH in the light stage. Only in the presence of SCN$^-$ above pH 7.0 did the pH gradient rise above 2.1–2.2 pH units and for optimal yields of ATP such a pH gradient was required in addition to the presence of a sufficient number of protons inside the chromatophores. Below this threshold level of 2.1 pH units no ATP could be synthesized in the postillumination stage even when a sufficient number of protons were taken up by the chromatophores (LEISER and GROMET-ELHANAN, 1975).

A similar dependence of postillumination ATP synthesis on a pH gradient above a ceratin threshold value as well as on a sufficient concentration of protons has been demonstrated in chloroplasts by SCHULDINER et al. (1973). These authors have also demonstrated that under suboptimal conditions postillumination or acid-base phosphorylation could be stimulated by a K$^+$ diffusion potential. Imposition of a K$^+$ diffusion potential by adding valinomycin+KCl to the dark stage was also found to stimulate postillumination ATP synthesis in chromatophores (GRO-MET-ELHANAN and LEISER, 1975).

For demonstration of acid-base phosphorylation in chromatophores imposition of K$^+$ or Na$^+$ diffusion potential was rather an obligate requirement (LEISER and GROMET-ELHANAN, 1974a, b). In the absence of the ionophore no acid-base phosphorylation could be obtained even under conditions which were found to be optimal for acid-base phosphorylation in chloroplasts (see this vol. Chap. III.9).

The demonstration of acid-base phosphorylation in chromatophores suggested that these particles as well as chloroplasts can tranduce pH gradient energy into ATP without passing through electron transport. They do not, however, provide unequivocal evidence that the light-induced high energy state is identical with the electrochemical proton gradient or that this gradient is indispensable for ATP synthesis.

V. ATPase, Pyrophosphatase and Exchange Reactions

Chromatophores, unlike chloroplasts (see this vol. Chap. III.5), exhibit a highly active dark ATPase (BOSE and GEST, 1965) and pyrophosphatase (BALTSCHEFFSKY et al., 1966) without need for a special activation step. Both activities are stimulated by uncoupling agents. Evidence that two different enzymes are responsible for these activities came from a number of observations:

1. ATPase was inhibited completely by oligomycin (BOSE and GEST, 1965) and partially by aurovertin (GROMET-ELHANAN, 1974a), which was shown also in mitochondria to inhibit the synthesis of ATP much more than its hydrolysis. Pyrophosphatase, on the other hand, was not affected by oligomycin (BALTSCHEFFSKY et al., 1966) and PP_i formation was rather stimulated by it. Two inhibitors which acted specifically on PP_i synthesis and breakdown, NaF and methylene diphosphate (see Fig. 1), were reported by KEISTER and MINTON (1971).

2. FISHER and GUILLORY (1969a) were able to remove ATPase and pyrophosphatase activities separately and selectively. (The first with LiCl and the second with butanol). The treated particles retained the alternative activity in each case.

3. KEISTER and MINTON (1971) have demonstrated a PP_i-driven ATP synthesis in *R. rubrum* chromatophores. Labeling experiments indicated that the reaction did not go through a phosphorylated intermediate and it seems, therefore, to involve the high energy state, possibly through the proton translocation (see below). A very active pyrophosphatase has recently been reported also in *Chromatium* chromatophores (HOCHMAN and CARMELI, 1973) but it was not affected by any reagent even not by the uncouplers which stimulated the *R. rubrum* pyrophosphatase.

A large number of energy linked reactions were found to be driven in the dark by ATP in a number of photosynthetic bacteria, and by PP_i in *R. rubrum* only. This situation is different than that in chloroplasts where the use of ATP as an energy donor was rather limited to uptake of cations, and only lately has been extended to reverse electron transport (see this vol. Chap. III.5). The energy linked reactions operating with ATP or PP_i in chromatophores include succinate linked NAD^+ reduction (see Sect. B.V); energy linked transhydrogenase (KEISTER and YIKE, 1966; 1967b); reduction of cytochrome b by cytochrome c (see Sect. B.IV); induction of proton translocation either by ATPase (SCHOLES et al., 1969) or by pyrophosphatase (MOYLE et al., 1972). All these reactions were inhibited by uncouplers and the ATP-induced ones were also inhibited by oligomycin while the light-induced and PP_i-induced reactions were rather stimulated by oligomycin.

ATP and PP_i were also shown to be direct energy donors for the creation of a high energy state. This was illustrated by their effect on carotenoid band shifts (BALTSCHEFFSKY, 1969) and on enhancement of ANS fluorescence (AZZI et al., 1971). In both cases the effect of ATP in the dark was similar to the light-induced effect (see Sect. C.II). This effect of ATP was not restricted to probes of the membrane potential (see Sect. C.II) but has also been obtained with the quenching of atebrin fluorescence (MELANDRI et al., 1972) which has been shown to reflect the light induced pH gradient in *R. rubrum* chromatophores (GROMET-EL-HANAN, 1971).

A large number of exchange reactions have also been recorded in chromato-phores, including P_i-ATP and ADP-ATP, both of which were stimulated by light (HORIO et al., 1965). It was not possible to define whether P_i-ATP exchange was dependent on ADP or ADP-ATP on P_i (see this vol. Chap. III.15), since, because of the active particle-bound dark ATPase, both ADP and P_i were always present. The P_i-ATP exchange was found to be inhibited by an antibody prepared against purified coupling factor (see Sect. C.I) from both *Rps. capsulata* (MELANDRI and BACCARINI-MELANDRI, 1971) and *R. rubrum* (JOHANSSON, 1975). It appears, there-fore, to report part of the coupling factor dependent overall phosphorylation reac-tion. A very slow P_i-ATP exchange has recently been observed also in *Chromatium* chromatophores (HOCHMAN and CAMELI, 1973). It was inhibited by uncouplers and by Dio-9 and phlorizin which in *Chromatium*, unlike in *R. rubrum* (GROMET-EL-HANAN, 1969) acted as energy transfer inhibitors. Oligomycin, on the other hand, was inactive in *Chromatium*.

KEISTER and MINTON (1971) have observed a two fold stimulation of the P_i-ATP exchange by PP_i. They have also reported on the activity of a dark P_i-PP_i exchange which, unlike the P_i-ATP exchange, was inhibited by the inhibitors of pyrophospha-tase and by uncouplers but stimulated by oligomycin (KEISTER and RAVEED, 1974). This exchange reaction appears, therefore, to be catalysed by the pyrophosphatase.

VI. Coupling Factors

Coupling factors have been removed from a number of photosynthetic bacteria, but a different method was required for each species. In *Chromatium* a 10-min incubation with tricine buffer at low ionic strength was sufficient to remove practi-cally all of the phosphorylation activity (HOCHMAN and CARMELI, 1971). In *Rps. cap-sulata* a similar treatment, even in the presence of EDTA, decreased phosphoryla-tion by only 50%, and for an 80 to 90% removal of coupling factor a 90-sec sonication period in the presence of EDTA was required (BACCARINI-MELANDRI et al., 1970). This treatment left, however, up to 50% of the phosphorylation activity in *R. rubrum* chromatophores (JOHANSSON, 1972; GROMET-ELHANAN, 1974b) and complete removal of phosphorylation was achieved only by extraction with 2 M LiCl (GUILLORY and FISHER, 1972; GROMET-ELHANAN, 1974b). The deple-tion of phosphorylation was accompanied in all three species by a similar decrease in ATPase but not in pyrophosphatase (see Sect. C.V). The different treatment required for removal of the coupling factor in each specie suggested differences in the binding of the coupling factor to the membrane and/or differences in the composition and structure of the membranes. All these aspects certainly deserve more intensive investigation.

The light-induced proton uptake as well as the light-induced quenching of atebrin fluorescence were not appreciably affected by the removal of the coupling factor from either *Rps. capsulata* (MELANDRI et al., 1970, 1972) or *R. rubrum* (GRO-MET-ELHANAN, 1974b; BINDER and GROMET-ELHANAN, 1974). Moreover, the noncyclic phosphorylating electron-transport from ascorbate-DCIP to oxygen (see Sect. B.IV) was not affected by the removal of the coupling factor and could still be stimulated by uncouplers (GROMET-ELHANAN, 1974b). This is in contrast with the loss of light-induced proton uptake and the general uncoupling of electron transport reported in coupling factor depleted chloroplasts (see this vol. Chap. III.7). These results were interpreted as indicating that the coupling fctor removed from chromatophores is functioning specifically in the terminal step of ATP synthesis and does not have any additional structural role.

Reconstitution of more than 80% of photophosphorylation by the supernatant containing the removed coupling factor was observed in *Rps. capsulata* (BACCARINI-MELANDRI et al., 1970) in *Chromatium* (HOCHMAN and CARMELI, 1971) and in *R. rubrum* provided that the LiCl extraction was carried out in the presence of ATP (BINDER and GROMET-ELHANAN, 1974). In *R. rubrum* a similar extent of depletion and restoration of phosphorylation was obtained in the endogenous system as well as in the PMS and DAD by-passes. It was, therefore, concluded that there was no release of electron carriers from the depleted chromatophores and that the removed coupling factor was participating in all sites of phosphorylation (GROMET-ELHANAN, 1974b).

The coupling factors prepared from acetone powder of *Rps. Capsulata* and *R. rubrum* chromatophores (BACCARINI-MELANDRI and MELANDRI, 1971; JOHANSSON et al., 1973) as well as the coupling factor present in the LiCl+ATP extract of *R. rubrum* (BINDER and GROMET-ELHANAN, 1974) were extensively purified. The molecular weight reported for the first two coupling factors (between 280,000 to 350,000) is similar to that reported for the chloroplast coupling factor (see this vol. Chap. III.7) but that of the last preparation appears to be different (PHILOSOF et al., unpublished observations). The *R. rubrum* acetone powder coupling factor was recently dissociated into five subunits with molecular weights of 54,000; 50,000; 32,000; 13,000 and 7,500 (JOHANSSON and BALTSCHEFFSKY, 1975) which are in close agreement with those reported in chloroplasts.

When the ATPase activity of the soluble coupling factor preparations was tested a surprising collection of contradictory results was obtained. In *Rps. capsulata* Mg^{2+} ATPase was observed (MELANDRI and BACCARINI-MELANDRI, 1971). Its activity increased during purification but even the most pure preparation had very low activity, < 10 μmoles of ATP hydrolysed per mg protein per h, and it could not be further activated by trypsin treatment or by any of the other methods used to activate chloroplast coupling factor ATPase (cf. Sect. III.7). In *R. rubrum* a Ca^{2+} ATPase, which was rather inhibited by Mg^{2+}, was observed (JOHANSSON et al., 1973). Its activity also inreased during purification of the coupling factor, but it was at least 20 fold more active than the *Rps. capsulata* enzyme. The coupling factor isolated by the LiCl+ATP extraction of *R. rubrum* chromatophores had, on the other hand, no ATPase activity even in the most purified stage but in a triton-extracted coupling factor high activity of both Ca^{2+} and Mg^{2+} ATPase was observed (OREN and GROMET-ELHANAN, unpublished observations). In the soluble crude *Chromatium* coupling factor a low activity of both Ca^{2+} and Mg^{2+} ATPase was reported and this could be increased several-fold after activation by trypsin (GEPSHTEIN and CARMELI, 1974).

This diversity is especially surprising when compared with the active particle-bound ATPase which was found in all three species to be both a Ca^{2+} and Mg^{2+} dependent enzyme. Further experiments will be required to clarify whether the differences between the soluble and bound enzyme are due to allotopy (see this vol. Chap. III.7) or result from the fact that optimal conditions for measuring the soluble enzyme have not been worked out yet.

Rps. capsulata is a facultative aerobe (see Table 1) and can be grown either photosynthetically or heterothrophically. Chromatophores prepared from such cells carry out photophosphorylation or oxidative phosphorylation. Melandri et al. (1971) have shown that coupling factor molecules derived from each of these particles could rebind and function in phosphorylation by the other. Moreover, the coupling factor proteins isolated from both types of cells could not be distinguished from one another and appear to be either very similar or identical (Baccarini-Melandri and Melandri, 1972; Lien and Gest, 1973).

D. Concluding Remarks

In the preceding sections an overall view of the present state of knowledge of the light-induced electron transport and phosphorylation in photosynthetic bacteria has been outlined. Many problems have been solved but many more still await critical investigation. New ideas and approaches will certainly be needed. A very important approach is the use of mutants. Surprisingly little has been done up to now in this field except for the isolation of a large number of mutants defective in various stages of pigment biosynthesis (see Lascelles, 1968). In this respect the purple nonsulfur bacteria have a distinct experimental advantage over the other groups because many of them can be grown either anaerobically in the light or aerobically in the dark. They, therefore, provide the possibility of isolation and characterization of mutants with specific genetic deletions in the photosynthetic apparatus.

The first such mutant has been isolated a decade ago in *Rps. spheroides* (Sistrom and Clayton, 1964). It could not grow photosynthetically although it was shown to possess a near-normal complement of antenna bacteriochlorophyll when grown in the dark under limited aeration. It was found to lack both the RC Bchl (Sistrom and Clayton, 1964) and the characteristic reaction center proteins (Clayton and Haselkorn, 1972).

Mutants depleted in any other possible electron transport carriers have not yet been reported, but this approach will certainly be used more extensively in the future. It has become especially attractive since the discovery of a genetic exchange system in *Rps. capsulata* (Marrs, 1974) and its reported operation in a large number of *Rps. capsulata* isolates (Wall et al., 1975). Bacteriophages have also been recently demonstrated in a number of Rhodopseudomonads (Bosecker et al., 1972; Abeliovich and Kaplan, 1974; Schmidt et al., 1974; Mural and Friedman, 1974) but no gene transduction by them has as yet been found. The reported transfer of genetic information might enable future studies on the control of synthesis and the regulation of function of the photosynthetic apparatus and its relation to respiration.

The correlation between respiration and photosynthesis has recently been tested by reconstitution studies in which membranes active in oxidative phosphorylation from bacteriochlorophyll-less mutants of *Rps. capsulata* and *Rps. spheroids,* were reconstituted with reaction centers from photosynthetically reactive strains (GARCIA et al., 1974, 1975; JONES and PLEWIS, 1974). These reconstituted systems carried out light-induced ATP synthesis, photooxidation of cytochrome c and photoreduction of cytochrome b. The exact extent of the association between these two modes of energy conservation in photosynthetic bacteria will certainly be the subject of extensive investigations, because it might determine the feasibility of isolating mutants defective in photosynthetic electron transport components other than the photosynthetic pigments.

References

Abeliovich, A., Kaplan, S.: J. Virol. **13**, 1392–1399 (1974)
Arnon, D.I., Allen, M.B., Whatley, F.R.: Nature (London) **174**, 394–396 (1954)
Azzi, A., Baltscheffsky, M., Baltscheffsky, H., Vainio, H.: FEBS Lett. **17**, 49–51 (1971)
Baccarini-Melandri, A., Gest, H., San Pietro, A.: J. Biol. Chem. **234**, 1224–1226 (1970)
Baccarini-Melandri, A., Melandri, B.A.: Methods Enzymol. **23**, 556–561 (1971)
Baccarini-Melandri, A., Melandri, B.A.: FEBS Lett. **21**, 131–134 (1972)
Baltscheffsky, H., Arwidsson, B.: Biochim. Biophys. Acta **65**, 425–428 (1962)
Baltscheffsky, H., Baltscheffsky, M.: Acta Chem. Scand. **12**, 1333–1335 (1958)
Baltscheffsky, H., Baltscheffsky, M., Thore, A.: Current Topics Bioenerg. **4**, 273–325 (1971)
Baltscheffsky, H., von Stedingk, L.V.: Biochim. Biophys. Res. Commun. **22**, 722–728 (1966)
Baltscheffsky, M.: Nature (London) **216**, 241–243 (1967)
Baltscheffsky, M.: Arch. Biochem. Biophys. **130**, 646–652 (1969)
Baltscheffsky, M., Baltscheffsky, H., von Stedingk, L.V.: Brookhaven Symp. Biol. **19**, 246–253 (1966)
Bartsch, R.G., Kakuno, T., Horio, T., Kamen, M.D.: J. Biol. Chem. **246**, 4489–4496 (1971)
Binder, A., Gromet-Elhanan, Z.: In: Proc. 3rd Intern. Congr. Photosynthesis. Avron, M. (ed.). Amsterdam: Elsevier, 1975, Vol. II, pp. 1163–1170
Bose, S.K., Gest, H.: Proc. Nat. Acad. Sci. U.S. **49**, 337–345 (1963)
Bose, S.K., Gest, H.: Biochim. Biophys. Acta **96**, 159–162 (1965)
Bosecker, K., Drews, G., Jank-Ladwig, R.: Arch. Mikrobiol. **86**, 69–81 (1972)
Briller, S., Gromet-Elhanan, Z.: Biochim. Biophys. Acta **205**, 263–272 (1970)
Buchanan, B.B., Evans, M.C.W.: Biochim. Biophys. Acta **180**, 123–129 (1969)
Casadio, R., Baccarini-Melandri, A., Zannoni, D., Melandri, B.A.: FEBS Lett. **49**, 203–207 (1974)
Case, G.D., Parson, W.W.: Biochim. Biophys. Acta **325**, 441–453 (1973)
Case, G.D., Parson, W.W., Thornber, J.P.: Biochim. Biophys. Acta **223**, 122–128 (1970)
Chance, B., Hollunger, G.: J. Biol. Chem. **236**, 1534–1543 (1961)
Chance, B., Nishimura, M.: Proc. Nat. Acad. Sci. U.S. **46**, 19–24 (1960)
Chance, B., Williams, G.R., Holmes, W.F., Higgins, J.: J. Biol. Chem. **217**, 439–451 (1955)
Clayton, R.K.: Ann. Rev. Biophys. Bioeng. **2**, 131–156 (1973)
Clayton, R.K., Haselkorn, R.: J. Mol. Biol. **68**, 97–105 (1972)
Cost, K., Bolton, J., Frenkel, A.: Photochem. Photobiol. **5**, 823–826 (1966)
Crofts, A.R., Prince, R.C., Holmes, N.G., Crowthers, D.: In: Proc. 3rd Intern. Congr. Photosynthesis. Avron, M. (ed.). Amsterdam: Elsevier, 1975, Vol. II, pp. 1131–1146
Cusanovich, M.A., Bartsch, R.G., Kamen, M.D.: Biochim. Biophys. Acta **153**, 397–417 (1968)
Cusanovich, M.A., Kamen, M.D.: Biochim. Biophys. Acta **153**, 376–396 (1968a)
Cusanovich, M.A., Kamen, M.D.: Biochim. Biophys. Acta **153**, 418–426 (1968b)
Dutton, P.L.: Biochim. Biophys. Acta **226**, 63–80 (1971)
Dutton, P.L., Jackson, J.B.: Europ. J. Biochem. **30**, 495–510 (1972)

Dutton, P.L., Kihara, T., McCray, J.A., Thornber, J.P.: Biochim. Biophys. Acta **226**, 81–87 (1971)
Duysens, L.N.M.: Transfer of Exicitation Energy in Photosynthesis. Thesis, Utrecht, 1952
Duysens, L.N.M.: Nature (London) **173**, 692–693 (1954)
Duysens, L.N.M., Huiskamp, W.J., Vos, J.J., van der Hart, J.M.: Biochim. Biophys. Acta **19**, 188–190 (1956)
Engelmann, Th.W.: Botan. Zh. **46**, 661–669 (1888)
Evans, E.H., Crofts, A.R.: Biochim. Biophys. Acta **333**, 41–51 (1974a)
Evans, E.H., Crofts, A.R.: Biochim. Biophys. Acta **357**, 89–102 (1974b)
Feldman, N., Gromet-Elhanan, Z.: In: Proc. 2nd Intern. Congr. Photosynthesis Research. Forti, G., Avron, M., Melandri, B.A. (eds.). The Hague: Dr. Junk, 1972, Vol. II, pp. 1211–1220
Fisher, R.R., Guillory, R.J.: FEBS Lett. **3**, 27–30 (1969a)
Fisher, R.R., Guillory, R.J.: J. Biol. Chem. **244**, 1078–1079 (1969b)
Fisher, R.R., Guillory, R.J.: J. Biol. Chem. **246**, 4687–4693 (1971)
Fleischman, D.E., Clayton, R.K.: Photochem. Photobiol. **8**, 287–298 (1968)
Fowler, C.F.: Biochim. Biophys. Acta **257**, 327–331 (1974)
Fowler, C.F., Nugent, N.A., Fuller, R.C.: Proc. Nat. Acad. Sci. U.S. **68**, 2278–2282 (1971)
Fowler, C.F., Sybesma, C.: Biochim. Biophys. Acta **197**, 276–283 (1970)
Frenkel, A.W.: J. Am. Chem. Soc. **76**, 5568–5569 (1954)
Frenkel, A.W.: J. Am. Chem. Soc. **80**, 3479–3480 (1958)
Frenkel, A.W.: Brookhaven Symp. Biol. **11**, 276–288 (1959)
Frenkel, A.W.: Biol. Rev. **45**, 569–616 (1970)
Garcia, A.F., Drews, G., Kamen, M.D.: Proc. Nat. Acad. Sci. U.S. **71**, 4213–4216 (1974)
Garcia, A.F., Drews, G., Kamen, M.D.: Biochim. Biophys. Acta **387**, 129–134 (1975)
Geller, D.M., Lipmann, F.: J. Biol. Chem. **235**, 2478–2484 (1960)
Gepshtein, A., Carmeli, C.: Europ. J. Biochem. **44**, 593–602 (1974)
Gest, H.: Nature (London) **209**, 879–882 (1966)
Gest, H.: Advan. Microbiol. Physiol. **7**, 243–282 (1972)
Govindjee, R., Smith, W.R., Govindjee: Photochem. Photobiol. **20**, 191–199 (1974)
Gray, B.H., Fowler, C.F., Nugent, N.A., Rigopoulos, W., Fuller, R.C.: Intern. J. Sys. Bacteriol. **23**, 256–264 (1973)
Gromet-Elhanan, Z.: Arch. Biochem. Biophys. **131**, 299–305 (1969)
Gromet-Elhanan, Z.: Biochim. Biophys. Acta **223**, 174–182 (1970)
Gromet-Elhanan, Z.: FEBS Lett. **13**, 124–126 (1971)
Gromet-Elhanan, Z.: Biochim. Biophys. Acta **275**, 125–129 (1972a)
Gromet-Elhanan, Z.: Europ. J. Biochem. **25**, 84–88 (1972b)
Gromet-Elhanan, Z.: In: Proc. 3rd Intern. Congr. Photosynthesis. Avron, M. (ed.). Amsterdam: Elsevier, 1975, Vol. I, pp. 791–797
Gromet-Elhanan, Z.: J. Biol. Chem. **249**, 2522–2527 (1974)
Gromet-Elhanan, Z., Briller, S.: Biochim. Biophys. Res. Commun. **37**, 261–265 (1969)
Gromet-Elhanan, Z., Leiser, M.: Arch. Biochem. Biophys. **159**, 583–589 (1973)
Gromet-Elhanan, Z., Leiser, M.: J. Biol. Chem. **250**, 90–93 (1975)
Guillory, R.J., Fisher, R.R.: Biochem. J. **129**, 471–481 (1972)
Hochman, A., Carmeli, C.: FEBS Lett. **13**, 36–40 (1971)
Hochman, A., Carmeli, C.: Photosynthetica **7**, 238–245 (1973)
Horio, T., Kamen, M.D.: Ann. Rev. Microbiol. **24**, 399–428 (1970)
Horio, T., Nishikawa, K., Katsumata, M., Yamashita, J.: Biochim. Biophys. Acta **94**, 371–382 (1965)
Horio, T., Yamashita, J.: Biochim. Biophys. Acta **88**, 237–250 (1964)
Isaev, P.I., Liberman, E.A., Samuilov, V., Skulachev, V.P., Tsofina, L.M.: Biochim. Biophys. Acta **216**, 22–29 (1970)
Jackson, J.B., Crofts, A.R.: FEBS Lett. **4**, 185–189 (1969)
Jackson, J.B., Crofts, A.R., von Stedingk, L.V.: Europ. J. Biochem. **6**, 41–54 (1968)
Jagendorf, A.T.: In: Bioenergetics of Photosynthesis. Govindjee (ed.). New York-San Francisco-London: Academic Press, 1975, pp. 413–492
Jensen, A., Aasmundrud, O., Eimhjellen, K.E.: Biochem. Biophys. Acta **88**, 466–479 (1964)
Johansson, B.C.: FEBS Lett. **20**, 339–340 (1972)

Johansson, B.C.: Thesis, Stockholm, 1975
Johansson, B.C., Baltscheffsky, M.: FEBS Lett. **53**, 221–224 (1975)
Johansson, B.C., Baltscheffsky, M., Baltscheffsky, H.: Europ. J. Biochem. **40**, 109–117 (1973)
Jones, O.T.G., Plewis, K.M.: Biochim. Biophys. Acta **357**, 204–214 (1974)
Kakuno, T., Bartsch, R.G., Nishikawa, K., Horio, T.: J. Biochem. **70**, 79–97 (1971)
Kamen, M.D., Horio, T.: Ann. Rev. Biochem. **39**, 673–700 (1970)
Keister, D.L., Minton, N.J.: Biochemistry **8**, 167–173 (1969)
Keister, D.L., Minton, N.J.: Arch. Biochem. Biophys. **147**, 330–338 (1971)
Keister, D.L., Raveed, N.J.: J. Biol. Chem. **249**, 6454–6458 (1974)
Keister, D.L., Yike, N.J.: Biochem. Biophys. Res. Commun. **24**, 519–525 (1966)
Keister, D.L., Yike, N.J.: Arch. Biochem. Biophys. **121**, 415–422 (1967a)
Keister, D.L., Yike, N.J.: Biochemistry **6**, 3847–3857 (1967b)
Kihara, T., Chance, B.: Biochim. Biophys. Acta **189**, 116–124 (1969)
Klemme, J.H.: Z. Naturforsch. **24B**, 67–76 (1969)
Klemme, J.H., Schlegel, H.G.: Arch. Mikrobiol. **63**, 154–169 (1968)
Knaff, D.B., Buchanan, B.B.: Biochim. Biophys. Acta **376**, 549–560 (1975)
Knaff, D.B., Buchanan, B.B., Malkin, R.: Biochim. Biophys. Acta **325**, 94–101 (1973)
Kuntz, I.D., Loach, P.A., Calvin, M.: Biophys. J. **4**, 227–249 (1964)
Lascelles, J.: The bacterial photosynthetic apparatus. In: Advances in Microbial Physiology. Rose, A.H., Wilkinson, J.F. (eds.). London-New York: Academic Press, 1968, Vol. II, pp. 1–42
Leiser, M., Gromet-Elhanan, Z.: FEBS Lett. **43**, 267–270 (1974)
Leiser, M., Gromet-Elhanan, Z.: In: Proc. 3rd Intern. Congr. Photosynthesis. Avron, M. (ed.). Amsterdam: Elsevier, 1975, Vol. II, pp. 941–949
Leiser, M., Gromet-Elhanan, Z.: J. Biol. Chem. **250**, 84–89 (1975)
Leiser, M., Gromet-Elhanan, Z.: Arch. Biochem. Biophys., in press (1976)
Lien, S., Gest, H.: Arch. Biochem. Biophys. **159**, 730–737 (1973)
Loach, P.A., Chu Kung, M., Hales, B.J.: Ann. N.Y. Acad. Sci. **244**, 297–319 (1975)
Marrs, B.: Proc. Nat. Acad. Sci. U.S. **71**, 971–973 (1974)
Melandri, B.A., Baccarini-Melandri, A.: In: Proc. 2nd Intern. Congr. Photosynthesis. Forti, G., Avron, M., Melandri, B.A. (eds.). The Hague: Dr. Junk, 1972, Vol. II, pp. 1169–1183
Melandri, B.A., Baccarini-Melandri, A., Crofts, A.R., Cogdell, R.J.: FEBS Lett. **24**, 141–145 (1972)
Melandri, B.A., Baccarini-Melandri, A., San Pietro, A., Gest, H.: Proc. Nat. Acad. Sci. U.S. **67**, 477–484 (1970)
Melandri, B.A., Baccarini-Melandri, A., San Pietro, A., Gest, H.: Science **174**, 514–516 (1971)
Mitchell, P.: Chemiosmotic Coupling and Energy Transduction. Bodmin: Glynn, 1968
Moyle, J., Mitchell, R., Mitchell, P.: FEBS Lett. **23**, 233–236 (1972)
Mural, R.J., Friedman, D.I.: J. Virol. **14**, 1288–1292 (1974)
Newton, J.W., Kamen, M.D.: Biochim. Biophys. Acta **25**, 462–474 (1957)
Niel, van, C.B.: Arch. Mikrobiol. **3**, 1–112 (1932)
Niel, van, C.B.: Rhodobacteriineae. In: Bergey's Manual of Determinative Bacteriology. 7th Edition. Breed, R.S., Murray, E.G.D., Smith, N.R. (eds.). Baltimore: Williams and Wilkins, 1957, pp. 35–67
Nishimura, M.: Biochim. Biophys. Acta **66**, 17–21 (1963)
Nozaki, M., Tagawa, K., Arnon, D.I.: Proc. Nat. Acad. Sci. U.S. **47**, 1334–1340 (1961)
Oelze, J., Drews, G.: Biochim. Biophys. Acta **265**, 209–239 (1972)
Okayama, S., Kakuno, T., Horio, T.: J. Biochem. **68**, 19–24 (1970)
Okayama, S., Yamamoto, N., Nishikawa, K., Horio, T.: J. Biol. Chem. **243**, 2995–2999 (1968)
Orlando, J.A., Sabo, A., Curnyn, C.: Plant Physiol. **41**, 937–945 (1966)
Parson, W.W.: Biochim. Biophys. Acta **131**, 154–172 (1967)
Parson, W.W.: Biochim. Biophys. Acta **153**, 248–259 (1968)
Parson, W.W.: Ann. Rev. Microbiol. **28**, 41–59 (1974)
Parson, W.W., Case, G.D.: Biochim. Biophys. Acta **205**, 232–245 (1970)
Parson, W.W., Cogdell, R.J.: Biochim. Biophys. Acta **416**, 105–149 (1975)
Pfennig, N., Trüper, H.G.: In: Bergey's Manual of Determinative Bacteriology. 8th Ed. Buchanan, R.E., Gibbons, W.E. (eds.). Baltimore: Williams and Wilkins, 1974, pp. 24–64

Pick, U., Avron, M.: Biochim. Biophys. Acta **440**, 189–204 (1976)
Reed, D.W., Clayton, R.K.: Biochem. Biophys. Res. Commun. **30**, 471–475 (1968)
Sauer, K.: In: Bioenergetics of Photosynthesis. Govindjee (ed.). New York-San Francisco-London: Academic Press, 1975, pp. 115–181
Schachman, H.K., Pardee, A.B., Stanier, R.Y.: Arch. Biochem. Biophys. **38**, 245–260 (1952)
Schmidt, G.L., Kamen, M.D.: Biochim. Biophys. Acta **234**, 70–72 (1971)
Schmidt, L.S., Yen, H.C., Gest, H.: Arch. Biochem. Biophys. **165**, 229–239 (1974)
Scholes, P., Mitchell, P., Moyle, J.: Europ. J. Biochem. **8**, 450–454 (1969)
Schuldiner, S., Padan, E., Rottenberg, H., Gromet-Elhanan, Z., Avron, M.: FEBS Lett. **49**, 174–177 (1974)
Schuldiner, S., Rottenberg, H., Avron, M.: Europ. J. Biochem. **39**, 455–462 (1973)
Seibert, M., de Vault, D.: Biochim. Biophys. Acta **253**, 396–411 (1971)
Shavit, N., Thore, A., Keister, D.L., San Pietro, A.: Proc. Nat. Acad. Sci. U.S. **59**, 917–922 (1968)
Silberstein, B.R., Gromet-Elhanan, Z.: FEBS Lett. **42**, 141–144 (1974)
Sistrom, W.R., Clayton, R.K.: Biochim. Biophys. Acta **88**, 61–73 (1964)
Smith, L., Baltscheffsky, M.: J. Biol. Chem. **234**, 1575–1579 (1959)
Stedingk, von, L.V.: Arch. Biochem. Biophys. **120**, 537–541 (1967)
Stedingk, von, L.V., Baltscheffsky, H.: Arch. Biochem. Biophys. **117**, 400–404 (1966)
Sybesma, C.: Biochim. Biophys. Acta **172**, 177–179 (1969)
Thore, A., Keister, D.L., Shavit, N., San Pietro, A.: Biochemistry **7**, 3499–3507 (1968)
Thornber, J.P.: Biochemistry **9**, 2688–2698 (1970)
Thornber, J.P., Olson, J.M., Williams, D.M., Clayton, M.L.: Biochim. Biophys. Acta **172**, 351–354 (1969)
Vernon, L.P.: Arch. Biochem. Biophys. **43**, 492–493 (1953)
Wall, J.D., Weaver, P.F., Gest, H.: Arch. Mikrobiol. in press, 1975
Wraight, C.A., Clayton, R.K.: Biochim. Biophys. Acta **333**, 246–260 (1974)
Yoch, D.C., Sweeney, W.V., Arnon, D.I.: J. Biol. Chem. in press (1975)
Yoch, D.C., Valentine, R.C.: Ann. Rev. Microbiol. **26**, 139–162 (1972)

Author Index

Baccarini-Melandri, A., Fabbri, E., Firstater, E., Melandri, B.A. 360, 361, *366*

Baccarini-Melandri, A., Gest, H., San Pietro, A. 656, 657, *659*

Baccarini-Melandri, A., Melandri, B.A. 657, 658, *659*

Baccarini-Melandri, A. see Casadio, R. 347, *348*, 364, 365, *367*, 382, 383, *390*, 653, *659*

Baccarini-Melandri, A. see Melandri, B.A. 328, *336*, 362, 364, 365, *367*, 384, 389, *391*, 656, 657, 658, *661*

Bachofen, R. 334, *334*

Bachofen, R., Arnon, D.I. 26, *51*

Bachofen, R., Lutz, H., Specht-Jurgensen, I. 380, *390*

Bachofen, R., Specht-Jurgensen, I. 323, *334*, 497, *499*

Bachofen, R. see Lewenstein, A. 453, *470*

Bachofen, R. see Lutz, H. 334, *336*, 351, *357*, 426, *428*

Baginsky, M. see Lozier, R.H. 98, *116*, 270, *281*

Bailey, J.L., Whyborn, A.G. 238, *245*

Bailey, J.L. see Döring, G. 75, 76, *89*, 137, 142, *146*

Bailey, J.L. see Thornber, J.P. 571, *573*, 574, *582*

Baker, R.A., Weaver, E.C. 139, *146*, 175, *177*

Bakker, E.P. see Fiolet, J.T.W. 85, *89*, 318, *335*, 386, *390*

Bakker-Grunwald, T. 372, *373*, 384, *390*, 497, *499*

Bakker-Grunwald, T., Van Dam, K. 346, *348*, 370, 371, 372, *373*, 384, *390*, 426, *428*

Baldry, C.W., Walker, D.A., Bucke, C. 38, *51*

Baldry, C.W. see Bucke, C. 38, *52*

Ball, E. see Lüttge, V. 595, *599*

Ballschmiter, K., Katz, J.J. 76, *88*

Ballschmiter, K. see Cotton, T.M. 64, 76, *89*

Ballschmiter, K. see Katz, J.J. 76, *90*

Baltscheffsky, H. 30, *52*, 452, 456, *468*

Baltscheffsky, H., Arwidsson, B. 645, *659*

Baltscheffsky, H., Baltscheffsky, M. 358, *366*, 401, *403*, 645, *659*

Baltscheffsky, H., Baltscheffsky, M., Thore, A. 347, *348*, 456, *468*, 650, *659*

Baltscheffsky, H., Stedingk, L.V. von 361, *366*, 647, *659*

Baltscheffsky, H. see Azzi, A. 655, 656, *659*

Baltscheffsky, H. see Baltscheffsky, M. 379, *390*, 655, *659*

Baltscheffsky, H. see Hall, D.O. 35, *53*

Baltscheffsky, H. see Horio, T. 379, *391*

Baltscheffsky, H. see Johansson, B.C. 657, *661*

Baltscheffsky, H. see Reeves, S.G. 308, *337*

Baltscheffsky, H. see Stedingk, L.V. von 317, *337*, 362, *368*, 381, *392*, 650, 651, *662*

Baltscheffsky, H. see Vainio, H. 385, *392*

Baltscheffsky, M. 362, *366*, 385, *390*, 452, 456, *468*, 645, 655, *659*

Baltscheffsky, M., Baltscheffsky, H., Stedingk, L.V. von 379, *390*, 655, *659*

Baltscheffsky, M., Hall, D.O. 385, *390*, 413, *415*

Baltscheffsky, M. see Azzi, A. 655, 656, *659*

Baltscheffsky, M. see Baltscheffsky, H. 347, *348*, 358, *366*, 401, *403*, 456, *468*, 645, 650, *659*

Baltscheffsky, M. see Dutton, P.L. 361, *367*

Baltscheffsky, M. see Johansson, B.C. 657, *661*

Baltscheffsky, M. see Smith, L. 643, 645, *662*

Baltscheffsky, M. see Vainio, H. 385, *392*

Bamberger, E., Gibbs, M. 38, *52*

Bamberger, E.S., Park, R.B. 508, *519*

Bamberger E.S., Rottenberg, H., Avron, M. 85, *88*, 318, *334*, 344, *348*, 361, *367*

Banai, M. 353, 354, *356*

Bannister, T.T. see Lien, S. 102, *116*, 256, *264*

Bannister, T.T. see Selman, B.R. 300, 301, *303*, 536, *542*

Bannister, T.T. see Teichler-Zallen, D. 459, *471*, 614, 615, *623*

Barber, J. 84, *88*, 320, *334*, 365, *367*, 408, *415*, 455, *468*, 476, 481, 482, 483, *489*

Barber, J., Kraan, G.B.P. 68, 83, *88*, 201, *203*, 406, 408, *415*, 481, 482, *489*

Barber, J., Telfer, A., Mills, J., Nicolson, J. *88*

Barber, J., Telfer, A., Nicolson, J. 69, 87, *88*

Barber, J., Varley, W.J. 320, *335*, 482, *489*

Barber, J. see Hopkins, M.E. 479, *490*

Barber, J. see Neumann, J. 483, *491*

Barbieri, G., Delosme, R., Joliot, P. 194, 197, *203*, 483, 484, *489*

Barbieri, G. see Joliot, P. 77, *90*, 191, 192, 193, 194, 197, *204*

Barbieri, G. see Lemasson, C. 193, *204*

Barcus, D.E. see Nobel, P.S. 455, *470*

Barfort, P., Arquilla, E.R., Vogelhut, P.O. 285, *295*

Barman, B.G., Tollin, G. 218, 220, *220*

Bar-Nun, S. see Ohad, I. 583, 599, 608, *622*

Barr, R., Crane, F.L. 238, *245*, 612, *618*

Barr, R., Crane, F.L., Giaquinta, R.T. 273, *279*

Barr, R., Hall, J.D., Baszynski, T., Brand, J., Crane, F.L. 525, *540*

Barr, R. see Baszynski, T. 525, *540*

Barr, R. see Giaquinta, R.T. 257, *263*, 271, *280*

Barr, R. see Hall, J.D. 525, *541*

Subject Index

Page numbers in *italics* refer to text parts mainly dealing with the corresponding notation.

Encyclopedia of Plant Physiology, New Series

Editors: A. Pirson, M.H. Zimmermann

Springer-Verlag
Berlin Heidelberg New York

European Journal of Biochemistry

Published on behalf of the Federation of European Biochemical Societies (FEBS)

As the first journal launched by the Federation of European Biochemical Societies, the EUROPEAN JOURNAL OF BIOCHEMISTRY continues the tradition of BIOCHEMISCHE ZEITSCHRIFT founded in 1906. The journal publishes important international contributions on the various aspects of biochemistry and molecular biology. Rather than preliminary reports, the editors select original papers on research that has been completed and can make a significant contribution, either experimental or theoretical, to our understanding of biological problems at the chemical or physical level. New methods for the study of biochemical problems are also covered.

Subcription information and sample copies upon request.

Springer-Verlag Berlin Heidelberg New York

96